Insect Virology

Edited by

Sassan Asgari and Karyn N. Johnson

School of Biological Sciences
The University of Queensland
St Lucia
Queensland
Australia

Caister Academic Press

Caister Academic Press
Norfolk, UK

www.caister.com

British Library Cataloguing-in-Publication Data
A catalogue record for this book is available from the British Library

ISBN 978-1-904455-71-4

Cover image was designed by Darren Underwood

Printed and bound in Great Britain

This book is dedicated to 'The family of the Society for Invertebrate Pathology'

Other Books of Interest

Streptomyces: Molecular Biology and Biotechnology 2011
Alphaherpesviruses: Molecular Virology 2011
Recent Advances in Plant Virology 2011
Vaccine Design: Innovative Approaches and Novel Strategies 2011
Salmonella: From Genome to Function 2011

Insect Virology 2010
Environmental Microbiology: Current Technology and Water Applications 2010
Sensory Mechanisms in Bacteria: Molecular Aspects of Signal Recognition 2010
Bifidobacteria: Genomics and Molecular Aspects 2010
Molecular Phylogeny of Microorganisms 2010
Nanotechnology in Water Treatment Applications 2010
Iron Uptake and Homeostasis in Microorganisms 2010
Caliciviruses: Molecular and Cellular Virology 2010
Epstein-Barr Virus: Latency and Transformation 2010
Lentiviruses and Macrophages: Molecular and Cellular Interactions 2010
RNA Interference and Viruses: Current Innovations and Future Trends 2010
Retroviruses: Molecular Biology, Genomics and Pathogenesis 2010
Frontiers in Dengue Virus Research 2010
Metagenomics: Theory, Methods and Applications 2010
Microbial Population Genetics 2010
Borrelia: Molecular Biology, Host Interaction and Pathogenesis 2010
Aspergillus: Molecular Biology and Genomics 2010
Environmental Molecular Microbiology 2010
Neisseria: Molecular Mechanisms of Pathogenesis 2010

Lab-on-a-Chip Technology: Biomolecular Separation and Analysis 2009
Lab-on-a-Chip Technology: Fabrication and Microfluidics 2009
Real-Time PCR: Current Technology and Applications 2009
ABC Transporters in Microorganisms 2009
Pili and Flagella: Current Research and Future Trends 2009
Bacterial Polysaccharides: Current Innovations and Future Trends 2009
Microbial Toxins: Current Research and Future Trends 2009
Acanthamoeba: Biology and Pathogenesis 2009
Bacterial Secreted Proteins: Secretory Mechanisms and Role in Pathogenesis 2009
Lactobacillus Molecular Biology: From Genomics to Probiotics 2009
Mycobacterium: Genomics and Molecular Biology 2009
Clostridia: Molecular Biology in the Post-genomic Era 2009
Plant Pathogenic Bacteria: Genomics and Molecular Biology 2009
Microbial Production of Biopolymers and Polymer Precursors 2009

Contents

Contributors

Adly M.M. Abd-Alla
Insect Pest Control Laboratory
Joint FAO/IAEA Division of Nuclear Techniques in Food and Agriculture
International Atomic Energy Agency
Vienna
Austria

a.m.m.abd-alla@iaea.org

Raul Andino
Department of Microbiology and Immunology
University of California
San Francisco, CA
USA

raul.andino@ucsf.edu

Basil Arif
Laboratory for Molecular Virology
Great Lakes Forestry Centre
Sault Ste Marie, ON
Canada

basil.arif@nrcan-rncan.gc.ca

Sassan Asgari
School of Biological Sciences
The University of Queensland
St Lucia
Australia

s.asgari@uq.edu.au

Manidipa Banerjee
School of Biological Sciences
Indian Institute of Technology, Delhi
Hows Khas
New Delhi
India

mbanerjee@bioschool.iitd.ac.in

Max Bergoin
Laboratoire de Pathologie Comparée
Université Montpellier 2
Montpellier
France

bergoinm@supagro.inra.fr

Dennis K. Bideshi
Department of Natural and Mathematical Sciences
California Baptist University
8432 Magnolia Avenue
Riverside, CA
USA

and

Department of Entomology
University of California
Riverside, CA
USA

dbideshi@calbaptist.edu

Yves Bigot
Université François Rabelais
UFR des Sciences et Techniques, Genetique, Immunotherapie, Chime et Cancer
UMR-CNRSA 6239
Tours
France

yves.bigot@univ-tours.fr

Bryony C. Bonning
Department of Entomology
Iowa State University
Ames, IA
USA

bbonning@iastate.edu

Drion G. Boucias
Entomology and Nematology Department
University of Florida
Gainesville, FL
USA

pathos@UFL.edu

Adam Chambers
School of Life Sciences and Oxford Expression Technologies Ltd
Oxford Brookes University
Oxford
UK

achambers@brookes.ac.uk

Rollie J. Clem
Molecular, Cellular, and Developmental Biology Program
Arthropod Genomics Center
Division of Biology
Kansas State University
Manhattan, KS
USA

rclem@ksu.edu

Jenny S. Cory
Department of Biological Sciences
Simon Fraser University
Burnaby, BC
Canada

jsc21@sfu.ca

John Danquah
School of Life Sciences
Oxford Brookes University
Oxford
UK

jdanquah@brookes.ac.uk

Rosemary A. Dorrington
Department of Biochemistry, Microbiology and Biotechnology
Rhodes University
Grahamstown
South Africa

r.dorrington@ru.ac.za

Brian A. Federici
Interdepartmental Programs in Microbiology, Genetics and Molecular Biology;
Department of Entomology;
Graduate Program in Cell, Molecular and Developmental Biology
University of California
Riverside, CA
USA

brian.federici@ucr.edu

Michelle L. Flenniken
Department of Microbiology and Immunology
University of California
San Francisco, CA
USA

michelle.flenniken@ucsf.edu

Caroline M. Griffiths
School of Life Sciences
Oxford Brookes University
Oxford
UK

cgriffiths@brookes.ac.uk

Richard B. Hitchman
Oxford Expression Technologies Ltd
Oxford Brookes University
Oxford
UK

r.hitchman@oetltd.com

Johannes A. Jehle
Institute for Biological Control
Julius Kühn-Institute
Federal Research Centre of Cultivated Plants
Darmstadt
Germany

jki.bund.edu

Ananya Jeshtadi
School of Life Sciences
Oxford Brookes University
Oxford
UK

ajeshtadi@brookes.ac.uk

John E. Johnson
Department of Molecular Biology
The Scripps Research Institute
La Jolla, CA
USA

jackj@scripps.edu

Karyn N. Johnson
School of Biological Sciences
The University of Queensland
St Lucia
Australia

karynj@uq.edu.au

Juan Jovel
Department of Molecular Biology
The Scripps Research Institute
La Jolla, CA
USA

jjovel@ucr.edu

Linda A. King
School of Life Sciences
Oxford Brookes University
Oxford
UK

laking@brookes.ac.uk

Mark Kunitomi
Department of Microbiology and Immunology
University of California
San Francisco CA
USA

mark.kunitomi@ucsf.edu

Zhen Li
Laboratory for Molecular Virology
Great Lakes Forestry Centre
Sault Ste Marie, ON
Canada

zhenli@nrcan.gc.ca

Peter Metcalf
School of Biological Sciences
University of Auckland
Auckland
New Zealand

peter.metcalf@auckland.ac.nz

Hajime Mori
Insect Biomedical Research Center,
Kyoto Institute of Technology,
Matsugasaki
Sakyo-ku
Kyoto
Japan

hmori@kit.ac.jp

Fernanda Murguia-Meca
School of Life Sciences
Oxford Brookes University
Oxford
UK

fmurguia@brookes.ac.uk

Lillian Pavlik
Laboratory for Molecular Virology
Great Lakes Forestry Centre
Sault Ste Marie, ON
Canada

lipavlik@nrcan.gc.ca

Srini Perera
Laboratory for Molecular Virology
Great Lakes Forestry Centre
Sault Ste Marie, ON
Canada

aperera@nrcan.gc.ca

Holly J.R. Popham
USDA
Agricultural Research Service
Biological Control of Insects Research Laboratory
Columbia, MO
USA

Holly.Popham@ars.usda.gov

Robert D. Possee
NERC Centre for Ecology and Hydrology
Oxford
UK

rdpo@ceh.ac.uk

Anette Schneemann
Department of Molecular Biology
The Scripps Research Institute
La Jolla, CA
USA

aschneem@scripps.edu

Kent S. Shelby
USDA
Agricultural Research Service
Biological Control of Insects Research Laboratory
Columbia, MO
USA

Kent.Shelby@ars.usda.gov

James R. Short
Department of Biochemistry, Microbiology and Biotechnology
Rhodes University
Grahamstown
South Africa

j.short@ru.ac.za

Tatsinda Spears
Graduate Program in Cell, Molecular and Developmental Biology
University of California
Riverside, CA
USA

tspea001@student.ucr.edu

Jeffrey A. Speir
Department of Molecular Biology
The Scripps Research Institute
La Jolla, CA
USA

speir@scripps.edu

Michael R. Strand
Department of Entomology
University of Georgia
Athens, GA
USA

mrstrand@uga.edu

Christopher S. Sullivan
Section of Molecular Genetics and Microbiology
The University of Texas at Austin
Austin, TX
USA

chris_sullivan@mail.utexas.edu

Michel Tassetto
Department of Microbiology and Immunology
University of California
San Francisco, CA
USA

michel.tassetto@ucsf.edu

Peter Tijssen
INRS-Institut Armand-Frappier
Université du Québec
Laval, QC
Canada

peter.tijssen@iaf.inrs.ca

Monique M. van Oers
Laboratory of Virology
Wageningen University
The Netherlands

monique.vanoers@wur.nl

P. Arno Venter
Department of Molecular Biology
The Scripps Research Institute
La Jolla, CA
USA

paventer@scripps.edu

Vernon K. Ward
Department of Microbiology and Immunology
University of Otago
Dunedin
New Zealand

vernon.ward@stonebow.otago.ac.nz

Trevor Williams
Instituto de Ecología AC
Xalapa
Veracruz
Mexico

trevor.williams@inecol.edu.mx

Preface

Insects are considered as one of the most successful group of animals on earth. Their prosperity is in part due to their incredible ability to defend their body from foreign invaders. However, some microorganisms, including viruses, cause chronic or lethal infections in their host insects. Entomopathogenic viruses fascinate many researchers around the world interested in microbial control, genetic engineering, gene therapy, protein expression and chemistry, cell biology, molecular virology and evolution of viruses. Certain groups of viruses continue to be problematic in beneficial insects and other arthropods causing millions of dollars of damage to industries such as sericulture, apiculture and aquaculture.

The focus of this book is on current advances and innovations in the molecular biology of insect viruses; providing information for graduate students, academics, research scientists and specialists. The book is divided into three parts starting with chapters on DNA viruses followed by RNA viruses and finally special topics in insect virology. The chapters cover all the major groups of insect pathogenic viruses, providing the latest information on each virus group and suggest future directions in research. A proposed new group of insect DNA viruses, Hytrosaviridae, is also covered in this book. *Polydnaviridae* associated with endoparasitoid wasps continue to be a controversial group violating a number of classical definitions in virology. Recent discovery of their association with nudiviruses has been a turning point in studies looking at the evolution of these virus-like particles, which provide an extraordinary example of endosymbiosis or domestication of ancestral viruses. In the last decade there has been an explosion of information on insect RNA viruses. This knowledge has implications for eukaryotic translation; virus replication, structure and assembly; invertebrate immunity and control of diseases of honey bees.

The special topics chapters cover exciting recent developments in insect virology. In the last decade, we have learned a great deal about antiviral responses to viruses with the discovery of RNAi responses from the host and mechanisms evolved by viruses to inhibit such responses. Whereas RNAi is an important antiviral response to RNA viruses, DNA viruses also face apoptosis and immune responses towards infected cells preventing spread of viral progeny. The role of non-coding small RNAs, such as microRNAs, in host–virus interactions has only recently been realized. There have been important advances in our understanding of virus particle structure and function arising from structural analysis of insect RNA viruses. Finally, compared to our knowledge on the molecular biology of various insect viruses, ecology of these viruses is certainly underexplored which will be looked at in the final chapter with the main focus on baculoviruses.

Our hope is that this book will provide a broad coverage of our current knowledge of insect viruses and also encourage future research in areas that require attention.

Sassan Asgari and Karyn Johnson

Part 1

DNA Viruses

Ascoviruses

1

Dennis K. Bideshi, Yves Bigot, Brian A. Federici and Tatsinda Spears

Abstract

The family *Ascoviridae* was erected almost a decade ago to accommodate a number of large double-stranded DNA viruses that are pathogenic to larvae and pupae of lepidopterous insects, primarily in the family *Noctuidae*. Ascoviruses are unique among members of known viral families with regard to their ultrastructure and pathobiology, and among entomoviruses, their mode of transmission is unusual, as they are vectored by parasitoid wasps. Ascovirus virions, for example, exhibit a reticulate pattern when negatively stained, and mature virions are produced by an elaborate process that combines virogenesis and the generation of virion-containing vesicles that accumulate in the blood (haemolymph) of the infected host. The virion vesicles are formed by a novel apoptotic process in which ascoviruses rescue and convert apoptotic bodies into infectious virion factories. These virion vesicles essentially serve as reservoirs for horizontal transmission of ascoviruses to susceptible hosts by female parasitoid wasps during oviposition. Though the study of ascovirus molecular biology is still in its early stages, recent advances in ascovirus molecular genetics and proteomics are beginning to reveal insights into biochemical processes related to their novel structure, pathobiology, and evolutionary origin. In this chapter, a brief historical perspective on the recognition of ascoviruses as a new family of viruses is presented, after which we discuss their evolutionary origin based on molecular profiling of genome sequences and proteomics, and possible mechanisms of virogenesis and the resulting pathology.

Introduction

In 2002 a new family of viruses (*Ascoviridae*) was erected to accommodate several species infectious to larvae and pupae of lepidopterous insects belonging in the family Noctuidae (Federici *et al.*, 2005). The genomes of these viruses are composed of large circular double-stranded DNA ranging from ~ 150 to 186 kilobasepairs, and are encapsidated in complex bacilliform, ovoidal, allantoid or reniform enveloped virions that measure 130 nm in diameter by 300–400 nm in length (Federici *et al.*, 2009). Compared to other known insect viruses, and in fact members of all vertebrate and invertebrate viral families, ascoviruses demonstrate unique virion ultrastructure, mechanism of transmission, gross pathology and cytopathology. Specifically, whereas virions of baculoviruses (family *Baculoviridae*; Theilmann *et al.*, 2005), entomopoxviruses (family *Poxviridae*; Buller *et al.*, 2005) and iridoviruses (family *Iridoviridae*; Chinchar *et al.*, 2005) are also large and structurally complex, they lack a reticulate surface pattern displayed by ascovirus virions when observed by transmission electron microscopy. Moreover, whereas other insect viruses are typically transmitted per os, ascoviruses are mechanically vectored by parasitoid wasps to lepidopteran larvae and pupae in which they cause a prolonged but fatal disease (Federici and Govindarajan, 1990; Tillman *et al.*, 2004). During the course of disease, larval growth and development is stunted, and a novel cytopathology occurs that is characterized by a modified form of virus-induced programmed cell death, i.e. apoptosis. Although apoptotic cascades often occur in the terminal stage of many viral infections, a distinct 'strategy' has evolved in ascoviruses in which apoptotic bodies are rescued as they develop, and subsequently converted into vesicles for production of infectious virions. The formation of these virion-containing vesicles, i.e. virion vesicles, is a hallmark of ascovirus pathobiology (Fig. 1.1A). After cleavage from the cell, virion vesicles dissociate and accumulate in the haemolymph, changing the colour from translucent green to opaque white (Federici, 1983; Federici *et al.*, 2009). Once in the haemolymph, maintenance and dissemination of ascoviruses in nature is perpetuated by female wasps as their ovipositors are contaminated with virions and virion vesicles that are subsequently transmitted horizontally to new hosts upon stinging or oviposition. At least one ascovirus, the type species *Spodoptera frugiperda ascovirus 1a* (SfAV1a), encodes a functional executioner caspase (Fig. 1.1B), which may be involved in mediating the apoptotic response leading to virion vesicle formation (Bideshi *et al.*, 2005). Caspase-like protein genes are also present in the genomes of *Heliothis virescens* ascovirus 3e (HvAV3e) and *Trichoplusia ni* ascovirus 6a (TnAV6a) (Asgari *et al.*, 2007; Wang *et al.*, 2006).

Although these distinguishing features of ascovirus structure and pathobiology are well documented (Federici, 1982, 1983; Bideshi *et al.*, 2005; Federici *et al.*, 2009), in contrast to the rich history of molecular data published on the biology of other insect pathogens, most

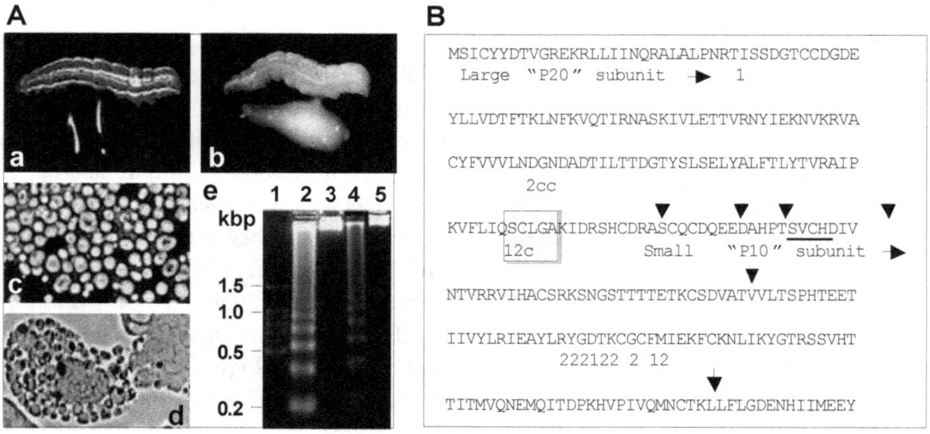

Figure 1.1 (A) Gross pathology and cytopathology induced by ascoviruses. (a) Healthy third instar larvae of *Trichoplusia ni* showing normal translucent haemolymph. (b) Third instar larvae of *T. ni* infected with ascovirus showing the diagnostic turbid, milky-white appearance of haemolymph due to the accumulation of virion containing vesicles (c) that are produced by a novel form of virus-regulated modified programmed cell death (modified apoptosis). (d) An example of an ascovirus-infected cell undergoing extensive hypertrophy to produce numerous virion-containing apoptotic vesicles that are reservoirs for horizontal transmission of the virus by parasitoid wasps. Ascovirus-induced apoptosis results in host nucleosomal fragmentation (e), a canonical response in programmed cell death; molecular mass standards in kilobasepairs (kbp) (lane 1), DNA isolated from *Spodoptera frugiperda ascovirus 1a* (SfAV1a)-infected (lane 2) and mock-infected (lane 3) insect cells (Sf21) of *Spodoptera frugiperda*, and SfAV1a-infected (lane 4) and mock-infected (lane 5) fat body of *Spodoptera exigua*. The terminal phase of apoptosis is mediated by executioner caspases-3/6/7, a functional homologue of which is encoded by SfAV1a. (B) (Bideshi *et al.*, 2005); potential cleavage sites after aspartate residues to generate the 'P20' and 'P10' active fragments are shown (arrowheads), as are residues known to be involved in catalysis (c) and which form the caspase substrate P1-binding pocket (1) that are adjacent to the substrate P2-P4 amino acids (2). The tetrad 'AHPT' (underlined) is a conserved motif known in inhibitor of apoptosis proteins (IAPs), and the conserved cysteine reside in the inhibitory allosteric site in the carboxy-terminal of caspase 3 is also shown (vertical arrow). The QACXG catalytic domain conserved in all members of the caspase superfamily is boxed in grey. [Adapted from Bideshi *et al.* (2005) Genes Dev. *19*, 1416–1421, with permission from Cold Spring Harbour Laboratory Press.]

notably baculoviruses, current understanding of the molecular biology of ascoviruses is in its early stages. A major contributing factor for the comparatively slow progress in ascovirus research is the lack of suitable *in vitro* systems that parallel infection as observed *in vivo*. Nevertheless, recent comparative analyses of the complete genome sequences of three ascoviruses, (Bideshi *et al.*, 2006; Wang *et al.*, 2006; Asgari *et al.*, 2007), and the *Diadromus pulchellus* AV 4a (DpAV4a), a symbiotic virus that apparently had a separate origin in the iridoviruses from pathogenic ascoviruses (Bigot *et al.*, 2009), are contributing to our knowledge of the molecular basis for the unique structural and pathological features displayed by ascoviruses. This chapter presents a brief historical perspective leading to the recognition of *Ascoviridae* as a unique family of viruses, describes the progression of ascovirus disease, and addresses the nomenclature and evolutionary origin of ascoviruses based on molecular phylogeny using partial and whole genome sequences. Finally, we propose hypotheses based on proteomics and the unique complement of genes shared among ascoviruses, which may help explain the novel cytopathology that characterizes ascovirus infections.

Historical perspective – ascoviruses as elusive pathogens

Although it now appears that ascoviruses are probably widely distributed in nature, as inferred from the ubiquity of their braconid (family *Braconidae*) and ichneumonid (family *Ichneumonidae*) endoparasitic wasp vectors and their noctuid hosts, these viruses were not known prior to the late 1970s. One reason diseased larvae went undetected in field surveys is because of the initial benign and chronic nature of ascovirus infection and slow progression of the disease. Even in advanced stages of disease just prior to death, overt gross pathology is difficult to detect, especially to the untrained eye, in comparison to other types of diseases. Thus, diseased larvae could be easily overlooked. Ascoviruses were originally discovered through the dissection of noctuid larvae in southern California, where the presence of disease was obvious by the white colour of haemolymph. The first infected larvae were those of *Scotogramma trifolii* collected in Fontana, California, in 1977 (Federici, 1983). Infected larvae of several other noctuids, including *Trichoplusia ni* and *Helicoverpa zea* (Federici, 1982; Browning *et al.*, 1982; Carner and Hudson, 1983; Hudson and Carner, 1981) were later identified. Initially, the pathology was so unusual that the cause of disease was not clear. Attempts to transmit the pathogen per os were unsuccessful. Subsequently, it was shown that the pathogen was an unusual virus that was readily transmitted by injecting the larvae with virions (Federici, 1983). These studies led to the identification of a new group of viruses that caused an opaque white discolouration of haemolymph, a pathology that contrasts markedly with the slightly green translucent haemolymph of healthy larvae (Fig. 1.1A). This pathology, not observed with any other virus, was shown to result from an accumulation of high concentrations of virion-containing vesicles, typically $>10^7$ vesicles per millilitre of haemolymph, and is diagnostic for the disease. Because the presence of virion vesicles is characteristic of ascovirus infection, the name 'ascovirus', derived from the Greek '*asco*' meaning 'sac', was chosen to refer to this group of viruses.

The initial description of the gross pathology induced by ascoviruses subsequently led to the identification of new isolates from lepidopterans collected in the United States, Europe (France), Australia, and Indonesia (Federici *et al.*, 1991; Hamm *et al.*, 1998; Cheng *et al.*, 2000, 2005; Newton, 2003). Though ascoviruses appear to occur most commonly in species of noctuid larvae, Bigot *et al.* (1997a,b) isolated an ascovirus (DpAV4a) that attacks the pupal stage of the leak moth, *Acrolepiopsis assectella*, a lepidopteran species belonging to the family Yponomeutidae. More recent studies, however, suggest that DpAV4a, though closely related to ascoviruses, may be the first member of a new family yet to be proposed (see below, Bigot *et al.*, 2009). Therefore, whether the natural host range of ascoviruses is limited to noctuids, as opposed to a wider spectrum of insect orders and families, remains to be resolved. However, it is likely that ascoviruses could have a broader host range, as it has been shown that under laboratory conditions, larvae of *Crocidolomia pavonana* (Lepidoptera: Crambidae) and *Plutella xyllostella* (Lepidoptera: Plutellidae) are semipermissive hosts for HvAV3e (Smede *et al.*, 2008; Furlong and Asgari, 2010).

Progression of ascovirus disease

Gross pathology

In both natural and laboratory environments, the initial signs of ascovirus disease in noctuid larvae are not obvious. However, upon careful examination and comparison with healthy larvae in the laboratory, within 24 hours a decline in the feeding rate of infected larvae can be observed. Infected larvae gain little or no weight, and may even lose weight. Moreover, anatomical and physiological changes become obvious, in which notably prolonged stunted larval growth accompanies an arrest of further development culminating in larval death. In comparison, healthy larvae forage actively, and within 3–4 days can gain as much as four times their weight and size (Govindarajan and Federici, 1990; Cheng *et al.*, 2000; Hamm *et al.*, 1998). Diseased larvae typically do not progress beyond the third instar and can survive about 14–34 days, a period well beyond the time in which healthy larvae have pupated (Federici, 1983; Govindarajan and Federici, 1990). As diseased larvae do not successfully moult to later instars and little or no apparent changes in physical appearance are obvious, ascovirus disease is easily missed in natural populations that typically comprise larvae in different stages of development.

Additional symptoms of ascovirus infection, though minor, include the inability to completely cast the moulted cuticle, a bloated thoracic region, and a white or creamy discoloration and hypertrophied appearance of the larval body especially at the terminal phase of the disease (Federici and Govindarajan, 1990; Cheng *et al.*, 2000; Newton, 2003). Interestingly, although these symptoms can be inapparent, ascovirus disease is paradoxical in the sense that it is probably the easiest of the viral diseases in lepidopterans to diagnose, even by the unaided eye. Visual examination of a drop of haemolymph will present as milky white in colour in contrast to the natural translucent green colour of haemolymph drawn from healthy larvae (Fig. 1.1A).

Tissue tropism, cytopathology, and biogenesis of virion-containing vesicles

Ascoviruses have been described from several noctuid larvae in which they are known to cause polyorganotropic infections or to have a more selective tissue-tropic preference. TnAV2 and HvAV3, as examples, are known to invade a variety of tissues including fat body, tracheal matrix and epidermis, whereas SfAV1 preferentially invades the fat body (Fig. 1.2), the only apparent site of productive pathogenesis (Federici and Govindarajan, 1990; Hamm *et al.*, 1998). Despite tissue tropic propensities, cellular invasion by ascoviruses invariably results in a novel program that leads to the formation and progressive accumulation of virion-containing vesicles in larval haemolymph.

A detailed investigation of the cytological changes induced by ascoviruses was reported soon after their discovery (Federici, 1983). All known ascoviruses initiate replication in the nucleus. The cytopathology (Fig. 1.3) is essentially identical in productive ascovirus infections regardless of host range and tissue tropism, and it explains the ultimate result of pathogenesis, i.e. the accumulation of virion-containing vesicles in haemolymph. After cellular invasion by infectious virions, the initial sign of disease is pronounced nuclear hypertrophy culminating in invagination of segments of the nuclear envelope. Following nuclear hypertrophy, the infected cell enlarges markedly and is easily distinguished as it grows from five to ten times the diameter of healthy cells. During this process, the nucleus increases in size, and the nuclear envelope ruptures and disintegrates into fragments. The plasma membrane also begins to invaginate along unusual planes composed of bilinear arrays of mitochondria. Between these planes, small vesicles of lipid membrane accrue, apparently synthesized *de novo*. The development of these mitochondria-lined planes varies among ascovirus species. For example, with SfAV1 and TnAV2 the planes of mitochondria and developing lipid membrane vesicles extend from the periphery of the hypertrophied cell almost to the centre, whereas in other species the planes first

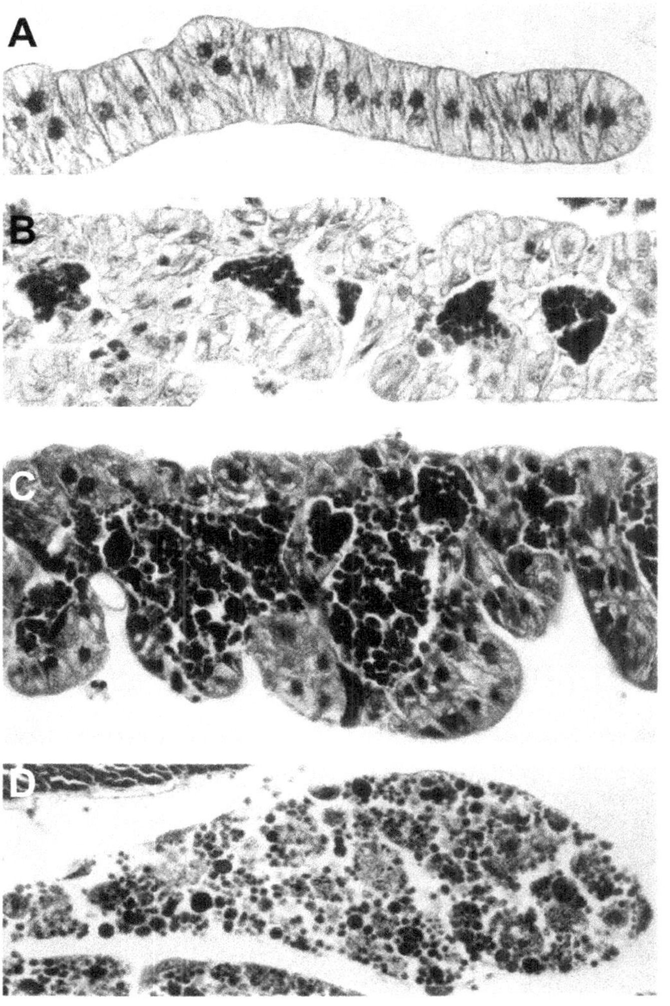

Figure 1.2 Progression of ascovirus disease in the fat body of fourth instars of *Spodoptera frugiperda* infected with the ascovirus from this species. (A) Lobe of fat body in a healthy late fourth instar. (B–D) Infected fat body at 3 (B), 7 (C), and 14 (D) days post infection. The dense masses located internally in the lobe are virion-containing vesicles and infected cells in the process of forming these (B and C). By 14 days post infection most fat body lobes have lysed completely, and lobes such as the one illustrated in (D), which is almost entirely filled with virion-containing vesicles and the remnants of cleaving cells, are uncommon. Brightfield, azan stain, X780. [From Federici, B.A. and Govindarajan, R. (1990) J. Invertebr. Pathol. *56*, 300–311. Used with permission from Elsevier BV.]

form near the cell periphery and cleave inward, progressively forming vesicles from the exterior to the interior regions of the cell. Regardless, these newly formed lipid membranes coalesce with the invaginating plasmalemma leading to cleavage of the cell into a cluster from 20 or 30 vesicles each ranging in size from 5 to 10 μm in diameter. These membrane bound viral vesicles are not absorbed in the tissue or neutralized by host defences. Instead, they are shed into the haemolymph as early as 2–3 days post infection and accumulate rapidly following deterioration and degeneration of the basement membrane. Within 3–4 days post infection, virion vesicle concentrations can be as high as 10^7–10^8 vesicles per millilitre of haemolymph.

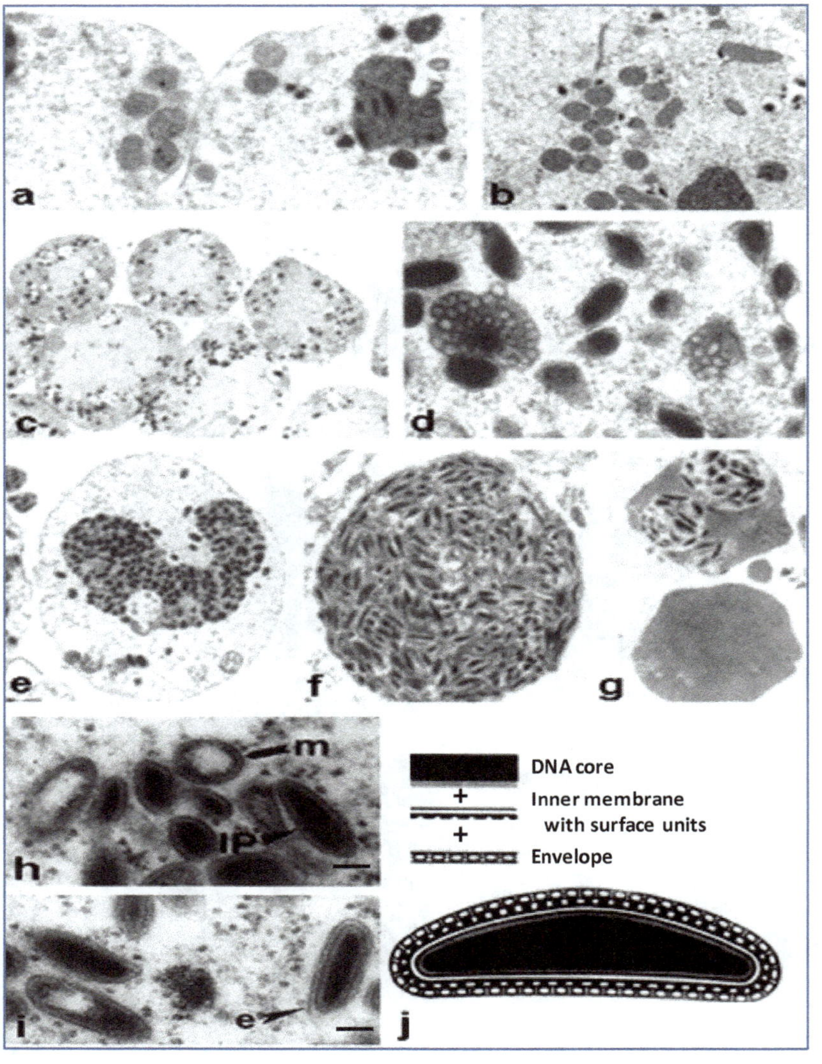

Figure 1.3 Cytopathology and virogenesis of ascoviruses. (a–g) Ultrastructural characteristics of ascovirus vesicle formation. (a) Plasmalemma invagination initiated by ascovirus formation. (b) Small lipid vesicle developing along a plane in a hypertrophied cell. These vesicles subsequently coalesce and together with the invaginating plasmalemma form the outer membrane that delimits ascovirus virion-containing vesicles. Clusters of mitochondria that occur near the developing lipid membranes are also seen in a and b. (c) Virion-containing vesicles that have cleaved from the fat body cells *in vivo* but are still contained within the fat body tissue. (d) Section through the portion of a SfAV virion-containing vesicle showing developing virions, occlusions of virions in the vesicular matrix characteristic of this ascovirus species, and virions at the vesicle periphery, apparently budding from the vesicle. (e, f) SfAV and HvAV vesicles, respectively, in the haemolymph of their hosts, *Spodoptera frugiperda* and *Heliothis virescens*. (g) Virions of TnAV in a protein granule in the fat body of a late fifth-instar larva of its host, *Trichoplusia ni*. (h, i) Virion assembly begins with the development of the outer membrane (m) of the inner particle (IP), which as it forms occludes DNA and proteins from the vesicle cytoplasmic matrix. After inner particle formation is complete, the virion is enveloped in a membrane (e), apparently formed *de novo* within the cytoplasmic matrix. (j) Schematic interpretation of ascovirus virion structure based on the appearance of virions in ultrathin sections and negatively stained preparations. The DNA core and inner membrane with surface units constitutes the inner particle. Bar = 100 nm. [From Federici, B.A. *et al.* (2009) Curr. Top. Microbiol. Immunol. *328*, 171–196. Used with permission from Springer.]

The novel cytopathology induced by ascovirus resembles apoptosis but has a peculiar hallmark that distinguishes it from the characteristic cellular suicide programme. Ascoviruses rescue apoptotic bodies that would normally be targeted for destruction, and re-engineer the programme to maximize viral proliferation in infected hosts. The evolutionary advantage to this strategy is that the prolonged process of infection and pathobiology provides a reservoir of virion-containing vesicles for dissemination among the larval host population by endoparasitic wasp vectors. The process is not a passive one, merely dictated by the host innate machinery to limit and contain viral propagation and systemic spread, as described for other viral agents of disease, including baculoviruses (Clem, 2007). In contrast, recent studies have shown that at least one ascovirus, SfAV1a, encodes a functional executioner caspase (Fig. 1.1B) synthesized within 9 hours post infection that apparently promotes the apoptotic program *in vitro* (Bideshi *et al.*, 2005). The function of this caspase and other putative proapoptotic proteins encoded by the SfAV1a and other ascovirus genomes indicate that these viruses have developed elaborate schemes to suppress antiviral countermeasures implemented by the host (see below).

Virion ultrastructure and assembly

Ultrastructural examination of ascovirus virions shows that they are enveloped and complex in symmetry, and depending on the species, vary in shape. The virions of SfAV1a, for example, are bacilliform, whereas those of TnAV2a and HvAV3a are allantoid (Fig. 1.3). Ascovirus virions typically range from 130 nm in diameter to 300–400 nm in length (Federici *et al.*, 2005). Although variations in virion morphology occur among various ascovirus species, the structural components, as observed by electron microscopy, appear to be essentially the same. The virion consists of a complex inner particle that contains a DNA/protein core confined by an apparent internal lipid bilayer surrounded by a distinctive layer of protein subunits (Fig. 1.3). Therefore, ascovirus virions appear to contain two lipid membranes, one associated with the inner particle and the other forming the lipid component of the external envelope. In negatively stained preparations, virions exhibit a unique reticulate pattern that distinguishes them from virions of other known viral families. It is thought that this unusual ultrastructural feature is due to the superimposition of protein subunits on the surfaces of the outer and inner membranes.

During virogenesis, mature virion progeny begin to appear about 12 hours after infection. Assembly and maturation of virions initiate after the nucleus ruptures, and prior to and during the cleavage of the cell into viral vesicles. The formation of the multilaminar layer of the inner particle, that is apparently composed of a unit membrane and an exterior layer of protein subunits, is the earliest recognizable structural component observed (Fig. 1.3). Concomitant with multilaminar assembly is the aggregation of a dense nucleoprotein core on the interior surface that develops until formation of the inner particle is complete. Subsequently, the inner particle is enveloped by membranes within the cell or vesicle that are thought to be synthesized *de novo*. A similar process of virion assembly and maturation is observed with other large dsDNA viruses, including herpesviruses, iridoviruses and poxviruses, where the virions differentiate after association of the precursors of virion structural components (Chinchar *et al.*, 2005; Buller *et al.*, 2005).

Although the cytopathology and virogenesis of different ascoviruses are essentially identical, both subtle and obvious differences are noted. In particular, for TnAV2 and HvAV3, virions aggregate and form inclusion bodies towards the periphery of the vesicle (Federici *et al.*, 1990; Asgari, 2006), whereas for SfAV1, occlusion bodies form a 'foamy' vesicular matrix that consists of a mixture of protein and minute spherical vesicles (Fig. 1.3) These occlusion bodies (OBs) are distinct from those produced by other types of viruses, including entomopoxvirus and the polyhedral or granular OBs of baculoviruses in which enveloped virions are embedded in rigid matrices composed of viral-encoded polyhedrin and granulin proteins, respectively (Buller *et al.*, 2005; Theilmann *et al.*, 2005).

Nomenclature of ascoviruses

Since the discovery and characterization of ascoviruses in the late 1970s and early 1980s, numerous isolates have been collected from diseased lepidopteran larvae in the field. However, detailed cytological and host-range studies of only a few have been reported (Federici *et al.*, 1991; Hamm *et al.*, 1998; Cheng *et al.*, 2000, 2005; Newton, 2003). In some of these studies, attempts were made to determine the relationships and nomenclature of ascoviruses by comparisons based on restriction fragment length polymorphisms (RFLPs), genomic (g) DNA hybridization, and protein profiles by sodium dodecylsulphate-polyacrylamide gel electrophoresis (SDS-PAGE). In addition to these methods, other criteria used in combination to distinguish species of ascoviruses were virion morphology, the presence or absence of occlusion bodies, tissue tropism, host of isolation and experimental host range, and association with specific hymenopteran parasites, if apparent. As with other viral families, for example *Baculoviridae*, the viral isolate was named according to the host from which it was isolated. However, it is clear that ascoviruses have both broad and narrow host ranges, for example, TnAV2a and HvAV3a have a broad experimental host range among noctuids, whereas SfAV1a is restricted primarily to *Spodoptera* species. Thus, the nomenclature based on host of isolation could be misleading. More recently, phylogenetic positions of genes encoding the major capsid protein, DNA polymerase and thymidine kinase have been added to the list of nomenclature criteria (Federici *et al.*, 2005).

With regard to hybridization, when the gDNA of a new isolate cross-hybridizes strongly with that of an existing species member suggesting that it belongs to one of the existing species, RFLPs have been used to distinguish variants within a species. Numerous ascovirus isolates, for example, have been obtained from larvae of different noctuid species, including *Heliothis virescens, Helicoverpa zea, Autographa precationis* and *Spodoptera exigua* in the US, as well as from *Helicoverpa* and *Spodoptera* species in Australia and Indonesia. The gDNA of many of these isolates show strong reciprocal hybridization with HvAV3a DNA under conditions of high stringency. RFLP profiles of these isolates, however, often show minor to major variations when compared with HvAV3a. Because these isolates cross-hybridize strongly with HvAV3a, they are considered variants of this viral species. Moreover, experimentally these isolates have been shown to have host ranges that overlap with HvAV3a, providing additional evidence that they are variants of the same species. Similar results are observed with isolates of TnAV2a and TnAV2b. To date, four species of ascoviruses have been characterized and three new species designations have been proposed (Table 1.1).

The taxonomic position of DpAV4 (Federici *et al.*, 2005) remains to be resolved, and, as noted above, it is likely a member of a new genus, or possibly even a new viral family. Robust evidence based on complete genome analyses of DpAV4a (GenBank accession number CU469068) indicates that this virus belongs to a different virus lineage having a distinct evolutionary origin among invertebrate iridoviruses (Bigot *et al.*, 2009), despite the structural similarities of its virion to those of other ascoviruses.

Ascovirus genomics and proteomics

The genomic sequences of three different ascoviruses, namely, the type species, *Spodoptera frugiperda ascovirus 1a* (SfAV1a; Bideshi *et al.*, 2006), *Trichoplusia ni ascovirus 2c* (TnAV2c, recently reclassified as TnAV6a; Wang *et al.*, 2006) and *Heliothis virescens ascovirus 3e* (HvAV3e; Asgari *et al.*, 2007), and the closely related *Diadromus pulchellus ascovirus* (DpAV4a; Bigot *et al.*, 2009), were reported recently. The genome size, G + C content, number of open reading frames (ORFs), and gene homologues and orthologues common to the ascoviruses are shown in Table 1.2. The variation in size of ascovirus genomes is due in part to the presence of repeated structural elements and multiple copies of a family of genes called *bro* (baculovirus repeated open reading frame) that encode nucleocytoplasmic shuttling proteins that utilize the CRM1-mediated nuclear export pathway (Kang *et al.*, 2006). The specific function of Bro protein in the biology of viruses is presently unknown. The *bro* genes and 'fossil' *bro*-like genes that are truncated variants interrupted by

Table 1.1 Recognized and proposed species of ascoviruses*

Species	Abbreviated name of the isolates	Accession number of the sequenced genome
Recognized species		
Spodoptera frugiperda ascovirus 1		
	SfAV1a	AM398843
	SfAV1b	–
	SfAV1c	–
Trichoplusia ni ascovirus 2		
	TnAV2a	–
	TnAV2b	–
Heliothis virescens ascovirus 3		
	HvAV3a	–
	HvAV3b	–
	HvAV3c	–
	HvAV3d	–
	HvAV3e	EF133465
	HvAV3f	–
	HvAV3g (ex-SeAV5a)	–
Trichoplusia ni ascovirus 6	TnAV6a**	DQ517337
Proposed species		
Helicoverpa armigera ascovirus 7		
	HaAV-7a	–
Helicoverpa punctigera ascovirus 8		
	HpAV-8a	–
Spodoptera exigua ascovirus 3		
	SeAV-9a	–

*DpAV4a is excluded as recent evidence indicates that it is not an ascovirus, but could be the first recognized member of a new family of viruses (Bigot *et al.*, 2009).

**Previous nomenclature for TnAV6a was TnAV2c.

one or several stop codons were first identified in baculoviruses and constitute a multigene family, typically with many copies per genome. These genes are also known to occur among other large insect dsDNA viruses and bacteriophages (Bideshi *et al.*, 2003). The SfAV1a genome contains two large tandemly repeated non-coding sequences of 2.9 kbp with inverted terminal repeats, and a fragment of the repeat, accounting for ~13.5 kbp or ~8.6% of the genome. Other *bro*-like genes occur in variable copy numbers. For example, 23 occur in HvAV3e, 7 in SfAV1a, 3 in TnAV6a, and 12 in DpAV4a.

More recent comparative genomic analyses have already contributed significantly to a greater understanding of the phylogeny and evolutionary origin of ascoviruses. Importantly, putative proteins encoded by ascovirus genomes, and proteomic analyses of SfAV1a and TnAV6a virions have already provided informed insights

Table 1.2 Characteristics of ascovirus genomes and selected ascovirus protein homologues assorted by putative functions

Characteristic	SfAV1a	HvAV3e	TnAV6a	DpAV4a
G + C content	49.2	45.8	35.4	49.7
Number of ORFs	123	180	165	119
Ascovirus protein homologues				
Nucleotide metabolism				
Delta DNA polymerase	001	001	001	001
DNA-dependent RNA polymerase	008	011	042	089
SNF2 DEAD-like helicase	009	015	161	020
RNAase III	022	027	008	003
DNA ligase	032	034/039	026	–
GIY-YIG endonuclease	037	–	–	–
Thymidine kinase	040	055	154	055
DNA-directed RNA polymerase, b subunit	067	082	110	070
DNA repair exonuclease/lipid metabolism,	059	071	121	026
FEN-1/FLAP, XPG RAD2-like endonuclease	066	080	112	–
S1/P1 nuclease	075	134	135	–
ATPase involved in DNA repair	103	118	077	033
ATPase	110	109	095	086
Poxvirus D5-like ATPase	099	119	078	093
Transcription factor TFIIS	113	105	100	108
Lipid metabolism				
Lipase/Juvenile hormone esterase-like	013	019	132	–
Sphingomyelinase/phosphodiesterase and nuclease-like protein	059	071	121	–
Fatty acid elongase	087	140	046	015
Patatin-like phospholipase (PLA2)	093	128	067	–
PlsC, phosphate acyltransferase	112	106	098	081
Apoptosis				
Inhibitor of apoptosis (IAP)-like	025	028	009	–
Cathepsin B	114	103	102	048
Caspase and caspase-like	073	165	072	–
Others				
Zinc-dependent metalloproteinase	014	020	158	–
Lipopolysaccharide modifying enzyme	090	131	058	–
Serine-threonine kinase	104	117	058	036

*Recent evidence suggests that DpAV4a is neither an ascovirus, nor a member of a separate new genus of the family *Ascoviridae* since it has a different origin. It has been proposed that DpAV4a represents the first recognized member of a new family of viruses (Bigot *et al.*, 2009).

into virions structural components and putative molecular processes and pathways leading to virogenesis and the unique modified apoptotic program induced by ascoviruses (Cui *et al.*, 2007; Tan *et al.*, 2009a,b; Bigot *et al.*, 2009).

Evolutionary origin of ascoviruses

Recent sequencing of the TnAV6a, SfAV1a, HvAV3e, and DpAV4a genomes in combination with earlier molecular phylogenetics (Stasiak *et al.*, 2000, 2003, 2005; Bigot *et al.*, 2009) and comparisons with vertebrate and invertebrate viruses (Eaton *et al.*, 2007) have further allowed for a more comprehensive evaluation of the evolutionary origin of ascoviruses. The number of homologues found between DpAV4a and TnAV6a, SfAV1a, HvAV3e, and iridoviruses (mosquito iridovirus, MIV or IIV3, and Chilo iridescent virus, CIV or IIV6), along with phylogenetic results, demonstrate that ascovirus and invertebrate iridoviruses have a common origin whereas the closest relative of DpAV4a is the invertebrate iridovirus CIV, rather than the other known ascoviruses (Bigot *et al.*, 2009). Interestingly, our current results reveal that invertebrate iridoviruses are genetically and evolutionarily closer to ascoviruses and DpAV4a, than to vertebrate iridoviruses (Bigot *et al.*, 2009).

Although the latest results strongly suggest that ascovirus and invertebrate iridoviruses have a common origin, further analyses with whole viral genome sequences, as they become available, are needed to definitively confirm the origin of ascoviruses from iridoviruses. At present, only two genera of invertebrate iridoviruses (*Iridovirus* and *Chloriridovirus*) have been recognized by the International Committee for Virus Taxonomy. Previous phylogenetic studies have proposed that there are at least three distinct iridovirus groups (Webby and Kalmakoff, 1998). CIV is currently a member of group II *Iridovirus* and MIV is the single member of *Chloriridovirus*. Thus, no genomic sequence data are currently available regarding the diversity within and among group I and group II iridoviruses. Sequencing and phylogenetic analyses of additional invertebrate iridovirus and ascovirus genomes will be required to more accurately define the evolutionary paths and relationships among these interesting large dsDNA viruses of invertebrates.

The recent results also provide new insights into the possible origins of various ascovirus-like particles (Avery *et al.*, 1977; Hess *et al* 1975; Stoltz and Faulkner, 1978; Stoltz *et al.* 1988; Barratt *et al.*, 1999). Importantly, the data suggest, for example, that some of these ascovirus-like particles might not be ascoviruses, but rather members of other virus families that could have originated from invertebrate iridoviruses.

Molecular modelling of ascovirus ultrastructure, virogenesis and cytopathology

The progression of virogenesis and virus-induced cytopathology as determined by electron microscopy is essentially identical among all known ascoviruses. As such, it is expected that these viruses share complements of gene homologues that function in virion assembly and maturation, nucleic acid metabolism including genome replication and gene transcription, and the modified apoptotic program unique to ascoviruses. As mentioned earlier, very little is known about the molecular mechanisms directly responsible for these processes, and indeed, an understanding of the molecular biology is still in its infancy when compared with, for example, the extensive knowledge gained from at least 30 years of work on the molecular biology of baculoviruses. However, comparative analyses of the available genome sequences of ascoviruses (Bideshi *et al.*, 2006; Wang *et al.*, 2006; Asgari *et al.*, 2007; Bigot *et al.*, 2009) and proteomics of ascovirus virions (Cui *et al.*, 2007; Tan *et al.*, 2009a) have provided insights that could uncover the molecular basis for their novel structural features and cytopathobiology. A list of gene homologues common among ascoviruses, and as they relate to biological and virion structural functions, are shown in Table 1.2 and Fig. 1.4, respectively.

Ascovirus virion proteins – structural and functional assignments

Although ultrastructural studies described above have revealed unique features of the ascovirus virion, little is known about the molecular basis for its biogenesis. The recent proteomic analyses

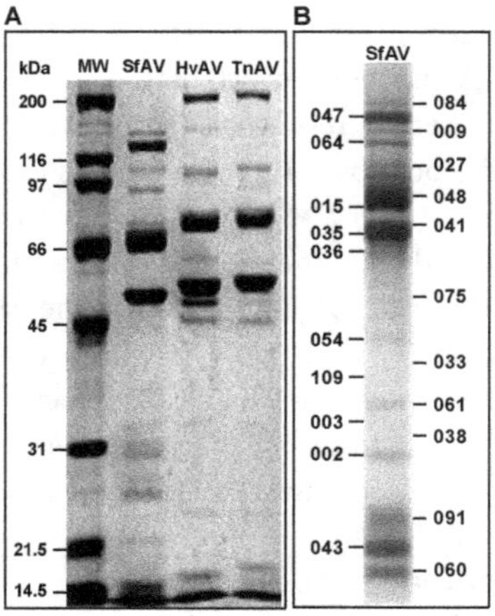

C

SfAV1a ORF	Putative Function	ORF homologue HvAV3e	TnAV6a	DpAV4a*
084	Dynein-like □ chain	146	043	085
009	SNF2 DEAD-like helicase	015	161	020
047	DNA puffC4B-like protein	060	-	-
064	Serine/threonine-like kinase	077	114	046
027	Unknown	030	011	-
048	DNA condensation/encapsidation	061	141	008
015	Inhibitor of apoptosis	022	006	-
041	Major capsid protein	056	145	019
054	Myristylated membrane-like protein	065	129	065
075	S1/P1 nuclease	134	135	-
033	Unknown	048	018	-
043	Unknown	058	149	-
035	Unknown	052	157	040
003	Unknown	004	164	-
061	Evr1/Alr thioloxidase	074	118	041
109	CTD phosphatase	110	093	117
038	Unknown	053	156	063
036	Unknown	052	157	-
091	HMG_Box/Yabby-like protein	130	059	022
002	Unknown	003	-	-
060	Unknown	072	-	-

D

```
MASKRKPARLNAEQCETFKRNKQAVSPLTNCPIDKFGRTAARFRKECDIASPPTRYTS 057
SVCKKFLANKTVSPYSGRPIKPGKKLYNDLEKHCSGRGTSPSRRSRSRSMSPRRRASP 114
ARRRASPNRSKPAKRTAANADERPDYCTNFHRDESRNPLTGKKLVPTSPIRKAWHKMC 171
SGTVQTRSTKCIAFDKNDKINPFTGRPINENNDTYRMIYSMCHGARYLPKKRSPRRKN 228
KSPARTVSFSPNRRSRSPSIGARRRPARPLRPSTSRSKTRSPSKSRSPSRRRSASKSRS 285
PSRRRSASKSRSPSRRRSASKSRSPSRRRSASKSRSPSRRRSASKSRSPSMRRSMSMAR 342
RSPSQRRSTSVARRSPSQRRSTSVARRSPSQRRSMSVARRSPSQRRSMSVARRSPSQRR 399
STSVARRSPSQRRSTSVARRSPSQSRRMTTPSRSPSRQRTSSSSRRMSARRSPSYSMSP 456
YRGRVLMTPMPEDAEPLSQDQLMRIASRMNVAQLRHVVTRNGFQPVDVAFNTNSSQLLN 513
LVRYHIREGNIKWLPANDQNVPNYRTTARPFMDRMKKN 565
```

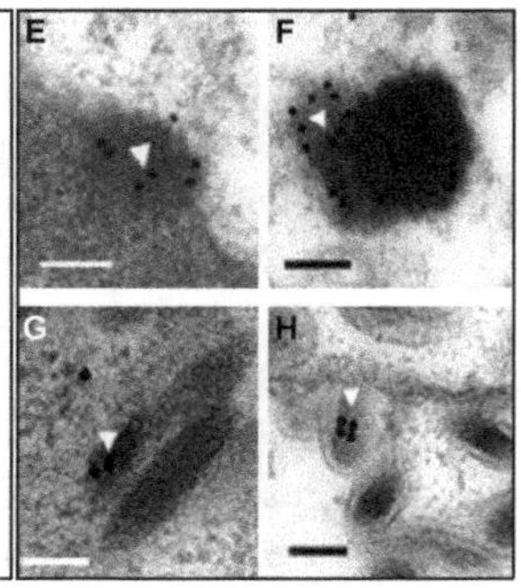

Figure 1.4 (A) Virion structural protein profiles of the *Spodoptera frugiperda ascovirus 1a* (SfAV1a, SAV), *Heliothis virescens ascovirus 3* (HAV), and *Trichoplusia ni ascovirus 2* (TAV) as determined by SDS-PAGE; MW, protein molecular mass standards in kilodaltons (kDa). (B) SDS-PAGE profile of SfAV used in virion proteomic analyses, (C) virion proteins encoded by SfAV1a are shown, as are homologues encoded by HvAV3e, TnAV6a, and *Diadromus pulchellus ascovirus 4a* (DpAV4a). (D) Sequence of the SfAV1a major virion protein P64 (ORF48) showing four copies of the virus-specific two-cysteine adaptor motifs (boxed) in the amino-terminal region and 14 copies of the repeat basic motif (consensus, SPSQRRSTS[V/K][A/S]RR) (boxed, light grey) in the carboxy-terminal region. (E, F) Immunogold labelling showing that P64 initially localizes in virogenic stroma. (G, H) Progressive incorporation of P64 into the SfAV1a virion core during virogenesis. [Adapted from Tan *et al.* (2009a) J. Gen. Virol. *90*, 359–365, with permission from the Society for General Microbiology; and Tan *et al.* (2009) J. Virol. *83*, 2708–2714].

of purified virions of TnAV6a and SfAV1a using nano-liquid chromatography/tandem (nano-LC-MS/MS) (Mueller *et al.*, 2007) identified several structural proteins encoded by the ascovirus genome and have already laid the foundation for the processes involved in the initial stages of virion assembly and maturation. Cui *et al.* (2007) identified seven virion proteins in TnAV6a (ORFs 2, 43, 115, 141, 142, 147 and 153), and in a later study Tan *et al.* (2009a) identified 21 proteins in the SfAV1a virion (Fig. 1.4). The molecular masses of a few of the virion proteins deviated from their predicted masses indicating that they are modified post-translationally by various processes including glycosylation, phosphorylation and proteolytic cleavage (Witze *et al.*, 2007).

Major virion proteins – capsid and P64, a novel DNA condensing protein

The two most abundant proteins in the SfAV1a and TnAV6a virions were homologues of the major capsid protein (MCP, Stasiak *et al.*, 2000, 2003) and P64 (Cui *et al.*, 2007; Tan *et al.*, 2009a,b). Given its function in forming the distinct protective layer in viral virions, the marked abundance of MCP relative to other virion proteins was predicted. To date, the SfAV1a

P64 protein (ORF048; Fig. 1.4D) has been one of the most studied of ascovirus proteins. P64 occurs in a similar proportion as MCP, and is an unusually large basic protein with a predicted p*I* of 12.2. The basic amino acids arginine (19.8%) and lysine (6.2%) constitute 26% of its residues, and an additional 24.1% of this protein is composed of serine (18.9%) and threonine (5.2%). Two domains, not known previously to occur together in other families of viral and cellular proteins, are found in P64. Thus, P64 represents a novel family of proteins known only in ascoviruses. The amino-terminal domain contains four copies of a well-conserved virus-specific 2-cysteine adaptor motif (protein family: pfam08793.1; Iyer *et al.*, 2006), and although this motif is known to be fused to ovarian tumour/A20-like peptidases and serine-threonine protein kinases, its biochemical function is unknown. However, the occurrence of the 2-cysteine adaptor in these proteins suggests that they could function as viral adaptors that facilitate SfAV1a virion morphogenesis.

The carboxy-terminal domain of P64 (Fig. 1.4D) shares no homology with virus proteins of known function. However, its residues from 269–365 shares a high level of identity (53%) and similarity (61%) with the sperm nuclear basic protein PL-1 isoform of the mollusc *Spisula solidassina*. Interestingly, this domain contains fourteen copies of a previously uncharacterized motif rich in arginine and serine residues (SPSQRRSTS[V/K][A/S]RR), and to a lesser extent, several lysine residues. It is the basic characteristic of P64 and its abundance in the SfAV1a virion that suggested it functions as a DNA-binding and genome-condensing protein. Indeed, studies by Tan *et al.* (2009b) showed that purified recombinant P64 has a strong non-specific DNA-binding activity, as it binds and condenses both SfAV1a and plasmid DNAs. Importantly, *in vivo* studies based on immunogold labelling electron microscopy showed that P64 initially localizes within virogenic stromae and is progressively packaged with SfAV1a gDNA into the virion core (Fig. 1.4E). Thus, the combination of data which includes its abundance in the virion, DNA-binding *in vitro*, and encapsidation *in vivo* led to the conclusion that P64 is a DNA-binding protein that packages viral DNA into the virion during assembly.

The involvement of a large protein such as P64 in viral genome condensation and encapsidation is unusual as other proteins with this function typically have low molecular masses (~5–11 kDa), and include histones, and polyvalent cations such as spermines, spermidine and protamines (Brewer *et al.*, 1999; Fuller *et al.*, 2007; Koltover *et al.*, 2000; Lanzer and Holowszak, 1975). These proteinaceous factors are capable of neutralizing the large negative electrostatic charge on phosphate groups of the nucleotide backbone that would otherwise resist condensation of viral genomes during encapsidation. For example, histones are involved in condensing viral genomes during virion assembly of several DNA viruses including polyoma, papovaviruses and herpes viruses (Chestier and Yaniv, 1979; Coca-Prados *et al.*, 1980; La Bella and Vesco, 1980; Milavetz, 2004; Maxwell and Frappier, 2007), and RNA retroviruses (Segura *et al.*, 2008), whereas the histone-like and protamine-like peptides are known to condense hepatitis B and baculovirus genomes prior to and during encapsidation (Gallina *et al.*, 1989; Wilson and Price, 1988; Wizemann and von Brunn, 1999).

Taken together with its abundance and its location in the virion core, the modular organization of P64 into the two distinct domains, i.e. the virus-specific 2-cysteine adaptor (pfam08793.1) (Iyer *et al.*, 2006) amino-terminus and protamine-like carboxy-terminus, suggests that it is a multifunctional protein that participates in configuring viral gDNA for assembly and recruiting other virion structural proteins during virogenesis. In this regard, the functions of P64 could be similar to those of the hepatitis B core antigen (HBcAg) protein (Galibert *et al.*, 1979). The virion of the hepatitis B virus (HBV) is composed of an outer envelope containing lipids and viral-encoded surface proteins that form a lipoprotein shell that encloses the nucleocapsid, or HBcAg, which contains the RNA genome of this virus. Interestingly, the capsid consists of dimers of a 183-residue protein that contains two domains with distinct functions, an amino-terminal assembly domain (residues 1–149) and a protamine-like carboxy-terminal domain (residues 150–183), responsible for polymerization into virion particles and RNA packaging, respectively (Gallina *et al.*, 1989).

Thus, by analogy, similar roles are anticipated for P64 in genome packaging, assembly, and maturation of the SfAV1a virion. This hypothesis finds support in the observation that P64 forms complexes with the SfAV1a major capsid protein *in vitro* (authors' unpublished data).

Lastly, the proportionally high level of serine and threonine in P64, collectively 25% (Fig. 1.4D), suggests that these amino acids potentially play crucial roles in its putative function, similar to those of the lower molecular mass histones and protamines. It is known that acylation, phosphorylation, and methylation of these polar residues in histones and protamines in the nucleus of host cells and viral chromatin result in epigenetic effects that influence processes such as chromatin structure and organization, gene transcription and silencing, and DNA replication and repair (Andrabi, 2007; Bartova *et al.*, 2008; Lieberman, 2008; Mikesh *et al.*, 2006; Milavetz, 2004). Interestingly, recombinant P64 produced in insect cells of *Spodoptera frugiperda* (Sf21) is phosphorylated, but is unphosphorylated in its native state in the SfAV1a virion (Tan *et al.*, 2009b). Thus, P64 could have additional physical and biochemical functions, such as the three-dimensional organization of viral genomic DNA and regulation of viral gene expression.

Virion proteins associated with nucleic acid metabolism

A number of proteins including enzymes that participate in viral nucleic acid metabolism are known to be part of the architecture of virions of many different viruses. These include replicative enzymes (DNA polymerases, RNA polymerases, and reverse transcriptase), nucleases, and transcription factors involved in immediate early and early gene expression (Braunagel *et al.*, 2003; Deng *et al.*, 2007; Resch *et al.*, 2007; Huang and Johnson, 2000; Ganser-Pornillos *et al.*, 2008). The SfAV1a virion contains at least five proteins (ORFs 009, 047, 075, 091 and 109) with conserved domains in protein families known to be involved in nucleic acid metabolism (Tan *et al.*, 2009a). These proteins potentially play crucial roles in the early phase of virogenesis following nuclear invasion by SfAV1a.

ORF009, a putative DNA helicase potentially involved in SfAV1a gDNA remodelling, contains conserved DEADc-box (cd00046) and SNF2_N (P00176) motifs in addition to sequences conserved in the HepA (COG0553) and SrmB (COG0513) protein superfamilies. Collectively, these proteins are known to function as DNA/RNA helicases, and participate in DNA replication, recombination and repair, transcription and translation. They are also known to be involved with pre-mRNA splicing, ribosome biogenesis, RNA decay and nucleocytoplasmic transport. In particular, its role could include viral chromatin remodelling as it is known that this function is mediated by protein complexes which typically consist of several subunits in which the core subunit is an ATPase of SWI/SNF2-subfamily of the DEAD/H superfamily of nucleic acid-stimulated ATPases (Boyer *et al.*, 2000).

The ORF047 protein also contains multiple conserved domains, including Smc (COG1196), HEC1 (COG5185) and MukB (COG3096). These proteins function as chromosome segregation and partitioning ATPases. ORF047 also contains the SMC_N, RecF/RecN/SMC amino-terminal (P02463) and SbcC (COG0419) domains, both known to be involved in DNA replication, recombination and repair, and a myosin tail 1 domain (P1576) with a calcium-binding coiled-coil domain (P07888) for which putative functions remain unassigned.

ORF075 contains a motif conserved in members of the S1/P1 nuclease family (P02265). The S1 and P1 nucleases (EC:3.1.30.1) are multifunctional enzymes ubiquitous in distribution, being found in microorganisms, plant and animals (Gite and Shankar, 1995). These enzymes are highly selective for single-stranded nucleic acids, degrading RNA and single-stranded DNA with no base specificity. In particular, they degrade altered structures defined by single-stranded regions in double-stranded nucleic acids, including unpolymerized or non-complementary regions present in dsDNA viral genomes (Chu *et al.*, 1990). Thus, the role of ORF075 could be both specific and non-specific in that it ensures the integrity of ascovirus dsDNA genomes and/or maintains and regulates appropriate turnover of both viral and host-encoded mRNA transcripts.

The ORF091 and ORF109 proteins are putative transcription factors. ORF091 contains a single high mobility group (HMG)-box/HMGB_UBF_HMG-box motif (P00505; cd01390) and an overlapping NHP6B motif (COG5648), both of which are present in a wide variety of eukaryotic chromosome-associated proteins and transcription factors. HMGs bind to the minor groove of DNA and can function as activators or repressors of viral and cellular promoters (Stros *et al.*, 2007). Three distinct classes of HMGs are recognized. Class I proteins bind DNA in a sequence-specific manner, and class II (e.g. non-histone chromosomal proteins HMG1 and HMG2) and class III proteins (HMGB-UBF, nucleolar and mitochondrial transcription factors) bind DNA non-specifically and contain two or more tandem HMG boxes. ORF109 contains a catalytic domain of CTD-like phosphatases (CPDc, smart00577), nuclear interacting factor-like phosphatases (NIF, NLI, P03031) and transcription factor TFIIF-interacting CTD-phosphatases (FCP1, COG5190). In addition to their roles in transcription, emerging roles of CTD-like phosphatases in DNA metabolism include chromatin remodelling and modification, DNA repair and mRNA packaging, editing and nuclear export (Meinhart *et al.*, 2005). Therefore, ORF109 could have multiple functions in the metabolism of nucleic acids. However, it is likely that its primary role involves mRNA biogenesis during the early phase of virogenesis.

ORF061 – a thiol oxidase/transferase involved with virion maturation

The ORF061 protein contains an Ervl/Alr conserved domain (COG5054) present in mitochondrial thiol oxidases involved in the biogenesis and maturation of cytosolic iron/sulphur proteins. The Evr1/Alr (Evr1, essential for respiration and vegetative growth; Alr, augmenter of liver regeneration) family of flavine adenine dinucleotide-containing sulphydryl oxidases uses oxygen as an electron acceptor (Senkevich *et al.*, 1997, 2002a,b). Representatives of the Evr1/Alr proteins include the *Saccharomyces cerevisiae* Evr1 protein required for mitochondrial development, and mammalian haematopoietin (ALR). These proteins function in a number of biochemical processes, including post-translational modification and protein turnover. Interestingly, ORF061 also shares 46% homology at residues 21–120 with the Kil-9GL protein of the African swine fever virus (ASFV). Experimental evidence suggests that Kil-9GL influences ASFV virion maturation and growth in macrophages, and plays a role in viral virulence in swine (Lewis *et al.*, 2000). The specific role of thiol transferases in virogenesis has been demonstrated for the E10R and G4L Erv1/Alr-like proteins in the vaccinia poxvirus. The G4L thiolredoxin and E10R oxidoreductase, also encoded by iridoviruses, poxviruses, and other large nuclear and cytoplasmic viruses (Gvakharia *et al*, 1996; Lu *et al.*, 2005; Senkevich *et al.*, 2002a,b), are required for formation of disulfide bonds between conserved cysteine residues in viral membrane-bound structural proteins, a process essential for poxvirus virion morphogenesis (Senkevich *et al.*, 2002a,b; White *et al.*, 2002). Thus, a similar role is anticipated for ORF061 in SfAV-1a virion maturation.

Virion proteins lacking conserved functional motifs

The remaining 13 SfAV1a-encoded virion structural proteins identified by nano-LC-MS/MS lacked conserved motifs and domains suggestive of their potential functions. Several of these proteins with low predicted molecular masses, for example, ORF002 (13.7 kDa) and ORF060 (Fig. 1.4A and B), could function as virion matrix components. ORF015 shares homology from 46% to 68% in different regions of a number of proteins that contain zinc-finger/RING finger motifs, with conserved cysteine and histidine residues at the corresponding positions. These include a cell growth regulator, mindbomb ubiquitin 3 ligase, a putative inhibitor of apoptosis protein (IAP) in *Arabidopsis thaliana*, and ORF084 of the *Spodoptera exigua* nucleopolyhedrovirus (SeMNPV) (Tan *et al.*, 2009a). In particular, the IAP superfamily of proteins, widespread in eukaryotes and their virus pathogens, were first discovered and characterized in baculoviruses where they function in suppressing or preventing apoptosis (Clem, 2007). Whether ORF015 functions as an IAP is unknown, but the prospect for this is intriguing as it provides a possible explanation for suppression of an apoptotic response elicited by ascovirus upon initial cellular invasion. Lastly, ORF084 and

ORF064 share low levels of sequence similarities with, respectively, dynein-like beta chain and serine-threonine kinase (Bideshi *et al.*, 2006).

Absence of DNA polymerase (ORF001) in the ascovirus virion

In many large dsDNA viruses such as the AcMNPV and *Helicoverpa armigera* NPV baculoviruses (Braunagel *et al.*, 2003; Deng *et al.*, 2007), vaccinia virus (Resch *et al.*, 2007) and human cytomegalovirus (Huang and Johnson, 2000), virus-encoded DNA polymerases are known to be present in their virions. These enzymes apparently function in replicating their genomes in the nucleus soon after uncoating. In contrast, no evidence has been provided to show that a DNA polymerase is present in ascovirus virions. In proteomic analyses of TnAV6a (Cui *et al.*, 2007) and SfAV1a (Tan *et al.*, 2009a), peptide sequences corresponding to ORF001, the putative ascovirus DNA polymerase, were not recovered and a protein band occurring at the predicted range (125.8 kDa) was not observed in SDS-PAGE (Fig. 1.4A and B). Whether ORF001 is present in ascovirus virions at quantities insufficient for detection by the methods used is not known. Further studies are required to determine whether ORF001 is required for SfAV1a genome replication or if its corresponding gene is transcribed early following infection of susceptible cells.

The differences in virion protein profiles among ascoviruses suggest that different species have different levels of structural complexity, which is not unexpected given that these viruses have different and overlapping host range and tissue tropism. The presence of at least 21 SfAV1a virion proteins compared with only seven for TnAV6a could indicate that SfAV1a is structurally more complex than TnAV6a, though further studies using identical experimental parameters are required to confirm this speculation. For example, the proteins surveyed for TnAV6a ranged from 31 to 220 kDa (Cui *et al.*, 2007), whereas the entire spectrum, from 6 to 220 kDa, was used in SfAV1a virion proteomics (Tan *et al.*, 2009a). In protein profile analyses of the SfAV1a virion, 10 of the 21 polypeptides migrated below 31 kDa and three of these (ORFs 043, 060 and 091) were prominent species of less than 14 kDa. Thus, whether homologues of these 10 proteins occur in TnAV6a virions is not known. Furthermore, aside from technical reasons, the distinctive vesicular occlusion body unique to SfAV1a (Federici *et al.*, 1990) could be responsible for some of these additional proteins.

Genes involved in cellular processes – virogenesis and cytopathology

DNA metabolism: replication, recombination and repair

The genomes of ascoviruses encode a DNA polymerase (ORF001) of the B family. ORF001 contains a nucleotide-polymerizing domain (II-VI-III-I-V) fused to an amino-terminal 3′ to 5′ exonuclease domain (Leipe *et al.*, 1999) and is similar to the principal replicative enzyme in both prokaryotes and eukaryotes. In eukaryotes, an associated processing factor called the proliferating cell nuclear antigen (PCNA)-like clamp, a ring-shaped protein known as a processing factor of DNA polymerase δ, is required for DNA replication. Several large dsDNA viruses including vertebrate poxviruses and baculoviruses (Iyer *et al.*, 2006; O'Reilly *et al.*, 1989) harbour genes that code for a PCNA-like clamp factor, but orthologues are absent in ascoviruses, insect iridoviruses and entomopoxviruses. In baculoviruses, a PCNA-like clamp protein is involved in viral genome replication. Viral PCNA homologues probably participate in replicating iridovirus (Chilo iridescent virus ORF436), entomopoxvirus, DpAV4a (ORF025), and ascovirus gDNA. Therefore, the absence of a PCNA-like clamp protein gene in these viruses suggests that a processivity factor function is provided by the host cell, as is the case for a mutant AcMNPV baculovirus lacking the PCNA-like factor (O'Reilly *et al.*, 1989). Also, associated with DNA polymerase function is ascovirus-encoded homologues of the poxvirus D5 family of proteins. The D5 family belongs to the helicase superfamily III within the AAA+ ATPase class that includes primary replicative helicases encoded by several DNA and RNA viruses (Gorbalenya *et al.*, 1989; Iyer *et al.*, 2005). The vaccinia virus D5 protein is a nucleic acid-independent nucleoside triphosphatase with a primase domain that is required for viral DNA replication (Evans *et al.*, 1995).

Several other proteins with presumed functions in viral DNA replication, recombination, and repair are encoded by ascoviruses. Orthologues are also present in other large dsDNA virus genomes which indicate they have conserved functions in DNA metabolism. These proteins include thymidine kinases, known to play a crucial role in synthesis of deoxyribonucleotides in cells with suboptimal pools of nucleotides (Efstathiou *et al.*, 1989; Field and Wildy, 1978), a topoisomerase I essential for maintaining the structural integrity and topology of dsDNA by transiently nicking and rejoining single DNA strands (Champoux, 1990; Shuman, 1991) and a DNA ligase that seals nicked duplex DNA substrates and Okazaki fragments produced during discontinuous DNA replication (Sriskanda and Shuman, 1998). Ascoviruses genomes also contain genes coding for ATPases, ssDNA and dsDNA nucleases, helicases, serine/threonine and tyrosine kinases, as well as members of the Rec/Sbc/Rad superfamilies, including RecD exonuclease V (ORF 95), SbcC-like and SbcD-like exonucleases, and a Rad50-like nuclease that participate in replication and repair (Ahmad *et al.*, 1998; Cromie *et al.*, 2001; Evans *et al.*, 1995; Liang and Rozman, 2006; Munter *et al.*, 2006; Okano *et al.*, 2006; Smith, 2001; Traktman and Boyle, 2004).

In particular, it is interesting to note that ascoviruses also encode a putative FEN-1/FLAP-like nuclease, an enzyme that is structurally and functionally related to eubacterial DNA polymerases, Rad2, the xeroderma pigmentosum G (XPG N and I regions) repair endonuclease and RNase H (Tian *et al.*, 2004). A T4 phage-encoded orthologue functions in RNA primer removal during viral replication, and furthermore, experimental evidence supports a role for the G5R orthologue in poxvirus in the early stage of viral morphogenesis and a regulatory role in herpesviruses where the enzyme inhibits cellular gene expression in infected cells thereby destabilizing pre-existing host mRNAs and ensuring rapid turnover of viral mRNA (da Fonseca *et al.*, 2004; Iyer *et al.*, 2006). Such functions could be conserved and essential in the biology of large nuclear and cytoplasmic DNA viruses, the so-called NCLDV family of dsDNA viruses, since orthologues occur in iridoviruses, poxvirus, *Emiliania huxleyi* virus, and mimivirus (Iyer *et al.*, 2006).

RNA metabolism

Several ascovirus encoded proteins that potentially participate in mRNA biogenesis include the DNA-dependent RNA polymerase subunit C and RNA polymerase II subunits 1 and 2, respectively, and transcription factor (TF)-like proteins, TFIIF, VLTF2 (virus-like transcription factor 2), and Yabby-like TF. The Yabby-like TF in plants constitute a group of proteins with a Cys-containing zinc finger and a DNA-binding domain related to the HMG (high-mobility group) family of TF (Meister *et al.*, 2005). Yabby proteins participate in abaxial–adaxial identity determination in lateral organs in *Arabidopsis thaliana*, whereas HMG proteins modulate gene expression through interactions with nucleosomes, transcription factors, nucleosomes-remodelling machines, and histone H1 (Bianchi and Agresti, 2005). The ascovirus orthologues of the FEN-1/FLAP-like nuclease family noted above, also contain a motif within the XPG N-I region that corresponds to sequences found in various virion host shut-off (VHS) proteins. VHS proteins encoded by herpesviruses, varicella-zoster virus, and pseudorabies virus inhibit cellular gene expression in infected cells and destabilize pre-existing host mRNAs, thereby promoting rapid turnover of virus-specific mRNAs (Lin *et al.* 2004; Smiley, 2004). Ascovirus genomes also code for an RNase III homologue with a well-conserved catalytic domain required for cleavage of dsRNA templates. RNase III mediates a number of functions in prokaryotes and eukaryotes such as mRNA degradation and processing of cellular and virus-encoded precursor RNAs into mRNAs, rRNAs, and tRNAs (Drider and Condon, 2004; Nicholson, 2003). The catalytic domain is also present in Dicer, a member of the RNase III protein superfamily, that cleaves precursor dsRNA into small temporal RNAs that function in control of developmental processes, and short interfering RNA (siRNA) of 22 nucleotides that participate in gene-specific inactivation (Aravind and Koonin, 2001). The presence of an RNase III gene in ascoviruses is unusual as homologues are absent in animal dsDNA viruses other than the *Paramecium bursaria* chlorella virus (PBCV) 1 (family *Phycodnaviridae*). The enzyme in PBCV1 is an active enzyme (Zhang *et al.*, 2003), but it lacks additional amino-terminal domains (ATPase/helicase, DUF283, and PAZ) present in Dicer (Nicholson, 2003; Macrae *et al.*, 2006). Among plant viruses, an RNase III (P22) encoded by the sweet potato chlorotic stunt virus, an RNA virus (family *Closteroviridae*), is known to enhance suppression of host-induced RNA silencing (Kreuze *et al.* 2005). Thus, it is possible that the RNase III encoded by the dsDNA viruses could function in a similar way to evade an anti-virogenic response elicited by host-directed siRNA. Interestingly, Hussain *et al.* (2008) have shown that HvAV3e encodes a micro (mi) RNA derived from its major capsid protein gene that targets the viral DNA polymerase thereby regulating virus replication (refer to Chapter 15). Whether the RNase III functions in this specific pathway or related pathways is unknown, but a more recent *in vitro* study (Hussain *et al.*, 2010) has shown that the HvAV3e RNase III is an active autoregulated enzyme involved in suppression of host cell defence.

Antiapoptotic proteins

As mentioned above, the most unusual cytopathological feature of ascovirus disease is the implementation of a novel program that leads to the formation of apoptotic bodies that are rescued from degradation by the virus. These preserved bodies serve as reservoirs for further development and maturation of ascovirus particles and are ultimately converted into infectious virion-containing vesicles, a virus-coded function that ensures transmission of ascoviruses among larval populations (Federici, 1983; Federici and Govindarajan, 1990; Bideshi *et al.*, 2005). It is likely that upon initial infection by ascoviruses the host cell mounts a response to suppress virogenesis (O'Brien, 1998; Roulston *et al.*, 1999; Barber, 2001; Clarke and Clem, 2003; Clem *et al.*, 1991; Imajoh *et al.*, 2004; Best and Bloom, 2004; Best, 2008). Indeed, the apoptotic response induced by SfAV1a *in vitro* in Sf21 cells is inhibited by the potent peptide inhibitor (z-DEVD-fmk) of executioner caspases-3, -6, and -7 (Bideshi *et al.*, 2005). However, many eukaryotic viruses have evolved molecular strategies that prolong cell viability to ensure efficient progeny production. Central to these strategies are virus genes that encode proteins in the serpin family (ex. CrmA/SPI-1 and SRP2; poxvirus), P35 family (P35, P49 and P33; baculoviruses and entomopoxvirus), FLIP family (vFLIP, MC159, γ-herpesviruses and Molluscum contagiosum virus), and IAP (inhibitor of apoptosis proteins, Op-IAP, AMV-IAP, A224L; baculoviruses, entomopoxvirus, and African swine fever virus) (Best, 2008). These proteins elaborate anti-apoptotic functions by targeting and inactivating caspases and many have at least two conserved domains (Dubrez-Daloz *et al.*, 2008). These are the BIR (baculovirus IAP repeat) domain that defines the IAP superfamily and is composed of about 70 amino acid residues organized in 4–5 α-helices and a three-stranded β-sheet stabilized by a zinc ion, and the RING (carboxy-terminal zinc finger) domain that behaves like an E3 ubiquitin ligase. The latter domain has been shown to mediate autoubiquitylation and ubiquitylation of caspase-3, -7 and -9 for proteosomal degradation. A few IAPs also contain CARD (caspase activation and recruitment domain), a domain involved in homotypic dimerization.

Ascoviruses lack genes encoding orthologues of the serpin, P35, and FLIP families of caspase inhibitors. However, they contain several copies of apparent full-length or fossil genes coding for proteins with degenerate and imperfect BIRs and E3 ubiquitin ligase-like domains. Based on sequence comparisons, ascoviruses apparently do not encode highly conserved orthologues of the baculovirus OpIAP3 and CpIAP anti-apoptotic proteins (Crooke *et al.*, 1993; Birnbaum *et al.*, 1994; Clem, 2001), and whether the function of these proteins is to suppress caspases by direct inactivation and ubiquitylation is unknown. A recent report (Hussain and Asgari, 2008) showed that the HvAV3e *ORF028* gene encoding a protein with an imperfect BIR and a RING domain, and a seemingly probable IAP candidate, was unable to block chemically induced apoptosis when expressed transiently and stably in insect cells. As chemically-induced apoptosis is blocked by HvAV3e, it is possible that genes other than *ORF028* could have anti-apoptotic functions in the biology of HvAV3e. Whether the ascovirus E3 ubiquitin ligase-like proteins have anti-apoptotic functions remains to be resolved. However, the ubiquitin-proteasome system is a well-conserved system in metazoans that is used for posttranslational modification and degradation of proteins. Three essential components are required in a cascade that targets proteins for ubiquitylation: E1 ubiquitin-activating enzyme, E2 ubiquitin-conjugating enzyme, and E3 ubiquitin ligase (Nandi *et al.*, 2006). E1 transfers ubiquitin, a protein of 76 amino acid residues, to E2 ubiquitin, which, together with E3, ligates ubiquitin to targets doomed for degradation by proteasomes. E3 ubiquitin ligase contains a zinc-binding RING motif that is essential for ubiquitination (Ardley and Robinson, 2005). Ascoviruses, as well as other members of NCLDV encode proteins with conserved RING domains that are characteristic of the E3 ubiquitin ligase (Iyer *et al.*, 2006). For example, the baculovirus Op-IAP3 contains a RING (C3H4) domain (Vaux and Silke, 2005) that functions as an E3 ubiquitin protein ligase (Muro *et al.*, 2005). Ascovirus E3 ubiquitin-like proteins, such as the SfAV1a ORF097 shares high homology with cellular proteins including breast- and ovarian-specific tumour suppressor-like protein (BRCA-1) which together with another protein (BARD1) functions as an E3 ubiquitin ligase (Starita and Parvin, 2006). It is becoming increasingly clear that ubiquitin posttranslational protein modification is involved in numerous cellular processes, including signal transduction, the cell cycle, and programmed cell death, in which the E3 ubiquitin ligase targets include the proapoptotic proteins, such as Hid, Reaper, Grim, and caspases (Muro *et al.*, 2005). Interestingly, more recent data suggest that an E3 ubiquitin ligase encoded by the white spot syndrome virus functions as an inhibitor of apoptosis to ensure efficient virus propagation (He *et al.*, 2006) and it is known that the E3 ubiquitin ligase of Op-IAP3 is required for autoubiquitylation and ubiquitylation of Hid (Muro *et al.*, 2005; Green *et al.*, 2004). Thus, it is possible that ascovirus E3 ubiquitin-like ligases play a similar role in efficient virus propagation.

Apoptosis, lipid metabolism and vesicle formation

Following cellular invasion by ascoviruses, the rapid progression of apoptosis and subsequent formation of specialized virion-containing vesicles (Bideshi *et al.*, 2005) indicate that after induction of cellular suicide, ascoviruses rapidly engineer pathways to promote these processes. Thus, a threshold is maintained to ensure that host-induced apoptotic bodies are not degraded but remain biochemically viable, in essence rescued, to allow efficient maturation of virions and vesicles. Undoubtedly, this process is complex, and not only encompasses virogenesis but the biogenesis of new vesicle membranes to allow for expansion of these virion-containing reservoirs. As host chromosomes are degraded in the apoptotic response, host nuclear gene products synthesized *de novo* are unlikely to participate in the biogenesis of virions and vesicles, although mitochondrial gene function could remain active, presumably to provide the energy source and other functions to drive reactions essential for the terminal phases of ascovirus cytobiology. Thus, ascoviruses apparently have the capability of promoting and regulating apoptosis while simultaneously synthesizing vesicle membrane *de novo*. Though likely, the participation of preformed cellular proteins in these processes is unknown, but ascoviruses encode a cassette of proteins, not known to occur in other known viruses, which could participate in a triad of functions, apoptosis, lipid metabolism and vesicle formation.

Proapoptotic proteins

The pathways leading to genetically-programmed cellular demise have been studied extensively in invertebrates and vertebrates (Duprez *et al.* 2009). Apoptosis is dependent upon the activation of procaspases, a family of cysteine-aspartic acid-specific proteases. Activated caspases then cleave and activate a number of protein targets, including other proteases and endonucleases that participate in chromatin condensation, nuclear fragmentation, membrane blebbing, and cytoskeletal rearrangements (Ghavami *et al.*, 2009), all characteristic of ascovirus cytopathology (Federici, 1983; Bideshi *et al.*, 2005). Apoptotic pathways to cellular death can be initiated by at least two different stimuli; i.e. at the level of the plasma membrane by ligand binding to death receptors followed by activation of initiator procaspases such as procaspase-8 (extrinsic pathway), or at the mitochondrial level following permeabilization of the outer mitochondrial membrane and release of proapoptotic factors, particularly cytochrome *c*, that mediates activation of procaspase-9 (intrinsic pathway). Regardless of the pathway, executioner procaspases (-3, -6, -7) are ultimately cleaved by upstream caspases into fragments containing 'P20' and 'P10' domains (Cohen, 1997; Lamkanfi *et al.*, 2002) that dimerize to generate the active enzyme. Active executioner caspases in turn activate proapoptotic factors that lead to cellular suicide. It is unlikely that ascoviruses propagate apoptosis via the extrinsic response, although they may be capable of suppressing this response through interactions with IAP-like proteins. It is more likely that cellular disintegration initiates from the intrinsic pathway as the cell enters disarray as virogenesis proceeds. Interestingly, a number of enzymes that can lead to mitochondrial destabilization are encoded by ascoviruses (discussed below). However, ascoviruses encode at least two proteins that could participate directly in initiating apoptosis. These are caspase and caspase-like proteins and cathepsin B. Indeed, ascoviruses are the only viruses known to encode caspases, and very few viruses, most notably baculoviruses and iridoviruses, encode cathepsins.

Caspases

SfAV1a encodes a novel caspase (SfAV-Casp or ORF073; 32.6 kDa) (Fig. 1.1B) that contains the two highly conserved and characteristic caspase-1 'P20' and 'P10' domains, along with the binding and catalytic residues (QSCLG) typical of the executioner caspase subfamily, and a putative IAP-binding motif similar to that present in caspase-9 (Bideshi *et al.*, 2005; Liston *et al.*, 2003). Synthesis of SfAV-Casp in *Escherichia coli* results in autocleavage of the proenzyme into fragments with the P20 and P10 domains and dimerization to produce an active enzyme that cleaves an executioner caspase substrate, z-DEVD-aminoluciferin (Bideshi *et al.*, 2005). Transient expression of SfAV-*casp* in insect cells induces apoptosis in the absence of extrinsic or intrinsic induction, and silencing of the gene by RNA interference (RNAi) in SfAV1a-infected Sf21 cells markedly inhibited formation of viral-containing vesicles from an average of 67% to 25% (Bideshi *et al.*, 2005). Interestingly, the caspase gene is expressed after 9 hours post infection, shortly after which the cardinal signs of apoptosis and virion vesicle formation are evident. Thus, it is tempting to speculate that the SfAV-Casp could be a contributor to SfAV1a-induced cytopathology that leads to virion vesicle formation. In this regard, it is possible that the timed accumulation of SfAV-Casp ultimately results in its autocleavage and activation of other proapoptotic enzymes that lead to a cascade of nuclear hypertrophy and degeneration, nucleosomal fragmentation, and membrane blebbing, concomitant with *de novo* synthesis of membranes needed to construct emerging viral vesicles. In addition, as executioner caspases also are known to be involved in the maturation of viral proteins (Best and Bloom, 2004; Best, 2008), the SfAV-Casp could participate in ascovirus virion maturation.

The scenario for the involvement of a virus-coded caspase in ascovirus cytopathology is intriguing, and even among known ascoviruses could be unique to SfAV1a. This is because apparent functional gene orthologues of executioner caspases are absent in the HvAV3e and TnAV6a. The HvAV3e encodes a protein (ORF165; HvAV-Casp) that is 37.4% identical to the SfAV-Casp and 23% identical to human caspase-7 (Asgari, 2007). The HvAV-Casp lacks the conserved cysteine residue in the catalytic motif (QT<u>N</u>RG vs. QA<u>C</u>XG) and apparently lacks an autocleavage site. It is not cleaved following overproduction in *E. coli*. Moreover, the HvAV-Casp does not induce apoptosis *in vitro* nor does inactivation of its transcripts by RNAi suppress apoptosis in HvAV3e-infected cells. Thus, its function in the apoptotic program is unclear (Asgari, 2007). However, in the same study, RNAi suppression of HvAV-Casp activity in cells resulted in a reduction in the viral MCP, which suggests that the HvAV-Casp could play a role in virion propagation.

The TnAV6a encodes a caspase (TnAV-Casp2; ORF072) with homology to caspase-2

(Wang *et al.*, 2006). The biochemical and physiologic functions of the TnAV-Casp2 has not been demonstrated, but what is known about caspase-2 suggests that this viral caspase could function in the apoptotic program. Caspase-2 belongs to the Ich-1 subfamily of caspases and has homology to initiator caspases, including caspases-1, -4, -5 and -9. It has a long prodomain that is similar to that of caspase-9 and contains a CARD (caspase recruitment domain) sequence that associates with several proteins including caspase-8 and BH3-interacting domain death agonists (Ghavami *et al.*, 2009). It must be noted that a CARD domain and the canonical caspase catalytic residues (QACRG) are absent in the TnAV-Casp2 (GEYRS), so it is possible that the putative protein lacks the conventional proapoptotic functions ascribed to caspase-2.

Cathepsin B

Cathepsins compose a family of cysteine proteases that belong to the C1 family of enzymes, the so-called papain family (Conus and Simon, 2008). They are synthesized as proenzymes that are glycosylated post-translationally, after which they are compartmentalized in lysosomes. Activated cathepsins released from lysosomes through various stimuli, including sphingomyelin diphosphodiesterase (SMase) activity (see below), direct many cellular pathways such as protein degradation, proenzyme activation, antigen processing, metabolism, and apoptosis. Its role in apoptosis is unclear, but recent evidence suggests that cathepsins B, D and L could regulate apoptosis by attacking the mitochondria and destabilizing its outer membrane to cause a release of cytochrome *c*, or by mobilizing proapoptotic proteins of the Bcl-2 superfamily such as Bid and Bax (Chwieralski *et al.*, 2006) and initiator caspases that activate cellular executioner caspases and possibly the SfAV-Casp. Unlike the unique occurrence of caspase genes in ascoviruses, cathepsin gene orthologues are also known in Chilo iridescent virus and baculoviruses (Jakob *et al.*, 2001; Slack *et al.*, 1995), and although the function of the cathepsin B homologue in CIV is unknown, the baculovirus cathepsin L orthologue is thought to be linked to degradation and liquefaction of host tissues during virus pathogenesis (Hawtin *et al.*, 1997; Slack *et al.*, 1995).

Enzymes involved in lipid metabolism

Growth, development and differentiation of ascovirus virion-containing vesicles require *de novo* lipid synthesis for extension of pre-existing membranes. Whereas synthesis of new membranes could be explained through cellular-directed processes (Bishop, 1988) during the initial stages of virogenesis, following the induction of apoptosis it is likely that ascovirus-encoded enzymes partially or completely supplement processes involved in lipid biogenesis. Ascoviruses code for several enzymes with conserved catalytic domains present in protein families known to play roles in lipid metabolism. Moreover, while a few of these enzyme genes occur individually in other viral genomes, the full cassette is unique to ascoviruses (Bideshi *et al.*, 2006). Moreover, these enzymes have putative roles in multiple lipid metabolic pathways which suggest that ascoviruses utilize diverse mechanisms to ensure efficient production of essential lipid ingredients required to manufacture virion-containing vesicles. The core lipid metabolizing enzymes of ascoviruses are the fatty acid elongase, sphingomyelin phosphodiesterase, phosphate acyltransferase, patatin-like phospholipase/alpha/beta hydrolase (phospholipase A_2, PLA_2), and a lipase-like carboxylesterase/juvenile hormone esterase. The cooperative activities of these enzymes appear likely to result in the production of raw materials required for vesicle lipid scaffolding and proliferation. Moreover, these products and by-products could have direct functions in propagating and/or amplifying the modified apoptotic program. The putative functions of these lipid-modifying enzymes and their contributions to the novel cytopathology induced by ascoviruses are presented below.

Ascovirus phosphate acyltransferases belong to the PlsC family of proteins. Members of this family, which include 1-acylglycerol-3-phosphate acyltransferase, function in *de novo* biosynthesis of cell membrane phospholipids and have glycerophosphate, 1-acylglycerolphosphate, or 2-acylglycerolphosphoethanolamine acyltransferase activities (Cullinane *et al.*, 2005;

Agarwal *et al.*, 2006). The carboxylesterase/JHE-like lipases of ascoviruses contain a well-conserved carboxylesterase domain found in the alpha/beta hydrolase fold lipase family and is also related to the juvenile hormone esterase (JHE) of *Apis mellifera*. Lipases catalyse the hydrolysis of a wide variety of ester- and amide-containing molecules, including short- and long-chain acylglycerols, long-chain acylcarnitine, and long-chain acyl-coenzyme A esters (Satoh and Hosokawa, 1998). JHE degrades juvenile hormone and functions in diverse roles in the life cycle of insects, including development, morphogenesis, reproduction, diapause, and metabolism (Gilbert *et al.*, 2000; Kamita *et al.*, 2003). Ascovirus encoded fatty acid elongases (ELO) belong to the GSN1/SUR4 enzyme family (Baudry *et al.*, 2001) and could participate in diverse processes involving lipid intermediates. In *Drosophila*, as well as in mammals and yeast, the majority of fatty acids are composed of species with carbon chain lengths of 14–18, whereas a minority of fatty acids is composed of 20 or more carbon atoms. The latter group is generally referred to as 'very long-chain fatty acids'. ELOs catalyse the biosynthesis of long-chain (20 or more) fatty acid precursors that are used in ceramide and other sphingolipid lipogenesis. In yeast, allelic variants of ELO2 and ELO3 (VBM1 and VBM2, respectively), also involved in long-chain fatty acid elongation and sphingolipid synthesis, appear to play an important role in protein sorting and trafficking of secretory vesicles, signal transduction, cell growth and differentiation, and cell death (Hannun, 1994; Jakobsson *et al.*, 2006).

The ascovirus sphingomyelin phosphodiesterase-like proteins (SMases), which also contain an SbcD DNA repair exonuclease domain, have a well-conserved overlapping calcineurin-like phosphoesterase domain found in a wide variety of phosphoesterases, including protein phosphoserine phosphatases, nucleotidases, 2′,3′-cyclic AMP phosphodiesterase, and SMases (Andrieu-Abadie and Lavade, 2002). SMases cleave sphingomyelin to release phosphocholine and ceramide. The patatin-like phospholipase/alpha/beta hydrolases of ascoviruses contain a FabD/lysophospholipase-like domain related to eukaryotic patatin-like phospholipases (Hirschberg *et al.*, 2001; Rydel *et al.*, 2003). These enzymes contain a patatin domain that overlaps with a highly conserved domain found in members of the alpha/beta hydrolase fold superfamily and are known to participate in a myriad of catalytic functions, including lipolytic, thioesterase, and peptidase activities. Patatin, a storage protein in potato tubers, also has an intrinsic lipid acylhydrolase activity that catalyses the cleavage of fatty acids from membrane lipids. Interestingly, the active site in patatin contains a Ser-Asp dyad similar to those of cytosolic phospholipase A_2 (PLA_2) (Hirschberg *et al.*, 2001) that is included in a family of enzymes involved in several cellular processes, including phospholipid digestion and metabolism, host defence, and signal transduction (Hirabayashi *et al.*, 2004; Zhao *et al.*, 2002). PLA_2 is also implicated in the induction of apoptosis via ceramide biogenesis (Atsumi *et al.*, 2000; Jayadev *et al.*, 1997) and nuclear shrinkage in caspase-independent cell death (Shinzawa and Tsujimoto, 2003). It is also interesting to note that the parvovirus capsid protein has an intrinsic PLA_2-like activity required for endocytic entry of the virus (Canaan *et al.*, 2004; Suikkanen *et al.*, 2003), a possible function for the ascovirus patatin PLA_2-like enzyme. However, the SfAV1a patatin-like protein (ORF093) was not detected in proteomic analysis of purified virions. Whether it occurs at a low level, undetectable by virion proteomic analyses, is unknown, but its presence could in part explain one mechanism responsible for invasion of target cells by ascoviruses. Finally, patatin was shown to inhibit growth of the maize root worm (*Diabrotica* sp.; Strickland *et al.*, 1995). As mentioned above, ascovirus-diseased larvae exhibit decreased feeding and stunted growth. Thus, it is intriguing to speculate that the ascovirus patatin-like PLA_2 contributes to the abnormal larval development.

Mechanistic interactions: apoptosis, lipid metabolism and virion vesicle formation

Ascovirus genomes are unusual among viruses with regard to a collection of genes that putatively mediate a number of diverse and independent processes, but that culminate in a modified apoptotic response directed towards formation of virion-containing vesicles. A simplified model that connects many ascovirus coded enzymes to the common purpose of virogenesis and virion vesicle maturation is shown in Fig. 1.5. The ascovirus proapoptotic proteins, particularly ascovirus caspase and cathepsin B, could directly elicit apoptosis, or more likely amplify a host-initiated death response, as observed with SfAV1a (Bideshi *et al.*, 2005). Whether the SfAV-Casp requires host factors such as initiator caspases for its activation is unknown. However, relative to indigenous executioner procaspases, a higher level of ascovirus caspases could be produced because, presumably, several gene copies will be present following viral DNA replication. It is known that autocleavage and activation of cellular caspases, and the SfAV1-Casp (Bideshi *et al.*, 2005), occur when overproduced in *E. coli*. Thus, the accumulation and timed-activation of ascovirus caspases could be a host-independent response utilized to amplify apoptosis, with concomitant generation of raw ingredients required for infectious virion-vesicle biogenesis.

Whereas the presence of an active SfAV1a executioner caspase provides one reasonable explanation for the rapid and coordinated degradation of the cell, the activities of ascovirus-encoded lipid metabolizing enzymes in both apoptosis and subsequent or concomitant virion vesicle biogenesis are less presumptive. However, it is also reasonable to speculate that the unusual complement of lipid enzymes encoded by ascoviruses collectively have tripartite activities: (1) lipogenesis required for virogenesis, (2) additive roles in amplifying the apoptotic programme and (3) virion-vesicle morphogenesis and maturation.

As described above, the ascovirus-encoded lipid enzymes could participate in generating precursors and terminal lipid components required for constructing virions and virion-containing vesicles. For example, the patatin-like phospholipase and coesterase/JHE hydrolases could have multiple functions such as *de novo* synthesis of double membranes and the lipase activities could assist in viral invasion and egress from infected cells, as have been observed with the viral p37 lipase, an envelope protein encoded by the vaccinia virus (Baek *et al.*, 1997). Interestingly, a recent study on the lipase-like gene (*ORF019*) in HvAV3e has shown that it is required for virus replication and vesicle formation (Smede *et al.*, 2009). Although lipase activity was not demonstrated for the ORF019 gene product, and it has not been demonstrated that it is directly involved in vesicle biogenesis, the results show that the putative lipid modifying enzyme plays a crucial role in the biology of ascoviruses.

In addition to lipid membrane biogenesis, at least three ascovirus lipid enzymes could play additional roles in apoptosis. These enzymes are the fatty acid elongase, sphingomyelin phosphodiesterase and the patatin-like PLA_2 phospholipase. In particular, cellular orthologues of these enzymes function in the metabolism of sphingolipids (ceramide and sphingosine), phospholipids (lysophospholipids) and arachidonic acid, all of which have diverse cellular roles, including their function as signalling molecules involved in apoptosis.

Ceramide, sphingosine and arachidonic acid

In recent years, it has been well established that ceramide is an endogenous regulator of programmed cell death and cellular stress (Schenk *et al.*, 2007; Taha *et al.*, 2006; Birbes *et al.*, 2001; Kolesnik and Hannun, 1999). The accumulation of ceramide is mediated by at least four mechanisms: (1) *de novo* synthesis by serine-palmitoyl-CoA transferase and ceramide synthase (Chalfant *et al.*, 2002; Perry *et al.*, 2000), (2) hydrolysis of pre-existing membrane sphingomyelin by enzymatic cleavage with sphingomyelinases (Lucci *et al.*, 1999; Strum *et al.*, 1994), (3) inhibition of ceramide hydrolysis, and (4) stimulation of glucosylceramide hydrolysis or inhibition of its synthesis (Senchenkov *et al.*, 2001). Interestingly, the generation and accumulation of ceramide via sphingomyelin hydrolysis has been shown to confer an immediate stress response leading to apoptosis while its *de novo* synthesis seems to elicit a delayed response (Gulbins and Kolesnick, 2003). Although the roles of ceramide are pleiotropic, in mammalian systems ceramide can activate proteases, such as lysosomal cathepsins, which in turn can cleave and activate or regulate the Bcl-2 family of proteins including the proapoptotic proteins, Bid and Bax. Indeed, localized synthesis of ceramide

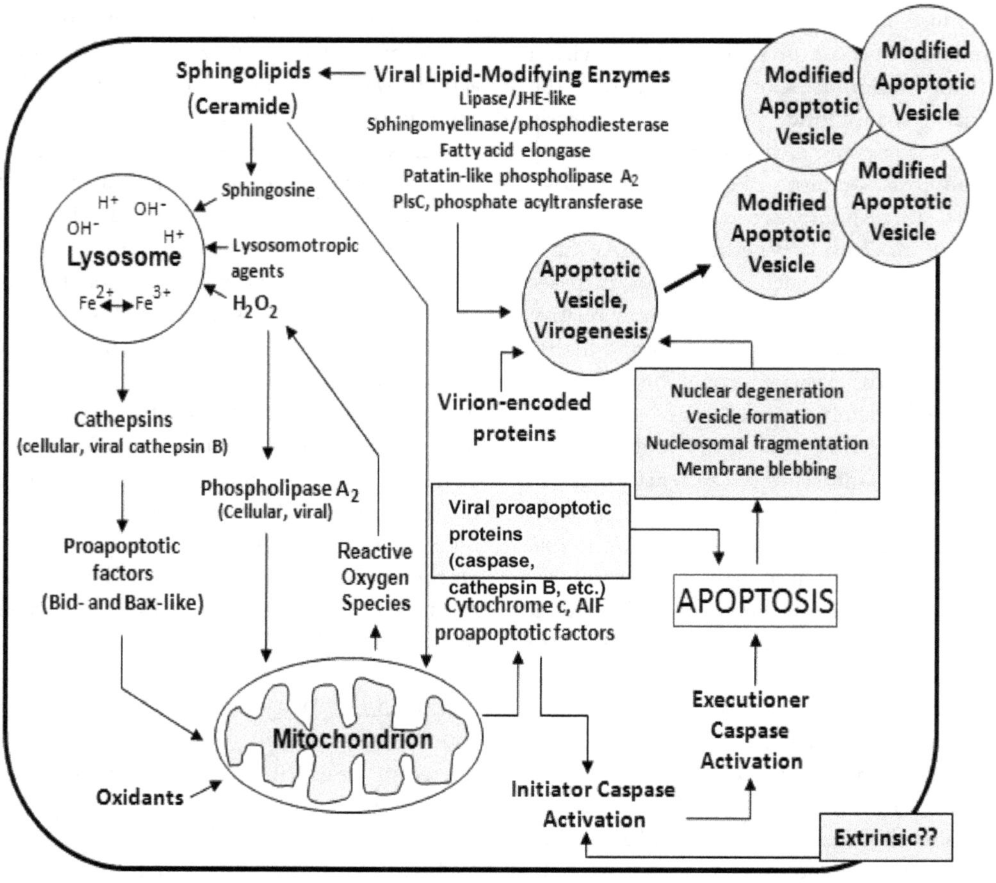

Figure 1.5 Molecular modelling of simplified pathways potentially involved in the production of ascovirus virion-containing vesicles based on complements of proapoptotic and lipid-modifying enzymes encoded by ascoviruses. Whether or not ascoviruses elicit an extrinsic apoptotic cascade upon viral invasion of host cells is unknown. However, it is likely that ascoviruses trigger a rapid intrinsic apoptotic response following cellular invasion. This response, as observed *in vitro*, occurs after 9 hours post infection with *Spodoptera frugiperda ascovirus 1a* and could involve direct participation of a virus-encoded executioner caspase (Bideshi *et al.*, 2005) and virus orthologues of the cathepsin B, D, and L families (Chwieralski *et al.*, 2006). The process of virion vesicle production is complex and apparently involves *de novo* synthesis of lipids presumably catalysed primarily by virus-encoded lipid modifying enzymes. Interestingly, putative byproducts of lipid metabolism could exacerbate a proapoptotic response, thereby escalating efficient production of virion vesicles. For example, sphingolipids, particularly sphingosine and ceramide, are known to permeabilize, respectively, lysosomes and mitochondria. Lysosomal leakage can result in the release of cathepsins which in turn can activate proapoptotic factors that further destabilize mitochondria resulting in the release of cytochrome *c* and activation of an initiator caspase that subsequently activates executioner caspases. Importantly, ceramide-induced destabilized mitochondria elicit the same cascade. Moreover, degradation of mitochondrial membranes results in the release of reactive oxygen species which creates positive feedback loops that intensify the apoptotic response by further destabilizing the integrity of lysosomal and mitochondrial membranes. Furthermore, it is known that even at low concentrations, H_2O_2 can activate phospholipase A_2, a protein with an enzymatic activity that could cause progressive membrane destabilization of organelles, including lysosomes and mitochondria (Zhao *et al.*, 2001). In addition, lysosomes are highly sensitive to oxidative stress due to the presence of oxidized iron that accumulates as a result of the degradation of metalloproteinases. Despite the rapid progression of the apoptotic response, it is clear that ascoviruses amplify the cellular suicide program and rescue apoptotic bodies to complete virogenesis (refer to Fig. 1.3). Whereas the typical apoptotic bodies are degraded by host defences and absorbed, ascovirus-rescued vesicles are manipulated and modified by viral gene products and are subsequently converted to highly-modified infectious virion-containing vesicles that are efficacious reservoirs for horizontal transfer of the virus by parasitoid wasps.

in the mitochodrion induces Bax oligomerization, thereby propagating the apoptotic response (Birbes *et al.*, 2005). This observation correlates with studies showing that ceramide treatment of isolated mitochondrion activates Bax-mediated induction of mitochondrial permeability transition (Pastorino *et al.*, 1999) and can inhibit oxidative phosphorylation by interfering with the electron transport chain at the level of complex III (Gudz *et al.*, 1997). A recent report showed that ceramide can enhance permeability of the inner mitochondrial membrane (Novgorodov *et al.*, 2005), but with regard to ascovirus cytopathology, its ability to increase permeability of the outer mitochondrial membrane is most intriguing as this response leads to the release of proapoptotic factors, particularly cytochrome *c*, that is directly involved in the intrinsic apoptotic response (Siskind *et al.*, 2002; Di Paola *et al.*, 2004; Siskind, 2005).

Sphingosine, produced from the degradation of ceramide by ceramidases, has also been shown to induce proapoptotic effects through the inhibition of pro-survival factors, including protein kinase C (PKC) (McCracken *et al.*, 2003) and Akt/protein kinase B (Chang *et al.*, 2001). In addition, sphingosine can induce the cleavage of Bid and activate downstream executioner caspases (Cuvillier *et al.*, 2001; Hung *et al.*, 1999). Apart from these protein targets, sphingosine is known to alter the integrity of intracellular membrane systems. Specifically, it increases the permeability of lysosomes causing the release of lysosomal cysteine proteases such as cathepsins B, D and L into the cytoplasm. These cathepsins can activate the lysosomal pathway of apoptosis (Kagedal *et al.*, 2001; Chwieralski *et al.*, 2006). In addition, sphingosine destabilizes the mitochondrial membrane resulting in the release of cytochrome *c*, which in turn activates the intrinsic apoptotic pathway (Cuvillier *et al.*, 2001).

In comparison to ceramide and sphingosine, the role of arachidonic acid in the apoptotic response is less clear. Arachidonic acid, as well as lysophospholipids and free fatty acids are generated through the hydrolysis of *sn*-2 fatty acid acyl bond of phospholipids by the PLA_2 family of lipases (Dennis *et al.*, 1994; Leslie, 1997). Whereas lysophosphatidic acid is implicated in diverse functions such as cellular proliferation and differentiation (Goetzl *et al.*, 2000), arachidonic acid is known to mediate the release of cytochrome *c* from the mitochondrial intermembrane space and to induce the activation of caspase to initiate apoptotic cell death (Di *et al.*, 2000; Garrido *et al.*, 2001). Moreover, arachidonic acid has been reported to mediate the biogenesis of ceramide through a sphingomyelinase-dependent pathway (Jayadev *et al.*, 1994). Interestingly, it has been shown recently that apoptosis induced by endoplasmic reticulum stress in insulin-secreting cells is amplified by the overexpression of $iPLA_2$ (Ramanadhan *et al.*, 2004).

Lysosomal cathepsin

Whether the lipid enzymes function as postulated in the cytopathology of ascovirus remains a central focus of current research. However, their putative functions seemingly coincide with cytological observations. As it is known that the synthesis and accumulation of ceramide and arachidonic acid incur stress on the cell, particularly mitochondrial stress, such a response could trigger or enhance the apoptotic programme. In this regard, the generation of reactive oxygen species (ROS) by aberrant mitochondria could have direct proapoptotic effects on stressed organelles, including the endoplasmic reticulum and lysosomes. For example, it is established that lysosomal permeabilization and cathepsin release are initiated by a number of extra- and intra-lysosomal stimuli, most notably by oxidative stress induced by ROS and by sphingolipid derivatives, especially ceramide and sphingosine, produced by sphingomyelinase and ceramidase catalysis (Andrieu-Abadie and Lavade, 2002; Chwieralski *et al.*, 2006). In addition, lysosomal membrane destabilization is induced by hydrogen peroxide which apparently activates PLA_2 that cleaves organelle membrane phospholipids (Zhao *et al.*, 2001). Regardless, the binding of ceramide to cathepsin D activates this enzyme to cleave and activate other lysosomal proteases including cathepsins B, L and D (Heinrich *et al.*, 1999), thereby providing a positive feedback loop in which proapoptotic factors further destabilize mitochondria to enhance the apoptotic response.

Concluding remarks

Ascoviruses are intriguing for experimental study from various perspectives that encompass their structural biology, molecular phylogeny and evolutionary origin. But perhaps most importantly, studies focused on elucidating molecular mechanisms related to the unique cytopathology that results from a novel program in which ascoviruses reengineer apoptotic bodies to manufacture infectious virion-containing vesicles for dissemination could yield insights into novel atypical pathways in cellular biology. Indeed, it is likely that advanced knowledge in this area will provide new models to explain unusual host–pathogen interactions. The maturation of virions in acellular compartments, i.e. virion vesicles, is also a unique phenomenon in the life cycle of ascoviruses as it is well established that viruses are obligate intracellular molecular parasites that rely on intact cellular functions for efficient propagation. Exploration of this phenomenon could provide new insight on how pathogens preserve cellular and acellular environments, including protracted mitochondrial activities, for their propagation. In this regard, ascoviruses are unusual among viruses as their pathobiology suggests that during evolution they have acquired genetic functions that allow for the maintenance of a viable physiologic environment in vesicles to ensure productive virogenesis. Undoubtedly, the specific roles of ascovirus proapoptotic proteins (caspases and cathepsins) and various lipid-modifying enzymes must be established to decipher complex interactions that generate infectious virions and virion-containing vesicles that are perpetually vectored by parasitic wasps among insect larval populations. Finally, our observations (unpublished) that various antioxidants and inhibitors of executioner caspases, sphingomyelinases, and cathepsin B adversely affect ascovirus-induced apoptosis *in vitro* suggest that the corresponding pathways influence various stages of virion vesicle morphogenesis and maturation. Thus, it is anticipated that future studies to decipher these complex interacting pathways will reveal previously undescribed mechanisms of cell biology that result in the novel cytopathobiology characteristic of ascoviruses.

References

Agarwal, A.K., Barnes, R.A., and Garg, A. (2006). Functional characterization of human 1-acylglycerol-3-phosphate acyltransferase isoform 8: Cloning, tissue distribution, gene structure, and enzyme activity. Arch. Biochem. Biophy. *449*, 64–76.

Ahmad, S.I., Kirk, S.H., and Eisenstark, A. (1998). Thymine metabolism and thymineless death in prokaryotes and eurkaryotes. Annu. Rev. Microbiol. *52*, 591–625.

Andrabi, S.M.H. (2007). Mammalian sperm chromatin structure and assessment for DNA fragmentation. J. Assist. Reprod. Gen. *24*, 561–569.

Andrieu-Abadie, N., and Lavade, T. (2002). Sphingomyelin hydrolysis during apoptosis. Biochim. Biophys. Acta *1585*, 126–134.

Aravind, L., and Koonin, E.V. (2001). A natural classification of ribonucleases. Meth. Enzymol. *341*, 3–28.

Ardley, H.C., and Robinson, A.A. (2005). E3 ubiquitin ligase. Essays Biochem. *41*, 15–30.

Asgari, S. (2006). Replication of *Heliothis virescens* ascovirus in insect cell lines. Arch. Virol. *151*, 1689–1699.

Asgari, S. (2007). A caspase-like gene from the *Heliothis virescens* ascovirus (HvAV3e) is not involved in apoptosis but is essential for virus replication. Virus Res. *128*, 99–105.

Asgari, S., Davis, J., Wood, D., Wilson, P., and McGrath, A. (2007). Sequence and organization of the *Heliothis virescens* ascovirus genome. J. Gen. Virol. *88*,1120–1132.

Atsumi, G., Murakami, M., Kojima, K., Hadano, A., Tajima, M., and Kudo, I. (2000). Distinct roles of two intracellular phospholipase A_2s in fatty acid release in the cell death pathway. J. Biol. Chem. *24*, 18248–18258.

Avery, S.W., Jouvenaz, D.P., Banks, W.A., and Anthony, D.W. (1977). Virus-like particles in a fire ant, *Solenopsis* sp., (Hymenoptera: Formicidae) from Brazil. Florida Entomol. *60*, 17–20.

Baek, S.H., Kwak, J.Y., Lee, S.H., Lee, T., Ryu, S.H., Uhlinger, D.J., and Lambeth, J.D. (1997). Lipase activities of p37, the major envelope protein of vaccinia virus. J. Biol. Chem. *272*, 32042–32049.

Barber, G.N. (2001). Host defence, viruses and apoptosis. *Cell Death Differ. 8*, 113–126.

Barratt, B.I.P., Evans, A.A., Stoltz, D.B., Vinson, S.B., and Easingwood, R. (1999). Virus-like particles in the ovaries of *Microctonus aethiopoides* loan (Hymenoptera: Braconidae), a parasitoid of adult weevils (Coleoptera: Curculionidae). J. Invertebr. Pathol. *73*, 182–188.

Bartova, E., Krejci, J., Harnicarova, A., Galiova, G., and Kozubek, S. (2008). Histone modifications and nuclear architecture: A review. J. Histochem. Cytochem. *56*, 711–721.

Baudry, K., Swain, E., Rahier, A., Germann, M., Batta, A., Rondet, S., Mandala, S., Henry, K., Tint, G.S., Edlind, T., Kurtz, M., and Nickels, Jr. T.J. (2001). The effects of the erg26-1 mutation on the regulation of lipid metabolism in *Saccharomyces cerevisiae*. J. Biol. Chem. *16*, 12702–12711.

Best, S.M. (2008). Viral subversion of apoptotic enzymes: Escape from death row. Annu. Rev. Microbiol. *62*, 171–192.

Best, S.M., and Bloom, M.E. (2004). Caspase activation during virus infection: More than just a kiss of death? Virology *320*, 191–194.

Bianchi, M.E., and Agresti, A. (2005). HMG proteins: Dynamic players in gene regulation and differentiation. Curr. Opin. Genet. Dev. *15*, 496–506.

Bideshi, D.K., Demattei, M.V., Rouleux-Bonnin, F., Stasiak, K., Tan, Y., Bigot, S., Bigot, Y., and Federici, B.A. (2006). Genomic sequence of *Spodoptera frugiperda ascovirus* 1a, an enveloped, double-stranded DNA insect virus that manipulates apoptosis for viral reproduction. J. Virol. *80*, 11791–11805.

Bideshi, D.K., Renault, S., Stasiak, K., Federici, B.A., and Bigot, Y. (2003). Phylogenetic analysis and possible function of *bro*-like genes, a multigene family widespread among large double-stranded DNA viruses of invertebrates and bacteria. J. Gen. Virol. *84*, 2531–2544.

Bideshi, D.B., Tan, Y., Bigot, Y., and Federici, B.A. (2005). A viral caspase contributes to modified apoptosis for virus transmission. Genes Dev. *19*, 1416–1421.

Bigot, Y., Rabouille, A., Doury, G., Sizaret, P.-Y., Delbost, F., Hamelim, M.-H., and Periquet, G. (1997a). Particle and genome characterization of a new member of Ascoviridae, *Diadromus pulchellus* ascovirus. J. Gen. Virol. *78*, 1139–1147.

Bigot, Y., Rabouille, A., Doury, G., Sizaret, P-Y., Delbost, F., Hamelim, M.-H., and Periquet, G. (1997b). Biological and molecular features of the relationships between *Diadromus pulchellus* ascovirus, a parasitoid hymenopteran wasp (*Diadromus pulchellus*) and its lepidopteran host, *Acrolepiopsis assectella*. J. Gen. Virol. *78*, 1149–1163.

Bigot, Y., Renault, S., Nicolas, J., Moundras, C., Demattei, M-V., Samain, S., Bideshi, D.K., and Federici, B.A. (2009). Symbiotic virus at the evolutionary intersection of three types of large DNA viruses: Iridoviruses, Ascoviruses, and Ichnoviruses. PLoS ONE *4*, 1–9.

Birbes, H.S., Bawab, S., Hannun, Y., and Obied, L.M. (2001). Selective hydrolysis of a mitochondrial pool of spingomyelin induces apoptosis. FASEB J. *14*, 2669–2679.

Birbes, H., Luberto, C., Hsu, Y.T., El Bawab, S., Hannun, Y.A., and Obeid, L.M. (2005). A mitochondrial pool of sphingomyelin is involved in TNFalpha-induced Bax translocation to mitochondria. Biochem. J. *386*, 445–451.

Birnbaum, M.J., Clem, R., and Miller, L.K. (1994). An apoptosis-inhibiting gene from a nuclear polyhedrosis virus encoding a peptide with cys/his sequence motifs. J. Virol. *68*, 2521–2528.

Bishop, W.R. (1988). Assembly of phospholipids into cellular membranes: biosynthesis, transmembrane movement and intracellular translocation. Annu. Rev. Cell. Biol. *4*, 579–610.

Boyer, L.A., Logie, C., Bonte, E., Becker, P.B., Wade, P.A., Wolffe, A.P., Wu, C., Imbalzano, A.N., and Peterson, C.L. (2000). Functional delineation of three groups of the ATP-dependent family of chromatin remodeling enzymes. J. Biol. Chem. *275*, 18864–18870.

Braunagel, S.C., Russell, W.K., Rosas-Acosta, G., Russell, D.H., and Summers, M.D. (2003). Determination of the protein composition of the occlusion-derived virus of *Autographa californica* nucleopolyhedrovirus. Proc. Natl. Acad. Sci. U.S.A. *100*, 9797–9802.

Brewer, L.R., Corzett, M., and Balhorn, R. (1999). Protamine-induced condensation and decondensation of small DNA molecules. Science *286*, 120–123.

Browning, H.W., Federici, B.A., and Oatman, E.R. (1982). Occurrence of a disease caused by a rickettsia-like organismin a larval population of the cabbage looper, *Trichoplusia ni*, in southern California. Environ. Entomol. *11*, 550–554.

Buller, R.M., Arif, B.M., Black, D.M., Dumbell, K.R., Espisito, J.J., Lefkowitz, E.J., McFadden, G., Moss, B., Mercer, A.A., Moyer, R.W., Skinner, M.A., and Tripathy, D.N. (2005). *Family Poxviridae*. In: Virus Taxonomy, Eighth Report of the International Committee on Taxonomy of Viruses. Eds. C.M. Fauquet, M.A. Mayo, J. Maniloff, U. Desselberger, L.A. Ball. (Elsevier, Academic Press, San Diego, CA). pp.117–143.

Canaan, S., Zadori, Z., Ghomashchi, F., Bollinger, J., Sadilek, M., Moreau, M.E., Tijssen, P., and Gelb, M.H. (2004). Interfacial enzymology of parvovirus phospholipase A_2. J. Biol. Chem. *279*, 14502–14508.

Carner, G.R., and Hudson, J.S. (1983). Histopathology of virul-like particles in *Heliothis* spp. J. Invertebr. Pathol. *41*, 238–249.

Chalfant, C.E., Rathman, K., Pinkerman, R.L., Wood, R.E., Obeid, L.M., Ogretmen, B., and Hannun, Y.A. (2002). De novo ceramide regulates the alternative splicing of caspase-9 and Bcl-x in A549 lung adenocarcinoma cells. Dependence on protein phosphatase-1. J. Biol. Chem. *277*, 12587–12595.

Champoux, J.J. (1990). DNA topology and its biological effects. Eds. Cozzarelli N.R., and Wang, J.C (Cold Spring Harbor Laboratory, Cold Spring Harbor, New York), pp. 217–242

Chang, H.C., Tsai, L.H., Chuang, L.Y., and Hung, W.C. (2001). Role of AKT kinase in sphinosine-induced apoptosis in human hepatoma cells. J. Cell. Physiol. *188*, 188–193.

Cheng, X.W., Carner, G.R., and Arif, B.M. (2000). A new ascovirus from *Spodoptera exigua* and its relatedness to the isolate from *Spodoptera frugiperda*. J. Gen. Virol. *81*, 3083–3092.

Cheng, X.Y., Wang, L., Carner, G.R., and Arif, B.M. (2005). Characterization of three ascovirus isolates from cotton insects. J. Invertebr. Pathol. *89*, 193–202.

Chestier, A., and Yaniv, M. (1979). Rapid turnover of acetyl groups in the four core histones of simian virus 40 minichromosomes. Proc. Natl. Acad. Sci. U.S.A. *76*, 46–50.

Chinchar, V.G., Essbauer, S., He, J.G., Hyatt, A., Miyazaki, T., Seligy, V., and Williams, T. (2005). *Family Iridoviridae*. In: Virus Taxonomy, Eighth Report of the International Committee on Taxonomy of Viruses. Eds. C.M. Fauquet, M.A. Mayo, J. Maniloff, U. Desselberger, L.A. Ball (Elsevier, Academic Press, San Diego, CA), pp 145–162.

Chu, Y., Huang, T., and Hsu, M-T. (1990). P1 nuclease defines a subpopulation of active SV40 chromatin – a new nuclease hypersensitivity assay. Nucleic Acids Res. *18*, 3705–3711.

Chwieralski, C.E., Welte, T., and Buhling, F. (2006). Cathepsin-regulated apoptosis. Apoptosis *11*, 143–149.

Clarke, T.E., and Clem, R.J. (2003). *In vivo* induction of apoptosis correlating with reduced infectivity during baculovirus infection. *J. Virol. 77*, 2227–2232.

Clem, R.J. (2001). Baculovirus and apoptosis: the good the bad, and the ugly. *Cell Death Differ. 8*, 137–143.

Clem, R.J. (2007). Baculoviruses and apoptosis: a diversity of genes and responses. Curr. Drug Targets *8*, 1069–1074.

Clem, R.J., Fechheimer, M., and Miller, L.K. (1991). Prevention of apoptosis by a baculovirus gene during infection of insect cells. *Science 254*, 1388–1390.

Coca-Prados, M., Vidali, G., and Hsu, M.T. (1980). Intracellular forms of simian virus 40 nucleoprotein complexes. III. Study of histone modifications. J. Virol. *36*, 353–360.

Cohen, G.M. (1997). Caspases: the executioners of apoptosis. Biochem. J. *326*, 1–16.

Conus, S., and Simon, H.U. (2008). Cathepsins: key modulators of cell death and inflammatory responses. Biochem. Pharmacol. *76*, 1374–1382.

Cromie, G.A., Connelly, J.C., and Leach, D.R. (2001). Recombination at double-strand breaks and DNA ends: conserved mechanisms from phage to humans. Mol. Cell. *8*, 1163–1174.

Crooke, E., Clem, R.J., and Miller, L.K. (1993). An apoptosis-inhibiting baculovirus gene with a zinc finger-like motif. *J. Virol. 67*, 2168–2174.

Cui, L., Cheng, X., Li, L, and Li, J. (2007). Identification of *Trichoplusia ni* ascovirus 2c virion structural proteins. J. Gen. Virol. *81*, 3083–3092.

Cullinane, M., Baysse, C., Morrissey, J.P., and O'Gara, F. (2005). Identification of two lysophosphatidic acid acyltransferase genes with overlapping functions in *Pseudomonas fluorescens*. Microbiology *151*, 3071–3080.

Cuvillier, O., Nava, V.E., Murthy, S.K., Edsall, L.C., Lavade, T., Milstien, S., and Spiegel, S. (2001). Sphingosine generation, cytochrome c release, and activation of caspase-7 in doxorubicin-induced apoptosis of MCF7 breast adenocarcinoma cells. Cell Death Differ. *8*, 162–172.

da Fonseca, F.G., Weisberg, A.S., Caeiro, M.F., and Moss, B. (2004). Vaccinia virus murants with alanine substitutions in the conserved G5R gene fail to initiate morphogenesis at the nonpermissive temperatiue. J. Virol. *78*, 10238–10248.

Deng, F., Wang, R., Fang, M., Jiang, Y., Xu, X., Wang, H., Chen, X., Arif, B.M., Guo, L., Wang, H., and Hu, Z. (2007). Proteomics analysis of *Helicoverpa armigera* single nucleocapsid nucleopolyhedrovirus identified two new occlusion derived virus-associated proteins, HA44 and HA100. J. Virol. *81*, 9377–9385.

Dennis, E.A. (1994). Diversity of group types, regulation, and function of phospholipase A2. J. Biol. Chem. *269*, 13057–13060.

Di, P.M., Cocco, T., and Lorusso, M. (2000). Arachidonic acid causes cytochrome c release from heart mitochondria. Biochem. Biophys. Res. Commun. *277*, 128–133.

Di Paola, M., Zaccagnino, P., Montedoro, G., Cocco, T., and Lorusso, M. (2004). Ceramide induces release of proapoptotic proteins from mitochondria by either a Ca^{2+}-dependent or a Ca^{2+}-independent mechanism. J. Bioenerg. Biomembr. *36*, 165–170.

Drider, D., and Condon, C. (2004). The continuing story of endoribonuclease III. J. Mol. Microbiol. Biotechnol. *8*, 195–200.

Dubrez-Daloz, L., Dupoux, A., and Cartier, J. (2008). IAPs: more than just inhibitors of apoptosis proteins. Cell Cycle 7, 1036–1046.

Duprez, B.A., Wirawan, E., Berghe, T.V., and Vandenabeele, P. (2009). Major cell death pathways at a glance. Microbes Infect. *11*, 1050–1062.

Eaton, H.E., Metcalf, J., Penny, E., Tcherepanov, V., and Upton, C. (2007). Comparative genomic analysis of the family *Iridoviridae*: re-annotating and defining the core set of iridovirus genes. Virol. J. *4*, 11.

Efstathiou, S., Kemp, D., Darby, G., and Minson, A.C. (1989). The role of herpes simplex virus type 1 thymidine kinase in pathogenesis. J. Gen. Virol. *70*, 869–879.

Evans, E., Klemperer, N., Ghosh, R., and Traktman, P. (1995). The vaccinia virus D5 protein, which is required for DNA replication, is a nucleic acid-independent nucleoside triphosphate. J. Virol. *69*, 5353–5361.

Federici, B.A. (1982). A new type of insect pathogen in larvae of the clover cutworm, *Scotogramma trifolii*. J. Invertebr. Pathol. *40*, 41–54.

Federici, B.A. (1983). Enveloped double-stranded DNA insect virus with novel structure and cytopathology. Proc. Natl. Acad. Sci. U.S.A. *80*, 7664–7668.

Federici, B.A., Bideshi, D.K., Tan, Y., Spears, T., and Bigot, Y. (2009). Ascoviruses: superb manipulators of apoptosis for viral replication and transmission. Curr. Top. Microbiol. Immunol. *328*, 171–196.

Federici, B.A., Bigot, Y., Granados, R.R., Hamm, J.J., Miller, L.K., Newton, I., Stasiak, K., and Vlak, J.M. (2005). Family *Ascoviridae*. In: Virus Taxonomy, Eighth Report of the International Committee on Taxonomy of Viruses. Eds, C.M. Fauquet, M.A. Mayo, J. Maniloff, U. Desselberger, L.A. Ball. (Elsevier, Academic Press, San Diego, CA), pp 269–274.

Federici, B.A., and Govindarajan, R. (1990). Comparative histopathology of three ascovirus isolates in larval noctuids. J. Invertebr. Pathol. *56*, 300–311.

Federici, B.A., Hamm, J.J., and Styer, E.L. (1991). Ascoviruses. In: An Atlas of Invertebrate Viruses. Eds. Adams J., and Bonami, G. (CRC Press, Boca Raton, Florida), pp 339–349.

Federici, B.A., Vlak, J.M., and Hamm, J.J. (1990). Comparison of virion structure, protein composition, and genomic DNA of three ascovirus isolates. J. Gen. Virol. 171, 1661–1668.

Field, H.J., and Wildy, P. (1978). The pathogenecity of thymidine kinase-deficient mutants of herpes simplex virus in mice. J. Hyg. (London) *81*, 267–277.

Fuller, D.N., Rickgauer, J.P., Hardine, P.J., Grimes, S., Anderson, D.L., and Smith, D.E. (2007). Ion effects on viral DNA packaging and portal motor function in bacteriophage φ29. Proc. Natl. Acad. Sci. U.S.A. *104*, 11245–11250.

Furlong, M.J., and Asgari, S. (2010). Effects of an ascovirus (HvAV-3e) on diamondback moth, *Plutella xylostella*, and evidence for virus transmission by a larval parasitoid. J. Invertebr. Pathol. *103*, 89–95.

Galibert, F., Mandart, E., Fitoussi, F., Tiollais, P., and Charnay, P. (1979). Nucleotide sequence of the hepatitis B virus genome (subtype ayw) cloned in *E. coli*. Nature *281*, 646–650.

Gallina, A., Bonelli, F., Zentilin, L., Rindi, G., Muttini, M., and Milanesi, M. (1989). A recombinant hepatitis B core antigen polypeptide with the protamine-like domain deleted self-assembly into capsid particles but fails to bind nucleic acids. J. Virol. *63*, 4645–4652.

Ganser-Pornillos, B.K., Yeager, M., and Sundquist, W.I. (2008). The structural biology of HIV assembly. Curr. Opin. Struct. Biol. *18*, 203–217.

Garrido, E., Matson, M.P., Hennig, B., and Toborek, M.J. (2001). Nicotine protects against arachidonic-acid-induced caspase activation, cytochrome c release and apoptosis of spinal cord neurons. Neurochemistry *76*, 1395–1403.

Ghavami, S., Hashemi, M., Ande, S.R., Yeganeh, B., Xiao, W., Eshraghi, M., Bus, C.J., Kadkhoda, K., Wiechec, E., Halayko, A.J., and Los, M. (2009). Apoptosis and cancer: mutations within caspase genes. J. Med. Genet. *46*, 497–510.

Gilbert, L., Granger, L.N.A., and Roe, R.M. (2000). The juvenile hormones, historical facts and speculations on future research directions. Insect Biochem. Mol. Biol. *20*, 617–644.

Gite, S.U., and Shanker, V. (1995). Single-strand-specific nucleases. Crit. Rev. Microbiol. *21*, 101–122.

Goetzl, E.J., Lee, H., Dolezalova, H., Kalli, K.R., Conover, C.A., Hu, Y.L., Azuma, T., Srossel, T.P., Karlina, J.S., and Jaffe, R.B. (2000). Mechanisms of lysolipid phosphate effects on cellular survival and proliferation. Ann. N.Y. Acad. Sci. *905*, 177–187.

Gorbalenya, A.E., Koonin, E.V., Donchenko, A.P., and Blinov, V.M. (1989). Two related superfamilies of putative helicases involved in replication, recombination, repair and expression of DNA and RNA genomes. Nucleic Acids Res.*17*, 4713–30.

Govindarajan, R., and Federici, B.A. (1990). Ascovirus infectivity and the effects of infection on the growth and development of noctuid larvae. J. Invertebr. Pathol. *56*, 291–299.

Green, M.C., Monser, K.P., and Clem, R.J. (2004). Ubiquitin protein ligase activity of the anti-apoptotic baculovirus protein Op-IAP3. Virus Res. *105*, 89–96.

Gudz, T.I., Tserng, K.Y., and Hoppel, C.L. (1997). Direct inhibition of mitochondrial respiratory chain complex III by cell-permeable ceramide. J. Biol. Chem. *24*, 24154–24158.

Gulbins, E., and Kolesnick, R. (2003). Raft ceramide in molecular medicine. Oncogene *22*, 7070–7077.

Gvakharia, B.O., Koonin, E.K., and Mathews, C.K. (1996). Vaccinia virus G4l gene encodes a glutaredoxin. Virology *226*, 408–411.

Hamm, J.J., Styer, E.L., and Federici, B.A. (1998). Comparison of field-collected ascovirus isolates by DNA hybridization, host range, and histopathology. J. Invertebr. Pathol. *72*, 138–146.

Hannun, Y.A. (1994). The sphingomyelin cycle and the second messenger function of ceramide. J. Biol. Chem. *269*, 3125–3128.

Hawtin, R.F., Zarkowska, T., Arnold, K., Thomas, C.J., Gooday, G.W., King, L.A., Kuzio, J.A., and Possee, R.D. (1997). Liquefaction of *Autographa californica* nucleopolyhedrovirus-infected insects is dependent on the integrity of virus-encoded chitinase and cathepsin genes. Virology *24*, 243–253.

He, F., Fenner, B.J., Godwin, A.K., and Kwang, J. (2006). White spot syndrome virus open reading frame 222 encodes a viral E3 ubiquitin ligase and mediates degradation of host tumor suppressor via ubiquitination. J. Virol. *80*, 3884–3892.

Heinrich, M., Wickel, M., Schnieder-Brachert, W. *et al.* (1999). Cathepsin D targeted by acid sphingomyelinase-derived ceramide. EMBO J. *18*, 5252–5263.

Hess, R.T., Benham, G.S., and Poinar, G.O. (1975). The ultrastructure of microorganisms in the tissues of *Casenaria infesta* (Hymenoptera: Ichneumonidae). J. Invertebr. Pathol. *26*, 181–191.

Hirabayashi, T., Murayama, T., and Shimizu, T. (2004). Regulatory mechanism and physiological role of cytosolic phospholipase A_2. Biol. Phar. Bull. *27*, 1168–1173.

Hirschberg, H.J.H.B., Simons, J.W.F.A., Dekker, N., and Egmond, M.R. (2001). Cloning, expression, purification and characterization of patatin, a novel phospholipase A. Eur. J. Biochem. *268*, 5037–5044.

Huang, E.-S., and Johnson, R.A. (2000). Human cytomegalovirus-no longer just a DNA virus. Nat. Med. *6*, 863–864.

Hudson, J.S., and Carner, G.R. (1981). Histopathology of an unidentified virus of *Heliothis zea* and *Heliothis virescens*. Proc. S. E. Electron Microsc. Soc. *4*, 27.

Hung, W.C., Chang, H.C., and Chuang, L.Y. (1999). Activation of caspase-3-like proteases in apoptosis induced by sphingosine and other long-chain bases in Hep3B hepatoma cells. *338*, 161–166.

Hussain, M., Abraham, A.M., and Asgari, S. (2010). An ascovirus encoded ribonuclease III autoregulates its expression and suppresses RNAi-mediated gene silencing. J. Virol. doi:10.1128/JVI.02362-09

Hussain, M., and Asgari, S. (2008). Inhibition of apoptosis by *Heliothis virescens* ascovirus (HvAV-3e): characterization of *orf28* with structural similarity to inhibitor of apoptosis proteins. *Apoptosis 13*, 1417–1426.

Hussain, M., Taft, R.J., and Asgari, S. (2008). An insect virus-encoded microRNA regulates viral replication. J. Virol. *82*, 9164–9170.

Imajoh, M., Sugiura, H., and Oshima, S. (2004). Morphological changes contribute to apoptotic cell death and are affected by caspase-3 and caspase-6 inhibitors during red sea bream iridovirus permissive replication. *Virology 322*, 220–230.

Iyer, L.M., Balaji, S., Koonin, E.V., and Aravind, L. (2006). Evolutionary genomics of nucleo-cytoplamsmic viruses. Virus Res. *111*, 156–184.

Iyer, L.M., Koonin, E.V., Leipe, D.D., and Aravind, L. (2005). Origin and evolution of the archaeo-eukaryotic primase superfamily and related palm-domain proteins: structural insights and new members. Nucleic Acids Res. *33*, 3875–3896.

Jakob, N.J., Muller, K., Bahr, U., and Darai, G. (2001). Analysis of the first complete DNA sequence of an invertebrate iridovirus: coding strategy of the genome of chilo iridescent virus. Virology *286*, 182–196.

Jakobsson, A., Westerberg, R., and Jacobsson, A. (2006). Fatty acid elongases in mammals: their regulation and roles in metabolism. Prog. Lipid Res. *45*, 237–249.

Jayadev, S., Hayter, H.L., Andrieu, N., Gamard, C.J., Liu, B., Balu, R., Hayakawa, M., Ito, F., and Hannun, Y.A. (1997). Phospholipase A_2 is necessary for tumor necrosis factor α-induced ceramide generation in L929 cells. J. Biol. Chem. *272*, 17196–17203.

Jayadev, S., Linardic, C.M., and Hannun, Y.A. (1994). Identification of arachidonic acid as a mediator of sphingomyelin hydrolysis in response to tumor necrosis factor alpha. J. Biol. Chem. *269*, 5757–5763.

Kagedal, K., Zhou, M., Svensson, I., and Brunk, U.T. (2001). Sphingosine-induced apoptosis is dependent on lysosomal proteases. Biochem. J. *359*, 335–343.

Kamita, S.G., Hinton, A.C., Wheelock, C.E., Wogulis, M.D., Wilson, D.K., Wolf, N.M., Stok, J.E., Hock, D., and Hammock, B.D. (2003). Juvenile hormone (JH) esterase: why are you so JH specific? Insect Biochem. Mol. Biol. *22*, 1261–1273.

Kang, W., Kurihara, M., and Mstsumoto, S. (2006). The BRO proteins of *Bombyx mori* nucleopolyhedrovirus are nucleopcytoplasmic shuttling proteins that utilize the CRM1-mediated export pathway. Virology *350*, 184–191.

Koltover, I., Wagner, K., and Safinya, C.R. (2000). DNA condensation in two dimension. Proc. Natl. Acad. Sci. U.S.A. *97*, 14046–14051.

Kolesnick, R., and Hannun, Y.A. (1999). Ceramide and apoptosis. Trends Biochem. Sci. *24*, 224–227.

Kreuze, J.F., Savenkov, E.I., Cuellar, W., Li, X., and Valkonen, J.P. (2005). Viral class 1 RNase III involved in suppression of RNA silencing. J. Virol. *79*, 7227–7238.

La Bella, F., and Vesco, C. (1980). Late modifications of simian virus 40 chromatin during the lytic cycle occur in an immature form of virion. J. Virol. 33, 1138–1150.

Lamkanfi, M., Declercq, W., Kalai, M., Saelens, X., and Vandenabeele, P. (2002). Alice in caspase land: A phylogenetic analysis of caspases from worm to man. *Cell Death Differ. 9*, 358–361.

Lanzer, W., and Holowszak, J.A. (1975). Polyamines of vaccinia virions and polypeptides released from viral cores by acid extraction. J. Virol. *16*, 1254–1264.

Leipe, D.D., Aravind, L., and Koonin, E.V. (1999). Did DNA replication evolve twice independently? Nucleic Acids Res. *27*, 3389–3401.

Leslie, C.C. (1997). Properties and regulation of cytosolic phospholipase A_2. J. Biol. Chem. *272*, 16709–16712.

Lewis, T., Zsak, K., Burrage, T.G., Lu, Z., Kutish, G.F., Neilan, J.G., and Rock, D.L. (2000). Maturation of viral growth in macrophages and viral virulence in swine. J. Virol. *74*, 1275–1285.

Liang, Y., and Roizman, B. (2006). State and role of SRC family kinases in replication of herpes simplex virus 1. J. Virol. *80*, 3349–3359.

Lieberman, P.M. (2008). Chromatin organization and virus gene expression. J. Cell. Physiol. *216*, 295–302.

Lin, H.W., Chang, Y.Y., Wong, M.L., Lin, J.W., and Chang, T.J. (2004). Functional analysis of virion host shutoff protein of pseudorabies virus. Virology *324*, 412–418.

Liston, P., Fong, W.G., and Korneluk, R.G. (2003). The inhibitors of apoptosis: there is more to life than Bcl2. *Oncogene 22*, 8568–8580.

Lu, L., Zhou, S.Y., Chen, C., Weng, S.P., Chan, S.M., and He, J.G. (2005). Complete genome sequence of an iridovirus isolated from the orange-spotted grouper, *Epinephelus coiodides*. Virology *339*, 81–100.

Lucci, A., Han, T.Y., Liu, Y.Y., Giuliano, A.E., and Cabot, M.C. (1999). Modification of ceramide metabolism increases cancer cell sensitivity to cytotoxics. Int. J. Oncol. *15*, 541–546.

Macrae, I.J., Li, F., Zhou, K., Cande, W.Z., and Doudna, J.A. (2006). Structure of Dicer and mechanistic implications for RNAi. Cold Spring Harb. Symp. Quant. Biol. *71*, 73–80.

Maxwell, K.L., and Frappier, L. (2007). Viral proteomics. Microbiol. Mol. Biol. Rev. *71*, 398–411.

McCracken, M.A., Miraglia, L.J., McKay, R.A., and Strobl, J.S. (2003). Protein kinase C δ is a prosurvival factor in human breast tumour cell lines. Mol. Cancer Ther. *2*, 273–281.

Meinhart, A., Kamenski, T., Hoeppner, S., Baumli, S., and Cramer, P. (2005). A structural perspective of CTD function. Genes Dev. *19*, 1401–1415.

Meister, R.J., Oldenhof, H., Bowman, J.L., and Gasser, C.S. (2005). Multiple protein regions contribute to differential activities of YABBY proteins in reproductive development. Plant Physiol. *137*, 651–662.

Mikesh, L., Ueberheide, M.B., Chi, A., Coon, J.H., Syka, J.E.P., Shabanowitz, J., and Hunt, D.F. (2006). The utility of ETD mass spectrometry in proteomic analysis. Biochem. Biophys. Acta *1764*, 1811–1822.

Milavetz, B. (2004). Hyperacetylation and differential deacetylation of histones H4 and H3 define two distinct classes of acetylated SV40 chromosomes early in infection. Virology *319*, 324–336.

Mueller, D.R., Voshol, H., Waldt, A., Wiedmann, B., and Van Oostrum, J. (2007). LC-MALDI MS and MS/MS – an efficient tool in proteome analysis. Subcell. Biochem. *43*, 355–380.

Munter, S., Way, M., and Frischknecht, F. (2006). Signaling during pathogen infection. Sci. STKE. *16*, 335:re5

Muro, I., Means, J.C., and Clem, R. (2005). Cleavage of the apoptosis inhibitor DIAP1 by the apical caspase DRONC in both normal and apoptotic *Drosophila* cells. J. Biol. Chem. *280*, 18683–18688.

Nandi, D., Tahiliani, P., Kumar, A., and Chandu, D. (2006). The ubiquitin-proteasome system. J. Biosci. *31*, 137–155.

Newton, I.R. (2003). The biology and characterisation of the ascoviruses (Ascoviridae: Ascovirus) of *Helicoverpa armigera* Hubner and *Helicoverpa punctigera* Wallengren (Lepidoptera: Noctuidae) in Australia. Ph.D. Dissertation, University of Queensland, Queensland, Australia.

Nicholson, A.W. (2003). The ribonuclease III superfamily: forms and function in RNA maturation, decay, and gene silencing. In: Hannon, G. (ed.), RNAi: A guide to gene silencing. Ed. G. Hannon, (Cold Spring Harbor Laboratory Press, Cold Spring Harbor, NY) pp. 149–174.

Novgorodov, S.A., Szulc, Z.M., Luberto, C., Jones, J.A., Bielawski, J., Bielawska, A., Hannun, Y.A., and Obeid, L.M. (2005). Positively charged ceramide is a potent inducer of mitochondrial permeabilization. J. Biol. Chem. *280*, 16096–16105.

O'Brien, V. (1998). Viruses and apoptosis. *J. Gen. Virol.* *79*, 1833–1845.

Okano, K., Vanarsdall, A.L., Mikhailov, V.S., and Rohrmann, G.F. (2006). Conserved molecular systems of the *Baculoviridae*. Virology *344*, 77–87.

O'Reilly, D.R., Crawford, A.M., and Miller, L.K. (1989). Viral proliferating cell nuclear antigen. Nature 337, 606.

Pastorino, J.G., Tafani, M., Rothman, R.J., Marcinkeviciute, A., Hoek, J.B., and Farber, J.L. (1999). Functional consequences of the sustained or transient activation by Bax of the mitochondrial permeability transition pore. J. Biol. Chem. *274*, 31734–31739.

Perry, D.K., Carton, J., Shah, A.K., Meredith, F., Uhlinger, D.J, and Hannun, Y.A. (2000). Serine palmitolytransferase regulates de novo ceramide generation during etoposide-induced apoptosis. J. Biol. Chem. *274*, 9078–9084.

Ramanadhan, S., Hus, F.F., Zhang, S., Jin, C., Bohrer, A., Song, H., Bao, S., Ma, Z., and Turk, J. (2004). Apoptosis of insulin-secreting cells induced by endoplasmic reticulum stress is amplified by overexpression of Group VIA calcium-independent phospholipase A_2 (iPLA$_2$B) and suppressed by inhibition of iPLA$_2$B. Biochemistry *43*, 918–930.

Resch, W., Hixson, K.K., Moore, R.J., Lipton, M.S., and Moss, B. (2007). Protein composition of the vaccinia virus mature virion. Virology *358*, 233–247.

Roulston, A., Marcellus, R.C., and Branton, P.E. (1999). Viruses and apoptosis. *Annu. Rev. Microbiol.* *53*, 577–628.

Rydel, T.J., Williams, J.M., Krieger, E., Moshiri, F., Stallings, W.C., Brown, S.M., Pershing, J.C., Purcell, J.P., and Alibhai, M.F. (2003). The crystal structure, mutagenesis, and activity studies reveal that patatin is a lipid acyl hydrolase with a ser-asp catalytic dyad. Biochemistry *42*, 6696–6708.

Satoh, T., and Hosokawa, M. (1998). The mammalian carboxylesterases: from molecules to functions. Annu. Rev. Pharmacol. Toxicol. *38*, 257–288.

Segura, M., Garnier, M.A., Di Falco, M.R., Whisell, G., Meneses-Acosta, A., Archand, N., and Kamen, A. (2008). Identification of host proteins associated with retroviral vector particles by proteonomic analysis of highly purified vector preparations. J. Virol. *82*, 1107–1117.

Schenk, M., Carpinteiro, A., Grassme, H., Lang, F., and Gulbins, E. (2007). Ceramide: physiologic and pathophysiological aspects. Arch. Biochem. Biophys. *462*, 171–175.

Senchenkov, A., Litvak, D.A., and Cabot, M.C. (2001). Targeting ceramide metabolism – a strategy for overcoming drug resistance J. Natl. Cancer Inst. *93*, 347–357.

Senkevich, T.G., Koonin, E.V., Bugert, J.J., Darai, G., and Moss, B. (1997). The genome of *Molluscum contagiosum* virus: analysis and comparison with other poxviruses. Virology *233*, 19–42.

Senkevich, T.G., White, C.L., Koonin, E.V., and Moss, B. (2002a). Complete pathway for protein disulfide bond formation encoded by poxvirus. Proc. Natl. Acad. Sci. U.S.A. *99*, 6667–6672.

Senkevich, T.G., White, C.L., Weisberg, A., Granek, J.A., Wolffe, E.J., Koonin, E.V., and Moss, B. (2002b). Expression of the vaccinia virus A2.5L redox protein is required for virion morphogenesis. Virology *300*, 296–303.

Shinzawa, K., and Tsujimoto, Y. (2003). PLA$_2$ activity is required for nuclear shrinkage in caspase-independent cell death. J. Cell. Biol. *163*, 1219–1230.

Shuman, S. (1991). Site-specific interaction of vaccinia virus topoisomerase I with duplex DNA. J. Biol. Chem. *17*, 11372–11379.

Siskind, L.J. (2005). Mitochondrial ceramide and the induction of apoptosis. J. Bioenerg. Biomembr. *37*, 143–153.

Siskind, L.J., Kolesnick, R.N., and Colombini, M (2002). Ceramide channels increase the permeability of the mitochondrial outer membrane to small proteins. J. Biol. Chem. *277*, 26796–267803.

Smede, M., Furlong, M.J., and Asgari, S. (2008). Effects of *Heliothis virescens* ascovirus (HvAV-3e) on a novel host, *Crocidolomia pavonana* (Lepidoptera: Crambidae). J. Invertebr. Pathol. *99*, 281–285.

Smede, M., Hussain, A., and Asgari, S. (2009). The lipase-like gene from *Heliothis virescens* ascovirus (HvAV-3e) is essential for replication and cell cleavage. Virus Genes 39, 409–417.

Smiley, J.R. (2004). Herpes simplex virus virion host shutoff protein: immune evasion mediated by viral RNase? J. Virol. *78*, 1063–1068.

Smith, G.R. (2001). Homologous recombination near and far from DNA breaks: alternative roles and contrasting views. Annu. Rev. Genet. 25, 243–274

Slack, J.M., Kuzio, and J., and Faulkner, P. (1995). Chatacterization of *v-cath*, a cathepsin L-like proteinase expressed by the baculovirus *Autographa californica* nuclear polyhedrosis virus. J. Gen. Virol. *76*, 1091–1098.

Sriskanda, V., and Shuman, S. (1998). Specificity and fidelity of strand joining by *Chlorella* virus DNA ligase. Nucleic Acids Res. *26*, 3536–3541.

Starita, L.M., and Parvin, J.D. (2006). Substrates of the BRCA1-dependent ubiquitin ligase. Cancer Biol. Ther. *5*, 137–41.

Stasiak, K., Demattei, M.V., Federici, B.A., and Bigot, Y., (2000). Phylogenetic position of the *Diadromus pulchellus* ascovirus DNA polymerase among viruses with large double-stranded DNA genomes. J. Gen. Virol. *81*, 3059–3072.

Stasiak, K., Renault, S., Demattei, M.V., Bigot, Y., and Federici, B.A. (2003). Evidence for the evolution of ascoviruses from iridoviruses. J. Gen. Virol. *84*, 2999–3009.

Stasiak, K., Renault, S., Federici, B.A., and Bigot, Y. (2005). Characteristics of pathogenic and mutualistic relationships of ascoviruses in field populations of parasitoid wasps. J. Insect Physiol. *51*, 103–115.

Stoltz, D.B., and Faulkner, G. (1978). Apparent replication of an unusual virus-like particle in both parasitoid wasp and its host. Can. J. Microbiol. *24*, 1509–1514.

Stoltz, D.B., Krell, P., Cook, D., MacKinnon, Z.A., and Lucarotti, C.J. (1988). An unusual virus from the parasitic wasp *Cotesia melanoscela*. Virology *162*, 311–320.

Strickland J.A., Orr, G.L., and Walsh, L.A. (1995). Inhibition of *Diabrotica* larval growth by patatin, the lipid acylhydrolase from potato tubers. Plant Physiol. *109*, 667–674.

Stros, M, Launholt, D., and Grasser, K.D. (2007). The HMG-box: a versatile protein domain occurring in a wide variety of DNA-binding proteins. Cell. Mol. Life Sci. *63*, 2590–2606.

Strum, J.C., Small, G.W., Pauig, S.B., and Daniel, L.W. (1994). 1-beta-D-arabinofuranosylcytosine stimulates ceramide and diglyceride formation in HL-60 cells. J. Biol. Chem. *269*, 15493–15497.

Suikkanen, S., Antila, M., Jaatinen, A., Vihinen-Ranta, A., and Vuento, M. (2003). Release of canine provirus from endocytic vesicles. Virology *316*, 267–280.

Taha, T.A., Mullen, T.D., and Obeid, L.M. (2006). A house divided: ceramide, sphingosine, and sphingosine-1-phosphate in programmed cell death. Biochem. Biophys. Acta *1758*, 2027–2036.

Tan, Y., Bideshi, D.K., Johnson, J.J., Bigot, Y., and Federici, B.A. (2009a). Proteomic analysis of the *Spodoptera frugiperda ascovirus 1a* virion reveals 21 proteins. J. Gen. Virol. *90*, 359–365.

Tan, Y., Spears, T., Bideshi, D.K., Johnson, J.J., Hice, R., Bigot, Y., and Federici, B.A. (2009b). P64, a novel major virion DNA-binding protein potentially involved in condensing the *Spodoptera frugiperda ascovirus 1a* genome. J. Virol. *83*, 2708–2714.

Theilmann, D.A., Blissard, G.W., Bonning, B., Jehle, J., O'Reilly, D.R., Rohrmann, G.F., Thiem, S., and Vlak, J. (2005). Family Baculoviridae. In: Virus Taxonomy, Eighth Report of the International Committee on Taxonomy of Viruses. Eds. C.M. Fauquet, M.A. Mayo, J. Maniloff, U. Desselberger, L.A. Ball. (Elsevier, Academic Press, San Diego, CA), pp177–185.

Tian, M., Jones, D.A., Smith, M., Shinkura, R., and Alt, F.W. (2004). Deficiency in nuclease activity of *Xeroderma pigmentosum* G in mice leads to hypersensitivity to UV irradiation. Mol. Cell. Biol. *24*, 2237–2242.

Tillman, P.G., Styer, E.L., and Hamm, J.J. (2004). Transmission of an ascovirus from *Heliothis virescens* (Lepidoptera, Noctuidae) and effects of the pathogen on survival of a parasitoid, *Cardiochiles nigriceps* (Hymenoptera: Braconidae). Environ. Entomol. 33, 633–643.

Traktman, P., and Boyle, K. (2004). Methods for analysis of poxvirus DNA replication. Meth. Mol. Biol. *269*, 169–186.

Vaux, D.L., and Silke, J. (2005). IAPs, RINGs and ubiquitylation. Nat. Rev. Mol. Cell. Biol. *6*, 287–297.

Wang, L., Xue, J., Seaborn, C.P., Arif, B.M., and Cheng, X.W. (2006). Sequence and organization of the *Trichoplusia ni* ascovirus 2c (*Ascoviridae*) genome. Virology *354*, 167–177.

Webby, R., and Kalmakoff, J. (1998). Sequence comparison of the major capsid protein gene from 18 diverse iridoviruses. Arch. Virol. *143*, 1949–1966.

White, C.L., Senkevich, T.G., and Moss, B. (2002). Vaccinia virus G4L glutaredoxin is an essential intermediate of a cytoplasmic disulfide bond pathway required for virion assembly. J. Virol. *76*, 467–472.

Wilson, M.E., and Price, K.H. (1988). Association of *Autographa californica* nuclear polyhedrosis virus with nuclear matrix. Virology *167*, 233–341.

Witze, E.S., Old, W.M., Resing, K.A., and Ahn, N.G. (2007). Mapping protein post-translation modification with mass spectrometry. Nat. Meth. *4*, 798–806.

Wizemann, H., and von Brunn, A. (1999). Purification of *E. coli*-expressed his-tagged hepatitis B core antigen by Ni^{2+}-chelate affinity chromatography. J. Virol. Meth. *77*, 189–197.

Zhang, Y., Calm-Jageman, I., Guron, J.R., Choi, T.J., Adams, B., Nicholson, A.W., and Van Etten, J.L. (2003). Characterization of a chlorella virus PBCV-1 encoded ribonuclease III. Virology *317*, 73–83.

Zhao, M., Brunk, U.T, and Eaton, J.W. (2001). Delayed oxidant-induced cell death involves activation of phospholipase A_2. FEBS Lett. *509*, 399–404.

Zhao, S., Du, X.Y., Chai, M.Q., Chen, J.S., Zhou, Y.C., and Song, J.G. (2002). Secretory phospholipase A_2 induces apoptosis via a mechanism involving ceramide generation. Biochim. Biophys. Acta *1581*, 75–88.

Baculoviruses: Biology, Replication and Exploitation

2

Robert D. Possee, Caroline M. Griffiths, Richard B. Hitchman, Adam Chambers, Fernanda Murguia-Meca, John Danquah, Ananya Jeshtadi and Linda A. King

Abstract

This chapter reviews recent progress to improve our understanding of baculovirus biology and replication at the cellular and whole insect levels, as well as providing an update on the exploitation of these viruses as expression and gene delivery vectors in both insect and mammalian cells. It does not discuss the ecology of baculoviruses, which is reviewed in Chapter 18, nor the use of these viruses as biocontrol agents.

Baculovirus taxonomy

Baculovirus taxonomy has kept pace with advances in our understanding of both the genetics and biology of members of the family. However, what has remained unchanged for many years is the convention of naming new virus isolates after the first insect in which they were found. While simple, this does cause problems if a virus is isolated subsequently in different host insects. The most notable example is the prototype baculovirus from *Autographa californica*, which was isolated from the Alfalfa looper. Very similar viruses were found in *Trichoplusia ni*, *Rachiplusia ou* and *Plutella xylostella*. Genetically, these baculoviruses are almost identical. They also have a wide host range, infecting 39 Lepidoptera species from 13 families (Bonning and Hammock, 1992). Theoretically, other baculoviruses similar to these could be isolated from many more species and thus complicate the taxonomy further. A similar observation was made when a baculovirus was isolated from *Mamestra brassicae*, which was more closely related to the baculovirus normally associated with *Panolis flammea* (Burden *et al.*, 2006). It remains to be seen how this problem can be resolved in future revisions of the taxonomy. Other members of the *Baculoviridae* demonstrate considerably more genetic diversity.

One common feature of all baculoviruses is the production of occlusion bodies, which contain infectious, lipid enveloped rod-shaped nucleocapsids called virus particles or virions (Fig. 2.1). These bacilliform structures suggested the name baculoviruses for members of the family. Although structurally similar, two major divisions can be recognized in baculovirus isolates. These are the nucleopolyhedroviruses (NPVs) and granuloviruses (GVs). The prime difference between NPVs and GVs is that the former have multiple virions packaged within the occlusion bodies to form polyhedra, whereas GVs usually package single virions in granules. Polyhedrin and granulin proteins are virus-encoded and genetically related. Other insect viruses that produce occlusion bodies, such as the entomopoxviruses and cypoviruses encode occlusion proteins that are genetically unrelated to each other or to polyhedrin or granulin. The virus particles in NPVs may envelope single or multiple nucleocapsids structures. At one time this was used as a taxonomic character but fell into disuse when sequence data did not support such partitions. Ironically, current nomenclatures of some baculoviruses reflect this feature. For example, *Autographa californica* multiple nucleopolyhedrovirus (AcMNPV) includes M in the abbreviation to indicate that multiple nucleocapsids may be enveloped by

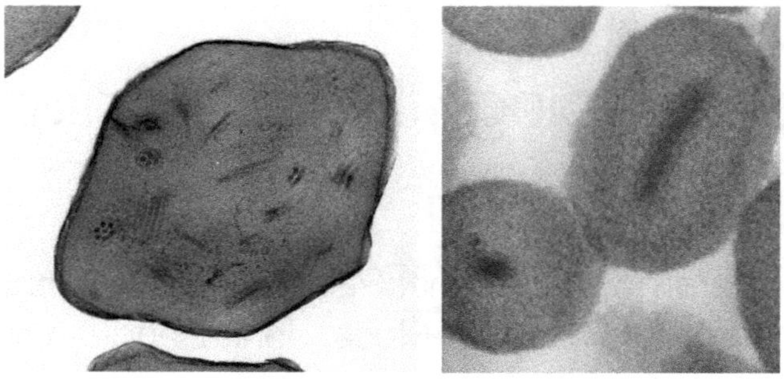

Figure 2.1 Typical cross-sections through polyhedra from *Autographa californica* MNPV (a) and granules from *Plodia interpunctella* GV (b) are shown.

a virus particle. The abbreviation of the name is also counter to recommendations that it should include the first and second letter of the insect genus and species. Other baculoviruses such as those from *P. flammea* are designated PaflNPV. This failure to conform for viruses such as AcMNPV owes much to historical convention. Baculoviruses have been isolated from Lepidoptera (butterflies and moths), Hymenoptera (wasps) and Diptera (mosquitoes).

A proposed classification of baculoviruses would recognize four major groups: alphabaculoviruses (lepidopteran NPVs), betabaculoviruses (lepidopteran GVs), gammabaculoviruses (hymenopteran NPVs) and deltabaculoviruses (dipteran NPVs) (Jehle *et al.*, 2006). This classification resolves the problem of a very wide genetic diversity in NPVs that is partly a reflection of the diverse insect species that they infect. It is also an acknowledgement of the contribution made by sequencing of whole baculovirus genomes to resolving their taxonomy.

Baculovirus genomes

Baculoviruses have double-stranded, circular, supercoiled DNA as genomes. The proportion of the DNA that is supercoiled is uncertain, as after isolation much of it presents as the open circular form. Currently, there are 50 complete baculovirus genomes available in GenBank. These range in size from 82,000 bp (*Neodiprion lecontei* NPV; Lauzon *et al.*, 2004) to 179,000 bp (*Xestia c-nigrum* GV; Hawakaya *et al.*, 1999), highlighting the considerable diversity in gene content of different baculovirus isolates. Only 31 genes appear to be common to all baculovirus genomes (Rohrmann, 2008). These appear to encode virus particle structural proteins, DNA polymerase functions and transcription factors; including an RNA polymerase that is alpha amanitin resistant. It is suggested that baculoviruses have co-evolved with their hosts and thus diverged genetically over time (Zanotto *et al.*, 1993; Herniou *et al.*, 2004).

The availability of so much sequence data has revolutionized the classification of baculoviruses. It was instrumental in the proposal for the four genera described above. Analysis of complete sequences also helped to subdivide further the alphabaculoviruses into Group I and II isolates. This subdivision is based on gene content, in particular the nature of a fusion protein required for entry of enveloped viruses into host cells. Group I NPVs use GP64 whereas Group II NPVs encode a protein designated F. Although this is the major feature used to separate these two groups, other genes are exclusive to Group I viruses. Interestingly, AcMNPV also contains an F homologue, although it is not active as a fusion protein. The GP64 and F proteins play key roles in virus replication but are not found in all structural forms of baculoviruses.

Baculovirus structure

The baculoviruses produce two structurally distinct forms of virus particle, each with a separate role in virus replication or survival outside of the insect host. These are budded virus (BV) and occlusion-derived virus (ODV); occlusion bodies have already been described above. The genomes of BV and ODV are identical, their nucleocapsids (NC) possess many identical proteins but both ODV and BV envelopes show key differences of protein and lipid content. Recent proteomic investigations of ODVs from different baculoviruses have revealed a range of conserved and novel proteins (Braunagel *et al.*, 2003; Deng *et al.*, 2007; Perera *et al.*, 2007). In addition to essential conserved protein functions, the virulence and specificity of viruses in the environment are likely to be represented by proteins within the ODV structures. Detailed reviews of baculovirus particle structure were recently published by Slack and Arif (2007) and Rohrmann (2008), so only a brief overview is provided here.

A structure common to each form of the virus is a rod-shaped nucleocapsid containing genomic DNA associated with a highly basic, arginine/serine/threonine-rich protein called P6.9 (Wilson *et al.*, 1987). The condensation of DNA into the nucleocapsids is thought to be a consequence of its association with P6.9. The major component of the nucleocapsids itself is VP39 (Thiem and Miller, 1989). The nucleocapsids are 40–70 nm in diameter and 250–400 nm in length (Bourcias and Pendland, 1998), depending on the size of the baculovirus genome. The nucleocapsid has an apical cap at one end and a base at the other. The base comprises PP78/83, which is encoded by ORF1629 in AcMNPV (Possee *et al.*, 1991; Ayres *et al.*, 1994). The gene appears critical for the production of infectious virus (Kitts and Possee, 1993). The BV form is produced when virus nucleocapsids formed in the nucleus bud through the plasma membrane and exit the cell. During the latter part of this process the nucleocapsids acquire a lipoprotein envelope containing a virus-encoded fusion protein; either GP64 or F.

Unlike BV, the ODV are formed in the nucleus of the virus-infected cell. The nucleocapsids of ODV also have a lipoprotein envelope but owing to its acquisition in the nucleus it does not contain either fusion protein associated with BV. It has a more complex structure than BV, with at least five proteins recognized as envelope components and other proteins designated as factors important for infectivity of the virus *in vivo*. In AcMNPV five *per os* infectivity factors (PIFs) have been identified: P74 or PIF0 (Kuzio *et al.*, 1989; Haas-Stapleton *et al.*, 2004), PIF-1 (Kikhno *et al.*, 2002), PIF-2 (Pijlman *et al.*, 2003), PIF-3 (Ohkawa *et al.*, 2005; Li *et al.*, 2007b) and PIF4 (Fang *et al.*, 2009b). Within the nucleus, ODVs are packaged further into occlusion bodies of which the major protein component is polyhedrin or granulin. Genes encoding these two proteins are remarkably conserved in the family, although a dipteran baculovirus has an occlusion body protein three times as large as any other and unrelated at the amino acid level (Afonso *et al.*, 2001; Perera *et al.*, 2006).

Surrounding the occlusions is a calyx/polyhedron envelope (Minion *et al.*, 1979). This is an electron-dense structure that appears to enhance stability of occlusion bodies. The envelope contains carbohydrate and a phosphorylated protein (PE), which if removed by manipulation of the virus genome results in less stable occlusion bodies (Whitt and Manning, 1988; Russell and Rohrmann, 1990). Also associated with occlusion bodies are the viral enhancing factors known as enhancins. These are a class of metalloproteases encoded by some lepidopteran NPVs and GVs, which can enhance virus infectivity *in vivo*, as described below.

Baculovirus replication *in vivo*

The biphasic replication of baculoviruses and the physical and functional separation of its two structural forms are most clearly appreciated during infection of the larval stages of lepidopteran hosts *in vivo*. The ODV, which initiates primary infection, gains entry to larvae as they consume OBs, the large proteinaceous structures that contain the enveloped virions. The number of ODV within an OB ranges from just one for GVs, to many for NPVs, as described above. Following primary infection, the cellular events of baculovirus replication proceed in a similar manner to those elucidated from infection of cultured cells although differences are inevitable. Replication yields the extracellular BV that spreads the virus between cells and tissues to achieve secondary infection.

Foraging larvae consume OBs, the protein component of which (polyhedrin or granulin) is dissolved by the alkaline pH within the midgut, releasing the ODV. Primary infection targets the columnar (and occasionally regenerative) cells lining the midgut, the only region of the insect gut not protected by a cuticular layer. Columnar cells are shielded by another chitinous barrier, the peritrophic matrix, but their microvillar tips can penetrate this matrix to be exposed to the midgut lumen, and hence to baculovirus ODVs. Many GVs (and some NPVs) increase their access to columnar cells through the action of enhancins, metalloproteases that degrade the major mucin protein of the peritrophic matrix (Lepore *et al.*, 1996; Wang and Granados, 1997; Peng *et al.*, 1999). Enhancins are structural components of the ODV (Slavicek and Popham, 2005) and act to permeabilize the peritrophic matrix, increasing the *per os* infectivity of baculoviruses that encode them (Popham *et al.*, 2001).

Primary infection of midgut cells is dependent on the PIFs that are found among the conserved core genes of all sequenced baculoviruses (Herniou *et al.*, 2003). The earliest PIF identified was P74, a membrane protein of ODVs that has been shown to bind the plasma membrane of columnar cells (Kuzio *et al.*, 1989; Haas-Stapleton *et al.*, 2004). Proteolytic cleavage of P74 in the alkaline environment of the midgut appears to be linked to enhanced virus infectivity (Slack and Lawrence, 2005; Slack *et al.*, 2008), possibly through increased binding of P74 to the cell receptor. The receptor is as yet unidentified, although a protein of unknown function isolated in brush border membrane vesicles from columnar cells (Zhou *et al.*, 2005) may prove to be a candidate. P74 is a key protein mediating viral attachment of OV during primary infection, although two other recently identified OV envelope proteins, PIF-1 and PIF2, have also been shown to be essential for virus binding to columnar cells (Kikhno *et al.*, 2002; Pijlman *et al.*, 2003; Ohkawa *et al.*, 2005). It is unclear if P74, PIF-1 and PIF-2 are collaborative partners in binding, but the loss of oral infectivity in mutants lacking even one of these proteins demonstrates they are not functionally redundant and therefore not likely to be linked in host range expansion of the baculoviruses. The functions of PIF-3 (Ohkawa *et al.*, 2005; Li *et al.*, 2007b) and PIF-4 (Fang *et al.*, 2009b) are not known but these proteins are also essential for oral infectivity. In addition to these vital proteins, a number of other structural proteins are found in ODV, such as ODV-E18, ODV-EC27, ODV-E35, ODV-E25 and ODV-E66, but the function of these proteins remains uncharacterised and they appear to play no essential part in *per os* infectivity (Slack *et al.*, 2001).

Membrane binding and fusion deliver viral nucleocapsids (NCs) into the columnar cells, which migrate through the length of the microvilli into the cytoplasm proper and thence to the nucleus, initiating genome uncoating, gene expression and replication. Movement of NCs through microvilli is as yet unexplained, although polymerization of actin may be involved in the nuclear targeting of NCs to the cell nucleus. Uniquely for MNPVs, a subset of NCs entering columnar cells may bypass the nucleus and travel intact through the cytoplasm (Washburn *et al.*, 2003a,b). These NCs associate with regions of the basal membrane enriched with the BV envelope glycoprotein GP64, expressed during early replication in the cell, and exit the cell by budding. Rapid egress of NCs from midgut cells is advantageous to MNPVs, ensuring spread to other body tissues before the sloughing of midgut can act to eliminate the virus from the host (Washburn *et al.*, 2003a,b).

Exit of assembled progeny NCs from the nucleus of infected cells involves the viral EXON0 protein, probably mediating the interaction of NCs with kinesin motors on microtubule networks within the nucleus and cytoplasm (Fang *et al.*, 2007, 2009a). Studies *in vitro* have also revealed a central role for actin in the assembly and morphogenesis of NCs (Volkman *et al.*, 1987; Ohkawa and Volkman, 1999; Goley *et al.*, 2006). GP64 itself co-localizes with f-actin and with lipid rafts in the membrane of AcMNPV-infected cells in culture, which polarize to form a GP64-enriched membrane patch (Haines *et al.*, 2009). These mechanisms focus the accumulation of GP64 in, and directed targeting of progeny NCs to, specific areas of the infected cell membrane, but it is not known if the same processes promote the rapid transit and egress of parental NCs from midgut cells seen for MNPVs.

Release of parental and/or progeny enveloped virions from the midgut signals the initiation of secondary infection, which results in further BV production throughout the infected host tissues. The GP64 envelope proteins of group I baculoviruses are essential for BV attachment, fusion and budding, promoting virus spread between cells during secondary infection (Hefferon *et al.*, 1999; Oomens and Blissard, 1999). Group II baculoviruses lack GP64 and have instead a functional homologue, the F (fusion) protein but despite their equivalent roles in BV spread through infected hosts, the proteins are not wholly comparable (Li and Blissard, 2008; Westenberg and Vlak, 2008) and bind to different cellular receptors (Westenberg *et al.*, 2007). Group I baculoviruses also possess F-like protein homologues that play no direct role in virus entry or budding, but have been suggested as accessory proteins for GP64 binding (Zhou and Blissard, 2008; Wang *et al.*, 2008b). The F-like protein of AcMNPV has been shown to increase BV production *in vitro* and the speed of kill of the virus *in vivo* (Lung *et al.*, 2003; Wang *et al.*, 2008b). Many viral proteins are essential for BV (and ODV) production within infected cells, including key structural proteins and the late expression factors or LEFs (reviewed in Lu and Miller, 1995; Kelly *et al.*, 2007; Passarelli and Guarino, 2007).

During the very late stage of replication, NCs remain within the nucleus after assembly, gaining an envelope prior to encapsulation into OBs. Little is known about the *de novo* synthesis of the ODV membrane, although recent work has confirmed the NC protein Ac142 and ODV envelope protein Ac143 (ODV-E18) as essential for envelopment of NCs prior to occlusion (McCarthy *et al.*, 2008; McCarthy and Theilmann, 2008). Nuclear localization of polyhedrin, its association with both NCs and the polyhedral membrane (calyx), and assembly into OB structures rely partly on inherent characteristics of the polyhedrin protein itself (Jarvis *et al.*, 1991; Carstens *et al.*, 1992) but other viral proteins contribute to the efficient assembly of OBs. The precise role of the viral P10 protein, produced to high levels during infection, has been difficult to elucidate but evidence from several sources reveals a key function in the successful assembly of OBs (reviewed in Carpentier and King, 2009).

An important maturation event is enclosure of the assembled OB within the calyx, a step that appears to depend on the close interaction of the major polyhedrin envelope protein (PEP, pp34) with P10 polymers (Williams *et al.*, 1989). P10 forms polymers or fibrils (van Oers *et al.*, 1994; Carpentier *et al.*, 2008) that interact with microtubules to form large fibrillar structures passing through the cell cytoplasm and nucleus (Patmanidi *et al.*, 2003; Carpentier *et al.*, 2008), forming extensive fibres and tubules within infected cells, including a cage-like structure lining the nuclear membrane (Carpentier *et al.*, 2008; Carpentier and King, 2009). This novel enclosure may assist during the assembly of OBs or it may serve to maintain the integrity of the nucleus until OB synthesis is complete. Observations *in vitro* have identified a small virus protein, FP25, involved in the synthesis and nuclear targeting of polyhedrin (Jarvis *et al.*, 1992) and loss of FP25 function leads to reduced ODV envelopment and occlusion (Katsuma *et al.*, 1999).

Following extensive OB assembly, nuclear disintegration and cell lysis liberate OBs into the host body cavity. Infected larval tissues become enlarged and lose integrity as the basal lamina weakens and the viral chitinase and cathepsin enzymes degrade the insect exoskeleton (Ohkawa *et al.*, 1994; Slack *et al.*, 1995; Hawtin *et al.*, 1995, 1997; Lanier *et al.*, 1996). Chitinase is produced at late times in infected cells, but liquefaction of tissues is delayed, most likely due to an endoplasmic reticulum (ER) retention motif within chitinase that retains chitinase within cells (Hawtin *et al.*, 1995) until OB synthesis has peaked and cellular lysis occurs (Hom *et al.*, 2002; Saville *et al.*, 2002). P10 appears to be influential in cell lysis and the release of chitinase from cells (Williams *et al.*, 1989; van Oers *et al.*, 1994; Thomas *et al.*, 1998) whilst FP25 similarly promotes secretion (Katsuma *et al.*, 2004) and prolonged synthesis (Katsuma *et al.*, 2009) of cathepsin at late times post infection. Degradation of the exoskeleton results in the 'melting' of infected larvae after death, such that a liquid virus inoculum is released over the soil and foliage.

The primary route of horizontal transmission for most baculoviruses is through this release of OBs from infected cadavers. Between hosts the OB provides limited protection for the ODV against UV-induced degradation (Behle *et al.*, 2000; Fuxa *et al.*, 2001) and recent work has shown that melanin (produced at increased levels following death of baculovirus-infected larvae) decreases OB inactivation by UV irradiation (King, F.L, Griffith, C.M. and King,

L.A., unpublished data). Susceptible hosts can be infected through OB-contaminated frass or regurgitate (Vasconcelos, 1996) and in crowded populations virus released from spiracles in the cuticle of infected larvae can also serve as a source of infection. Many baculoviruses do not induce host liquefaction, making these and cannibalism the major routes for transmission (Dhandapani *et al.*, 1993; Boots, 1998). Transmission has also been documented during oviposition by parasitoid wasps, either by surface contamination or by passing virus-contaminated haemolymph between larvae (reviewed by Sait *et al.*, 1996). Contamination of egg surfaces, or of egg contents, can cause vertical transmission (Murray and Elkington, 1989; Tanada and Kaya, 1993) but in the environment baculoviruses must develop mechanisms to reduce cytopathic effect, evade immune clearance and persist during periods of low host density. The ability of baculoviruses to infect hosts persistently provides an important and effective route for vertical transmission (Kukan, 1999; Fuxa *et al.*, 1999, 2002; Myers *et al.*, 2000; Burden *et al.*, 2002; Matthews *et al.*, 2002).

Replication in cultured cells

Baculoviruses, like other intracellular pathogens, have co-evolved with their hosts to efficiently produce BV progeny to spread infection between cells *in vivo* and in insect cells cultured *in vitro*. The clever adaptation of viruses to survive is rooted in the very nature of virus co-evolution and exploitation of the complex cellular machinery of host cells, and often the study of virus replication reveals knowledge of fundamental cellular mechanisms. Essential for virus replication is the ability of the virus particles, whether as incoming NCs or as newly assembled NCs destined to become BV, to negotiate and move within the crowded cytoplasm. As with all particles of over 500 kDa, the cytoplasm is too crowded for NC to move freely and therefore baculoviruses, as with other large DNA viruses, have evolved mechanisms to exploit the intracellular highways using actin filaments and the microtubular networks. In recent times, with increasing interest in baculoviruses as vectors capable of delivering genes into mammalian cells (reviewed by Hu, 2008; He *et al.*, 2009; Wu *et al.*, 2009; see also Baculovirus expression vectors section below), there have been a number of studies that have started to elucidate the mechanisms by which baculovirus NCs move into the nucleus of transduced cells (Laakkonen *et al.*, 2009; Tsai *et al.*, 2009).

Understanding the intracellular trafficking of baculoviruses, both in insect and mammalian cells, is therefore important because these viruses are no longer just the work horses of the insect cell expression system, they are being actively considered as promising candidates for gene therapy and vaccine vectors in mammalian cells. For example, previous studies on the role, and subsequent deletion of, the chitinase gene from the AcMNPV genome has helped improve the production of recombinant secretory and membrane-targeted proteins in insect cells (Hitchman *et al.*, 2009), because chitinase accumulates in, and compromises the function of, the ER during virus infection (Thomas *et al.*, 1998; Saville *et al.*, 2002, 2004). Similarly, it is believed that investigating and understanding the intracellular trafficking of baculoviruses can help in producing improved vectors for gene delivery into both insect and mammalian cells (as discussed in Baculovirus expression vectors section below).

Baculovirus entry into cells

Viruses use normal cellular processes such as clathrin-dependent and -independent endocytosis (Matilainen *et al.*, 2005; Long *et al.*, 2006; Laakkonen *et al.*, 2009) and macropinocytosis (reviewed by Kalia and Jameel, 2009) to ensure an efficient entry into host cells. Baculovirus BV enter cultured insect cells (Fig. 2.2) and a broad spectrum of mammalian cells; reviewed by Condreay and Kost, 2007). BV comprise single NCs enwrapped by host plasma membrane that is enriched with glycoprotein, GP64 (or F protein in alphabaculoviruses; Pearson *et al.*, 2000, 2001) at one end of the nucleocapsid (Summers and Volkman, 1976), which is responsible for entry and exit of baculovirus (Monsma *et al.*, 1996; Oomens and Blissard, 1999; Spenger *et al.*, 2002). Recently the post-fusion structure of GP64 was solved (Kadlec *et al.*, 2008), showing that GP64 is a type III virus fusion protein, with structural similarities to

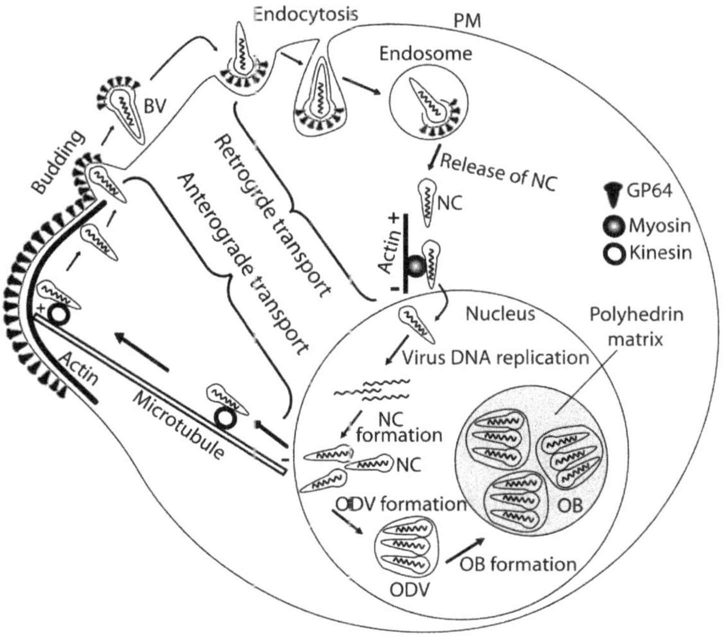

Figure 2.2 Life cycle of baculovirus *Autographa californica* multinucleopolyhedrovirus (AcMNPV). The virus produces two distinct virus forms, budded virus (BV) and occlusion derived virus (ODV). The BV enters cells through receptor-mediated endocytosis following interaction between the BV surface glycoprotein, GP64, and an unknown cell surface receptor. BV-containing endosomes release nucleocapsids (NC) into cytoplasm, which are then transported into the nucleus using actin microfilaments, possibly in conjunction with myosin. Following virus DNA replication, newly formed NC at first move from the nucleus into the cytoplasm, where they are transported, most likely using microtubules, to the cell surface where BV buds from the plasma membrane gaining a GP64-enriched envelope. Actin may be involved in the final stages of budding. At later times in infection, NCs remain in the nucleus and are enveloped to form ODV, which then become embedded in polyhedrin protein to form occlusion bodies (OB). BV is required for cell-to-cell spread within the insect hosts and within cell culture whereas ODV and OB are required to spread virus infection to new insect hosts.

vesicular stomatitis virus G and herpes simplex virus type 1 gB proteins.

During BV entry, GP64 binds to as yet unknown host receptors resulting in the fusion of both GP64 and the plasma membrane, and recently we came a step closer to determining the nature of the receptor when it was demonstrated that a GP64 domain comprising amino acids 21–159 was responsible for receptor binding (Zhou and Blissard, 2008) and that the length of the hydrophobic transmembrane domain was critical for fusion activity (Li and Blissard, 2009a). Li and Blissard (2009b) have also demonstrated that GP64 contains a 23 amino acid pre-transmembrane domain (PTM) that appears analogous to the PTM of other viral fusion proteins and may be important in the formation of the fusion pore. Several binding factors appear to influence the effectiveness of GP64, including carbohydrates, phospholipids and electrostatic forces. Duisit *et al.* (1999) reported that adsorptive binding of baculovirus onto the surface of human 293 cells was promoted by electrostatic forces. Phospholipids have also been shown to play a role in the interaction between GP64 and the surface of mammalian cells (Tani *et al.*, 2001).

Volkman and Goldsmith (1985) were the first to indicate that BV may enter host insect cells via receptor-mediated endocytosis. More recent studies have shown that baculovirus employs clathrin-mediated endocytosis when entering both insect (Long *et al.*, 2006) and mammalian cells (Matilainen *et al.*, 2005). It has also been shown that clathrin-independent endocytosis may be important in entry into mammalian cells, in which RhoA-regulated actin ruffling of plasma membrane rafts lead to baculovirus entry (Laakkonen *et al.*, 2009). After penetrating the plasma membrane via endocytosis, baculovirus particles are internalized within endosomes within 10–20 minutes (Hefferon *et al.*, 1999), and have two possible fates. Firstly, the endosomes may fuse with lysosomes leading to the destruction of baculovirus, a common defensive mechanism through which organisms remove pathogens. Secondly, the endosomes may be processed resulting in their maturation and eventual breakdown to release the endocytic contents. It has been suggested that in insect cells, baculoviruses evade the former fate through GP64 fusion with the endosome membrane in a low-pH milieu, at a precise time in the late endocytic stage, resulting in the release of naked NCs into the cytoplasmic environment (Hefferon *et al.*, 1999). In mammalian cells, early endosomes are formed within 15 minutes, which then mature into late endosomes, releasing naked NCs into the cytoplasm within 45 minutes after transduction (Salminen *et al.*, 2005).

Retrograde transport of nucleocapsids

Naked NCs that are released from late endosomes are transported into the nucleus using active, energy-dependent mechanisms because free diffusion of molecules larger than about 500 kDa is highly restricted in the tightly packed cytoplasm (Fig. 2.2). In both insect and mammalian cells, a number of indirect studies have shown that the actin cytoskeleton is the main transport mechanism used by baculoviruses to move towards the nucleus (Charlton and Volkman, 1991, 1993; van Loo *et al.*, 2001). The basal capsid protein ORF1629 (PP78/83) is a Wiskott–Aldrich syndrome protein (WASP) homologue that has been shown to be important in the nucleation of actin prior to its polymerization (Machesky *et al.*, 2001; Goley *et al.*, 2006) and it is therefore possible that this interaction facilitates NC transport along actin microfilaments. Depolymerization of actin with cytochalasin D and/or inhibition of myosin function with butanedione monoxime (DBM) prevents naked NCs from reaching the nucleus in insect cells (Lanier and Volkman, 1998). The effect of DBM, a drug that prevents the release of the phosphate ion after ATP hydrolysis in the globular head of myosins, but does not affect actin structural integrity, suggests that both actin and myosins are required for the retrograde transport of baculoviruses in insect cells (Fig. 2.2). However, it is not clear how actin polymerization and myosins interplay to ensure that naked NCs are delivered into the nucleus. Interestingly, disruption of microtubules in human hepatocytes led to an increase in the number of baculovirus NCs entering the nucleus of transduced cells, suggesting that microtubules may actually hinder the retrograde trafficking of NCs in mammalian cells (Salminen *et al.*, 2005).

Nuclear import of baculovirus nucleocapsids

Our understanding of the mechanisms involved in NC movement into the nucleus is very limited. Most of what we do know is based on the early microscopy studies of Summers (1971) and Granados (1978). These studies, together with Granados and Lawler (1981), showed the presence of both DNA-filled and empty NPV NCs within the nucleus of infected insect cells, suggesting that the NCs were delivered intact into the nucleus, where subsequent unpackaging of the viral genome occurs prior to DNA replication (Fig. 2.2). However, it has been suggested that GV NCs dock at nuclear pores using the basal capsid protein and then inject their DNA into the nucleus leaving the empty NC shell within the cytoplasm (Summers, 1971; Raghow and Grace, 1974; Granados, 1980). Once the DNA is released within the nucleus, the transcription cascade and DNA replication processes are initiated leading to the synthesis and nuclear assembly of progeny NCs. These processes have recently been the subject of an excellent review (Rohrmann, 2008) and will not be covered here.

A capsid protein C42, which is found in both BV and ODV, has been implicated in NC assembly and is believed to bind to, and transport, the p78/83 capsid protein into the nucleus via the nuclear pores (Wang *et al.*, 2008a). Although C42 appears responsible for the translocation of PP78/83 capsid protein into the nucleus, it is unlikely that C42 is responsible for the nuclear transport of NCs during retrograde transport. This is because even though C42 is conserved among all the *Baculoviridae* family, including GVs, GV NCs do not enter the nucleus but dock at the nuclear pore.

An important feature of baculovirus infection of insect cells is the gradual accumulation of actin within the nucleus, which subsequently polymerizes from the globular (g) form to filamentous (f) actin. By the late stages of infection, most of the cellular actin has been concentrated within the nucleus (reviewed by Volkman, 2007). The WASP-homologue basal capsid protein (PP78/83, ORF1629), which is conserved amongst all NPVs, has been shown to promote f-actin formation by interacting with g-actin and the host cell Arp2/3 complex (Goley *et al.*, 2006). Mutation analysis of PP78/83 has suggested that nuclear f-actin formation is important for proper NC and ODV formation, possibly by forming a scaffold upon which NCs are assembled (Goley *et al.*, 2006).

Newly synthesized NCs, destined to become BV, leave the nucleus and move towards the plasma membrane, but after DNA replication ceases, there appears to be a switch, which remains to be understood, and NCs stay within the nucleus where they become enveloped to form ODV. A real time PCR study (Rosinski *et al.*, 2002) has quantified the number of virus DNA molecules within infected cells and has suggested that whilst up to 2000 NCs may go on to be released from the cell as BV, more than 84,000 genome copies are made with the vast majority remaining in the nucleus, with some destined to become ODV. Rohrmann (2008) suggests that these large numbers, which amount to ten times the normal host cell DNA content, might explain the enormous expansion of the nucleus that is characteristic of virus-infected cells and may also help explain how the virus is able to hyper-express the very late genes. High-level gene expression in the very late phase of gene expression may, therefore, be partly dependent on the very high gene copy number present in infected nuclei.

In NPV-infected cells, NCs remaining in the nucleus are believed to become enveloped with membranes taken from microvesicles derived from the inner nuclear membrane that have become enriched with ODV-specific proteins (Braunagel *et al.*, 2003; reviewed by Braunagel and Summers, 2007). Hyperexpression of the very late proteins polyhedrin and P10 lead to the formation of occlusion bodies and fibrillar/tubular structures, respectively. Polyhedrin accumulates forming a lattice-like structure that surrounds and occludes the ODV. P10 initially forms long fibrillar-like structures, whose formation is dependent on the integrity of microtubules (van Oers *et al.*, 1994; Patmanidi *et al.*, 2003; Carpentier *et al.*, 2008), however, in the very late stages of infection hollow, tubular-like structures are formed that pass through both the cytoplasm and nucleus but which eventually forms a 3D cage-like structure encircling the nucleus (Carpentier *et al.*, 2008). As the OB develop, P10 fibrils can be observed closely associated with their surfaces and may be involved in the final stage of OB maturation, the assembly of the polyhedral envelope (Minion *et al.*, 1979; Lee, 1996; Williams, 1989; van Oers, 1993). Although the role of P10 remains elusive and intriguing, its possible functions have recently been reviewed (Carpentier and King, 2009) and will not be further discussed here.

Anterograde transport of baculovirus

Anterograde transport is the movement of substances including viruses from the nucleus to the cell periphery. Our knowledge of the cellular processes involved in the anterograde transport of baculovirus NCs from the nucleus to the site of budding at the plasma membrane remains limited but is thought to involve both microtubules and f-actin (Fig. 2.2). There are two possible pathways in which NCs may be transported out of the nucleus; by passing through nuclear pores or by budding through the nuclear membrane (NM). The former will result in the ejection of naked NCs into the cytoplasm and the latter will result in NCs enveloped with nuclear membrane. It seems unlikely that NCs move through the nuclear pore as the typical size of a NC is 300–250 nm × 30–25 nm (Bourcias and Pendland, 1998), much larger than the diameter of the nuclear pore complex, which is about 120 nm with sieve-like properties of 2.6 nm wide (Mohr *et al.*, 2009). However, the nuclear pore has been shown to dilate to allow macromolecules of diameters up to 39 nm to translocate in certain circumstances (Pante and Kann, 2002). Because infected cells do not divide, it is also not possible for the NCs to rely on cell division and NM disintegration as a mechanism for release into the cytosol.

Some human viruses such as herpes viruses modify the NM by inserting virus encoded glycoproteins and then bud through acquiring a NM envelope, which is subsequently lost by passage through the Golgi apparatus before finally budding through the plasma membrane (Stannard *et al.*, 1996; Mettenleiter, 2004; Baines *et al.*, 2007). Using herpes viruses as a model, the NM of AcMNPV-infected insect cells may also have a similar modification to permit NCs to bud through into the cytoplasm. Early electron microscopy studies have provided some evidence to support this hypothesis, where virions were observed to be budding from the NM (Knudson and Harrap, 1975). Since BV has a single envelope derived from the plasma membrane, it suggests that if the budding hypothesis is correct, the NM-derived envelope must be removed en route to the plasma membrane, presumably by interacting with membranous bodies such as ER and the Golgi apparatus. Neither is it yet clear how such enveloped NCs may move using the host cell cytoskeleton and motor proteins, although it seems accepted that both actin and microtubules are almost certainly involved. Microtubule-based transport is directional and requires motor proteins such as dynein, which carries cargos towards the nucleus, and kinesin, which carries cargos towards the plasma membrane (reviewed by Hirokawa *et al.*, 2009; Verhey and Hammond, 2009). Actin-based motility is non-directional, and could be based on either actin polymerization alone or in cooperation with actin motor protein, the myosins.

The first attempt to dissect anterograde transportation of baculovirus NCs was carried out by Volkman and Zaal (1990). In that study, microtubules of baculovirus-infected Sf9 cells (derived from pupal ovarian tissue of the fall armyworm *Spodoptera frugiperda*) were treated with colchicine and taxol, which depolymerize and stabilize microtubules, respectively. Depolymerization of microtubules with colchicine did not show any significant decrease in the titre of the BV compared to the control cells. However, the stabilization of microtubules with taxol resulted in a very significant reduction in the titre of BV. In a more recent study depolymerising microtubules using colchicine, there was a marked decrease (90%) of BV in drug-treated Sf9 cells at 72 hpi (Fang *et al.*, 2009a). Although colchicine treatment experiments are not conclusive but rather indicative, these studies do suggest that microtubules play a role in the anterograde transport of NCs. To determine whether the baculovirus NCs interact with microtubules, Fang *et al.* (2009a) performed a yeast-two-hybrid experiment using tubulin and the baculovirus capsid protein EXON0 since confocal microscopy studies had showed a strong colocalization between EXON0 and tubulin. Although the yeast-two-hybrid result was negative, demonstrating no direct interaction between tubulin and EXON0, it was suggested that an intermediate protein, in this case kinesin, might be sandwiched between EXON0 and tubulin, since kinesin is known to be involved in the anterograde movement of cargos on microtubules. The final step in the replication cycle of the virus is the egress of budded virus from the plasma membrane enriched in the virus glycoprotein GP64. Haines *et al.* (2009) have suggested that virus egress may take place in cholesterol-enriched membrane domains, also known as lipid rafts. Confocal microscopy showed that actin colocalized both with the cholesterol-enriched domains and with GP64, suggesting that, as with other viruses that bud from the plasma membrane, f-actin may be involved in the final stages of egress.

Baculovirus expression vectors

The baculovirus expression vector system (BEVS) has been widely used for the production of recombinant protein in insect cells since its development in the mid-1980s (Summers, 2006). Cells are infected with a recombinant BV containing a foreign gene coding region usually under the control of the polyhedrin (*polh*) promoter. During the later stage of virus infection large quantities of recombinant protein are synthesized, effectively turning the cells into protein production factories. These cells have the capacity to produce high yields of authentically processed proteins, including signal peptide recognition and secretion, assembly of oligomeric proteins, proper post-translational modifications such as glycosylation, phosphorylation and disulfide bond formation. In recent years, several systems have been developed to allow easy insertion of foreign genes into baculovirus expression vectors. These have been reviewed elsewhere and will not be described extensively here (King *et al.*, 2007). Baculovirus expression vectors have also been modified to remove genes encoding proteins such as chitinase and cathepsin that are deleterious to recombinant protein production, intracellular transport and stability (Hitchman *et al.*, 2009). These improved vectors have greatly increased recombinant protein yields and quality. Coupled with the authentic post-translational modification and high yields of recombinant proteins is the ability of recombinant baculovirus-infected cells to process foreign proteins further into tertiary and quaternary structures. One application that has exploited this feature of the system is the assembly of virus-like particles (VLPs) in insect cells.

VLPs are built from proteins that form a virus outer shell with surface proteins, sometimes embedded within a lipid bilayer. However, the particles do not contain the nucleic acid required for replication, meaning that they are not infectious. This makes VLPs excellent vaccine candidates as they are not infectious and their size and structure have good immunostimulating properties. The first VLP-based vaccine, produced in yeast cells, was successfully introduced in 1987 against hepatitis B virus, containing the surface antigen of the hepatitis B virus that co-assembles with cellular membranes into spherical particles (17–25 nm diameter) (Gavilanes *et al.*, 1982). The first VLP to be produced by baculovirus vectors was in the early 1990s using the human immunodeficiency virus (HIV) group-specific antigen protein (gag), which is normally processed into the basic physical infrastructure of HIV. It was observed forming retroviral core like particles within the cytoplasm of baculovirus-infected insect cells (Morikawa *et al.*, 1991; Lerch and Friesen, 1992; Royer *et al.*, 1992). Since this time, VLPs from several different viruses have been produced using BEVS including rotavirus, influenza, orbivirus, Norwalk virus and others (Roy and Noad, 2008). More recently a VLP-based vaccine (Cervarix, GlaxoSmithKline) was approved against human papillomavirus (HPV) that contains the major HPV structural protein L1 (Grgacic and Anderson, 2006). Cervarix is currently produced using the baculovirus expression system and is a bivalent vaccine against HPV types 16 and 18. This section outlines the growing use of the BEVS in VLP production and describes the strategies employed to meet the demand for VLP-based vaccines.

Why use baculoviruses to produce VLPs?

There are a number of advantages of using the BEVS for VLP production, compared to other expression systems. Traditionally, VLPs that can be assembled from a single protein were produced in bacteria or yeast. However, baculovirus vectors are able to incorporate multiple, large DNA inserts allowing them to express simultaneously more than one recombinant protein during infection. These VLPs are also produced in cells which are free of contaminants from mammalian sources. Insect cells are now routinely cultured in serum-free medium, which avoids the risk of prion-like particles. The insect-derived products have very similar properties to mammalian-derived VLPs and can be easily made in large quantities and at a lower cost compared to mammalian systems. Baculovirus-derived VLPs are also highly stable and can induce similar immune responses to mammalian-derived VLPs. This coupled with the relative ease of making viruses and their high yield of protein makes the BEVS an ideal choice of system for VLP production.

Optimizing VLP production in insect cells

Prior to using the BEVS to produce VLPs an infection strategy will need to be determined and the system must be optimized to maximize yields and increase process efficiency. Several different recombinant proteins can be produced at the same time by either infecting insect cells with multiple recombinant baculoviruses each containing a single recombinant gene (monocistronic) or by infecting cells with a single recombinant baculovirus containing multiple recombinant genes (multicistronic). The latter of these two strategies has several advantages including fewer amplifications and viral stocks to handle and also ensures that each recombinant protein is produced in an infected cell, which is a requirement for some VLPs to assemble. The disadvantage of this strategy is that recombinant baculoviruses that contain more than one recombinant protein are often unstable and manipulation of individual recombinant protein production is not possible. For example, by using several viruses yields of individual recombinant proteins can be regulated by adjustment of their multiplicity of infection.

There are a number of different methods available for constructing recombinant baculoviruses, the most commonly used being recombination *in vitro* (BaculoDirect™, Invitrogen), transposition in bacteria (Bac-to-Bac®, Invitrogen) and homologous recombination in insect cells (Possee *et al.*, 2008; flashBAC, OET Ltd). A novel system called MultiBac (Redbiotec) allows insertion of multiple genes using both recombination and transposition sites in the virus genome. Senger *et al.* (2009) demonstrated up to 40-fold improvement for VLP production for several difficult to express papillomavirus-like particles using this method. The homologous recombination system was used by Stewart *et al.* (2010) to construct a baculovirus that expressed all four of the major capsid proteins of Bluetongue virus. Recently, Noad *et al.* (2009) demonstrated that the sites for insertion of VLP genes into the virus genome are not limited to the *polh* locus. They used the lambda red and bacteriophage P1 Cre-recombinase system to carry out multiple, single-locus insertions of foreign genes into the baculovirus genome at seven different loci. Using this bipartite selection system they showed that successful expression and correct assembly of VLPs could be achieved for two different viruses, a two protein complex Influenza A virus (H7N7) and a four protein complex Bluetongue virus (Noad *et al.*, 2009).

The two most commonly used insect cell lines for VLP production are *Spodoptera frugiperda* (Sf9) or *Trichoplusia ni* cells (i.e. High Five®, Invitrogen). In many cases both cell lines will be tested at the optimization stage before one is chosen for VLP production. For example, Ebola and Marburg VLP production (VP40 matrix protein and nucleoprotein) were compared in Sf9 and High Five® cells and yields were found to be higher in the *T. ni* cells. High Five® cells have also been used for papillomavirus-like particle production (Senger *et al.*, 2009) and to produce chimeric HIV-1 VLPs (GagRT and GagTN) (Pillay *et al.*, 2009). High Five® cells may produce more recombinant protein and provide a more suitable environment for certain VLP assembly than Sf cells. However, Sivakumar *et al.* (2009) found that Sf21 cells resulted in the highest yield of *Pea enation* mosaic virus VLP when compared to both High Five® and Sf9 cells. This highlights the need to test different insect cell lines when expressing VLPs. There are also now concerns over the use of High Five® cells since it was discovered that some stocks of these cells harbour a nodavirus (Li *et al.*, 2007a). However, another cell line MSU-TnT4 derived from *T. ni* tested negative for nodavirus (Zhang and Thiem, 2009).

The multiplicity of infection (MOI) at the time of infection is an important factor in determining recombinant protein expression yield and quality, and is usually chosen on the basis of kinetics and site of accumulation. In VLP production, this is often more complex due to several different recombinant proteins being expressed at the same time. Hence, these conditions are normally optimized for VLP assembly, not recombinant protein production. This is due to evidence suggesting that concentration of recombinant protein does not necessarily result in a change of VLP concentration, and the optimum conditions for recombinant protein production may not necessarily be the same for VLP production (Cruz *et al.*, 1998). Generally, the MOI is calculated by the Poisson distribution method which can predict the fraction of the population that will be infected during primary infection. However, the Poisson distribution is only followed during infection of insect cells with a single recombinant baculovirus and is not applicable when insect cells are infected with multiple viruses. The reason for this discrepancy remains to be investigated, but it does require careful optimization to be carried out for VLP production.

Generally, higher MOI appears to yield higher levels of VLP formation, but this is likely to be VLP specific and also depend on the culture conditions used (i.e. cell type, stage of infection etc). For example, Clark *et al.* (2009) used monocistronic recombinant baculoviruses to express three major capsid proteins for rotavirus in Sf9 cells using different MOIs (0.1, 0.2, 1 and 2) for each and showed higher yields of VLP formation with increasing MOI. Mena *et al.* (2006) and Vieira *et al.* (2005) used a relatively high MOI 5 when expressing rotavirus major capsid proteins in High Five® (Invitrogen) cells. However, low MOI has been used with success for the production of recombinant proteins and may have a beneficial effect for VLP production, reducing the number of virus amplification steps and minimizing virus passage effect (Maranga *et al.*, 2003).

Generally the optimum time of VLP harvest is between 48 and 72 hours post infection (hpi), although in some cases it may be extended to 96 hpi. However, at later time points the cells start to break apart releasing proteases into the extracellular environment and this will influence the optimal time of harvest. The baculovirus itself also contains the cysteine protease cathepsin (*v-cath*), which will degrade recombinant proteins. The use of deletion vectors that do not contain *v-cath* (i.e. *flash*BACGOLD and Multibac) can greatly improve the final yield and quality of VLP obtained (Senger *et al.*, 2009).

One of the advantages of using the BEVS is that it facilitates the scale up to large-batch manufacturing. Insect cells can be easily scaled up from 3 L Erlenmyer flasks to large (>10 L) agitated bioreactors and can be cultivated in most culture systems from batch to perfusion (Maranga *et al.*, 2002). However, since the baculovirus infection induces cell lysis, for the production of VLPs the batch mode or eventually, fed-batch, is generally used (Maranga *et al.*, 2004). Baculovirus infections impose a high metabolic burden upon the cells and there is a much greater requirement for oxygen and glucose, compared with uninfected cells. Insect cells are also susceptible to high shear stress, usually in the form of vortex formation and bubble bursting and entrapment. These variables can be controlled by the optimization of conditions, including vessels, impellers and control modules for pH, temperature, nutrients and oxygenation and the inclusion of Pluronic F-68 in the media. Infecting cells at half of the maximum cell yield attained in uninfected control cultures has also been used to avoid oxygen limitations when producing HIV–VLPs (Cruz *et al.*, 1998). Recently, single-use disposable systems have become popular, usually supplied pre-sterilized by gamma irradiation and with controls for sensing critical growth parameters. Such technologies offer rapid turnaround between runs, reduce risk of cross-contamination, reduce validation requirements and minimize initial start-up costs.

Removal of residual trace amounts of baculovirus or insect cells within VLP preparations is important to provide clinical grade material. The standard technique for VLP purification from infected cell culture is isopycnic gradients, for example using caesium chloride for non-enveloped icosahedral viruses. Enveloped viruses (i.e. influenza virus) are purified on sucrose gradients. However, there are a number of disadvantages in using isopycnic gradients, including the requirement of high amounts of culture sample, variable purification yields, destabilization of VLPs and low purity of the final product. Isopycnic gradients also lack the sensitivity to separate unassembled and incomplete intermediaries of VLPs, leading to unknown quality of the final purified VLP. This has led to the development of a number of different strategies for VLP purification including ion exchange, affinity and size exclusion chromatography. High yields of VLP can be obtained by first removing the baculovirus with an aqueous two-phase system (sucrose gradient) followed by size exclusion chromatography, allowing up to recovery of ~65% of VLP with ~95% purity overall (Palomares and Ramírez, 2009). The typical yield of purified assembled VLPs obtained from baculovirus-infected cells is in the range of 0.4 to 15 mg/L.

More recently, the modification of baculoviruses to harbour mammalian promoters has been described and these BacMams have been used for the expression of VLPs in mammalian cells. For example, Chen *et al.* (2005) reported the use of BacMams to express the two protein complex hepatitis delta VLP driven by the CMV promoter in BHK cells (baby hamster kidney cells). The idea of using baculoviruses to deliver recombinant genes/proteins to mammalian cells has promoted the intriguing prospect of this system being used as vectors for human gene therapy.

Baculoviruses for human gene therapy and vaccination

Cumulative evidence underlines the potential of gene transfer and an effective strategy to replace the defective/faulty genes and cure a variety of monogenic diseases. Moreover, gene therapy has the potential of re-establishing the function of group of cells, tissues or organs and improves the condition of polygenic diseases such as cancer, diabetes mellitus and cardiovascular diseases among many others (Gillet *et al.*, 2009). Having evolved to deliver their genes to target cells, viruses are currently one of the most effective means of gene delivery. Viral vectors based on adenovirus, adenovirus-related virus, retrovirus, vaccinia virus, and herpes simplex virus have been used in the majority of gene therapy clinical trials. Unfortunately, these vectors remain of limited utility in gene therapy because of their pathogenicity, immunogenicity, insertional mutagenesis and the possibility of reversion to wild type Sakurai *et al.*, 2008; Nayak and Herzog, 2009; Zaiss *et al.*, 2009).

Immune responses to mammalian vectors can cause severe side effects and malignant complications in humans. Moreover, the deaths associated with the use of adenoviral vectors in human phase I/II gene therapy clinical trials call for a new perspective and radical thinking about future vector development for gene therapy in humans (Tan *et al.*, 2008; Wilson, 2008). Baculoviruses offer an alternative that is non-pathogenic to humans but able to enter human cells, uncoat and deliver transgenes to the nuclei (Hu, 2008). The safety of insect baculoviruses is well documented. There is a long track record of their safe use in insect biocontrol (Black *et al.*, 1997). Moreover, because the natural host of baculovirus is phylogenetically distant from mammals, AcMNPV-based vectors elicit much lower innate immune responses in comparison with conventional viral vectors and do not face a pre-existing host memory immune response (Georgopoulos *et al.*, 2009). The lack of specific mammalian promoters in the baculoviral DNA avoids the activation of cellular oncogenes, or silencing the expression of endogenous tumour suppressor genes, which, in turn, may trigger malignant cell transformation in the host.

Another major advantage of baculoviruses as human gene therapy vectors is their capacity for recombinant DNA; the size of the rod-shaped virus particles appears to be determined by the size of the genome to be packaged (Bourcias and Pendland, 1998). Against their use is the fact that transduction efficiency can be low, requiring the production of virus stocks with high titres, which usually means that BV must be purified from the virus-infected cell growth medium. Typically, MOIs of 20–00 pfu/cell are used to ensure the human cell lines are transduced with the baculovirus vector (Boyce and Bucher, 1996; Shoji *et al.*, 1997). A further problem may be the nature of the BV envelope, which is derived from insect cells. This may induce a host immunogenic response, such as complement, which can inactivate the virus particles very quickly and hinder systemic use. However, the problem has been at least partially overcome by treating BVs with an antibody against a complement component or cobra venom factor (Hoffman and Strauss, 1998). Complement-resistant BVs were also derived by inserting the gene for complement-regulatory protein human decay accelerating factor into the virus genome so that this protein was incorporated into the baculovirus envelope (Hüser *et al.*, 2001).

Although recombinant baculoviruses were originally used to transduce mammalian cell cultures (Boyce and Butcher, 1996; Hoffman *et al.*, 1995), they were soon used with primary hepatic cells *in vitro* and human livers *ex vivo* (Sandig *et al.*, 1996). More recently, transduction of de-differentiated stem cells *in vitro* and the *ex vivo* modification of human cell progenitors has been achieved stressing the potential of these vectors for cell therapy or tissue engineering (Chuang *et al.*, 2007, 2009).

The use of stem cells represents the main thrust of human regenerative medicine and a promising strategy to treat a wide range of diseases because of their properties of self-renewal, pluripotency and undifferentiated state. However, to achieve specific functions the stem cells are usually genetically modified. Human germinal cells and de-differentiated pluripotential stem cells have been successfully transduced with recombinant baculovirus with high levels of transgene expression and no evidence of cytotoxicity or malignant transformation (Ho *et al.*, 2006; Chuang *et al.*, 2007, 2009). Baculovirus can be used effectively to transduce mouse, rat, porcine, rhesus, and human bone marrow-derived mesenchymal stem cells (BMSCs) *in vitro* with transduction efficiencies ranging from 20% to 80%. Interestingly, comparatively more rhesus and human BMSCs were successfully transduced and showed higher mean fluorescence intensity than BMSCs from mouse, rats and pigs (Zeng *et al.*, 2009).

Similarly, using a recombinant baculovirus vector with the woodchuck hepatitis virus post-transcriptional regulatory element (WPRE), cultured human embryonic stem cell-derived neurons (HESCN) showed early transgene expression after baculovirus transduction with an efficiency of up to 80%. Strong expression was observed for more than 1 month post-transduction, and the signal was still detectable after 3 months. In addition, areas in which baculovirus-transduced HESCN were transplanted into the brain of nude mice, were integrated into the mouse brain and maintained transgene expression for at least 4 weeks post-transduction. These encouraging results highlight that due to the high transduction efficiency for primate and human BMSCs, baculovirus is possibly a more suitable gene vector to genetically modify BMSCs for preclinical and clinical human gene therapy trials. Another example of where baculovirus gene therapy vectors have been used effectively *in vivo* is in the central nervous system to inhibit glioma cell growth by expressing the A-chain of the diphtheria toxin (Wang *et al.*, 2006).

In related studies, baculoviruses have been shown to represent a viable option as a recombinant vaccine delivery system in mammals. The molecular flexibility of recombinant baculovirus allows an efficient modification of the vector phenotype for tissue and cell targeting, resistance to innate immunity and vaccine design. It has been shown that one of the major AcMNPV surface glycoproteins, GP64, can be modified to display complex foreign eukaryotic proteins on the surface of the viral envelope (Boublik *et al.*, 1995). More recently, *in vitro* studies have shown that baculovirus vectors can transduce professional antigen-presenting cells and trigger their maturation, which is a prerequisite for efficient antigen presentation. Moreover, recombinant baculovirus vectors specifically designed to express foreign antigens on the viral surface are able to induce strong antigen-specific CD4(+) and CD8(+) T-cell responses (Strauss *et al.*, 2007).

Insect cell technology is a promising platform for mass-production of influenza vaccine and might be a practical alternative to conventional vaccines. A baculovirus cell-derived influenza vaccine has been shown to provide equivalent or better immunogenicity than an established surrogate marker for protection in clinical trials Meghrous *et al.*, 2009). In another study baculovirus pseudotyped with vesicular stomatitis virus glycoprotein was used as a vector to express the haemagglutinin (HA) protein of highly pathogenic H5N1 avian influenza virus. The resultant recombinant baculovirus (BV-G-HA) mediated gene delivery and HA was expressed efficiently in mammalian cells. In a subsequent step, mice immunized with 1×10^9 pfu of BV-G-HA developed significantly higher levels of H5-specific antibodies and cellular immunity than those that received 100 μg of DNA vaccines expressing HA. In consequence, these mice were completely protected from lethal challenge with H5N1 avian influenza virus. Furthermore, different vaccination doses were further tested in chickens, and these experiments demonstrated that 1×10^8 pfu of BV-G-HA offered complete protection from challenge with 100 LD50 of HB/327. These data indicate that the pseudotype baculovirus-mediated vaccine could be utilized as an alternative strategy against the pandemic spread of H5N1 influenza virus (Meghrous *et al.*, 2009). The major advantage of the insect cell/baculovirus expression system (BEVS)-derived subunit vaccine is that the manufacturing of the HA proteins does not require the cultivation and subsequent inactivation of live viruses.

Other problems remaining in the use of baculoviruses as gene therapy vectors concern the lack of detailed information about the transduction process, including efficiency of virus uptake, transport of nucleocapsids to the nucleus, uncoating and release of DNA. Integration of baculovirus DNA into the chromosome has been observed, probably a consequence of illegitimate recombination (Merrihew *et al.*, 2001). If these stages in the transduction process can be made more efficient then the application of baculoviruses as gene therapy vectors will be more promising.

Concluding remarks

Baculoviruses were one of the first pathogens to be recognized in insects from their often devastating effects on the silkworm industry in China and Europe. Their large occlusion bodies were also readily visible when the first light microscopes were employed to study the pathology of this disease. Their deadly effect in high density insect populations was later exploited in biological control of pests in agriculture. This led to detailed studies of the ecology of baculoviruses and their hosts (see Chapter 18). With the development of modern techniques in virology and molecular biology much has also been learned of their replication strategies in insect cells, particularly the biphasic production of budded and occluded viruses. In turn, this led to the development of baculovirus expression systems, which despite the complexity of the virus, has proven to be widely utilized in acàdemia and industry alike. Although initially used for production of single foreign gene products, developments in multiple expression vectors have allowed the synthesis of more than one protein so that complex structures can be assembled in recombinant baculovirus-infected cells. These have enabled ambitious projects to assemble virus-like particles as candidates for vaccines. Free of the original virus genetic material, they offer very safe alternatives to other vaccines derived from the wild type pathogen. VLPs are also rapidly expanding beyond vaccines, as new applications begin to appear. They have the potential to be convenient delivery agents to target organs or cells for therapeutic applications, for use in diagnostic assays, structural studies for the determination of protein function and nanotechnology, where VLPs can have new functionality conferred to them by coating their surface with metal particles. Given the capacity of baculovirus genomes and virus particles to accommodate additional genetic material there seems no limitation to what can be made using this system for recombinant protein production. Finally, in a twist to the well-accepted safety of baculoviruses as biopesticides, they are also being developed as vectors for human gene therapy. Baculoviruses can be used to transduce human cells, thus delivering genes to combat genetic diseases, without the risks posed by using mammalian virus vectors. Baculoviruses therefore offer a versatile and well accepted system for making recombinant proteins in both insect and mammalian cells.

References

Afonso, C.L., Tulman, E.R., Lu, Z., Balinsky, C.A., Moser, B.A., Becnel, J.J., Rock, D.L., and Kutish, G.F. (2001). Genome sequence of a baculovirus for *Culex nigripalpus*. J. Virol. *75*, 11157–11165.

Ayres, M.D., Howard, S.C., Kuzio, J., Lopez-Ferber, M., and Possee, R.D. (1994). The complete nucleotide sequence of *Autographa californica* nuclear polyhedrosis virus. Virology *202*, 586–605.

Baines, J.D., Wills, E., Jacob, R.J., Pennington, J., and Roizman, B. (2007). Glycoprotein M of herpes simplex virus 1 is incorporated into virions during budding at the inner nuclear membrane. J. Virol. *81*, 800–812.

Behle, R.W., McGuire, M.R., and Tamez-Guerra, P. (2000). Effect of light energy on alkali-released virions from *Anagrapha falcifera* nucleopolyhedrovirus. J. Invertebr. Pathol. *76*, 120–126.

Black, B.C., Brennan, L.A., Dierks, P.M., and Gard, I.E. (1997). Commerialization of baculoviral insecticides. In The Baculoviruses, L.K. Miller, ed. (New York and London: Plenum Press), pp.341–381.

Bonning, B.C., and Hammock, B.D. (1992). Development and potential of genetically engineered viral insecticides. Biotech. Genet. Eng. Rev. *10*, 455–489.

Boots, M. (1998). Cannibalism and the stage dependant transmission of a viral pathogen of the Indian meal moth, *Plodia interpunctella*. Ecol. Entomol. *23*, 118–122.

Boublik, Y., Di Bonito, P., and Jones, I.M. (1995). Eucaryotic virus display: engineering of the major surface glycoprotein of the *Autographa californica* nuclear polyhedrosis virus (AcNPV) for the presentation of foreign proteins on the virus surface. Bio/Technology *13*, 1079–1084.

Bourcias, D.G., and Pendland, J.C. (1998). Baculoviruses. In Principles of Insect Pathology, Norwell: Kluwer Academic Publishers), pp 111–146.

Boyce, F.M., and Bucher, N.L. (1996). Baculovirus-mediated gene transfer into mammalian cells. Proc. Natl. Acad. Sci. U.S.A. *93*, 2348–2352.

Braunagel, S.C., Russell, W.K., Rosas-Acosta, G., Russell, D.H., and Summers, M.D. (2003). Determination of the protein composition of the occlusion-derived virus of *Autographa californica* nucleopolyhedrovirus. Proc. Natl. Acad. Sci. U.S.A. *100*, 9797–9802.

Braunagel, S.C., and Summers, M.D. (2007). Molecular biology of the baculovirus occlusion-derived virus envelope. Curr. Drug Targets *8*, 1084–1095.

Burden, J.P., Griffiths, C.M., Cory, J.S., Smith, P., and Sait, S.M. (2002). Vertical transmission of sublethal granulovirus infection in the Indian meal moth, *Plodia interpunctella*. Mol. Ecol. *11*, 547–555.

Burden, J.P., Possee, R.D., Sait, S.M., King, L.A., and Hails, R.S. (2006). Phenotypic and genotypic characterisation of persistent baculovirus infections in populations of the cabbage moth (*Mamestra brassicae*) within the British Isles. Arch. Virol. *151*, 635–649.

Carpentier, D.C., Griffiths, C.M., and King, L.A. (2008). The baculovirus P10 protein of *Autographa californica* nucleopolyhedrovirus forms two distinct cytoskeletal-like structures and associates with polyhedral occlusion bodies during infection. Virology *371*, 278–291.

Carpentier, D.C., and King, L.A. (2009). The long road to understanding the baculovirus P10 protein. Virol. Sinica *24*, 227–242.

Carsten, E.B., Williams, G.V., Faulkner, P., and Partington, S. (1992). Analysis of polyhedra morphology mutants of *Autographa californica* nuclear polyhedrosis virus: molecular and ultrastructural features. J. Gen. Virol. *73*, 1471–1479.

Charlton, C.A., and Volkman, L.E. (1991). Sequential rearrangement and nuclear polymerization of actin in baculovirus-infected *Spodoptera frugiperda* cells. J. Virol. *65*, 1219–1227.

Charlton, C.A., and Volkman, L.E. (1993). Penetration of *Autographa californica* nuclear polyhedrosis virus nucleocapsids into IPLB Sf21 cells induces actin cable formation. Virology *197*, 245–254.

Chen, Y.H., Wu, J.C., Wang, K.C., Chiang, Y.W., Lai, C.W., Chung, Y.C., and Hu, Y.C. (2005). Baculovirus-mediated production of HDV-like particles in BHK cells using a novel oscillating bioreactor. J. Biotechnol. *118*, 135–147.

Chuang, C.K., Sung, L.Y., Hwang, S.M., Lo, W.H., Chen, H.C., and Hu, Y.C. (2007). Baculovirus as a new gene delivery vector for stem cell engineering and bone tissue engineering. Gene Ther. *14*, 1417–1424.

Chuang, C.K., Wong, T.H., Hwang, S.M., Chang, Y.H., Chen, G.Y., Chiu, Y.C., Huang, S.F., and Hu, Y.C. (2009). Baculovirus transduction of mesenchymal stem cells: *in vitro* responses and *in vivo* immune responses after cell transplantation. Mol. Ther. *17*, 889–896.

Clark, K.B., Lin, S.C., Humphrey, C., Foytich, K., Esona, M., Wang, Y., Liu, M., and Jiang, B. (2009). Expression and characterization of human group C rotavirus virus-like particles in insect cells. Virology *387*, 267–272.

Condreay, J.P., and Kost, T.A. (2007). Baculovirus expression vectors for insect and mammalian cells. Curr. Drug Targets *8*, 1126–1131.

Cruz, P.E., Cunha, A., Peixoto, C.C., Clemente, J., Moreira, J.L., and Carrondo, M.J. (1998). Optimization of the production of virus-like particles in insect cells. Biotechnol. Bioeng. *60*, 408–418.

Deng, F., Wang, R., Fang, M., Jiang, Y., Xu, X., Wang, H., Chen, X., Arif, B.M., Guo, L., Wang, H., and Hu, Z. (2007). Proteomics analysis of *Helicoverpa armigera* single nucleocapsid nucleopolyhedrovirus identified two new occlusion-derived virus-associated proteins, HA44 and HA100. J. Virol. *81*, 9377–9385.

Dhandapani, N.,Jayaraj, S., and Rabindra, R.J. (1993). Cannibalism on nuclear polyhedrosis virus infected larvae by *Heliothis armigera* (Hubner) and its effect on viral infection. Insect Sci. Appl. *14*, 427–430.

Duisit, G., Saleun, S., Douthe, S.G., and Moullier, P. (1999). Baculovirus vector requires electrostatic interactions including heparan sulphate for efficient gene transfer in mammalian cells. J. Gene. Med. *1*, 93–102.

Fang, M., Dai, X., and Theilmann, D.A. (2007). *Autographa californica* multiple nucleopolyhedrovirus core EXON0 (ORF141) is required for efficient egress of nucelocapsids from the nucleus. J. Virol. *81*, 9859–9869.

Fang, M., Nie, Y., and Theilmann, D.A. (2009a). AcMNPV EXON0 (Ac141) which is required for the efficient egress of budded virus nucleocapsids interacts with b-tubulin. *Virology 385*, 496–504.

Fang, M., Nie, Y., Harris, S., Erlandson, M.A., and Theilmann, D.A. (2009b). *Autographa californica* multiple nucleopolyhedrovirus core gene *Ac96* encodes a per os infectivity factor (*pif-4*). J. Virol. *83*, 12569–12578.

Fuxa, J.R., Sun, J.S., Weidner, E.H., and LaMotte, M.R. (1999). Stressors and rearing densities of *Trichoplusia ni*: evidence of vertical transmission of NPV and CPV. J. Invertebr. Pathol. *74*, 149–155.

Fuxa, J.R., Matter, M.M., Abdel-Rahman, A., Micinski, S., Richter, A.R., and Flexner, J.L. (2001). Persistence and distribution of wild type and recombinant nucleopolyhedrovirus in soil. Microb. Ecol. *41*, 222–231.

Fuxa, J.R., Richter, A.R., Ameen, A.O., and Hammock, B.D. (2002). Vertical transmission of TnSNPV, TnCPV, AcMNPV and possibly recombinant NPV in *Trichoplusia ni*. J. Invertebr. Pathol. *79*, 44–50.

Gavilanes, F., Gonzalez-Ros, J.M., and Peterson, D.L. (1982). Structure of hepatitis B surface antigen. Characterization of the lipid components and their association with the viral proteins. J. Biol. Chem. *257*, 7770–7777.

Georgopoulos, L.J., Elgue, G., Sanchez, J., Dussupt, V., Magotti, P., Lambris, J.D., Totterman, T.H., Maitland, N.J., and Nilsson, B. (2009). Preclinical evaluation of innate immunity to baculovirus gene therapy vectors in whole human blood. Mol. Immunol. *46*, 2911–2917.

Gillet, J.P., Macadangdang, B., Fathke, R.L., Gottesman, M.M., and Kimchi-Sarfaty, C. (2009). The development of gene therapy: from monogenic recessive disorders to complex diseases such as cancer. Methods Mol. Biol. 542, 5–54.

Goley, E.D., Ohkawa, T., Mancuso, J., Woodruff, J.B., D'Alessio, J.A., Cande, W.Z., Volkman, L.E., and Welch, M.D. (2006). Dynamic nuclear actin assembly by Arp2/3 complex and a baculovirus WASP-like protein. Science *314*, 464–467.

Granados, R.R. (1978). Early events in the infection of *Heliothis zea* midgut cells by a baculovirus. Virology *90*, 170–174.

Granados, R.R. (1980). Infectivity and mode of action of baculoviruses. Biotechnol. Bioeng. *22*, 1377–1405.

Granados, R.R., and Lawler, K.A. (1981). *In vivo* pathways of *Autographa californica* baculovirus invasion and infection. Virology *108*, 297–308.

Grgacic, E.V., and Anderson, D.A. (2006). Virus-like particles: passport to immune recognition. Methods *40*, 60–65.

Haas-Stapleton, E.J., Washburn, J.O., and Volkman, L.E. (2004). P74 mediates specific binding of *Autographa californica* M nucleopolyhedrovirus occlusion-derived virus to primary cellular targets in the midgut epithelia of Heliothis virescens. J. Virol. *78*, 6786–6791.

Haines, F.J., Griffiths, C.M., Possee, R.D., Hawes, C.R., and King, L.A. (2009). Involvement of lipid rafts and cellular actin in AcMNPV GP64 distribution and virus budding. Virol. Sinica *24*, 333–349.

Hawtin, R.E., Arnold, K., Ayres, M.D., Zanotto, P.M., Howard, S.C., Gooday, G.W., Chappell, L.H., Kitts, P.A., King, L.A., and Possee, R.D. (1995). Identification and preliminary characterisation of a chitinase gene in *Autographa californica* nuclear polyhedrosis virus genome. Virology *212*, 673–685.

Hawtin, R.E., Zarkowska, T., Arnold, K., Thomas, C.J., Gooday, C.W., King, L.A., Kuzio, J.A., and Possee, R.D. (1997). Liquefaction of *Autographa californica* nucleopolyhedrovirus-infected insects is dependent on the integrity of virus-encoded chitinase and cathepsin. Virology *238*, 243–253.

Hayakawa, T., Ko, R., Okano, K., Seong, S., Goto, C., and Maeda, S. (1999). Sequence analysis of the *Xestia c-nigrum* granulovirus genome, Virology *262*, 277–297.

He, F., Madhan, F., and Kwang, J. (2009). Baculovirus vector as a delivery vehicle for influenza vaccines. Expert Rev. Vaccines *8*, 455–467.

Hefferon, K.L., Oomens, A.G., Monsma, S.A., Finnerty, C.M., and Blissard, G.W. (1999). Host cell receptor binding by baculovirus GP64 and kinetics of virion entry. Virology *258*, 455–468.

Herniou, E., Olszewski, J.A., Cory, J.S., and O'Reilly, D. (2003). The genome sequence and evolution of baculoviruses. Annu. Rev. Entomol. *48*, 211–234.

Herniou, E.A., Olszewski, J.A., O'Reilly, D.R., and Cory, J.S. (2004). Ancient coevolution of baculoviruses and their insect hosts. J. Virol. *78*, 3244–3251.

Hirokawa, N., Noda, Y., Tanaka, Y., and Niwa, S. (2009). Kinesin superfamily motor proteins and intracellular transport. Nat. Rev. Mol. Cell Biol. *10*, 682–696.

Hitchman, R.B., Possee, R.D., Crombie, A.T., Chambers, A., Ho, K., Siaterli, E., Lissina, O., Sternard, H., Novy, R., Loomis, K., Bird, L.E., Owens, R.J., and King, L.A. (2009). Genetic modification of a baculovirus vector for increased expression in insect cells. Cell Biol. Toxicol. DOI 10.1007/s10565-009-9133-y

Ho, Y.C., Lee, H.P., Hwang, S.M., Lo, W.H., Chen, H.C., Chung, C.K., and Hu, Y.C. (2006). Baculovirus transduction of human mesenchymal stem cell-derived progenitor cells: variation of transgene expression with cellular differentiation states. Gene Ther. *13*, 1471–1479.

Hoffman, C., Sandig, V., Jennings, G., Rudolph, M., Schlag, P., and Strauss, M. (1995). Efficient gene transfer into human heptocytes by baculovirus vectors. Proc. Natl. Acad. Sci. U.S.A. *92*, 10099–10103.

Hoffman, C., and Strauss, M. (1998). Baculovirus-mediated gene transfer in the presence of human serum or blood facilitated by inhibition of the complement system. Gene Ther. *5*, 531–536.

Hom, L.G., Ohkawa, T., Trudeau, D., and Volkman, L.E. (2002). *Autographa californica* M nucleopolyhedrovirus ProV-CATH is activated during infected cell death. Virology *296*, 212–218.

Hu, Y.C. (2008). Baculoviral vectors for gene delivery: a review. *Curr. Gene Ther. 8*, 54–65.

Hüser, A., Rudolph, M., and Hofmann, C. (2001). Incorporation of decay-accelerating factor into the baculovirus envelope generates complement-resistant gene transfer vectors. Nat. Biotech. *19*, 451–455.

Jarvis, D.L., Bohlmeyer, D.A., and Garcia Jr., A. (1991). Requirements for nuclear localization and supramolecular assembly of a baculovirus polyhedrin protein. Virology. *186*, 795–810.

Jarvis, D.L., Bohlmeyer, D.A., and Garcia Jr,. A. (1992). Enhancement of polyhedrin nuclear localization during baculovirus infection. J. Virol. *66*, 6903–6911.

Jehle, J.A., Blissard, G.W., Bonning, B.C., Cory, J.S. Herniou, E.A., Rohrmann, G.F. Theilmann, D.A., Thiem, S.M., and Vlak, J.M. (2006). On the classification and nomenclature of baculoviruses: A proposal for revision. Arch. Virol. *151*, 1257–1266.

Kadlec, J., Loureiro, S., Abrescia, N.G., Stuart, D.I., and Jones, I.M. (2008). The postfusion structure of baculovirus gp64 supports a unified view of viral fusion machines. Nat. Struct. Mol. Biol. *15*, 1024–1030.

Kalia, M., and Jameel, S. (2009). Virus entry paradigms. Amino acids DOI 10.1007/s00726-009-0363-3.

Katsuma, S., Noguchi, Y., Zhou, C.L., Kobayashi, M., and Maeda, S. (1999). Characterization of the 25K FP gene of the baculovirus *Bombyx mori* nucleopolyhedrovirus: implications for post-mortem host degradation. J. Gen. Virol. *80*, 783–791.

Katsuma, S., Tanaka, S., Shimada, T., and Kobayashi, M. (2004). Reduced cysteine protease activity of the haemolymph of *Bombyx mori* larvae infected with *fp25k*-inactivated *Bombyx mori* nucleopolyhedrovirus results in the reduced postmorten host degradation. Arch. Virol. *149*, 1773–1782.

Katsuma, S., Nakanishi, T., and Shimada, T. (2009). *Bombyx mori* nucleopolyhedrovirus FP25K is essential for maintaining a steady-state level of v-cath expression throughout the infection. Virus Res. *140*, 155–160.

Kelly, B.J., King, L.A., and Possee, R.D. (2007). Introduction to baculovirus molecular biology. In

Baculovirus and Insect Cell Expression Protocols, D. W. Murhammer, ed. (Totowa, New Jersey: Humana Press), pp. 25–53.

Kikhno, I., Gutiérres, S., Croizier, L., Croizier, G., and López Ferber, M. (2002). Characterisation of *pif*, a gene required for the per os infectivity of *Spodoptera littoralis* nucleopolyhedrovirus. J. Gen. Virol. *83*, 3013–3022.

King, L.A., Hitchman, R.B., and Possee, R.D. (2007). Recombinant baculovirus isolation. In Baculovirus and Insect Cell Expression Protocols, D. W. Murhammer, ed. (Totowa, New Jersey:Humana Press), pp 54–76.

Kitts, P.A., and Possee, R.D. (1993). A method for producing recombinant baculovirus expression vectors at high frequency. *Biotechniques 14,* 810–817.

Knudson, D.L., and Harrap, K.A. (1975). Replication of nuclear polyhedrosis virus in a continuous cell culture of *Spodoptera frugiperda*: microscopy study of the sequence of events of the virus infection. J. Virol. *17,* 254–268.

Kukan, B. (1999). Vertical transmission of nucleopolyhedrovirus in insects. J. Invertebr. Pathol. *74*, 103–111.

Kuzio, J., Jaques, R., and Faulkner, P. (1989). Identification of p74, a gene essential for virulence of baculovirus occlusion bodies. Virology *173*, 759–763.

Laakkonen, J.P., Mäkelä, A.R., Kakkonen, E., Turkki, P., Kukkonen, S., Peränen, J., Ylä-Herttuala, S., Airenne, K.J., Oker-Blom, C., Vihinen-Ranta, M., and Marjomäki, V. (2009). Clathrin-independent entry of baculovirus triggers uptake of *E. coli* in non-phagocytic human cells. PLoS ONE *4*, e5093.

Lanier, L.M., Slack, J.M., and Volkman, L.E. (1996). Actin binding and proteolysis by the baculovirus AcMNPV: The role of the virion associated V-CATH. Virology *216,* 380–388.

Lanier, L.M., and Volkman, L.E. (1998). Actin binding and nucleation by *Autographa californica* M nucleopolyhedrovirus. Virology *243*, 167–177.

Lauzon, H.A.M., Lucarotti, C.J., Krell, P.J., Feng, Q., Retnakaran, A., and Arif, B.M. (2004). Sequence and organization of the *Neodiprion lecontei* nucleopolyhedrovirus genome. J. Virol. *78,* 7023–7035.

Lee, S.Y. (1996). A common pathway for p10 and calyx proteins in progressive stages of polyhedron envelope assembly in AcMNPV-infected *Spodoptera frugiperda* larvae. Arch. Virol. *141*, 1247–1258.

Lepore, L.S., Roelvink, P.R., and Granados, R.R. (1996). Enhancin, the granulovirus protein that facilitates nucleopolyhedrovirus (NPV) infections, is a metalloproteinase. J. Invertebr. Pathol. *68*, 131–140.

Lerch, R.A., and Friesen, P.D. (1992). The baculovirus-integrated retrotransposon TED encodes gag and pol proteins that assemble into viruslike particles with reverse transcriptase J. Virol. *66*, 1590–601.

Li, T.C., Scotti, P.D., Miyamura, T., and Takeda, N. (2007a). Latent infection of a new Alphanodavirus in an insect cell line. J. Virol. *81*, 10890–10896.

Li, X., Song, J., Jiang, T., Liang, C., and Chen, X. (2007b). The N-terminal hydrophobic sequence of *Autographa californica* nucleopolyhedrovirus PIF-3 is essential for oral infection. Arch. Virol. *152*, 1851–1858.

Li, Z., and Blissard, G.W. (2008). Functional analysis of the transmembrane (TM) domain of the *Autographa californica* multicapsid nucleopolyhedrovirus GP64 protein: Substitution of heterologous TM domains. J. Virol. *82*, 3329–3341.

Li, Z., and Blissard, G.W. (2009a). The *Autographa californica* multicapsid nucleopolyhedrovirus GP64 protein: analysis of transmembrane domain length and sequence requirements. J. Virol. *83*, 4447–4461.

Li, Z., and Blissard, G.W. (2009b). The pre-transmembrane domain of the *Autographa californica* multicapsid nucleopolyhedrovirus GP64 protein is critical for membrane fusion and virus infectivity. J. Virol. *83,* 10993–11004.

Long, G., Pan, X., Kormelink, R., and Vlak, J.M. (2006). Functional entry of baculovirus into insect and mammalian cells is dependent on clathrin-mediated endocytosis. J. Virol. *80*, 8830–8833.

Lu, A., and Miller, L.K. (1995). The roles of eighteen baculovirus late expression factor genes in transcription and DNA replication. J. Virol. *69*, 975–982.

Lung, O.Y., Cruz-Aolvarez, M., and Blissard, G.W. (2003). Ac23, an envelope fusion protein homolog in the baculovirus *Autographa californica* multicapsid nucleopolyhedrovirus, is a viral pathogenicity factor. J. Virol. *77,* 328–339.

Machesky, L.M., Insall, R.H., and Volkman, L.E. (2001). WASP homology sequences in baculoviruses. Trends Cell Biol. *11*, 286–287

Maranga, L., Cruz, P.E., Aunins, J.G., and Carrondo, M.J. (2002). Production of core and virus-like particles with baculovirus infected insect cells. Adv. Biochem. Eng. Biotechnol. *74*, 183–206.

Maranga, L., Brazao, T.F., and Carrondo, M.J. (2003). Virus-like particle production at low multiplicities of infection with the baculovirus insect cell system. Biotechnol. Bioeng. *84*, 245–253.

Maranga, L., Cunha, A., Clemente, J., Cruz, P., and Carrondo, M.J. (2004). Scale-up of virus-like particles production: effects of sparging, agitation and bioreactor scale on cell growth, infection kinetics and productivity. J. Biotechnol. *107*, 55–64.

Matilainen, H., Rinne, J., Gilbert, L., Marjomäki, V., Reunanen, H., and Oker-Blom, C. (2005). Baculovirus entry into human hepatoma cells. J. Virol. *79*, 15452–15459.

Matthews, H.J., Smith, I., and Edwards, J.P. (2002). Lethal and sublethal effects of a granulovirus on the tomato moth *Lacanobia oleracea*. J. Invertebr. Pathol. *80*, 73–80.

McCarthy, C.B., Dai, X., Donly, C., and Theilmann, D.A. (2008). *Autographa californica* multiple nucleopolyhedrovirus *ac142*, a core gene that is essential for BV production and ODV envelopment. Virology *372*, 325–339.

McCarthy, C.B., and Theilmann, D.A. (2008). AcMNPV *ac143* (odv-e18) is essential for mediating budded virus production and is the 30th baculovirus core gene. *Virology 375*, 277–291.

Meghrous, J., Mahmoud, W., Jacob, D., Chubet, R., Cox, M., and Kamen, A.A. (2009). Development of a simple and high-yielding fed-batch process for the production of influenza vaccines. *Vaccine 28*, 309–316.

Mena, J.A., Ramirez, O.T., and Palomares, L.A. (2006). Intracellular distribution of rotavirus structural proteins and virus-like particles expressed in the insect cell-baculovirus system. J. Biotechnol. *122*, 443–452.

Merrihew, R.V., Clay, W.C., Condreay, J.P., Witherspoon, S.M., Dallas, W.S., and Kost, T.A. (2001). Chromosomal integration of transduced recombinant baculovirus DNA in mammalian cells. J. Virol. *75*, 903–909.

Mettenleiter, T.C. (2004). Budding events in herpesvirus morphogenesis. Virus Res. *106*, 167–180.

Minion, F.C., Coons, L.B., and Broome, J.R. (1979). Characterisation of the polyhedral envelope of the nuclear polyhedrosis virus of *Heliothis virescens*. J. Invertebr. Pathol. *34*, 303–307.

Mohr, D., Frey, S., Fischer, T., Guttler, T., and Gorlich, D. (2009). Characterisation of the passive permeability barrier of nuclear pore complexes. EMBO J. *28*, 2541–2553.

Monsma, S.A., Oomens, A.G., and Blissard, G.W. (1996). The GP64 envelope fusion protein is an essential baculovirus protein required for cell-to-cell transmission of infection. J. Virol. *70*, 4607–4616.

Morikawa, S., Booth, T.F., and Bishop, D.H.L. (1991). Analysis for the requirements for the synthesis of virus like particles by feline immunodeficiency virus gag using baculovirus vectors. Virology *183*, 288–297.

Murray, K.D., and Elkington, J.S. (1989). Environmental contamination of egg masses as a major component of transgenerational transmission of the gypsy moth nuclear polyhedrosis virus (LdMNPV). J. Invertebr. Pathol. *53*, 324–334.

Myers, J.H., Malakar, R., and Cory, J.S. (2000). Sublethal nucleopolyhedrovirus effects on female pupal weight, egg mass size and vertical transmission in gypsy moth (Lepidoptera: Lymantriidae). Environ. Entomol. *29*, 1268–1272.

Nayak, S., and Herzog, R.W. (2009). Progress and prospects: immune responses to viral vectors. Gene Ther. doi:10.1038/gt.2009.148.

Noad, R.J., Stewart, M., Boyce, M., Celma, C.C., Willison, K.R., and Roy, P. (2009). Multigene expression of protein complexes by iterative modification of genomic Bacmid DNA. BMC Mol. Biol. *10*, 87 doi:10.1186/1471-2199-10-87

Ohkawa, T., Majima, K., and Maeda, S. (1994). A cysteine protease encoded by the baculovirus *Bombyx mori* nuclear polyhedrosis virus. J. Virol. *68*, 6619–6625.

Ohkawa, T., and Volkman, L.E. (1999). Nuclear F-actin is required for AcMNPV nucleocapsid morphogenesis. Virology *264*, 1–4.

Ohkawa, T., Washburn, J.O., Sitapara, R., Sid, E., and Volkman, L.E. (2005). Specific binding of *Autographa californica* nucleopolyhedrovirus occlusion-derived virus to midgut cells of *Heliothis virescens* larvae is mediated by products of *pif* genes *Ac119* and *Ac022* but not by *Ac115*. J. Virol. *79*, 15258–15264.

Oomens, A.G.P., and Blissard, G.W. (1999). Requirement of GP64 to drive efficient budding of *Autographa californica* multicapsid nucleopolyhedrovirus. Virology *254*, 297–314.

Palomares, L.A., and Ramírez, O.T. (2009). Challenges for the production of virus-like particles in insect cells: The case of rotavirus-like particles. Biochem. Eng. J. *45*, 158–167.

Pante, N., and Kann, M. (2002). Nuclear pore complex is able to transport macromolecules with diameters of about 39 nm. Mol. Biol. Cell *13*, 425–434.

Passarelli, L., and Guarino, L.A. (2007). Baculovirus late and very late gene regulation. Curr. Drug Targets *8*, 1103–1115.

Patmanidi, A.L., Possee, R.D., and King, L.A. (2003). Formation of P10 tubular structures during AcMNPV infection depends on the integrity of host cell microtubules. Virology *317*, 318–320.

Pearson, M.N., Groten, C., and Rohrmann, G.F. (2000). Identification of the *Lymantria dispar* nucleopolyhedrovirus envelope fusion protein provides evidence for a phylogenetic division of the Baculoviridae. J. Virol. *74*, 6126–6131.

Pearson, M.N., Russell, R.L., and Rohrmann, G.F. (2001). Characterization of a baculovirus-encoded protein that is associated with infected-cell membranes and budded virions. Virology *291*, 22–31.

Peng, J., Zhong, J.R., and Granados, R.R. (1999). A baculovirus enhancing alters the permeability of a mucosa peritrophic matrix from Lepidopteran larvae. J. Insect Physiol. *45*, 159–166.

Perera, O.P., Valles, S.M., Green, T.B., White, S., Strong, C.A., and Becnel, J.J. (2006). Molecular analysis of an occlusion body protein from *Culex nigripalpus* nucleopolyhedrovirus (CuniNPV). J. Invertebr. Pathol. *91*, 35–42.

Perera, O., Green, T.B., Stevens, Jr. S.M., White, S., and Becnel, J.J. (2007). Proteins associated with *Culex nigripalpus* nucleopolyhedrovirus occluded virions. J. Virol. *81*, 4585–4590.

Pijlman, G.P., Pruijssers, A.J.P., and Vlak, J.M. (2003). Identification of *pif-2*, a third conserved baculovirus gene required for per os infection of insects. J. Gen. Virol. *84*, 2041–2049.

Pillay, S., Meyers, A., Williamson, A.L., and Rybicki, E.P. (2009). Optimization of chimeric HIV-1 virus-like particle production in a baculovirus-insect cell expression system. Biotechnol. Prog. *25*, 1153–1160.

Popham, H.J., Bischoff, D.S., and Slavicek, J.M. (2001). Both *Lymantria dispar* nucleopolyhedrovirus enhancing genes contribute to viral potency. J. Virol. *75*, 8639–8648.

Possee, R.D., Sun, T.-P., Howard, S.C., Ayres, M.D., Hill-Perkins, M., and Gearing, K.L. (1991). Nucleotide sequence of the *Autographa californica* nuclear polyhedrosis virus 9.4 kbp EcoRI-I and –R (polyhedrin gene) region. Virology *185*, 229–241.

Possee, R.D., Hitchman, R.B., Richards, K.S., Mann, S.G., Siaterli, E., Nixon, C.P., Irving, H., Assenberg, R., Alderton, D., Owens, R.J., and King, L.A. (2008). Generation of baculovirus vectors for the high-throughput production of proteins in insect cells. Biotechnol. Bioeng. *101*, 1115–1122.

Raghow, R., and Grace, T.D. (1974). Studies on a nuclear polyhedrosis virus in *Bombyx mori* cells *in vitro*. 1. Multiplication kinetics and ultrastructural studies. J. Ultrastruct. Res. *47*, 384–399.

Rohrmann, G.F. (2008). Baculovirus molecular biology. www.ncbi.nlm.nih.gov/bookshelf/br.fcgi?book = bacvir

Rosinski, M., Reid, S., and Nielson, L.K. (2002). Kinetics of baculovirus replication and release using real time quantitative polymerase chain reaction. Biotechnol. Bioeng. *77*, 476–480.

Roy, P., and Noad, R. (2008). Virus-like particles as a vaccine delivery system: myths and facts. Hum. Vaccines *4*, 5–12.

Royer, M., Hong, S.S., Gay, B., Cerutti, M., and Boulanger, P. (1992). Expression and extracellular release of human immunodeficiency virus type 1 Gag precursors by recombinant baculovirus-infected cells. J. Virol. *66*, 3230–3235.

Russell, R.L.Q., and Rohrmann, G.F. (1990). A baculovirus polyhedral envelope protein: Immunogold localization in infected cells and mature polyhedra. Virology *174*, 177–184.

Sait, S.M., Begon, M., Thompson, D.J., and Harvey, J.A. (1996). Parasitism of baculovirus-infected, *Plodia interpunctella* by *Venturia canescens* and subsequent virus transmission. Funct. Ecol. *10*, 586–591.

Sakurai, H., Kawabata, K., Sakurai, F., Nakagawa, S., and Mizuguchi, H. (2008). Innate immune response induced by gene delivery vectors. Int. J. Pharm. *354*, 9–15.

Salminen, M., Airenne, K.J., Rinnankoski, R., Reimari, J., Valilehto, O., Rinne, J., Suikkanen, S., Kukkonen, S., Yla-Herttuala, S., Kulomaa, M.S., and Vihinen-Ranta, M. (2005). Improvement in nuclear entry and transgene expression of baculoviruses by disintegration of microtubules in human hepatocytes. J. Virol. *79*, 2720–2728.

Sandig, V., Hofmann, C., Steinert, S., Jennings, G., Schlag, P., and Strauss, M. (1996). Gene transfer into hepatocytes and human liver tissue by baculovirus vectors. Hum. Gene Ther. *7*, 1937–1945.

Saville, G.P., Thomas, C.J., Possee, R.D., and King, L.A. (2002). Partial redistribution of the *Autographa californica* nucleopolyhedrovirus chitinase in virus-infected cells accompanies mutation of the carboxyl-terminal KDEL ER-retention motif. J. Gen. Virol. *83*, 685–694.

Saville, G.P., Patmanidi, A.L., Possee, R.D., and King, L.A. (2004). Deletion of the *Autographa californica* nucleopolyhedrovirus chitinase KDEL motif and *in vitro* and *in vivo* analysis of the modified virus. J. Gen. Virol. *85*, 821–831.

Senger, T., Schadlich, L., Gissmann, L., and Muller, M. (2009). Enhanced papillomavirus-like particle production in insect cells. Virology *388*, 344–353.

Shoji, I., Aizaki, H., Tani, H., Ishii, K., Chiba, T., Saito, I., Miyamura, T., and Matsuura, Y. (1997). Efficient gene transfer into various mammalian cells, including non-hepatic cells, by baculovirus vectors. J Gen. Virol. *78*, 2657–2664.

Sivakumar, S., Wang, Z., Harrison, R.L., Liu, S., Miller, W.A., and Bonning, B.C. (2009). Baculovirus-expressed virus-like particles of *Pea enation mosaic virus* vary in size and encapsidate baculovirus mRNAs. Virus Res. *139*, 54–63.

Slack, J.M., Kuzio, J., and Faulkner, P. (1995). Characterisation of v-cath, a cathepsin L-like proteinase expressed by the baculovirus *Autographa californica* multiple nuclear polyhedrosis virus. J. Gen. Virol. *76*, 1091–1097.

Slack, J.M., Dougherty, E.M., and Lawrence, S.D. (2001). A study of the *Autographa californica* nucleopolyhedrovirus multiple nucleopolyhedrovirus ODV envelope protein p74 using a GFP tag. J. Gen. Virol. *82*, 2279–2287.

Slack, J.M., and Lawrence, S.D. (2005). Evidence for proteolytic cleavage of the baculovirus occlusion-derived virion envelope protein P74. J. Gen. Virol. *86*, 1637–1643.

Slack, J., and Arif, B.M. (2007). The baculovirus occlusion-derived virus: virion structure and function. Adv. Virus Res. *69*, 99–165.

Slack, J.M., Lawrence, S.D., Krell, P.J., and Arif, B.M. (2008). Trypsin cleavage of the baculovirus occlusion-derived virus attachment protein P74 is a prerequisite in per os infection. J. Gen. Virol. *89*, 2388–2397.

Slavicek, J.M., and Popham, H.J.R. (2005). The *Lymantria dispar* nucleopolyhedrovirus enhancins are components of the occlusion-derived virus. J. Virol. *79*, 10578–10588.

Spenger, A., Grabherr, R., Tollner, L., Katinger, H., and Ernst, W. (2002). Altering the surface properties of baculovirus *Autographa californica* NPV by insertional mutagenesis of the envelope protein gp64 Eur. J. Biochem. *269*, 4458–4467.

Stannard, L.M., Himmelhoch, S., and Wynchank, S. (1996). Intra-nuclear localization of two envelope proteins, gB and gD, of herpes simplex virus. Arch. Virol. *141*, 505–524.

Stewart, M., Bhatia, Y., Athmaran, T.N., Noad, R., Gastaldi, C., Dubois, E., Russo, P., Thiery, R., Sailleau, C., Breard, E., Zientara, S., and Roy, P. (2010). Validation of a novel approach for the rapid production of immunogenic virus-like particles for bluetongue virus. Vaccine in press.

Strauss, R., Huser, A., Ni, S., Tuve, S., Kiviat, N, Sow, P. S., Hofmann, C., and Lieber, A. (2007). Baculovirus-based vaccination vectors allow for efficient induction of immune responses against *Plasmodium falciparum* circumsporozoite protein. Mol. Ther. *15*, 193–202.

Summers, M.D. (1971). Electron microscopic observations on granulosis virus entry, uncoating and replication processes during infection of the midgut cells of *Trichoplusia ni*. J. Ultrastruct. Res. *35*, 606–625.

Summers, M.D. (2006). Milestones leading to the genetic engineering of baculoviruses as expression vector systems and viral pesticides. Adv. Virus Res. *68*, 3–73.

Summers, M.D., and Volkman, L.E. (1976). Comparison of biophysical and morphological properties of occluded and extracellular nonoccluded baculovirus from *in vivo* and *in vitro* host systems. J. Virol. *17*, 962–972.

Tan, P.H., Janes, S.E., Handa, A.I., and Friend, P.J. (2008). More trouble ahead; is gene therapy coming of age? Expert Opin. Biol. Ther. *8*, 561–567.

Tanada, Y., and Kaya, H.K. (1993). Insect pathology (San Diego: Academic Press).

Tani, H., Nishijima, M., Ushijima, H., Miyamura, T., and Matsuura, Y. (2001). Characterization of cell-surface determinants important for baculovirus infection. Virology *279*, 343–353.

Thomas, C.J., Brown, H.L., Hawes, C.R., Lee, B.Y., Min, M.K., and King, L.A. (1998). Localization of a baculovirus-induced chitinase in the insect cell endoplasmic reticulum. J. Virol. *72*, 10207–10212.

Thiem, S.M., and Miller, L.K. (1989). Identification, sequence, and transcriptional mapping of the major capsid protein gene of the baculovirus *Autographa californica* nuclear polyhedrosis virus. J. Virol *63*, 2008–2018.

Tsai, C.T., Chuang, C.K., and Hu, Y.C. (2009). Baculovirus-mediated gene transfer into mesenchymal stem cells. Methods Mol. Biol. *515*, 339–351.

van Loo, N.D., Fortunati, E., Ehlert, E., Rabelink, M., Grosveld, F., and Scholte, B.J. (2001). Baculovirus infection of nondividing mammalian cells: mechanisms of entry and nuclear transport of capsids. J. Virol. *75*, 961–970.

van Oers, M.M. (1993). Functional domains of the p10 protein of *Autographa californica* nuclear polyhedrosis virus. J. Gen. Virol. *74*, 563–574.

van Oers, M.M., Flipsen, J.T.M., Reusken, C.B.E., and Vlak, J.M. (1994). Specificity of baculovirus p10 functions. Virology *200*, 513–523.

Vasconcelos, S.D. (1996). Alternative routes for the horizontal transmission of a nucleopolyhedrovirus. J. Invertebr. Pathol. *68*, 269–274.

Verhey, K.J., and Hammond, J.W. (2009). Traffic control: regulation of kinesin motors. Nat. Rev. Mol. Cell Biol. *10*, 765–777.

Vieira, H.L., Estevao, C., Roldao, A., Peixoto, C.C., Sousa, M.F., Cruz, P.E., Carrondo, M.J., and Alves, P.M. (2005). Triple layered rotavirus VLP production: kinetics of vector replication, mRNA stability and recombinant protein production. J. Biotechnol. *120*, 72–82.

Volkman, L.E., and Goldsmith, P.A. (1985). Mechanism of neutralization of budded *Autographa californica* nuclear polyhedrosis virus by a monoclonal antibody: Inhibition of entry by adsorptive endocytosis. Virology *143*, 185–195.

Volkman, L.E., Goldsmith, P.A., and Hess, R.T. (1987). Evidence for microfilament involvement in budded *Autographa californica* nuclear polyhedrovirus virus production. Virology *156*, 32–39.

Volkman, L.E., and Zaal, K.J. (1990). *Autographa californica* M nuclear polyhedrosis virus: microtubules and replication. Virology *175*, 292–302.

Volkman, L.E. (2007). Baculovirus infectivity and the actin cytoskeleton. Curr. Drug Targets *8*, 1075–1083.

Wang, P., and Granados, R.R. (1997). An intestinal mucin is the target substrate for a baculovirus enhancin. Proc. Natl. Acad. Sci. U.S.A. *94*, 6977–6982.

Wang, C.-Y., Li, F., Yang, Y., Guo, H.-Y., Wu, C.-X., and Wang, S. (2006). Recombinant baculovirus containing the Diphtheria toxin A gene for malignant glioma therapy. Cancer Res. *66*, 5798–5806.

Wang, Y., Wang, Q., Liang, C., Song, J., Li, N., Shi, H., and Chen, X. (2008a). *Autographa californica* multiple nucleopolyhedrovirus nucleocapsid protein BV/ODV-C42 mediates the nuclear entry of P78/83. J. Virol. *82*, 4554–4561.

Wang, M., Tan, Y., Yin, F., Deng, F., Vlak, J.M., Hu, Z., and Wang, H. (2008b). The F-like protein Ac23 enhances the infectivity of the budded virus of *gp64*-null *Autographa californica* multinucleocapsid nucleopolyhedrovirus pseudotyped with baculovirus envelope fusion protein F. J. Virol. *82*, 9800–9804.

Washburn, J.O., Chan, E.Y., Volkman, L.E., Aumiller, J.J., and Jarvis, D.L. (2003a). Early synthesis of budded virus envelope fusion protein GP64 enhances *Autographa californica* multicapsid nucleopolyhedrovirus virulence in orally infected *Heliothis virescens*. J. Virol. *77*, 280–290.

Washburn, J.O., Trudeau, D., Wong, J.F., and Volkman, L.E. (2003b). Early pathogenesis of *Autographa californica* multiple nucleopolyhedrovirus and *Helicoverpa zea* single nucleopolyhedrovirus in *Heliothis virescens*: a comparison of the 'M' and 'S' strategies for establishing fatal infections. J. Gen. Virol. *84*, 343–351.

Westenberg, M., Uijtdewilligen, P., and Vlak, J.M. (2007). Baculovirus envelope fusion proteins F and GP64 exploit different receptors to gain entry into cultured insect cells. J. Gen. Virol. *88*, 302–306.

Westernberg, M., and Vlak, J.M. (2008). GP64 of group I nucleopolyhedroviruses cannot readily rescue infectivity of group II *f*-null nucleopolyhedroviruses. J. Gen. Virol. *89*, 424–431.

Whitt, M.A., and Manning, J.S. (1988). A phosphorylated 34-kDa protein and a sub-population of polyhedrin are thiol-linked to the carbohydrate layer surrounding a baculovirus occlusion body. Virology *163*, 33–42.

Williams, G.V. (1989). A cytopathological investigation of *Autographa californica* nuclear polyhedrosis virus p10 gene function using insertion/deletion mutants. J. Gen. Virol. *70*, 187–202.

Williams, G.V., Rohel, D.Z., Kuzio, J., and Faulkner, P. (1989). A cytopathological investigation of *Autographa californica* nuclear polyhedrosis virus p10 gene function using insertion/deletion mutants. J. Gen. Virol. *70*, 187–202.

Wilson, M.E., Mainprize, T.H., Friesen, P.D., and Miller, L.K. (1987). Location, transcription and sequence of a baculovirus gene encoding a small arginine-rich polypeptide. J. Virol. *61*, 661–666.

Wilson, J. M. (2008). Adverse events in gene transfer trials and an agenda for the new year. Hum. Gene Ther. *19*, 1–2.

Wu, Q., Fang, L., Wu, X., Li, B., Luo, R., Yu, Z., Jin, M., Chen, H., and Xiao, S. (2009). A pseudotype baculovirus-mediated vaccine confers protective immunity against lethal challenge with H5N1 avian influenza virus in mice and chickens. Mol. Immunol. *46*, 2210–2217.

Zaiss, A.K., Machado, H.B., and Herschman, H.R. (2009). The influence of innate and pre-existing immunity on adenovirus therapy. J. Cell Biochem. *108*, 778–790.

Zanotto, P.M., Kessing, B.D., and Maruniak, J.E. (1993). Phylogenetic interrelationships among baculoviruses: evolutionary rates and host associations. J. Invertebr. Pathol. *62*, 147–164.

Zeng, J., Du, J., Lin, J., Bak, X.Y., Wu, C., and Wang, S. (2009). High-efficiency transient transduction of human embryonic stem cell-derived neurons with baculoviral vectors. Mol. Ther. *17*, 1585–1593.

Zhang, F., and Thiem, S.M. (2009). The *Trichoplusia ni* cell line MSU-TnT4 does not harbour a latent nodavirus. *In Vitro* Cell Dev. Biol. DOI 10.1007/s11626-009-9241-2.

Zhou, J., and Blissard, G.W. (2008). Display of heterologous proteins on *gp64*-null baculovirus virions and enhanced budding mediated by a vesicular stomatitis virus G-stem construct. J. Virol. *82*, 1368–1377.

Zhou, W., Yao, L., Xu, H., Yan, F., and Qi, Y. (2005). The function of envelope protein P74 from *Autographa californica* multiple nucleopolyhedrovirus in primary infection to host. Virus Genes *30*, 139–150.

3

Densoviruses: a Highly Diverse Group of Arthropod Parvoviruses

Max Bergoin and Peter Tijssen

Abstract

Densoviruses (DNVs) are defined as small (25 nm), non-enveloped viruses with an icosahedral symmetry containing an unsegmented single-stranded linear DNA genome, 4–6 kb in length, which terminates in short duplex hairpin telomeres involved in DNA replication. According to these properties, DNVs are bona fide members of the family *Parvoviridae,* along with vertebrate parvoviruses. The genome of DNVs contains two sets of genes encoding non-structural (NS) and capsid (VP) proteins. The manner their coding sequences are organized and transcribed as well as the structure of their non-coding 3′ and 5′ extremities appear very diversified. Some DNVs have a monosense organization, i.e. their gene products are encoded in tandem from a single DNA strand. Others have an ambisense organization, i.e. their NS and VP coding sequences are located in the 5′ half on both complementary strands. Most DNVs cause fatal diseases in their hosts. However, mortality is less common for shrimp DNVs. Owing to their specificity and high virulence, they have been considered for the biological control of some major insect pests. The ability of recombinant DNV genomes to integrate into host cell DNA has been exploited for stable transformation of insect cells allowing constitutive expression of foreign genes and somatic transformation of insects.

Introduction

The description in the early 1960s by Vago and coworkers of a pathological condition in the greater wax moth, *Galleria mellonella,* under the name 'densonucléose' is at the origin of the discovery of densoviruses (DNVs) (Meynadier *et al.*, 1964; Vago *et al.*, 1964). Examination of ultrathin sections from tissues of heavily infected larvae revealed that the enlarged, strongly Feulgen-positive nuclei observed in histological sections (at the origin of the name 'densonucléose') corresponded to electron-dense viral factories producing thousands of small isometric particles (Amargier *et al.*, 1965). Icosahedral, non-enveloped virus particles 20–22 nm in diameter were then isolated and purified from dead larvae and their chemical analysis revealed that they contained DNA and proteinaceous moieties.

Further biochemical and biophysical analyses of the virion components showed that the capsid was made up of 4 proteins and contained a single-stranded linear DNA molecule about 6 kb in size. This virus thus appeared to share the main biological (nuclear site of replication), structural and biochemical features of vertebrate parvoviruses and was accordingly classified as a new genus, *Densovirus,* within the family *Parvoviridae.* Since then, several DNVs have been isolated not only from different insect orders, including Lepidoptera, Hemiptera, Orthoptera, Diptera and Dictyoptera, but also from Crustacea, providing evidence of the ubiquity of DNVs in arthropods. Presently, some thirty DNVs have been identified, a figure very low compared to the one million-odd insect species, suggesting that many more remain to be discovered. Different aspects of arthropod parvoviruses, including their associated pathologies, have recently been reviewed (Bergoin and Tijssen, 1998, 2008; Fédière, 2000; Flegel, 2006; Tijssen *et al.*, 2006; Tijssen and Bergoin, 1995).

The cloning and complete sequencing of some twenty DNV genomes during the last two decades provided significant information on the biology of DNVs at the molecular level and revealed the great diversity of their genomic structure and organization. Furthermore, the discovery of highly conserved sequence motifs in both invertebrate and vertebrate parvoviruses demonstrated that they very likely originate from common ancestors.

We have divided this chapter into four major parts, the first dealing with the biological properties of DNVs, the second presenting the overall properties of DNV virus particles and their components, the third presenting the molecular biology of DNVs and the fourth the applications of DNVs as expression vectors or as biological control agents.

Biological properties of DNVs

Most DNVs are responsible for fatal diseases of their insect host in contrast to those that have crustacean hosts. In insects, they infect essentially larval stages, although infections at pupal or imaginal stages were also observed in natural populations or following experimental contamination at late larval instars or nymphs (Boemare and Brès, 1969; Suto *et al.*, 1979).

Symptoms

In most Lepidoptera, symptoms associated with DNV infections start with anorexia and lethargy followed by flaccidity and inhibition of moulting or metamorphosis. Larvae become whitish and progressively paralysed (Chao *et al.*, 1985; Kelly *et al.*, 1980; Meynadier *et al.*, 1964; Vago *et al.*, 1966). Death occurs in 2–20 days depending on the DNV isolate, larval stage, virus concentration, and the way of infection: natural (epidemic) or experimental (per os or inoculation) transmission. The cockroach *Periplaneta fuliginosa* infected with the PfDNV displays a typical symptom of hind legs paralysis and uncoordinated movements (Suto *et al.*, 1979). Similar symptoms are observed in the house cricket *Acheta domesticus* infected with AdDNV (Meynadier *et al.*, 1977c).

Some DNVs induce tumour lesions in the intestine of their hosts. More than half of the infected *P. fuliginosa* cockroaches develop ulcers in the hindgut as a result of accumulation of haemocytes around injured hindgut epithelial cells (Suto, 1979). Typical tumours generated by the intensive mitotic activity of regenerative cells were observed in *Leucophaea maderae* midgut or gastric caeca (Boufersaoui and Matz, 1974). These tumours which have a densoviral aetiology can be induced by severance of the recurrent nerve and are transmissible by injection of their filtered homogenates (Matz and Bergoin, 1984). Tumour lesions in the intestine have also been observed in slug caterpillars *Sibine fusca* and *Casphalia extranea* heavily infected with SfDNV and CeDNV, respectively (Amargier *et al.*, 1979; Fédière, 1983; Meynadier *et al.*, 1977a). The intensive proliferation of the midgut epithelial cells leads to the thickening of the gut wall and almost complete obstruction of the gut lumen. Mosquito larvae infected with DNVs exhibit symptoms of paralysis. Despite its lack of cytopathic effect in the mosquito C6/36 cell line (Barreau *et al.*, 1996), the AalDNV proved to be highly pathogenic for first instar *Aedes aegypti* larvae by per os infection. Affected larvae with distorted and curved bodies lost their mobility and hung near to the water surface.

So far, three different shrimp parvoviruses have been found involved in infections (Flegel, 2006); infectious hypodermal and haematopoietic necrosis virus (IHHNV, renamed as *Penaeus stylirostris* DNV, PstDNV), hepatopancreatic parvoviruses (*Penaeus monodon* DNV, PmDNV, *Penaeus merguiensis* DNV, PmergDNV, and *Penaeus chinensis* DNV, PchinDNV, with sequence identities of about 70%; Flegel, 2006) and the little characterized spawner-isolated mortality virus (SMV). PstDNV-infected juvenile or sub-adult *P. stylirostris* show a high mortality whereas PstDNV-infected *P. vannamei* present cuticular deformities (specially deformed rostra) and stunted growth referred as runt-deformed syndrome (Bell and Lightner, 1984, 1987). Infected adults seldom show this syndrome, although they may carry the virus for life and transmit it both vertically and horizontally (Lightner, 1996). PmDNV-infected shrimp show the highest density of lesions (basophilic inclusions in enlarged nuclei) in dividing epithelial cells at the distal ends of hepatopancreatic tubules. No distinctive pathology has been described for SMV, other than the original outbreak of high mortality.

Tissue tropism

Most DNVs known so far present a wide tissue tropism. Thus, in *G. mellonella*, the virus replicates in almost all tissues, except midgut epithelium. Fat body, hypodermis, muscles, tracheal cells, Malpighian tubules and haemocytes are the most severely infected tissues (Amargier, 1966; Amargier *et al.*, 1965; Bergoin and Brès, 1968), but infection of ovaries and the central nervous system have also been reported (Bergoin and Brès, 1968; Garzon and Kurstak, 1968). A similar tissue polytropism was observed in other DNV-infected Lepidoptera such as *Spodoptera littoralis* infected with *Junonia coenia* DNV, JcDNV (Diallo, 1978), or *Agraulis vanillae* (Kelly *et al.*, 1980), *Diatarea saccharalis* (Meynadier *et al.*, 1977b), *Pieris rapae* (Sun *et al.*, 1981) and *Pseudoplusia includens* (Chao *et al.*, 1985) infected with their native DNVs. Similarly, DNVs infecting mosquitoes and crickets appear to have a wide tissue tropism (Barreau *et al.*, 1996; Buchatsky, 1989; Meynadier *et al.*, 1977c). In contrast, the DNVs belonging to the genus *Iteravirus* (see Table 3.1) replicate either exclusively (*Bombyx mori* DNV, BmDNV-1) or predominantly (SfDNV and CeDNV) in the midgut epithelium cells of their hosts (Amargier *et al.*, 1979; Fédière, 1996; Maeda and Watanabe, 1978; Watanabe and Kurihara, 1988; Watanabe *et al.*, 1976). PfDNV appears to replicate also exclusively in the hindgut epithelial cells of its native host (Suto *et al.*, 1979). The tropism of shrimp DNVs extends to most tissues of ectodermal and mesodermal origin (Lightner and Redman, 1985; Lightner *et al.*, 1983).

Histopathology and ultrastructure of DNV-infected cells

As mentioned above, the hypertrophy of nuclei which stain intensively with methyl green or Feulgen reagent or exhibit green fluorescence after staining with acridine orange at acidic pH is the salient histopathological feature of all DNV-infected cells (Garzon and Kurstak, 1968; Vago *et al.*, 1964). Other cytochemical staining reactions such as immunofluorescent or immunoperoxidase-labelled antibodies have been used to localize viral antigen synthesis and to follow the time course analysis of DNV infections. GmDNV viral antigens were detected about 4–7 hours post inoculation at 32°C in the cytoplasm of *G. mellonella* haemocytes and only 4 to 6 hours later in the nucleus (Kurstak and Stanislawski-Birencwajg, 1968). The same technique was used by (Maeda and Watanabe, 1978) to detect the first appearance of BmDNV-1 antigens in columnar cells of *B. mori* at 24 hours post infection (p.i.). The principal feature of PstDNV lesions is the presence of large eosinophilic intranuclear Cowdry type A inclusions in hypertrophied nuclei (Lightner *et al.*, 1983).

Ultrastructural studies performed in the 1960–1970s using different tissues of infected *G. mellonella* larvae and different cytochemical approaches have led to a comprehensive description of the pathogenesis of DNV infection. The first ultrastructural changes in GmDNV infection occur during the first 6 hours p.i. with the rapid disappearance of polyribosomes and formation of cytoplasmic microbody-like structures arising from the accumulation of spherical bodies 17–20 nm inside of vesicles. This step could correspond to accumulation and transport of viral capsid proteins from the cytoplasm to the nucleus. The main changes in the nucleus concern the hypertrophy of the nucleolus along with the segregation of its fibrillar and granular components. Simultaneously, a virogenic stroma corresponding to accumulation of double-stranded DNA of the replicative form appears in close vicinity to the nucleolus. The rapid growth of the virogenic stroma is accompanied by the hypertrophy of the nucleus and pushing of the chromatin and the nucleolus to the nuclear periphery. The virus morphogenesis occurs inside the virogenic stroma and mature virions progressively form aggregates which fuse to give rise to large areas filled with viral particles. At the final step of infection, the nuclear envelope is disrupted releasing virions in the cytoplasm where they often appear as paracrystalline arrays. In several DNV-infected insects, the formation of cytoplasmic paracrystalline virions arrays occurs prior to or without disruption of the nuclear membrane (Chao *et al.*, 1985; Charpentier, 1979;

Diallo, 1978). Ultrathin sections of shrimp tissues infected with PstDNV or the hepatopancreatic PmDNV revealed very similar type of virus morphogenesis. Virions appear in electron-dense granular or fibrillar material (virogenic stroma) in hypertrophied nuclei (Lightner and Redman, 1985; Lightner *et al.*, 1983). The early steps in the replication of DNVs: adsorption, penetration and transport of the virion to the nuclear membrane have long been ignored. Recently, a precise analysis of the entry of JcDNV virions in *Lymantria dispar* LD 652 cells has been reported (Vendeville *et al.*, 2009). Following their attachment to the cell membrane, JcDNV particles are rapidly internalized into clathrin-coated vesicles and slowly traffic within early and late endocytic compartments. The mechanism of entry of DNVs seems thus similar to that observed for vertebrate parvoviruses (Bartlett *et al.*, 2000; Linser *et al.*, 1979).

Host range

Investigations on the host range of DNVs indicate that it varies considerably from one isolate to another. While some DNVs such as GmDNV, CeDNV and AdDNV have a host range apparently restricted to their original hosts (Fédière, 1996; Giran, 1966; Jousset *et al.*, 1986), others such as JcDNV and *Mythimna loreyi* DNV (MlDNV) have a broader host range. The JcDNV can replicate in *Aglais urticae, B. mori, Chrysodexis chalcites, L. dispar, Mamestra brassicae, M. oleracea, Scotia ipsilon, Spodoptera littoralis, S. exigua* but not in *G. mellonella*. Similarly, MlDNV is infectious for *Chilo agamemnon, G. mellonella, Ostrinia nubilalis, Pectinophora gossypiella, S. cretica* and *S. littoralis* (Fédière, 1996). The host range of PfDNV extends to four other species of the genus *Periplaneta*: *P. americana, P. australasiae, P. brunnea* and *P. japonica* (Suto, 1979). The host range of mosquito DNVs, *Aedes albopictus* DNV (AalDNV) and *A. aegypti* DNV (AeDNV), extend to different species. Larvae of *A. albopictus, A. cantans, A. caspius, A. geniculatus, A. vexans, Culex pipiens,* and *Culiseta annulata* are all susceptible to per os infection with AeDNV (Lebedeva *et al.*, 1973). In contrast, *Culex pipiens* DNV (CpDNV) does not replicate in *A. aegypti* (Jousset *et al.*, 2000). In the case of BmDNV-1, the only available information about its host spectrum concerns its ability to infect *Glyphodes pyloalis*, a pyralid infesting mulberry plantations of sericultural farms. Of practical interest for sericulture was the demonstration that the susceptibility of silkworm to BmDNV-1 varied from one strain to another and resistant strains could be selected. The mode of inheritance of the resistance to BmDNV-1 has been investigated and it was established that the non-susceptibility is determined by two genes, *nsd-1* (non-susceptibility to BmDNV-1) and *nid-1* (non-infection to BmDNV-1), located at chromosome 21 and 17, respectively (Eguchi *et al.*, 2007). Finally it is worth mentioning that despite their high virulence for their hosts, DNVs do not appear to be able to replicate in vertebrates, including humans (see section 'Potential of DNVs as biopesticides to control insect pests').

DNV virions

Classification of DNVs

Densoviruses are defined as small isometric, non-enveloped viruses containing an unsegmented single-stranded, linear DNA genome. This definition excludes viruses such as BmDNV-2 from this virus family. BmDNV-2 has also a genome of linear single-stranded DNA, but unlike parvoviruses, it is segmented and twice as long and codes for proteins with a polymerase motif (Hayakawa *et al.*, 2000). Like their hosts, ranging from various insect orders to Crustacea, parvoviruses of arthropods (subfamily *Densovirinae*) form a heterogeneous group. Overall, two types of DNVs can be distinguished: the ambisense DNVs that encode open reading frames (ORFs) on both complementary strands and the monosense DNVs that use only one strand for this purpose. The ambisense DNVs package both single-stranded DNAs in separate capsids. Upon DNA extraction under high-salt conditions, double-stranded genomes are obtained. In contrast, monosense DNVs tend to package mainly one type of DNA. Both the ambisense and monosense DNVs have been subdivided into several genera (Tattersall, 2006; Tattersall *et al.*, 2005). There are also a large number of unclassified DNVs due to insufficient physicochemical and molecular biology data.

Two recognized genera, *Densovirus* and *Pefudensovirus*, exist for the ambisense DNVs but two other candidate genera, 'Aphidensovirus' and 'Cupidensovirus', can also be recognized. For the monosense DNVs, two genera, *Brevidensovirus* and *Iteravirus*, have been recognized but a third candidate genus, 'Hepandensovirus', can equally be recognized. The characteristics of these genera, as well as the DNVs belonging to them, are summarized in Table 3.1.

All members of the *Densovirus* genus have lepidopteran hosts and have high sequence identities of 85–95% (Dumas *et al.*, 1992; Fediere *et al.*, 2004; Tijssen *et al.*, 2003). However, recently a second DNV was isolated from *G. mellonella*, GmDNV-2, with an identity of about 50%, but overall properties similar to those of other viruses from this genus (Tijssen, unpublished results). Moreover, GmDNV-2 lacks the significant serological cross-reactivity that exists among the established members of this genus.

The members of the *Pefudensovirus* genus have low sequence identities (often less than 30%), but retain a striking uniformity genome regarding structure and expression strategy (Guo *et al.*, 2000; Kapelinskaya *et al.*, 2008; Thao *et al.*, 2001; Yamagishi *et al.*, 1999; Yang *et al.*, 2008). The low sequence identity may be a result of the diversity of the hosts who belong to diverse orders of insects. Recent results in our laboratories have shown that these viruses have the unique property among parvoviruses of having two sets of structural proteins that make up the capsid (VP2, VP3, VP4 and VP1, VP3, VP4; Bergoin and Tijssen, unpublished results). The expression strategy of the NS proteins of this genus is identical to that of the DNVs of the *Densovirus* genus.

The proposed genus Aphidensovirus contains aphid DNVs that have a high sequence identity. However, the distribution of NS-ORFs is different in the published data (Ryabov *et al.*, 2009; van Munster *et al.*, 2003a,b). In contrast, the VP-ORFs are very similar, both in structure and sequence identity.

So far, CpDNV is the only member for the proposed Cupidensovirus genus. Its expression strategy for the NS proteins is very different from the other ambisense DNVs.

Like the *Densovirus* genus, all members of the *Iteravirus* genus have lepidopteran hosts. The iteraviruses have high sequence and genome structure identities (Fediere *et al.*, 2002; Li *et al.*, 2001; Wang *et al.*, 2005). They are the only monosense DNVs that use a phospholipase A2 in their largest VP to enter cells, like ambisense DNVs and virtually all vertebrate parvoviruses (Zádori *et al.*, 2001).

Members of the Brevidensoviruses are found in mosquitoes and shrimp that have either a partial or complete part of their life cycle in water (Boublik *et al.*, 1994; Ren *et al.*, 2008; Shike *et al.*, 2000). Parts of their NS1 proteins (Walker motifs) have a significant identity with other parvovirus NS proteins. The capsid protein is much smaller than the capsid proteins of other parvoviruses and shows very little sequence identity.

The proposed Hepandensovirus genus has so far three members (PmDNV, PmergDNV and PchinDNV) that differ in genome length but have 70% sequence identity (Flegel, 2006). PmDNV has a typical parvovirus genome structure, although it is the longest among all parvoviruses (Sukhumsirichart *et al.*, 2006). Although the VP-ORF is very long, with a coding potential of a 92-kDa protein, only a 57-kDa capsid protein was observed.

Physicochemical properties of DNV virions

Most of the data available on the physicochemical properties of DNV virions were essentially obtained for members of the first discovered two genera *Densovirus* and *Iteravirus* in the 1960s and 1970s, i.e. before the development of recombinant DNA technology in the late 1970s. At that time, virus particles were mainly characterized by their hydrodynamic properties and biochemical analysis of their components. Later, the newly discovered DNVs have been characterized essentially by cloning and sequencing their genomes, sometimes even without thoroughly purifying the virus particles.

Like most of the vertebrate parvoviruses, DNV virions are very resistant to extreme environmental conditions, including a wide pH range and temperature, especially in an unpurified state (dead larvae). However, as a purified virus suspension or in dry state, the infectivity tends to decrease rapidly (Boemare *et al.*, 1970; Odier and Boemare, 1972).

Table 3.1 Isolates and classification of densoviruses

Ambisense densoviruses			
Genus *Densovirus*, 6-kb genome with 0.55-kb ITRs, 3/1 ORFs for NS/VP			
JcDNV	*Junonia coenia*	Lepidoptera	NC_004284
GmDNV	*Galleria mellonella*	Lepidoptera	NC_004286
DsDNV	*Diatraea saccharalis*	Lepidoptera	NC_001899
HaDNV	*Helicoverpa armigera*	Lepidoptera	Sequenced, not submitted
MlDNV	*Mythimna loreyi*	Lepidoptera	NC_005341
PiDNV	*Pseudoplusia includens*	Lepidoptera	Sequenced, not submitted
Genus *Pefudensovirus*, 5.5-kb genome with 0.2-kb ITRs, 3/2 ORFs for NS/VP			
AdDNV	*Acheta domesticus*	Orthoptera	Sequenced, not submitted
BgDNV	*Blattella germanica*	Orthoptera	NC_005041
PcDNV	*Planococcus citri*	Hemiptera	NC_004289
PfDNV	*Periplaneta fuliginosa*	Dictyoptera	AB028936
Proposed genus Aphidensovirus, 5.5-kb genome, 0.1-kb ITRs, 2/2 ORFs for NS/VP			
MpDNV	*Myzus persicae*	Hemiptera	NC_005040
DplDNV	*Dysaphis plantaginea*	Homoptera	EU851411
Proposed genus Cupidensovirus, 5.8-kb genome, 0.3-kb ITRs, 6/1 ORFs for NS/VP			
CpDNV	*Culex pipiens*	Diptera	FJ810126
Monosense densoviruses			
Genus *Iteravirus*, 5-kb genome, 0.25-kb ITRs, phospholipase A2 motif in VP, 2/1 ORFs for NS/VP			
CeDNV	*Casphalia extranea*	Lepidoptera	AF375296
BmDNV-1	*Bombyx mori*	Lepidoptera	AY033435
DpDNV	*Dendrolimus punctatus*	Lepidoptera	NC_006555
Genus *Brevidensovirus*, 4-kb genome, no ITRs but terminal hairpins, no phospholipase A2 motif in VP, 2/1 ORFs for NS/VP			
AeDNV	*Aedes aegypti*	Diptera	M37899
AalDNV-1	*Aedes albopictus* (cell line)	Diptera	NC_004285
AalDNV-2	*Aedes albopictus* (cell line)	Diptera	AY095351
AgDNV	*Anopheles gambiae*	Diptera	NC_011317
PmoDNV	*Penaeus monodon*	Decapoda	AF499102
PstDNV	*Penaeus stylirostris*	Decapoda	AF273215

Table 3.1 continued

Proposed genus Hepandensovirus, 5.7–6.3-kb genome, 0.2-kb terminal hairpins, no phospholipase A2 motif in VP 2/1 ORFs for NS/VP			
PmDNV	*Penaeus monodon*	Decapoda	EU247528
PmergDNV	*Penaeus merguienses*	Decapoda	NC_007218
Non-characterized densoviruses			
AvDNV	*Agraulis vanillae*	Lepidoptera	Kelly *et al.* (1980)
EaDNV	*Euxoa auxilliaris*	Lepidoptera	Sutter (1973)
LduDNV	*Leucorrhinia dubia*	Odonata	Charpentier (1979)
LdiDNV	*Lymantria dispar*	Lepidoptera	Grignon (1982)
PrDNV	*Pieris rapae*	Lepidoptera	Sun (1981)
SfDNV	*Sibine fusca*	Lepidoptera	Meynadier *et al.* (1977a)
HeDNV	*Haemagogus equinus*	Diptera	O'Neill *et al.* (1995)
TaDNV	*Toxorhynchites amboinensis*	Diptera	O'Neill *et al.* 1995)
SvDNV	*Simulium vittatum*	Diptera	Federici (1976)

LdiDNV, HeDNV, TaDNV were isolated from cell lines.

The total molecular mass of DNVs is about 5–6 × 10^6 Da of which the DNA represents 20–30%. DNVs do not contain lipids but several of them have been shown to contain the three polyamines putresceine, spermidine (predominant) and spermine comprising about 1.4% of the virus molecular mass. These polyamines are sufficient to neutralize 26% and about 15% of the ssDNA encapsidated in the virion of GmDNV and BMDNV-1, respectively (Bando *et al.*, 1983; Kelly and Elliott, 1977).

Two types of infectious virus particles have been reported for GmDNV (Tijssen *et al.*, 1977), PfDNV (Suto, 1979) and BmDNV-1 (Nakagaki and Kawase, 1980). The two virus populations differ in both their buoyant density and sedimentation coefficient. The light virus particles (DNV-I) are predominant and band in equilibrium CsCl gradients at a buoyant density of 1.40 g/ml, whereas the heavy particles (DNV-II) concentrate as a thin band at a density of 1.44 g/ml. DNV-I virions sediment at about 110S, whereas DNV-II sediment at about 90S. Empty capsids have a sedimentation coefficient of 60S and a buoyant density of 1.30 g/ml (Longworth *et al.*, 1968). The almost complete conversion of DNV-II to DNV-I in the presence of bivalent cations and its reversibility in the presence of a chelating agent (EDTA) strongly supported the role of these cations in the quaternary structure of the virus capsid (Tijssen *et al.*, 1977). The demonstration of a Ca^{2+} binding site in the phospholipase A2 region of VP1 and of the critical role played by bivalent cations for PLA2 activity provided a good explanation of these two forms (Zadori *et al.*, 2001).

Owing to their high DNA content and their low light scattering due to their small size, highly purified virus suspensions of DNVs have a UV absorption spectrum close to that of a nucleoprotein complex with a peak around 260 nm and a 260/280 absorbance ratio of about 1.50–1.60. The extinction coefficient of DNVs virus particles varies from 6.9 to 9.5 cm^2/mg at 260 nm.

The organization of the viral DNA molecule inside the capsid of DNVs and its interactions with internal VP sequences or polyamines are not known. However, the single-strandedness of the encapsidated viral DNA has been demonstrated by the flame-red fluorescence of virus suspensions stained with acridine orange and the hyperchromicity (around 10%) and shift in absorption maximum (from 260 to 264) following 24 hours exposure of the virus to 1.8% formaldehyde (Barwise and Walker, 1970; Kelly *et al.*, 1977; Suto, 1979).

Structure and morphogenesis of DNV virions

The X-ray crystallographic structures of parvoviruses have been reviewed previously (Chapman and Agbandje-McKenna, 2006; Chapman and Rossmann, 1993). Among the DNVs, only the structure of GmDNV has so far been established (Simpson *et al.*, 1998), although the PstDNV structure to 2.5 Å resolution has recently been obtained by X-ray crystallography (Rossmann, unpublished). In comparison, the structure of seven parvoviruses of vertebrates has been established. Unfortunately, these viruses have very low sequence identity with DNVs making it impossible to extrapolate.

All parvoviruses have a diameter of about 250 Å and a T-1 structure (60 subunits in an icosahedron with a 2–3–5 point symmetry (Fig. 3.1). Therefore, 31 rotational symmetry axes intersect at the centre of the virus particle. The PDB databank (www.rcsb.org) contains the coordinates of a GmDNV subunit, accession number 1DNV. Oligomers or the complete capsid can be generated at the Viperdb databank (viperdb.scripps.edu) using the Oligomer Generator (in the Utilities tab) and can then be visualized by many different molecular viewing programs. A common problem of these programs is that the location and extent of the secondary structure elements in cartoons do not represent the helix and β-strand notations in the original file. The Cn3D 4.1 visualization program (NCBI) does not take the data directly from the PDB databank and uses the MMDB database (parsed versions of the PDB files) but does not accurately assign secondary structures in cartoons. We use the popular PyMol program (version 0.99 is free but supported versions are relatively cheap). To make subunits in an oligomer individually accessible, one should first modify the pdb file (or output from Viperdb) by a software text editor [e.g. ConTEXT 0.98 (free)]. GmDNV pdb files for PyMol corrected for secondary structures, or how-to methods, are available from one of the authors (P.T.).

Structures are much more conserved than sequences. For instance, when the subunits of minute virus of mice (MVM) and Porcine Parvovirus (PPV) are superimposed, remarkably 96% of the Cαs of the amino acids have virtually identical positions (root mean square deviation of 1.0 Å) despite having only 52% sequence identity. A high degree of structure conservation is often observed to below 30% sequence identity, i.e. between distinct parvovirus genera.

All parvoviruses have a 'jellyroll' fold (β-barrel or β-sandwich) consisting of two curved sheets of four antiparallel β-strands (Fig. 3.1A1 and A2). The strands participating in this fold are labelled A, B, C, to F from the N- to the C-terminus and can be folded as CHEF and BIDG sheets (Fig. 3.1A2). The BIDG sheet is near the inner surface and the CHEF is shorter and near the viral surface. The B and D β-strands point to the 5-fold axis and have a right-handed twist to the viral surface. An important difference between vertebrate parvoviruses and DNVs is the position of the A β-strand and its role in dimeric interactions ('domain swapping' see below and Fig. 3.1D).

The loops between the β-strands are named after the β-strands they bridge (e.g. loop GH is loop 3 in Fig. 3.1A1). These loops are prominent in the viral surface and the interactions of subunits, particularly in the trimeric and pentameric oligomers (Fig. 3.1B and 3.1E). The outside of densovirus particles is much smoother than that of vertebrate parvoviruses, mainly due to the absence of loop 4. Instead of a 3-fold spike, DNVs have a β-annulus at the 3-fold axis (Fig. 3.1B). A clear depression is seen in the structure at the 2-fold axis, like for vertebrate parvoviruses. A roadmap of the surface amino acids has been shown elsewhere (Tijssen *et al.*, 2003), but can also be analysed in the Viperdb database (under Data>Family Index>Parvoviridae>Densovirus to Biodata>φ-ψ explorer; contact tables are found in Annotations).

Assembly of the virion for MVM (Lombardo *et al.*, 2002; Riolobos *et al.*, 2006) and PPV (Tijssen, unpublished) starts with the formation of trimers (Fig. 3.1C), which, if folded correctly,

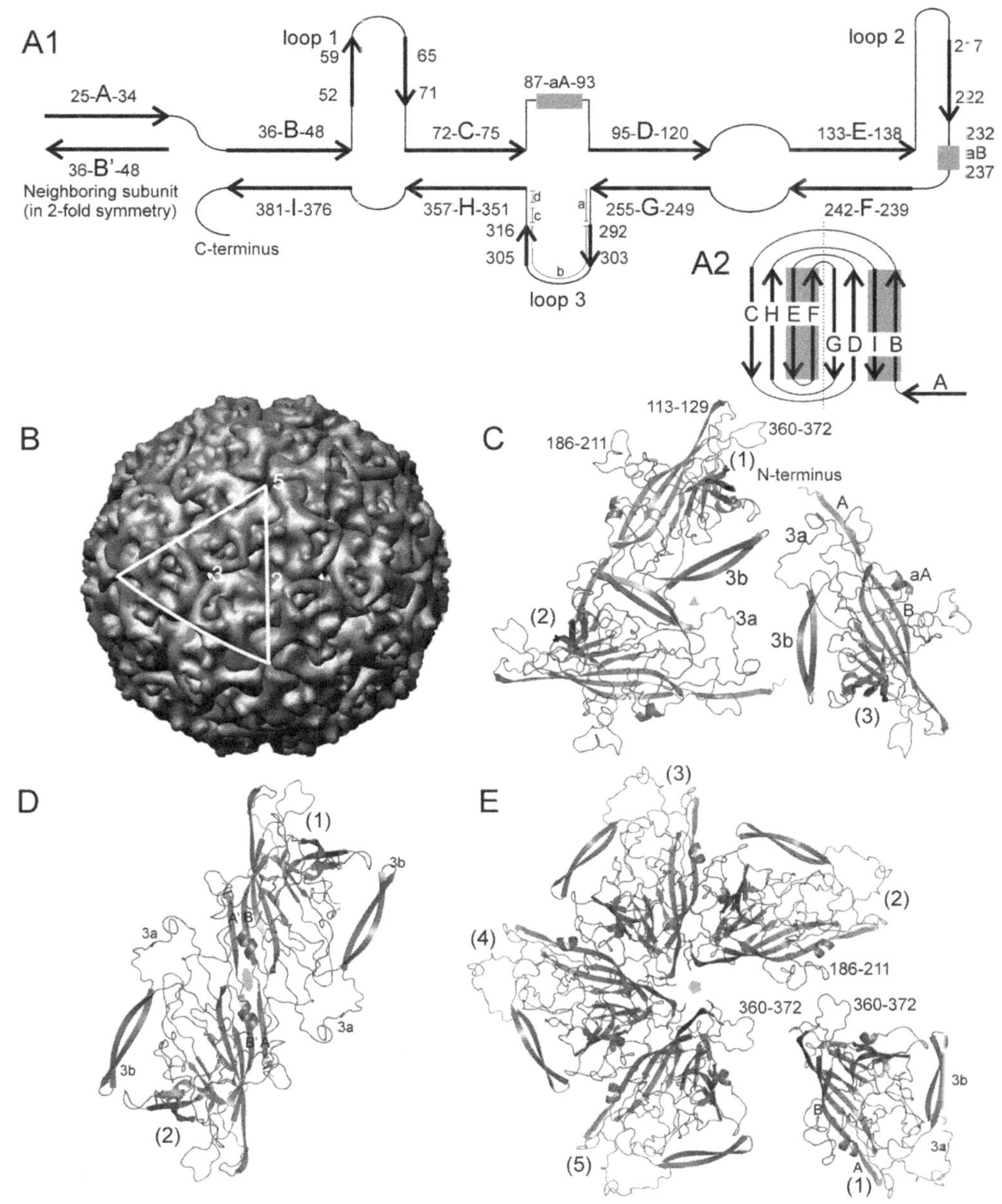

Figure 3.1 X-ray crystallographic structure of GmDNV. A1 shows the position of the secondary structures of the capsid proteins (arrows: β-strands; gray rectangles: α-helices A and B; thin line: random coil). The antiparallel β-strands can also be represented as in A2, whereby BIDG forms one leaflet and CHEF the other leaflet that can be folded along the dotted line to form a β-barrel that forms the inner core and inner surface of the structural protein. (B) GmDNV particle rendered to 5 Å 'resolution' by the Chimera program (Viperdb. scripps.edu). The vertexes of the triangle are at the 5-fold symmetry axes. Twenty triangles (trimers of the structural protein) cover the whole particle. In contrast to vertebrate parvoviruses, this virus has a channel at the 3-fold axis (indicated by 3). The 2-fold axes on the three edges go through structural depressions. (C) Trimer of capsid proteins. Subunit (3) has been translated 30 Å to the right to emphasize the large number of intertwining contacts between the subunits. The gray triangle in the centre indicates the position of the 3-fold axis and the β-annulus channel. (D) Dimer with grey oval indicating 2-fold axis. There is very little intertwining between the 2 subunits. (E) Pentamer of capsid protein with subunit (1) translated 25 Å to the right and downwards to demonstrate the small number of interwoven loops near the 5-fold channel (penton indicating 5-fold axis). C, D, and E were obtained with the PyMol software after correction for secondary structures and individualizing the subunits in the oligomers.

acquire a nuclear localization signal and thus are allowed to participate in the formation of empty capsids in the nucleus. Likewise, intertwining of loop 3 domains to create densovirus trimer intermediates is much more complex than the succeeding steps (Fig. 3.2). The 3b β-stranded loop interweaves with the 3a and 3c domains of the neighbouring subunit. Additional contacts are made with loops 2, 3d, 5a and 5b (Fig. 3.2) to yield very tight trimers.

The next step is probably the formation of pentamers of these trimers and includes simultaneously the 2-fold interactions. The 5-fold interactions are relatively simple, involving part of loops 1, 2 and HI (Fig. 3.1E), whereas the 2-fold interaction involves hardly any intertwining (Fig. 3.1D). In the latter case, the A and the linearly extended B β-strands interact in an antiparallel fashion with the A′ and B′ β-strands of the neighbouring subunits at the inner surface of the subunits. The α-helix A and part of β-strand D on top of these antiparallel β-strands interact with amino acids 381–400 in the C-terminus of the neighbouring subunit.

Both vertebrate parvoviruses, such as canine parvovirus (CPV) and PPV, and GmDNV have a 5-fold channel with a similar diameter. However, the channel is much longer for the vertebrate parvoviruses (e.g. G37, V138, L169, N166, V164, S156, T158 and Q159 for CPV) than for GmDNV (L127, F115, T117) and serves as an entry portal of the viral DNA into the provirion (probably using the viral helicase as a motor) and externalization of the phospholipase domain during cell entry. These aspects have not yet been investigated for DNVs. The role of the β-annulus at the 3-fold axis of GmDNV (Fig. 3.1B) instead of a 3-fold spike seen for vertebrate parvoviruses, is not clear.

The 3D structure of capsid of *A. albopictus* C6/36 cells densovirus was also determined to 14-Å resolution by electron cryomicroscopy and computer reconstruction (Cheng *et al.*, 2004). Like for GmDNV, there is a central hole in each triangular face, but unlike GmDNV there is a spike on each 5-fold vertex. Structure comparisons between the C6/36 DNV and other parvoviruses (Cheng *et al.*, 2007) suggested that the degree of structural similarity between C6/36 DNV and the human parvovirus B19 is higher than that between C6/36 DNV and GmDNV. Moreover, the amino acid sequence comparisons of structural and non-structural proteins, surprisingly, suggested higher levels of similarity between C6/36 DNV and parvovirus B19 than those between C6/36 DNV and GmDNV. These results await confirmation by X-ray crystallographic studies.

Molecular biology of DNVs

Overview

The genome of all vertebrate parvoviruses consists of a small (4.5–5.5 kb) linear, single-stranded DNA molecule containing two major genes organized in a monosense manner, i.e. non-structural (NS) and structural (VP) genes located on the same strand. The NS gene occupying the 5′ half (by convention designated 'left' half) of the coding sequence encodes a small number of replication proteins whereas the structural (VP) gene occupies the 'right' half of the coding sequence and encodes two or three capsid proteins. The coding sequence is flanked by short (120–400 nt) imperfect terminal palindromes able by folding back on themselves to form duplex telomeres. These terminal hairpins render the genome self-priming for complementary strand synthesis by host cell DNA polymerase (Cotmore and Tattersall, 2006). Alternative splicing is the mechanism usually employed by vertebrate parvoviruses to regulate the expression of their overlapping genes.

Like their vertebrate counterparts, and in accordance with their small size (4–6 kb), DNV genomes are limited in their coding capacities. However, the manner their two sets of NS and VP genes are organized and transcribed as well as the structure of their non-coding 3′ and 5′ extremities appear very diversified. The structural criteria retained for their classification by the International Committee on Taxonomy of Viruses (ICTV) take into account the number and size of ORFs encoding NS and VP proteins, their location on the same strand (monosense) or both strands (ambisense) of the genome as well as the size, sequence and folding properties of the 5′ and 3′ extremities. The availability of molecular biology data, as outlined below, on more DNVs justifies in our opinion the creation of new

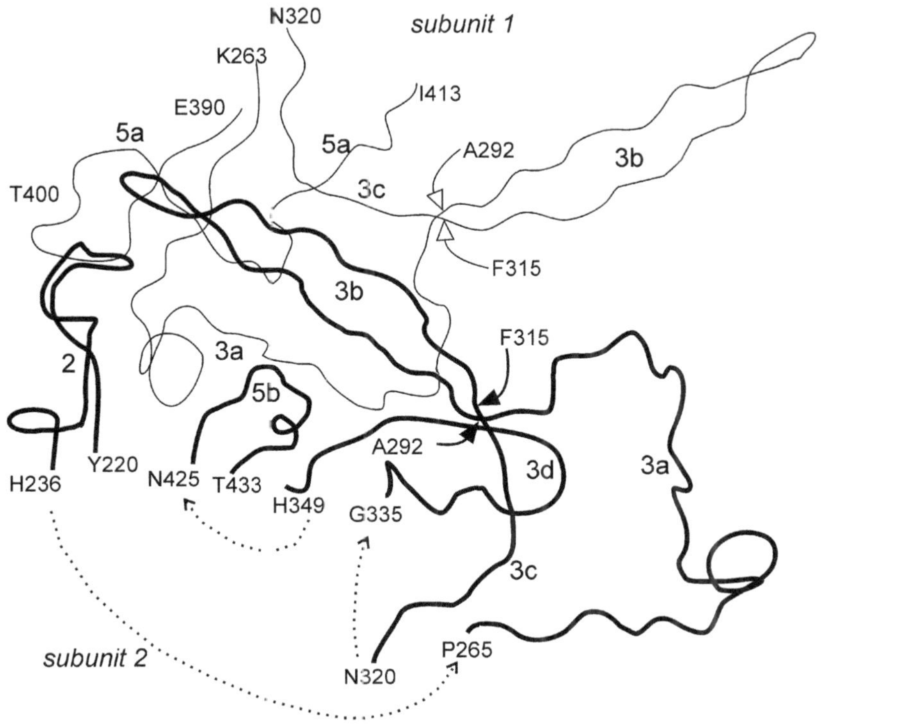


```
Major contacts between subunits 1 and 2

Loop 2
2-TYR220 -> 1-GLN273,PRO274,GLY278
2-GLN221 -> 1-ILE271
2-GLN222 -> 1-LYS277,GLY278
2-GLN225 -> 1-ILE280
2-GLY226 -> 1-ILE270,LYS277
2-THR227 -> 1-ILE271
2-GLY228 -> 1-ILE271,PHE396,SER399
2-GLY229 -> 1-ILE271,PHE396,PRO397

Loop 3
1-ASN267 -> 2-GLU306
1-TYR268 -> 2-GLU306,THR308
1-LYS269 -> 2-ALA305,GLU306
1-ILE270 -> 2-GLY226,GLN302,GLU306,GLN221
            THR227,GLY228,GLY229
1-GLN273 -> 2-LEU68,GLY145,ILE146
            MET219,TYR220,ASN425
1-PRO274 -> 2-ILE146,TYR220,PHE424,ASN425
1-THR275 -> 2-PHE424,THR433
1-VAL276 -> 2-LEU427
1-LYS277 -> 2-GLN222,GLY226
1-GLY278 -> 2-ASN66, TYR220
3a
1-THR279 -> 2-GLN222
1-ILE280 -> 2-GLN225,ALA300,GLN302
1-ASP284 -> 2-PRO297,ARG313
1-ASN285 -> 2-ASN296,LEU420,LEU427,GLY431
1-LEU286 -> 2-THR295,PHE315,ILE413,LEU415
            SER419,LEU420,ALA432
1-VAL287 -> 2-VAL294,ASN296
1-ASN288 -> 2-VAL293,VAL294,PHE315
1-MET289 -> 2-VAL294,ILE344,PRO346
2-ARG290 -> 1-ARG290
1-ARG290 -> 2-VAL294,THR312,ASN314
1-GLY291 -> 2-THR312
1-ALA292 -> 2-THR312

2-VAL293 -> 1-ASN288
2-VAL294 -> 1-VAL287,ASN288,MET289,ARG290
2-THR295 -> 1-LEU286
2-ASN296 -> 1-ASN285,VAL287
2-PRO297 -> 1-ASP284
2-ALA300 -> 1-ILE280
2-GLN302 -> 1-ILE270,ILE280
2-ALA305 -> 1-LYS269
3b
2-GLU306 -> 1-ASN267,TYR268,LYS269,ILE270
2-SER307 -> 1-ASP318
2-THR308 -> 1-TYR268,ASP318,TYR411
2-ASN310 -> 1-ALA317,HSD339,PRO342,TYR411
2-LEU311 -> 1-PRO316
2-THR312 -> 1-ARG290,GLY291,ALA292
2-ARG313 -> 1-ASP284
2-ASN314 -> 1-ARG290
2-PHE315 -> 1-LEU286,ASN288
3c
1-PRO316 -> 2-LEU311
1-ALA317 -> 2-ASN310
1-ASP318 -> 2-SER307,THR308

Loop 3d
1-HSD339 -> 2-ASN310
1-PRO342 -> 2-ASN310
2-ILE344 -> 1-MET289

C-terminus ("Loop 5")
1-HSD395 -> 2-GLU235
1-PHE396 -> 2-TRP144,GLY228,GLY229,PRO231
5a
1-PRO397 -> 2-GLY229
1-SER399 -> 2-GLY228
1-PRO406 -> 2-LEU427
1-TYR411 -> 2-THR308,ASN310
2-ASN425 -> 1-GLN273,PRO274
2-LEU427 -> 1-VAL276,ASN285,PRO406
5b
2-GLY431 -> 1-ASN285
2-ALA432 -> 1-LEU286
2-THR433 -> 1-THR275
```

Figure 3.2 Three-fold symmetry interactions in GmDNV trimers. Wire diagram showing the major 3-fold contacts between subunits (1; thin line) and (2; bold line) and interaction of critical amino acids. For loop 2, for instance, 2-Tyr220 (#220 in subunit (2)) interacts with 1-Gln273, 1-Pro274 and 1-Gly278 in subunit (1). Similarly, 1-Asn267 in loop 3a of subunit (1), interacts with 2-Glu306 in loop 3b of subunit (2) (2-Asn267 would interact with 3-Glu306 that is not shown in this diagram). The most striking feature of these interactions is the three 3b β-stranded hairpins that are held between the 3a and 3c loops of the neighbouring subunit. Moreover, there is a significant interaction of loop 3a with loops 2 and 5b.

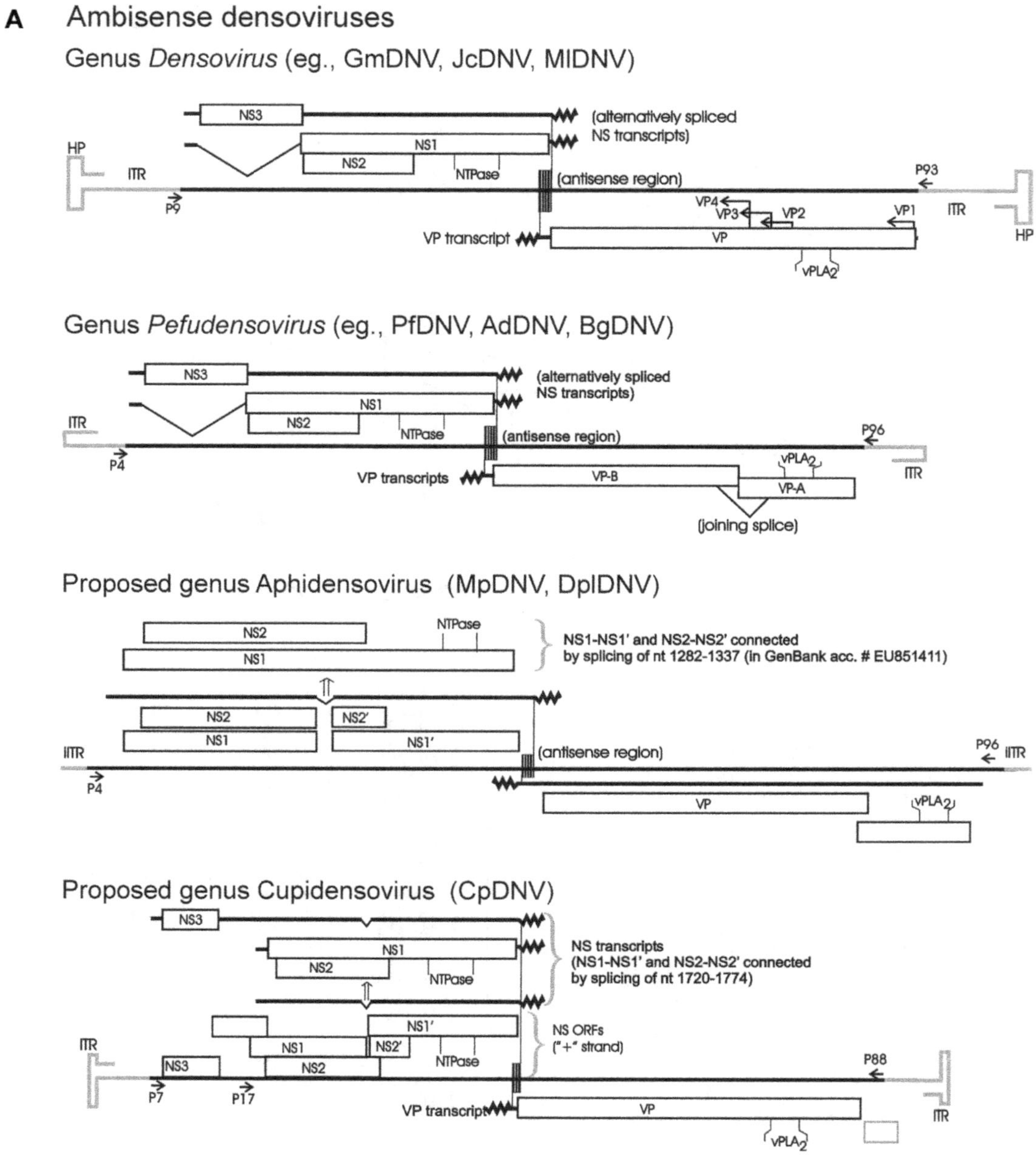

Figure 3.3 Expression strategies of ambisense and monosense densoviruses. (A) The ambisense densoviruses can be divided into 4 genera, two of which have split ORFs for the gene of the structural proteins and two of which have split ORFs for the non-structural proteins (four combinations). The *Densovirus* genus members have in general a high sequence identity (about 85%) and identical genome organization. The sequences of the pefudensoviruses is very diverse, perhaps due to the diversity of their hosts, but their genome organization and expression strategy is identical. The sequences of MpDNV and DplDNV are incomplete but have very high sequence identities and show a unique expression strategy. A new genus, Aphidensovirus, is proposed for these densoviruses. CpDNV cannot be fitted into any of these genera and a new genus, Cupidensovirus, is proposed. All ambisense densoviruses as well as the iteraviruses have 'flip' and 'flop' orientations in their inverted terminal repeats (ITRs), although those of the 'Aphidensovirus' have not been fully sequenced (incomplete ITR or iITR).

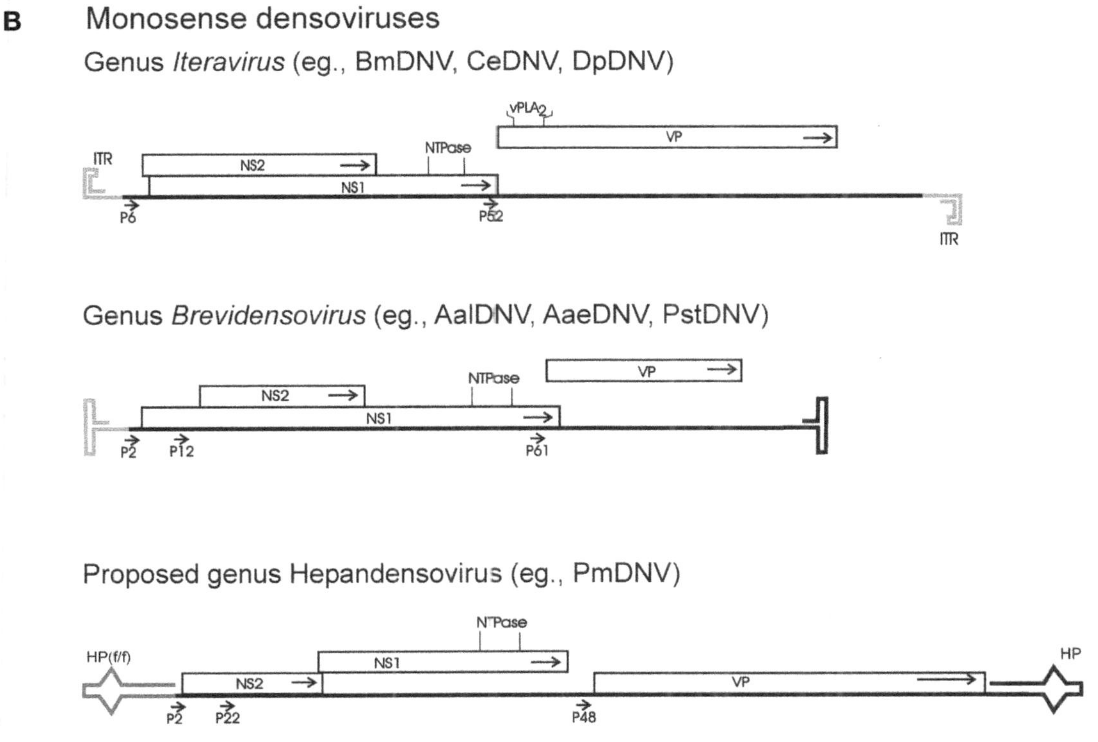

Figure 3.3 (continued) (B) The monosense densoviruses are, with respect to sequence identity, more diverse than the ambisense densoviruses and are so far divided into 2 genera and an additional proposed genus. Only members of the *Iteravirus* genus carry the phospholipase A2 motif which is a hallmark enzyme in parvovirus capsids. The brevidensoviruses have the smallest genome (4 kb) among parvoviruses, whereas the members of the newly proposed hepandensoviruses have the largest genome (> 6 kb). Transcription strategies of the monosense densoviruses, except their promoters for use in expression vectors, have not yet been investigated. HP represents hairpin, f/f flip-flop sequences and P# the promoters.

genera. A schematic representation of the various strategies is shown in Fig. 3.3.

Genus Densovirus

Members of this genus have a genome of about 6 kb in length possessing long (>500 nt) inverted terminal repeats (ITRs) (Dumas *et al.*, 1992; Fediere *et al.*, 2004; Tijssen *et al.*, 2003). Their genome exhibits an ambisense organization, with sequences encoding NS and VP proteins being located in the 5′ half on the complementary strands. During the course of virus morphogenesis both strands are encapsidated separately in equimolecular ratio. The sequences encoding NS polypeptides are organized in three ORFs on the NS strand designated by convention as the positive strand. The left-most ORF encoding NS3 is separated only by a stop codon (TAA) from the main ORF encoding NS1. A third ORF overlapping the N-terminal region of NS1, encodes NS2 in a frame different from NS1.

The NS genes are under the control of the P7 promoter whose TATA box and upstream promoter elements are located in the 3 terminal sequence of the left ITR. Two transcripts sharing the same transcription start have been identified for NS genes. An unspliced 2.5 kb mRNA directs the translation of NS3 polypeptide from the 5′ proximal initiation codon to generate a protein of about 20 kDa. The splicing of the 2.5 kb mRNA from a position just upstream of the NS3 AUG codon to an acceptor site just upstream of NS1 initiation codon generates a 1.8 kb mRNA deleted of the NS3 coding sequence. This transcript allows the translation of the NS1/NS2 cassette to generate two polypeptides of about 30 and 60 kDa by a leaky scanning mechanism. The initiation codons of NS1 and NS2 are very close to each other, but the rather unfavourable context of NS1 AUG codon very likely facilitates the leaky scanning to the second (NS2) AUG codon.

The VP gene is located in a single large ORF occupying the 5′ half of the complementary strand. This gene is under the control of the P93 promoter located in an equivalent position as the P7 NS promoter in the left-most sequence of the right ITR. A single 2.5 kb VP mRNA is transcribed from this promoter and four structural polypeptides VP1–4 shown to be N-extended isoforms are generated by a leaky scanning mechanism.

The 550 nucleotides of GmDNV and MlDNV as well as the 517 nt of JcDNV ITRs exhibit a complex secondary structure when analysed through the Fold program. The folding of the distal 100 nucleotides of each extremity is able to generate a Y-shaped structure serving as primer for replication. The sequence of ITR is highly conserved among members of the *Densovirus* genus and the region forming the two arms of the Y-shaped structure occurs in two orientations called 'flip' and 'flop' as a result of hairpin transfer during the rolling circle mechanism of replication.

Genus Pefudensovirus

Members of this genus have a smaller genome than those of the *Densovirus* genus since their ITRs are much smaller at about 150 nt (Kapelinskaya *et al.*, 2008; Mukha *et al.*, 2006; Yamagishi *et al.*, 1999). These hairpins do not fold into Y-shaped telomeres but in regular fold-back hairpins. The strategy of their NS expression is identical to that of the *Densovirus* genus. However the VP gene cassette is different and has some unique features. The VP coding sequence is split into 2 ORFs, of which the smaller upstream ORF contains the phospholipase A2 domain and the larger downstream ORF contains the complete VP2 coding sequence. The VP1 protein is generated by splicing of the two ORFs. Unlike all other parvoviruses, the N-terminus of VP2 is therefore not contained within VP1. Unspliced VP transcripts yield proteins coded by the small ORF that do not participate in capsid formation and may have a, yet unknown, role in the viral cycle.

Proposed genus 'Aphidensovirus'

Only two densoviruses, with high sequence identity, have been described for this proposed genus (Ryabov *et al.*, 2009; van Munster *et al.*, 2003a). Although the sequences for *Dysaphis plantaginea* DNV (DplDNV) and *Myzus persicae* DNV (MpDNV) are still incomplete, a unique expression strategy is emerging for these viruses. Like for the *Pefudensovirus* members, the VP coding sequence is split into 2 ORFs. However, The NS1 and NS2 ORFs are split (as for the Cupidensovirus, see below) and spliced to yield in-frame coding sequences.

Proposed genus 'Cupidensovirus'

With a large (6kb), ambisense genome possessing an inverted repeated sequence at each extremity, the CpDNV genome shares the main characteristics of lepidopteran DNV genomes belonging to the genus *Densovirus*, such as a single ORF for VP. However, it deviates from this model in two structural domains (Baquerizo-Audiot *et al.*, 2009). The first concerns the structure of ITRs, which are significantly shorter (285 bases), lack the GAC repeats assumed to function as an NS1 binding site and their 68 terminal nucleotides fold back to generate a J-shaped structure. The second concerns the split of NS1 and NS2 coding sequences each in two ORFs, implying a splicing of the mRNA to put in frame their N-terminal and C-terminal sequences. A third significant difference lies in the mode of expression of NS genes, which requires two specific promoters, one (P7) to drive expression of NS3 and one (P17) to drive expression of NS1 and NS2. Both promoters have their TATA boxes located downstream of the ITR sequence

Genus Iteravirus

These viruses have been so far described only in three species of Lepidoptera: *B. mori*, *Casphalia extranea* and *Dendrolimus punctatus* (Fédière *et al.*, 2002; Li *et al.*, 2001; Wang *et al.*, 2005). As examplified by BmDNV and CeDNV, members of the *Iteravirus* genus possess a monosense genome of only 5.1 kb in size. It contains two overlapping ORFs in the left half encoding NS1 and NS2 polypeptides (no NS3) and a right half ORF encoding the four or five VP proteins.

So far, no transcription map of iteravirus coding sequences has been published. However, it has been clearly demonstrated by peptide mapping that both the four BmDNV-1 and the five CeDNV structural polypeptides are N-extended isoforms, a result strongly supporting the hypothesis that, like members of the genus *Densovirus*, the iteravirus structural proteins are translated by leaky scanning of a single mRNA (Bando *et al.*, 1984; Li *et al.*, 2001; Fédière *et al.*, 2002).

The coding sequences of iteraviruses are flanked at each extremity by a 230 nt ITR, the first 160 of which are able to fold to form a J-shaped terminal hairpin. As for the ITRs within the genus *Densovirus*, this hairpin is present in the two possible flip/flop orientations.

Genus Brevidensovirus

This genus encompasses DNVs with the shortest (4kb) genome (Afanasiev *et al.*, 1991; Boublik *et al.*, 1994; Shike *et al.*, 2000; Zhai *et al.*, 2008). They have been described in several mosquitoes (including mosquito cell lines) and shrimp species. In addition to their small size, their genome differs from that of other genera of DNVs by the absence of ITRs, which are replaced by short (130–150) disparate hairpin structures at each extremity. The folding of these hairpins generates T-shaped structures of about 60 nucleotides very similar to those described in most vertebrate parvoviruses. The coding sequence of their monosense genome is organized in three overlapping ORFs, a large left ORF encoding NS1 and nested in its 5′ half, the ORF encoding NS2, and a short right ORF encoding the two VP capsid proteins with sizes of about 40 kDa only, that lack the phospholipase A2 domain.

Proposed genus 'Hepandensovirus'

The 6.3 kb genome of hepatopancreatic parvovirus (HPV), a ssDNA virus isolated from infected *Penaeus monodon* in Thailand contains three large ORFs and two non-coding termini (Sukhumsirichart *et al.*, 2006). The left (ORF1), mid (ORF2), and right (ORF3) ORFs may code for 50, 68, and 92 kDa proteins, respectively. The 5′ and 3′ ends of viral genome lacked ITRs, but contain hairpin-like structures of approximately 220 nts, respectively. The ORF2 contained conserved replication initiator motif, NTP-binding and helicase domain similar to NS1 of other parvoviruses. The ORF1 most probably codes for NS2 of unknown function. The ORF3 of the HPV genome has a coding potential for VP of about 92 kDa, lacking the phospholipase A2 motif, but by SDS-PAGE a single 57 kDa band was observed (maybe a trace of 92 kDa).

Viral proteins

Structural proteins

Whereas the capsid of vertebrate parvoviruses is usually made up of only two or three proteins with molecular masses in the range of 60–90 kDa, SDS-PAGE analysis of most of densovirus capsids reveals four polypeptides with molecular masses ranging from 40 to 90 kDa. Notable exceptions are the capsids of brevidensoviruses which contain only two major polypeptides with molecular masses not exceeding 50 kDa. Far before the sequencing of its genome, peptide mapping of the four GmDNV structural proteins revealed that they are closely related and are encoded by the same gene (Tijssen and Kurstak, 1981). This was confirmed more recently by showing that the four structural polypeptides are N-extended isoforms proteins translated by leaky scanning from a single mRNA. A common observation regarding DNV capsid proteins is the overestimation of about 10% of their molecular masses by SDS-PAGE analysis compared with their amino acid sequence. This discrepancy is very likely due to the fact that they seem to be phosphorylated.

Alignments of VP sequences revealed the presence of two conserved sequences in most vertebrate and invertebrate parvoviruses: a short stretch of amino acids (S/T-G) in the N-terminal sequences of densovirus VP4 and parvovirus VP2 proteins and a larger region in the unique sequence of VP1. This last region contains in its N-terminal sequence the structural calcium-binding loop (GPGN) and the phospholipase A2 (PLA2) motif (DxxAxxHDxxY). This enzymatic activity is essential for virus infectivity and is present in all genera of DNVs except for brevidensoviruses and hepandensoviruses. Expression of GmDNV, JcDNV, MlDNV and PstDNV VP sequences in a baculovirus expression system showed that the smallest VP is enough to generate virus-like particles.

Non-structural proteins

Compared to vertebrate parvoviruses, little is still known about the function(s) of the non-structural proteins of DNVs. Sequence alignments revealed that NS1 of all DNVs share homologies with their vertebrate counterparts. These homologies concern motifs involved in DNA replication including the RCR (rolling circle replication) nuclease and Walker A and B motif of ATPase, both characteristic of the NS1 proteins of vertebrate parvoviruses. In contrast, NS2 and NS3 proteins share very low homology among the different genera of DNVs.

By using a recombinant NS1 protein of JcDNV fused to a maltose binding tag, it has been shown that this protein specifically binds the four GAC tandem repeat sequence of the hairpin stem and nicks single-stranded forms of the hairpin at two specific sites (Ding *et al.*, 2002). The organization of the three NS proteins is conserved among DNVs with an ambisense genome. NS3 is always in the leftmost position and is translated separately by the 5′ extremity of an about 2.4 kb transcript bearing the NS1/NS2 coding sequences. The roles played by NS2 and NS3 are so far not yet elucidated, but it has been shown by deletion and rescue experiments that NS3 protein of JcDNV is essential for DNA replication (Abd-Alla *et al.*, 2004).

Practical applications of DNVs

The use of DNVs as expression vectors

The demonstration in 1990 that the transfection to *Spodoptera littoralis* larvae of as few as 30 ng of pBRJ DNA, a plasmid encompassing an almost complete viral sequence of the JcDNV genome, resulted in an infection in all respects identical to injection of a purified virus suspension (Jourdan *et al.*, 1990) prompted researches to investigate the potentialities of DNVs as expression vectors. Virions extracted from transfected larvae were undistinguishable from wild type virions with regard to their biophysical and biological properties. In particular, the DNA extracted from rescued virions and restricted with enzymes cutting specifically in the ITRs had exactly the same profile as wt DNA, showing that the 88 nt missing at the 3 termini of the viral DNA cloned in pBRJ was rescued during the process of DNA replication. This rendered possible the construction of recombinant viruses expressing reporter genes to analyse the structure, strength and mode of regulation of viral promoters. Furthermore, the presence at each extremity of the viral sequences of JcDNV and GmDNV genomes of large ITRs containing short direct repeats suggested that these structures could play a role similar to the extremities of AAVs genomes or transposable elements in the integration in host cell DNA. This in turn suggested their possible use as tools for stable transformation of insect cells, allowing constitutive expression of a foreign gene, and possibly the production of transgenic insects.

Originally, all JcDNV vectors were derived from pBRJ by inserting different reporter (*LacZ*, *cat*) genes in frame with the coding sequence of VPs or NS1 or with their own ATG initiation codons. These constructs were used for transient expression in *S. littoralis* SPCSL-52 cells (Giraud *et al.*, 1992). The highest levels of expression were achieved by inserting the reporter gene in frame and downstream of VP4 ATG codon, so that the four chimeric proteins were produced at the same time by the leaky scanning translation mechanism of VP mRNA. The observation that some blue cells could be detected after several subcultures led to the development of a second set of constructs expressing either the dominant selectable marker *neo*^{r} gene or the easily *in vivo* detectable marker green fluorescent protein (*gfp* gene) to analyse the persistence of expression of these reporter genes. Following transfections with different constructs expressing neomycin phosphotransferase; several G418-resistant cell colonies were obtained which successfully underwent 50 subcultures in the presence or absence of the antibiotic, indicating stable expression of the foreign gene. In cell clones originally transfected with constructs that expressed NS genes, the recombinant viral genome was detected as integrated into the cellular DNA at low copy number (one or few), whereas in cells transfected with constructs deleted of the NS genes (ΔNS), the viral DNA was maintained under a multicopy episomal form (Rolling, 1992).

A more precise study to elucidate the genomic status of JcDNV-derived vectors was undertaken by using constructs expressing the GFP reporter gene fused in frame into the major open reading frame encoding VP proteins under the control of the P9 promoter. A series of constructs that included either fully functional (pJGFPH), partially (pJGFPΔNS3), or entirely (pJGFPΔNSH) deleted NS genes were generated to assess the role of non-structural proteins in the persistence of JcDNV vectors (Bossin *et al.*, 2003) (Fig. 3.4). Transfection of *Spodoptera frugiperda* cells (Sf9) with each of the constructs conferred a persistent GFP phenotype to a high proportion of cells, exceeding at least 10 times the frequency of random insertion events of plasmids that did not contain densovirus DNA. This allowed direct isolation of GFP-fluorescent clones in the absence of selective pressure. Structural analyses of cell DNA extracted from different clones by pulsed-field gel electrophoresis, inverted PCR, Southern blotting and metaphase chromosome *in situ* hybridization revealed that the majority of the JcDNV plasmid sequence was integrated into the genome of the fluorescent clones. Integration was observed whether or not NS proteins were expressed but the presence of NS genes in the constructs greatly influenced the number of integrated copies and their distribution in the host genome. In cells transfected with pJGFPH, i.e. expressing the three NS proteins, integration was always found at low copy number and primed *in situ* synthesis analysis of metaphase chromosome spreads consistently revealed a single spot, indicating that pJGFPH had inserted in this fluorescent clone at a single site in the Sf9 polyploid genome. In contrast, integration of plasmid pJGFPΔNS3 occurred at low copy numbers in several sites into the host cell genome, and deletion of all NS genes (pJGFPΔNSH) resulted in integration of up to 60 copies of the vector in head-to-tail concatemers. Most important, sequence analysis of integration junctions by inverse PCR demonstrated that the *cis* JcDNV 5′ ITR region was the primary site of recombination. These findings clearly established that JcDNV vectors integrate into insect cells in a manner similar to AAV plasmids in mammalian cells. JcDNV-derived vectors thus appeared as appropriate tools for producing stably transformed insect cell lines able to express constitutively proteins of interest. Constructs expressing recombinant human erythropoietin, interleukin 2 and γ interferon have been constructed and expression of these proteins in stably transformed Sf9 cell clones was achieved (M. B., unpublished results). The potential of JcDNV-derived vectors for stable production of transgenic insects has also been explored. A first attempt was made by microinjecting *Drosophila melanogaster* syncytial embryos with plasmids containing the *lacZ* gene coding sequence inserted in frame with the capsid protein gene. High expression of β-galactosidase chimeric proteins was detected throughout ontogenesis from larvae to adult flies in different tissues including Malpighian tubules, tracheal cells, adipose cells and epidermis (Royer *et al.*, 2001). However, attempts to detect expression in G1 progeny repeatedly failed. The recombinant genomes thus appeared as efficient vectors for somatic transformation, but are apparently unable to integrate in the germ line cells to generate transgenic flies

Using the pJDsRedH plasmid, a pBRJ-derived construct where the GFP gene of pJGFPH was replaced by the Ds*Red* coding sequence to express the four chimeric VP-Ds*Red* proteins, Bossin and co-workers (Bossin *et al.*, 2007) obtained a high rate of somatic transformation by microtransfecting syncytial *Drosophila* eggs and eggs from three Lepidoptera species, *Ephestia kuehniella, Plodia interpunctella* and *Trichoplusia ni*. The somatic transformation efficiencies were higher using JcDNV vectors than using piggyBac vectors. Deletion of viral sequences expressing all NS proteins did not block somatic transformation in either *D. melanogaster* or *P. interpunctella* embryos. Similarly, insertion of a second reporter cassette expressing GFP under the control of 3xP3 promoter within or outside the densoviral sequence did not affect the rate of somatic transformation. However, unlike the transfections with piggyBac vectors, none of the JcDNV vectors were able to transform germline cells and no expression was detected in G1 progeny.

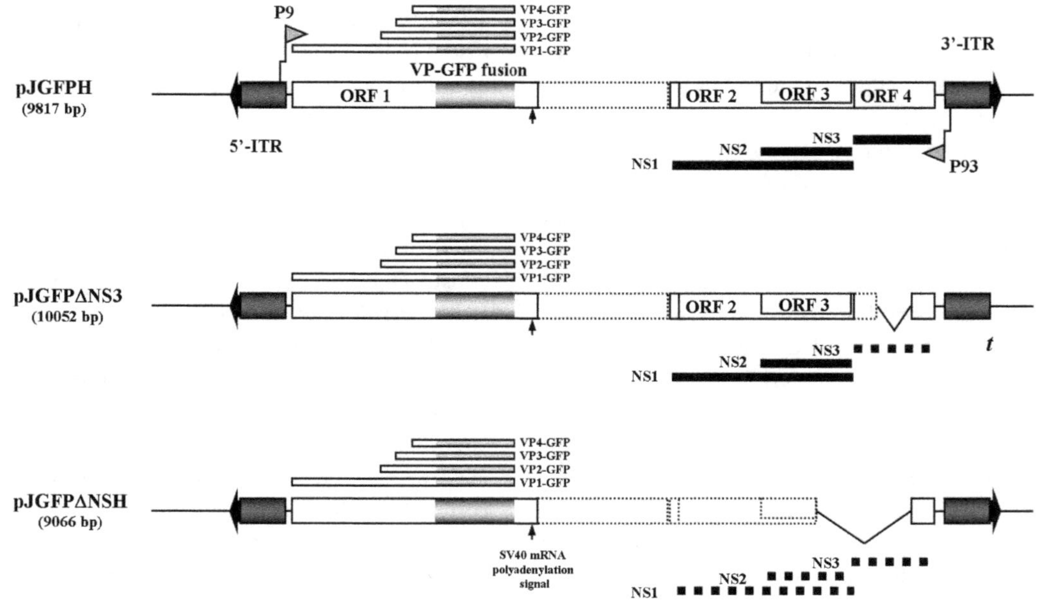

Figure 3.4 Densovirus expression vectors. Schematic organization of recombinant JcDNV plasmids expressing GFP-VP chimeric proteins. All plasmids derive from pBRJ, a plasmid encompassing an infectious sequence of the JcDNV genome (Jourdan *et al.*, 1990). The P9 and P93 viral promoters (dotted arrows) drive the expression of the VP-GFP gene fusion and of the NS genes, respectively. Plain thick arrows represent sequences that form self-priming hairpins at the end of the viral genome. *t* indicates the location of the 3′ ITR truncation on plasmid pJGFPDNS3, which is restored in pJGFPH and pJGFPΔNSH. VP–GFP fusion proteins are shown above their coding sequence. Dotted boxes show JcDNV regions that are no longer transcribed. Plain thick lines represent the NS proteins, and dark dotted lines show NS proteins that are no longer expressed (after Bossin *et al.*, 2003).

Parallel to JcDNV vectors, a plasmid carrying an infectious sequence of the *Aedes aegypti* DNV, a prototype of the *Brevidensovius* genus was constructed (Afanasiev *et al.*, 1994). When transfected to C6/36 cell line, this plasmid (pUCA) produced the same amount of the monomeric replicative form of viral DNA as did infection with wild-type virions. This plasmid was used to analyse the three putative P0.5, p7 and p61 promoters located upstream of the left, mid and right ORFs, respectively. These promoters were assumed to drive expression of NS1, NS2 and VP genes, respectively. The coding sequence of lacZ, GFP, hygromycin and zeocine resistance genes were inserted in frame with the coding sequence of the left (NS1) mid (NS2) and right (VP) ORFs and the constructs were used to transfect C6/36 cells. High expression levels of these reporter genes were achieved in cells transfected with the two pUCAP7 and pUCAP61 constructs. A 15-fold increase of expression from both promoters was achieved by cotransfecting the cells with a helper plasmid expressing the NS proteins. In contrast, attempts to express foreign genes inserted as non-fusion proteins with their own translational initiation codons have failed. Detailed analysis of the P7 promoter showed that the 35 nucleotides sequence immediately downstream of the TATA box are crucial for the expression of this promoter and that the sequences located on the 5′ terminus of NS transcript are necessary for the expression from the P7 promoter and are involved at the level of translation initiation of the NS mRNA (Kimmick *et al.*, 1998). Expression of foreign proteins under the control of the P61 promoter resulted in a comparable level of expression as P7 promoter, but did not necessitate that the foreign gene be in frame with the VP coding sequence.

Several AeDNV-derived constructs have been developed to produce transducing virus particles able to deliver a foreign gene in mosquito larvae. Owing to their small size, the capsid of densovirus virions can only package DNA with a limited size. As previously demonstrated for AAVs transducing particles, the recombinant parvovirus genome is most efficiently encapsidated when its size is close to that of wt virus DNA. The highest transduction efficiency was achieved with a construct expressing the GFP gene fused to the C-terminal sequence of NS1. This construct retained the NS functions but lacked VP gene. Transducing virions were thus obtained by using a helper plasmid providing *in trans* the expression of VP gene (Afanasiev *et al.*, 1999). To avoid the recombination events leading to the production of wt virions, Allen-Muria and coworkers (Allen-Miura *et al.*, 1999) developed a system to package transducing genome expressing NS1-GFP gene via a Sindbis-derived virus expressing *in trans* AeDNV VP proteins.

More recently, transducing virus particles of the *Anopheles gambiae* DNV (AgDNV) have been produced using a two-plasmid helper-transducer system. EGFP-transducing virions were highly infectious to *An. gambiae* larvae, persisted in adults and were transmitted to subsequent mosquito generations (Ren *et al.*, 2008). Such transducing particles are promising tools for the control of malaria, and more largely of medically important mosquito vectors

Finally, Hu and coworkers (Hu *et al.*, 2007) recently reported the cloning of an infectious sequence of PfDNV and the construction of PfDNV-derived vectors expressing GFP able to integrate into the host cell genome. Infectious recombinant virions were produced from plasmids expressing GFP fused downstream of and in frame with NS1 and NS2 coding sequences.

Potential of DNVs as biopesticides to control insect pests

So far, the only viruses which received official authorizations to be put on the market as biopesticides belong to the family *Baculoviridae* (nucleopolyhedroviruses and granuloviruses). Their lack of morphological resemblance and of sequence homology of their genome with viruses infecting mammals and humans as well as their incapacity to replicate in mammalian cells greatly facilitated their acceptance as biopesticides.

Despite their small size, the lack of specific protective forms such as polyhedra of baculoviruses and cypoviruses, or spherules of entomopoxviruses, DNVs are at the origin of important epizootics in natural populations of insect pests (oil palm defoliators, sugarcane borer) and in mass rearing of insects used as baits or pet food (crickets, greater wax moth). Most DNVs have a high virulence for their host and have a rather limited host range. In this respect, they appear as good candidates for the biological control of their hosts, most of them belonging to economically or medically important insect pests.

Owing to sequence homologies between DNV and vertebrate parvovirus genomes, a concern was raised about the safety of using them as biocontrol agents. Several arguments pledge in favour of their inocuity to vertebrates and man, such as their incapacity to replicate in vertebrate and human cell lines (El-Far *et al.*, 2004; Jousset *et al.*, 1993), the lack of any pathogenic effect when GmDNV, AalDNV or AeDNV are inoculated intracerebrally into suckling mice (Giran, 1966; Jousset *et al.*, 1993; Lebedinets and Kononko, 1989) or following injection of rabbits, guinea pigs or mice with most of DNVs for raising antibodies and the lack of any unusual pathological condition reported in human populations living in close contact with naturally infected insects such as farmers, or workers in industrial oil palm tree plantations or in insect mass rearing facilities. Of course further safety tests need to be thoroughly performed to assess their inocuity and before the approval of a DNV as a pesticide is granted. Preliminary tests have been performed to estimate the efficiency of some DNVs to control their host. As early as 1965, the introduction into beehives infested by *G. mellonella* of cadavers of larvae which had died of GmDNV infection was successful to get rid of this pest (Lavie *et al.*, 1965). SfDNV and CeDNV have been used at large scales to control two major pests in industrial palm trees plantations, *Sibine fusca* in Colombia and *Casphalia extranea* in Ivory Coast (Fedière *et al.*, 1986; Genty and Mariau, 1975).

The spreading by aeroplane over several hundreds of hectares of a suspension containing as little as the equivalent of 20 g/ha of a homogenate of heavily infested *S. fusca* larvae resulted in over

90% mortality after 20 days. In Europe, AeDNV has been used to control mosquito larvae in the former Soviet Union and a commercial product (Viroden) was developed (Buchatsky *et al.*, 1987). Finally, a commercial product based on PfDNV suspension was developed in China to control the smoky-brown cockroach *Periplaneta fuliginosa* (Jiang *et al.*, 2007). These preliminary reports clearly demonstrate the potential of DNVs as biocontrol agents and more attention should be paid in the future to integrate their use in integrated pest management programs against some major insect pests.

Concluding remarks

Densoviruses have shown a striking diversity that may reflect host differences among arthropods. Only about 25 DNVs have been isolated and even fewer have been characterized in some detail when compared to the large number of arthropods that exist. Still, it is surprising that the number of genera that have been distinguished is rather limited. Molecular study of these viruses not only sheds light on the properties of DNVs but also on *Parvoviridae* overall. For example, the few conserved sequences that exist between vertebrate and invertebrate parvoviruses led to the discovery of viral phospholipase A_2 and its role in cell entry. Likewise, there are similar examples with regard to viral structure and non-structural proteins.

The extremely high stability and high pathogenicity of DNVs make them eminent candidates for biological control agents but also feared contaminants in monocultures (e.g. shrimp, wax moths and crickets) in farms.

Like vertebrate parvoviruses, densovirus genomes can also be exploited for use as expression vectors and occupies its own niche with specific properties.

Acknowledgements

We would like to thank François Cousserans for his help in preparing the manuscript and the two reviewers for critical reading and suggestions.

References

Abd-Alla, A., Jousset, F.X., Li, Y., Fediere, G., Cousserans, F., and Bergoin, M. (2004). NS-3 protein of the *Junonia coenia* densovirus is essential for viral DNA replication in an Ld 652 cell line and *Spodoptera littoralis* larvae. J. Virol. *78*, 790–797.

Afanasiev, B.N., Galyov, E.E., Buchatsky, L.P., and Koslov, Y.V. (1991). Nucleotide sequence and genomic organization of *Aedes* densonucleosis virus. Virology *185*, 323–336.

Afanasiev, B.N., Kozlov, Y.V., Carlson, J.O., and Beaty, B.J. (1994). Densovirus of *Aedes aegypti* as an expression vector in mosquito cells. Exp. Parasitol. *79*, 322–339.

Afanasiev, B.N., Ward, T.W., Beaty, B.J., and Carlson, J.O. (1999). Transduction of *Aedes aegypti* mosquitoes with vectors derived from *Aedes* densovirus. Virology *257*, 62–72.

Allen-Miura, T.M., Afanasiev, B.N., Olson, K.E., Beaty, B.J., and Carlson, J.O. (1999). Packaging of AeDNV-GFP transducing virus by expression of densovirus structural proteins from a sindbis virus expression system. Virology *257*, 54–61.

Amargier, A. (1966). Action de la densonucléose du lépidoptère *Galleria mellonella* sur la sécrétion et l'émission de substance soyeuse. Arch. Ges. Virusforsch. *19*, 13.

Amargier, A., Vago, C., Duthoit, J.L., and Meynadier, G. (1979). Formation tumorale d'origine parvovirale chez *Sibine fusca* (Lep. Limacodidae). Entomophaga *24*, 259–271.

Amargier, A., Vago, C., and Meynadier, G. (1965). Etude histopathologique d'un nouveau type de virose mise en évidence chez le lépidoptère *Galleria mellonella.* Arch. Ges. Virusforsch. *15*, 659–667.

Bando, H., Kondo, N., and Kawase, S. (1984). Molecular homology among the structural proteins of densonucleosis virus from silkworm, *Bombx mori.* Arch. Virol. *80*, 209–218.

Bando, H., Nakagaki, M., and Kawase, S. (1983). Polyamines in densonucleosis virus from the silkworm, *Bombyx mori.* J. Invertebr. Pathol. *42*, 264–266.

Baquerizo-Audiot, E., Abd-Alla, A., Jousset, F.X., Cousserans, F., Tijssen, P., and Bergoin, M. (2009). Structure and expression strategy of the genome of *Culex pipiens* densovirus, a mosquito densovirus with an ambisense organization. J. Virol. *83*, 6863–6873.

Barreau, C., Jousset, F.X., and Bergoin, M. (1996). Pathogenicity of the *Aedes albopictus* parvovirus (AaPV), denso-like virus, for *Aedes aegypti* mosquitoes. J. Invertebr. Pathol. *68*, 299–309.

Bartlett, J.S., Wilcher, R., and Samulski, R.J. (2000). Infectious entry pathway of adeno-associated virus and adeno-associated virus vectors. J. Virol. *74*, 2777–2785.

Barwise, A.H., and Walker, I.O. (1970). Studies on the DNA of a virus from *Galleria mellonella.* FEBS Lett. *6*, 13–16.

Bell, T.A., and Lightner, D.V. (1984). IHHN virus: infectivity and pathogenicity studies in *Penaeus stylirostris* and *Penaeus vannamei.* Aquaculture *38*, 185–194.

Bell, T.A., and Lightner, D.V. (1987). IHHN disease of *Penaeus stylirostris*: effects of shrimp size on disease expression. J. Fish Dis. *10*, 165–170.

Bergoin, M., and Brès, N. (1968). Lésions tissulaires chez la larve du lépidoptère *Galleria mellonella* L. atteinte du virus de la densonucléose. Bull. Apicole *11*, 5–12.

Bergoin, M., and Tijssen, P. (1998). Biological and molecular properties of densoviruses and their use in protein expression and biological control. In The insect viruses, L.K. Miller, and L.A. Ball, eds. (New York and London, Plenum Press), pp. 141–169.

Bergoin, M., and Tijssen, P. (2008). Parvoviruses of arthropods. In Encyclopedia of Virology, B. Mahy, and M. van Regenmortel, eds. (Oxford, Elsevier), pp. 76–85.

Boemare, N., and Brès, N. (1969). Le développement du virus de la densonucléose au cours de la métamorphose de son hôte *Galleria mellonella* L. (Lepidoptera). Ann. Zool. Ecol. Anim. *1*, 309–320.

Boemare, N., Croizier, G., and Veyrunes, J.C. (1970). Contribution à la connaissance des propriétés du virus de la densonucléose. Entomophaga *15*, 327–332.

Bossin, H., Fournier, P., Royer, C., Barry, P., Cerutti, P., Gimenez, S., Couble, P., and Bergoin, M. (2003). *Junonia coenia* densovirus-based vectors for stable transgene expression in Sf9 cells: influence of the densovirus sequences on genomic integration. J. Virol. *77*, 11060–11071.

Bossin, H., Furlong, R.B., Gillett, J.L., Bergoin, M., and Shirk, P.D. (2007). Somatic transformation efficiencies and expression patterns using the JcDNV and piggyBac transposon gene vectors in insects. Insect Mol. Biol. *16*, 37–47.

Boublik, Y., Jousset, F.X., and Bergoin, M. (1994). Complete nucleotide sequence and genomic organization of the *Aedes albopictus* parvovirus (AaPV) pathogenic for *Aedes aegypti* larvae. Virology *200*, 752–763.

Boufersaoui, A., and Matz, G. (1974). Augmentation de l'activité mitotique dans l'intestin moyen après section du nerf récurrent chez l'imago de *Leucophaea maderae* Fabr. (Insecte, Dictyoptère). C. R. Acad. Sci. D *279*, 1889.

Buchatsky, L.P. (1989). Densonucleosis of bloodsucking mosquitoes. Dis. Aquat. Org. *6*, 145–150.

Buchatsky, L.P., Kuznetsova, M.A., Lebedinets, N.N., and Kononko, A.G. (1987). Development and basic properties of a virus preparation viroden. Vopr.Virusol. *6*, 729–733.

Chao, Y.C., Young, S.Y.I., Kim, K.S., and Scott, H.A. (1985). A newly isolated densonucleosis virus from *Pseudoplusia includens* (Lepidoptera: Noctuidae). J. Invertebr. Pathol. *46*, 70–82.

Chapman, M.S., and Agbandje-McKenna, M. (2006). Atomic structure of virus particles. In Parvoviruses, J.R. Kerr, Cotmore, S. F., Bloom, M. E., Linden, R. M., Parrish, C.R., ed. (London, Hodder Arnold), pp. 107–123.

Chapman, M.S., and Rossmann, M.G. (1993). Structure, sequence, and function correlations among parvoviruses. Virology *194*, 491–508.

Charpentier, R. (1979). A nonoccluded virus in nymph of the dragonfly *Leucorrhinia dubia* (Odonata, Anisoptera). J. Invertebr. Pathol. *34*, 95–98.

Cheng, L., Chen, S., Brannan, J.M., Jakana, J., Zhang, Q., Zhou, Z.H., and Zhang, J. (2004). Three-dimensional structure determination of capsid of *Aedes albopictus* C6/36 cell densovirus. Sci. China C. Life Sci. *47*, 224–228.

Cheng, L., Chen, S., Zhou, Z.H., and Zhang, J. (2007). Structure comparisons of *Aedes albopictus* densovirus with other parvoviruses. Sci. China C. Life Sci. *50*, 70–74.

Cotmore, S.F., and Tattersall, P. (2006). Structure and organization of the viral genome. In Parvoviruses, J.R. Kerr, Cotmore, S. F., Bloom, M. E., Linden, R. M., Parrish, C. R., ed. (London, Hodder Arnold), pp. 73–94.

Diallo, B. (1978). Etude de l'infection à densovirus chez le lépidoptère *Spodoptera littorallis* BOISDUVAL. (Thèse 3e cycle Montpellier, Univ. Sci. Techn. Languedoc), pp. 141.

Ding, C., Urabe, M., Bergoin, M., and Kotin, R.M. (2002). Biochemical characterization of *Junonia coenia* densovirus nonstructural protein NS-1. J. Virol. *76*, 338–345.

Dumas, B., Jourdan, M., Pascaud, A.M., and Bergoin, M. (1992). Complete nucleotide sequence of the cloned infectious genome of *Junonia coenia* densovirus reveals an organization unique among parvoviruses. Virology *191*, 202–222.

Eguchi, R., Nagayasu, K.I., Ninagi, O., and Hara, W. (2007). Genetic analysis on the dominant non-susceptibility to densonucleosis virus type 1 in the silkworm, *Bombyx mori*. J. Insect Biotechnol. Sericol. *76*, 159–163.

El-Far, M., Li, Y., Fediere, G., Abol-Ela, S., and Tijssen, P. (2004). Lack of infection of vertebrate cells by the densovirus from the maize worm *Mythimna loreyi* (MlDNV). Virus Res. *99*, 17–24.

Federici, B.A. (1976). Pathology and histochemistry of a densonucleosis virus in larvae of the blackfly *Simulium vittatum*. Proc. 1st Intern. Coll. Invertebr. Pathol. (Kingston, Canada) *p* 341–342.

Fédière, G. (1983). Recherches sur des viroses épizootiques de lépidoptères limacodidae ravageurs de palmacées (Thèse 3e cycle Montpellier, Univ. Sci. Techn. Languedoc), pp. 103.

Fédière, G. (1996). Recherches sur des viroses de lépidoptères ravageurs de cultures pérennes en Côte d'Ivoire et de cultures annuelles en Egypte. (Thèse d'Etat Montpellier, Université Montpellier II), pp. 204.

Fédière, G. (2000). Epidemiology and pathology of Densovirinae. In Parvoviruses From molecular biology to pathology and therapeutic uses, S. Faisst, and J. Rommelaere, eds. (Basel, Karger), pp. 1–11.

Fédière, G., El-Far, M., Li, Y., Bergoin, M., and Tijssen, P. (2004). Expression strategy of densonucleosis virus from *Mythimna loreyi*. Virology *320*, 181–189.

Fédière, G., Li, Y., Zadori, Z., Szelei, J., and Tijssen, P. (2002). Genome organization of *Casphalia extranea* densovirus, a new iteravirus. Virology *292*, 299–308.

Fédière, G., Monsarrat, P., Mariau, D., and Bergoin, M. (1986). A densovirus of *Casphalia extranea* (Lepidoptera: Limacodidae): characterization and use for biological control. In Fundamental and Applied Aspects of Invertebrate Pathology, R.A. Samson, J.M. Vlak, D. Peters ed. (Wageningen) p705.

Flegel, T.W. (2006). Shrimp parvoviruses. In Parvoviruses, J.R. Kerr, Cotmore, S. F., Bloom, M. E., Linden, R. M., Parrish, C. R., ed. (London, Hodder Arnold), pp. 487–493.

Garzon, S., and Kurstak, E. (1968). Infection des cellules des gonades et du système nerveux de *Galleria mellonella* par le virus de la densonucléose. Naturalist Can. *95*, 1125.

Genty, P., and Mariau, D. (1975). Utilisation d'un germe entomopathogène dans la lutte contre *Sibine fusca* (Limacodidae). Oléagineux *30*, 349–354.

Giran, F. (1966). Action de la 'densonucléose' de lépidoptères sur les mammifères. Entomophaga *11*, 405–407.

Giraud, C., Devauchelle, G., and Bergoin, M. (1992). The densovirus of *Junonia coenia* (JcDNV) as an insect cell expression vector. Virology *186*, 207–218.

Grignon, N. (1982). Recherches sur une infection chronique dans une lignée cellulaire de lépidoptères (Montpellier, Université de Montpellier), pp. 102.

Guo, H., Zhang, J., and Hu, Y. (2000). Complete sequence and organization of *Periplaneta fuliginosa* densovirus genome. Acta Virol. *44*, 315–322.

Hayakawa, T., Kojima, K., Nonaka, K., Nakagaki, M., Sahara, K., Asano, S., Iizuka, T., and Bando, H. (2000). Analysis of proteins encoded in the bipartite genome of a new type of parvo-like virus isolated from silkworm – structural protein with DNA polymerase motif. Virus Res. *66*, 101–108.

Hu, L., Zhang, L., Shen, C., Lu, J., Zhang, J., and Hu, Y. (2007). The densovirus of *Periplaneta fuliginosa* (PfDNV) as an insect vector for persistent foreign gene expression *in vivo*. Biochem. Biophys. Res. Commun. *358*, 976–982.

Jiang, H., Zhang, J.M., Wang, J.P., Yang, B., Liu, C.F., Lu, J., and Hu, Y.Y. (2007). Genetic engineering of *Periplaneta fuliginosa* densovirus as an improved biopesticide. Arch. Virol. *152*, 383–394.

Jourdan, M., Jousset, F.X., Gervais, M., Skory, S., Bergoin, M., and Dumas, B. (1990). Cloning of the genome of a densovirus and rescue of infectious virions from recombinant plasmid in the insect host Spodoptera littoralis. Virology *179*, 403–409.

Jousset, F.X., Baquerizo, E., and Bergoin, M. (2000). A new densovirus isolated from the mosquito Culex pipiens (Diptera: Culicidae). Virus Res. *67*, 11–16.

Jousset, F.X., Barreau, C., Boublik, Y., and Cornet, M. (1993). A parvo-like virus persistently infecting a C6/36 clone of *Aedes/Albopictus* mosquito cell-line and pathogenic for *Aedes-Aegypti* larvae. Virus Res. *29*, 99–114.

Jousset, F.X., Compagnon, B., and Bergoin, M. (1986). Comparison of the restriction map and infectivity of the genomes of three densoviruses (Wageningen, Fundation IVth Int.Coll.Invertebr.Pathol. Wageningen, The Netherlands).

Kapelinskaya, T.V., Martynova, E.U., Korolev, A.L., Schal, C., and Mukha, D.V. (2008). Transcription of the German cockroach densovirus BgDNV genome: alternative processing of viral RNAs. Dokl. Biochem. Biophys. *421*, 176–180.

Kelly, D.C., Ayres, M.D., Spencer, L.K., and Rivers, C.F. (1980). Densonucleosis virus 3: a recent insect parvovirus isolated from *Agraulis vanillae* (Lepidoptera: Nymphalidae). Microbiologia *3*, 455–460.

Kelly, D.C., Barwise, A.H., and Walker, I.O. (1977). DNA contained by two densonucleosis viruses. J. Virol. *21*, 396–407.

Kelly, D.C., and Elliott, R.M. (1977). Polyamines contained by two densonucleosis viruses. J. Virol. *21*, 408–410.

Kimmick, M.W., Afanasiev, B.N., Beaty, B.J., and Carlson, J.O. (1998). Gene expression and regulation from the p7 promoter of *Aedes* densonucleosis virus. J. Virol. *72*, 4364–4370.

Kurstak, E., and Stanislawski-Birencwajg, M. (1968). Localisation du matériel antigénique du virus de la densonucléose au cours de l'infection chez *Galleria mellonella* L. Can. J. Microbiol. *14*, 1350–1352.

Lavie, P., Fresnaye, J., and Vago, C. (1965). L'action de la virose à noyaux denses sur les larves de *Galleria mellonella* dans les ruches. Ann. Abeille *8*, 321–323.

Lebedeva, O.P., Kuznetsova, M.A., Zelenko, A.P., and Gudz-Gorban, A.P. (1973). Investigation of a virus disease of the densonucleosis type in a laboratory culture of *Aedes aegypti*. Acta Virol. *17*, 253–256.

Lebedinets, N.N., and Kononko, A.G. (1989). Experimental study of the pathways of densovirus infection transmission in a population of blood-sucking mosquitoes. Med. Parazitol. Parazit. Bolezni *2*, 79–83.

Li, Y., Zadori, Z., Bando, H., Dubuc, R., Fediere, G., Szelei, J., and Tijssen, P. (2001). Genome organization of the densovirus from *Bombyx mori* (BmDNV-1) and enzyme activity of its capsid. J. Gen. Virol. *82*, 2821–2825.

Lightner, D.V. (1996). Epizootiology, distribution and the impact on international trade of two penaeid shrimp viruses in the Americas. Rev. Sci. Techn. Off. Int. Epiz. *15*, 579–601.

Lightner, D.V., and Redman, R.M. (1985). A parvo-like virus disease of penaeid shrimp. J. Invertebr. Pathol. *45*, 47–53.

Lightner, D.V., Redman, R.M., and Bell, T.A. (1983). Infectious hypodermal and hematopoietic necrosis, a newly recognized virus disease of penaeid shrimp. J. Invertebr. Pathol. *42*, 62–70.

Linser, P., Bruning, H., and Armentrout, R.W. (1979). Uptake of minute virus of mice into cultured rodent cells. J. Virol. *31*, 537–545.

Lombardo, E., Ramirez, J.C., Garcia, J., and Almendral, J.M. (2002). Complementary roles of multiple nuclear targeting signals in the capsid proteins of the parvovirus minute virus of mice during assembly and onset of infection. J. Virol. *76*, 7049–7059.

Longworth, J.F., Tinsley, T.W., Barawise, A.H., and Walker, I.O. (1968). Purification of a non-occluded virus of *Galleria mellonella*. J. Gen. Virol. *3*, 167–174.

Maeda, S., and Watanabe, H. (1978). Immunofluorescence observation of the infection of densonucleosis virus in the silkworm, *Bombyx mori*. Japan J. Appl. Entomol. Zool. 22, 98–101.

Matz, G., and Bergoin, M. (1984). Experimental tumors in *Leucophaea maderae* (Dictyoptera): a viral etiology. J. Invertebr. Pathol. 43, 424426.

Meynadier, G., Amargier, A., and Genty, P. (1977a). Une virose de type densonucléose chez le lépidoptère *Sibine fusca* STOLL. Oléagineux 32, 357–361.

Meynadier, G., Galichet, P.F., Veyrunes, J.C., and Amargier, A. (1977b). Mise en évidence d'une densonucléose chez *Diatraea saccharalis* (Lep. Pyralidaea). Entomophaga 22, 115–120.

Meynadier, G., Matz, G., Veyrunes, J.C., and Brès, N. (1977c). Virose de type densonucléose chez les orthoptères. Ann. Soc. Entomol. Fr. 13, 487–493.

Meynadier, G., Vago, C., Plantevin, G., and Atger, P. (1964). Virose d'un type inhabituel chez le lépidoptère *Galleria mellonella* L. Rev. Zool. Agric. Appl. 63, 207–208.

Mukha, D.V., Chumachenko, A.G., Dykstra, M.J., Kurtti, T.J., and Schal, C. (2006). Characterization of a new densovirus infecting the German cockroach, *Blattella germanica*. J. Gen. Virol. 87, 1567–1575.

Nakagaki, M., and Kawase, S. (1980). DNA of a new parvo-like virus isolated from the silkworm, *Bombyx mori*. J. Invertebr. Pathol. 35, 124–133.

O'Neill, S.L., Kittayapong, P., Braig, H.R., Andreadis, T.G., Gonzalez, J.P., and Tesh, R.B. (1995). Insect densoviruses may be widespread in mosquito cell lines. J. Gen. Virol. 76, 2067–2074.

Odier, F., and Boemare, N. (1972). Investigations on the preservation of the virus causing densonucleosis. Entomophaga 17, 197–202.

Ren, X., Hoiczyk, E., and Rasgon, J.L. (2008). Viral paratransgenesis in the malaria vector *Anopheles gambiae*. PLoS Pathog. 4, e1000135.

Riolobos, L., Reguera, J., Mateu, M.G., and Almendral, J.M. (2006). Nuclear transport of trimeric assembly intermediates exerts a morphogenetic control on the icosahedral parvovirus capsid. J. Mol. Biol. 357, 1026–1038.

Rolling, F. (1992). Vecteurs dérivés du densovirus de Junonia coenia (JcDNV) exprimant sous forme épisomale ou intégrée des gènes étrangers dans des cellules d'insectes. Thèse de Doctorat de Biologie Cellulaire et Microbiologie, Université Aix-Marseille II, 153 *p*.

Royer, C., Bossin, H., Romane, C., Bergoin, M., and Couble, P. (2001). High amplification of a densovirus-derived vector in larval and adult tissues of *Drosophila*. Insect Mol. Biol. 10, 275–280.

Ryabov, E.V., Keane, G., Naish, N., Evered, C., and Winstanley, D. (2009). Densovirus induces winged morphs in asexual clones of the rosy apple aphid, *Dysaphis plantaginea*. Proc. Natl. Acad. Sci. U.S.A. 106, 8465–8470.

Shike, H., Dhar, A.K., Burns, J.C., Shimizu, C., Jousset, F.X., Klimpel, K.R., and Bergoin, M. (2000). Infectious hypodermal and hematopoietic necrosis virus of shrimp is related to mosquito brevidensoviruses. Virology 277, 167–177.

Simpson, A.A., Chipman, P.R., Baker, T.S., Tijssen, P., and Rossmann, M.G. (1998). The structure of an insect parvovirus (*Galleria mellonella* densovirus) at 3.7 A resolution. Structure 6, 1355–1367.

Sukhumsirichart, W., Attasart, P., Boonsaeng, V., and Panyim, S. (2006). Complete nucleotide sequence and genomic organization of hepatopancreatic parvovirus (HPV) of *Penaeus monodon*. Virology 346, 266–277.

Sun, F.L., Ma, G.H., and Chen, M.S. (1981). A new insect virus of *Pieris rapae*. Acta Microbiol. Sin. 21, 41.

Suto, C. (1979). Characterization of a virus newly isolated from the smoky-brown cockroach, *Periplaneta fuliginosa* (Serville). Nagoya J. Med. Sci. 42, 13–25.

Suto, C., Kawamoto, F., and Kumada, N. (1979). A new virus isolated from the cockroach, *Periplaneta fuliginosa* (Serville). Microbiol. Immunol. 23, 207–211.

Sutter, G.R. (1973). A nonoccluded virus of the army cutworm. J. Invertebr. Pathol. 21, 62–70.

Tattersall, P. (2006). The evolution of parvovirus taxonomy. In Parvoviruses, J.R. Kerr, Cotmore, S.F., Bloom, M.E., Linden, R. M., Parrish, C. R., ed. (London, Hodder Arnold), pp. 5–14.

Tattersall, P., Bergoin, M., Bloom, M.E., Brown, K.E., Linden, R.M., Muzyczka, N., Parrish, C R., and Tijssen, P. (2005). Parvoviridae. In Virus Taxonomy VIIIth Report of the ICTV, C.M. Fauquet, M.A. Mayo, J. Maniloff, U. Desseloberger, and L.A. Ball, eds. (London, Elsevier/Academic Press).

Thao, M.L., Wineriter, S., Buckingham, G., and Baumann, P. (2001). Genetic characterization of a putative Densovirus from the mealybug *Planococcus citri*. Curr. Microbiol. 43, 457–458.

Tijssen, P., Bando, H., Li, Y., Jousset, F.X., Zadori, Z., Fédière, G., El-Far, M., Szelei, J., and Bergoin, M. (2006). Evolution of densoviruses. In Parvoviruses, J.R. Kerr, S.F. Cotmore, M.E. Bloom, R.M Linden, and P.C. R., eds. (London, UK, Hodder Arnold), pp. 55–68.

Tijssen, P., and Bergoin, M. (1995). Densonucleosis viruses constitute an increasingly diversified subfamily among the parvoviruses. Sem.Virol. 6, 347–355.

Tijssen, P., and Kurstak, P. (1981). Biochemical, biophysical, and biological properties of densonucleosis virus (parvovirus). III. Common sequences of structural proteins. J. Virol. 37, 17–23.

Tijssen, P., Li, Y., El-Far, M., Szelei, J., Letarte, M., and Zadori, Z. (2003). Organization and expression strategy of the ambisense genome of densonucleosis virus of *Galleria mellonella*. J. Virol. 77, 10357–10365.

Tijssen, P., Tijssen-van der Slikke, T., and Kurstak, E. (1977). Biochemical, biophysical, and biological properties of densonucleosis virus (paravovirus). II. Two types of infectious virions. J. Virol. 21, 225–231.

Vago, C., Duthoit, J.L., and Delahaye, F. (1966). Les lésions nucléaires de la 'virose à noyaux denses' du lépidoptère *Galleria mellonella*. Arch. Ges. Virusforsch. 18, 344–349.

Vago, C., Meynadier, G., and Duthoit, J.L. (1964). Etude d'un nouveau type de maladie à virus chez les lépidoptères. Ann. Epiphyties 15, 475–479.

van Munster, M., Dullemans, A.M., Verbeek, M., van den Heuvel, J.F., Reinbold, C., Brault, V., Clerivet, A., and van der Wilk, F. (2003a). Characterization of a new densovirus infecting the green peach aphid *Myzus persicae*. J. Invertebr. Pathol. *84*, 6–14.

van Munster, M., Dullemans, A.M., Verbeek, M., van den Heuvel, J.F., Reinbold, C., Brault, V., Clerivet, A., and van der Wilk, F. (2003b). A new virus infecting *Myzus persicae* has a genome organization similar to the species of the genus *Densovirus*. J. Gen. Virol. *84*, 165–172.

Vendeville, A., Ravallec, M., Jousset, F.X., Devise, M., Mutuel, D., Lopez-Ferber, M., Fournier, P., Dupressoir, T., and Ogliastro, M. (2009). Densovirus infectious pathway requires clathrin-mediated endocytosis followed by trafficking to the nucleus. J. Virol. *83*, 4678–4689.

Wang, J., Zhang, J., Jiang, H., Liu, C., Yi, F., and Hu, Y. (2005). Nucleotide sequence and genomic organization of a newly isolated densovirus infecting *Dendrolimus punctatus*. J. Gen. Virol. *86*, 2169–2173.

Watanabe, H., and Kurihara, Y. (1988). Comparative histopathology of two densonucleoses in the silkworm, *Bombyx mori*. J. Invertebr. Pathol. *51*, 287–290.

Watanabe, H., Maeda, S., Matsui, M., and Shimizu, T. (1976). Histopathology of the midgut epithelium of the silkworm, *Bombyx mori*, infected with a newly isolated virus from the flacherie diseased larvae. J. Sericult. Sci. Japan *45*, 29–34.

Yamagishi, J., Hu, Y., Zheng, J., and Bando, H. (1999). Genome organization and mRNA structure of *Periplaneta fuliginosa* densovirus imply alternative splicing involvement in viral gene expression. Arch. Virol. *144*, 2111–2124.

Yang, B., Dong, X., Cai, D., Wang, X., Liu, Z., Hu, Z., Wang, H., Cao, X., Zhang, J., and Hu, Y. (2008). Characterization of the promoter elements and transcription profile of *Periplaneta fuliginosa* densovirus nonstructural genes. Virus Res. *133*, 149–156.

Zádori, Z., Szelei, J., Lacoste, M.C., Li, Y., Gariépy, S., Raymond, P., Allaire, M., Nabi, I.R., and Tijssen, P. (2001). A viral phospholipase A2 is required for parvovirus infectivity. Dev. Cell *1*, 291–302.

Zhai, Y.G., Lv, X.J., Sun, X.H., Fu, S.H., Gong, Z.D., Fen, Y., Tong, S.X., Wang, Z.X., Tang, Q., Attoui, H., *et al.* (2008). Isolation and characterization of the full coding sequence of a novel densovirus from the mosquito *Culex pipiens pallens*. J. Gen. Virol. *89*, 195–199.

Entomopoxviruses

4

Srini Perera, Zhen Li, Lillian Pavlik and Basil Arif

Abstract

Like other poxviruses, entomopoxviruses (EPVs) replicate in the cytoplasm of infected cells but form proteinic bodies (occlusion bodies) in which virions are occluded. Unlike baculoviruses, EPVs appear to utilize haemocytes to disseminate infection to permissive larval cells and tissues. Our observations showed that the tracheal system is not the conduit to spread virus infection in larval tissues. Poxviruses contain a large, linear double-stranded DNA genome but in the case of EPVs, the genome is extremely AT-rich. To date, only two EPV genomes have been fully sequenced. Indeed, the very high AT content makes sequencing and assembly a bit of a problem. Comparative analyses of all sequenced poxviruses revealed 49 genes conserved among all poxviruses. The conserved genes are involved in basic functions such as transcription, DNA replication and virion assembly.

Introduction

Although poxviruses have been known to man for decades and some of their diseases have been observed for centuries, the entomopoxviruses (EPVs) have been relatively recently discovered by Vago in 1963. He reported that EPVs have close structural similarities to orthopoxviruses except that EPV virions were distinguished by being occluded in a spheroidal proteinic body we now term the spheroid. The spheroid functions much like the baculovirus occlusion bodies providing certain protection to the virions against environmental inactivating agents such as UV light, heat, desiccation, etc. The spheroid also dissolves in the alkaline midgut of larvae and delivers the virions to permissive midgut cells. Therefore, the early stages of infection of larvae are much like those of baculoviruses. Once the virions are released from the spheroids with the aid of spheroid-associated alkaline protease (Bilimoria and Arif, 1979), infection of midgut cells is initiated. The virions enter the midgut epithelial cells by fusion of the viral outer envelope with the plasma membrane allowing entry of the viral core and lateral bodies to the cell cytoplasm. Eventually, evidence of virus replication in the cytoplasm is indicated by the accumulation of electron dense material, also known as viroplasms. Through a series of processes, the viroplasm forms the immature virion cores (Granados, 1973a; Arif and Kurstak, 1991). In some hosts, virions undergo maturation after occlusion into the spheroids. Similarly, extracellular virus may mature after budding out of the infected cell to spread the infection to other susceptible tissues.

Entomopoxviruses infect four orders of insects, Lepidoptera, Orthoptera, Diptera and Coleoptera (Granados, 1973b) and, while the virus is infective to larvae, studies have shown that an EPV of the bark beetle, *Ips typographus*, appears to target only adult insects (Wegensteiner and Weiser, 1995). Host range, morphology and the size of the virus are the criteria for the different genera. An illustrated review of viral morphology in the different genera has already been well illustrated (Kurstak and Garzon, 1997).

Relatively recently, the International Committee on the Taxonomy of Viruses has divided the family *Poxviridae* into two subfamilies of *Chordopoxvirinae* and *Entomopoxvirinae*. The first subfamily encompasses several genera of viruses that infect Chordates. The *Entomopoxvirinae* are poxviruses that infect insects and contain three genera:

- *Alphaentomopoxvirus,* of which the *Melolontha melolontha entomopoxvirus* (MMEV) is the type species and members of this genus infect coleopteran insects (beetles). Virions are 450 × 250 nm in size and ovoid in shape. They contain a unilateral concave core and a single lateral body. The genome size of these viruses is 260–370 kb.
- *Betaentomopoxvirus,* of which the *Amsacta moorei entomopoxvirus 'L'* (AMEV) is the type species. Viruses of this genus infect lepidopteran insects (butterflies and moths). The virions contain a cylindrical core and a 'sleeve-shaped' lateral body. The virions are similar in shape to alphaentomopoxviruses (ovoid) but are smaller in size (250 × 350 nm).
- *Gammaentomopoxvirus,* of which the *Chironomus luridus entomopoxvirus* (CLEV) is the type species. Members of this genus are brick shaped (320 × 230 × 110 nm) and infect flies and mosquitoes. The virions contain two lateral bodies and a biconcave core.

Interactions with the larval host

Unlike baculoviruses, to date, EPVs have not been used extensively in pest control for a variety of reasons. One, they take a long time (3–5 weeks) to kill the host. Interestingly, many EPVs encode a protein that forms spindle-shaped structures and, except for some like MMEV, *Anomala cuprea* EPV (ACEV) and *Pseudaletia separata* EPV (PSEV), the spindles are occluded alongside the virions into the spheroid. The spindle structure is composed primarily of a single protein that has been termed fusolin and appears to share high amino acid identity with fusolins from other EPVs and, to a certain extent, to GP37 of baculoviruses (Dall *et al.,* 1993). Like GP37, fusolin has a 20 amino acid leader sequence that is cleaved post-translationally (Dall *et al.,* 1993). It was demonstrated that EPV spindles enhanced the infectivity of nucleopolyhedrovirus (NPV) (Mitsuhashi *et al.,* 1998; Wijonarko and Hukuhara, 1998). It was observed that the spindles, but not the spheroids, of ACEV caused the disintegration of the peritrophic matrix in the silkworm larvae, thus allowing greater numbers of NPV virions to enter microvilli (Mitsuhashi and Miyamoto, 2003). Mitsuhashi *et al.* (2007) inoculated ACEV spindles into the host and thereafter observed the effect in the midgut area by scanning electron microscopy. A significant disruption of the peritrophic matrix by the spindles was observed. Further experiments using quantitative PCR (qPCR) on viral DNA showed that the disruption of the peritrophic matrix facilitated the entry of ACEV into the midgut and to the proximity of permissive cells.

Takemoto *et al.* (2008) constructed a series of deletion mutants in the gene encoding fusolin, expressed in a baculovirus expression system and assayed for enhancement of BmNPV in the silkworm. Deletion mutagenesis revealed that the N-terminus (aa 1–253) is essential for fusolin function, while the carboxy terminus is dispensable. The amino terminus appears to have a chitin binding domain needed for the degradation of the peritrophic matrix. They also observed that site directed mutagenesis of the glycosylation site at N191 caused significant reduction or total abolishment of the enhancing activity. The authors concluded that the N-terminus and the carbohydrate side chain were essential for function and stability of fusolin in midgut juices.

Baculoviruses encode a protein, P35, which is a potent inhibitor of apoptosis, specifically targeting caspases. AMEV encodes a protein, P33, that has been shown to inhibit apoptosis in human and insect cell lines and has low but observable identity to baculovirus P35 (Means *et al.,* 2007). The authors found that a recombinant P33 was cleaved specifically by effector caspases and the cleavage fragments became associated with caspases. Thus P33 is specific for effector caspases but not initiator caspases. Means *et al.* (2007) went on to show that site directed mutagenesis of the caspase cleavage site abrogated cleavage and anti-apoptotic activity. This was the first report to show that a homologue of the baculovirus P35 is present in another virus family.

EPVs carry other inhibitors of apoptosis genes (*iap*) but their anti-apoptotic function has not been totally demonstrated.

It has been observed by a number of investigators that EPV infected larvae do not undergo metamorphosis but continue to feed and eventually die. A certain percentage of EPV infected larvae die as larval/pupal intermediates or pupal/adult intermediates similar to phenotypic observations on larvae treated with a juvenile hormone (JH) (Palli *et al.*, 2000). Various assays on the levels of JH and ecdysteroids in infected versus control larvae showed that the virus causes an elevation in the JH titre and a lowering of the ecdysteroid titre. Therefore, it appears that EPV interferes with metamorphosis by modulating the levels of ecdysteroids and JH in the haemolymph (Palli *et al.*, 2000).

Pathology

There has been some correlation observed between the host and the amount of time it takes to kill the larvae after infection with EPV. EPVs infecting Coleoptera may take up to 37 weeks, depending on the temperature (Hurpin, 1968). In contrast, infection of Lepidoptera requires approximately one month. In addition, the symptoms vary from host to host. The *Elasmopalpus lignosellus* larvae become sluggish and unresponsive just prior to death. Their normally brown cuticle changes to red and infected haemolymph turns light blue or white (Mitchell *et al.*, 1985). Virus-infected *Othnonius batesi* have a distinct white-spotted or mottled exterior caused by the accumulation of infected fat body in the last segment that surrounds the rectal sac (Goodwin *et al.*, 1969). Lepidopteran larvae such as *Euxoa auxiliaries* and *Operophtera brumata* become white in colour but larvae of *Estigmene acrea* remain dark after infection (Weiser and Vago, 1966; Sutter, 1972; Granados, 1973b). Larvae infected with *Choristoneura biennis* EPV double in size and pupation is normally delayed (Bird *et al.*, 1971). This is due to the modulation of hormonal levels in infected larvae (Palli *et al.*, 2000).

The primary site of replication is the fat body (Roberts *et al.*, 1968), but other tissues are infected. In some larvae, such as those of *Malanoplus sanguinipes*, viral infection is restricted to fat body cells and haemocytes. An AMEV-*gfp*, constructed in the laboratory of Dr Richard Moyer at the University of Florida, was used (in the laboratory of Basil Arif) to infect larvae of the gypsy moth, *Lymantria dispar*. The gene for the green fluorescent protein (*gfp*) was under the control of the very late promoter of the spheroidin gene. Infection was followed by observing GFP expression every 8–12 hours. The first signs of fluorescence were in the haemocytes. Unlike baculoviruses, that use the trachea as a conduit to spread infection within larval tissues, overt infection of tracheoblasts with AMEV was not observed (Fig. 4.1). The fat body tissue appeared to be the main site of infection (Fig. 4.2). However a number of other tissues were also infected including the silk gland (Fig. 4.3), hindgut and midgut (data not shown). Fig. 4.4 shows the accumulation of virions in the silk gland. The Malpighian tubules were also refractory to infection.

Genome comparison

Poxvirus genomes

Poxviruses contain large, linear, double-stranded DNA genomes with a high A+T content (Arif, 1995). As shown in Table 4.1, over 100 poxvirus genomes have been completely sequenced: two complete sequences from the entomopoxviruses (EPV), one sequence from an unclassified poxvirus, and the nine chordopoxvirus (ChPV) genera each represented by one or more sequences. The length of poxvirus genomes can vary in size from 134 kb in bovine papular stomatitis virus (BPSV) to 360 kb in canarypox virus (CNPV), while the number of proteins encoded is also variable ranging from 130 (Orf virus, ORFV) to 328 (CNPV). The G+C content of the majority of poxvirus genomes is 25–40%, while some [BPSV, ORFV, molluscum contagiosum virus (MOCV), and crocodilepox virus (CRV)] may be as high as approximately 60%. The lowest G+C content is reported in the subfamily *Entomopoxvirinae* with 17% and 18% for AMEV and the unclassified *Melanoplus sanguinipes* entomopoxvirus (MSEV), respectively.

A typical poxvirus genome contains a conserved central region bordered by more

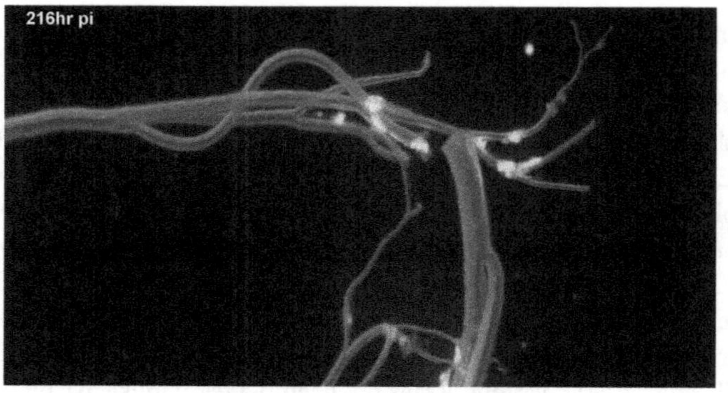

Figure 4.1 Infected haemocytes in the proximity of uninfected trachea. Gypsy moth larvae were inoculated with infected *Lymantria dispar* (Ld) cells and at various times thereafter, tissues were dissected out and examined for GFP expression from the spheroidin promoter, which indicated infection. p.i., post infection. A colour version of this figure is located in the plate section at the back of the book.

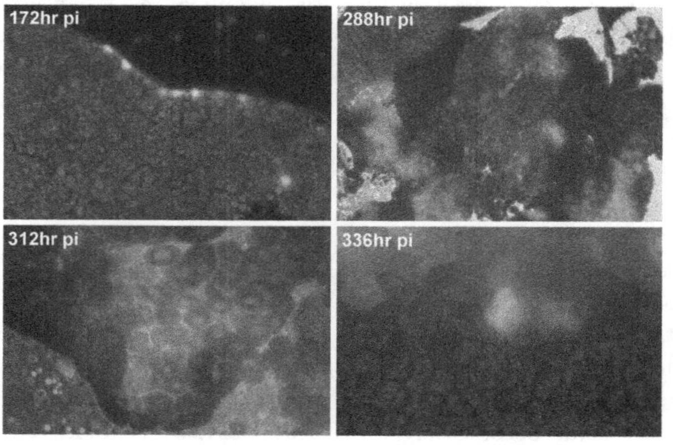

Figure 4.2 Infection of fat body tissue at various times after inoculation. Same procedure as in Fig. 4.1. p.i., post infection. A colour version of this figure is located in the plate section at the back of the book.

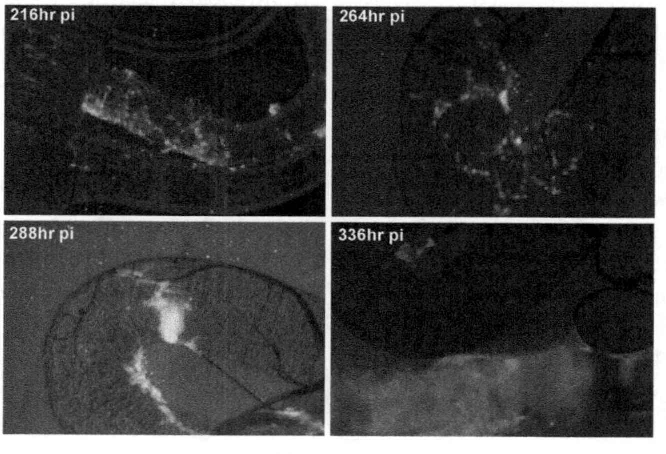

Figure 4.3 Infection in the silk gland. Same procedure as in Fig. 4.1. pi, post infection. A colour version of this figure is located in the plate section at the back of the book.

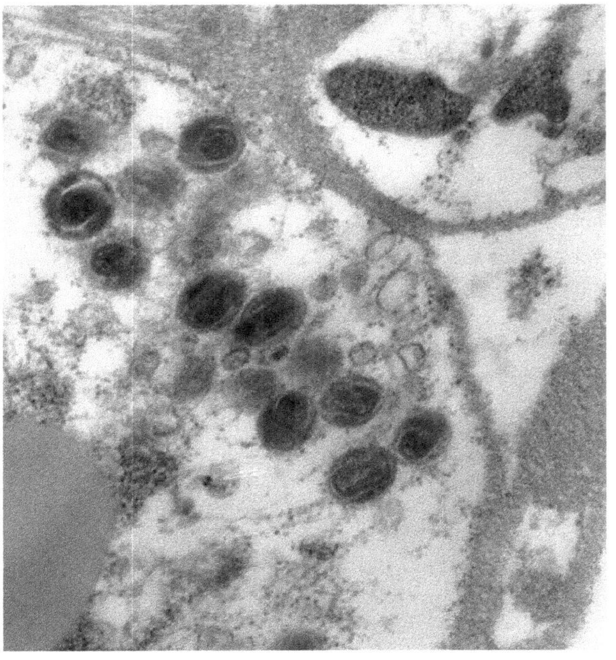

Figure 4.4 Virus particles in the silk gland. Infected silk glands were isolated, embedded, thin sectioned and examined for virus infection by transmission electron microscopy.

variable terminal regions. The extreme termini of the genome contain inverted terminal repeat (ITR) regions which range in length from < 100 bp [variola virus strains: Germany 1958 Heidelberg, Somalia 1977 (V77-1252), Somalia 1977 (V77-1605), United Kingdom 1947 Higgins (Staffordshire), India 3 Major 1967; Garcia 1966] to 16 kb (vaccinia virus strain, Acambis, 2000; http://www.poxvirus.org/). There are, however, some exceptions. The ITR of the cowpox virus strain Brighton red (CPXV-BR) is 10 kb, but some of its variants contain ITR regions as long as 50 kb as a result of sequence duplications (Pickup *et al.*, 1984). The ITRs often consist of tandemly repeated sequences interspersed with non-repetitive spacer regions. At the left and right termini, the two DNA strands of the genome are covalently linked by an A+T rich, incompletely basepaired hairpin loop. The hairpin loop and the adjacent highly conserved concatemer resolution motif play an important role in poxvirus genome replication (Moss, 2001).

In most ChPV genomes, the genes are arranged in groups that are transcribed in the same direction; with the genes at the left and right termini of the genome transcribed towards their respective ends, and most of the genes in the central region transcribed towards the centre, giving a characteristic 'W' appearance to a plot of transcriptional orientation (Lefkowitz *et al.*, 2006). This arrangement of gene transcription is assumed to reduce collisions between transcription complexes. No such pattern of transcriptional orientation is evident in the betaentomopoxvirus AMEV. However, in MSEV, the unclassified member of the subfamily *Entomopoxvirinae*, most ORFs in the terminal 50 kb of the genome are transcribed towards their respective termini while no discernible pattern with regard to the direction of transcription is seen in the central region of the genome.

Conserved and shared genes

Comparison of sequenced poxvirus genomes revealed 49 genes that are completely conserved among members of both subfamilies. The conserved genes are involved in key functions such as transcription, replication of the DNA genome and virion assembly. As such, these genes are essential for the survival of the virus

Table 4.1 Completely sequenced poxvirus genomes

Genus	Species	Abbreviation	Strains/ isolates*	Representative strain/ isolate
Subfamily: Entomopoxvirinae				
Betaentomopoxvirus	*Amsacta moorei* entomopoxvirus	AMEV	1	Strain: Moyer
Unclassified	*Melanoplus sanguinipes* entomopoxvirus	MSEV	1	Strain: Tucson
Subfamily: Chordopoxvirinae				
Orthopoxvirus	Camelpox virus	CMLV	2	Isolate: M-96
	Cowpox virus	CPXV	3	Strain: Brighton Red
	Ectromelia virus	ECTV	2	Strain: Moscow
	Monkeypox virus Zaire-96-I-16	MPXV	9	Strain: Zaire-96-I-16
	Taterapox virus	TATV	1	Strain: Dahomey 1968
	Vaccinia virus	VACV	16	Strain: WR (Western Reserve)
	Variola virus	VARV	49	Strain: India-1967, ssp. major; isolate: Ind3
Parapoxvirus	Bovine papular stomatitis virus	BPSV	1	Strain: BV-AR02; isolate: ORFB
	Orf virus	ORFV	4	Strain: OV-SA00; isolate: ORFD
Avipoxvirus	Canarypox virus	CNPV	1	Strain: ATCC VR-111; isolate: Wheatley C93
	Fowlpox virus	FWPV	2	Strain: Iowa
Capripoxvirus	Goatpox virus Pellor	GTPV	2	Strain: Pellor
	Lumpy skin disease virus NI-2490	LSDV	3	Isolate: Neethling 2490
	Sheeppox virus 17077–99	SPPV	3	Strain: TU-V02127
Leporipoxvirus	Myxoma virus	MYXV	2	Strain: Lausanne
	Rabbit fibroma virus (Shope fibroma virus)	RFV (SFV)	1	Strain: Kasza
Suipoxvirus	Swinepox virus	SWPV	1	Isolate: 17077–99
Molluscipoxvirus	Molluscum contagiosum virus	MOCV	1	Strain: subtype 1
Yatapoxvirus	Tanapox virus	TANV	3	Isolate: TPV-Kenya
	Yaba monkey tumor virus	YMTV	1	
Cervidpoxvirus	Deerpox virus W-848–83	DPV	2	Isolate: W-848–83
Unclassified members of family Poxviridae				
	Crocodilepox virus	CRV	1	Strain: Zimbabwe

*Number of strains/isolates sequenced from each species.

Sources: http://www.ncbi.nlm.nih.gov/sites/entrez?db = Genome&itool = toolbar, http://www.ncbi.nlm.nih.gov/ICTVdb/, http://www.poxvirus.org/ and cited references.

Genome (bp)	No. of ORFs	GC content	Accession no.	Reference
232,392	294	17%	NC_002520	Bawden *et al.* (2000)
236,120	267	18%	NC_001993	Afonso *et al.* (1999)
205,719	211	33%	NC_003391	Afonso *et al.* (2002b)
224,499	233	33%	NC_003663	–
209,771	173	33%	NC_004105	Chen *et al.*, (2000, 2003)
196,858	191	33%	NC_003310	Shchelkunov *et al.* (2001)
198,050	225	33%	NC_008291	Esposito *et al.*, 2006)
194,711	223	33%	NC_006998	–
185,578	197	32%	NC_001611	Shchelkunov *et al.* (1994, 1996), Shchelkunov and Totmenin (1995)
134,431	131	64%	NC_005337	Delhon *et al.* (2004)
139,962	130	63%	NC_005336	Delhon *et al.* (2004)
359,853	328	30%	NC_005309	Tulman *et al.* (2004)
288,539	261	30%	NC_002188	Afonso *et al.* (2000)
149,599	150	25%	NC_004003	Tulman *et al.* (2002)
150,773	156	25%	NC_003027	Tulman *et al.*, 2001)
149,955	148	25%	NC_004002	Tulman *et al.* (2002)
161,773	170	43%	NC_001132	Cameron *et al.* (1999)
159,857	165	39%	NC_001266	Willer *et al.* (1999)
146,454	150	27%	NC_003389	Afonso *et al.* (2002a)
190,289	163	63%	NC_001731	Senkevich *et al.* (1996)
144,565	156	26%	NC_009888	Nazarian *et al.* (2007)
134,721	140	29%	NC_005179	Brunetti *et al.* (2003)
166,259	169	26%	NC_006966	Afonso *et al.* (2005)
190,054	173	61%	NC_008030	Afonso *et al.* (2006)

and can be considered as the minimum poxvirus genome. A total of 90 genes were found to be conserved among chordopoxvirus genera, including the 49 genes common to all poxviruses. In chordopoxviruses, the essential genes are clustered in the central region of the genome while the genes that are genus-specific are found near the termini. Many of the terminal genes encode virulence proteins that are involved in counteracting the host immune responses (Gubser *et al.*, 2004; Upton *et al.*, 2003). Poxviruses that belong to different genera have widely different immune evasion genes suggesting that these were acquired during the course of poxvirus host adaptation (Senkevich *et al.*, 1996).

In chordopoxviruses, the conserved genes located in the central region share a co-linear arrangement, as shown in the gene parity plot between the two orthopoxvirus genomes vaccinia virus (VACV) and variola virus (VARV) (Fig. 4.5A). A close inspection of the two sequenced EPV genomes reveal that this is not the case in entomopoxviruses; the core genes do not show a co-linear arrangement with chordopoxviruses (Fig. 4.5B) or with each other (Fig. 4.5C). It should be noted that AMEV and MSEV do not belong to the same genus; the former is a betaentomopoxvirus (lepidopteran entomopoxvirus), while the latter is an unclassified entomopoxvirus isolated from the migratory grasshopper. A comparison of AMEV with the partially sequenced genome of another betaentomopoxvirus, *Heliothis armigera entomopoxvirus* (HAEV), shows that there is some degree of conservation in gene arrangement within this genus (Bawden *et al.*, 2000). More sequence data from betaentomopoxviruses as well as the other two EPV genera are needed before meaningful genome comparisons can be made among entomopoxviruses and between ChPV and EPVs.

Table 4.2 shows genes shared by the two poxvirus subfamilies. They include the 49 genes that are conserved among all poxviruses (Upton *et al.*, 2003) and an additional 10 ChPV genes that share identity with entomopoxviruses. As the last 10 are not found in all poxviruses, they are not considered conserved genes. PSI-BLAST searches were used to construct the list of EPV orthologues in Table 4.2. In the case of the 49 conserved genes, the AMEV and MSEV ORFs with the best match were selected as the corresponding orthologues regardless of the blast expect (E) value. The 10 non-conserved genes were selected based on a threshold E value of 1.0×10^{-5}.

Some of the ChPV genes listed in Table 4.2 (VACV F2L, J2R, H1L and CNPV085) have orthologues in AMEV but not in the unclassified entomopoxvirus MSEV. Similarly, the VACV B2R gene has an MSEV orthologue (MSV237) but shares no identity with AMEV.

Vaccinia virus poly(A) polymerase and methyltransferase gene (J3R) has two AMEV orthologues, AMV060 and AMV115. AMV060 has been found to be an active (nucleoside-2′-O-) methyltransferase (Becker *et al.*, 2008), while the function of AMV115 is not clear. In contrast, MSV041 serves as the only J3 orthologue in MSEV.

VACV B1 and B12 are structurally related proteins with significant sequence similarity to known protein kinases (Howard and Smith, 1989). However, only B1 has been shown to possess enzymatic activity (Lin *et al.*, 1992), while B12 which lacks some of the key residues, was reported to be non-functional (Banham and Smith, 1993). AMEV too codes for two structurally similar protein kinases (AMV197 and AMV084) both of which are listed as B1 and B12 orthologues in Table 4.2. Of these two, AMV197, which contains all the residues required for enzymatic activity, is most likely a functional protein kinase. AMV084, like B12, lacks some of the key amino acids in its sequence and is probably not a functional enzyme. B12 and its orthologues have been placed in the 'unknown function' category as their function has not yet been elucidated.

The 59 EPV orthologues identified in Table 4.2 can be grouped into several functional categories such as those involved in virion structure and morphogenesis, transcription and RNA modification, genome replication, other known functions and genes with unknown functions. Genes involved in structure and virion morphogenesis (37%) and transcription (34%) comprise the largest groups followed by those involved in other functions (14%) and replication

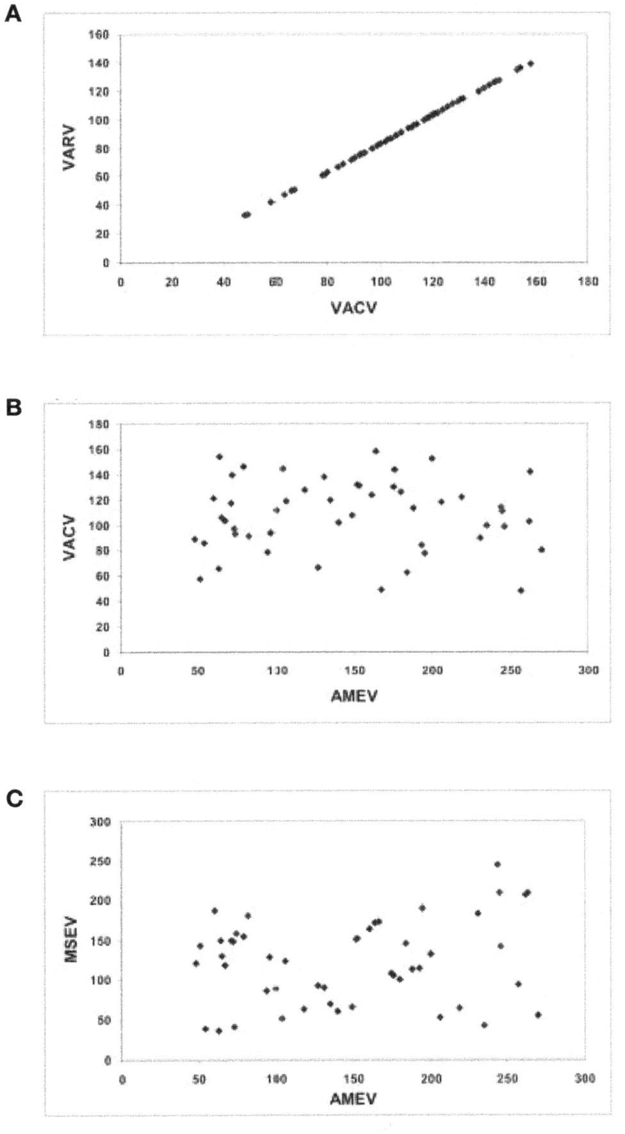

Figure 4.5 Gene parity plots of (A) vaccinia virus (VACV) versus variola virus (VARV), (B) *Amsacta moorei entomopox-virus* (AMEV) versus VACV and (C) AMEV versus *Melanoplus sanguinipes entomopoxvirus* (MSEV). Each graph shows a comparison of shared orthologues between two poxvirus genomes. Each pair of orthologues is plotted based on their location in their respective genomes. Orthologues with co-linear arrangement appear as straight lines on the graph while those with no co-linearity are shown as scattered diamonds. Only the 49 conserved poxvirus genes are shown.

(12%). Only 3% of the genes, namely the B2R and B12R, are yet to be characterized.

Genomes of entomopoxviruses

The first entomopoxvirus genome to be completely sequenced was the 236-kb MSEV genome containing 267 ORFs (Afonso *et al.*, 1999). In 2000, the sequence of AMEV (232-kb) which was presumed to encode 294 ORFs became available (Bawden *et al.*, 2000). As in other poxviruses, both entomopoxvirus genomes contain inverted repeat regions (ITR) which are

Table 4.2 Chordopoxvirus (ChPV) orthologues in entomopoxviruses *Amsacta moorei* EPV (AMEV) and *Melanoplus sanguinipes* EPV (MSEV)

ChPV ORF*	AMEV ORF	MSEV ORF	Comments
Structure/morphogenesis			
A3L	AMV147	MSV164	Precursor p4b of core protein 4b
A9L	AMV161	MSV108	Late virion membrane protein (MP), essential
A10L	AMV139	MSV152	Precursor p4a of core protein 4a, complexes with A4L
A11R	AMV138	MSV151	Virion membrane formation
A16L	AMV118	MSV090	Membrane protein involved in fusion and entry
A21L	AMV249	MSV209	Membrane protein involved in fusion and entry
A28L	AMV186	MSV132	Membrane protein involved in fusion and entry
A32L	AMV150	MSV171	DNA packaging; morphogenesis
D13L	AMV122	MSV069	Rifampicin target; associates with inner surface of immature virus membrane
E6R	AMV170	MSV145	Core protein; involved in morphogenesis
F9L	AMV243	MSV094	Membrane protein associated with entry–fusion complex
F10L	AMV153	MSV173	Major vaccinia core protein kinase; involved in morphogenesis
G1L	AMV256	MSV056	Core metalloproteinase; essential for virus maturation
G5R	AMV179	MSV115	Core protein; involved in morphogenesis
G9R	AMV035	MSV121	Membrane protein involved in fusion and entry
H2R	AMV127	MSV060	Membrane protein involved in fusion and entry
H3L	AMV248	MSV206	Membrane protein; MV morphogenesis
I7L	AMV181	MSV189	Core cysteine proteinase; required for virion morphogenesis
J5L	AMV232	MSV142	Membrane protein involved in fusion and entry
L1R	AMV217	MSV183	Membrane protein associated with entry–fusion complex
L4R	AMV061	MSV158	Core protein; microtubule associated; ssDNA/ssRNA binding protein
L5R	AMV083	MSV129	Membrane protein involved in fusion and entry
Transcription/RNA modification			
A1L	AMV047	MSV187	Late gene transcription factor VLTF-2
A2L	AMV205	MSV065	Late gene transcription factor VLTF-3
A5R	AMV166	MSV100	DNA-dependent RNA polymerase subunit rpo19
A7L	AMV105	MSV063	82K large subunit of early gene transcription factor VETF
A18R	AMV059	MSV148	DNA helicase; transcript release factor
A23R	AMV091	MSV052	Component of intermediate transcription factor VITF-3
A24R	AMV066	MSV155	DNA-dependent RNA polymerase subunit rpo132
A29L	AMV051	MSV149	DNA-dependent RNA polymerase rpo35
D1R	AMV135	MSV067	Large subunit of mRNA capping enzyme, transcriptioN-termination factor VTF

ChPV ORF*	AMEV ORF	MSEV ORF	Comments
D6R	AMV174	MSV113	70K small subunit of early gene transcription factor VETF
D7R	AMV230	MSV245	DNA-dependent RNA polymerase subunit rpo18
D11L	AMV192	MSV053	ATPase, nucleoside triphosphate phosphohydrolase-I, NPH-I transcription elongation, termination, release factor
D12L	AMV093	MSV124	Small subunit of mRNA capping enzyme, transcriptioN-termination factor VTF
E1L	AMV038	MSV143	Poly(A) polymerase catalytic subunit VP55
H4L	AMV054	MSV119	Rap94; tightly associated with DNA-dependent RNA polymerase, aids early-stage transcription, preinitiation, and termination
H6R	AMV052	MSV130	DNA topoisomerase type 1B
I8R	AMV081	MSV086	DExH RNA/DNA helicase stimulated by L4R (NPH-II); required for early transcription
J3R	AMV060 AMV115	MSV041	Poly(A) polymerase and RNA methyltransferase
J6R	AMV221	MSV043	DNA-dependent RNA polymerase subunit rpo147
L3L	AMV069	MSV180	Involved in early gene transcription
Replication			
A22R	AMV162	MSV106	Holliday Junction Resolvase; resolves viral DNA concatemers into unit length genomes
B1R	AMV197 AMV084	MSV154	Early Ser/Thr Kinase essential for ongoing DNA replication
D4R	AMV231	MSV208	Uracil DNA glycosylase, Component of the processive DNA polymerase
D5R	AMV087	MSV089	Nucleic acid-independent nucleoside triphosphatase involved in DNA replication.
E9L	AMV050	MSV036	Catalytic DNA polymerase component of the trimeric processive polymerase complex
F2L	AMV002		Deoxyuridine triphosphatase
J2R	AMV016		Thymidine kinase
Other functions			
A45R	AMV255		Superoxide dismutase
D10R	AMV058	MSV150	Contains mutT-like motif of nudix hydrolases, involved in regulating RNA levels
E10R	AMV114	MSV093	Disulphide bond formation pathway protein
G6R	AMV041	MSV039	*In vivo* virulence
H1L	AMV078		Dual specificity phosphatase; Ser/Thr and Tyr protein phosphatase
CNPV206	AMV025	MSV235	CPD photolyase
CNPV096	AMV167	MSV144	Ubiquitin
CNPV085	AMV246		Protein tyrosine phosphatase

Table 4.2 continued

ChPV ORF*	AMEV ORF	MSEV ORF	Comments
Unknown functions			
B2R		MSV237	Hypothetical protein
B12R	AMV197 AMV084	MSV154	Serine/Threonine kinase-like protein

Proteins that are not conserved in all poxviruses are shaded.

*ChPV ORFs are from vaccinia virus (VACV) unless otherwise stated.

7-kb and 9-kb for MSEV and AMEV, respectively. As mentioned earlier, EPV genomes contain the lowest G+C content (17–18%) of all sequenced poxviruses.

Although AMEV genome is expected to code for 294 ORFs, later studies indicate that the actual number of ORFs may be less than initially predicted. Da Silva and Upton (2005) described a method to predict functional ORFs in A+T rich poxvirus genomes based on their purine content and the average aspartate-glutamate-serine composition. When this method was applied to the AMEV genome, approximately 45 previously predicted ORFs were found to be non-coding and thus not to be 'real' genes. Two other studies using a different method of predicting ORFs (Z-curve method) reported that 38–48 of previously annotated AMEV ORFs do not code for proteins (Guo and Zhang, 2006; Guo and Yu, 2007). Based on these reports, the AMEV genome codes for 246–256 functional proteins, which is considerably less than the previously reported 294 ORFs.

Entomopoxviruses encode a number of genes that can be grouped into seven gene families (Table 4.3). Of these, AMEV genome contains 23 genes in six families and MSEV codes for 43 genes in five families. None of these genes have been functionally characterized. However, it is likely that the members of the ALI family and 17K ORF family play a role in gene transcription and DNA replication. The ALI (alanine-leucine-isoleucine) gene family is based on an amino-terminal motif which contains invariant alanine, leucine, and isoleucine residues. All but one member (AMV257) of the ALI motif family contain the Bro-N domain [N-terminus of baculovirus repeated open reading frame (BRO) proteins]. Although the function of BRO proteins is unknown, it has been suggested that in baculoviruses they act as DNA binding proteins and involved in regulating host DNA replication and/or transcription (Zemskov *et al.*, 2000). The 17K ORF gene family consists of five AMEV proteins with homology to the 17K ORF of HAEV. This family also contains a putative DNA binding domain (KilA-N domain) found in a wide range of proteins in bacterial and eukaryotic DNA viruses which suggests a possible role in transcription and/or replication.

Phylogeny and evolution

Phylogeny is the inference of evolutionary relationships, which traditionally was assessed by comparing morphological features between organisms from a variety of species (Pevsner, 2003). As techniques progress and data accumulate, molecular sequence data are now widely used in phylogenetic analysis. Because of viruses' relatively high speed of evolution and mutation at the molecular level, phylogenetic analysis are now widely used in studying their evolution and epidemiology.

Poxviruses have linear double-strand DNA genomes that vary from 130 kb to 350 kb. The two ends of genomes are covalently closed and adjoining small inverted terminal repetitions (ITRs) region. The coding region sequences (CRS) are between these two ITRs (Moss, 2001; Esposito *et al.*, 2006).

Table 4.3 Entomopoxvirus gene families. *Amsacta moorei* EPV (AMEV); *Melanoplus sanguinipes* EPV (MSEV)

Family	AMEV ORF	MSEV ORF
Leucine-rich repeat (LRR)	AMVITR01, AMV005, AMV014, AMV076, AMV134	MSV008, MSV009, MSV010, MSV011, MSV013, MSV014, MSV015, MSV016, MSV018, MSV186, MSV227, MSV228, MSV239, MSV240, MSV241, MSV253, MSV254, MSV255, MSV257, MSV260, MSV261
ALI (alanine-leucine-isoleucine) motif		
Subgroup 1	AMV257	MSV024,MSV026, MSV196, MSV204
Subgroup 2	AMV055, AMV057, AMV175, AMV177	MSV023, MSV194, MSV195
Tryptophan (W) repeat	AMV029, AMV254	MSV027, MSV029, MSV034, MSV197, MSV205, MSV252
Methionine–threonine–glycine (MTG)	AMV194, AMV207, AMV209	MSV021, MSV191, MSV198, MSV199
AMV176	AMV056, AMV176, AMV178	
17K ORF (KilA-N domain-containing proteins)	AMV024, AMV100, AMV110, AMV112, AMV132	
Serine–cysteine–glycine (SCG)		MSV062, MSV214, MSV215, MSV216, MSV217

Entomopoxvirus phylogeny

Different methods have been applied to phylogeny analysis, such as morphological, individual genes, gene contents and gene order and arrangements in the genome. Recently, alignment-free methods and single nucleotide polymorphism (SNP) have been applied to phylogeny analysis of poxviruses (Esposito *et al.*, 2006; Gao and Qi, 2007; Li *et al.*, 2007; Wu *et al.*, 2009). The phylogeny of entomopoxviruses based on spheroidin is illustrated in Fig. 4.6 and shows that the betaentomopoxviruses of AMEV group together.

Among comparable genes, MSEV and AMEV have a relatively low degree of identity (Afonso *et al.*, 1999; Bawden *et al.*, 2000). For example, the MSEV spheroidin protein exhibits only about 20% amino acid identity to spheroidins of AMEV and other betaentomopoxviruses (HAEV and *Choristoneura biennis entomopoxvirus*, CBEV) as well as the spheroidin of the alphaentomopoxvirus MMEV. This is in contrast to the 76–92% amino acid identity observed among spheroidins of betaentomopoxviruses (Afonso *et al.*, 1999). Phylogenetic analysis of spheroidins shows that MSEV is divergent to AMEV (Fig. 4.6) (Hernández-Crespo *et al.*, 2000).

Furthermore, the gene order and organization are conserved among the lepidopteran betaentomopoxviruses. We sequenced the central part of other betaentomopoxviruses. CBEV and *Choristoneura rosaceana* entomopoxvirus (CREV), and found that this part of the two betaentomopoxviruses are highly conserved (data not shown). This conservation is likely to apply to other betaentomopoxviruses, such as HAEV and *Choristoneura fumiferana* entomopoxvirus (CFEV) (Sriskantha *et al.*, 1997; Li and Arif, unpublished data). No such conservation was found between MSEV and AMEV, the two entomopoxviruses that have been fully sequenced (Bawden *et al.*, 2000; McLysaght *et al.*, 2003).

Based on the total sequence, the number of homologues between genomes of AMEV and MSEV is about 120 ORFs. Based on our partial sequencing of the genome of CBEV, there are no less than 200 homologous ORFs in the AMEV genome.

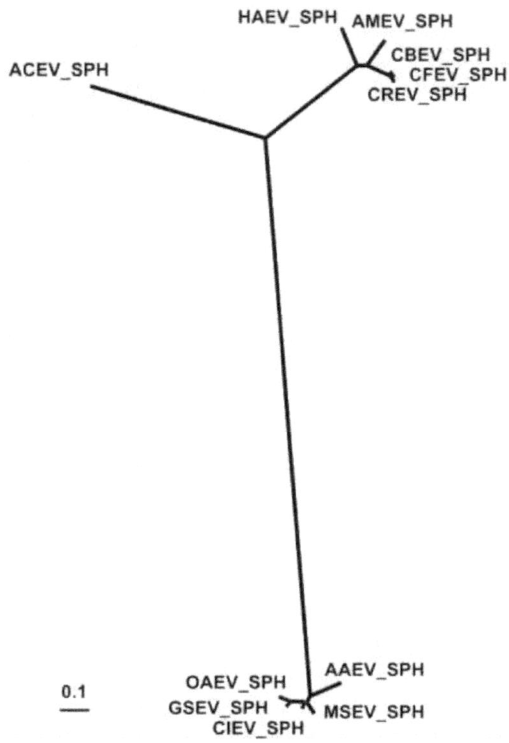

Figure 4.6 Phylogeny of entomopoxviruses based on spheroidin. Entomopoxviruses (EPVs) were from *Oedaleus asiaticus* (OAEV), *Gomphocerus sibiricus* (GSEV), *Calliptamus italicus* (CIEV), *Anacridium aegyptium* (AAEV), *Melanoplus sanguinipes* (MSEV), *Anomala cuprea* (ACEV), *Heliothis armigera* (HAEV), *Amsacta moorei* (AMEV), *Choristoneura biennis* (CBEV), *Choristoneura fumiferana* (CFEV) and *Choristoneura rasaceana* (CREV).

The molecular data show that AMEV and MSEV are divergent, and led the International Committee on Taxonomy of Viruses (ICTV) to remove the latter from the genus *Betaentomopoxvirus* and reclassify it as an undesignated member of the subfamily *Entomopoxvirinae*.

Relationship to chordopoxviruses

As a subfamily of *Poxviridae*, entomopoxviruses share a number of characters with their vertebrate counterparts. However, a striking difference is their nucleotide constitutions; entomopoxviruses being extremely rich in A + T residues with content greater than 80%. (Arif, 1976; Langridge *et al.*, 1977). It also means that the codon usage is strongly bias to A/T. By investigating 21 completely sequenced poxvirus genomes, Upton and his colleagues (2003) identified a set of 49 gene families that are conserved throughout the entire *Poxviridae* family. Among these families; 25 are known to be involved in DNA replication and transcription, and 15 are associated with virions, virions assembly, or maturation, while 12 are of unknown function (three gene families are involved in both transcription and morphogenesis) (Upton *et al.*, 2003). Recently, investigators have shown that 7 of 12 gene families of unknown function are related to entry/fusion step (Senkevich *et al.*, 2004, 2005; Senkevich and Moss, 2005; Townsley *et al.*, 2005a,b; Brown *et al.*, 2006; Izmailyan *et al.*, 2006; Ojeda *et al.*, 2006). These conserved gene families indicate that both entomopoxviruses and chordopoxviruses inherited these genes vertically from a common ancestor before their divergence and, they keep these genes in their genomes not only for virus replication but also for virion transmission. Phylogeny analysis showed that the topologies of these genes are quite similar and coincident with the topologies generated with concatenated protein sequences of all conserved genes. It means these conserved genes are under strict selective pressure, and can be used to define a poxvirus (Upton *et al.*, 2003).

On the other hand, entomopoxviruses may be the most 'distant relatives' of the vertebrate poxviruses known (Becker and Moyer, 2007). Using different phylogeny analyses, the *Poxviridae* are always divided into two subfamilies of entomopoxviruses and chordopoxviruses (Lefkowitz *et al.*, 2006; Xing *et al.*, 2006). Similar results were obtained by using concatenated protein sequences of conserved genes and by studying gene gain and loss (McLysaght *et al.*, 2003; Hughes and Friedman, 2005; Xing *et al.*, 2006), which are consistent with the proteome trees constructed by composition vector and feature frequency profiles methods (Gao and Qi, 2007; Wu *et al.*, 2009). For gene contents, there are 49 homologous genes between MSEV and chordopoxviruses, and 52 homologues between AMEV and chordopoxviruses; but it is predicted that entomopoxviruses encode more than 250 genes and approximately 200 genes by chordopoxviruses. Homologues take only a small part of all genes. The biggest difference is at the gene organization level within the genomes. For chordopoxviruses, the overall gene order and content are well conserved but such conservation was not found between chordopoxvirus and entomopoxviruses, or even between two genome-sequenced entomopoxviruses, MSEV and AMEV (McLysaght *et al.*, 2003; Gubser *et al.*, 2004).

Among the 250 predicted genes in the two sequenced entomopoxvirus genomes, AMEV and MSEV, many may have been acquired by horizontal transfer, as in the case of baculoviruses (Hughes, 2002a,b; Hughes and Friedman, 2003; McLysaght *et al.*, 2003). McLysaght *et al.* (2003) proposed that there have been frequent events of gene gain and loss over the evolution of poxviruses. These authors also suggested that much of the gene acquisition was by particular horizontal transfer events and the host was an important reservoir in these events.

Phylogenetic methods have also been used to test the hypothesis of a recent horizontal transfer from host to virus (Hughes, 2002a; Hughes and Friedman, 2003, 2005). For example, viruses have evolved various strategies to inhibit apoptosis, thereby allowing productive virus replication. In the case of chordopoxviruses, apoptosis is controlled in part by serpins (serine protease inhibitors) (Turner and Moyer, 1998; Taylor and Barry, 2006). The entomopoxviruses clearly control the host immune response, but probably in a fashion different from that of the chordopoxviruses. AMV021, an ORF in AMEV genome, encodes an IAP protein. It shows significant identity to the IAP of *Cydia pomonella* granulovirus (47%), and contains two typical baculovirus IAP repeats and a C-terminal RING-finger motif. The AMV021 ORF has been shown to be functionally active (Li *et al.*, 2005). IAP homologues are also found in MSEV. AMV10, which is homologous to P35, also has the function of inhibiting apoptosis in insect cells. IAP and P35, which are both inhibitors of apoptosis, were first reported from baculoviruses (see Chapter 17). IAPs have been found to exist widely in cells, but P35 has only been found in some baculoviruses and entomopoxviruses (Deveraux and Reed, 1999; Miller, 1999; Afonso *et al.*, 1999; Bawden *et al.*, 2000; Means *et al.*, 2007).

Relationship to other viruses

The origin of poxviruses is not very well understood and many investigators have proposed different hypotheses. Some investigators suggested that poxviruses with other nucleocytoplasmic large DNA virus (NCLDV) are all divergent from a common ancestor. Iyer and his colleagues (2001) found nine genes that are shared by four viruses (poxviruses, asfarviruses, iridoviruses, and phycodnaviruses), and 22 more genes that are present in at least three of the four viral families. These four viruses are monophyletic by cladistic analysis (Iyer *et al.*, 2001). Later, the latter authors revised their data after analysing more virus sequences, specially whole genomes of mimiviruses (Iyer *et al.*, 2006). In this system, two principal lineages of NCLDVs were the poxviruses/asfarviruses group and the iridoviruses/phycodnaviruses-mimiviruses group. Recently, alignment-free methods were used to delineate the phylogeny of viruses. Gao and Qi (2007) used composition vector methods to analyse the protein sequences and topology of viruses. Poxviruses first grouped with *Nimaviridae*, then grouped with another cluster containing *Adenoviridae*, *Phycodnaviridae* and *Mimiviridae*. Another proteome tree, constructed by feature frequency profile methods, showed that *Poxviridae* was initially grouped with *Asfarviridae*, then another comprising the *Iridoviridae* and *Ascoviridae* (Wu *et al.*, 2009). These results confirmed those of Iyer *et al.* (2001, 2006) on relationships between poxviruses and other large DNA viruses.

As a subfamily, entomopoxvirus not only inherited genes from a common poxvirus ancestor by vertical transfer, but also obtained genes from other sources to adapt to the insect host and environment. Further analysis of phylogeny can provide insight into virus–host interaction and generate more knowledge on virus evolution and epidemiology.

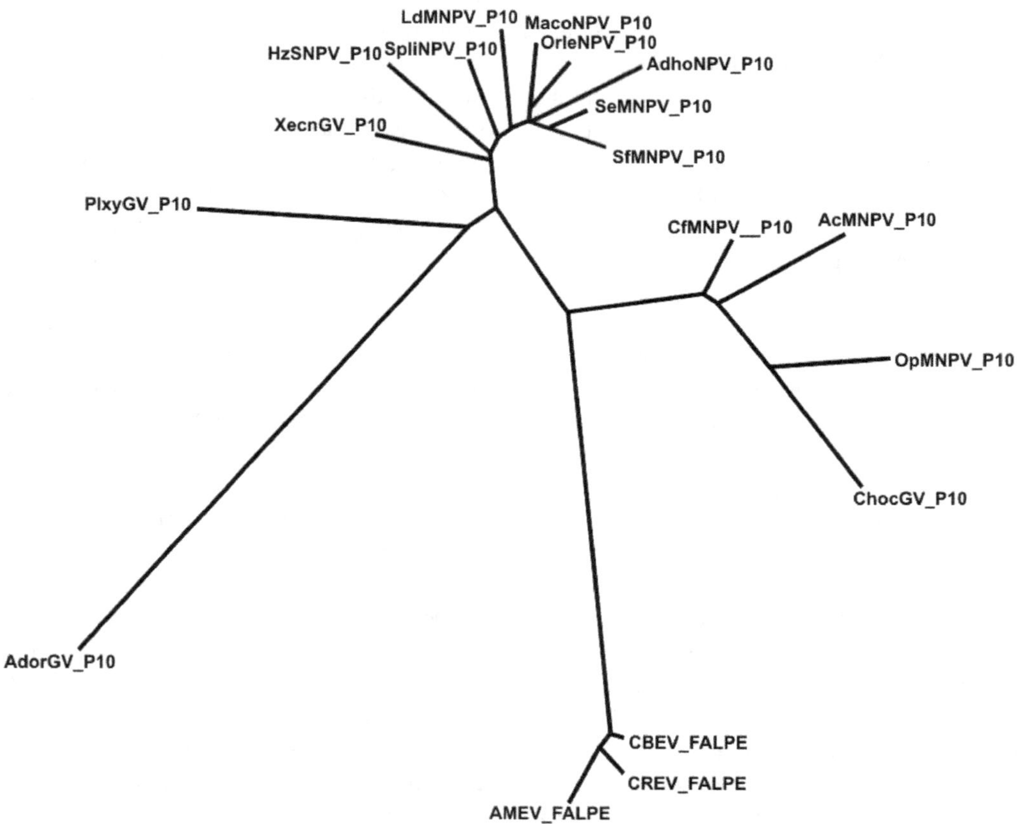

Figure 4.7 Phylogenetic relationship between P10 and FALPE proteins among baculoviruses and entomopoxviruses. AcMNPV, *Autographa californica* MNPV; AdhoNPV, *Adoxophyes honmai* NPV; CfMNPV, *Choristoneura fumiferana* MNPV; ChocGV, *Choristoneura occidentalis* GV; HzSNPV, *Helicoverpa zea* single NPV; LdMNPV, *Lymantria dispar* MNPV; MacoNPV, *Mamestra configurata* NPV; MaviMNPV, *Maruca vitrata* MNPV; OpMNPV, *Orgyia pseudotsugata* MNPV; OrleNPV, *Orgyia leucostigma* NPV; PlxyNPV, *Plutella xylostella* MNPV; RoMNPV, *Rachiplusia ou* MNPV; SeMNPV, *Spodoptera exigua* MNPV; SfMNPV, *Spodoptera frugiperda* MNPV; SpliNPV, *Spodoptera litura* NPV; XecnGV, *Xestia c-nigrum* GV. Entomopoxviruses: AMEV, *Amsacta moorei* EPV; AAEV, *Anacridium aegyptium* EPV; ACEV, *Anomala cuprea* EPV; CIEV, *Calliptamus italicus* EPV; CBEV, *Choristoneura biennis* EPV; CFEV, *Choristoneura fumiferana* EPV; CREV, *Choristoneura rosaceana* EPV; GSEV, *Gomphocerus sibiricus* EPV; HAEV, *Heliothis armigera* EPV; MSEV, *Melanoplus sanguinipes* EPV; OAEV, *Oedaleus asiaticus* EPV.

Apart from the anti-apoptotic proteins P33/P35 and the GP37/Fusolin proteins, one other protein has been found in EPVs with a homologue in baculoviruses. The baculovirus P10 has been well characterized with a number of functions ascribed to it. A homologue of P10 in EPVs is a filament associated late protein designated as FALPE (Alaoui-Ismaili and Richardson, 1996). FALPE forms large cytoplasmic fibrils that are associated with the viral spheroids. Much like P10, the amino-terminal region is predicted to be an amphipathic helix. While the amino acid identity between P10 and FALPE is low, the structural similarities are rather clear (van Oers and Vlak, 1997). Phylogenetic analysis reveals that FALPE is distantly related to baculovirus P10s (Fig. 4.7).

Concluding remarks

It is ironic that at the time when sequencing has become very cost effective, we only have two totally sequenced EPV genomes. Long stretches of AT-rich sequences have been problematic. Once a reasonable number of fully sequenced EPV genomes are available, we should be able to classify existing and new viruses on the basis of molecular data rather than structure and the family of the host insect. Clearly, much research is yet to be conducted on EPVs not only to understand their mode of action but because they represent a unique niche of host–parasite interactions. We have a glimpse on how EPVs interfere with larval metamorphosis by modulating the levels of juvenile hormone and ecdysteroids in the haemolymph. This is a fascinating area which deserves much attention because this type of interaction must have arisen through evolution when the virus began to acquire genes to maximize its replication potential in infected hosts. Another area that is equally fascinating is how the virus circumvents the larval immune system. Genomic approaches showed that the virus possesses genes needed to evade larval immunity and facilitate replication in the insect. The apparent little interest in EPVs stems from the fact that they may not represent a viable tool in the control of insect pests. However, research on EPV biology, interactions with the host, potential as protein expression systems that can accommodate large transgenes as well as possible use in gene therapy are areas that should lead to bountiful scientific and economic returns.

References

Afonso, C.L., Tulman, E.R., Lu, Z., Oma, E., Kutish, G.F., and Rock, D.L. (1999). The genome of *Melanoplus sanguinipes* entomopoxvirus. J. Virol. *73*, 533–552.

Afonso, C.L., Tulman, E.R., Lu, Z., Zsak, L., Kutish, G.F., and Rock, D.L. (2000). The genome of fowlpox virus. J. Virol. *74*, 3815–3831.

Afonso, C.L., Tulman, E.R., Lu, Z., Zsak, L., Osorio, F.A., Balinsky, C., Kutish, G.F., and Rock, D.L. (2002a). The genome of swinepox virus. J. Virol. *76*, 783–790.

Afonso, C.L., Tulman, E.R., Lu, Z., Zsak, L., Sandybaev, N.T., Kerembekova, U.Z., Zaitsev, V.L., Kutish, G.F., and Rock, D.L. (2002b). The genome of camelpox virus. Virology *295*, 1–9.

Afonso, C.L., Delhon, G., Tulman, E.R., Lu, Z., Zsak, A., Becerra, V.M., Zsak, L., Kutish, G.F., and Rock, D.L. (2005). Genome of deerpox virus. J. Virol. *79*, 966–977.

Afonso, C.L., Tulman, E.R., Delhon, G., Lu, Z, Viljoen, G.J., Wallace, D.B., Kutish, G.F., and Rock, D.L. (2006). Genome of crocodilepox virus. J. Virol. *80*, 4978–4991.

Alaoui-Ismaili, M.H., and Richardson, C.D (1996). Identification and characterization of a filament associated protein encoded by the *Amsacta moorei* entomopoxvirus. J. Virol. *70*, 2697–2705.

Arif, B.M. (1976). Isolation of an entomopoxvirus and characterization of its DNA. Virology *69*, 626–634.

Arif, B.M. (1995). Recent advances in the molecular biology of entomopoxviruses. J. Gen. Virol. *75*, 1–13.

Arif, B.M., and Kurstak, E. (1991). The Entomopoxviruses. In Virus of Invertebrates, Edouard Kurstak, ed. (Montreal, Canada: Marcel Dekker inc.), pp. 179–195.

Banham, A.H., and Smith, G.L. (1993). Characterization of vaccinia virus gene B12R. J. Gen. Virol. *74*, 2807–2812.

Bawden, A.L., Glassberg, K.J., Diggans, J., Shaw, R., Farmerie, W., and Moyer, R.W. (2000). Complete genomic sequence of the *Amsacta moorei* entomopoxvirus: analysis and comparison with other poxviruses. Virology *274*, 120–139.

Becker, M., and Moyer, R. (2007). Subfamily *Entomopoxvirinae*. In Poxviruses, A.A. Mercer, A. Schmidt and O. Weber, eds. (Basel, Switerland: Birkhauser Verlag), pp. 253–271.

Becker, M.N., Todd, T.M., and Moyer, R.W. (2008). An *Amsacta moorei* entomopoxvirus orthologue of the poly(A) polymerase small subunit exhibits methyltransferase activity and is non-essential for virus growth. Virology *375*, 624–636.

Bilimoria, S., and Arif, B.M. (1979). Subunit protein and alkaline protease of entomopoxvirus spheroids. Virology *96*, 596–603.

Bird, F.T., Saunders, C.J., and Burke, J.M. (1971). A newly discovered virus disease of the spruce budworm *Choristonuera biennis* (Lepidoptera: Tortricidae). J. Invertebr. Pathol. *18*, 159–161.

Brown, E., Senkevich, T.G., and Moss, B. (2006). Vaccinia virus F9 virion membrane protein is required for entry but not virus assembly, in contrast to the related L1 protein. J. Virol. *80*, 9455–9464.

Brunetti, C.R., Amano, H., Ueda, Y., Qin, J., Miyamura, T., Suzuki, T., Li, X., Barrett, J.W., and McFadden, G. (2003). Complete genomic sequence and comparative analysis of the tumorigenic poxvirus Yaba monkey tumor virus. J. Virol. *77*, 13335–13347.

Cameron, C., Hota-Mitchell, S., Chen, L., Barrett, J., Cao, J.X., Macaulay, C., Willer, D., Evans, D., and McFadden, G. (1999). The complete DNA sequence of myxoma virus. Virology *264*, 298–318.

Chen, N., Buller, R.M.L., Wall, E.M., and Upton, C. (2000). Analysis of host response modifier ORFs of ectromelia virus, the causative agent of mousepox. Virus Res. *66*, 155–173.

Chen, N., Danila, M.I., Feng, Z., Buller, R.M., Wang, C., Han, X., Lefkowitz, E.J., and Upton, C. (2003). The genomic sequence of ectromelia virus, the causative agent of mousepox. Virology *317*, 165–186.

Da Silva, M., and Upton, C. (2005). Using purine skews to predict genes in AT-rich poxviruses. BMC Genomics *6*, 22.

Dall, D., Sriskantha, A., Vera, A., Lai-Fook, J., and Symonds, T. (1993). A gene encoding a highly expressed spindle body protein of *Heliothis armigera* entomopoxvirus. J. Gen. Virol. *74*, 1811–1818.

Delhon, G., Tulman, E.R., Afonso, C.L. Weiser, Lu, Z., de la Concha-Bermejillo, A., Lehmkuhl, H.D., Piccone, M.E., Kutish, G.F., and Rock, D.L. (2004). Genomes of the parapoxviruses ORF virus and bovine papular stomatitis virus. J. Virol. *78*, 168–177.

Deveraux, Q.L., and Reed, J.C. (1999). IAP family proteins--suppressors of apoptosis. Genes Dev. *13*, 239–252.

Esposito, J.J., Sammons, S.A., Frace, A.M., Osborne, J.D., Olsen-Rasmussen, M., Zhang, M., Govil, D., Damon, I.K., Kline, R., Laker, M., *et al.* (2006). Genome sequence diversity and clues to the evolution of variola (smallpox) virus. Science *313*, 807–812.

Gao, L., and Qi, J. (2007). Whole genome molecular phylogeny of large dsDNA viruses using composition vector method. BMC Evol. Biol. *7*, 41.

Goodwin, R.H., and Filshe, B.K. (1969). Morphology and development of an occluded virus from the blacksoil scarab, *Othnonius bastesi*. J. Invertebr. Pathol. *13*, 317–329.

Granados, R.R. (1973a). Entry of an insect poxvirus by fusion of the virus envelope with host cell membrane. Virology *52*, 305–309.

Granados, R.R. (1973b). Insect Poxviruses: In Pathology, morphology and development, Some Recent Advances in Insect Pathology, D.W. Roberts and W.G. Yendol, eds. (Miscellaneous Publications of Entomological Society of America), pp. 73–94.

Gubser, C., Hue, S., Kellam, P., and Smith, G.L. (2004). Poxvirus genomes: a phylogenetic analysis. J. Gen. Virol. *85*, 105–117.

Guo, F.B., and Yu, X.J. (2007). Re-prediction of protein-coding genes in the genome of *Amsacta moorei* entomopoxvirus. J. Virol. Meth. *146*, 389–392.

Guo, F.B., and Zhang, C.T. (2006). ZCURVE_V: a new self-training system for recognizing protein-coding genes in viral and phage genomes. BMC Bioinformatics *7*, 9.

Hernández-Crespo, P., Veyrunes, J.C., Cousserans, F., and Bergoin, M. (2000). The spheroidin of an entomopoxvirus isolated from the grasshopper *Anacridium aegyptium* (AaEPV) shares low homology with spheroidins from lepidopteran or coleopteran EPVs. Virus Res. *67*, 203–213.

Howard, S.T., and Smith, G.L. (1989). Two early vaccinia virus genes encode polypeptides related to protein kinases. J. Gen. Virol. *70*, 3187–3201.

Hughes, A.L. (2002a). Evolution of inhibitors of apoptosis in baculoviruses and their insect hosts. Infect. Genet. Evol. *2*, 3–10.

Hughes, A.L. (2002b). Origin and evolution of viral interleukin-10 and other DNA virus genes with vertebrate homologues. J. Mol. Evol. *54*, 90–101.

Hughes, A.L., and Friedman, R. (2003). Genome-wide survey for genes horizontally transferred from cellular organisms to baculoviruses. Mol. Biol. Evol. *20*, 979–987.

Hughes, A.L., and Friedman, R. (2005). Poxvirus genome evolution by gene gain and loss. Mol. Phylogenet. Evol. *35*, 186–195.

Hurpin, B. (1968). The influence of temperature and larval stage on certain diseases of *Melolontha melolontha*. J. Invertebr. Pathol. *10*, 252–262.

Iyer, L.M, Aravind, L., and Koonin, E.V. (2001). Common origin of four diverse families of large eukaryotic DNA viruses. J. Virol. *75*, 11720–11734.

Iyer, L.M., Balaji, S., Koonin, E.V., and Aravind, L. (2006). Evolutionary genomics of nucleo-cytoplasmic large DNA viruses. Virus Res. *117*, 156–184.

Izmailyan, R.A., Huang, C.Y., Mohammad, S., Isaacs, S.N., and Chang, W. (2006). The envelope G3L protein is essential for entry of vaccinia virus into host cells. J. Virol. *80*, 8402–8410.

Kurstak, E., and Garzon, S. (1997). Entomopoxviruses. In Atlas of Insect and Plant Viruses, K. Maramorosch, ed. (San Francisco, New York, London: Academic Press), pp. 29–66.

Langridge, W.H., Bozarth, R.F., and Roberts, D.W. (1977). The base composition of entomopoxvirus DNA. Virology *76*, 616–620.

Lefkowitz, E.J., Wang, C., and Upton, C. (2006). Poxviruses: past, present and future. Virus Res. *117*, 105–118.

Li, Y., Carroll, D.S., Gardner, S.N., Walsh, M.C., Vitalis, E.A., and Damon, I.K. (2007). On the origin of smallpox: correlating variola phylogenics with historical smallpox records. Proc. Natl. Acad. Sci. U.S.A. *104*, 15787–15792.

Li, Q., Liston, P., and Moyer, R. W. (2005). Functional analysis of the inhibitor of apoptosis (iap) gene carried by the entomopoxvirus of *Amsacta moorei*. J. Virol. *79*, 2335–2345.

Lin, S., Chen, W., and Broyles, S.S. (1992). The vaccinia virus B1R gene product is a serine/threonine protein kinase. J. Virol. *66*, 2717–2723.

Means, J.C., Penabaz, T., and Clem, R.J. (2007). Identification and functional characterization of AMVp33, a novel homolog of the baculovirus caspase inhibitor p35 found in *Amsacta mocrei* entomopoxvirus. Virology *358*, 436–447.

McLysaght, A., Baldi, P.F., and Gaut, B.S. (2003). Extensive gene gain associated with adaptive evolution of poxviruses. Proc. Natl. Acad. Sci. U.S.A. *100*, 15655–15660.

Miller, L.K. (1999). An exegesis of IAPs: salvation and surprises from BIR motifs. Trends Cell Biol *9*, 323–328.

Mitchell, F.L., and Smith, J.W. (1985). Pathology and bioassays of the lesser cornstalk borer (*Elasmopalus lignosellus*) entomopoxvirus. J. Invertebr. Pathol. *45*, 75–80.

Mitsuhashi, W., Furuta, Y., and Sato, M. (1998). The spindle of an entomopoxvirus of Coleoptera (*Anomala cuprea*) strongly enhance the infectivity of a nucleopolyhedrovirus in Lepidoptera (*Bombyx mori*). J. Invertebr. Pathol. *71*, 186–188.

Mitsuhashi, W., and Miyamoto, K. (2003). Disintegration of the peritrophic membrane of silkworm larvae due to spindles of an entomopoxvirus. J. Invertebr. Pathol. *82*, 34–40.

Mitsuhashi, W., Kawakita, H., Murakami, R., Takemoto, Y., Saiki, T., Miyamoto, K., and Wada, S. (2007). Spindles of an entomopoxvirus facilitate its infection of the host insect by disrupting the peritrophic membrane. J. Virol. *81*, 4235–4243.

Moss, B. (2001). Poxviridae: The viruses and their replication. In Fields Virology, B.N Fields, D.M. Knipe, and P.M. Howley, eds. (Philadelphia: Lippincott-Raven Publishers) pp. 2849–2883.

Nazarian, S.H., Barrett, J.W., Frace, A.M., Olsen-Rasmussen, M., Khristova, M., Shaban, M., Neering, S., Li, Y., Damon, I.K., Esposito, J.J. *et al.* (2007). Comparative genetic analysis of genomic DNA sequences of two human isolates of Tanapox virus. Virus Res. *129*, 11–25.

Ojeda, S., Senkevich, T.G., and Moss, B (2006). Entry of vaccinia virus and cell–cell fusion require a highly conserved cysteine-rich membrane protein encoded by the A16L gene. J. Virol. *80*, 51–61.

Palli, S.R., Ladd, T.R., Tomkins, W.L., Shu, S., Ramaswamy, S.B., Tanaka, Y. Arif, B., and Retnakaran, A. (2000). *Choristoneura fumiferana* entomopoxvirus prevents metamorphosis and modulates juvenile hormone and ecdysteroid titers. Insect Biochem. Mol. Biol. *30*, 869–876.

Pevsner, J. (2003). Bioinformatics and functional genomics. (Hoboken, NJ: John Wiley & Sons, Inc.).

Pickup, D.J., Ink, B.S., Parsons, B.L., Hu, W., and Joklik, W.K. (1984). Spontaneous deletions and duplications of sequences in the genome of cowpox virus. Proc. Natl. Acad. Sci. U.S.A. *81*, 6817–6821.

Roberts, D.W., and Granados, R.R. (1968). A poxlike virus from *Amsacta moorei* (Lepidoptera: Arctiidae). J. Invertebr. Pathol. *12*, 141–143.

Senkevich, T.G., Bugert, J.J., Sisler, J.R., Koonin, E.V., Darai, G., and Moss, B. (1996). Genome sequence of a human tumorigenic poxvirus: prediction of specific host response-evasion genes. Science 273, 813–816.

Senkevich, T.G., and Moss, B. (2005). Vaccinia virus H2 protein is an essential component of a complex involved in virus entry and cell–cell fusion. J. Virol. *79*, 4744–4754.

Senkevich, T.G., Ojeda, S., Townsley, A., Nelson, G.E., and Moss, B. (2005). Poxvirus multiprotein entry–fusion complex. Proc. Natl. Acad. Sci. U.S.A. *102*, 18572–18577.

Senkevich, T.G., Ward, B.M., and Moss, B. (2004). Vaccinia virus entry into cells is dependent on a virion surface protein encoded by the A28L gene. J. Virol. *78*, 2357–2366.

Shchelkunov, S.N., Blinov, V.M., Resenchuk, S.M., Totmenin, A.V., Olenina, L.V., Chirikova, G.B., and Sandakhchiev, L.S. (1994). Analysis of the nucleotide sequence of 53 kbp from the right terminus of the genome of variola major virus strain India-1967. Virus Res. *34*, 207–236.

Shchelkunov, S.N., and Totmenin, A.V. (1995). Two types of deletions in orthopoxvirus genomes. Virus Genes *9*, 231–245.

Shchelkunov, S.N., Totmenin, A.V., Babkin, I.V., Safronov, P.F., Ryazankina, O.I., Petrov, N.A., Gutorov, V.V., Uvarova, E.A., Mikheev, M.V., and Sisler, J.R. (2001). Human monkeypox and smallpox viruses: genomic comparison. FEBS Lett. *509*, 66–70.

Shchelkunov, S.N., Totmenin, A.V., and Sandakhchiev, L.S. (1996). Analysis of the nucleotide sequence of 23.8 kbp from the left terminus of the genome of variola major virus strain India-1967. Virus Res. *40*, 169–183.

Sriskantha, A., Osborne, R.J., and Dall, D.J. (1997). Mapping of the *Heliothis armigera* entomopoxvirus (HaEPV) genome, and analysis of genes encoding the HaEPV spheroidin and nucleoside triphosphate phosphohydrolase I proteins. J. Gen. Virol. *78*, 3115–31123.

Sutter, G.R. (1972). A poxvirus of the army cutworm. J. Invertebr. Pathol. *19*, 375–382.

Takemoto, Y., Mitsuhashi, W., Murakami, R., Konishi, H., and Miyamoto, K. (2008). The N-terminal region of an entomopoxvirus fusolin is essential for the enhancement of peroral infection, whereas the C-terminal region is eliminated in digestive juice. J. Virol. *82*, 12406–12415.

Taylor, J.M., and Barry, M. (2006). Near death experiences: poxvirus regulation of apoptotic death. Virology *344*, 139–150.

Townsley, A., Senkevich, T.G., and Moss, B. (2005a). The product of the vaccinia virus L5R gene is a fourth membrane protein encoded by all poxviruses that is required for cell entry and cell–cell fusion. J. Virol. *79*, 10988–10998.

Townsley, A., Senkevich, T.G., and Moss, B. (2005b). Vaccinia virus A21 virion membrane protein is required for cell entry and fusion. J. Virol. *79*, 9458–9469.

Tulman, E.R., Afonso, C.L., Lu, Z., Zsak, L., Kutish, G.F., and Rock, D.L. (2001). Genome of lumpy skin disease virus. J. Virol. *75*, 7122–7130.

Tulman, E.R., Afonso, C.L., Lu, Z., Zsak, L., Kutish, G.F., and Rock, D.L. (2004). The genome of canarypox virus. J. Virol. *78*, 353–366.

Tulman, E.R., Afonso, C.L., Lu, Z., Zsak, L., Sur, J.H., Sandybaev, N.T., Kerembekova, U.Z., Zaitsev, V.L., Kutish, G.F., and Rock, D.L. (2002). The genomes of sheeppox and goatpox viruses. J. Virol. *76*, 6054–6061.

Turner, P.C., and Moyer, R.W. (1998). Control of apoptosis by poxviruses. Semin. Virol. *8*, 453–469.

Upton, C., Slack, S., Hunter, A.L., Ehlers, A., and Roper, R.L. (2003). Poxvirus orthologueous clusters: toward defining the minimum essential poxvirus genome. J. Virol. *77*, 7590–7600.

Vago, C. (1963). A new type of insect virus. J. Insect Pathol. *5*, 275–276.

van Oers, M.M., and Vlak, J.M. (1997). The baculovirus 10-kDa protein. J. Invertebr. Pathol. *70*, 1–17.

Wegensteiner, R., and Weiser, J. (1995). A new entomopoxvirus in the bark beetle *Ips typographus* (Coleoptera: Scolytidae). J. Invertebr. Pathol. *65*, 203–205.

Weiser, J., and Vago, C. (1966). A newly described virus of the winter moth, *Operophtera brumata* Hübner (Lepidoptera: Geometridae). J. Invertebr. Pathol. *8*, 314–319.

Willer, D.O., McFadden, G., and Evans, D.H. (1999). The complete genome sequence of shope (rabbit) fibroma virus. Virology *264*, 319–343.

Wijonarko, A., and Hukuhara, T. (1998). Detection of a virus factor in the spheroid, spindle and virion of an entomopoxvirus. J. Invertebr. Pathol. 72, 82–86.

Wu, G.A., Jun, S.R., Sims, G.E., and Kim, S.H. (2009). Whole-proteome phylogeny of large dsDNA virus families by an alignment-free method. Proc. Natl. Acad. Sci. U.S.A. *106*, 12826–12831.

Xing, K., Deng, R., Wang, J., Feng, J., Huang, M., and Wang, X. (2006). Genome-based phylogeny of poxvirus. Intervirology. *49*, 207–214.

Zemskov, E.A., Kang, W., and Maeda, S. (2000). Evidence for nucleic acid binding ability and nucleosome association of *Bombyx mori* nucleopolyhedrovirus BRO proteins. J. Virol. *74*, 6784–6789.

Hytrosaviruses: Structure and Genomic Properties

5

Adly M.M. Abd-Alla, Drion G. Boucias and Max Bergoin

Abstract

Hytrosaviruses are the only insect virus group that elicit salivary gland hypertrophy symptoms in their respective host. Salivary gland hypertrophy viruses (SGHVs) have been identified from several dipteran species, including the tsetse fly *Glossina pallidipes* (GpSGHV), the housefly *Musca domestica* (MdSGHV), and the narcissus bulb fly *Merodon equestris* (MeSGHV). The main characteristics of the hytrosaviruses are: (i) they produce non-occluded, enveloped, rod-shaped virions that measure 550–1000 nm in length and 80–100 nm in diameter; (ii) they possess a large circular double-stranded DNA (dsDNA) genome ranging in size from 120 to 190 kbp and having G+C ratios ranging from 28% to 44%; (iii) they cause overt salivary gland hypertrophy symptoms, testicular degeneration and ovarian abnormalities in dipteran adults. Although sharing several important features with other large DNA viruses (baculoviruses, nudiviruses, ascoviruses, and nimaviruses), such as producing rod-shaped, enveloped virions replicating in the nucleus, and possessing a large, circular, double-stranded DNA genome, GpSGHV and MdSGHV (and MeSGHV) differ significantly in several respects from these already-classified viruses. In this chapter, the discovery of the hytrosaviruses and their pathological impact on their host will be discussed and their genomic organization and relation to the other large DNA viruses will be reviewed.

Introduction

The hytrosaviruses constitute a unique group of entomopathogenic viruses that are as yet unclassified. These viruses possess a large (approximately 120–190 kbp), circular, double-stranded DNA genome that is packaged in long, enveloped rod-shaped virions. The salient feature of hytrosavirus infection is the hypertrophy of adult salivary glands, which can increase four-fold in size when compared to the uninfected salivary glands (Fig. 5.1A and B). This syndrome is therefore designated as 'salivary gland hypertrophy' (SGH) in the literature (Burtt, 1945). The proposed family name Hytrosaviridae is derived from *Hytrosa,* a sigla from the Greek '<u>Hy</u>per<u>tro</u>phia' for 'hypertrophy', and '<u>s</u>ialo<u>a</u>denitis' for 'salivary gland inflammation' (Abd-Alla *et al.*, 2009). Although numerous viruses have been reported to replicate in insect salivary glands (Mellor, 2000), only three viruses are known to induce overt SGH symptoms (Abd-Alla *et al.*, 2008; Garcia-Maruniak *et al.*, 2008). To date, these viruses have been detected only in adult dipteran flies. These include salivary gland hypertrophy viruses (SGHVs) detected in various tsetse fly species including *Glossina pallidipes* (family: Glossinidae, GpSGHV), in the narcissus bulb fly *Merodon equestris* (family: Syrphidae, MeSGHV) and in the housefly *Musca domestica* (family Muscidae, MdSGHV). The limited number of hytrosaviruses reported to date may reflect in part: the absence of external symptoms of the infected adult flies; the fact that the virus does not cause rapid mortality of the infected flies; the relatively low incidence of this

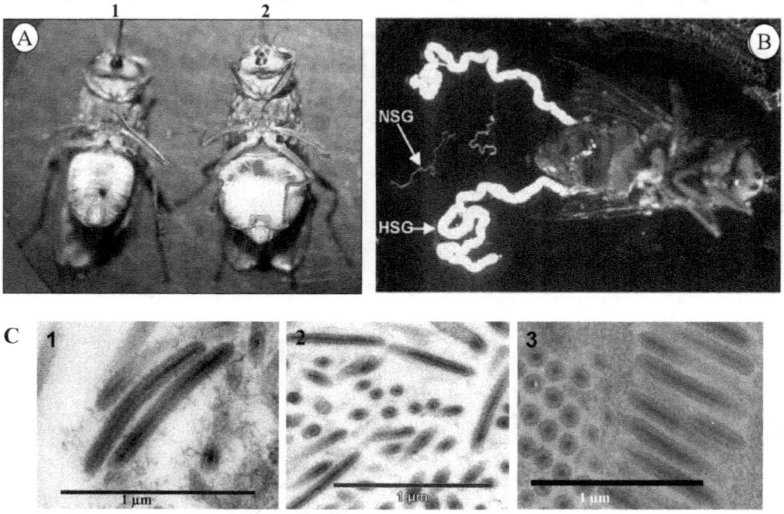

Figure 5.1 Symptoms and ultrastructure of hytrosaviruses. A, Male *Glossina pallidipes* with (1) asymptomatic and (2) symptomatic salivary glands. B, Normal (NSG) and hypertrophied salivary glands (HSG) of *G. pallidipes* (Abd-Alla *et al.*, 2007, with permission). C, Ultrastructure of various members of the Hytrosaviridae: 1, thin section of an infected salivary gland of *G. pallidipes* with GpSGHV. 2, thin section showing enveloped virus particles in the lumen of an infected salivary gland of *Musca domestica*, and 3, thin section of an infected salivary gland of *Merodon equestris* (Abd-Alla *et al.*, 2009, with permission).

virus in field populations; and/or the paucity of disease surveys on adult dipteran populations. It should be noted, that all three reported viruses were detected accidentally during research addressing other objectives. It is likely that many more SGHVs are associated with other dipterans in nature. Most of the information presented in this review concerns two viruses detected in the calyptrate flies, the haematophagous tsetse flies and the synanthropic housefly.

Discovery and identification

The tseste hytrosavirus was discovered by Jenni in 1973 at the Swiss Tropical Institute, Basel, Switzerland during a histological study of *Trypanosoma brucei* development in the salivary gland of adult *G. morsitans centralis* collected from Singida, United Republic of Tanzania. At that time, the virus-like particles in the salivary glands were speculated to be an arbovirus (Jenni, 1973; Jenni and Steiger, 1974a,b). Five years later, Jaenson at Uppsala, Sweden, reported a virus associated with the hypertrophied salivary glands in adult *G. pallidipes* collected from Kibwezi Forest, Kenya (Jaenson, 1978). In this study, the presence of the virus associated with the enlarged or hypertrophied salivary glands of the flies caught in the wild was correlated with both reduced female fertility and male sterility. Jaenson stated that the virus shared morphological properties with the non-occluded baculoviruses (formerly subgroup C Baculoviruses, presently nudiviruses). He speculated that the virus could play an important role as a biological control agent against tsetse flies, an important group of trypanosome vectors (Jaenson, 1978). Following this study, viruses inducing similar symptoms were observed in multiple populations of different tsetse species collected from different African countries (Otieno *et al.*, 1980; Odindo, 1982; Ellis and Maudlin, 1987; Gouteux, 1987; Jura *et al.*, 1988, 1989; Minter-Goedbloed and Minter, 1989; Shaw and Moloo, 1993; Abd-Alla *et al.*, 2007).

MdSGHV was first discovered in 1993 by Coler and co-authors (Coler *et al.*, 1993) in adult houseflies. During surveys of parasitic nematodes in housefly populations in Florida, the authors observed that some dissected flies possessed grossly enlarged salivary glands. These glands were whitish-blue in colour and contained a viscous milky fluid which leaked rapidly from the lumen of damaged glands. Electron microscopy of these glands revealed enveloped, rod-shaped virus particles in the cells and luminal fluid. Like GpSGHV, infection by MdSGHV induced female sterility. The prevalence of MdSGHV in field housefly populations is higher in males than females (Geden *et al.*, 2008). MdSGHV has a global distribution and has been detected in housefly populations in the USA (FL, CA, MA, UT, and KA), Europe (Denmark), Asia (Thailand) and South America (Brazil) (Prompiboom *et al.*, 2010).

The third hytrosavirus was detected in the adult narcissus fly *M. equestris* (Amargier *et al.*, 1979). Long rod-shaped virus particles similar to those previously described in tsetse flies were isolated from hypertrophied salivary glands. Symptoms in the narcissus fly were associated with hosts displaying similar phenotypic characters (i.e. black band in the abdominal region) in two varieties of the *Merodon* flies, *M. equestris* var. *nobilis* and *M. equestris* var. *transversalis* (Lyon, 1973). The field incidence of this virus was very high in certain fly populations. The survey conducted in early 1970s in southern France demonstrated that 88% of males and 16% of females of *M. equestris* var. *nobilis* displayed the hypertrophied salivary gland syndrome. In *M. equestris* var. *transversalis* 93% of males and 30% of females showed hypertrophied salivary glands (Iyon, 1973).

Virus morphology and protein composition

Transmission electron microscope examination of ultrathin sections of hypertrophied salivary glands dissected from the different host insects revealed the presence of numerous rod-shaped virus particles in the nucleus, cytoplasm, intracellular spaces and gland lumen (Jaenson, 1978). Inside the nucleus, the viral nucleocapsids are long and slender with a diameter of ~30 nm. The lengths of the different SGHVs ranging from 500 to 1000 nm, are significantly longer than the enveloped forms of the non-occluded nudiviruses, for example the *Gryllus bimaculatus* nudivirus (GbNV) and *Oryctes rhinoceros* nudivirus (OrNV) virions measure 100 and 200 nm, respectively (Wang *et al.*, 2007a). SGHV nucleocapsids exit the nuclei via nuclear pores and acquire the outer membrane in the cytoplasm of salivary gland cells. Enveloped virus particles consist of an inner membrane which encloses the nucleocapsid, and an outer membrane separated from the inner membrane by a narrow space (Fig. 5.1C). The enveloped virus particles range from 60–80 nm in diameter (Jaenson, 1978). The length of nucleocapsid and enveloped particles varies among the different hytrosaviruses. The tsetse SGHV virus particles could be grouped in median (870 nm) and long virus particles (1175 nm). These virus particles are characterized by the presence of rounded and thickened caps (Odindo *et al.*, 1986). The virus rod shows a regular arrangement of the structural units in the capsid consisting of a series of bands about 7 to 10 nm wide (Odindo *et al.*, 1986). The enveloped particles of MdSGHV (Coler *et al.*, 1993) and MeSGHV (Amargier *et al.*, 1979) (length of 550 nm) are significantly shorter than those of GpSGHV (Abd-Alla *et al.*, 2009). Negative staining of Nycodenz-purified MdSGHV revealed enveloped particles that possessed a unique braided, bead-like surface topography not observed with other rod-shaped insect viruses (Garcia-Maruniak *et al.*, 2008). Furthermore, the ends of MdSGHV particles are round and not blunt-ended as observed in nudiviruses.

Analysis of the protein composition of the SGHVs has demonstrated that these viruses are complex containing dozens of different structural peptides. Following the first report of the structural polypeptides of a Kenyan isolate of GpSGHV (Odindo *et al.*, 1986), recent SDS-PAGE analysis of the Ugandan GpSGHV isolate revealed more than 35 protein bands ranging in size from 10 to 220 kDa (Abd-Alla *et al.*, 2009). At least six bands were larger than 100 kDa, with the major bands having a molecular mass (M_w) of 39 and 40 kDa (Fig. 5.2).

Analysis of MdSGHV produced a protein profile similar to that of the GpSGHV. Initial analysis of sucrose gradient purified MdSGHV, using non-gradient 12% sodium dodecyl sulphate polyacrylamide gel electrophoresis (SDS-PAGE), revealed the presence of two major bands (40 and 148 kDa) and eight minor bands (Coler *et al.*, 1993). More recent analysis of virus purified on Nycodenz gradients in 5–15% gradient SDS-PAGE gels demonstrated the presence of at least 35 polypeptides ranging from 10 to 200 kDa (Fig.

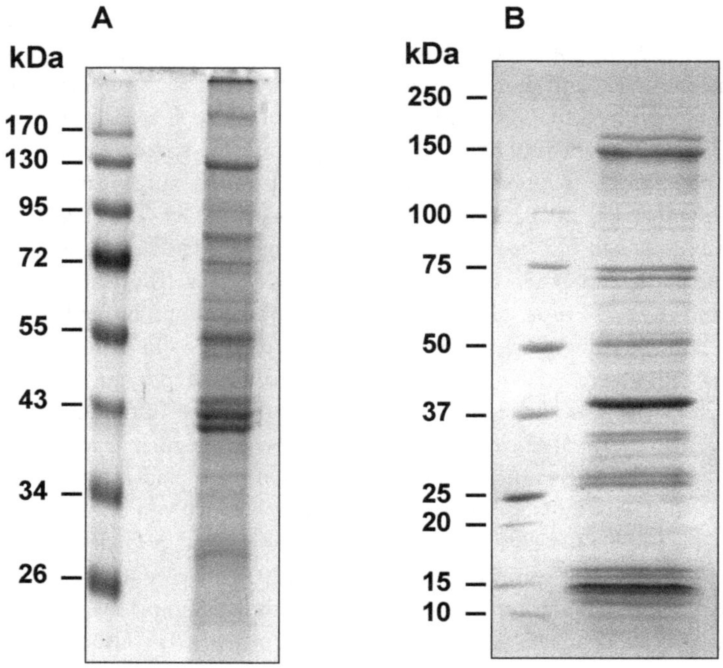

Figure 5.2 Comparative pattern of the structural proteins of GpSGHV and MdSGHV virions. SDS-PAGE of (A) GpSGHV and (B) MdSGHV stained with Coomassie Brilliant Blue. Molecular masses (kDa) of markers are denoted on the left side of the gel (Garcia-Maruniak *et al.*, 2009, with permission).

5.2). The presence and distribution of two major bands and 14 moderately stained bands was in agreement with the protein profile previously reported (Coler *et al.*, 1993). Follow-up analysis using GeLC-MS/MS analysis of the trypsin digests of different slices of the gradient gel confirmed the presence of 29 bona fide structural proteins based on unique peptide fragments ranging from 6 to 27 amino acids sharing 100% match with 29 MdSGHV open reading frames (ORFs) (Garcia-Maruniak *et al.*, 2008). Only five ORFs (13, 16, 39, 50, and 91) had less than 10% coverage. Fourteen ORFs had 10–50% coverage, and 10 ORFs had coverage greater than 50%. The structural ORFs appeared to be clustered loosely on the genome; a total of 17 structural ORFs were located between MdSGHV013 and MdSGHV052, and eight structural ORFs were located between MdSGHV084 and MdSGHV097 (Fig. 5.3B).

Genome organization of GpSGHV and MdSGHV

GpSGHV

The long GpSGHV virus particle contains a double-stranded circular DNA molecule of 190,032 bp. This genome contains a total of 322 potential (ORFs) with a methionine start codon that codes for proteins composed of at least 50 amino acids (Abd-Alla *et al.*, 2008). Among these, 160 ORFs with none or minimal overlap (< 82 nt) were assumed to encode putative proteins (Abd-Alla *et al.*, 2008) (Fig. 5.3A). The GpSGHV genome has a high A+T content of 72%. The low G+C content of the GpSGHV genome (28%) differs significantly from those reported for most invertebrate viruses with large circular dsDNA genomes, including baculoviruses (33% to 58% G+C) (Jehle *et al.*, 2006), several nudiviruses (42% G+C) (Cheng *et al.*, 2002; Wang *et al.*, 2007b) (an exception is the cricket nudivirus GbNV; 28% G+C, Wang *et al.*, 2007b), nimaviruses (41% G+C) (van Hulten *et al.*, 2001;

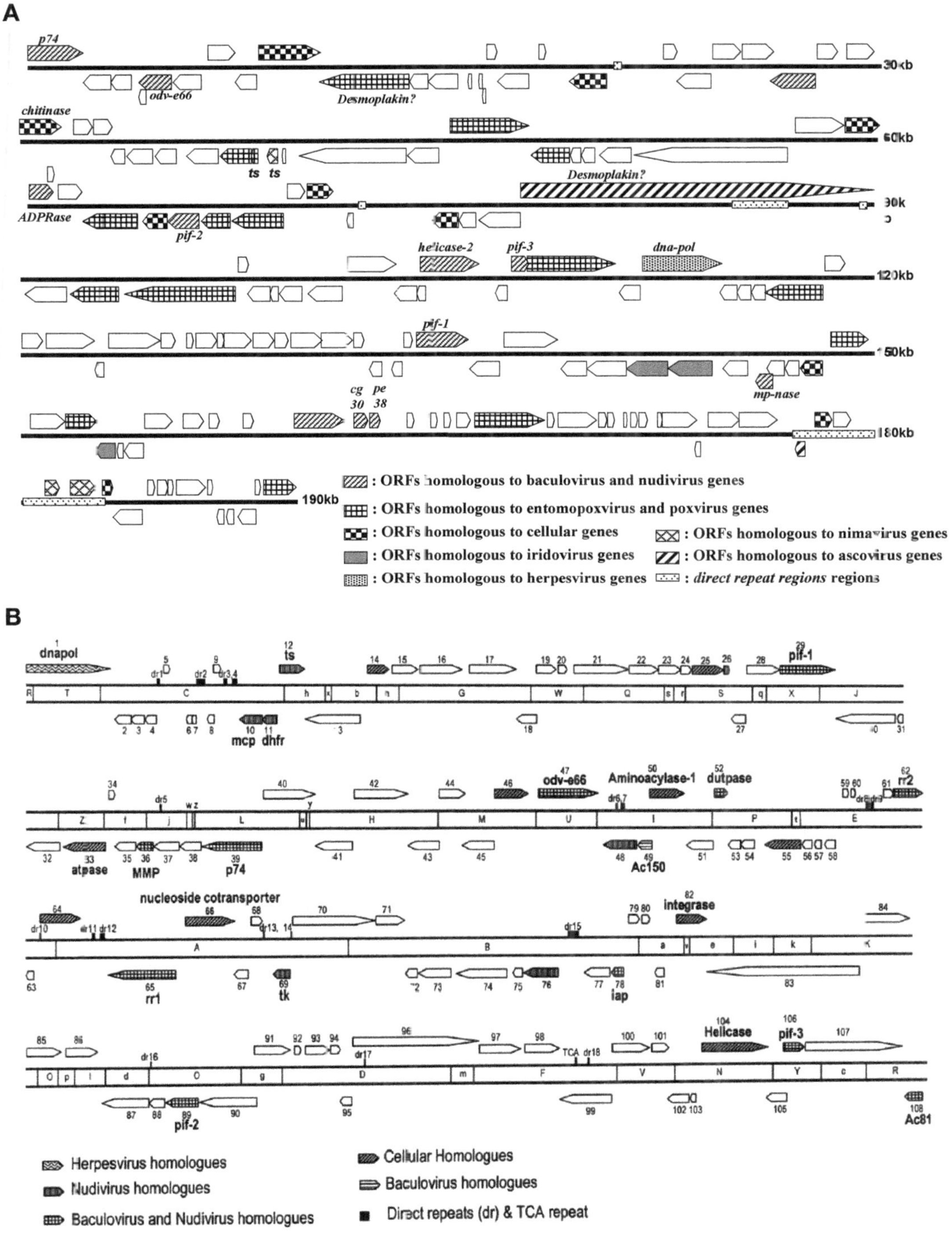

Figure 5.3 Genome organization of GpSGHV and MdSGHV. A, Linearized representation of the GpSGHV genome. The genome was linearized at the ATG start codon of *p74* (= SGHV001). Arrows indicate the position and direction of transcription for potential ORFs, which are represented according to suggested homology to those of other viruses (Abd-Alla *et al.*, 2008, with permission). (B) Linear representation of the MdSGHV genome and *Eco*RI physical map. The genome location, relative size and transcriptional direction of each putative ORF are indicated by the arrows. Arrow patterns and colour indicate homology to organisms found in the current protein database. (Garcia-Maruniak *et al.*, 2008, with permission).

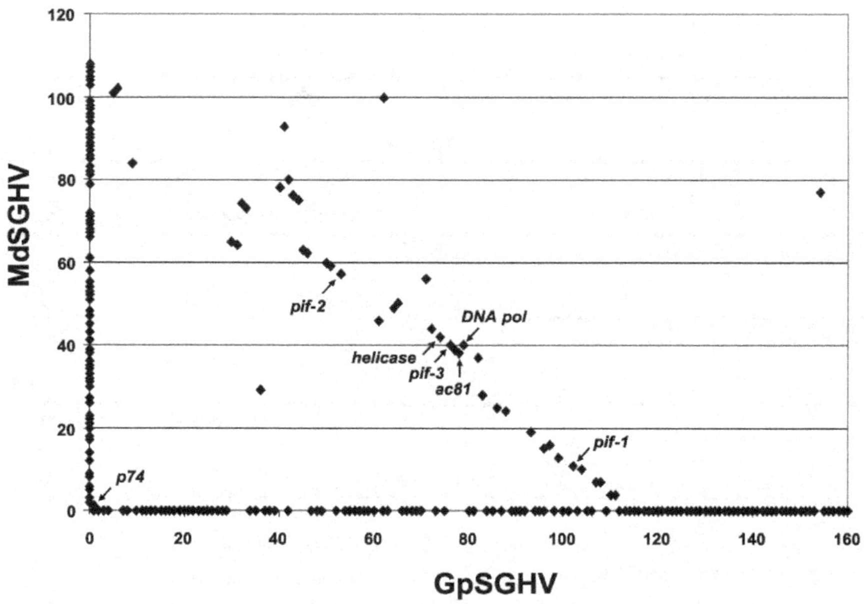

Figure 5.4 Gene parity plot comparing gene order of MdSGHV and GpSGHV. Gene order was 'aligned' by choosing the *p74* gene homologues as the starting point for the plot. Unique ORFs of GpSGHV and MdSGHV are on the x- and y-axes, respectively (modified from Garcia-Maruniak *et al.*, 2009).

Yang *et al.*, 2001), and ascoviruses (35.5% to 49% G+C) (Bideshi *et al.*, 2006; Wang *et al.*, 2006; Asgari *et al.*, 2007). In contrast, the G+C content of GpSGHV is similar to that of the large linear dsDNA entomopoxviruses genomes (18%) (Afonso *et al.*, 1999; Bawden *et al.*, 2000) and *Chilo* iridescent virus (29%) (Jakob *et al.*, 2001).

The predicted 160 ORFs are evenly distributed on both strands (51% forward and 49% reverse); many are arranged in unidirectional gene clusters (Fig. 5.3A). The ORFs represent 86% of the genome, with an average density of one gene per 1.2 kb. Position 1 of the GpSGHV genome sequence was arbitrarily assigned at the A of the ATG start codon of ORF SGHV001, encoding a homologue of the baculovirus *p74* gene encoding an occlusion-derived virion envelope protein. BLAST searches revealed that 47 of the 160 predicted GpSGHV ORFs have putative homologues in the NCBI protein database. These include 28 homologues to putative or known genes from other large dsDNA insect viruses, essentially baculoviruses (12 ORFs) and entomopoxviruses (16 ORFs), and 10 ORFs homologous to putative or known genes from cellular organisms. The GpSGHV genome encodes homologues to the four baculovirus per os infectivity factors *p74* [*pif-0*], *pif-1*, *pif-2*, and *pif-3* (Abd-Alla *et al.*, 2008). The DNA polymerase encoded by GpSGHV is of type B and appears to be phylogenetically distant from all DNA polymerases encoded by large double-stranded DNA viruses (Fig. 5.4). It should be noted that the level of amino acid identity between the GpSGHV putative ORFs and those of other viruses is usually low, ranging from 13.8% to 37.4% with an average of around 22%. The remaining 113 predicted ORFs produced no significant BLASTP hits. Surprisingly, no homologues to DNA-dependent RNA polymerase subunits were detected.

There are 14 direct repeats (*drs*) with head-to-tail tandem repeat sequences (TRS) and one inverted repeat sequence distributed throughout the GpSGHV genome, representing ~3% of the genome sequence. The repeated sequences are 171 to 556 nt in length and consist of 78 TRS minifragments. Most of these TRS are highly homologous to each other and are clustered in two genome areas located in the 86- to 88-kb and 179- to 183-kb regions. The sizes of the TRS vary from 52 to 246 nt, and the number of TRS per *dr* vary from 2.7 to 14.5. Within a particular *dr*, the identity of TRS is over 80%, whereas, the sequence identity among the different *drs* vary from 21% to 96%. Repeat sequences have been observed in all large dsDNA viruses, e.g. baculoviruses, entomopoxviruses, insect iridoviruses, ascoviruses, herpesviruses, and adenoviruses (Jancovich *et al.*, 2003). In baculoviruses, direct repeats occur at multiple locations along the genome and are believed to serve as origins of DNA replication (Kool *et al.*, 1995) and/or as enhancers of transcription (Guarino *et al.*, 1986; Guarino and Summers, 1986).

MdSGHV

The shorter MdSGHV genome consists of a 124,279 bp circular dsDNA molecule (Garcia-Maruniak *et al.*, 2008). Examination of end-labelled fragments of the restriction endonucleases (REN) digested DNA showed the sequenced MdSGHV2005 isolate (derived from a single hypertrophied gland; Garcia-Maruniak *et al.*, 2008) to be more homogenous than the DNA of the MdSGHV1992 isolate that was extracted from pooled glands preparation (Coler *et al.*, 1993). However, analysis of the data derived from Sanger and pyrophosphate-based (454) sequencing confirmed that MdSGHV2005 viral preparation was still polymorphic. It should be noted that there is no known cell culture system that supports SGHV replication to generate clonal preparations. The MdSGHV genome is significantly smaller than the GpSGHV genome (Abd-Alla *et al.*, 2008) and its G+C content is 43.5%, a ratio similar to that found in several nudiviruses (*Heliothis zea* nudivirus 1 (HzNV-1) and OrNV), but significantly higher than those of GbNV and GpSGHV contents (Cheng *et al.*, 2002; Wang *et al.*, 2006, 2007b).

The physical map of MdSGHV is depicted in Fig. 5.3B. Nucleotide 1 of the MdSGHV genome was designated to be the A of the ATG initiation codon for the putative DNA polymerase gene (*dnapol*). This gene was chosen as the start site in an effort to find a common gene for comparison to other DNA viruses. A total of 108 putative ORFs were identified *in silico* in the MdSGHV genome (Fig. 5.3B). There are 53 putative ORFs (49%) transcribed in the clockwise direction and 55 (51%) in the opposite direction. Interestingly, there are several clusters of ORFs transcribing in one direction. A similar balance in transcriptional direction is often found in baculoviruses and nudiviruses. A total of 22 ORFs overlap adjacent ORFs with a maximum of 158 nucleotide overlaps between ORFs 105 and 106. These two ORFs have opposite transcriptional orientations and code for >240 aa proteins. No biological activity could be assigned to ORF 105, since there was no protein homology, nor known motifs, whereas ORF 106 had limited homology to the baculovirus *pif-3* gene. Of the putative 108 ORFs, only 30 had significant homology to proteins from the current database (Fig. 5.3B). Eleven had homology to proteins encoded by both baculovirus and nudivirus genes, seven to nudivirus, one to baculovirus, one to herpesvirus, and ten to cellular genes. Of the 12 ORFs homologous to baculovirus proteins, only five corresponded to conserved core baculovirus genes (Garcia-Maruniak *et al.*, 2004).

Eighteen tandem direct repeats (*drs*) were found distributed throughout the genome (Garcia-Maruniak *et al.*, 2008). The size of the repeated sequences ranged from 149 bp (*dr*15) to only 9 bp long (*dr*18), and the number of copies of the *drs* ranged from 1.9 to 7.4, making the total sizes of the *drs* from 30 bp to 380 bp long. Unlike the homologous regions found in large, circular invertebrate dsDNA viruses that have imperfect palindromic sequences within the repeated region, (Cochran and Faulkner, 1983; Garcia-Maruniak *et al.*, 1996), no palindromes were found in the MdSGHV direct repeats. Seven *drs* (*dr*5, 6, 7, 10, 16, 17 and 18) were located inside putative ORFs. None of the 18 MdSGHV *drs*, had significant homology to each other.

Promoter and transcript analysis

The relative abundance of putative regulatory elements in the promoter regions of the ORFs was studied by computational analysis of the GpSGHV genome (Abd-Alla *et al.*, 2008). This analysis revealed that the 4-mer motif TAAG has the highest relative enrichment (3 times) in the 100-nt promoter compared to the whole GpSGHV genome. Additionally, the 5-mer motifs, ATAAG and TAAGA are highly enriched (4.9 and 3.7 times, respectively). These findings suggest that the TAAG motif may have a regulatory function in transcription. The 4-mer TAAG motif is also strongly over-represented in the 100- and 200-bp sequences upstream of the MdSGHV ORFs (Garcia-Maruniak *et al.*, 2009). This TAAG motif is not enriched in upstream regions of ORFs identified from either the insect iridovirus or nudviruses (Garcia-Maruniak *et al.*, 2009). Analysis of 5-mer sequences, a consensus motif [(A/G/T) TAAG (A/G/C)] could be detected in MdSGHV, a situation reminiscent of what is found in baculoviruses (Marks *et al.*, 2006). The same motif is part of the major late transcription initiation site of baculoviruses, and this may suggest an ancient regulatory relationship between GpSGHV and baculoviruses. To date, no transcriptional analysis has been done to support the hypothesis that TAAG plays a role in SGHV transcription. Further analyses are needed to address the role(s) of the SGHV ORF motifs in transcriptional regulation and to confirm that the TAAG motif is indeed a crucial element of SGHV viral promoters

A search (TBLASTX) for putative GpSGHV mRNAs in the *G. morsitans morsitans* expressed sequence tag (EST) database (http://old.genedb.org/genedb/glossina/blast.jsp) revealed significant identities (64–100%) to twelve GpSGHV ORFs. These ESTs were derived from cDNA synthesized from mRNA from tsetse flies asymptomatically infected with GpSGHV (S. Aksoy, personal communication). The homologies included SGHV010, SGHV044, SGHV045, SGHV060, SGHV061, SGHV062, SGHV064, SGHV074, SGHV107, SGHV108, SGHV111, and SGHV112. Six of these ORFs (SGHV010, SGHV062, SGHV064, SGHV074, SGHV107, and SGHV108) had hits to viral genes, whereas six (SGHV044, SGHV045, SGHV060, SGHV061, SGHV111, and SGHV112) had no hit with any of the NCBI protein databases.

The transcriptome of the MdSGHV was analysed by rapid amplification of cDNA 3′-ends (3′-RACE) and RT-PCR (Salem *et al.*, 2009). Direct sequencing of 3′-RACE products revealed 78 poly(A) transcripts that contained 95 of the 108 putative ORFs. Six ORFs, not amplified by 3′-RACE, were detected by RT-PCR. A series of 5′-RACE reactions were conducted on selected ORFs that were identified by 3 -RACE to be transcribed in tandem (tandem transcripts). In the majority of cases, the downstream ORFs were detected as single transcripts as well as components of the tandem transcripts, whereas the upstream ORFs were found only in tandem transcripts. In addition to the production of tandem transcripts, a single splicing event was detected in the 3′ untranslated region (3′-UTR) of MdSGHV045. The 3′-RACE sequence of this ORF revealed that the initial 335 nucleotides matched perfectly the last 322 bases of the genomic MdSGHV045, plus an additional 13 bases downstream of the stop codon. However, the following 262 bases of the MdSGHV045 3′- RACE product aligned perfectly with the beginning of MdSGHV043, an ORF located 847 bases downstream of MdSGHV045. The sequence obtained from the MdSGHV045 3′-RACE product indicated that the majority of MdSGHV044 had been spliced out from the MdSGHV045 mRNA, and that the splicing occurred in the 3′-UTR. Sequence analysis of the spliced ends revealed the presence of the dinucleotides GU at the 5′- and AG at the 3′-splice sites, which were homologous to the consensus splice sites (Mount, 1982). In general, most of the splicing events found in dsDNA viruses are located in exon–intron borders or at 5′-ends (Berget *et al.*, 1977; Nasseri *et al.*, 1982; Rixon and Clements, 1982; Chisholm and Henner, 1988; Hilger *et al.*, 1991). Although some of the reported viral 3′-UTRs possess a sequence homologous to 5′ splice sites, none of these viral transcripts show deletions in their 3′-UTRs (Furth and Baker, 1991; Furth *et al.*, 1994; Barksdale and Baker, 1995).

Extensive heterogeneity was observed in the polyadenylation signals (PSs) and cleavage sites (CSs) of MdSGHV transcripts (Salem *et al.*, 2009). Thirty-one of the 77 transcripts analysed showed CS heterogeneity and contained two to four potential CSs. In general, the 3′-UTRs were relatively short; 63 MdSGHV transcripts less than 100 bases, with only three having 3′-UTRs longer than 400 bases. Eighteen transcripts had stop codons located within their PS and seven transcripts had their stop codons located downstream of the PS. The majority of transcripts (53) had multiple PSs, only 24 transcripts had a single PS. The canonical PS (AAUAAA) was found 78 times in the 3′-UTRs of 65 transcripts. Less frequent and rare PSs were also found in many transcripts. It is interesting to note that some viral AAUAAA transcripts are known to use other PS variants such as AGUAAA, UAUAAA, and CAUAAA (Klemenz *et al.*, 1981; Simonsen and Levinson, 1983; Nasseri *et al.*, 1987; Hilger *et al.*, 1991; Andrews and DiMaio, 1993; Silver Key and Pagano, 1997).

Sets of transcript pairs, distributed throughout the MdSGHV genome, were found to overlap. The most common pattern was the convergent (3′–3′ overlap, 17 pairs), followed by unidirectional (12 pairs), and then the divergent (5′–5′ overlap, four pairs) pattern. The presence of overlapping ORFs in the MdSGHV genome may be caused by the close proximity of the putative ORFs (Garcia-Maruniak *et al.*, 2008). Such gene overlap, or 'overprinting', maximizes the coding information within genomes of limited size (Lamb and Horvath, 1991; Fukuda *et al.*, 2003). In the MdSGHV genome, the most frequent pattern was 3′ to 3′ overlap of adjacent genes; this bias may reflect the lack of 5′-UTR sequence information or a preference of 3′ to 3′ overlapping sense–antisense pairs (Sun *et al.*, 2005). The presence of both divergent and convergent gene pairs in MdSGHV suggests the potential involvement of *cis*-encoded natural antisense transcripts (*cis*-NATs) in gene expression. It should be pointed out that 7 out of 21 of the 3′ to 3′ and 5′ to 5′ overlapped regions in MdSGHV extended more than 100 bases. In addition to the *cis*-NATs identified in this study, several pre-miRNA sequences have been predicted in the MdSGHV genome (Garcia-Maruniak *et al.*, 2009).

Relation between hytrosavirus and the other large DNA viruses: basis for classification as a new virus family

Although SGHVs share several important features with other large DNA viruses (baculoviruses, nudiviruses, ascoviruses, and nimaviruses) such as rod-shaped, enveloped virions, a nuclear site of replication, a large, circular, double-stranded DNA genome and the presence of PIF genes, they differ significantly in several respects from these already-classified viruses. From a biological viewpoint, their host range is, so far, limited to the order Diptera, and they are the only insect viruses to induce symptoms of salivary gland hypertrophy in adult insects (Fig. 5.1B). Salivary gland hypertrophy, although reducing fertility, does not significantly impact fly survival. The general morphology of the long, enveloped, rod-shaped nucleocapsids is distinctly different from that of rod-shaped baculoviruses, nudiviruses, and ovoid-shaped ascoviruses or short rod-shaped nimaviruses (Fig. 5.1C). The differences include the length/diameter ratio and flexibility of the nucleocapsid as well as the structure of the envelope (Abd-Alla *et al.*, 2009). They are also excluded from the other large dsDNA virus families such as *Poxviridae* and *Iridoviridae* that harbour non- rod-shaped viruses with linear genomes (Table 5.1). Most importantly, genome analysis of GpSGHV and MdSGHV revealed that they not only differ from baculoviruses, nudiviruses, ascoviruses, and nimaviruses, but that they cannot be assigned to any of the large dsDNA virus families known to infect invertebrates or vertebrates (Abd-Alla *et al.*, 2008; Garcia-Maruniak *et al.*, 2009).

The similarity between SGHVs genomes and those of other large DNA viruses is limited to a very small number of genes with well-known functions. Only 29 out of 108 MdSGHV putative genes, and 37 out of 160 GpSGHV putative genes show slight homology to viral proteins in the databases (see Genome organization of GpSGHV and MdSGHV section).

The genome of all large DNA viruses contains a set of genes with enzymatic functioning as a DNA-dependent RNA polymerase necessary to express their late genes. Surprisingly, analysis of GpSGHV and MdSGHV genomes failed to reveal any clear evidence of genes homologous to DNA-dependent RNA polymerase subunits of baculoviruses (Lu and Miller, 1997), nudiviruses (Wang and Jehle, 2009), ascoviruses (Asgari *et al.*, 2007), and iridoviruses (Eaton *et al.*, 2007). As previously mentioned (Promoter and transcript analysis section), the 100- and 200-bp sequences upstream of both GpSGHV and MdSGHV ORFs

Table 5.1 Comparison between Hytrosaviridae and large dsDNA invertebrate viruses (after Abd-Alla *et al.*, 2009, with permission)

Virus family	Ascoviridae	Baculoviridae	Iridoviridae	Entomopoxvirinae	Nimaviridae	Nudiviruses	Hytrosaviridae
Morphology	Enveloped ovoid	Enveloped rod	Non-enveloped isometric	Enveloped ovoid, pleomorphic	Enveloped rod	Enveloped rod	Enveloped rod
Size (nm) (L × D)	220–240 × 130	200–450 × 30–100	160–350	300–400 × 170–250	270–290 × 1 20–150	20–250 × 50–70	700–1000 50
Genome (kb)	Circular (160–190)	Circula (90–180)	Linear (150–280)	Linear (130–300)	Circular (300)	Circular (125–220)	Circular (120–190)
Replication	Nucleus	Nucleus	Nucleus	Cytoplasm	Nucleus	Nucleus	Nucleus
Occlusion body	No	Yes	No	Yes	No	Yes/No	No
Symptoms/ histopathology	Impairment of development, vesicle formation	Nuclear and cellular hypertrophy	Iridescence	Spheroid (and spindle) formation, impairment of motility and development	White spot syndrome	Nuclear and cellular hypertrophy	Hypertrophied Salivary gland
Host	Lepidoptera Hymenoptera	Lepidoptera, Hymenoptera Diptera	Lepidoptera Diptera Coleoptera	Lepidoptera Orthoptera Coleoptera Diptera	Crustacea	Lepidoptera, Coleoptera Orthoptera	Diptera

contain at high frequency the TAAG motif typical of the major late transcription initiation site of baculoviruses. This motif is very likely involved in the transcription of hytrosavirus late genes but the genes encoding the subunits of their specific RNA polymerase remain to be decrypted.

Comparative analyses between GpSGHV and MdSGHV genome sequences (Abd-Alla *et al.*, 2008; Garcia-Maruniak *et al.*, 2008) have shown that only 37 MdSGHV ORFs have homology to 42 GpSGHV ORFs. This discrepancy is due to the fact that five MdSGHV ORFs (ORFs 025, 033, 036, 074 and 082) have homology to five pairs of GpSGHV ORFs (096/97, 107/108, 110/111, 032/033 and 030/031), respectively (Garcia-Maruniak *et al.*, 2009). However comparison of the gene order of GpSGHV and MdSGHV by gene parity plots (Fig. 5.4) and syntenic maps revealed two major regions of collinear ORFs, suggesting a common ancestry of the two SGHVs. These two regions are interspersed by a large gap of non-related genes and a distinct inversion. This collinearity is comparable to that observed when several representatives of the baculovirus NPVs group II were compared (Hu *et al.*, 1998; van Oers and Vlak, 2007; Wolff *et al.*, 2008). In contrast, using this approach to compare the gene order of each of the SGHVs to two nudiviruses, HzNV-1 and GbNV failed to reveal any collinearity between them. This indicated that SGHVs and nudiviruses are clearly separated from one another (Garcia-Maruniak *et al.*, 2009).

MdSGHV001 and GpSGHV079 both encode a protein homologous to the delta catalytic subunit of DNA polymerase from several eukaryotic organisms (Abd-Alla *et al.*, 2008; Garcia-Maruniak *et al.*, 2008). The best match for both ORFs to viral genes was to herpesvirus DNA polymerase with 22.6% and 24.9% amino acid identity for MdSGHV001 and GpSGHV079, respectively. A phylogenetic analysis of the DNA polymerase gene from all large dsDNA viruses failed to assign MdSGHV and GpSGHV to any of the presently recognized families (Fig. 5.5A). The two SGHVs are in the same clade, but they branched distantly from the baculoviruses, nudiviruses and nimaviruses. Since this gene is considered as significant to infer relationships of dsDNA viruses, this analysis clearly indicated the lack of close relationship of hytrosaviruses to any other known virus family.

The identification in GpSGHV and MdSGHV genomes of genes encoding proteins homologous to per os infectivity factors (pifs) and p74 and odv-e66 structural proteins (Abd-Alla *et al.*, 2008; Garcia-Maruniak *et al.*, 2008), common to all sequenced nudiviruses and baculoviruses, and believed to be essential for infectivity (Kuzio *et al.*, 1989; Faulkner *et al.*, 1997; Kikhno *et al.*, 2002; Pijlman *et al.*, 2003; Ohkawa *et al.*, 2005; Song *et al.*, 2008) suggests a common mechanism of virus entry via the insect alimentary tract and a possible common origin of these viruses. These genes are presumed to encode envelope proteins typical of the occlusion-derived baculovirus phenotype, which are important for the entry process into the midgut epithelial cells (Slack and Arif, 2007). The phylogenetic analysis of these proteins showed a very close relationship with structural proteins in MdSGHV and GpSGHV. These proteins shared the same clade but branched distantly from baculoviruses and nudiviruses (Fig. 5.5B). This tree was strongly supported by bootstrap analysis. The phylogenetic trees of two other proteins of GpSGHV and MdSGHV found to be homologous to conserved core proteins of baculoviruses and nudiviruses illustrated in Fig. 5.5B. Both MdSGHV108 and GpSGHV078, homologues of Ac81 baculovirus and nudivirus proteins, and GpSGHV74 and MdSGHV104, homologues of baculovirus helicase-2 proteins, branched very close to each other in the same clade and separated from the corresponding conserved proteins in baculoviruses and nudiviruses.

Impact of hytrosaviruses on host insects

To date, the hytrosaviruses have been isolated from dipterans that have distinct biology and unrelated phylogenies, the viviparous hematophagous tsetse flies, the housefly, and the phytophagous narcissus bulb fly. This host diversity suggests that other dipterans may harbour novel hytrosaviruses. Unlike many other insect viruses, the hytrosaviruses replicate and cause pathological symptoms in only the adult stage. All known SGHV isolates induce an enlargement or hypertrophy of infected

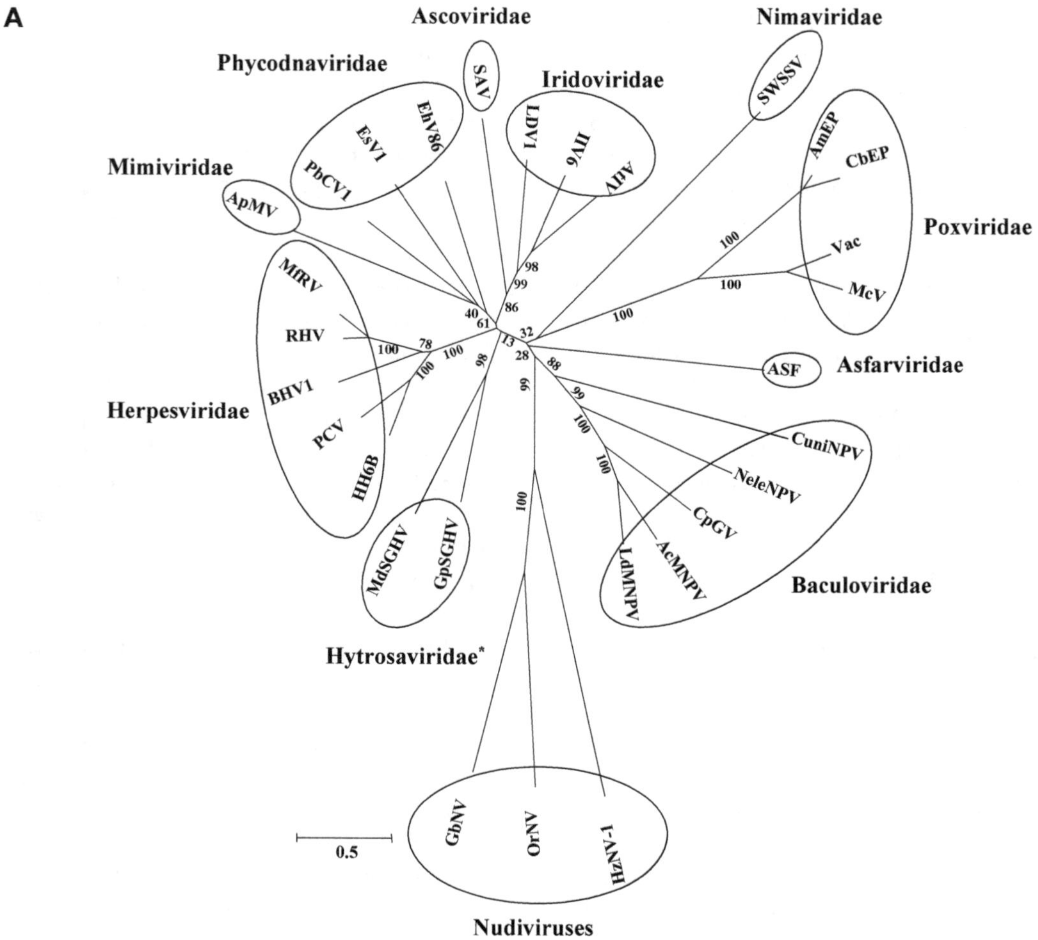

Figure 5.5 (A) Phylogenetic analysis of selected GpSGHV and MdSGHV proteins. (A) Neighbour-joining (NJ) trees of predicted amino acid sequence of the DNA polymerase genes. The phylogenetic tree of DNA polymerase and its homologues is based on 2374 sites of DNA polymerase of 29 viruses from various families. Distances were calculated using Poisson correction. The robustness of the tree was tested using bootstrap analysis (500 replicates). Numbers on the nodes indicate bootstrap values. The names of the selected virus families are indicated on the tree. The asterisk indicates the proposed family name for the SGHVs. DNA polymerases used and their accession numbers in brackets: ASF: African swine fever virus [NP_042783], GpSGHV: *Glossina pallidipes* salivary gland hypertrophy virus [ABQ08852], MdSGHV: *Musca domestica* salivary gland hypertrophy virus [YP_001883329], HHB6: Human herpesvirus 6B [NP_050219], PCV: Porcine cytomegalovirus [AAF80107], BHV1: Bovine herpesvirus 1 [NP_045328], RHV: Retroperitoneal fibromatosis-associated herpesvirus [AAC57976], MfRV: *Macaca fuscata* rhadinovirus [YP_238317], ApMV: *Acanthamoeba polyphaga* mimivirus [YP_142676], PbCV1: *Paramecium bursaria* Chlorella virus 1 [AAC00532], EsV1: *Ectocarpus siliculosus* virus 1 [NP_077578], EhV86: *Emiliania huxleyi* virus 86 [AAL58859], SAV: *Spodoptera* ascovirus [AAC54632], LDV1: Lymphocystis disease virus 1 [NP_078724], IIV6: Invertebrate iridescent virus 6 [CAC19195], AtIV: *Aedes taeniorhynchus* iridescent virus [ABF82150], AmEPV: *Amsacta moorei* entomopoxvirus [NP_064832], CbEP: *Choristoneura biennis* entomopoxvirus [CAA40566], Vac: Vaccinia virus [NP_063712], McV: *Molluscum contagiosum* virus [AAL40129], WSSV: shrimp white spot syndrome virus [AAK77696], CuniNPV: *Culex nigripalpus* nucleopolyhedrovirus (NPV) [AAK13281], NeleNPV: *Neodiprion lecontei* NPV [YP_025217], CpGV: *Cydia pomonella* granulovirus [NP_148895], AcMNPV: *Autographa californica* NPV [NP_054095], LdMNPV: *Lymantria dispar* NPV [NP_047720], HzNV-1: *Heliothis zea* virus 1 [NP_690550], OrNV: *Oryctes rhinoceros* virus [ABF93350], GbNV: *Gryllus bimaculatus* nudivirus [YP_001111279] (Garcia-Maruniak *et al.*, 2009).

B

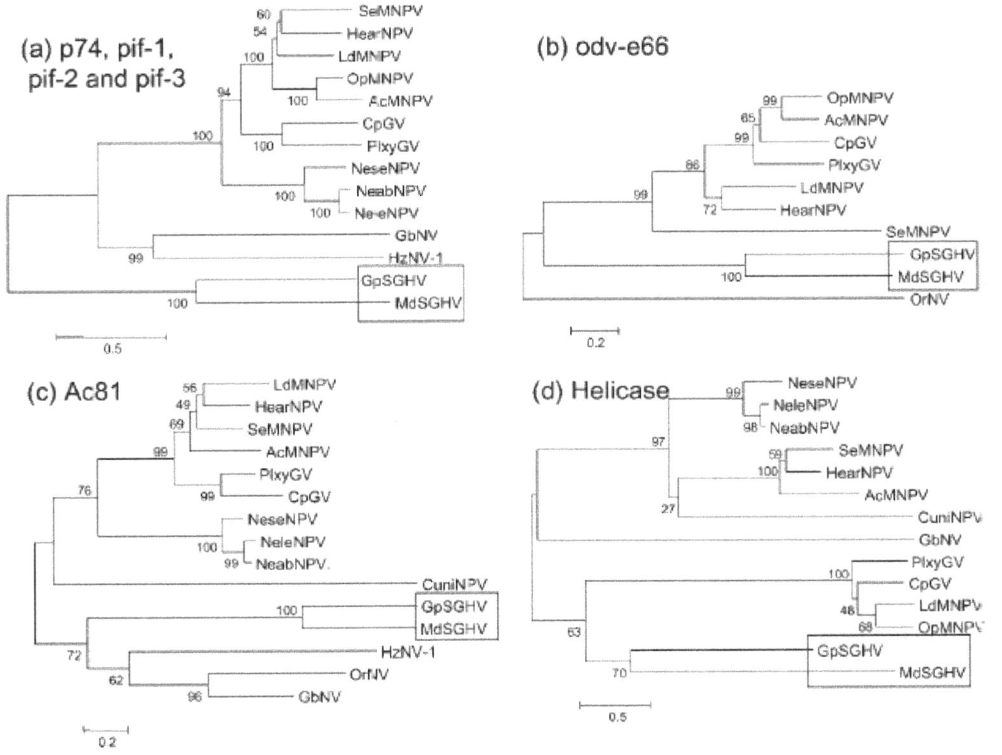

Figure 5.5 (B) Neighbour-joining phylogenetic amino acid trees of (a) combined p74, pif-1, pif-2 and pif-3, (b) odv-e66, (c) Ac81 and (d) helicase proteins and their homologues in baculoviruses and nudiviruses. The following viruses were included: Alphabaculoviruses: *Autographa californica* (Ac) MNPV [NC_001623], *Orgyia pseudotsugata* (Op) MNPV [NC_001875], *Lymantria dispar* (Ld) MNPV [NC_001973], *Spodoptera exigua* (Se) MNPV [NC_002169], *Helicoverpa armigera* (Hear) NPV [NC_002654]; Betabaculoviruses: *Cydia pomonella* (Cp) GV [NC_002816], *Plutella xylostella* (Plxy) GV [NC_002593]; Hymenopteran-specific NPVs: *Neodiprion sertifer* (Nese) NPV [NC_005905], *N. lecontei* (Nele) NPV [NC_005906], *N. abietis* (Neab) NPV [NC_008252]; Dipteran-specific NPV: *Culex nigripalpus* (Cuni) NPV [NC_003084]; and Nudiviruses: *Heliothis zea* (Hz) NV-1 [NC_004156], *Gryllus bimaculatus* (Gb) NV [NC_009240], *Oryctes rhinoceros* (Or) NV[DQ665871, DQ665870). Distances were calculated using Poisson correction. Homogeneous substitution pattern among lineages with gamma distributed rate among sites (gamma parameter 2.25) was employed for reconstruction of the trees. The robustness of the tree was tested using bootstrap analysis (500 replicates). Numbers on the nodes indicate bootstrap values (Garcia-Maruniak *et al.*, 2009, with permission).

salivary glands; this adult tissue appears to be the major site of virus replication. Available assay data suggest that hytrosaviruses are host specific. Attempts to infect various insect species, including adult dipterans (*Sarcophaga crassipalpis, Haematobia irritans,* and *Stomoxys calcitrans*) and the larvae of *Diaprepes abbreviatus* (Coleoptera) and *Helicoverpa zea* (Lepidoptera) by injecting filter-sterilized virus preparations that induce 100% SGH in houseflies failed to induce SGH symptoms. In addition, bioassays involving challenge of *G. pallipides* and *M. domestica* with their heterologous SGHV failed to produce hypertrophy symptoms. The specificity displayed by these viruses is very likely related to the unique biology of their insect hosts.

Blood-feeding adult tsetse flies are responsible for the transmission of two major diseases in sub-Saharan Africa, human African trypanosomiasis (HAT or sleeping sickness), and animal African trypanosomiasis (AAT or nagana) (Steelman, 1976; WHO, 2001). Both diseases have profound impacts on the health of livestock and humans in tropical Africa (Jordan, 1986). In the absence of effective vaccines and inexpensive drugs to control HAT, and the development of drug resistance of the agent of AAT (Aksoy and Rio, 2005), vector control remains the most viable strategy for the sustainable management of these diseases. Following the successful eradication of *Glossina austeni* from the island of Unguja, United Republic of Tanzania, using an area-wide integrated pest management approach (Vreysen *et al.,* 2000; Hendrichs *et al.,* 2007), similar strategies have been attempted on the African mainland. Such a programme against *G. pallidipes* was initiated in 1996 in the Southern Rift Valley of Ethiopia (Feldmann *et al.,* 2005). Within the framework of this project, rearing of an Ethiopian strain of *G. pallidipes* was established at the Insect Pest Control Laboratory (former Entemology Unit) of the Joint FAO/IAEA Programme of Nuclear Technique in Food and Agriculture, Seibersdorf, Austria, to provide a seed colony and a back-up for the Ethiopian mass rearing colony. However, following its successful establishment, the colony experienced a steady decline over 2 years and finally collapsed. Investigations to identify the cause of this colony decline led to the discovery that up to 85% of adults had SGH symptoms. The onset of SGH symptoms causes dramatic host changes including testicular degeneration and ovarian abnormalities (Jura *et al.,* 1988; Kokwaro *et al.,* 1990; Sang *et al.,* 1998, 1999) as well as impacting the development, survival, fertility and fecundity of both naturally or experimentally infected flies (Feldmann *et al.,* 1992a; Jura *et al.,* 1993; Sang *et al.,* 1997).

Importantly, the SGH syndrome was described initially in wild populations of *G. pallidipes* (Whitnall, 1934; Burtt, 1945), but later detected in many tsetse species from different African countries (Otieno, 1980; Odindo 1982; Ellis and Maudlin, 1987; Gouteux, 1987; Jura *et al.,* 1988, 1989; Minter-Goedbloed and Minter, 1989; Shaw and Moloo, 1993). In field tsetse populations, the SGHV operates as an enzootic agent and is found in only a small percentage of trapped adults. Due to the relative low density of tsetse flies and their blood-feeding behaviour, transmission in nature among conspecifics is believed to occur by either transovum transmission or to developing larvae through infected milk glands (Jura *et al.,* 1989; Sang *et al.,* 1996, 1998). It should be noted that tsetse flies exhibit viviparity, give birth to fully developed larvae, that upon larviposition immediately transform into the pupal stage. The high incidence of SGH observed in the declining Ethiopian *G. pallidipes* colony at the Insect Pest Control Laboratory was most likely due to the protocols used in colony maintenance. In these colonies, caged flies are kept at high densities and are communally fed processed animal blood placed under membranes (Feldmann *et al.,* 1992b). Under this scenario, a single infected fly could deposit virus via salivary secretions into blood under the membrane during the feeding event that could be easily acquired by healthy flies. This type of horizontal transfer is unlikely to occur when adults acquire their blood meal from live animals.

In order to monitor the presence and the level of virus in these colonies, a sensitive, non-destructive, diagnostic PCR assay was developed by doing the PCR on the DNA extracted from an intermediate excised leg (Abd-Alla *et al.,* 2007). Interestingly, the implementation of this specific PCR method demonstrated a very high (up to 100%) prevalence of the virus in the colony of the Uganda (Tororo) strain of *G. pallidipes.* However, dissection of assayed flies from this colony revealed that only 3.1% of females and 3.8% of males displayed the SGH symptoms (Abd-Alla *et al.*, 2007). The high percentage of asymptomatic flies in colonized tsetse flies may be due to (1) a lower pathogenicity of this virus as compared to the Ethiopian strain; (2) an acquired resistance of the flies in this colony to the virus; and/or (3) that the virus exists in a latent or integrated state in tsetse flies. Significantly, the presence of an asymptomatic infection in this insect suggests that factors regulating viral activation, such as stress, may play a role in the colony collapse observed in *G. pallidipes* of Ethiopian origin.

The second SGHV host is the housefly, *M. domestica* (L.) (Diptera: Muscidae), a cosmopolitan insect, recognized as both nuisance pest of humans as well as both a mechanical vector of dysentery, typhoid and polio and a biological vector of various protozoan, cestode and nematode parasites (Coler *et al.*, 1993). Additionally, *M. domestica* may oviposit on wounds and/or infections of man and animals resulting in myiasis. Housefly populations, reaching high densities in poultry and dairy operations, often require management with insecticides at the rural/urban interface. Due to the high costs associated with insecticide use and application, the increasing resistance of housefly pests to insecticides, and the increasing environmental concern of both producers and consumers, there has been concerted interest in cultural and biological control methods as alternative management strategies. In fact, MdSGHV was discovered initially during a survey of the natural enemy complex of the housefly (Coler *et al.*, 1993). Bioassays demonstrated that the MdSGHV was infectious per os to healthy adult male and female *M. domestica*. Like GpSGHV, infection with MdSGHV suppressed egg development in female flies, which makes this virus a potential candidate as a biological control (birth control) of *M. domestica* populations (Lietze *et al.*, 2007).

A series of studies have examined the transmission of MdSGHV and have demonstrated that it is not sexually transmitted from infected to healthy adults (Lietze *et al.*, 2007; Geden *et al.*, 2008). Additionally, there has been no indication of vertical transmission of MdSGHV and this virus is thought to be acquired only by the adult stage. Attempts to infect the egg or larval stages via injection of the virus have failed to produce adults displaying SGH symptoms. Experiments have demonstrated that this virus is transmitted orally via deposition and consumption of saliva, which is a composite of salivary gland secretions and crop contents (Geden *et al.*, 2008; Lietze *et al.*, 2009). The lumen of both the salivary gland and crops of infected flies contain large numbers of enveloped virus particles. By quantitative PCR, MdSGHV can be detected initially in salivary secretions at two days post infection (dpi) with copy numbers increasing exponentially until 4 dpi. Between 5 and 21 dpi, infected flies released an average of 1.0×10^6 MdSGHV copies per feeding event (Geden *et al.*, 2008). The virus that was deposited on food via oral secretions was infectious when ingested by newly emerged adult flies. In addition to saliva, the excreta collected overnight at 5 dpi from individual infected flies contained detectable levels of MdSGHV. Low infection rates were produced when newly emerged flies were challenged with filtered excreta by either oral or injection routes. Evaluation of the quantity and infectivity of MdSGHV released by individual infected houseflies showed that salivation onto a shared food substrate is the most likely route of natural MdSGHV transmission among adult houseflies. The MdSGHV ingested by healthy conspecifics replicates quickly, causing SGH and a down-regulation of egg formation in female flies. Unlike the situation with GpSGHV, MdSHGV does not sterilize the infected males. In addition to suppressing egg development in females, the onset of SGH suppresses the female responsiveness to male copulation attempts and reduces the avidity of males to induce courtship behaviours (Lietze *et al.*, 2007).

It should be noted that the host insect *M. domestica* has a global distribution excluding polar regions and that recent surveys have shown that populations throughout its distribution harbour MdSGHV (Prompiboon *et al.*, 2009). The relative incidence of MdSGHV in specific locations has been found to vary throughout the fly season but does in part seem to be correlated with fly density (Geden *et al.*, 2008). In several North Florida dairies, the incidence of detectable SGH symptoms ranged from 0% to 40% over a 2-year period. Although the MdSGHV incidence is variable over the season, it does appear to persist at particular locations over time. For example, the dairies sampled in 2005–2006 were negative while they are the same dairies that were positive in the initial surveys conducted in the early 1990s. Limited sampling has detected the MdSGHV in fly populations in North America and South America, Europe, and Asia. The relative incidence of SGH in collected flies at these locations ranged from 0–15%. In all cases, the viruses isolated from these populations display similar pathology and possess high genetic similarity (Prompiboon *et al.*, 2010).

Concluding remarks

Available data on the biology, the structure of virions and the genome organization of GpSGHV and MdSGHV indicate that these two viruses (and probably also MeSGHV) have many common characteristics. Many of their shared properties differ enough from other large vertebrate and invertebrate DNA viruses to warrant their grouping in a new virus family for which the name Hytrosaviridae has been proposed (Abd-Alla *et al.*, 2009). The current ultrastructural studies on SGHV architecture and morphogenesis will certainly highlight their originality among other rod-shaped insect DNA viruses. Likewise, deciphering of the modalities of expression of their genome and the identification of genes involved in the transcription of SGHV late genes will better define the salient features of this new group of viruses.

Acknowledgements

We would like to thank Andrew Parker, Marc Vreysen and Verena Leitze for reviewing the manuscript.

References

Abd-Alla, A., Bossin, H., Cousserans, F., Parker, A., Bergoin, M., and Robinson, A. (2007). Development of a non-destructive PCR method for detection of the salivary gland hypertrophy virus (SGHV) in tsetse flies. J. Virol. Meth. *139*, 143–149.

Abd-Alla, A.M.M., Cousserans, F., Parker, A.G., Jehle, J.A., Parker, N.J., Vlak, J.M., Robinson, A.S., and Bergoin, M. (2008). Genome analysis of a *Glossina pallidipes* salivary gland hypertrophy virus (GpSGHV) reveals a novel large double-stranded circular DNA virus. J. Virol. *82*, 4595–4611.

Abd-Alla, A.M.M., Vlak, J.M., Bergoin, M., Maruniak, J.E., Parker, A.G., Burand, J.P., Jehle, J.A., and Boucias, D.G. (2009). Hytrosaviridae: a proposal for classification and nomenclature of a new insect virus family. Arch. Virol. *154*, 909–918.

Afonso, C.L., Tulman, E.R., Lu, Z., Oma, E., Kutish, G.F., and Rock, D.L. (1999). The genome of *Melanoplus sanguinipes* entomopoxvirus. J. Virol. 73, 533–552.

Aksoy, S., and Rio, R.V.M. (2005). Interactions among multiple genomes: Tsetse, its symbionts and trypanosomes. Insect Biochem. Mol. Biol. *35*, 691–698.

Amargier, A., Lyon, J.P., Vago, C., Meynadier, G., and Veyrunes, J.C. (1979). Discovery and purification of a virus in gland hyperplasia of insects. Study of *Merodon equestris* F. (Diptera, Syrphidae). C. R. Acad. Sci. D *289*, 481–484.

Andrews, E.M., and DiMaio, D. (1993). Hierarchy of polyadenylation site usage by bovine papillomavirus in transformed mouse cells. J. Virol. *67*, 7705–7710.

Asgari, S., Davis, J., Wood, D., Wilson, P., and McGrath, A. (2007). Sequence and organization of the *Heliothis virescens* ascovirus genome. J. Gen. Virol. *88*, 1120–1132.

Barksdale, S.K., and Baker, C.C. (1995). The human immunodeficiency virus type 1 Rev protein and the Rev-responsive element counteract the effect of an inhibitory 5′ splice site in a 3′ untranslated region. Mol. Cell. Biol. *15*, 2962–2971.

Bawden, A.L., Glassberg, K.J., Diggans, J., Shaw, R., Farmerie, W., and Moyer, R.W. (2000). Complete genomic sequence of the *Amsacta moorei* entomopoxvirus: analysis and comparison with other poxviruses. Virology *274*, 120–139.

Berget, S.M., Moore, C., and Sharp, P.A. (1977). Spliced segments at the 5′ terminus of adenovirus 2 late mRNA. Proc. Natl. Acad. Sci. U.S.A. *74*, 3171–3175.

Bideshi, D.K., Demattei, M.V., Rouleux-Bonnin, F., Stasiak, K., Tan, Y., Bigot, S., Bigot, Y., and Federici, B.A. (2006). Genomic sequence of *Spodoptera frugiperda ascovirus 1a*, an enveloped, double-stranded DNA insect virus that manipulates apoptosis for viral reproduction. J. Virol. *80*, 11791–11805.

Burtt, E. (1945). Hypertrophied salivary glands in *Glossina*: evidence that *G. pallidipes* with this abnormality is particularly suited to trypanosome infection. Ann. Trop. Med. Parasitol. *39*, 11–13.

Cheng, C.H., Liu, S.M., Chow, T.Y., Hsiao, Y.Y., Wang, D.P., Huang, J.J., and Chen, H.H. (2002). Analysis of the complete genome sequence of the Hz-1 virus suggests that it is related to members of the *Baculoviridae*. J. Virol. *76*, 9024–9034.

Chisholm, G.E., and Henner, D.J. (1988). Multiple early transcripts and splicing of the *Autographa californica* nuclear polyhedrosis virus IE-1 gene. J. Virol. *62*, 3193–3200.

Cochran, M.A., and Faulkner, P. (1983). Location of homologous DNA sequences interspersed at five regions in the baculovirus AcMNPV genome. J. Virol. *45*, 961–970.

Coler, R.R., Boucias, D.G., Frank, J.H., Maruniak, J.E., Garcia-Canedo, A., and Pendland, J.C. (1993).

Characterization and description of a virus causing salivary gland hyperplasia in the housefly, *Musca domestica* Med. Vet. Entomol. 7, 275–282.

Eaton, H.E., Metcalf, J., Penny, E., Tcherepanov, V., Upton, C., and Brunetti, C.R. (2007). Comparative genomic analysis of the family *Iridoviridae*: re-annotating and defining the core set of iridovirus genes. Virol. J. *4*, 1–17.

Ellis, D.S., and Maudlin, I. (1987). Salivary gland hyperplasia in wild caught tsetse from Zimbabwe. Entomol. Exp. Appl. *45*, 167–173.

Faulkner, P., Kuzio, J., Williams, G.V., and Wilson, J.A. (1997). Analysis of p74, a PDV envelope protein of *Autographa californica* nucleopolyhedrovirus required for occlusion body infectivity *in vivo*. J. Gen. Virol. *78*, 3091–3100.

Feldmann, H.U., Barnor, H., and Acs, R. (1992a). Abweichungen in der Reproduktion von *Glossina morsitans submorsitans* Newstead (Diptera: Glossinidae): Untersuchungen zur Behebung eines gestörten Geschlechterverhältnisses und zum Übertragungsweg von Fertilitäts-reduzierenden Viren an die Nachkommen. Mitt. Dtsch. Ges. Allg. Angew. Ent. *8*, 248–251.

Feldmann, U., Dyck, V.A., Mattioli, R.C., and Jannin, J. (2005). Potential impact of tsetse fly control involving the sterile insect technique. In Sterile Insect Technique. Principles and Practice in Area-Wide Integrated Pest Management, V.A. Dyck, J. Hendrichs, and A.S. Robinson, eds. (Dordrecht, The Netherlands: Springer), pp. 701–723.

Feldmann, U., Luger, D., Barnor, H., Dengwat, L., Ajagbonna, B., Vreysen, M.J.B., and Van Der Vloedt, A.M.V. (1992b). Tsetse fly mass rearing: Colony management, deployment of sterile flies, related research and development. In Tsetse control, diagnosis and chemotherapy using nuclear techniques. Proceedings of a seminar jointly organized by the International Atomic Energy Agency and the Food and Agriculture Organization of the United Nations and held in Muguga, Kenya, 11–15 February 1991, (Vienna, Austria: IAEA), pp. 167–180.

Fukuda, Y., Nakayama, Y., and Tomita, M. (2003). On dynamics of overlapping genes in bacterial genomes. Gene 323, 181–187.

Furth, P.A., and Baker, C.C. (1991). An element in the bovine papillomavirus late 3′ untranslated region reduces polyadenylated cytoplasmic RNA levels. J. Virol. *65*, 5806–5812.

Furth, P.A., Choe, W.T., Rex, J.H., Byrne, J.C., and Baker, C.C. (1994). Sequences homologous to 5′ splice sites are required for the inhibitory activity of papillomavirus late 3′ untranslated regions. Mol. Cell. Biol. *14*, 5278–5289.

Garcia-Maruniak, A., Abd-Alla, A.M.M., Salem, T.Z., Parker, A.G., van Oers, M.M., Maruniak, J.E., Kim, W., Burand, J.P., Cousserans, F., Robinson, A.S. *et al.* (2009). Two viruses that cause salivary gland hypertrophy in *Glossina pallidipes* and *Musca domestica* are related and form a distinct phylogenetic clade. J. Gen. Virol. *90*, 334–346.

Garcia-Maruniak, A., Maruniak, J.E., Farmerie, W., and Boucias, D.G. (2008). Sequence analysis of a non-classified, non-occluded DNA virus that causes salivary gland hypertrophy of *Musca domestica*, MdSGHV. Virology *377*, 184–196.

Garcia-Maruniak, A., Maruniak, J.E., Zanotto, P.M.A., Doumbouya, A.E., Liu, J.-C., Merrit, M., and Lanoie, J.S. (2004). Sequence analysis of the genome of the *Neodiprion sertifer* nucleopolyhedrovirus. J. Virol. *78*, 7036–7051.

Garcia-Maruniak, A., Pavan, O.H., and Maruniak, J.E. (1996). A variable region of *Anticarsia gemmatalis* nuclear polyhedrosis virus contains tandemly repeated DNA sequences. Virus Res. *41*, 123–132.

Geden, C.J., Lietze, V.U., and Boucias, D.G. (2008). Seasonal prevalence and transmission of salivary gland hypertrophy virus of houseflies (Diptera: Muscidae). J. Med. Entomol. *45*, 42–51.

Gouteux, J.P. (1987). Prevalence of enlarged salivary glands in *Glossina palpalis*, *G. pallicera*, and *G. nigrofusca* (Diptera: Glossinidae) from the Vavoua area, Ivory Coast. J. Med. Entomol. *24*, 268.

Guarino, L.A., Gonzalez, M.A., and Summers, M.D. (1986). Complete sequence and enhancer function of the homologous DNA regions of *Autographa californica* nuclear polyhedrosis virus. J. Virol. *60*, 224–229.

Guarino, L.A., and Summers, M.D. (1986). Interspersed homologous DNA of *Autographa californica* nuclear polyhedrosis virus enhances delayed-early gene expression. J. Virol. *60*, 215–223.

Hendrichs, J., Kenmore, P., Robinson, A.S., and Vreysen, M.J.B. (2007). Area-wide integrated pest management (AW-IPM): Principles, practice and prospects. In Area-wide control of insect pests. From research to field implementation, M.J.B. Vreysen, A.S. Robinson, and J. Hendrichs, eds. (Dordrecht, The Netherlands: Springer), pp. 3–33.

Hilger, C., Velhagen, I., Zentgraf, H., and Schroder, C.H. (1991). Diversity of hepatitis B virus X gene-related transcripts in hepatocellular carcinoma: a novel polyadenylation site on viral DNA. J. Virol. *65*, 4284–4291.

Hu, Z.H., Arif, B.M., Jin, F., Martens, J.W., Chen, X.W., Sun, J.S., Zuidema, D., Goldbach, R.W., and Vlak, J.M. (1998). Distinct gene arrangement in the *Buzura suppressaria* single-nucleocapsid nucleopolyhedrovirus genome. J. Gen. Virol. *79*, 2841–2851.

Jaenson, T.G.T. (1978). Virus-like rods associated with salivary gland hyperplasia in tsetse, *Glossina pallidipes*. Trans. R. Soc. Trop. Med. Hyg. 72, 234–238.

Jakob, N.J., Müller, K., Bahr, U., and Darai, G. (2001). Analysis of the first complete DNA sequence of an invertebrate iridovirus: coding strategy of the genome of *Chilo* iridescent virus. Virology *286*, 182–196.

Jancovich, J.K., Mao, J., Chinchar, V.G., Wyatt, C., Case, S.T., Kumar, S., Valente, G., Subramanian, S., Davidson, E.W., Collins, J.P. *et al.* (2003). Genomic sequence of a ranavirus (family Iridoviridae) associated with salamander mortalities in North America. Virology *316*, 90–103.

Jehle, J.A., Blissard, G.W., Bonning, B.C., Cory, J.S., Herniou, E.A., Rohrmann, G.F., Theilmann, D.A., Thiem, S.M., and Vlak, J.M. (2006). On the classification and nomenclature of baculoviruses: a proposal for revision. Arch. Virol. *151*, 1257–1266.

Jenni, L. (1973). Virus-like particles in a strain of *G. morsitans centralis,* Machado 1970. Trans. R. Soc. Trop. Med. Hyg. *67*, 295.

Jenni, L., and Steiger, R. (1974a). Viruslike particles of *Glossina fuscipes fuscipes* Newst. 1910. Acta Trop. 177–180.

Jenni, L., and Steiger, R.F. (1974b). Virus like particles in the tsetse fly, *Glossina morsitans* sspp. Preliminary results. Rev. Suisse Zool. *81*, 663–666.

Jordan, A.M. (1986). Trypanosomiasis control and African rural development. (London (UK): Longman).

Jura, W.G.Z.O., Odhiambo, T.R., Otieno, L.H., and Tabu, N.O. (1988). Gonadal lesions in virus-infected male and female tsetse, *Glossina pallidipes* (Diptera: Glossinidae). J. Invertebr. Pathol. *52*, 1–8.

Jura, W.G.Z.O., Otieno, L.H., and Chimtawi, M.M.B. (1989). Ultrastructural evidence for trans-ovum transmission of the DNA virus of tsetse, *Glossina pallidipes* (Diptera: Glossinidae). Curr. Microbiol. *18*, 1–4.

Jura, W.G.Z.O., Zdarek, J., and Otieno, L.H. (1993). A simple method for artificial infection of tsetse, *Glossina morsitans morsitans* larvae with the DNA virus of *G. pallidipes*. Insect Sci. Appl. *14*, 383–387.

Kikhno, I., Gutierrez, S., Croizier, L., Croizier, G., and Ferber, M.L. (2002). Characterization of *pif*, a gene required for the per os infectivity of *Spodoptera littoralis* nucleopolyhedrovirus. J. Gen. Virol. *83*, 3013–3022.

Klemenz, R., Reinhardt, M., and Diggelmann, H. (1981). Sequence determination of the 3′ end of mouse mammary tumor virus RNA. Mol. Biol. Rep. *7*, 123–126.

Kokwaro, E.D., Nyindo, M., and Chimtawi, M. (1990). Ultrastructural changes in salivary glands of tsetse, *Glossina morsitans morsitans*, infected with virus and rickettsia-like organisms. J. Invertebr. Pathol. *56*, 337–346.

Kool, M., Ahrens, C.H., Vlak, J.M., and Rohrmann, G.F. (1995). Replication of baculovirus DNA. J. Gen. Virol. *76*, 2103–2118.

Kuzio, J., Jaques, R., and Faulkner, P. (1989). Identification of p74, a gene essential for virulence of baculovirus occlusion bodies. Virology *173*, 759–763.

Lamb, R.A., and Horvath, C.M. (1991). Diversity of coding strategies in influenza viruses. Trends Genet. *7*, 261–266.

Lietze, V.U., Geden, C.J., Blackburn, P., and Boucias, D.G. (2007). Effects of salivary gland hypertrophy virus on the reproductive behavior of the housefly, *Musca domestica*. Appl. Environ. Microbiol. 73, 6811–6818.

Lietze, V.U., Sims, K.R., Salem, T.Z., Geden, C.J., and Boucias, D.G. (2009). Transmission of MdSGHV among adult houseflies, Musca domestica (Diptera: Muscidae), occurs via oral secretions and excreta. J. Invertebr. Pathol. *101*, 1–7.

Lu, A., and Miller, L.K. (1997). Regulation of baculovirus late and very late gene expression. In The baculoviruses, L.K. Miller, ed. (NY and London: plenum Press), pp. 193–216.

Lyon, J.P. (1973). La mouche des Narcisses (*Merodon equestris* F., Diptere *Syrphidae*). I. Identification de l'insecte et de ses degats et biologie dans le sud-est de la France. Rev. Zool. Agric. Pathol. Veg. *72*, 65–92.

Marks, H., Ren, X.Y., Sandbrink, H., van Hulten, M.C.W., and Vlak, J.M. (2006). *In silico* identification of putative promoter motifs of White Spot Syndrome Virus. BMC Bioinformatics *7*, 309.

Mellor, P.S. (2000). Replication of arboviruses in insect vectors. J. Comp. Pathol. *123*, 231–247.

Minter-Goedbloed, E., and Minter, D.M. (1989). Salivary gland hyperplasia and trypanosome infection of *Glossina* in two areas of Kenya. Trans. R. Soc. Trop. Med. Hyg. *83*, 640–641.

Mount, S.M. (1982). A catalogue of splice junction sequences. Nucleic Acids Res. *10*, 459–472.

Nasseri, M., Hirochika, R., Broker, T.R., and Chow, L.T. (1987). A human papilloma virus type 11 transcript encoding an E1-E4 protein. Virology *159*, 433–439.

Nasseri, M., Wettstein, F.O., and Stevens, J.G. (1982). Two colinear and spliced viral transcripts are present in non-virus-producing benign and malignant neoplasms induced by the shope (rabbit) papilloma virus. J. Virol. *44*, 263–268.

Odindo, M.O. (1982). Incidence of salivary gland hypertrophy in field populations of the tsetse *Glossina pallidipes* on the South Kenya coast. Insect Sci. Appl. *3*, 59–64.

Odindo, M.O., Payne, C.C., Crook, N.E., and Jarret, P. (1986). Properties of a novel DNA virus from the tsetse fly, *Glossina pallidipes*. J. Gen. Virol. *67*, 527–536.

Ohkawa, T., Washburn, J.O., Sitapara, R., Sid, E., and Volkman, L.E. (2005). Specific binding of *Autographa californica M* nucleopolyhedrovirus occlusion-derived virus to midgut cells of *Heliothis virescens* larvae is mediated by products of pif genes Ac119 and Ac022 but not by Ac115. J. Virol. *79*, 15258–15264.

Otieno, L.H., Kokwaro, E.D., Chimtawi, M., and Onyango, P. (1980). Prevalence of enlarged salivary glands in wild populations of *Glossina pallidipes* in Kenya, with a note on the ultrastructure of the affected organ. J. Invertebr. Pathol. *36*, 113–118.

Pijlman, G.P., Pruijssers, A.J., and Vlak, J.M. (2003). Identification of *pif-2*, a third conserved baculovirus gene required for per os infection of insects. J. Gen. Virol. *84*, 2041–2049.

Prompiboon, P., Lietze, V.-U., Denton, J.S.S., Geden, C.J., Steenberg, T., and Boucias, D.G. (2010). The *Musca domestica* salivary gland hypertrophy virus: an insect virus that globally infects and sterilizes female houseflies. Appl. Environ. Microbiol. *74*, 994–998.

Rixon, F.J., and Clements, J.B. (1982). Detailed structural analysis of two spliced HSV-1 immediate-early mRNAs. Nucleic Acids Res. *10*, 2241–2256.

Salem, T.Z., Garcia-Maruniak, A., Lietze, V.U., Maruniak, J.E., and Boucias, D.G. (2009). Analysis of transcripts from predicted open reading frames of *Musca domestica* salivary gland hypertrophy virus. J. Gen. Virol. *90*, 1270–1280.

Sang, R.C., Jura, W.G.Z.O., Otieno, L.H., and Mwangi, R.W. (1998). The effects of a DNA virus infection on the reproductive potential of female tsetse flies, *Glossina morsitans centralis* and *Glossina morsitans morsitans* (Diptera: Glossinidae). Mem. Inst. Oswaldo Cruz *93*, 861–864.

Sang, R.C., Jura, W.G.Z.O., Otieno, L.H., Mwangi, R.W., and Ogaja, P. (1999). The effects of a tsetse DNA virus infection on the functions of the male accessory reproductive gland in the host fly *Glossina morsitans centralis* (Diptera; Glossinidae). Curr. Microbiol. *38*, 349–354.

Sang, R.C., Jura, W.G.Z.O., Otieno, L.H., and Ogaja, P. (1996). Ultrastructural changes in the milk gland of tsetse *Glossina morsitans centralis* (Diptera; Glissinidae) female infected by a DNA virus. J. Invertebr. Pathol. *68*, 253–259.

Sang, R.C., Jura, W.G.Z.O., Otieno, L.H., Tukei, P.M., and Mwangi, R.W. (1997). Effects of tsetse DNA virus infection on the survival of a host fly *Glossina morsitans centralis* (Diptera: Glossinidae). J. Invertebr. Pathol. *69*, 253–260.

Shaw, M.K., and Moloo, S.K. (1993). Virus-like particles in Rickettsia within the midgut epithelial cells of *Glossina morsitans centralis* and *Glossina brevipalpis*. J. Invertebr. Pathol. *61*, 162–166.

Silver Key, S.C., and Pagano, J.S. (1997). A noncanonical poly(A) signal, UAUAAA, and flanking elements in Epstein–Barr virus DNA polymerase mRNA function in cleavage and polyadenylation assays. Virology *234*, 147–159.

Simonsen, C.C., and Levinson, A.D. (1983). Analysis of processing and polyadenylation signals of the hepatitis B virus surface antigen gene by using simian virus 40-hepatitis B virus chimeric plasmids. Mol. Cell. Biol. *3*, 2250–2258.

Slack, J., and Arif, B.M. (2007). The baculoviruses occlusion-derived virus: virion structure and function. Adv. Virus Res. *69*, 99–165.

Song, J., Wang, R., Deng, F., Wang, H., and Hu, Z. (2008). Functional studies of per os infectivity factors of Helicoverpa armigera single nucleocapsid nucleopolyhedrovirus. J. Gen. Virol. *89*, 2331–2338.

Steelman, C.D. (1976). Effects of external and internal arthropod parasites on domestic livestock production. Annu. Rev. Entomol. *21*, 155–178.

Sun, M., Hurst, L.D., Carmichael, G.G., and Chen, J. (2005). Evidence for a preferential targeting of 3′-UTRs by cis-encoded natural antisense transcripts. Nucleic Acids Res. *33*, 5533–5543.

van Hulten, M.C.W., Witteveldt, J., Peters, S., Kloosterboer, N., Tarchini, R., Fiers, M., Sandbrink, H., Lankhorst, R.K., and Vlak, J.M. (2001). The white spot syndrome virus DNA genome sequence. Virology *286*, 7–22.

van Oers, M.M., and Vlak, J.M. (2007). Baculovirus genomics. Curr. Drug Targets *8*, 1051–1068.

Vreysen, M.J.B., Saleh, K.M., Ali, M.Y., Abdulla, A.M., Zhu, Z.-R., Juma, K.G., Dyck, V.A., Msangi, A.R., Mkonyi, P.A., and Feldmann, H.U. (2000). *Glossina austeni* (Diptera: Glossinidae) eradicated on the island of Unguja, Zanzibar, using the sterile insect technique. J. Econ. Entomol. *93*, 123–135.

Wang, L., Xue, J., Seaborn, C.P., Arif, B.M., and Cheng, X.W. (2006). Sequence and organization of the *Trichoplusia ni* ascovirus 2c (*Ascoviridae*) genome. Virology *354*, 167–177.

Wang, Y., Burand, J.P., and Jehle, J.A. (2007a). Nudivirus genomics: diversity and classification. Virol. Sin. *22*, 128–136.

Wang, Y., and Jehle, J.A. (2009). Nudiviruses and other large, double-stranded circular DNA viruses of invertebrates: new insights on an old topic. J. Invertebr. Pathol. *101*, 187–193.

Wang, Y., Kleespies, R.G., Huger, A.M., and Jehle, J.A. (2007b). The genome of *Gryllus bimaculatus* nudivirus indicates an ancient diversification of baculovirus-related nonoccluded nudiviruses of insects. J Virol. *81*, 5395–5406.

Wang, Y., van Oers, M.M., Crawford, A.M., Vlak, J.M., and Jehle, J.A. (2006). Genomic analysis of *Oryctes rhinoceros* virus reveals genetic relatedness to *Heliothis zea* virus 1. Arch. Virol. *152*, 519–531.

Whitnall, A.B.M. (1934). The trypanosome infections of *Glossina pallidipes* in the Umfolosi Game Reserve, Zululand. Onderstepoort. J. Vet. Sci. Anim. Ind. *2*, 7–21.

WHO (2001). Report on African trypanosomiasis (sleeping sickness). (Geneva, Switzerland: World Health Organization).

Wolff, J.L.C., Valicente, F.H., Martins, R., Oliveira, J.V., and Zanotto, P.M. (2008). Analysis of the genome of *Spodoptera frugiperda* nucleopolyhedrovirus (SfMNPV-19) and of the high genomic heterogeneity in group II nucleopolyhedroviruses. J. Gen. Virol. *89*, 1202–1211.

Yang, F., He, J., Lin, X., Li, Q., Pan, D., Zhang, X., and Xu, X. (2001). Complete genome sequence of the shrimp white spot bacilliform virus. J. Virol. *75*, 11811–11820.

Iridoviruses

6

Trevor Williams and Vernon K. Ward

Abstract

Invertebrate iridescent viruses (IIVs) of the family *Iridoviridae* are icosahedral dsDNA viruses with large circularly permuted and terminally redundant genomes that infect a number of agricultural pests, medically important insect vectors and terrestrial isopods that live in damp or aquatic habitats. IIVs currently figure among the most neglected groups of entomopathogenic viruses, although they present a number of unusual and intriguing aspects in their physical properties, replication strategies and host–virus interactions. We present an overview of this group of viruses with emphasis on recent advances and on-going lines of research focussing on the current classification and progress in understanding the structure of IIV virions. We then outline the genomic characteristics and replication process followed by a brief description of their pathology and the ecological factors that determine the distribution and abundance of these viruses. Finally, we summarize future lines for research and identify potential novel applications in biotechnology and materials science.

Introduction

Invertebrate iridescent viruses (IIVs) are non-occluded icosahedral dsDNA viruses of the family *Iridoviridae* that infect invertebrates, particularly insects and terrestrial isopods that inhabit damp or aquatic environments. There can be little doubt that IIVs currently figure among the most neglected groups of entomopathogenic viruses. Generally perceived as being of little value as biocontrol agents and of minor economic importance, they have attracted the attention of virologists and insect pathologists more for their curiosity value and their intriguing relationships to other groups of vertebrate and invertebrate viruses rather than because of an inherent interest to understand IIV biology and ecology *per se*.

Whether or not that mindset is set to undergo a marked change remains to be seen, although as usually happens when working on viruses, recent advances in IIV research have generated a number of novel avenues of study that have the potential to significantly transform our view of these pathogens.

The *Iridoviridae* family has been the subject of two recent reviews (Williams *et al.*, 2005; Chinchar *et al.*, 2009). The older literature can be accessed through a number of reviews published in the 1980s and 1990s (Anthony and Comps, 1991; Devauchelle *et al.*, 1985a,b; Hall, 1985; Kelly, 1985; Darai, 1990; Ward and Kalmakoff, 1991; Williams, 1996, 1998). In the present work, we present an overview of this group with emphasis on recent advances and on-going lines of research. For this, we start by reviewing the current classification of the group, and then describe significant advances in understanding the structure of IIV virions. We then outline the genomic characteristics and replication process followed by a brief overview of pathology and an examination of the ecological factors that determine the distribution and abundance of these viruses. Finally, we summarize future lines for research and identify key issues that merit additional attention.

Classification

The classification of these viruses has been severely hampered by a lack of comparative molecular studies and a paucity of genome sequence information. The family *Iridoviridae* is currently divided into five genera (Table 6.1), of which three infect poikilothermic vertebrates (*Lymphocystivirus*, *Megalocytivirus* and *Ranavirus*) and two infect invertebrates, particularly insects in humid or aquatic habitats (*Chloriridovirus* and *Iridovirus*). The prototype species of the family is the ranavirus *Frog virus 3* (FV-3) which represents the model virus for our understanding of key aspects of the replication process in these viruses (Chinchar *et al.*, 2005).

The IIVs, represented by the *Iridovirus* and *Chloriridovirus* genera, were originally named according to the host species and assigned a type number according to the sequence of their discovery (Tinsley and Kelly, 1970). However, a number of problems have been recognized with this system including a widespread lack of characterization data for many of the viruses, the absence of reference material for many of the isolates described several decades ago, and different type numbers assigned to the same virus isolated from different host species.

An alternative system of nomenclature was proposed for isolates for which some characterization information and reference material was available (Williams and Cory, 1994). This system involved assigning geographical descriptors to each virus based on the nearest large town to the site of isolation. More recently, an alternative nomenclature has been proposed to resolve the confusion that arises when a given virus displays a range of pathological symptoms, a broad host range and wide geographical distribution (Chinchar *et al.*, 2009). This approach is aimed at systematic naming of isolates rather than virus species. In the case of IIVs, this would involve indicating the host from which the isolate was obtained, the location of the isolate, the isolate number, and the year of isolation, e.g. (*Spodoptera frugiperda*/Tapachula/3/2002) for the third isolate from *S. frugiperda* larvae collected close to the town of Tapachula (Mexico) in 2002. Once the virus has been clearly identified, it could be referred to in an abbreviated form such as SF/T/3/02. However, these alternative proposals for nomenclature have yet to be adopted by researchers so that virus type numbers and host species names continue to be used.

The genus *Iridovirus* is the largest of the genera that infect invertebrates and currently comprises two species and 11 tentative species that require additional characterization in order to determine their species status (Table 6.2). The two virus species are *Invertebrate iridescent virus 1* (IIV-1) isolated from soil-dwelling larvae of the cranefly *Tipula paludosa* in Great Britain (Xeros, 1954), and the type species of the genus, *Invertebrate iridescent virus 6* (IIV-6), originally isolated from larvae of the rice stemborer moth, *Chilo suppressalis*, in Japan (Fukaya and Nasu, 1966). The members of this genus are sometimes referred to as the small IIVs, having a dehydrated virion diameter of 110–160 nm, more typically in the range 120–130 nm. The viruses in this genus can be further categorized into one of three complexes based on serological, DNA hybridization and partial major capsid protein

Table 6.1 Genera in the family *Iridoviridae*

Genus	Hosts infected	Type species (abbreviation)
Chloriridovirus	Diptera	*Invertebrate iridescent virus 3* (IIV-3)
Iridovirus	Insects and terrestrial isopods	*Invertebrate iridescent virus 6* (IIV-6)
Lymphocystivirus	Flat fish (worldwide)	*Lymphocystis disease virus 1* (LCDV-1)
Megalocytivirus	Fish (SE Asia)	*Infectious spleen and kidney necrosis virus* (ISKNV)
Ranavirus	Cold blooded vertebrates (mainly amphibians, fish and reptiles)	*Frog virus 3* (FV-3)

Table 6.2 Species and tentative species in the *Iridovirus* genus

Virus name (strain common names)	Accession numbers	Hosts infected
I. Species		
Invertebrate iridescent virus 1	M33542, M62953	*Tipula paludosa* (Diptera)
Tipula iridescent virus		
Invertebrate iridescent virus 6	AF003534, M99395	*Chilo suppressalis* (Lepidoptera)
Chilo iridescent virus		
Gryllus iridovirus		*Gryllus* spp. (Orthoptera)
II. Tentative species		
Invertebrate iridescent virus 2	AF042335	*Sericesthis pruinosa* (Coleoptera)
Sericesthis iridescent virus		
Invertebrate iridescent virus 9	AF025774	*Wiseana cervinata* (Lepidoptera)
Invertebrate iridescent virus 10		*Eudonia (Witlesia)* sp. (Lepidoptera)
Invertebrate iridescent virus 18		*Opogonia* sp. (Lepidoptera)
Invertebrate iridescent virus 16	AF025775	*Costelytra zealandica* (Coleoptera)
Costelytra zealandica iridescent virus		
Invertebrate iridescent virus 21		*Helicoverpa armigera* (Lepidoptera)
Invertebrate iridescent virus 28		*Hydrocyrius columbiae columbiae*[1] (Hemiptera)
Invertebrate iridescent virus 22	AF042341, M32799	*Simulium variegatum* (Diptera)
Simulium sp. iridescent virus		
Invertebrate iridescent virus 23	AF042342	*Heteronychus arator* (Coleoptera)
Heteronychus arator iridescent virus		
Invertebrate iridescent virus 24	AF042340	*Apis cerana* (Hymenoptera)
Apis iridescent virus		
Invertebrate iridescent virus 29	AF042339	*Tenebrio molitor* (Coleoptera)
Tenebrio molitor iridescent virus		
Invertebrate iridescent virus 30	AF042336	*Helicoverpa zea* (Lepidoptera)
Helicoverpa zea iridescent virus		
Invertebrate iridescent virus 31	AF042337, AJ279821	*Armadillidium vulgare*[2] (Isopoda)
Invertebrate iridescent virus 32		*Porcellio dilatatus* (Isopoda)
Anticarsia gemmatalis iridescent virus	AF042343	*Anticarsia gemmatalis* (Lepidoptera)

[1]Host identity originally reported as *Lethocerus columbiae* but more likely to be *Hydrocyrius columbiae columbiae* Spinola or *Lethocerus niloticus* Stal. Williams (2008).

[2]Described by Federici (1980) but may also infect *Porcellio laevis,* Grosholz (1992), *Porcellio scaber*, Cole and Morris (1980), and possibly a number of other isopod species. Wijnhoven and Berg (1999); Schultz *et al.* (1982), although this requires confirmation.

(MCP) gene sequence information (Williams and Cory, 1994; Webby and Kalmakoff, 1998). Group I comprises IIV-6 and possibly IIVs –21 and -28, although the identity of the two latter viruses requires confirmation. Group II comprises IIVs-1, -2, -9, -16, -22, -23, -24, -29, -30 and *Anticarsia gemmatalis* IIV that lacks a type number. Group III comprises the terrestrial isopod virus IIV-31 and a closely related isolate from the soil-dwelling scarab beetle *Phyllophaga anxia*.

The genus *Chloriridovirus* currently comprises a single species, *Invertebrate iridescent virus 3* (IIV-3), also known as Mosquito iridescent virus. Isolates from mosquitoes and midges tend to be larger than those of the *Iridovirus* genus. Dehydrated virion diameters of the so-called large IIVs range from 170 to 200 nm, although the true significance of virion size in defining chloriridoviruses is uncertain. Sequencing of the IIV-3 genome has revealed that the 190,132 bp genome has no obvious collinearity with IIV-6 (Delhon *et al.*, 2006). Of the 126 predicted genes, 52 had homologues in IIV-6 and 27 had homologues in all currently sequenced members of the family, although with low levels of deduced amino acid sequence identity. Phylogenetic analysis of a concatenated set of 11 conserved proteins indicated that the genetic distance between IIV-3 and IIV-6 was similar to that of members of other genera present within the family. However, as IIV infections have been reported in a total of 32 species of mosquitoes and midges (Williams, 2008), none of which have been characterized, the scope and validity of the genus is clearly in need of revision.

Relationships to other families

Iridoviruses are members of a monophyletic clade of nucleocytoplasmic large DNA viruses (NCLDVs) that includes the *Phycodnaviridae*, *Asfarviridae*, *Poxviridae* and the giant icosahedral *Mimiviridae* (Iyer *et al.*, 2001; Claverie *et al.*, 2009). These viruses encode their own machinery for transcription and RNA modification including RNA polymerase and transcription factors, many of which appear to have been acquired from their eukaryotic hosts (Tidona and Darai, 2000; Filée *et al.*, 2008), whereas in the case of the protist-infecting phycodnaviruses and mimivirus, most genes seem to have been acquired from bacterial sources (Filée *et al.*, 2008). Phycodnaviruses lack RNA polymerase genes and therefore depend on host transcriptional mechanisms in the nucleus (Wilson *et al.*, 2009). The clade shares a total of 31 ancestral genes including three genes for structural proteins: the capsid protein and two lipid membrane localized proteins. Of these ancestral genes, nine are core genes shared by all the families of the clade.

Based on the analysis of LCDV-1 and IIV-6, iridoviruses appear to have lost two ancestral genes: an ATP dependent DNA ligase and a capping enzyme required for RNA capping in the cytoplasm (Iyer *et al.*, 2001), but these viruses also differed from the ancestral group by possessing seven additional genes including thymidylate synthase, a cathepsin B-like cysteine protease and RNase III. This cladistic analysis did not include the more recently sequenced IIV-3 genome.

The similarities between iridoviruses, African swine fever virus (ASFV) and the algal phycodnaviruses extend beyond the architecture of the icosahedral capsid, a common inner lipid membrane and an electron dense core. Phycodnaviruses have a rather less prominent inner lipid membrane, a right-handed (dextro) skewed capsid lattice comprising doughnut-shaped capsomers with a smooth outer surface (Nandhagopal *et al.*, 2002), a high degree of similarity in the MCP primary and secondary structure, and instead of the fibrillar structures seen in iridoviruses, a single vertex spike has been observed recently in Paramecium bursaria Chlorella virus 1 (PBCV-1) (Cherrier *et al.*, 2009). In contrast, structural similarities between iridoviruses and the allantoid virions of ascoviruses are not immediately apparent whereas molecular evidence indicates a phylogenetic relationship between these viruses.

Ascoviruses differ from IIVs in having bacilliform virions and a circular genome with a high level of methylation of deoxycytidines, mostly at CpT and CpC residues (Cheng *et al.*, 1999; Bigot *et al.*, 2000). Evolutionary relationships were first proposed based on sequence information of the δ DNA polymerases of IIVs and the ascoviruses of lepidopteran insects (Knopf, 1998; Stasiak *et al.*, 2000), although caution is warranted when interpreting the relatedness of morphologically distinct viruses based on the phylogeny of their replication proteins (Krupovic and Bamford, 2009). More recently, Bigot *et al.* (2009) identified 28 core genes that are shared by the symbiotic ascovirus DpAV-4a, three pathogenic ascoviruses and the invertebrate iridoviruses IIV-6 and IIV-3. Previous findings involving the analysis of multiple phylogenetic trees based on a selection of different genes from ascoviruses, iridoviruses, phycodnaviruses and ASFV had postulated that ascoviruses had evolved from iridoviruses (Federici and Bigot, 2003; Stasiak *et al.*, 2003). For the moment, it appears that ascoviruses and IIVs share a common origin but confirming that ascoviruses originated from IIVs will require additional IIV genome sequence information. However, based on the number of genes shared between DpAV-4a and IIV-6, Bigot *et al.* (2009) have suggested that DpAV-4a may have had a separate origin in the iridoviruses from that of the pathogenic ascoviruses. These relationships are discussed in detail in Chapter 1 of this book.

The virion

The icosahedral virions of IIVs are structurally complex and comprise an hydrated electron-dense core of DNA and associated proteins, an internal lipid membrane, an outer capsid and an external fringe of fibrils. Virions that bud out of cells may acquire an outer envelope, but this is not essential for infectivity.

One-dimensional SDS-PAGE of IIV-6 virions typically resolves 20–32 polypeptides in the range of 11 to 200 kDa (Krell and Lee, 1974; Barray and Devauchelle, 1979; Cole and Morris, 1980; Black *et al.*, 1981; Tajbakhsh and Seligy, 1990), whereas two-dimensional PAGE indicates the presence of more than 30 entities (Cerutti and Devauchelle, 1985). Most of this diversity is associated with the core and lipid membrane, although one surface labelling study on IIV-1 has indicated that the surface structure may be quite complex, with 10 or 11 polypeptides labelled under non-reducing and reducing conditions, respectively (Watson and Seligy, 1997).

The virion core comprises a 12.5 kDa major component and at least five additional polypeptides ranging in size up to a maximum of 81 kDa (Cerutti and Devauchelle, 1985). Freeze-etching studies of FV-3 have indicated that the DNA–protein complex may be organized as rods of ~10 nm in width arranged in a long coiled filament (Tripier-Darcy and Nermut, 1983), whereas the core structure of a mosquito isolate was described as bundles of parallel threads (Buchatsky, 1977). Lipid enveloped cores of IIV-3 were not infectious in cell culture or in mosquito larvae (Wagner *et al.*, 1975).

The internal lipid layer, 4 nm thick, constitutes 5–9% of the virion dry weight, of which 44–75% is phospholipid (Kalmakoff and Tremaine, 1968; Kelly and Vance, 1973; Balange-Orange and Devauchelle, 1982). The absence of microscopy evidence of virion cores budding through intracellular membranes (Stoltz, 1971) and the observations that the fatty acid and phospholipid composition of IIVs differ from that of the host cell (Balange-Orange and Devauchelle, 1982; Williams and Thompson, 1995) lead to the suggestion that the lipid layer may be generated *de novo* during the infection process. However, studies on other members of the NCLDV clade have revealed that their internal lipid envelopes are acquired from intracellular compartments (Schmelz *et al.*, 1994; Andrés *et al.*, 1998; Rouiller *et al.*, 1998), in a process facilitated by vimentin filaments concentrated in viral assembly sites (Risco *et al.*, 2002). To what degree IIV virion morphogenesis and lipid acquision mirrors that of poxviruses remains to be seen. As a case of history repeating itself, initial assertions that the lipid membrane was part of a double-layered structure (Stoltz, 1971; Yan *et al.*, 2000) were later rectified to indicate the single bilayer nature of this component (Stoltz, 1973; Yan *et al.*, 2009).

The formative electron microscope (EM) studies on IIV structure were performed by Wrigley (1969, 1970) and Stoltz (1971, 1973), and their findings have been broadly confirmed by the current model for IIV structure, based on detailed cryo-EM and image reconstruction studies on IIV-6 (Yan *et al.*, 2000, 2009). The vitrified IIV-6 virion has diameters of 162, 165 and 185 nm along the two, three and five fold axes of symmetry, respectively (Fig. 6.1A and B). Each particle is formed of 12 pentasymmetrons and 20 trisymmetrons arranged in an icosahedral, quasi-equivalent symmetry with a triangulation number of $T = 147$ ($h = 7$, $k = 7$). Both types of

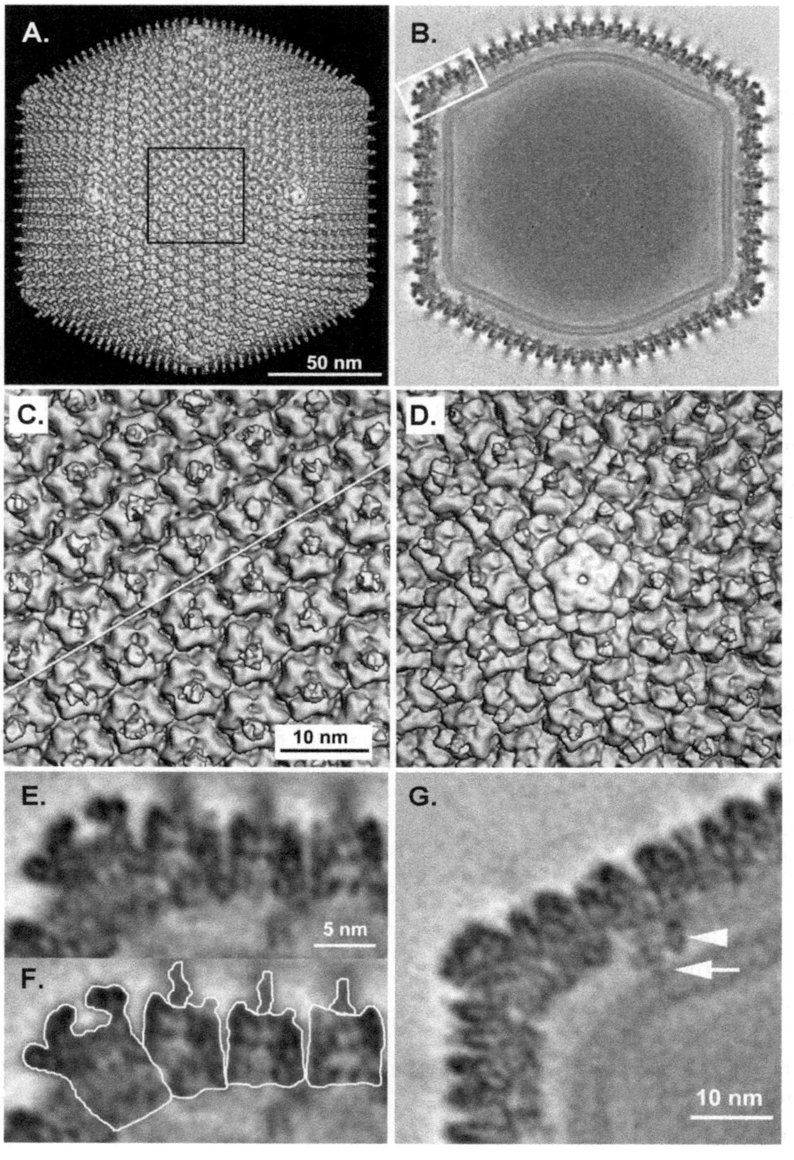

Figure 6.1 Three dimensional reconstruction of an IIV-6 particle at a resolution of 1.3 nm. (A.) 3-D density map of particle viewed along an icosahedral two-fold axis. The external fringe of filaments has been blurred away due to their varying position in relation to the symmetry of the capsid. (B) Central cross section, one pixel thick. A lipid bilayer follows the inner contour and icosahedral symmetry of the capsid shell whereas the core appears to lack any structures that are arranged following icosahedral symmetry. (C) Magnified view of the central region outlined in (A). The diagonal white line indicates the cleavage plane between adjacent trisymmetrons. (D) Magnified view of the propeller-like pentamer complex at the five-fold vertex. This complex differs from the trimeric capsomers in that it is larger, has a small axial hole, and lacks an external fibre. (E) Magnified view of boxed region in (B) showing transverse section of the pentamer complex and adjacent trisymmetrons. (F) Same as (E) but with the pentamer complex, three capsomers, and their fibres individually outlined. (G) Transmembrane anchor proteins beneath the pentasymmetrons showing two stick-like entities. The longer of the two (long arrow) crosses both leaflets of the bilayer, whereas the other (short arrow) stops at the outer leaflet of the internal lipid membrane. (Images reproduced from Yan *et al.* 2009 by permission of Elsevier.)

structure predominantly comprise hexavalent capsomers, a total of 1460 per virion, that are composed of the highly conserved MCP (Tidona *et al.*, 1998), that in IIV-6, comprises 467 amino acids (51.4 kDa) and represents ~40% of the total virion polypeptide (Moore and Kelly, 1980; Stohwasser *et al.*, 1993). Each hexavalent capsomer is formed by a non-covalent MCP trimer on the outer surface and a second MCP trimer linked by disulphide bonds on the inner surface (Cerutti and Devauchelle, 1985, 1990). Each MCP monomer contains two β-barrel domains, of the viral jelly-roll type, which are an ancient feature of icosahedral virus shell proteins (Khayat *et al.*, 2005; Krupovic and Bamford, 2008). The 55 capsomers in a trisymmetron are uniformly packed with an intercapsomer distance of 7.5 nm and in a single shared orientation, which is rotated by ~60° compared to the capsomers of the neighbouring trisymmetron (Fig. 6.1C). In addition, each pentasymmentron comprises 30 hexavalent (trimeric) capsomers and a single pentavalent capsomer at its centre, at the vertex of each pentasymmetron, a total of 12 in each virion. This pentavalent capsomer is significantly larger than the trimeric capsomers and has a five-bladed propeller-shaped external appearance and a small central pore below which a flask-shaped cavity is located (Fig. 6.1D–F).

A number of additional proteins have recently been identified in the capsid shell and in association with the lipid membrane (Yan *et al.*, 2009). These have been named zip monomers, zip dimers, finger proteins and anchor proteins (Fig. 6.2). The zip dimers appear as two halves of a clasp connecting the trimer capsomers along the edges of adjacent trisymmetrons, whereas zip monomers appear to be involved in linking trisymmetron capsomers with those of neighbouring pentasymmetrons. Three sets of nine inward-pointing finger proteins bind the capsomers along the first, second and third rows from the edge of each trisymmetron. Finally, an anchor protein connects each pentasymmetron with the lipid membrane at a distance of two capsomers from the pentavalent vertex (Figs. 6.1G and 6.2). It seems likely that this protein stabilizes the structure of the virion close to the vertices. Many other transmembrane proteins are present, but only the anchor protein can be visualized in image reconstruction studies due to its invariable position with respect to the icosahedral symmetry of the particle. Volume estimates, using the MCP as a reference, were used to estimate the molecular masses of these minor capsid proteins at 11.9, 19.7 and 32.4 kDa for the zip, finger and anchor proteins, respectively.

The outer surface of the capsid is covered by flexible fibrils or fibres. Conventional EM studies have suggested that these fibrils are often rather short (~2.5 nm in length) (Cole and Morris, 1980; Black *et al.*, 1981; Devauchelle *et al.*, 1985b) and may have terminal knobs (Stoltz, 1971, 1973), but this is clearly not the case in IIV-6 where a single fibril of 2 nm width and ~35 nm in length extends outwards from the centre of each trimeric capsomer (Figs. 6.1F and 6.3). The fragility of these structures is uncertain and they may have been damaged during sample preparation procedures involving dehydration and epoxy embedding for conventional EM studies. Cryo-EM studies that maintain the samples in a hydrated state have shown the presence of fibrils in IIV-6 and IIV-9 (Yan *et al.*, 2000, 2009; Juhl *et al.*, 2006) suggesting that this may be a feature of many IIVs. The presence of fibrils would be expected to affect the interparticle distance in quasi-crystalline arrays, which is influential in determining the iridescent phenomenon. Indeed, an isolate from the midge, *Chironomus plumosus*, was notable for the length of its well preserved fibrils that increased interparticle distance to such an extent that the isolate did not cause visible iridescence of infected host tissues (Stoltz *et al.*, 1968). The role of the fibrils remains elusive, but interactions with the host or improved stability in the environment have been postulated. Interestingly, the icosahedral mimivirus of amoeba is also surrounded by a long fringe of external fibrils that are partially digested by the host to achieve infection (Claverie *et al.*, 2009).

The genome

The IIV genome is a linear molecule of dsDNA. However, these genomes are characterized by being circularly permuted when mapped by restriction endonuclease analysis due to packaging of concatemeric DNA in a headful packaging system. The circular permutation and terminal redundancy of iridovirus genomes

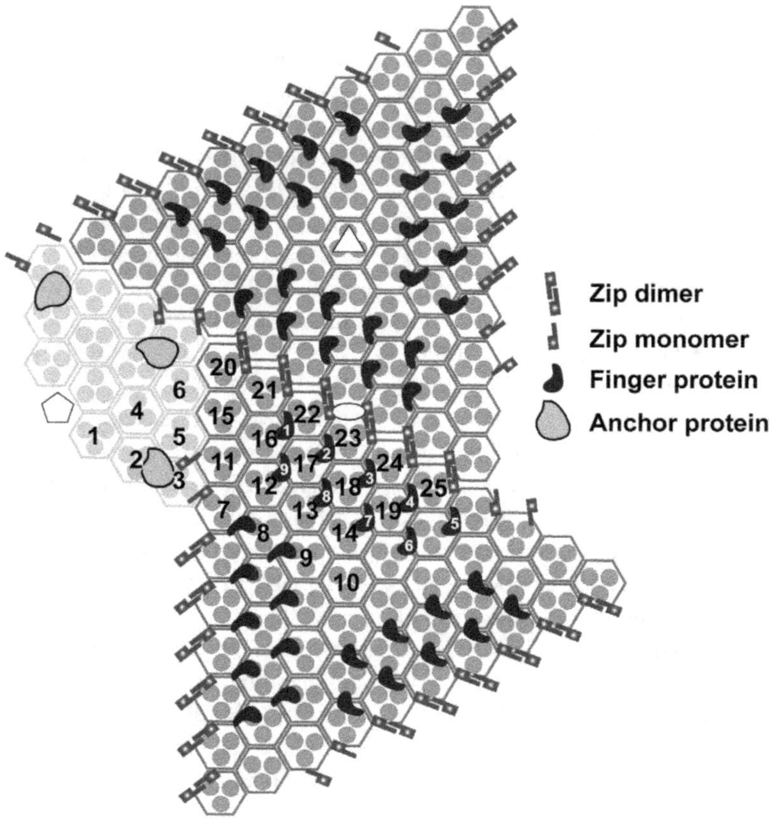

Figure 6.2 Schematic planar diagram of the IIV-6 capsid structure with the central pentamer complex (white pentagon) omitted for clarity. The centrally located white ellipse, triangle, and pentagon symbols highlight the positions of 2-, 3-, and 5-fold icosahedral axes, respectively. All major capsid protein trimers are shown as three small grey disks enclosed by a hexagon. The icosahedral asymmetric unit contains 24 and one third of these capsomers (numbered 1–25). Each trisymmetron contains 55 capsomers, all oriented similarly and rotated by 60° relative to those in the adjacent trisymmetron. Finger proteins bind to each trisymmetron (nine within one asymmetric unit are numbered in white), a total of 27 in each trisymmetron. A total of 18 Zip dimers are present at the interface between one trisymmetron and its adjacent trisymmetrons. Zip monomers are located at the interface between a trisymmetron and its neighbouring pentasymmetrons. The transmembrane anchor proteins (shown in Fig. 6.1G) are located under capsomers 2 and 3 beneath the pentasymmetron. (Image adapted from Yan *et al.* 2009, with permission from Elsevier)

was first proposed for FV-3 (Goorha and Murti, 1982), and subsequently confirmed by analysis of the genomes of IIV-6, IIV-9 and IIV-16 (Delius *et al.*, 1984; McMillan *et al.*, 1990; Ward and Kalmakoff, 1987; Webby and Kalmakoff, 1999). The terminal redundancy for IIV varies between species with pulse-field gel electrophoresis producing estimates ranging from 0% terminal redundancy for IIV-1 to 5.6% for IIV-6 and 7.8% for IIV-22 (Webby and Kalmakoff, 1999), which is somewhat lower than previous estimates for IIV-6 based on EM observations of terminal annealing (Delius *et al.*, 1984). The terminal redundancy is randomly distributed over the genome with all regions of the genome found at the termini of purified DNA (Delius *et al.*, 1984; McMillan *et al.*, 1990; Webby and Kalmakoff, 1999). This contrasts with FV-3 in which the termini are believed to be derived from a limited subset of the viral genome (Murti *et al.*, 1982).

To date, a total of 14 genomes have been fully sequenced in the family *Iridoviridae* (Table 6.3), but of these, the only viruses from invertebrates are those of IIV-3 (Delhon *et al.*, 2006) and

Table 6.3 Fully sequenced genomes from vertebrate and invertebrate iridoviruses

Genus, virus[a]	Size (bp)	ORF no.[b]	%GC	Accession no.	Reference
Chloriridovirus					
IIV-3	191,132	126	48	DQ643392	Delhon *et al.* (2006)
Iridovirus					
IIV-6	212,482	243[d]	29	AF303741	Jakob *et al.* (2001)
Lymphocystivirus					
LCDV-1	102,653	110	29	L63545	Tidona and Darai (1997)
LCDV-C	186,250	240	27	AY380826	Zhang *et al.* (2004)
Ranavirus					
TFV	105,057	105	55	AF389451	He *et al.* (2002)
ATV	106,332	96	54	AY150217	Jancovich *et al.* (2003)
FV-3	105,903	98	55	AY548484	Tan *et al.* (2004)
STIV	105,890	105	55	EU627010	Huang *et al.* (2009)
SGIV	140,131	162	49	AY521625	Song *et al.* (2004)
GIV	139,793	120	49	AY666015	Tsai *et al.* (2005)
Megalocytivirus					
ISKNV	111,362	124	55	AF371960	He *et al.* (2001)
RBIV	112,080	118	53	AY532606	Do *et al.* (2004)
RSIV	112,414	93[c]	53	BD143114	Kurita *et al.* (2002)
OSGIV	112,636	121	54	AY894343	Lu *et al.* (2005)

[a]IIV-3, invertebrate iridescent virus 3; IIV-6, invertebrate iridescent virus 6; LCDV-1, lymphocystis disease virus 1; LCDV-C, lymphocystis disease virus, China strain; TFV, tiger frog virus; ATV, *Ambystoma tigrinum* virus; FV-3, frog virus 3; STIV, soft-shelled turtle iridovirus; SGIV, Singapore grouper iridovirus; GIV, grouper iridovirus; ISKNV, infectious spleen and kidney necrosis virus; RBIV, rock bream iridovirus; RSIV, Red Sea bream iridovirus; OSGIV, orange spotted grouper iridovirus.

[b]Essentially non-overlapping ORFs encoding a minimum length of 40–62 amino acids.

[c]Minimum ORF of 86 amino acids.

[d]Revised annotation of Eaton *et al.* (2007).

IIV-6 (Jakob *et al.*, 2001), while the genome sequence of IIV-9 (genus *Iridovirus*) has been determined and is currently undergoing analysis (Wong, 2009). Genomes from different genera show substantial differences with no apparent collinearity between their genes unless they are very closely related (Jancovich *et al.*, 2003). The genomes have widely different GC contents at 48% and 29% for IIV-3 and IIV-6, respectively. Individual IIV genomes display less than 10% homology between genomes, reflecting the lack of collinearity and diversity of sequences found in these genomes (Webby and Kalmakoff, 1999). The codon bias of IIVs is thought to reflect this GC content (Tsai *et al.*, 2007). As mentioned previously, phylogenetic analyses of the MCP, for which many more sequences are available, suggests that IIV-3 may in fact be more closely related to the major complex of IIVs than IIV-6 (Delhon *et al.*, 2006).

IIV genomes possess extensive repeat sequences with 20% and 39% of repetitive DNA observed for IIV-3 (Delhon *et al.*, 2006) and IIV-16 (McMillan *et al.*, 1990), respectively. The characterization of a repeat region from IIV-6 revealed a complex repeat structure including overlapping and interspersed repeats (Fischer *et al.*, 1988). The role of these highly repetitive sequences is unknown, although a role in genome replication through branched chain concatemer formation is one possibility. The analysis of the genome of IIV-9 has shown a similar complex arrangement of repeats suggesting that this is a common feature of IIVs (Wong, 2009). One of the hallmarks of vertebrate iridovirus genomes is the high degree of methylation with a viral encoded DNA methyltransferase that, with one exception (Song *et al.*, 2004), results in the methylation of approximately 20% of the cytosines in the genome (Willis and Granoff, 1980). In contrast, IIV genomes are not methylated (<5% methylation at CpG residues).

The genomes of these viruses encode a range of genes with a surprising diversity. Analysis across all sequenced iridoviruses shows that only 26 genes are conserved across all members of the family, and when comparing IIV-3 and IIV-6, only 52 of the 126 IIV-3 genes are conserved in IIV-6 (Delhon *et al.*, 2006). Twelve of the 26 core genes are associated with DNA replication or transcription, indicating that the replication of the genome and viral gene transcription share a common basis between these viruses that are currently classified in different genera. Of the 52 genes shared between IIV-3 and IIV-6, only 12 have a predicted function, underlining the differences between these two viruses (Delhon *et al.*, 2006). The recent sequencing of IIV-9, which is in the main phylogenetic cluster of IIVs in which IIV-3 clusters by MCP analysis, may indicate whether IIV-3 or IIV-6 is the outlier of this group and if the current separation of *Iridovirus* and *Chloriridovirus* genera is appropriate (Wong, 2009).

The genes encoded by vertebrate and invertebrate iridoviruses have been described in detail in the respective publications (Table 6.3) and will not be covered in this chapter. However, a clear feature is the lack of co-linearity in gene order between the viruses unless they are essentially strains of the same species. This remarkable diversity in gene order indicates that there is no apparent requirement for clusters of genes with complimentary functions or clustering of transcripts. Whether this is a consequence of the replication strategy of hybridizing concatemers driving a high rate of recombination (Chinchar and Granoff, 1986), and hence a need for the control of individual genes to be independent of their neighbours, is unknown.

A number of interesting elements are being identified in iridovirus genomes. For example, an intein has been identified in the ribonucleotide reductase gene of IIV-6, representing the first report of such an element in an insect virus (Pietrokovski, 1998). Mini-inteins have also been identified in IIV-9 and IIV-16 (Goodwin *et al.*, 2006) suggesting that this may be a common feature of IIV genomes. In addition, the genome of the soft-turtle iridovirus was recently found to contain 12 putative precursor RNAs encoding 20 predicted microRNAs (miRNAs) (Huang *et al.*, 2009). Whether these miRNAs are produced was not determined, however, it offers an intriguing possibility for interaction with and control of host or viral gene expression (Cullen, 2009; also see Chapter 15). Whether these elements are also present in IIVs is not known, but given the high incidence of complex repeats in these genomes combined with the nuclear site of early gene expression, it would seem likely that candidates exist on these genomes.

Replication

Iridoviruses show very broad host ranges provided the temperature is kept relatively low (<30°C), indicating that they bind non-specifically to a wide range of cell surface receptors, or to a limited number of common receptor molecules (Williams, 1996). Both the naked and enveloped virions are infectious, and each presumably recognizes different receptors, although this has not been confirmed. Both types of virions appear to enter cells by receptor-mediated endocytosis (Hall, 1985), although the mechanisms by which these events occur are unclear. Once in the cytoplasm, particles may be surrounded by a membrane, in lysosome-like structures or as free virions that undergo uncoating. The DNA core then migrates to the nucleus and initiates a two-stage replication cycle.

The replication of the genome is presumed to be essentially similar to that of FV-3, although the information supporting this is limited. The unit length genome of the infecting virus enters the nucleus where it is replicated into approximately unit or twice unit length progeny genomes in the first stage of genome replication (Goorha, 1982). The involvement of the nucleus is essential (Goorha *et al.*, 1977), emphasizing the importance of this step in viral replication. These nuclear genomes are the templates for early mRNA production and are transcribed by host RNApol II. The essentially unit length genomes are transported to the cytoplasm where the second stage of DNA replication occurs (Goorha *et al.*, 1978). Cytoplasmic replication leads to the formation of concatemeric replication complexes that are over 10 unit lengths in size (Goorha, 1982). Evidence for the formation of cross-hybridizing complexes includes the detection of single-stranded regions through the use of nuclease S1 and their subsequent repair (Goorha, 1982). Whether this process is the same in IIVs is not clear. Studies on IIV-9 showed that large complexes are indeed formed, although nuclear and cytoplasmic separation was not clearly defined as in FV-3 (Watson, 1997). In contrast, Cerutti *et al.* (1989) reported evidence for recombination in IIV-6 during replication, indicating that a similar process was occurring to that elucidated in FV-3. Further research is needed to confirm the presence and nature of the cytoplasmic stage of genome replication in IIVs.

Late in the infection cycle particle maturation occurs at cytoplasmic assembly sites. The exact relationship between assembly site formation, recruitment of concatemeric DNA, late gene transcription and assembly of virus particles is largely unknown. However, the accumulation of DNA and virion components leads to virion assembly and the presumed packing of DNA by a headful mechanism from concatemeric DNA that includes more than genome length DNA, thus leading to circular permutation and terminal redundancy (Goorha, 1982). An ATPase is one of the 26 core genes identified in all iridoviruses and may be a packaging ATPase involved in DNA packaging. A terminal redundancy of 10% through this mechanism would mean an extra ~20 kbp packaged into each particle and hence a significant gene duplication. Exactly how the network of cross-hybridizing concatemers is resolved into unit length genomes is unknown and the potential role(s) of these gene duplications representing all regions of the genome remain to be elucidated.

Iridovirus assembly sites are perinuclear electron translucent structures surrounded by rough ER membranes and polyribosomes. Netherton *et al.* (2007) have argued that these cytoplasmic inclusions resemble aggresomes as they sequester intermediate filaments and mitochondria that often appear damaged. In FV-3, vimentin filaments are phosphorylated and form cages around each assembly site that appear to act as a scaffold for virus replication and that simultaneously exclude cytoplasmic components (Chen *et al.*, 1986; Murti *et al.*, 1988). Consistent with these findings, polyribosomes and most newly synthesized viral proteins associate with intermediate filaments during infection (Murti and Goorha, 1989). A 108 kDa early protein is also recruited to viral assembly sites during FV-3 infection, although its role remains unclear (Chinchar *et al.*, 1984).

The processes that control virion assembly are also unclear. Proteins such as the MCP are required but what other factors are critical is largely unstudied. One potentially important gene is the myristylated membrane protein found in all sequenced iridoviruses, which may be analogous to the myristylated membrane protein of vaccinia virus (Huang *et al.*, 2009) that has been shown to be critical for virion assembly (Ravanello and Hruby, 1994). Once the virions have assembled there are two possible routes of final maturation for the virus particles. The majority exit the assembly site and accumulate a regular paracrystalline arrays, whereas some virions bud from the cell to become enveloped. The factors that define which pathway virions mature by are not known.

Gene expression

As for many aspects of IIV replication, transcript synthesis is presumed to be essentially similar to that of FV-3 and many similarities have been supported through studies of IIV-6 and IIV-9.

Parallel to the first stage of genome replication, viral gene transcription commences in the nucleus with the production of immediate early (IE) transcripts that initiate a temporal cascade characterized by delayed early (DE) and late (L) phases of gene expression. IE gene expression is defined as the events that occur prior to *de novo* protein synthesis and are expressed in the presence of the protein synthesis inhibitor cycloheximide. Subsequent DE genes are dependant upon one or more factors encoded by IE genes and are expressed in the presence of DNA synthesis inhibitors such as aphidicolin or AraC. Late gene expression is dependent upon factors encoded by DE genes and is also blocked by inhibitors of DNA synthesis (Barray and Devauchelle, 1985; Elliott and Kelly, 1980; D'Costa *et al.*, 2001; Nalcacioglu *et al.*, 2003).

Early gene transcription requires a functional nucleus and the use of α-aminitin during FV-3 infection has shown that host RNApol II is required (Goorha, 1981), whereas later phases of the transcription cascade are presumed to involve virus encoded subunit homologues of RNApol II designated vpol IIα and vpol IIβ. Indeed, the targeting of FV-3 vpol IIα with morpholino antisense oligonucleotides inhibits late gene transcripts (Sample *et al.*, 2007; Chinchar *et al.*, 2009). The transcription complex required for IIV transcription, particularly in the cytoplasm, is unknown. One feature of the large subunit of viral RNApol is the absence of a region termed the carboxyl-terminal domain that is associated with most RNA polymerases. This domain has been linked to transcription termination and polyadenylation as well as to interactions with the cap-synthesizing complex. It is interesting to note that this correlates with the absence of polyadenylation on iridovirus transcripts. It is not known what viral factors are associated with host RNApol II for IE gene expression other than they are virion associated transcriptional transactivators and that these transcripts are also non-polyadenylated (Chinchar *et al.*, 2009). The lack of a polyA tail combined with the absence of a domain usually associated with 5′ cap synthesis suggests that viral factors may be involved in translation of viral proteins.

Detailed analysis of IIV-6 transcripts by northern blotting using plasmid clones of the genome identified 137 transcripts ranging from 0.2 to 8.3 kb, of which 38 were classified as IE, 34 were DE and 65 designated as L (D'Costa *et al.*, 2001). The IE genes could be detected as early as 0.5 hours post infection, DE transcripts from 3 hours and L transcripts from 6 hours. This analysis complements earlier studies using ^{35}S methionine labelling to separate IIV-6 proteins produced in *Choristoneura fumiferana* cells into early and late phases, with early proteins detected as soon as 1 hour post infection and later proteins not detected before 5 hours post infection (Barray and Devauchelle, 1985). Many genes are transcribed across multiple phases (D'Costa *et al.*, 2004) and it is not clear if the switch to late gene expression is limited purely to the cytoplasmic concatemeric DNA generated during the second stage of replication, thus linking subcellular replication sites with the temporal control of replication, or if transcription continues within the nucleus, which is likely to require the nuclear translocation of IE and/or DE viral transcription factors from the cytoplasm. In this context, it is interesting to note that *bro*-like genes (baculovirus repeated ORF) have been identified in IIV-6 and IIV-31, as well as a number of other members of the NCLDV clade (Bideshi *et al.*, 2003). The BRO proteins have now been identified as nucleocytoplasmic shuttling proteins that use the CRM1(exportin 1/Xpo1p)-mediated nuclear export pathway in nucleopolyhedroviruses (Kang *et al.*, 2006).

Approximately 90% of the genome of IIV-6 is transcriptionally active and while many IE and DE genes appeared to occur in four and three physical clusters, respectively, the late gene transcripts are scattered across the entire genome (D'Costa *et al.*, 2004). A similar clustering of gene classes was also suggested for IIV-9, although the clustering was much less apparent (McMillan *et al.*, 1990). There are clear clusters of early and late genes in vertebrate iridoviruses (Lua *et al.*, 2005). Of particular interest to the observation of gene clustering is the apparent lack of conservation of gene order between disparate iridoviruses, which raises the question of the importance of clustering and whether temporal clusters are retained even when gene order is not. Combined with the observation that many transcripts are very large and that co-terminal 5′ and 3′ transcripts can be identified during infection (D'Costa *et al.*, 2004), it is apparent that individual gene transcript analysis will provide valuable information on gene expression in these viruses.

Analysis of the early DNA polymerase (DNApol) and helicase genes and the late MCP gene was undertaken to identify promoter elements associated with different phases of the expression cascade in IIV-6 (Nalcacioglu *et al.*, 2003). As shown previously for FV-3, the promoters were only active in the presence of IIV-6 infection, presumably mediated by one or more proteins in the virus particle given that UV-irradiated virus can provide the necessary function (Cerutti *et al.*, 1989). A TATA-like element was identified, however, a subsequent study on the DNApol gene showed it to be a DE gene that was not produced in the presence of cycloheximide and an essential AAAAT promoter sequence was identified upstream of 20 DE genes and upstream of DNApol in eight iridoviruses for which sequence information was available, including IIV-3 (Nalcacioglu *et al.*, 2007). This provides the first information of a promoter sequence in IIV that may define a class of transcript. Intriguingly, the AAAAT promoter has also been identified in a single additional member of the NCLDV virus clade; namely the giant mimivirus where it appears as an octomer (AAAATTGA) identical to that of the IIV-6 DNApol promoter element, and is located 50 to 80 nucleotides upstream of almost half the mimivirus ORFs where it is believed to function as a signal for transcription of late-early genes by virus encoded transcription machinery present in the virion (Suhre *et al.*, 2005). How relevant the mimivirus model is to IIVs remains to be seen.

Recent studies on vertebrate iridoviruses using microarrays have shown similar results, with a classic temporal gene cascade, clusters of early and late genes from around the genome, the general separation of regulatory and nucleic acid metabolism associated genes occurring during the early phases and a link to DNA replication and virion proteins associated with late genes, together with a number of genes being expressed across the different temporal phases (Dang *et al.*, 2007; Lua *et al.*, 2005; Majji *et al.*, 2009; Teng *et al.*, 2008). In addition, the slow pace of IIV infection in insects compared to the rapid replication in cell culture has been echoed in the vertebrate viruses with a much slower progression of replication *in vivo* than that observed *in vitro* and distinct differences in gene expression in different cells and tissues (Dang *et al.*, 2007; Teng *et al.*, 2008).

Concomitant with viral macromolecular synthesis is the shutdown of host macromolecular synthesis (Cerutti and Devauchelle, 1980). There is a decrease in host RNA, DNA and protein synthesis and this effect is associated with the virion. UV-treated IIV-6 is able to inhibit host cell macromolecule production and this effect is also observed in cells treated with virion extracts. What factor(s) are causing these effects are unknown, although viral replication can continue, indicating a fundamental difference between viral and host cell macromolecule synthesis.

Apoptosis: induction and inhibition

The interactions between iridoviruses and the cells within which they replicate have received limited attention. An exception is in the induction and inhibition of apoptosis within infected cells. It is interesting to note that while the general principles of apoptosis are the same between vertebrates and invertebrates, they differ in a number of important details. This necessitates care when comparing vertebrate to invertebrate iridoviruses in the area of apoptosis. We start by considering apoptosis modulated by the vertebrate viruses and then signal how this process differs in IIV-infected cells.

Vertebrate iridoviruses and apoptosis

It is well known that infection by a number of vertebrate iridoviruses leads to apoptosis in cultured cells (Essbauer and Ahne, 2002; Chinchar *et al.*, 2003; Hu *et al.*, 2004; Imajoh *et al.*, 2004; Huang *et al.*, 2007; Lai *et al.*, 2008), following the well-characterized hallmarks of cellular apoptosis, including cell blebbing, chromatin condensation, DNA fragmentation, relocalization of phosphatidylserine to the cell surface and activation of caspases. The induction of apoptosis by FV-3 occurs following treatment with heat or UV-inactivated virus, indicating that gene expression is not required. This activation also occurs in BHK cells that do not support viral replication (Chinchar *et al.*, 2003).

The induction of apoptosis by other ranaviruses has been shown to involve mitochondrial morphological and distribution changes with a clear indication of a loss of mitochondrial membrane function as indicated by depolarization of membrane potential in the late stages of viral infection. The induction of apoptosis has been associated with an LPS-induced TNFα factor that has been linked to a collapse in mitochondrial membrane potential and consequent effects upon caspase-3 activity and Nf-κb dependant transcription (Huang *et al.*, 2008). Apoptosis induction has also been linked in part to the activation of a protein kinase in another fish iridovirus (Essbauer and Ahne, 2002). In all cases, central caspase-3 activity has been shown to increase during vertebrate iridovirus replication and where studied, the pancaspase inhibitor Z-VAD-FMK is able to block iridovirus-induced apoptosis, thus confirming the central role of caspase-dependant apoptosis in response to infection (Chinchar *et al.*, 2003; Imajoh *et al.*, 2004; Huang *et al.*, 2007, 2008; Chiou *et al.*, 2009).

The observation that high doses of virions induced rapid apoptosis whereas lower doses did not induce apoptosis until late in infection suggested the presence of an apoptosis inhibitor in vertebrate iridoviruses. The induction of apoptosis in fish cells exposed to UV light was rapidly inhibited by infection with a fish iridovirus (Lin *et al.*, 2008). Expression of the inhibitory factor occurred in the presence of the protein synthesis inhibitor cycloheximide and the DNA synthesis inhibitor aphidicolin, indicating that the anti-apoptotic protein was an IE gene. This was subsequently shown to be a virally encoded Bcl2-like protein that contained the signature BH1, 2, 3 and 4 motifs of this group of proteins. Furthermore, this protein was located in the virion, indicating that it is delivered as part of the infection process. Heat or UV-inactivation of the virus did not prevent the inhibition of apoptosis induced by UV-treatment of host cells (Lin *et al.*, 2008). Moreover, vertebrate iridoviruses encode initiation factor eIF2α homologues (Essbauer *et al.*, 2001) that are known to function as pseudo-substrate inhibitors of protein kinase R (PKR) in poxviruses. In the absence of the inhibitor, PKR phosphorylates host eIF2α leading to translational shut-off, but this mechanism has not been fully characterized in vertebrate iridoviruses and candidate eIF2α genes have not been identified in IIV-3 or IIV-6.

Overall then, studies on a variety of vertebrate iridoviruses have identified a combination of an antiapoptotic factor present in the virion, a Bcl2-like IE gene and a putative pseudosubstrate inhibitor of PKR. However, differences in the biology of vertebrate and invertebrate cells and the gene content of the viruses mean that some of the observations on vertebrate iridoviruses will not translate to IIVs, emphasizing that caution is required when extrapolating between these groups of viruses.

Invertebrate iridoviruses and apoptosis

Studies on apoptosis in IIVs are largely limited to IIV-6. Paul *et al.* (2007) investigated the observation that IIV-6 infections induce apoptosis in insect cells as indicated by cell blebbing and DNA fragmentation. Treatment of cells with a virion protein extract resulted in shutdown of host cell protein production and apoptosis, indicating that the inducing factor was associated with the virion. In addition, the use of the pancaspase inhibitor Z-VAD-FMK prevented the induction of apoptosis, indicating that a caspase dependant response was involved, as observed for vertebrate iridoviruses. However, this did not affect protein synthesis, suggesting that these two responses are not linked in IIV-6 infection.

In a follow up study, the delivery of a high dose of virus to cells was observed to trigger a massive apoptotic response; however, at lower doses the apoptotic response was inhibited, confirming that IIV-6 causes apoptosis, and that IIV-6 must possess an antiapoptotic factor that blocks apoptosis (Chitnis *et al.*, 2008). The entry of virions or virion proteins induced apoptosis and blocking this entry through the use of ammonium chloride to prevent endosome acidification, bafilomycin to block endocytosis, or attachment to polystyrene beads to prevent uptake, blocked apoptosis induction. Therefore, the uptake of virus or viral proteins is a critical step in induction of apoptosis by IIV-6. Furthermore, the use of cycloheximide to block *de novo* protein synthesis shows that protein expression is not required to induce apoptosis and hence the inducing factor must be part of the virion.

Of particular interest was the observation that the Janus Kinase (JNK) signalling pathway was activated by IIV-6 infection and that the SP600125 inhibitor of JNK blocked IIV-6 apoptosis induction; the first indication that JNK is involved in an insect virus infection. Consistent with this was the identification of a Dronc-like apical caspase designated Cf-caspase-I in *C. fumiferana* cells that was activated by the JNK pathway, confirming a caspase-dependant apoptotic response for an IIV. UV inactivated virus, protein extracts and cycloheximide strongly induced apoptosis. In contrast, IIV-6 infection showed muted apoptosis and the use of aphidicolin to block late gene expression indicated that the antiapoptotic factor was an early gene (Chitnis *et al.*, 2008).

The apoptosis inducing polypeptide has now been identified as a 49-kDa serine-threonine kinase with an ATP binding motif (Bilimoria, 2009). This is likely to be the same virion-associated kinase that was characterized by Monnier and Devauchelle (1980) from solubilized extracts of IIV-6. A 37-kDa fragment derived from this protein has been patented for its apoptotic and insecticidal properties. The insecticidal fragment has been named iridoptin; no doubt the details of the function of this protein will appear shortly in the scientific literature.

In a separate study, Ince *et al.* (2008) analysed the genome of IIV-6 and identified three open reading frames with homology to the baculovirus inhibitors of apoptosis (IAP). Of these, only ORF193R contained the RING finger and baculovirus inhibitor repeat (BIR) motifs found in functional IAPs. This was the only iridovirus protein identified with a BIR motif. ORF193R was demonstrated to be an IE gene that was capable of blocking actinomycin D-induced apoptosis and represents the first functional *iap* gene identified in an IIV.

IIV apoptosis induction and inhibition has many parallels to vertebrate iridovirus interactions with this pathway. One or more inducing factors are clearly associated with the particle and an IE gene(s) limits the apoptotic response during infection. However, the specific details appear to be different. Studies of more IIV apoptosis interactions will be required to determine if the observations in IIV-6 are universal, particularly given that IIV-6 and its close relatives appear to be phylogenetically separated from the largest complex of IIVs. However, apoptosis is one area where the IIV investigations are in many respects equal with, or leading, our overall understanding of these viruses.

IIV disease: signs and pathology

Signs of disease

The most evident symptom of patent infection is the iridescent hues, which arise from the quasicrystalline arrangement of virus particles in host cells or purified pellets of virus. Light reflected from the planes of regularly packed particles causes interference with light in similar phase that has been reflected from a neighbouring plane, a phenomenon known as Bragg diffraction, that results in a range of colours from lavender blue, to turquoise or green (Klug *et al.*, 1959). Iridescent colours are highly directional so that changes in viewing geometry can dramatically alter the hue and/or intensity of the observed colours (Doucet and Meadows, 2009). The larger particle size and greater interparticle spacing of viruses from mosquitoes results in iridescent colours of yellow-green, orange and red (Hemsley *et al.*, 1994), although there are a number of records of inconsistencies between observed iridescent colours and particle sizes that may be related to the degree of hydration of the infected cell, the length of capsid fibrils or some other optical property of the virus arrays (Williams, 2008). The biological function or adaptive significance of iridescence is not known and may be a trivial characteristic that arises from the physical structure and assembly characteristics of IIV particles (Stoltz, 1971; Kelly, 1985).

Natural patent infections are mainly observed in immature stages (larvae, nymphs), although adults are susceptible to lethal infection in terrestrial isopods and bees. Infection by an IIV is believed to be the causative agent of 'Clustering disease' in the Asian bee, *Apis cerana*. Diseased colonies become inactive, form small groups of flightless individuals that crawl on the ground, and the colony quickly perishes (Bailey

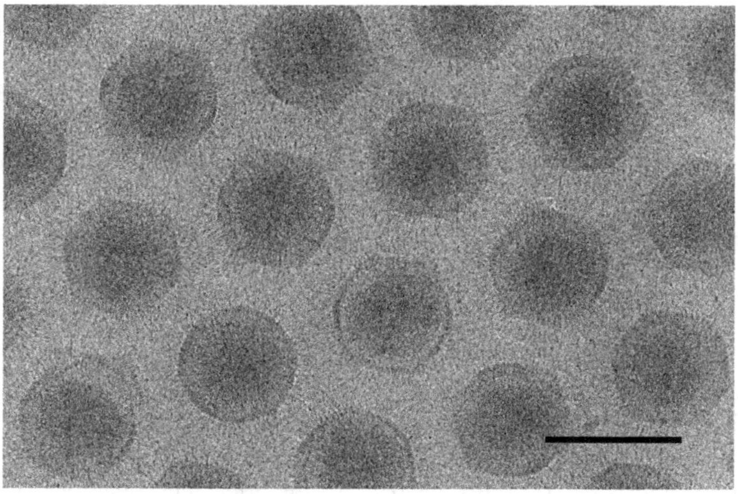

Figure 6.3 Photomicrograph of vitrified virions of IIV-6 showing external fringe of fibrils extending from the surface of the capsid (scale bar indicates 200 nm). Image courtesy of Xiaodong Yan and Tim Baker.

et al., 1976; Bailey and Ball, 1978; Singh, 1979; Mishra *et al.*, 1980). Reductions in fecundity and longevity have been reported in an aphid (*Toxoptera citricida*) that became patently infected after feeding on crude suspensions of IIV-6 (Hunter *et al.*, 2001b).

Sublethal (covert) infection is becoming increasingly recognized as a viral infection strategy with an important impact on the dynamics of certain insect populations (Sorrell *et al.*, 2009). Covert infections by IIVs have been observed in natural populations of the blackfly *Simulium variegatum* (Williams, 1993) and the mayfly *Ecdyonurus torrentis* (Tonka and Weiser, 2000), and laboratory populations of *Helicoverpa zea, Aedes aegypti* and *Galleria mellonella* (Sikorowski and Lawrence, 1997; Marina *et al.*, 1999; Constantino *et al.*, 2001). Covertly infected mayflies were observed to be infected by a low density of particles in the haemocytes and fat body with no obvious signs of disease. A large number of tubular structures typical of aberrant capsid formation were observed in cytoplasmic viral assembly sites (Tonka and Weiser, 2000), suggesting that *E. torrentis* may not be an optimal host for this virus.

In *Ae. aegypti*, adult mosquitoes that were covertly infected by IIV-6 in the larval stage suffered reduced body size (wing length of females) and reduced longevity with an early peak in the death rate and very few long-lived individuals. Covert infection was associated with a decrease in fecundity ranging from 22% on a single gonotrophic cycle to 36–39% on multiple (lifetime) gonotrophic cycles; the average number of female offspring produced by each mosquito (R_0) was reduced by 22–50%. No effects on egg fertility were detected (Marina *et al.*, 1999, 2003b).

The effects observed in mosquitoes may be due to the diversion of host resources to repair cells and tissues damaged by the pathological effects of sublethal disease, or the cost of mounting an immune response to control the infection. An alternative hypothesis, that sublethal effects were a consequence of a toxic virus protein associated with the internal lipid membrane (Cerutti and Devauchelle, 1980; Ohba *et al.*, 1990), was not supported in experimental studies; heat- or UV-inactivated IIV-6 did not significantly affect the reproductive capacity or longevity of mosquitoes and caused only minor changes in adult body size compared to controls (Marina *et al.*, 2003c).

Pathology

IIVs replicate extensively in the majority of host tissues, especially the fat body, haemocytes, and epidermis. Infection of the midgut epithelial cells is rarely observed and it seems that the midgut is an unlikely route of entry for these viruses, which may account for their low prevalence in most host populations. IIV-1 caused the formation of epidermal tumours in silkworm larvae (Hukuhara, 1964), but such pathology is not observed in other hosts. Patently infected individuals that survive to pupate often show deformations of the pupa (Smith et al., 1961; Carter, 1974; Stadelbacher et al., 1978).

IIVs can replicate *in vitro* in cell lines from many insects albeit with different degrees of efficiency (Cerutti et al., 1981; Charpentier et al., 1986; Constantino et al., 2001). Some insect cell lines appear not to support replication but suffer marked cytotoxic effects following inoculation (Cerutti and Devauchelle, 1979; Belloncik et al., 2007), that are related to the recognized toxic effects of virions and virion extracts (Ohba and Aizawa, 1979a; Cerutti and Devauchelle, 1982; Ohba et al., 1990). IIVs are also unusual in being capable of non-genetic reactivation. In this process, viral DNA alone is not infectious to cells but infection is achieved when cells are inoculated with mixtures of DNA and solubilized virion extract or UV-inactivated virus indicating that one or more inner membrane-associated components of the virion are required to initiate a productive infection (Cerutti et al., 1989).

Cytopathic effects associated with IIV infection include cell rounding and the appearance of extensive areas of finely granulated material devoid of cell organelles that are the cytoplasmic viral assembly sites. Large numbers of virions form quasicrystalline arrays in close association but not generally within this zone (Fig. 6.4). Marked cell contraction and detachment are also commonly observed. Blebbing of cell membranes, cell to cell fusion (syncytia) and related apoptotic changes are observed at high multiplicities of infection (Cerrutti and Devauchelle, 1979; Mathieson and Lee, 1981; Charpentier et al., 1986; Czuba et al., 1994). Changes in the position and morphology of mitochondria have been reported (Hess and Poinar, 1985), consistent with the aggresome-like nature of these cytoplasmic inclusions (Netherton et al., 2007).

Ecology

Host range

IIVs have been isolated from infected hosts on all continents except Antarctica. Examination of confirmed and suspected species records from

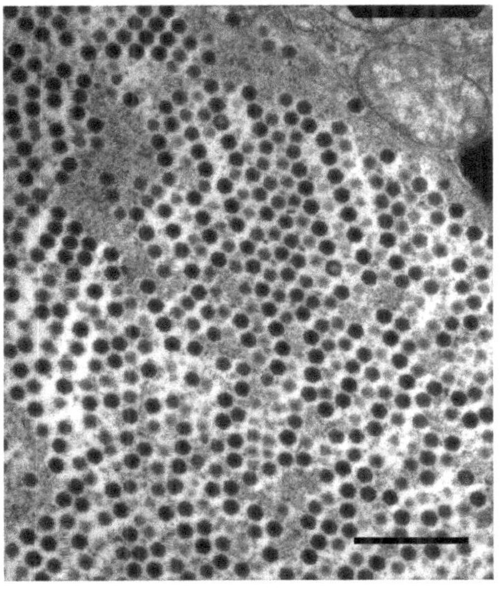

Figure 6.4 Quasicrystalline array of IIV-3 particles and granular material from the viral assembly site in infected cells of *Ochlerotatus taeniorhynchus* (scale bar indicates 1 μm). Image courtesy of James Becnel.

insects and isopods shows the greatest interest in these viruses in North America (41 species) and Europe (36 species) followed by Australia/Oceania (10 species), Africa (six species), South America (three species) and Asia (three species) (Williams, 2008). This represents a 50-fold difference between Europe and Asia in terms of host species per area and suggests that many host species remain to be discovered in Africa, South America and Asia. Indeed, this relationship may be set to change given the rapid growth in insect virology in China over the past decade.

IIVs have been reported to infect 99 species of terrestrial and aquatic invertebrates (Hall, 1985; Williams, 2008), although a number of these records are based on the characteristic signs of patent infection rather than the observation of virus particles or molecular evidence. Of these, 70 are insect species and 18 are species of terrestrial isopods commonly known as woodlice, sowbugs or pillbugs. The insect records are dominated by Diptera with aquatic immature stages (44 species), followed by Coleoptera (nine species, of which eight have soil-dwelling larvae), Lepidoptera (nine species), Orthoptera (four species), and four species from other orders. There are also nine species records from invertebrates such as daphnids, molluscs, crabs, and shrimp among others, in freshwater and marine habitats for which the relationships with IIVs are uncertain. A record from the marine mollusc *Nautilus* sp. suggested that this isolate was more closely related to vertebrate ranaviruses than to IIVs (Gregory *et al.*, 2006), whereas an iridovirus isolated from the sergestid shrimp *Acetes erythraeus* showed ~80% similarity in the MCP sequence to those of IIVs (Tang *et al.*, 2007). As mentioned at the beginning of this chapter, perhaps the principal feature that unites the large majority of these host species is the damp or aquatic nature of the habitats they occupy during part, or all, of their life cycle.

The host range of many IIVs in the laboratory depends on the route of inoculation. In general, the range of hosts infected by peroral inoculation is markedly more restricted than the host range defined by studies in which the inoculum is injected. For instance, the best studied example is that of IIV-6, that can infect numerous (>100) insect species of agricultural and medical importance including Lepidoptera, Coleoptera, Hymenoptera, Hemiptera, Homoptera, Diptera and Orthoptera (Mitsuhashi, 1967; Hama, 1968; Fukuda, 1971; Jensen *et al.*, 1972; McLaughlin *et al.*, 1972; Ohba, 1975; Kleespies *et al.*, 1999; Henderson *et al.*, 2001; Hunter *et al.*, 2001b) and non-insect species such as terrestrial isopods and a centipede (Ohba and Aizawa, 1979b). A strain of IIV-6 was also observed to cause lethal infection of hymenopteran endo- and ectoparasitoids that developed on infected lepidopteran larvae (López *et al.*, 2002).

The majority of these viruses can be easily grown in very large quantities by intrahaemocoelic inoculation of wax moth larvae, *G. mellonella*, which is the standard laboratory host. However, a few IIVs do not replicate in this host, including the mosquito virus IIV-3, IIV-16 from *Costelytra zealandica* and the bee virus, IIV-24. A few studies have attempted to determine host range given asymptomatic infection. Ohba (1975) observed an increase in the titre of IIV-6 in insects that showed no signs of patent disease. Ward and Kalmakoff (1991) reported that dot-blot DNA hybridization could be used to detect replication of IIV-9 in heterologous host species prior to, or in the absence of iridescence of host tissues. Similarly, an indirect fluorescent antibody technique was employed to demonstrate the presence of IIV-22 antigen in the tissues of mosquitoes, sandflies and a triatomine bug, with the quantity of antigen proportional to the interval between inoculation and testing (Tesh and Andreadis, 1992). As is the case with most insect viruses, it is now widely accepted that the appraisal of host range requires diagnosis of infection using techniques other than observation of the characteristic signs of patent disease.

A steadily growing body of evidence indicates that some IIVs naturally infect multiple species of hosts, sometimes across different orders or even different classes. Three of the best examples are as follows. First, strains of IIV-9, originally isolated from soil-dwelling larvae of *Wiseana cervinata* (Lepidoptera) (Fowler and Robertson, 1972; Ward and Kalmakoff, 1987) were also found infecting sympatric larvae of *Eudonia* (*Witlesia*) *sabulosella* (Lepidoptera) and *Opogonia* sp. (Coleoptera) in New Zealand (Williams and Cory, 1994; Webby and Kalmakoff, 1998). Second, strains of IIV-6 have been reported naturally infecting lepidopteran larvae of *C. suppressalis* and *S. frugiperda* (Williams and Hernández, 2006) and commercial colonies of the crickets *Gryllus bimaculatus, G. campestris* and *Acheta domesticus* (Kleespies *et al.*, 1999; Just and Essbauer, 2001). A record of IIV-6 naturally infecting whiteflies (Homoptera) requires confirmation (Hunter *et al.*, 2001a). Lastly, the woodlouse virus IIV-31 naturally infects at least four and probably more species of terrestrial isopod (Crustacea), a mermithid nematode (Hess and Poinar, 1985) and a closely related strain has been isolated from larvae of the soil-dwelling beetle *Popillia japonica,* possibly via the consumption of infected isopods by the beetle (Lacey and Adams, 1994; Webby and Kalmakoff, 1998).

Single host species may also be infected by different IIVs in different locations or at different times. A population of blackfly larvae (*S. variegatum*) was infected by three different IIVs in different months and at different sites along the River Ystwyth in Wales (U.K.), which lead to speculation concerning the potential role of alternative hosts in the lotic environment for the survival and transmission of each of these viruses (Williams, 1995).

Finally, IIV-6 can productively infect reptile cells *in vitro* (McIntosh and Kimura, 1974) and there is evidence that it may be able to infect reptiles (Just *et al.*, 2001), but not amphibians (Ohba and Aizawa, 1982). A chameleon believed to have consumed infected crickets died and post-mortem tissue samples were used to inoculate two reptilian cell lines that subsequently developed cytopathic effects. Injection of the cell culture supernatant resulted in patent disease in crickets, and the sequence of a PCR product from inoculated crickets was identical to the IIV-6 strain sequence reported from commercially reared crickets (Weinmann *et al.*, 2007). However, it remains to be seen whether the virus is capable of replication in the chameleon host and whether transmission occurs in natural reptile populations that prey on infected insects.

Transmission

In general, the mechanisms and route of transmission of these viruses are poorly understood. As stated previously, transmission is highly dependent on the route of infection because of the marked differences in IIV infectivity by ingestion compared to intrahaemocoelic injection. The reasons for the low *per os* infectivity may reside in the virion's susceptibility to host digestive enzymes. EM studies have shown that virions of IIV-3 are degraded shortly after entering the mosquito larval midgut (Stoltz and Summers, 1971), so that infection of midgut cells may only be possible at the anterior section of the midgut where the newly formed peritrophic matrix is relatively thin or fragile (Hall and Anthony, 1971; Hembree and Anthony, 1980), although this has yet to be demonstrated.

Mixing virus inoculum with silicon carbide fibres or ground sand has been shown to result in significant increases in the prevalence of infection of mosquito larvae, presumably due to physical damage to the peritrophic matrix by these substances that facilitated access to the midgut cells by virions (Undeen and Fukuda, 1994; Marina *et al.*, 1999). In contrast, mixtures of virus with a stilbene derived optical brightener, known to degrade the chitin structure of the peritrophic matrix, or small silicon carbide grains, did not result in a significant increase in infection of mosquito larvae (Marina *et al.*, 2003a).

It is clear that cannibalism or predation of infected individuals, involving the consumption of very large quantities of virus, is a viable route of transmission in isopods (Federici, 1984), tipulid larvae (Carter, 1973b), caterpillars (Williams and Hernández, 2006), mosquitoes (Linley and Nielsen, 1968a), and crickets (Fowler, 1989; Kleespies *et al.*, 1999). Similarly, density-modulated aggressive interactions, both intra and inter-specific, influence transmission rates in isopods (Grosholz, 1992, 1993) and mosquitoes (Marina *et al.*, 2005), presumably as wounding compromises the integrity of the cuticle and allows the entry of virions into the injured host. Rainfall and soil moisture positively influence the prevalence of infection in both isopods and tipulid larvae (Ricou, 1975; Grosholz, 1993). Few studies have considered the efficacy of transmission from alternative sources of IIV inoculum in the environment. Patently infected insects disseminate viable virions in their faeces but at concentrations insufficient to result in *per os* transmission (Carter, 1973a; Ohba, 1975; Williams and Hernandez, 2006).

Diseased insects are reported to be more sluggish than healthy conspecifics but may continue to live for several weeks or months (Hama, 1968; Linley and Nielsen, 1968a; Sikorowski and Tyson, 1984; Poprawski and Yule, 1990). This sluggish response means that infected hosts are more prone to become the victims of cannibals than their healthy counterparts (Kleespies *et al.*, 1999; Williams and Hernández, 2006). In addition, infected soil-dwelling larvae have been reported to move upwards, close to the soil surface prior to death, where the probability of transmission by cannibalism or predation may be enhanced (Fowler and Robertson, 1972).

IIV-3 survives in populations of the mosquito *Ochlerotatus taeniorhynchus* by alternating cycles of horizontal and vertical transmission. Infected larvae tend to die in the final instar and are cannibalized by early and late instar conspecifics. Individuals that become infected in the early instars develop patent disease and die prior to pupation whereas individuals that acquire infection in the late instars pupate and emerge as adults. A significant proportion of the adult females transmit the infection to their progeny. These progeny develop normally until the third or fourth instar when they die of patent disease, leading to a further cycle of transmission (Linley and Nielsen, 1968a,b; Woodard and Chapman, 1968; Fukuda and Clark, 1975). Serial passage and prolonged exposure experiments have indicated that the dynamics of the virus–host relationship are highly stable in laboratory systems (Woodard and Chapman, 1968; Undeen and Fukuda, 1994). Both the transmission and replication of IIV-3 were reported to be enhanced by the presence of a picornavirus that persisted as a low level inapparent infection in the laboratory mosquito colony (Wagner *et al.*, 1974). The advent of molecular techniques would now permit a thorough dissection of the transmission cycles reported in this pathosystem.

Persistence

The persistence of IIVs in the environment is not well understood. IIVs are highly sensitive to desiccation with complete loss of infectivity in dry habitats in periods of less than 24 hours (Linley and Nielsen, 1968b; Reyes *et al.*, 2004), or a few days (McLaughlin *et al.*, 1972; Grosholz, 1993). In contrast, IIVs gradually lose infectivity over periods of weeks or months when incubated in water in the laboratory (Day and Gilbert, 1967; Mullens *et al.*, 1999), or moist soil or aquatic environments (Marina *et al.*, 2000; Reyes *et al.*, 2004). IIV-6 binds strongly to certain clay minerals found in soils but not to the degree observed in the occlusion bodies of nucleopolyhedroviruses (Christian *et al.*, 2006). Virus persistence is reduced in the presence of leaf exudates on cotton (McLaughlin *et al.*, 1972), the presence of soil microbiota (Reyes *et al.*, 2004), and rapidly lost following exposure to sunlight (Linley and Nielsen, 1968b; Hernández *et al.*, 2005), or elevated temperatures (Day and Mercer, 1964; Marina *et al.*, 2000).

In addition to the persistence of IIV-3 in populations of *O. taeniorhynchus*, described above, there is steadily growing evidence that the infection strategies adopted by IIVs vary with the host-virus system. The great majority of infections of isopods and *S. frugiperda* larvae were patent and lethal (Grosholz, 1992; López *et al.*, 2002), whereas covert sublethal infections were common in laboratory studies on *Ae. aegypti* and *G. mellonella* and *H. zea* (Sikorowski and Lawrence, 1997; Marina *et al.*, 1999; Constantino *et al.*, 2001). Covert infections have also been detected in natural populations of the blackfly, *S. variegatum* and the mayfly, *E. torrentis* (Williams, 1993; Tonka and Weiser, 2000). In the case of the blackfly, the prevalence of covert infection was quantified by bioassay of larval homogenates in *G. mellonella* and confirmed by a nested PCR targeted at the MCP gene. Covert infection was highest in the spring and autumn and preceded that appearance of patent infections in the larval blackfly population (Williams, 1993, 1995). Sublethal infections would allow the virus to exploit vertical routes of transmission to host offspring, which appears common in baculoviruses (Vilaplana *et al.*, 2008), but this has not been studied in IIVs, except for IIV-3 and some evidence from aphids and weevils infected by IIV-6 in the laboratory (Hunter *et al.*, 2001b, 2003).

Dispersal

The mechanisms by which IIVs achieve dispersal and factors that determine the rate of that dispersal are largely speculative. It is likely that the flights of sublethally infected adult insects in search of oviposition sites are a major route of virus dispersal to new habitats, whereas infected aquatic insects in rivers may disperse downstream by drifting. The incidence of disease in terrestrial isopods was highest during the wetter months of the year and was positively correlated with their tendency to disperse and negatively correlated with the distance between suitable patches of habitat (Grosholz, 1993). Endoparasitic wasps were capable of disseminating virus between *S. frugiperda* larvae on adjacent plants in field cages (López *et al.*, 2002), and potentially could disperse the virus over much larger distances in their search for suitable hosts.

Genetic diversity in IIV populations

The factors that generate and maintain genetic diversity in insect virus populations, particularly in baculoviruses, have become the subject of considerable interest over the past decade, not simply due to an increasing awareness of the prevalence and magnitude of this diversity (Graham *et al.*, 2004; Cory *et al.*, 2005), but also because of the implications that this has on virus phenotype (Hodgson *et al.*, 2004), interactions between different genotypes (López-Ferber *et al.*, 2003; Simón *et al.*, 2008), and the efficacy of virus-based insecticides (Muñoz and Caballero, 2000). In the context of IIVs, high levels of genetic variation have been observed in strains isolated from blackflies and *S. frugiperda* larvae. In blackflies collected from a river in Wales, each host larva appeared to be infected by a slightly different strain of IIV-22, as revealed by restriction endonuclease analyses (Williams and Cory, 1993; Williams, 1995). Similarly, almost all the diseased fall armyworm larvae collected from different maize plants in a field in southern Mexico were found to be infected with slightly different strains of an IIV-6 like virus, while one strain was found to be present in several different individuals (N. Hernández and T. Williams, unpublished data). The mechanisms generating this variation and the selective advantages that may, or may not, accompany it have not been the subject of study, although similar heterogeneity has been reported in iridovirus pathogens of fish (Goldberg *et al.*, 2003), suggesting that variation may be an intrinsic component in the replication or infection strategies of this family of viruses.

Concluding remarks

Iridoviruses have provided many surprises over the years, starting with the observation of a cytoplasmic virus that caused iridescence in *Tipula* larvae (Xeros, 1954) and the subsequent observations on massive replication and the Bragg diffraction of light by viral arrays (Williams and Smith, 1957). The circular permutation and terminal redundancy of the genome was also a surprise in an animal virus (Murti *et al.*, 1985). Now, advances in nanotechnology are opening new approaches to the study of these unusual viruses. Of particular interest is how the biophysical properties of the virus, such as iridescence, can be used to develop new biomaterials. Recent studies have shown that the regular and repetitive structures of virus particle biocolloids offer advantages over other colloidal materials, with viruses used to construct multilayered polymers (Yoo *et al.*, 2006), or electronic materials (Tseng *et al.*, 2006).

IIVs offer a number of properties for biomaterial construction. The 140–180 nm iridovirus virion is large for a virus and it is this size, combined with a regular array of particles providing spaced refractive centres, which brings Bragg refraction of light into the visible spectrum. This opens the possibility of developing novel photonic materials from IIVs. For example, IIV-9 has been assembled into crystals with defined optical properties that are determined by the hydration state of the crystal affecting interparticle spacing (Juhl *et al.*, 2006). Furthermore, these particles have a range of chemical functionalities on their surface from the constituent amino acids and have been used as biotemplates to build complete metal nanoshells with defined sizes (Radloff *et al.*, 2005). Biophysical analysis of IIV virions using Brillouin light scattering has shown that the virions are surprisingly hard with the DNA core dominating the rigidity, reflecting a high compression state of the genome within the core (Hartschuh *et al.*, 2008). These studies indicate that IIVs have potential as alternatives to classic colloidal materials; they have low surface charges and multifunctionality that can be exploited in materials synthesis and prospective new technological applications.

The sequencing of additional IIV genomes will provide important information on the relationships among these viruses and to other virus families. At present, the analysis of the IIV-3 and IIV-6 genomes has supported the concept that these are very different viruses from different genera, yet MCP sequence analysis suggests that IIV-3 may be more similar to other members of the *Iridovirus* genus than IIV-6. The upcoming analysis of IIV-9 should help resolve this issue. Moreover, additional genome sequencing is urgently needed to support virus species definitions in the invertebrate genera.

The majority of IIV genes are of unknown function and even those that show domain homology to known proteins are largely based on theoretical activities with undefined roles in virus replication. Analysis of the viral proteome and its interaction with the host is an area that deserves considerable attention. Molecular tools for investigative genome manipulation are now widely employed and the application of a genetic modification system and the use of antisense or RNAi technologies offer a wealth of possibilities to improve our understanding of IIV infection and replication strategies.

IIVs naturally infect major insect pests and medically important vectors, but are not considered useful as biocontrol agents due to the often low prevalence of patent infections and their broad host range. However, IIVs have the potential to generate considerable scientific interest due to their potential biotechnological applications (Bilimoria, 2009), and their impact as pathogens of economically important invertebrates such as mopane worms (*Gonimbrasia belina*, Lepidoptera) that are a major food industry in southern Africa, and possibly oyster populations that periodically suffer putative iridovirus disease, although the proximity of the relationship between the iridoviruses of molluscs and the IIVs is unknown. Despite the enthusiasm of a small number of researchers, major advances in the study of these viruses will require a fundamental shift in our perception of the importance of IIVs as pathogens, or their potential applications in biotechnology or materials science.

Acknowledgements

We thank Xiaodong Yan and Tim Baker (University of California, San Diego) for their kind permission to use reconstruction images of IIV-6, and James Becnel (USDA-ARS, Gainesville, FL) for images of IIV-3.

References

Andrés, G., García-Escudero, R., Simón-Mateo, C., and Viñuela, E. (1998). African swine fever virus is enveloped by a two-membraned collapsed cisterna derived from the endoplasmic reticulum. J. Virol. 72, 8988–9001.

Anthony, D.W., and Comps, M. (1991). Iridoviridae. In Atlas of Invertebrate Viruses, J.R. Adams and J.R. Bonami eds. (Boca Raton, Florida: CRC Press), pp. 55–86.

Bailey, L., and Ball, B.V. (1978). *Apis* iridescent virus and 'clustering disease' of *Apis cerana*. J. Invertebr. Pathol. *31*, 368–371.

Bailey, L., Ball, B.V., and Woods, R.D. (1976). An iridovirus from bees. J. Gen. Virol. *31*, 459–461.

Balange-Orange, N., and Devauchelle, G. (1982). Lipid composition of an iridescent virus type 6 (CIV). Arch. Virol. *73*, 363–367.

Barray, S., and Devauchelle, G. (1979). Etude des polypeptides de structure du virus iridescent de *Chilo suppressalis* (iridovirus type 6). Can. J. Microbiol. *25*, 841–849.

Barray, S., and Devauchelle, G. (1985). Protein synthesis in cells infected by *Chilo* iridescent virus (iridovirus, type 6). Arch. Virol. *86*, 315–326.

Belloncik, S., Petcharawan, O., Couillard, M., Charpentier, G., Larue, B., Guardado, H., Chareonsak, S., and Imanishi, S. (2007). Development and characterization of a continuous cell line, AFKM-On-H, from hemocytes of the European corn borer *Ostrinia nubilalis* (Hübner) (Lepidoptera, Pyralidae). *In Vitro* Cell. Dev.-An. *43*, 245–254.

Bideshi, D.K., Renault, S., Stasiak, K., Federici, B.A., and Bigot, Y. (2003). Phylogenetic analysis and possible function of *bro*-like genes, a multigene family widespread among large double-stranded DNA viruses of invertebrates and bacteria. J. Gen. Virol. *84*, 2531–2544.

Bigot, Y., Renault, S., Nicolas, J., Moundras, C., Demattei, M.V., Samain, S., Bideshi, D.K., and Federici, B.A. (2009). Symbiotic virus at the evolutionary intersection of three types of large DNA viruses; iridoviruses, ascoviruses, and ichnoviruses. PLoS One *4*, e6397.

Bigot, Y., Stasiak, K., Rouleux-Bonnin, F., and Federici, B.A. (2000). Characterization of repetitive DNA regions and methylated DNA in ascovirus genomes. J. Gen. Virol. *81*, 3073–3082.

Bilimoria, S.L. (2009). Use of iridoptin to induce toxicity in insects. US Patent Application number US2009/0069239 A1.

Black, P.N., Blair, C.D., Butcher, A., Capinera, J.L., and Happ, G.M. (1981). Biochemistry and ultrastructure of iridescent virus type 29. J. Invertebr. Pathol. *38*, 12–21.

Buchatsky, L.P. (1977). An iridovirus from larvae of *Culiseta annulata* and *Culex territans*. Acta Virol. *21*, 85–86.

Carter, J.B. (1973a). The mode of transmission of *Tipula* iridescent virus. II Source of infection J. Invertebr. Pathol. *21*, 123–130.

Carter, J.B. (1973b). The mode of transmission of *Tipula* iridescent virus. II Route of infection. J. Invertebr. Pathol. *21*, 136–143.

Carter, J.B. (1974). *Tipula* iridescent virus infection in the developmental stages of *Tipula oleracea*. J. Invertebr. Pathol. *24*, 271–281.

Cerutti, M., and Devauchelle, G. (1979). Cell fusion induced by an invertebrate virus. Arch. Virol. *61*, 149–155.

Cerutti, M., and Devauchelle, G. (1980). Inhibition of macromolecular synthesis in cells infected with an invertebrate virus (iridovirus type 6 or CIV). Arch. Virol. *63*, 297–303.

Cerutti, M., and Devauchelle, G. (1982). Isolation and reconstitution of *Chilo* iridescent virus membrane. Arch. Virol. *74*, 145–155.

Cerutti, M., and Devauchelle, G. (1985). Characterisation and localisation of CIV polypeptides. Virology *145*, 123–131.

Cerutti, M., and Devauchelle, G. (1990). Protein composition of *Chilo* iridescent virus. In Molecular Biology of Iridoviruses, G. Darai, ed. (Boston: Kluwer), pp 81–112.

Cerutti, M., Cerutti, P., and Devauchelle, G. (1989). Infectivity of vesicles prepared from *Chilo* iridescent virus inner membrane: evidence for recombination between associated DNA fragments. Virus Res. *12*, 299–313.

Cerutti, M., Guerillon, J., Arella, M., and Devauchelle, G. (1981). La réplication de l'iridovirus de type 6 (CIV) dans différentes lignées cellulaires. C.R. Acad. Sci. *292*, 797–802.

Charpentier, G., Bordeleau-Cloutier, P., Massicotte-Roussea, C., Guerillon, J., and Devauchelle, G. (1986). *Chilo suppressalis* iridescent virus induced antigens in insect cells and tissues: localization and kinetics of appearance. Ann. Inst. Pasteur Virol. *137*, 37–49.

Chen, M., Goorha, R., and Murti, K.G. (1986). Interaction of frog virus 3 with the cytomatrix. IV. Phosphorylation of vimentin precedes the reorganization of intermediate filaments around the virus assembly sites. J. Gen. Virol. *67*, 915–922.

Cheng, X.W., Carner, G.R., and Brown, T.M. (1999). Circular configuration of the genome of ascoviruses. J. Gen. Virol. *80*, 1537–1540.

Cherrier, M.V., Kostyuchenko, V.A., Xiao, C., Bowman, V.D., Battisti, A.J., Yan, X., Chipman, P.R., Baker, T.S., Van Etten, J.L., and Rossmann, M.G. (2009). An icosahedral algal virus has a complex unique vertex decorated by a spike. Proc. Natl. Acad. Sci. U.S.A. *106*, 11085–11089.

Chinchar, V.G., and Granoff, A. (1986). Temperature-sensitive mutants of frog virus 3: biochemical and genetic characterization. J. Virol. *58*, 192–202.

Chinchar, V.G., Bryan, L., Wang, J., Long, S., and Chinchar, G.D. (2003). Induction of apoptosis in frog virus 3-infected cells. Virology *306*, 303–312.

Chinchar, V.G., Essbauer, S., He, J.G., Hyatt, A, Miyazaki, T., Seligy, V., and Williams, T. (2005). *Iridoviridae*. In Virus Taxonomy: Eighth Report of the International Committee on the Taxonomy of Viruses, C.M. Fauquet, M.A. Mayo, J. Maniloff, U. Desselberger, and L.A. Ball eds. (Elsevier, London), pp. 163–175.

Chinchar, V.G., Goorha, R., and Granoff, A. (1984). Early proteins are required for the formation of frog virus 3 assembly sites. Virology *135*, 148–156.

Chinchar, V.G., Hyatt, A., Miyazaki, T., and Williams, T. (2009). Family *Iridoviridae*: poor viral relations no longer. Curr. Top. Microbiol. Immunol. *328*, 123–170.

Chiou, P.P., Chen, Y.C., and Lai, Y.S. (2009). Caspase-dependent induction of apoptosis in barramundi, *Lates calcarifer* (Bloch), muscle cells by grouper iridovirus. J. Fish Dis. doi: 10.1111/j.1365-2761.2009.01083.x

Chitnis, N.S., D'Costa, S.M., Paul, E.R., and Bilimoria, S.L. (2008). Modulation of iridovirus-induced apoptosis by endocytosis, early expression, JNK, and apical caspase. Virology *370*, 333–342.

Christian, P.D., Richards, A.R., and Williams, T. (2006). Differential adsorption of occluded and non-occluded insect pathogenic viruses to soil forming minerals. Appl. Environ. Microbiol. *72*, 4648–4652.

Claverie, J.M., Abergel, C., and Ogata, H. (2009). Mimivirus. Curr. Top. Microbiol. Immunol. *328*, 89–121.

Cole, A., and Morris, T.J. (1980). A new iridovirus of two species of terrestrial isopods, *Armadillidium vulgare* and *Porcellio scaber*. Intervirology *14*, 21–30.

Constantino, M., Christian, P., Marina, C. F., and Williams, T. (2001). A comparison of techniques for detecting Invertebrate iridescent virus 6. J. Virol. Meth. *98*, 109–118.

Cory, J.S., Green, B.M., Paul, R.K., and Hunter-Fujita, F. (2005). Genotypic and phenotypic diversity of a baculovirus population within an individual insect host. J. Invertebr. Pathol. *89*, 101–111.

Cullen, B.R. (2009). Viral and cellular messenger RNA targets of viral microRNAs. Nature *457*, 421–425.

Czuba, M., Tajbakhsh, S., Walker, T., Dove, M.J., Johnson, B.F., and Seligy, V.L. (1994). Plaque assay and replication of *Tipula* iridescent virus in *Spodoptera frugiperda* ovarian cells. Res. Virol. *145*, 319–330.

Dang, T.L., Yasuike, M., Hirono, I., Kondo, H., and Aoki, T. (2007). Transcriptional profile of red seabream iridovirus in a fish model as revealed by viral DNA microarrays. Virus Genes *35*, 449–461.

Darai, G. (1990). Molecular Biology of Iridoviruses. (Boston: Kluwer).

Day, M.F., and Gilbert, N. (1967). The number of particles of *Sericesthis* iridescent virus required to produce infections of *Galleria* larvae. Aust. J. Biol. Sci. *20*, 691–693.

Day, M.F., and Mercer, E.H. (1964). Properties of an iridescent virus from the beetle *Sericesthis pruinosa*. Aust. J. Biol. Sci. *17*, 892–902.

D'Costa, S.M., Yao, H., and Bilimoria, S.L. (2001). Transcription and temporal cascade in *Chilo* iridescent virus infected cells. Arch. Virol. *146*, 2165–2178.

D'Costa, S.M., Yao, H.J., and Bilimoria, S.L. (2004). Transcriptional mapping in *Chilo* iridescent virus infections. Arch. Virol. *149*, 723–742.

Delhon, G., Tulman, E.R., Afonso, C.L., Lu, Z., Becnel, J.J., Moser, B.A., Kutish, G.F., and Rock, D.L. (2006). Genome of Invertebrate iridescent virus type 3 (mosquito iridescent virus). J. Virol. *80*, 8439–8449.

Delius, H., Darai, G., and Flugel, R.M. (1984). DNA analysis of Insect iridescent virus 6: evidence for circular permutation and terminal redundancy. J. Virol. *49*, 609–614.

Devauchelle, G., Attias, J., Monnier, C., Barray, S., Cerutti, M., Guerillon, J., and Orange-Balange, N. (1985a). *Chilo* iridescent virus. Curr. Top. Microbiol. Immunol. *116*, 37–48.

Devauchelle, G., Stoltz, D.B., and Darcy-Tripier, F. (1985b). Comparative ultrastructure of Iridoviridae. Curr. Top. Microbiol. Immunol. *116*, 1–21.

Do, J.W., Moon, C.H., Kim, H.J., Ko, M.S., Kim, S.B., Son, J.H., Kim, J.S., An, E.J., Kim, M.K., Lee, S.K., Han, M.S., Cha, S.J., Park, M.S., Park, M.A., Kim, Y.C., Kim, J.W., and Park, J.W. (2004). Complete genomic DNA sequence of rock bream iridovirus. Virology *325*, 351–363.

Doucet, S.M., and Meadows, M.G. (2009). Iridescence: a functional perspective. J. R. Soc. Interface *6*, S115-S132. doi: 10.1098/rsif.2008.0395.focus

Eaton, H.E., Metcalf, J., Penny, E., Tcherepanov, V., Upton, C., and Brunetti, C.R. (2007). Comparative genomic analysis of the family Iridoviridae: re-annotating and defining the core set of iridovirus genes. Virol. J. *4*, 11.

Elliott, R.M., and Kelly, D.C. (1980). Frog Virus 3 replication: induction and intracellular distribution of polypeptides in infected cells. J. Virol. 33, 28–51.

Essbauer, S., and Ahne, W. (2002). The epizootic haematopoietic necrosis virus (Iridoviridae) induces apoptosis *in vitro*. J. Vet. Med. B *49*, 25–30.

Essbauer, S., Bremont, M., and Ahne, W. (2001). Comparison of the eIF-2 alpha homologous proteins of seven ranaviruses (Iridoviridae). Virus Genes *23*, 347–359.

Federici, B.A. (1980). Isolation of an iridovirus from two terrestrial isopods, the pill bug, *Armadillidium vulgare*, and the sow bug, *Porcellio dilatatus*. J. Invertebr. Pathol. *36*, 373–381.

Federici, B.A. (1984). Diseases of terrestrial isopods. Symp. Zool. Soc. Lond. 53, 233–245.

Federici, B.A., and Bigot, Y. (2003). Origin and evolution of polydnaviruses by symbiogenesis of insect DNA viruses in endoparasitic wasps. J. Insect Physiol. *49*, 419–432.

Filée, J., Pouget, N., and Chandler, M. (2008). Phylogenetic evidence for extensive lateral acquisition of cellular genes by nucleocytoplasmic large DNA viruses. BMC Evol. Biol. *8*, 320.

Fischer, M., Schnitzler, P., Scholz, J.R., Rosen-Wolff, A., Delius, H., and Darai, G. (1988). DNA nucleotide sequence analysis of the *Pvu*II DNA fragment L of the genome of insect iridescent virus type 6 reveals a complex cluster of multiple tandem, overlapping, and interdigitated repetitive DNA elements. Virology *167*, 497–506.

Fowler, H.G. (1989). An epizootic iridovirus of Orthoptera (Gryllotalpidae: *Scaptericus borellii*) and its pathogenicity to termites (Isoptera: Cryptotermes). Rev. Microbiol. São Paulo *20*, 115–120.

Fowler, M., and Robertson, J.S., (1972). Iridescent virus infection in field populations of *Wiseana cervinata* (Lepidoptera: Hepialidae) and *Witlesia* sp. (Lepidoptera: Pyralidae) in New Zealand. J. Invertebr. Pathol. *19*, 154–155.

Fukaya, M., and Nasu, S. (1966). A *Chilo* iridescent virus from the rice stem borer, *Chilo suppressalis* Walker (Lepidoptera: Pyrallidae). Appl. Entomol. Zool. *1*, 69–72.

Fukuda, T. (1971). Per os transmission of *Chilo* iridescent virus to mosquitoes. J. Invertebr. Pathol. *18*, 152–153.

Fukuda, T., and Clark, T.B. (1975). Transmission of the mosquito iridescent virus (RMIV) by adult mosquitoes of *Aedes taeniorhynchus* to their progeny. J. Invertebr. Pathol. *25*, 275–276.

Goldberg, T.L., Coleman, D.A., Grant, E.C., Inendino, K.R., and Philipp, D.P. (2003). Strain variation in an emerging iridovirus of warm-water fishes. J. Virol. *77*, 8812–8818.

Goodwin, T.J., Butler, M.I., and Poulter, R.T. (2006). Multiple, non-allelic, intein-coding sequences in eukaryotic RNA polymerase genes. BMC Biol. *4*, 38.

Goorha, R. (1981). Frog virus 3 requires RNA polymerase II for its replication. J. Virol. *37*, 496–499.

Goorha, R. (1982). Frog virus 3 DNA replication occurs in two stages. J. Virol. *43*, 519–528.

Goorha, R., and Murti, K.G. (1982). The genome of frog virus 3, an animal DNA virus, is circularly permuted and terminally redundant. Proc. Natl. Acad. Sci. U.S.A. *79*, 248–252.

Goorha, R., Murti, G., Granoff, A., and Tirey, R. (1978). Macromolecular synthesis in cells infected by frog virus 3. VIII. The nucleus is a site of frog virus 3 DNA and RNA synthesis. Virology *84*, 32–50.

Goorha, R., Willis, D.B., and Granoff, A. (1977). Macromolecular synthesis in cells infected by frog virus 3. VI. Frog virus 3 replication is dependent on the cell nucleus. J. Virol. *21*, 802–805.

Graham, R.I., Tyne, W.I., Possee, R.D., Sait, S.M., and Hails, R.S. (2004). Genetically variable nucleopolyhedroviruses isolated from spatially separate populations of the winter moth *Operophtera brumata* (Lepidoptera: Geometridae) in Orkney. J. Invertebr. Pathol. *87*, 29–38.

Gregory, C.R., Latimer, K.S., Pennick, K.E., Benson, K., and Moore, T. (2006). Novel iridovirus in a nautilus (*Nautilus* spp.). J. Vet. Diagn. Invest. *18*, 208–211.

Grosholz, E.D. (1992). Interactions of intraspecific, interspecific and apparent competition with host-pathogen population dynamics. Ecology 73, 507–514.

Grosholz, E.D. (1993). The influence of habitat heterogeneity on host-pathogen population dynamics. Oecologia *96*, 347–353.

Hall, D.W. (1985). Pathobiology of invertebrate icosahedral cytoplasmic deoxyriboviruses (Iridoviridae). In Viral Insecticides for Biological Control, K. Maramorosh and K.E. Sherman, eds. (New York: Academic Press), pp. 163–196.

Hall, D.W., and Anthony, D.W. (1971). Pathology of a mosquito iridescent virus (MIV) infecting *Aedes taeniorhynchus*. J. Invertebr. Pathol. *18*, 61–69.

Hama, H. (1968). On the cross-inoculation of *Chilo* iridescent virus (CIV). Jap. J. Appl. Entomol. Zool. *12*, 34–39 (in Japanese).

Hartschuh, R.D., Wargacki, S.P., Xiong, H., Neiswinger, J., Kisliuk, A., Sihn, S., Ward, V., Vaia, R.A., and Sokolov, A.P. (2008). How rigid are viruses? Phys. Rev. E Stat. Nonlin. Soft Matter. Phys. *78*, 021907.

He, J.G., Deng, M., Weng, S.P., Li, Z., Zhou, S.Y., Long, Q.X., Wang, X.Z., and Chan, S.M. (2001). Complete genome analysis of the mandarin fish infectious spleen and kidney necrosis iridovirus. Virology *291*, 126–139.

He, J.G., Lu, L., Deng, M., He, H.H., Weng, S.P., Wang, X.H., Zhou, S.Y., Long, Q.X., Wang, X.Z., and Chan, S.M. (2002). Sequence analysis of the complete genome of an iridovirus isolated from the tiger frog. Virology *292*, 185–197.

Hembree, S.C., and Anthony, D.W. (1980). Possible site of entry of the regular mosquito iridescent virus (RMIV) in *Aedes taeniorhynchus* larvae. Mosq. News *40*, 449–451.

Hemsley, A.R., Collinson, M.E., Kovach, W.L., Vincent, B., and Williams, T. (1994). The role of self-assembly in biological systems: evidence from iridescent colloidal sporopollenin in *Selaginella* megaspore walls. Phil. Trans. R. Soc. Lond. B *345*, 163–173.

Henderson, C.W., Johnson, C.L., Lodhi, S.A., and Bilimoria, S.L. (2001). Replication of *Chilo* iridescent virus in the cotton boll weevil, *Anthonomus grandis*, and development of an infectivity assay. Arch. Virol. *146*, 767–775.

Hernández, A., Marina, C.F., Valle, J., and Williams, T. (2005). Persistence of Invertebrate iridescent virus 6 in artificial tropical aquatic environments. Arch. Virol. *150*, 2357–2363.

Hess, R.T., and Poinar Jr., G.O. (1985). Iridoviruses infecting terrestrial isopods and nematodes. Curr. Top. Microbiol. Immunol. *116*, 49–76.

Hodgson, D.J., Hitchman, R.B., Vanbergen, A.J., Hails, R.S., Possee, R.D., and Cory, J.S. (2004). Host ecology determines the relative fitness of virus genotypes in mixed-genotype nucleopolyhedrovirus infections. J. Evol. Biol. *17*, 1018–1025.

Hu, G.B., Cong, R.S., Fan, T.J., and Mei, X.G. (2004). Induction of apoptosis in a flounder gill cell line by lymphocystis disease virus infection. J. Fish Dis. *27*, 657–662.

Huang, X., Huang, Y., Gong, J., Yan, Y, and Qin, Q. (2008). Identification and characterization of a putative lipopolysaccharide-induced TNF-alpha factor (LITAF) homolog from Singapore grouper iridovirus. Biochem. Biophys. Res. Comm. *373*, 140–145.

Huang, Y., Huang, X., Liu, H., Gong, J., Ouyang, Z., Cui H., Cao, J., Zhao, Y., Wang, X., Jiang, Y., and Qin, Q. (2009). Complete sequence determination of a novel reptile iridovirus isolated from soft-shelled turtle and evolutionary analysis of Iridoviridae. BMC Genomics *10*, 224.

Huang, Y.H., Huang, X.H., Gui, J.F., and Zhang, Q.Y. (2007). Mitochondrion-mediated apoptosis induced by *Rana grylio* virus infection in fish cells. Apoptosis *12*, 1569–1577.

Hukuhara, T. (1964). Induction of an epidermal tumour in *Bombyx mori* (L.) with *Tipula* iridescent virus, J. Insect Pathol. *6*, 246–251.

Hunter, W.B., Lapointe, S.L., Sinisterra, X.H., Achor, D.S., and Funk, C.J. (2003). Iridovirus in the root weevil *Diaprepes abbreviatus*. J. Insect Sci. *3*, 9.

Hunter, W.B., Patte, C.P., Sinisterra, X.H., Achor, D.S, Funk, C.J., and Polston, J.E. (2001a). Discovering new insect viruses: whitefly iridovirus (Homoptera: Aleyrodidae: *Bemisia tabaci*). J. Invertebr. Pathol. *78*, 220–225.

Hunter, W.B., Sinisterra, X.H., McKenzie, C.L., and Shatters, Jr., R.G. (2001b). Iridovirus infection and vertical transmission in citrus aphids. Proc. Fl. State Hort. Soc. *114*, 70–72.

Imajoh, M., Sugiura, H., and Oshima, S. (2004). Morphological changes contribute to apoptotic cell death and are affected by caspase-3 and caspase-6 inhibitors during red sea bream iridovirus permissive replication. Virology *322*, 220–230.

Ince, I.A., Westenberg, M., Vlak, J.M., Demirbag, Z., Nalcacioglu, R., and van Oers, M.M. (2008). Open reading frame 193R of *Chilo* iridescent virus encodes a functional inhibitor of apoptosis (IAP). Virology *376*, 124–131.

Iyer, L.M., Aravind, L., and Koonin, E.V. (2001). Common origin of four diverse families of large eukaryotic DNA viruses. J. Virol. *75*, 11720–11734.

Jakob, N.J., Muller, K., Bahr, U., and Darai, G. (2001). Analysis of the first complete DNA sequence of an invertebrate iridovirus: coding strategy of the genome of *Chilo* iridescent virus. Virology *286*, 182–196.

Jancovich, J.K., Mao, J., Chinchar, V.G., Wyatt, C., Case, S.T., Kumar, S., Valente, G., Subramanian, S., Davidson, E.W., Collins, J.P., and Jacobs, B.L. (2003). Genomic sequence of a ranavirus (family Iridoviridae) associated with salamander mortalities in North America. Virology *316*, 90–103.

Jensen, D.D., Hukuhara, T., and Tanada, Y. (1972). Lethality of *Chilo* iridescent virus to *Colladonus montana* leafhoppers. J. Invertebr. Pathol. *19*, 276–278.

Juhl, S., Chan, E.P., Ha, Y.H., Maldovan, M., Brunton, J., Ward, V., Dokland, T., Kalmakoff, J., Farmer, B., Thomas, E.L., and Vaia, R.L. (2006). Assembly of *Wiseana* iridovirus: viruses for colloidal photonic crystals. Adv. Funct. Mat. *16*, 1086–1094.

Just, F., Essbauer, S., Ahne, W., and Blahak, S. (2001). Occurrence of an invertebrate iridescent-like virus (Iridoviridae) in reptiles. J. Vet. Med. B *48*, 685–694.

Just, F.T., and Essbauer, S.S. (2001). Characterization of an iridescent virus isolated from *Gryllus bimaculatus* (Orthoptera: Gryllidae). J. Invertebr. Pathol. *77*, 51–61.

Kalmakoff, J., and Tremaine, J.H. (1968). Physiochemical properties of *Tipula* iridescent virus. J. Virol. *2*, 738–744.

Kang, W.K., Kurihara, M., and Matsumoto, S. (2006). The BRO proteins of *Bombyx mori* nucleopolyhedrovirus are nucleocytoplasmic shuttling proteins that utilize the CRM1-mediated nuclear export pathway. Virology *350*, 184–191.

Kelly, D.C. (1985). Insect iridescent viruses. Curr. Top. Microbiol. Immunol. *116*, 23–35.

Kelly, D.C., and Vance, D.E. (1973). The lipid content of two iridescent viruses. J. Gen. Virol. *21*, 417–423.

Khayat, R., Tang, L., Larson, E.T., Lawence, C.M., Young, M., and Johnson, J.E. (2005). Structure of an archaeal virus capsid protein reveals a common ancestry to eukaryotic and bacterial viruses. Proc. Natl. Acad. Sci. U.S.A. *102*, 18944–18949.

Kleespies, R.G., Tidona, C.A., and Darai, G. (1999). Characterization of a new iridovirus isolated from crickets and investigations on the host range. J. Invertebr. Pathol. 73, 84–90.

Klug, A., Franklin, R.E., and Humphreys-Owen, S.P.F. (1959). The crystal structure of *Tipul*a iridescent virus as determined by Bragg reflection of visible light. Biochim. Biophys. Acta 32, 203–211.

Knopf, C.W. (1998). Evolution of viral DNA-dependent DNA polymerases. Virus Genes *16*, 47–58.

Krell, P., and Lee, P.E. (1974). Polypeptides in *Tipula* iridescent virus (TIV) and in TIV-infected hemocytes of *Galleria mellonella* (L.) larvae. Virology *60*, 315–326.

Krupovic, M., and Bamford, D.H. (2008). Virus evolution: how far does the double β-barrel viral lineage extend? Nat. Rev. Microbiol. *6*, 941–948.

Krupovic, M., and Bamford, D.H. (2009). Does the evolution of viral polymerases reflect the origin and evolution of viruses? Nat. Rev. Microbiol. 7, 250.

Kurita, J., Nakajima, K., Hirono, I., and Aoki, T. (2002). Complete genome sequencing of Red sea bream iridovirus (RSIV). Fish. Sci. *68*, 1113–1115.

Lacey, L.A., and Adams, J.R. (1994). An iridescent virus from *Popillia japonica* (Col., Scarabaeidae). Entomophaga 39, 131–136.

Lai, Y.S., Chiou, P.P., Chen, W.J., Chen, Y.C., Chen, C.W., Chiu, I.S., Chen, S.D., Cheng, Y.H., and Chang, C.Y. (2008). Characterization of apoptosis induced by grouper iridovirus in two newly established cell lines from barramundi, *Lates calcarifer* (Bloch). J. Fish Dis. *31*, 825–834.

Lin, P.W., Huang, Y.J., John, J.A., Chang, Y.N., Yuan, C.H., Chen, W.Y., Yeh, C.H., Shen, S.T., Lin, F.P., Tsui, W.H., and Chang, C.Y. (2008). Iridovirus Bcl-2 protein inhibits apoptosis in the early stage of viral infection. Apoptosis *13*, 165–176.

Linley, J.R., and Nielsen, H.T. (1968a). Transmission of a mosquito iridescent virus in *Aedes taeniorhynchus*. I. Laboratory experiments. J. Invertebr. Pathol. *12*, 7–16.

Linley, J.R., and Nielsen, H.T. (1968b). Transmission of a mosquito iridescent virus in *Aedes taeniorhynchus*. II. Experiments related to transmission in nature. J. Invertebr. Pathol. *12*, 17–24.

López, M., Rojas, J. C., Vandame, R., and Williams, T. (2002). Parasitoid-mediated transmission of an iridescent virus. J. Invertebr. Pathol. *80*, 160–170.

López-Ferber, M., Simón, O., Williams, T., and Caballero, P. (2003). Defective or effective? Mutualistic interactions between virus genotypes. Proc. R. Soc. B: Biol. Sci. *270*, 2249–2255.

Lu, L., Zhou, S.Y., Chen, C., Weng, S.P., Chan, S.M., and He, J.G. (2005). Complete genome sequence analysis of an iridovirus isolated from the orange-spotted grouper, *Epinephelus coioides*. Virology *339*, 81–100.

Lua, D.T., Yasuike, M., Hirono, I., and Aoki, T. (2005). Transcription program of red sea bream iridovirus as revealed by DNA microarrays. J. Virol. *79*, 15151–15164.

Majji, S., Thodima, V., Sample, R., Whitley, D., Deng, Y., Mao, J., and Chinchar, V.G. (2009). Transcriptome analysis of Frog virus 3, the type species of the genus *Ranavirus*, family Iridoviridae. Virology *391*, 293–303.

Marina, C.F., Arredondo-Jiménez, J., Castillo, A., and Williams, T. (1999). Sublethal effects of iridovirus disease in a mosquito. Oecologia *119*, 383–388.

Marina, C.F., Feliciano, J.M., Valle, J., and Williams, T. (2000). Effect of temperature, pH, ion concentration and chloroform treatment on the stability of Invertebrate iridescent virus 6. J. Invertebr. Pathol. *75*, 91–94.

Marina, C.F., Fernández-Salas, I., Ibarra, J. E., Arredondo-Jiménez, J. I., Valle, J., and Williams, T. (2005). Transmission dynamics of an iridescent virus in an experimental mosquito population: the role of host density. Ecol. Entomol. *30*, 376–382.

Marina, C.F., Arredondo-Jiménez, J.I., Ibarra, J.E., Fernández-Salas, I., and Williams, T. (2003a). Effects of an optical brightener and an abrasive on iridescent virus infection and development of *Aedes aegypti*. Entomol. Exp. Appl. *109*, 155–161.

Marina, C.F., Ibarra, J.E., Arredondo-Jiménez, J.I., Fernández-Salas, I., Liedo, P., and Williams, T. (2003b). Adverse effects of covert iridovirus infection on life history and demographic parameters of *Aedes aegypti*. Entomol. Exp. Appl. *106*, 53–61.

Marina, C.F., Ibarra, J.E., Arredondo-Jiménez, J.I., Fernández-Salas, I., Valle, J., and Williams, T. (2003c). Sublethal iridovirus disease of the mosquito *Aedes aegypti* is due to viral replication not cytotoxicity. Med. Vet. Entomol. *17*, 187–194.

Mathieson, W.B., and Lee, P.E. (1981). Cytology and autoradiography of *Tipula* iridescent virus infection of insect suspension cell cultures. J. Ultra. Res. *74*, 59–68.

McIntosh, A.H., and Kimura, M. (1974). Replication of the insect *Chilo* iridescent virus (CIV) in a poikilothermic vertebrate cell line. Intervirology *4*, 257–267.

McLaughlin, R.E., Scott, H.A., and Bell, M.R. (1972). Infection of the boll weevil by *Chilo* iridescent virus, J. Invertebr. Pathol. *19*, 285–290.

McMillan, N.A., Davison, S., and Kalmakoff, J. (1990). Comparison of the genomes of two sympatric iridescent viruses (types 9 and 16). Arch. Virol. *114*, 277–284.

Mishra, R.C., Dogra, G.S., and Gupta, P.R. (1980). Some observations on iridovirus of bees. Indian Bee J. *42*, 9–10.

Mitsuhashi, J. (1967). Infection of the green rice hopper, *Nephotettix cincticeps* by *Chilo* iridescent virus. J. Invertebr. Pathol. *9*, 432–436.

Monnier, C., and Devauchelle, G. (1980). Enzyme activities associated with an invertebrate iridovirus: protein kinase activity associated with iridescent virus type 6 (*Chilo* iridescent virus). J. Virol. *35*, 444–450.

Moore, N.F., and Kelly, D.C. (1980). A comparative study of the polypeptides of three iridescent viruses by N-terminal analysis and surface labelling. J. Invertebr. Pathol. *36*, 415–422.

Mullens, B.A., Velten, R.K., and Federici, B.A. (1999). Iridescent virus infection in *Culicoides variipennis* sonorensis and interactions with the mermithid parasite *Heleidomermis magnapapula*. J. Invertebr. Pathol. *73*, 231–233.

Muñoz, D., and Caballero, P. (2000). Persistence and effects of parasitic genotypes in a mixed population of the *Spodoptera exigua* nucleopolyhedrovirus. Biol. Control *19*, 259–264.

Murti, K., and Goorha, R. (1989). Synthesis of frog virus 3 proteins occurs on intermediate filament-bound polyribosomes. Biol. Cell *65*, 205–214.

Murti, K.G., Goorha, R., and Granoff, A. (1982). Structure of frog virus 3 genome: size and arrangement of nucleotide sequences as determined by electron microscopy. Virology *116*, 275–283.

Murti, K.G., Goorha, R., and Granoff, A. (1985). An unusual replication strategy of an animal iridovirus. Adv. Virus Res. *30*, 1–19.

Murti, K.G., Goorha, R., and Klymkowsky, M.W. (1988). A functional role for intermediate filaments in the formation of frog virus 3 assembly sites. Virology *162*, 264–269.

Nalcacioglu, R., Ince, I.A., Vlak, J.M., Demirbag, Z., and van Oers, M.M. (2007). The *Chilo* iridescent virus DNA polymerase promoter contains an essential AAAAT motif. J. Gen. Virol. *88*, 2488–2494.

Nalcacioglu, R., Marks, H., Vlak, J.M., Demirbag, Z., and van Oers, M.M. (2003). Promoter analysis of the *Chilo* iridescent virus DNA polymerase and major capsid protein genes. Virology *317*, 321–329.

Nandhagopol, N., Simpson, A.A., Gurnon, J.R., Yan, X., Baker, T.S., Graves, M.V., Van Etten, J.L., and Rossmann, M.G. (2002). The structure and evolution of the major capsid protein of a large lipid-containing DNA virus. Proc. Natl. Acad. Sci. U.S.A. *99*, 14758–14763.

Netherton, C., Moffat, K., Brooks, E., and Wileman, T. (2007). A guide to viral inclusions, membrane rearrangements, factories, and viroplasm produced during virus replication. Adv. Virus Res. *70*, 101–182.

Ohba, M. (1975). Studies on the pathogenesis of *Chilo* iridescent virus 3. Multiplication of CIV in the silkworm *Bombyx mori* L. and field insects. Sci. Bull. Fac. Agr. Kyushu Univ. *30*, 71–81 (in Japanese).

Ohba, M., and Aizawa, K. (1979a). *In vivo* insect hemocyte destruction by UV-irradiated *Chilo* iridescent virus. J. Invertebr. Pathol. *34*, 32–40.

Ohba, M., and Aizawa, K. (1979b). Multiplication of *Chilo* iridescent virus in noninsect arthropods. J. Invertebr. Pathol. *33*, 278–283.

Ohba, M., and Aizawa, K. (1982). Failure of *Chilo* iridescent virus to replicate in the frog *Rana limnocharis*. Proc. Assoc. Plant Protec. Kyushu *28*, 164–166 (in Japanese).

Ohba, M., Kanda, K., and Aizawa, K. (1990). Cytotoxicity of *Chilo* iridescent virus to *Antheraea eucalypti* cultured cells. Appl. Entomol. Zool. *25*, 528–531.

Paul, E.R., Chitnis, N.S., Henderson, C.W., Kaul, R.J., D'Costa, S.M., and Bilimoria, S.L. (2007). Induction of apoptosis by iridovirus virion protein extract. Arch. Virol. *152*, 1353–1364.

Pietrokovski, S. (1998). Identification of a virus intein and a possible variation in the protein-splicing reaction. Curr. Biol. *8*, R634–635.

Poprawski, T.J., and Yule, W.N. (1990). A new small iridescent virus from grubs of *Phyllophaga anxia* (LeConte) (Col.: Scarabidae). J. Appl. Entomol. *110*, 63–67.

Radloff, C., Vaia, R.A., Brunton, J., Bouwer, G.T., and Ward, V.K. (2005). Metal nanoshell assembly on a virus bioscaffold. Nano. Lett. *5*, 1187–1191.

Ravanello, M.P., and Hruby, D.E. (1994). Conditional lethal expression of the vaccinia virus L1R myristylated protein reveals a role in virion assembly. J. Virol. *68*, 6401–6410.

Reyes, A., Christian, P., Valle, J., and Williams, T. (2004). Persistence of Invertebrate iridescent virus 6 in soil. Biocontrol *49*, 433–440.

Ricou, G. (1975). Production de *Tipula paludosa* Meig. en prarie en function de l'humiditié du sol. Rev. Ecol. Biol. Sol. *12*, 69–89.

Risco, C., Rodriguez, J.R., Lopez-Iglesias, C., Carrascosa, J.L., Esteban, M., and Rodriguez, D. (2002). Endoplasmic reticulum-Golgi intermediate compartment membranes and vimentin filaments participate in vaccinia virus assembly. J. Virol. *76*, 1839–1855.

Rouiller, I., Brookes, S.M., Hyatt, A.D., Windsor, M., and Wileman, T. (1998). African swine fever virus is wrapped by the endoplasmic reticulum. J. Virol. *72*, 2373–2387.

Sample, R., Bryan, L., Long, S., Majji, S., Hoskins, G., Sinning, A., Olivier, J., and Chinchar, V.G. (2007). Inhibition of iridovirus protein synthesis and virus replication by antisense morpholino oligonucleotides targeted to the major capsid protein, the 18 kDa immediate-early protein, and a viral homolog of RNA polymerase II. Virology *358*, 311–320.

Schmelz, M., Sodeik. B., Ericsson, M., Wolffe, E.J., Shida, H., Hiller, G., and Griffiths, G. (1994). Assembly of vaccinia virus: the second wrapping cisterna is derived from the trans Golgi network. J. Virol. *68*, 130–147.

Schultz, G.A., Garthwaite, R.L., and Sassaman, C. (1982). A new family placement for *Mauritaniscus littorinus*

(Miller) N. Comb. from the west coast of North America with ecological notes (Crustacea: Isopoda: Oniscoidea: Bathytropidae). Wasmann J. Biol. *40*, 77–89.

Sikorowski, P.P., and Lawrence, A.M. (1997). Major diseases of *Heliothis virescens* and *Helicoverpa zea* in Mississippi fields and insectaries. Mississippi State University Extension Service, Technical Bulletin 218. Online: http://msucares.com/pubs/techbulletins/tb218.htm

Sikorowski, P.P., and Tyson, G.E. (1984). Per os transmission of iridescent virus of *Heliothis zea* (Lepidoptera: Noctuidae). J. Invertebr. Pathol. *44*, 97–102.

Simón, O., Williams, T., López-Ferber, M., Taulemesse, J.M., and Caballero, P. (2008). Population genetic structure determines the speed of kill and occlusion body production in *Spodoptera frugiperda* multiple nucleopolyhedrovirus. Biol. Control *44*, 321–330.

Singh, Y. (1979). Iridescent virus in the Indian honey bee *Apis cerana* F. Am. Bee J. *119*, 398–340.

Smith, K.M., Hill, G.J., and Rivers, C.F. (1961). Studies on the cross-inoculation of the *Tipula* iridescent virus. Virology *13*, 233–241.

Song, W.J., Qin, Q.W., Qiu, J., Huang, C.H., Wang, F., and Hew, C.L. (2004). Functional genomics analysis of Singapore grouper iridovirus: complete sequence determination and proteomic analysis. J. Virol. *78*, 12576–12590.

Sorrell, I., White, A., Pedersen, A.B., Hails, R.S., and Boots, M. (2009). The evolution of covert, silent infection as a parasite strategy. Proc. R. Soc. B Biol. Sci. *276*, 2217–2226.

Stadelbacher, E.A., Adams, J.R., Faust, R.M., and Tompkins, G.J. (1978). An iridescent virus of the bollworm *Heliothis zea* (Lepidoptera: Noctuidae). J. Invertebr. Pathol. *32*, 71–76.

Stasiak, K., Demattei, M.V., Federici, B.A., and Bigot, Y. (2000). Phylogenetic position of the *Diadromus pulchellus* ascovirus DNA polymerase among viruses with large double-stranded DNA genomes. J. Gen. Virol. *81*, 3059–3072.

Stasiak, K., Renault, S., Demattei, M.V., Bigot, Y., and Federici, B.A. (2003). Evidence for the evolution of ascoviruses from iridoviruses. J. Gen. Virol. *84*, 2999–3009.

Stohwasser, R., Raab, K., Schnitzler, P., Janssen, W., and Darai, G. (1993). Identification of the gene encoding the major capsid protein of insect iridescent virus type 6 by polymerase chain reaction. J. Gen. Virol. *74*, 873–879.

Stoltz, D.B. (1971). The structure of icosahedral cytoplasmic deoxyriboviruses. J. Ultra. Res. *37*, 219–239.

Stoltz, D.B. (1973). The structure of icosahedral cytoplasmic deoxyriboviruses II. An alternative model. J. Ultra. Res. *43*, 58–74.

Stoltz, D.B., and Summers, M.D. (1971). Pathway of infection of mosquito iridescent virus. I. Preliminary observations on the fate of ingested virus. J. Virol. *8*, 900–909.

Stoltz, D.B., Hilsenhoff, W.L., and Stich, H.F. (1968). A virus disease of *Chironomus plumosus*. J. Invertebr. Pathol. *12*, 118–126.

Suhre, K., Audic, S., and Claverie, J.M. (2005). Mimivirus gene promoters exhibit an unprecedented conservation among all eukaryotes. Proc. Natl. Acad. Sci. U.S.A. *102*, 14689–14693.

Tajbakhsh, S., and Seligy, V.L. (1990). Molecular biology of *Tipula* iridescent virus. In Molecular Biology of Iridoviruses, G. Darai, ed. (Boston: Kluwer), pp. 13–46.

Tan, W.G., Barkman, T.J., Chinchar, V.G., and Essani, K. (2004). Comparative genomic analyses of frog virus 3, type species of the genus *Ranavirus* (family Iridoviridae). Virology *323*, 70–84.

Tang, K.F., Redman, R.M., Pantoja, C.R., Groumellec, M.L., Duraisamy, P., and Lightner, D.V. (2007). Identification of an iridovirus in *Acetes erythraeus* (Sergestidae) and the development of in situ hybridization and PCR method for its detection. J. Invertebr. Pathol. *96*, 255–260.

Teng, Y., Hou, Z., Gong, J., Liu, H., Xie, X., Zhang, L., Chen, X., and Qin, Q.W. (2008). Whole-genome transcriptional profiles of a novel marine fish iridovirus, Singapore grouper iridovirus (SGIV) in virus-infected grouper spleen cell cultures and in orange-spotted grouper, *Epinephulus coioides*. Virology *377*, 39–48.

Tesh, R.B., and Andreadis, T.G. (1992). Infectivity and pathogenesis of Iridescent virus type 22 in various insect hosts. Arch. Virol. *126*, 57–65.

Tidona, C.A., and Darai, G. (1997). The complete DNA sequence of lymphocystis disease virus. Virology *230*, 207–216.

Tidona, C.A., and Darai, G. (2000). Iridovirus homologues of cellular genes – implications for the molecular evolution of large DNA viruses. Virus Genes *21*, 77–81.

Tidona, C.A., Schnitzler, P., Kehm, R., and Darai, G. (1998). Is the major capsid protein of iridoviruses a suitable target for the study of viral evolution? Virus Genes *16*, 59–66.

Tinsley, T.W., and Kelly, D.C. (1970). An interim nomenclature system for the iridescent group of viruses. J. Invertebr. Pathol. *12*, 66–68.

Tonka T., and Weiser, J. (2000). Iridovirus infection in mayfly larvae. J. Invertebr. Pathol. *76*, 229–231.

Tripier-Darcy, F., and Nermut, M.V. (1983). Cryodecapage d'un virus, le FV3 (frog virus 3): structure fine de la capside et organisation interne. Biol. Cell. *48*, 17a.

Tsai, C.T., Lin, C.H., and Chang, C.Y. (2007). Analysis of codon usage bias and base compositional constraints in iridovirus genomes. Virus Res. *126*, 196–206.

Tsai, C.T., Ting, J.W., Wu, M.H., Wu, M.F., Guo, I.C., and Chang, C.Y. (2005). Complete genome sequence of the grouper iridovirus and comparison of genomic organization with those of other iridoviruses. J. Virol. *79*, 2010–2023.

Tseng, R.J., Tsai, C., Ma, L., Ouyang, J., Ozkan, C.S., and Yang, Y. (2006). Digital memory device based on tobacco mosaic virus conjugated with nanoparticles. Nat. Nanotechnol. *1*, 72–77.

Undeen, A.H., and Fukuda, T. (1994). Effects of host resistance and injury on the susceptibility of *Aedes taeniorhynchus* to mosquito iridescent virus. J. Am. Mosquito Contr. *10*, 64–66.

Vilaplana, L., Redman, E.M., Wilson, K., and Cory, J.S. (2008). Density-related variation in vertical transmission of a virus in the African armyworm. Oecologia *155*, 237–246.

Wagner, G.W., Webb, S.R., Paschke, J.D., and Campbell, W.R. (1974). A picornavirus isolated from *Aedes taeniorhynchus* and its interaction with mosquito iridescent virus. J. Invertebr. Pathol. *24*, 380–382.

Wagner, G.W., Webb, S.R., Paschke, J.D., and Campbell, W.R. (1975). Production and characterization of the cores of the 'R' strain of mosquito iridescent virus. Virology *64*, 430–437.

Ward, V.K., and Kalmakoff, J. (1991). Invertebrate Iridoviridae. In Viruses of Invertebrates, E. Kurstak, ed. (New York: Marcel Dekker), pp. 197–226.

Ward, V.K., and Kalmakoff, J. (1987). Physical mapping of the DNA genome of insect iridescent virus type 9 from *Wiseana* spp. larvae. Virology *160*, 507–510.

Watson, D., and Seligy, V.L. (1997). Characterization of iridovirus IV1 polypeptides: mapping by surface labelling. Res. Virol. *148*, 239–250.

Watson, S.K. (1997). A study of *in vitro* replication of an iridovirus. Unpublished Ph.D. thesis, Department of Microbiology (New Zealand: Dunedin, University of Otago), pp. 172.

Webby, R., and Kalmakoff, J. (1998). Sequence comparison of the major capsid protein gene from 18 diverse iridoviruses. Arch. Virol. *143*, 1949–1966.

Webby, R.J., and Kalmakoff, J. (1999). Comparison of the major capsid protein genes, terminal redundancies, and DNA-DNA homologies of two New Zealand iridoviruses. Virus Res. *59*, 179–189.

Weinmann, N., Papp, T., Alves de Matos, A.P., Teifke, J.P., and Marschang, R.E. (2007). Experimental infection of crickets (*Gryllus bimaculatus*) with an invertebrate iridovirus isolated from a high-casqued chameleon (*Chamaeleo hoehnelii*). J. Vet. Diagn. Invest. *19*, 674–679.

Wijnhoven, H., and Berg, M.P. (1999). Some notes on the distribution and ecology of iridovirus (Iridovirus, Iridoviridae) in terrestrial isopods (Isopoda, Oniscidae). Crustaceana 72, 145–156.

Williams, R.C., and Smith, K.M. (1957). A crystallizable insect virus. Nature *179*, 119–120.

Williams, T. (1993). Covert iridovirus infection of blackfly larvae. Proc. R. Soc. B Biol. Sci. *251*, 225–230.

Williams, T. (1995). Patterns of covert infection by invertebrate pathogens: iridescent viruses of blackflies. Mol. Ecol. *4*, 447–457.

Williams, T. (1996). The iridoviruses. Adv. Virus Res. *46*, 347–412.

Williams, T. (1998). Invertebrate iridescent viruses. In The Insect Viruses. L.K. Miller and L.A. Ball, eds. (New York: Plenum), pp. 31–68.

Williams, T. (2008). Natural invertebrate hosts to iridoviruses (Iridoviridae). Neotrop. Entomol. *37*, 615–632.

Williams, T., and Cory, J.S. (1993). DNA restriction fragment polymorphism in iridovirus isolates from individual blackflies (Diptera: Simuliidae). Med. Vet. Entomol. 7, 199–201.

Williams, T., and Cory, J.S. (1994). Proposals for a new classification of iridescent viruses. J. Gen. Virol. *75*, 1291–1301.

Williams, T., and Hernández, O. (2006). Costs of cannibalism in the presence of an iridovirus pathogen of *Spodoptera frugiperda*. Ecol. Entomol. *31*, 106–113.

Williams, T., and Thompson, I.P. (1995). Fatty acid profiles of iridescent viruses. Arch. Virol. *140*, 975–981.

Williams, T., Barbosa-Solomieu, V., and Chinchar, V.G. (2005). A decade of advances in iridovirus research. Adv. Virus Res. *65*, 173–248.

Willis, D.B., and Granoff, A. (1980). Frog virus 3 DNA is heavily methylated at CpG sequences. Virology *107*, 250–257.

Wilson, W.H., Van Etten, J.J., and Allen, M.J. (2009). The Phycodnaviridae: the story of how tiny giants rule the world. Curr. Top. Microbiol. Immunol. *328*, 1–42.

Wong, C.K. (2009). Whole genome sequencing of *Wiseana* iridescent virus (WIV). Unpublished MSc. thesis, Department of Microbiology and Immunology, (New Zealand: Dunedin, University of Otago), pp. 94.

Woodard, D.B., and Chapman, H.C. (1968). Laboratory studies with the mosquito iridescent virus (MIV). J. Invertebr. Pathol. *11*, 296–301.

Wrigley, N.G. (1969). An electron microscope study of the structure of *Sericesthis* iridescent virus. J. Gen. Virol. *5*, 123–134.

Wrigley, N.G. (1970). An electron microscope study of the structure of *Tipula* iridescent virus. J. Gen. Virol. *6*, 169–173.

Xeros, N. (1954). A second virus disease of the leather jacket, *Tipula paludosa*. Nature *174*, 562–563.

Yan, X., Olson, N.H., Van Etten, J.L., Bergoin, M., Rossmann, M.G., and Baker, T.S. (2000). Structure and assembly of large lipid-containing dsDNA viruses. Nat. Struct. Mol. Biol. 7, 101–103.

Yan, X., Yu, Z., Zhang, P., Battisti, A.J., Holdaway, H.A., Chipman, P.R., Bajaj, C., Bergoin, M., Rossmann, M.G., and Baker, T.S. (2009). The capsid proteins of a large, icosahedral dsDNA virus. J. Mol. Biol. *385*, 1287–1299.

Yoo, P.J., Nam, K.T., Qi, J., Lee, S.K., Park, J., Belcher, A.M., and Hammond, P.T. (2006). Spontaneous assembly of viruses on multilayered polymer surfaces. Nat. Mater. *5*, 234–240.

Zhang, Q.Y., Xiao, F., Xie, J., Li, Z.Q., and Gui, J.F. (2004). Complete genome sequence of lymphocystis disease virus isolated from China. J. Virol. *78*, 6982–6994.

Nudiviruses: their Biology and Genetics

7

Johannes A. Jehle

Abstract

Nudiviruses are a diverse group of rod-shaped viruses with large circular closed dsDNA genomes, which have previously been considered as non-occluded baculoviruses, because of the lack of an occlusion body. In this chapter, the biology as well as recent advances in genome sequencing, analysis of gene function and phylogenetic analyses will be reviewed. These studies revealed that nudiviruses are indeed related to baculoviruses and share a set of 20 core genes involved in virus structural components, virus–host interaction, DNA replication, nucleotide metabolism and RNA replication. As shown in this chapter, the lack of an occlusion body is not a suitable demarcation between nudiviruses and baculoviruses, as some nudiviruses produce an occlusion body.

Introduction

For many decades, rod-shaped, enveloped, dsDNA viruses with large circular genomes and specific for arthropods have been considered as the so-called 'non-occluded baculoviruses' (Huger and Krieg, 1991). For some of these viruses, the name 'nudiviruses' (Latin *nudi* = bare, naked, uncovered) was used since the beginning of the 1990s (Burand, 1998), other authors also used the term 'intranuclear bacilliform viruses' (IBVs) (Evans and Edgerton, 2002). Some nudiviruses (NVs) have been considered as potential biocontrol agents of economically important arthropod pests, including Lepidoptera and Coleoptera. For example, already more than 40 years ago the palm beetle *Oryctes rhinoceros* was successfully controlled using a nudivirus (Burand, 1998; Huger, 1966).

Unlike baculoviruses, nudiviruses were thought to generally lack a proteinacous occlusion body (OB). Thus far, a variety of nudiviruses and nudivirus-like viruses have been reported from various host species belonging to Lepidoptera, Trichoptera, Diptera, Siphonaptera, Hymenoptera, Neuroptera, Coleoptera, Homoptera, Thysanura, Orthoptera, Acarina, Araneina, and Crustacea (Huger and Krieg, 1991). However, it is important to note that the similarity of these viruses is only based on structural characteristics and some similarities in replication. Owing to the lack of genetic data, however, these viruses were later removed from the family *Baculoviridae* and remained unassigned (Mayo, 1995). Today, it is apparent that the so-called 'non-occluded baculoviruses' are an assemblage of different viruses, which most likely comprises several diverged lineages of viruses. Their relationship to each other and to baculoviruses was not duly confirmed until the genome sequences of some of these viruses had been determined. From comparative phylogenetic analyses of gene and genome sequences, it is obvious now that at least five of these viruses form a monophyletic clade, which we consider as nudiviruses. These viruses comprise the *Oryctes rhinoceros* nudivirus (OrNV, also *O. rhinoceros* virus (= OrV)), the *Heliothis zea* nudivirus 1 (HzNV-1, also *H. zea* virus 1 = Hz-1V), the *H. zea* nudivirus 2 (HzNV-2, also *H. zea* virus 2 = Hz-2V, formerly known as gonad-specific virus (GSV), the *Gryllus bimaculatus* nudivirus (GbNV), and the so-called *Penaeus sp.* baculoviruses, which represent crustacean-specific occluded viruses, phylogenetically most closely related to HzNV-1 and HzNV-2. In recent years, some comprehensive reviews have been written on the discovery, biology and field application of nudiviruses (Huger and Krieg, 1991; Burand, 1991, 1998; Huger, 2005; Jackson *et al.*, 2005; Wang and Jehle, 2009). Therefore this chapter will primarily focus on the description of the diversity, molecular biology and phylogeny of these viruses. Recent work has provided clear evidence that polydnavirus of braconid wasps derived from an ancestral nudivirus and represents a virus-like agent domesticated by the braconids to manipulate the parasitoid host defence (Bézier *et al.*, 2009a,b). This unusual group of viruses is covered in Chapter 8 of this volume.

Oryctes rhinoceros nudivirus (OrNV)

The *Oryctes rhinoceros* nudivirus (OrNV), also known as *O. rhinoceros* virus (OrV), was discovered in 1963 in Malaysia, when a field survey of natural pathogens of the scarab beetle *O. rhinoceros* was performed in oil palm plantations (Huger, 1966, 2005). The rhinoceros beetle caused severe damage to coconut palms in the South Pacific Islands. Later OrNV was introduced into many South East and South Pacific countries, e.g. Samoa (Marshall, 1970), South Pacific Islands (Bedford, 1980), Papua New Guinea (Gorick, 1980), Mauritius (Monty, 1978), the Philippines (Zelazny and Alfiler, 1991), the Maldives and Indonesia (Zelazny *et al.*, 1992), and the Andaman Islands (Jacob, 1996; Shyam Prasad *et al.*, 2008). In recent years, research was initiated to control the rhinoceros beetle in palm oil plantations in Malaysia (Ramle *et al.*, 2005).

Both the larval stages and the adult beetles of *O. rhinoceros* become infected with OrNV. The viral disease is orally transmitted. Infected larvae succumb to a lethal infection within 1–4 weeks. The infection starts in the midgut epithelial cells and rapidly spreads to other tissues initiating a systemic infection. Infected larvae stop feeding and become lethargic (Huger and Krieg, 1991). While fat body cells increasingly disintegrate and vacuolate, an increase of haemolymph can be observed. At the final stages of infection, the larvae become turgid and vitreous, often beige and waxen. Also adult beetles may become lethally infected; however, most frequently they undergo a chronic infection without apparent symptoms. Here, the infection is initially restricted to the midgut epithelial cells. These cells become heavily packed with newly synthesized viruses and are sloughed off into the gut lumen (Huger and Krieg, 1991). In this stage, the midgut is heavily swollen and appears whitish. Since the infected epithelial cells are continuously regenerated, the midgut lumen becomes packed with virus infected cells excreted with their faeces. This results in an extremely efficient horizontal spread of OrNV in the beetle populations and virus contamination of nesting sites. These chronically infected adults have been considered as 'flying virus factories' (Huger, 2005). Infected adults stop feeding and egg lying, which contributes additionally to the control of *Oryctes* populations.

At the cellular level, replication of a cloned OrNV isolate, termed PV505, was studied in a cell line derived from the black beetle *Heteronychus arator* (Crawford and Shehan, 1985). It is thought that viruses enter the cell via pinocytosis. The first cytopathogenic effects of virus infection were nuclear hypertrophy and cell rounding of the fibroblast-like cells, followed by complete cell lysis about 72–96 hours post infection (hpi). First signs of virus replication occur 7 hpi, when infected nuclei show chromatin-free areas, which become filled with virion envelope material. Between 12 and 36 hpi, virus replication is fully switched on and OrNV virions assemble in dense arrays closed to the nuclear membrane (Fig. 7.1) (Crawford and Shehan, 1985). Finally, the virions cross the cytoplasm and bud through the cell membrane.

OrNV virions are enveloped and rod-shaped with a size of 220 × 120 nm (Payne, 1974, 1977). When treated with detergent, the envelope may detach and show an artificially mug-shaped distortion (Huger and Krieg, 1991). Purified nucleocapsids are 180 × 65 nm in size and consist of an electron-dense DNA protein core and a thin capsid with slightly thickened caps on their longitudinal ends (Fig. 7.2). One end of the nucleocapsid is linked to a flexuous, tail-like appendage of about 270 × 10 nm.

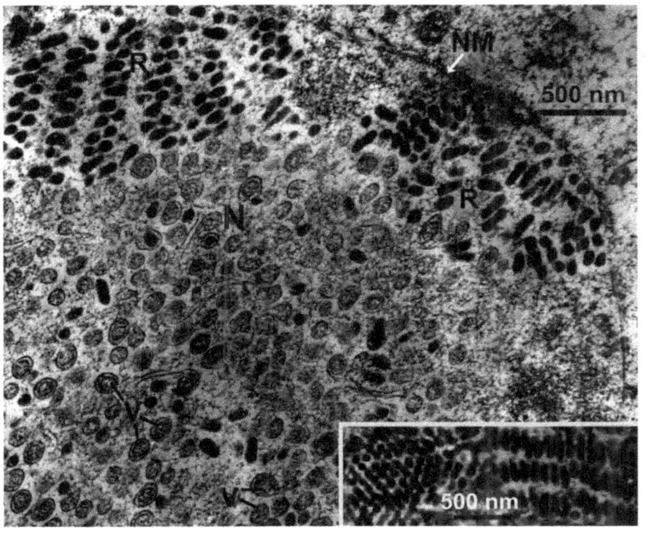

Figure 7.1 Electron micrograph of a thin section showing part of the nucleus (N) of a fat body cell of a third instar rhinoceros beetle larva with heavy virus infection. Note the accumulation of virus rods (R) at the nuclear periphery and the single- and double-membraned vesicles (V) in the nuclear centre. NM nuclear membrane. Inset: Virus rods in cross (left) and longitudinal (right) section arranged in a pseudocrystalline pattern. Reprinted from Huger (2005) with permission from Elsevier.

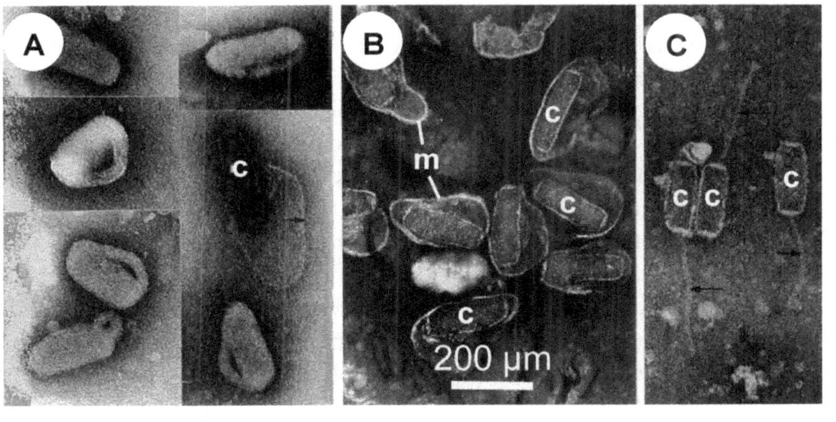

Figure 7.2 Electron micrographs with structural details of *Oryctes* virus rods, negatively stained with phosphotungstic acid. (A) Virions unpenetrated by stain, often being artificially mug-shaped; middle right: the virus membrane (arrow) is shed off from the capsid (c). (B) Virions with longer penetration by stain, thus displaying the capsids (c) and the surrounding viral membrane (m). (C) Three capsids (c) showing the typical thread-like appendix (arrows). Reprinted from Huger (2005) with permission from Elsevier.

Interestingly, similar tails can also be observed with the virion(!) of whispoviruses of shrimp (Wongteerasupaya *et al.*, 1995). However, whether this is a significant structural similarity or an artefact of nucleocapsid/virion preparation remains to be analysed.

The protein composition of the virion is not very well analysed. Proteomic data are completely missing until today. Using SDS-PAGE analysis, up to 27 virion structural proteins with a size of 10 to 215 kDa have been identified. Since 14 of these proteins could be removed by detergent treatment, 13 proteins may be associated with the nucleocapsids (Crawford and Sheehan, 1985). When pulse labelling with L-[^{35}S]methionine was applied to OrNV infected cell cultures only eight of these proteins could be identified within 36 hpi. Two proteins (p4.6 and p10) were identified within 4 hpi and six other proteins (p13, p22, p25, p27, p40, and p46.5) at 6 hpi. It was hypothesized that a possible reason for the difficulty in identifying newly synthesized virus proteins in pulsed cells was that, owing to a lack of a shut-off of host protein synthesis, the viral protein synthesis may have been obscured by the large number of host cell proteins.

The genome of OrNV is a circularly closed dsDNA molecule. Both the Philippines isolate PV505 and the Indian isolate KI were mapped by DNA restriction endonuclease analysis to 123–127 kb in length (Crawford *et al.*, 1985; Mohan and Gopinathan, 1991). Several restriction fragments were cloned and sequenced, allowing the development of PCR primers for detection of the virus (Richards *et al.*, 1999). Sequencing of the cloned *Pst*I-C (19,805 bp) and *Pst*I-D (17,146 bp) of PV505 covered about 30% of the viral genome and revealed a first glance on the genetic organization of the OrNV genome (Wang *et al.*, 2007a). Most interestingly, a significant number of the predicted open reading frames (ORFs) revealed homology to HzNV-1 as well as to some baculovirus core genes, suggesting a common ancestry of these viruses. After amplification of the spurious amounts of OrNV genomic DNA by applying the multiple displacement amplification method, it was possible to obtain the complete sequence of the 127,615 bp large genome of the OrNV isolate Ma07 (Wang *et al.*, 2008; Table 7.1). Based on the full genome sequence, the G+C content was determined to be 42% (Wang and Jehle, 2009). Most strikingly, OrNV and other sequenced nudiviruses (GbNV, HzNV-1; see below) encode 20 out of 30 core gene homologues identified in all of the 50 sequenced baculovirus genomes. These core genes are associated with late and very late gene transcription (*p47*, *lef-8*, *lef-9*, *lef-4*, *vlf-1*, and *lef-5*), DNA replication (*dnapol* and *helicase*), virus structure (*p74*, *pif-1*, *pif-2*, *pif-3*, *ac68*, *vp91*, *vp39*, *38K*, *19kDa*, and *odv-e56*), or have unknown functions (*ac81* and *p33*) (van Oers and Vlak, 2007). From the presence of these core gene homologues in nudiviruses and baculoviruses and from phylogenetic analyses it can be clearly concluded that nudiviruses and baculoviruses are two monophyletic sister groups which share a common ancestor (Wang *et al.*, 2007a; Wang and Jehle, 2009). Another surprising feature of the OrNV and other nudivirus genomes is the presence of a homologue of the *polyhedrin/granulin* (*polh/gran*) gene of alpha-, beta- and gammabaculoviruses, which typically form occlusion bodies, while the nudiviruses are generally deficient in those. However, a facultative occurrence of occlusion bodies in midgut cells of OrNV-infected rhinoceros beetles was reported previously (Huger and Krieg, 1991). It remains unknown whether the baculovirus *polh/gran* homologue of OrNV is involved in occlusion body formation or whether other genes are involved. Strikingly, atypical occlusion bodies can also occasionally be observed with HzNV-2 (Raina *et al.*, 2000; Rallis *et al.*, 2002b).

Gryllus bimaculatus nudivirus

The cricket specific nudivirus GbNV infects nymphs and adults of several field crickets, *Gryllus bimaculatus*, *G. campestris*, *Teleogryllus oceanicus* and *T. commodus* (Huger, 1985). When early nymphs of *G. bimaculatus* were orally infected, they died from infection after 3 to 12 weeks. Infected crickets are lethargic and smaller in size, sometimes even crippled (Huger and Krieg, 1991). GbNV has rod-shaped or irregularly ellipsoidal, enveloped virions of 90 × 180 nm (Fig. 7.3) (Huger, 1985). Similar to other nudiviruses, GbNV has a circular dsDNA genome and replicates in the nuclei of the infected fat body cells; infected cells become extremely hypertrophied. The GbNV is not very well studied: only a very few reports describe its biology and virion structure (Huger and Krieg, 1991).

After propagation of GbNV in a laboratory maintained *G. bimaculatus*, the complete genome of GbNV was cloned and sequenced by Wang *et al.* (2007b). The GbNV genome is 96,944 bp in length and contains 98 predicted ORFs (Fig. 7.4). Its G+C content is 28%, one of the lowest of any sequenced dsDNA insect virus. GbNV shares 66 gene homologues with OrNV but only 33 homologues with HzNV-1 (see below), underscoring the closer relationship between GbNV and OrNV. Beyond the homologues to 20 baculovirus core genes, six ORFs of GbNV are homologous to non-conserved baculovirus genes, which are *dnaligase*, *helicase 2*, *rr1*, *rr2*, *iap-3*, and

Table 7.1 Nudiviruses with completely sequenced genomes. Modified from Wang *et al.* (2007c)

			Size (nm)		Genome		
Virus	Host	Host stage and/or tissue tropism	Virus particles	Nucleocapsids	Size (bp)*	G+C (%)	ORFs
HzNV-1	*Heliothis zea* (cell line) (Lepidoptera: Noctuidae)	–	384–444 × 77–83	–	228,089	42	154
HzNV-2	*Heliothis zea* (Lepidoptera: Noctuidae)	Larvae and adults; reproductive tissues	415 × 80	500 × 25	231,621‡	42	113
OrNV	*Oryctes rhinoceros* (Coleoptera: Dynastidae)	Larvae and adults; midgut and fat body	200–235 × 100–120	160–180 × 50–65	127,615	42	139
GbNV	*Gryllus bimaculatus* (Orthoptera: Gryllidae)	Nymphs and adults; fat body	145–240 × 80–100	162 × 66	96,944	28	98

*Predicted/sequenced genomic DNA size. ‡Burand, unpubl.

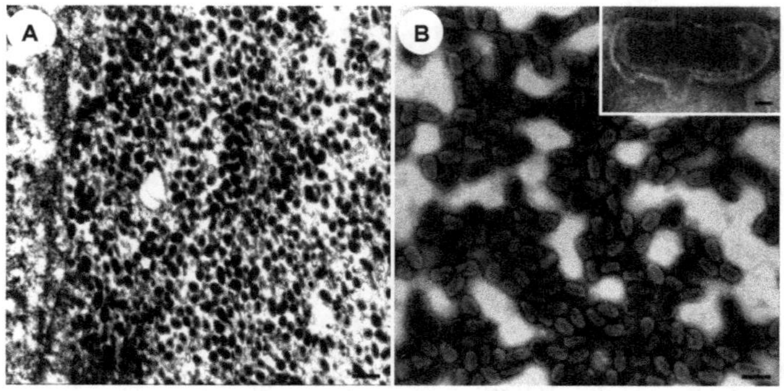

Figure 7.3 Electron micrographs of GbNV particles. (A) Thin section of infected cell nucleus. (B) Purified virions. Inset: Enlarged image of a typical virion showing the envelope and the rod-shaped nucleocapsid. Bar: 200 nm in (A) and (B) and 20 nm in inset. Courtesy of Dr Alois M. Huger, JKI Darmstadt. Reprinted from Wang and Jehle (2009) with permission from Elsevier.

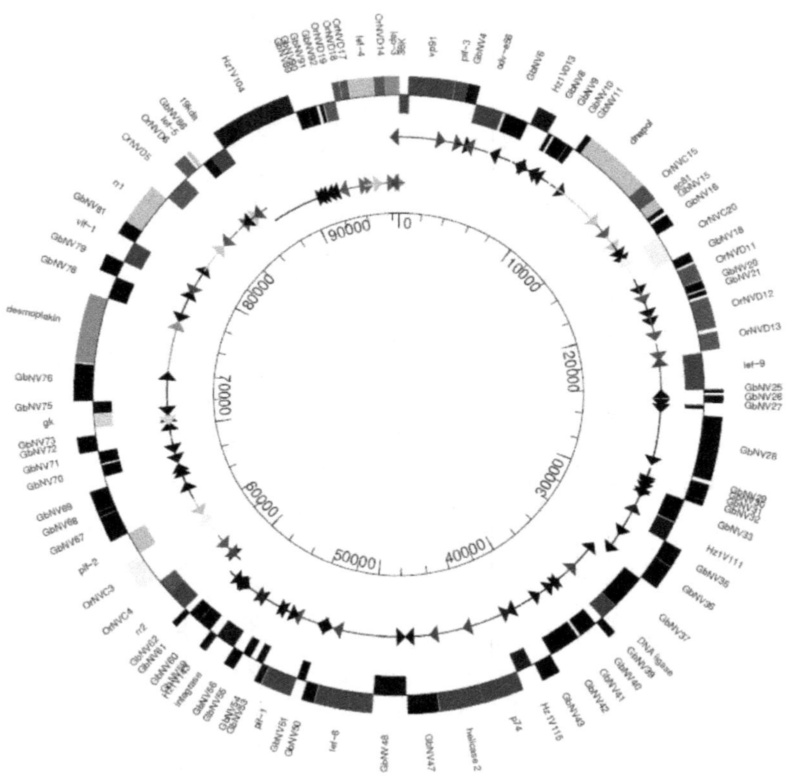

Figure 7.4 Circular map of the genome of *Gryllus bimaculatus* nudivirus based on the sequence published by Wang *et al.* (2007b) with permission from the American Society for Microbiology.

desmoplakin (Wang *et al.,* 2007b). Thirty-two GbNV ORFs showed no or only very weak similarity to other genes in the GenBank. The GbNV genome does not contain homologous repeat sequences (*hrs*), as they are typical for baculovirus genomes. However, fourteen short direct repeat regions (*drs*) distributed throughout the genome and constituting about 0.6% of the genome sequence were identified. These *drs* are extremely AT rich (up to 98%) and contain 2 or 3 copies of short reiterated sequences of 11 to 42 bp in length. Despite the considerable number of homologues between GbNV and OrNV, gene order is poorly conserved between the two genomes. They share only some short gene clusters of 2 to 7 collinearly arranged genes which appear randomly distributed throughout their genomes. Conservation between GbNV and HzNV-1 is even less; only two conserved clusters were identified. Interestingly, a gene cluster of *helicase*, *19kDa*, and/or *lef-5* is present in GbNV and all other sequenced nudivirus genomes, resembling the conserved gene cluster of *helicase*, *19kDa*, *38K* and *lef-5* identified in all sequenced baculovirus genomes (Herniou *et al.*, 2003; Jehle and Backhaus, 1994).

Heliothis zea nudiviruses 1 and 2

The *H. zea* nudivirus 1 (HzNV-1), also known as IMC-HZ-I-NOV, baculovirus X or Hz-1 virus, was first described as a persistent viral infection in the ovarian *H. zea* cell line IMC-Hz-1, of which only less than 1% showed cytopathogenic effects (Granados *et al.*, 1978). Plaque derived HzNV-1 virions measure 385–445 × 77–83 nm and 28 structural viral proteins ranging from 14 to 150 kDa were identified (Burand *et al.*, 1983a). Typically for nudiviruses, the genome of HzNV-1 consists of covalently closed dsDNA. A first physicochemical and biochemical characterization of the HzNV-1 genome was performed by Huang *et al.* (1982) and Langridge (1981). Using CsCl equilibrium centrifugation a buoyant density of 1.70 g/ml corresponding to a (G+C) content of 43% was predicted. A physical map was constructed by Chao *et al.* (1990) and its size was estimated to be 230 to 245 kb (Huang *et al.*, 1982; Burand *et al.*, 1983b). These findings matched very well with the G+C content (41.8%) and size (228 kb) determined later by full genome sequencing (Cheng *et al.*, 2002).

HzNV-1 develops persistent infections of several other lepidopteran cell lines, such as IPLB-1075 (*H. zea*), IPLB-SF-21 (*Spodoptera frugiperda*), IPLB-65Z (*Lymantria dispar*) and TN-368 (*Trichoplusia ni*) (Granados *et al.*, 1978; Kelly *et al.*, 1981; Ralston *et al.*, 1981). In contrast, the two lepidopterans, *Estigmene acrea* and *H. zea,* were not susceptible after intrahaemocoelic injection and/or per os feeding with HzNV-1 (Granados *et al.*, 1978). McIntosh *et al.* (2007) studied the susceptibility of 13 insect cell lines originating from Lepidoptera, Coleoptera, Diptera, and Homoptera. Though HzNV-1 showed a wide *in vitro* host cell spectrum, only lepidopteran cell lines supported HzNV-1 replication. Interestingly, HzNV-1 caused acute infection and massive cytopathic effects (CPE) of TN-368 cell cultures of TN-368. Cells surviving the acute infection could be re-established to cultures, which were persistently infected and resistant to HzNV-1 but remained susceptible to infection with *Autographa californica* multiple nucleopolyhedrovirus (AcMNPV) (Ralston *et al.*, 1981). As shown for *S. frugiperda* insects and cells, persistence is linked to apoptosis. Superinfection with HzNV-1 results in massive cell death accompanied by a significant reduction of virus progeny (Lee *et al.*, 1993). Apoptosis induced by superinfection could be effectively blocked in *S. frugiperda* cells, when the cells were stably transfected with the apoptosis inhibitor gene p35 construct of AcMNPV before or after the establishment of persistence. However, these cells showed a robust resistance when they were challenged with HzNV-1, suggesting that the observed virus interference did not solely depend on the induction of apoptosis but may be caused by an early block in virus infection (Lee and Chao, 1998). When transcription of productive and persistent infections was compared, 101 virus-specific transcripts of 0.8 kb to more than 9.5 kb in length belonging to three different temporal classes (<2 hours, 4–6 hours, 8 hours post infection) were observed in acute infections. In contrast, massive viral gene shutoff was observed in persistently infected cells and the only detectable viral RNA was a transcript of about 2.9 kb, termed persistency–associated transcript (PAT1). Notably, three transcripts of different size were produced from the same genome region as PAT1, one of these transcripts had similar size, initiation and 5′end (Chao *et al.*, 1992). PAT1 transcripts were exclusively located in the nucleus and were not associated with ribosomes in the cytoplasm. When the persistency-associated gene 1 (*pag1*), encoding PAT1 was cloned, no significant protein coding open reading frames, but some repeated motifs, could be detected (Chao *et al.*, 1998). However, there was clear experimental evidence that PAT1 functions in establishing persistent HzNV-1 infection.

Promoter analysis of the *pag1* gene revealed a negative regulatory sequence containing GATA elements (Chen *et al.*, 2001).

It appears that persistently infected cells contain various amounts of viral DNA comprising up to 16% of the cellular DNA in extreme cases. There is strong evidence that at least some viral DNA molecules are inserted into the host chromosomes (Lin *et al.*, 1999). Therefore, the persistent infection of HzNV-1 can be considered as a form of latency during which the virus genome is propagated with each host cell division and needs activation for becoming virulent again. Reactivation of HzNV-1 infection from persistent infections appears at a low rate of 0.2% over a 5-day period (Lin *et al.*, 1999).

A gene probably involved in reactivating latent HzNV-1 infection is *hhi1*. Whereas during the productive viral infection, *hhi1* is expressed as a 6.2 kb transcript as early as 0.5 hpi and reached a maximum at 2 hpi, its transcription was undetectable in cells latently infected with HzNV-1. However, it was observed that transient expression of *hhil* in latently infected cells resulted in an increased virus titre and DNA replication. In addition, *hhi1* induced expression of several early genes (Wu *et al.*, 2010). Expression of *hhi1* in persistently infected Sf21 cells can be activated by super-infection with AcMNPV. In particular, it was found that the AcMNPV ie1 and p35 were needed for *hhi1* promoter activation (Wu *et al.*, 2008). Taking together, it appears that *hhi1* may be important in the reactivation of virus infection.

The HzNV-1 genome is 228,089 bp in size and encodes 154 putative ORFs of 50 or more amino acids (Cheng *et al.*, 2002). Similar to other nudivirus genomes, no homologous repeat regions (*hrs*) were found. For HzNV-1, a number of short tandem repeats of 21 to 75 bp are distributed throughout its genome sequence (Cheng *et al.*, 2002). HzNV-1 harbours a significant number of ORFs which are homologous to baculovirus genes, including 20 baculovirus core genes (Cheng *et al.*, 2002; Wang *et al.*, 2007a). Ten ORFs are homologues of cellular genes coding for histidine kinase, dihydrofolate reductase, dUTP pyrophosphatase, matrix metalloproteinase, deoxynucleoside kinase, glycine hydroxymethyl-transferase, ribonucleotide reductase small subunit, thymidylate synthase, alt1, and carboxylesterase (Cheng *et al.*, 2002; Chen *et al.*, 2001). Interestingly, a continuous 9.6 kb locus containing five homologues of cellular genes (Hz1V069 to Hz1V073) was located between the nucleotide positions 99,357–108,961 (Cheng *et al.*, 2002).

Most strikingly, the genomic sequence and gene content of HzNV-1 is very similar to *H. zea* nudivirus-2 (HzNV-2), also known as gonad-specific virus (GSV), *H. zea* reproductive virus and Hz-2V (for reviews see Burand, 1998, 2009; Wang *et al.*, 2007c). HzNV-2 has a genome of 231,621 bp in size and is only slightly larger than that of HzNV-1 (228 kb) (Wang *et al.*, 2007c). The G+C content of the HzNV-2 genomic DNA is 41.8% and is identical to that of HzNV-1. HzNV-2 contains 113 predicted ORFs and shares a 94–95% sequence similarity with HzNV-1 (Wang *et al.*, 2007c, Burand, 2009). As other nudiviruses, 20 ORFs of HzNV-2 are homologous to baculovirus core genes. Only two segments of cloned HzNV-2 DNA has been deposited in the GenBank (Raina *et al.*, 2000): A 1,127 bp fragment with a more than 99% nucleotide identity to a HzNV-1 sequence containing the ORFs Hz1V134 and Hz1V135, which latter is similar to a baculovirus *iap*-3 gene. A second 2,295 bp fragment encodes two partial ORFs with more than 97% nucleotide identity to HzNV-1 ORFs Hz1V058 and Hz1V059, respectively (Wang *et al.*, 2007c). These highly significant similarities of gene homologues clearly indicate that HzNV-1 and HzNV-2 are genetically very closely related.

Despite this close genetic relationship, HzNV-1 and HzNV-2 show a very distinct biology and pathology. HzNV-1 infections have been demonstrated only from infected *in vitro* cell cultures but not from *in vivo* host insects, while HzNV-2 exhibit a highly unusual and unique pathology in adult *H. zea* males and females. HzNV-2 was discovered in a laboratory colony of *Helicoverpa zea* from Stoneville, Massachusetts (Fig. 7.5) (Raina and Adams, 1995). Later, it was found to replicate in insect cell lines of *Trichoplusia ni* (Tn-368) and *Lymantria dispar* (Ld652Y) (Burand and Lu, 1997; Lu and Burand, 2001). Tn-368 derived viruses produced the same symptoms in *H. zea* adults as observed with *in vivo* produced HzNV-2 (Burand and Lu, 1997).

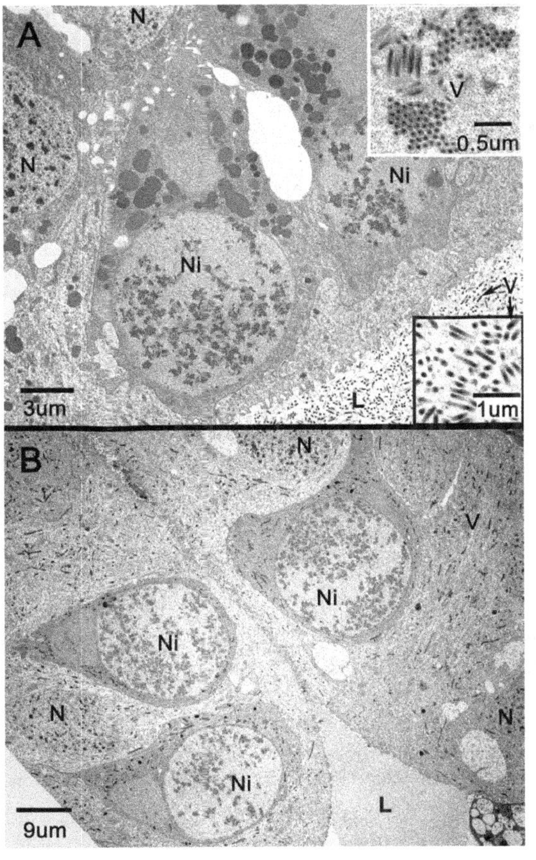

Figure 7.5 Electron micrographs of oviduct tissues from agonadal females. (A) Lateral oviduct of an agonadal female 10 days post pupation (dpp) showing numerous virus particles (V) filling the lumen (L) (inset, lower right) and organized in arrays in infected nuclei (Ni) (inset, upper right). Note the absence of a well-defined nuclear membrane surrounding the nucleus of the virus-infected cell on the right. (B) Common oviduct of a 10 dpp agonadal female showing cluster of cells containing infected nuclei (Ni) filled with virus particles. The infected cells are electron-dense and therefore stain more darkly than uninfected cells [cells with normal nuclei (N)]. Reprinted from Rallis and Burand (2002b) with permission from Elsevier.

One of the most striking features of HzNV-2 is that infection generally does not cause mortality of infected insects as observed with many other nudiviruses. But it results in severe deformations of the reproductive organs of infected adults leading to host sterility in both male and female moths (Raina and Adams, 1995; Hamm *et al.*, 1996; Burand and Rallis, 2004). It also leads to a dramatic change in the physiology and behaviour of the infected hosts favouring the vertical and horizontal transmission of the virus in the host population (Burand, 2009).

Infected moths are typically agonadal (AG) due to atrophic testicular and ovarian cells. Whereas infected males are sterile and AG, infected female moths can show two distinct phenotypes: sterile AG *and* fertile asymptomatic (AS) infected females (Raina and Adams, 1995; Hamm *et al.*, 1996). All three phenotypes play specific roles for the efficient cycling of HzNV-2 in the host population. AG males lack testes, accessory glands and other reproductive tissues. Their deformed reproductive tissues have a Y-like structure and are the site of virus replication (Fig. 7.6) (Rallis and Burand, 2002a; Burand, 2009).

AG females produce a waxy plug, also known as the 'virus plug', which covers their copulatory opening. On the other hand, they lack ovaries, the bursa copulatrix, accessory glands and spermatheca. Their oviducts are hypertrophic and

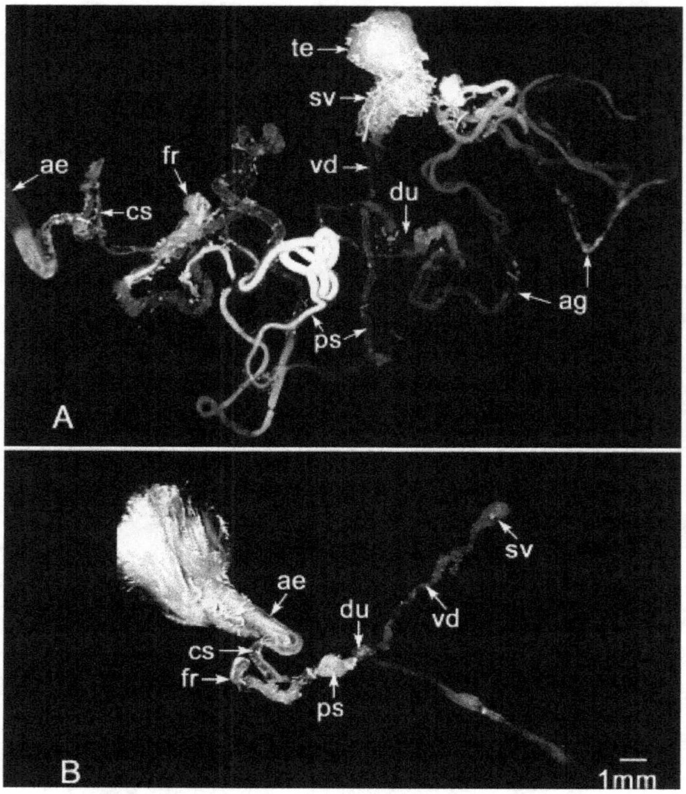

Figure 7.6 Normal and agonadal reproductive systems of the adult male *H. zea*. (A) Normal male, accessory glands (ag), testes (te), seminal vesicles (sv), vasa deferentia (vd), duplex (du), primary simplex (ps), cuticular simplex (cs), twisted area of the cuticular simplex where the frenum of the spermatophore is formed (fr), and the aedeagus (ae). (B) Agonadal male exhibiting grossly malformed reproductive system lacking testes and accessory glands, but exhibiting other tissues corresponding to the same tissues of normal male. Reprinted from Rallis and Burand (2002a) with permission from Elsevier.

deformed and show a 'Y-shaped' structure. Their cells are packed with newly synthesized virus (Fig. 7.7) (Rallis and Burand, 2002b). Burand (2009) developed a model for transmission of HzNV-2. When AG females mate with a non-infected male, the latter becomes contaminated from the 'virus plug' during the copulation attempt. Virus transmission to healthy males is facilitated by two physiological and behavioural changes in AG females: First, they produce more pheromones than healthy females and are therefore more attractive to males. Secondly, after mating they do not stop exhibiting a calling behaviour and continue attracting males for copulation and transmitting virus to other healthy males (Burand and Tan, 2006). In fertile AS females, the virus stays persistent and may replicate in reproductive tissues transmitting the virus offspring to the eggs. Progeny insects may become again AS females or AG females or AG males. Finally, when AG males copulate with fertile females the virus is transmitted to the females. However, the AG males are incapable of transferring a spermatophore and a pheromonostatic peptide (PSP), which causes a steep decline in female pheromone production and an arrest of their calling behaviour. Thus, females mated with AG males continue to call for and copulate with males until they are inseminated by a fertile male. These females may then produce healthy, AS or AG offspring (Burand, 2009). There appears to be a strong correlation between virus infection, deformations of reproductive tissues, alterations in the physiology and behaviour of

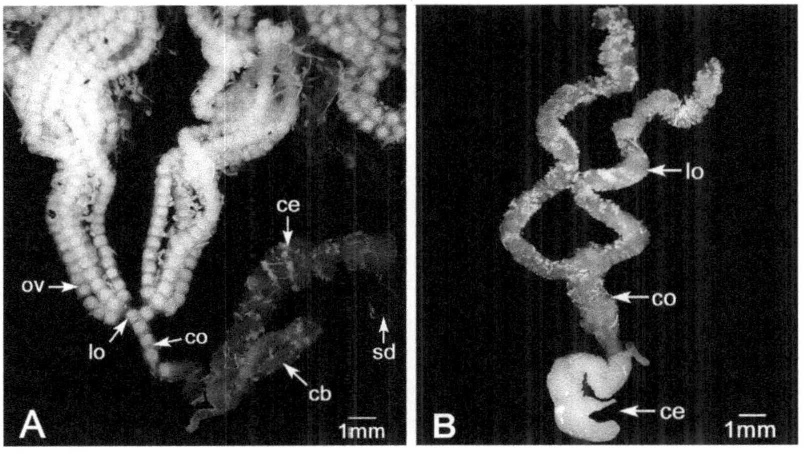

Figure 7.7 Reproductive systems of adult female *H. zea* showing normal, agonadal, and intermediate malformed tissues resulting from Hz-2V infection of different life stages. (A) Normal female ovarioles (ov), lateral oviduct (lo), common oviduct (co), cervix bursa (ce), corpus bursa (cb), and seminal duct (sd). (B) Grossly malformed reproductive tissues of a typical agonadal female adult moth arising from an egg laid by an infected adult female. Reprinted from Rallis and Burand (2002b) with permission from Elsevier.

infected insects and the vertical and horizontal transmission of virus. As pointed out, the primary mode of transmission is through mating by AS females. Recent studies, however, provided evidence that HzNV-2 can also be acquired by first instar larvae, through feeding on virus contaminated diet (Raina and Lupiani, 2006).

Nudiviruses of other arthropods

A number of poorly characterized viruses possibly related to nudiviruses have been described and were classified according to some biological and morphological data (Table 7.2) (Huger and Krieg, 1991; Wang *et al.*, 2007c). However, no molecular data, which would allow sequence comparison and phylogenetic analyses, are available for these viruses. Though not specific to insects, occluded viruses infecting penaeid shrimp, *Penaeus vannamei* and *P. monodon*, are referred to here, as they appear to belong also to nudiviruses. Discovered in the 1970s (Couch, 1974a,b; Lightner and Redman, 1981; Johnson and Lightner, 1988), these viruses produce occlusion bodies and were therefore considered as Crustacea-specific single nucleopolyhedroviruses and named as *Baculovirus penaei* (BP) or *P. vannamei* SNPV (Couch 1974a; Mayo, 1995) and *P. monodon* baculovirus (MBV) or *P. monodon* single nucleopolyhedrovirus (PmSNPV) (Lightner and Redman, 1981; Bonami *et al.*, 1997). However, recent sequence comparisons and phylogenetic analyses of this shrimp virus and nudivirus sequences clearly suggest that MBV is most closely related to HzNV-1 and HzNV-2 and therefore should be considered as an occluded nudivirus (Wang and Jehle, 2009). Interestingly, the occlusion body protein has no sequence similarity to Alpha-, Beta-, and Gammabaculovirus polyhedrins or granulin nor to the Deltabaculovirus polyhedrin (Chaivisuthankura *et al.*, 2008). Apparently, occlusion body formation appears to be a significant ecological and evolutionary advantage for viruses specific for insects and other arthropods and has been developed independently in several virus lineages such as Alpha-, Beta- and Gammabaculoviruses (Jehle *et al.*, 2006), in Deltabaculoviruses (Perera *et al.*, 2006), Shrimp nudiviruses (Chaivisuthankura *et al.*, 2008), Entomopoxviruses (Yuen *et al*, 1990; Banville *et al.*, 1992) and the Cypoviruses (Coulibaly *et al.*, 2007).

Nudivirus diversity and evolution

In summary, nudiviruses are a highly diverse group of arthropod viruses with a very wide

Table 7.2 The putative nudiviruses of insects and other arthropods. Adapted from Wang *et al*. (2007c), modified from Huger and Krieg (1991)

Host		Host stage and/or tissue tropism	Size (nm)		Molecular data	References
			Virus particles	Nucleocapsids		
Insects	*Chaoborus crystallinus* (Diptera: Chaoboridae)	Larvae; midgut	260–300 × 75–106	210–226 × 38–43	–	1
	Chaoborus astictopus (Diptera: Chaoboridae)	Larvae; midgut	220 × 90	150 × 60		2
	Pulex simulans (Siphonaptera: Pulicidae)	Midgut	120 × 50	90 × 30	–	3
	Microplitis croceipes (Hymenoptera: Braconidae)	Larvae and adults; fat body, midgut and haemocytes				4
	Gyrinus natator (Cleoptera: Gyrinidae)	Adults; midgut	ca. 160 × 75	150 × 35	DNA	5
	Diabrotica undecimpunctata (Coleoptera: Chrysomelidae)	Adults; haemocytes	ca. 295 × 95	230 × 52	–	6
	Aphis sp.; *Pentalonia nigronervosa* (Homoptera: Aphidae)	Fat body and muscle cells	200–300 × 50–60	180–200 × 40–50	–	7
	Bacillus rossius (Orthoptera: Phasmidae)	Midgut	250–300×100	210 × 50–60	DNA	8
	Acheta domesticus (Orthoptera: Gryllidae)	Haemocytes (?)	–	223 × 57	–	9

Table 7.2 continued

Host		Host stage and/or tissue tropism	Size (nm)		Molecular data	References
			Virus particles	Nucleocapsids		
	Gryllus rubens (Orthoptera: Gryllidae)	Nymphs; fat body	154 × 91	127 × 55	Circular dsDNA, 87 kb; six major and 11 minor polypeptides	10
Other arthropods	*Panonychus ulmi* (Acarina: Tetranychidae)	Fat body	200 × 90	150 × 38	–	11
	Panonychus citri (Acarina: Tetranychidae)	Midgut and hindgut	266 ×111	194 × 58		12
	Penaeus japonicus (Decapoda: Penaeidae)	Adults; gill, epidermal tissues, head soft tissues and haemolymph	322–378 × 130–159	316–350 × 65–66	Four structural proteins, ca. 183 kb of genome	13, 14
	Crangon crangon (Decapoda: Crangonidae)	Hepatopancreas and midgut	280 × 72	215 × 40		15
	Carcinus maenas (Decapoda: Portunidae)	Haemocytes and connective tissues	300–320 × 90–100			16
	Callinectes sapidus (Decapoda: Portunidae)	Haemocytes and hepatopancreas		300×50		17–18

1, Larsson (1984); 2, Federici (1986); 3, Beard *et al.* (1989); 4, Hamm *et al.* (1988); 5, Gouranton (1972); 6, Kim and Kitajima (1984); 7, Kitajima *et al.* (1978); 8, Scali *et al.* (1980); 9, Grégoire (1951); 10, Boucias *et al.* 1989; 11, Bird (1967); 12, Reed and Hall (1972); 13, Nadala *et al.* (1998); 14, Sano *et al.* (1981); 15, Stentiford *et al.* (2004); 16, Bazin *et al.* (1974); 17, Johnson (1978); 18, Johnson (1984).

Table 7.3 The demarcation criteria of the baculoviruses and the nudiviruses. Adapted from Wang *et al.* (2007c)

Demarcation criteria	Baculoviruses	Nudiviruses
Rod-shaped virion	+	+
Circular dsDNA genome	+	+
Replication in nucleus	+	+
Nucleus hypertrophy	+	+
Host range		
Holometabolous	+	+
Hemimetabolous	–	+
Host stage infected		
Larvae	+	+
Adults	–	+
Horizontal transmission		
Parenteral	–	+
Peroral	+	+
Vertical transmission	+/–	+/–
Occlusion body	+	+/–

host range covering species from Crustacea, Coleoptera, Orthoptera, Lepidoptera and possibly other arthropod orders. Some of them have evolved very close association with the reproductive system of their host (Table 7.3). So far, the complete genomes of four nudiviruses have been sequenced, which are HzNV-1 (Cheng *et al.*, 2002), HzNV-2 (Burand unpublished), GbNV (Wang *et al.*, 2007b), and OrNV (Wang *et al.*, 2008). Among all nudivirus genomes, gene order is poorly conserved. Most strikingly, all sequenced nudivirus genomes harbour a gene cluster of *helicase, 19kDa* and/or *lef-5* (Wang and Jehle, 2009). A similar cluster consisting of *helicase, 19kDa, 38K* and *lef-5* was identified in all sequenced baculovirus genomes (Herniou *et al.*, 2003; Jehle and Backhaus, 1994). The existence of homologous gene clusters alone would not be an indication of a common origin of baculoviruses and nudiviruses, as such clusters could be horizontally transferred among virus ancestors. However, the presence of 20 baculovirus core genes in nudivirus genomes can be hardly explained by horizontal gene transfer, especially as these genes are involved in fundamental virus functions, such as DNA replication, nucleotide metabolism, transcription, virus morphogenesis and structure as well as oral infectivity (Wang *et al.*, 2007b; Wang and Jehle, 2009).

Based on gene sequences as well as gene content, it became obvious that both nudiviruses and baculoviruses form two sister lineages of dsDNA viruses sharing a common ancestor. In addition, nudiviruses, baculoviruses and hytrosaviruses derived from a common ancestor and that bracoviruses evolved from an ancestral nudivirus (Bézier *et al.*, 2009a,b; Wang and Jehle, 2009, see also Chapters 5 and 2 in this volume). Based on phylogenetic analyses of conserved homologous genes, it was proposed that HzNV-1 and -2 are very closely related to each other and share a recent common ancestor with PmNV (Wang and Jehle, 2009). On the other hand, GbNV and OrNV were also clustered in a single clade and emerged from a common ancestor with the HzNV/PmNV clade. According to the current understanding of nudivirus evolution; both nudiviruses and baculoviruses are sister lineages, which are derived from a common ancestor with hytrosaviruses. Some gene homologies with shrimp whispoviruses even suggests that all these viruses may have derived from a circular dsDNA marine arthropod virus. It is important to note that the proposed phylogeny of nudiviruses and baculoviruses does not mirror the present-day picture of the phylogeny of their host species and that the common ancestor of nudiviruses and baculoviruses did not date back to the origin of arthropods (Wang *et al.*, 2007b; Wang and Jehle, 2009). Thus, not the evolution of the phylogeny of the host species rather than their specific ecological niches appears to be the main driving forces of nudivirus evolution.

Concluding remarks

Nudiviruses form a highly diverse group of dsDNA viruses. So far, only a very few nudiviruses have been described in more detail and at the genomic level; many more nudiviruses await for their identification and characterization. However, the few nudiviruses presented in this chapter make clear that nudiviruses show an extreme diversity on genetic level, in their biology, their host range, as well as their transmission strategies ranging from horizontal transmission to vertical transmission or even gonad specific replication. The latter links nudiviruses to bracoviruses, which genetic relationship to nudiviruses is meanwhile corroborated, as well as their relationship to baculoviruses and hytrosaviruses is without any doubt. Thus, nudiviruses form the missing link combining these different virus taxa to a super-group of phylogenetically related dsDNA viruses of arthropods. There is no other example in invertebrate virology that allows studying so many different mechanisms of virus–host interaction, virus evolution and diversification than present in nudiviruses. For that reason, it is time to shed more light and stronger scientific focus on this unique but mainly neglected group of viruses.

Acknowledgements

Parts of work reviewed in this chapter were funded by the Deutsche Forschungsgemeinschaft (DFG) to J.A.J (Je245-7).

References

Banville, M., Dumas, F., Trifiro, S., Arif, B., and Richardson, C. (1992). The predicted amino acid sequence of the spheroidin protein from *Amsacta moorei* entomopoxvirus: lack of homology between major occlusion body proteins of different poxviruses. J. Gen. Virol. 73, 559–566.

Bazin, F., Monsarrat, T., Bonami, J.R., Croizier, G., Meynadier, G., Quiot, J. M., and Vago, C. (1974). Particules virales de type baculovirus observées chez le crabe *Carcinus maenas*. Rev. Trav. Inst. Pêches Marit. *38*, 205–208.

Beard, C.B., Butler, J.F, and Maruniak, J.E. (1989). A baculovirus in the flea, *Pulex simulans*. J. Invertebr. Pathol. *54*, 128–131.

Bedford, G.O. (1980). Biology, ecology, and control of palm rhinoceros beetles. Annu. Rev. Entomol. 25, 309–339.

Bézier, A., Annaheim, M., Herbinière, J., Wetterwald, C., Gyapay, G., Bernard-Samain, S., Wincker, P., Roditi, I., Heller, M., Belghazi, M., Pfister-Wilhem, R., Periquet, G., Dupuy, C., Huguet, E., Volkoff, A.N., Lanzrein, B., and Drezen, J.M. (2009a). Polydnaviruses of braconid wasp derive from an ancestral nudivirus. Science 323, 926–930.

Bézier, A., Herbinière, J., Lanzrein, B., and Drezen, J.M. (2009b). Polydnavirus hidden face: the genes producing virus particles of parasitic wasps. J. Invertebr. Pathol. *101*, 194–203.

Bird, F.T. (1967). A virus disease of the European red mite *Panonychus ulmi* (Koch). Can. J. Microbiol. *13*, 1131.

Bonami, J.R., Aubert, H., Mari, J., Poulos, B.T., and Lightner, D.V. (1997). The polyhedra of the occluded baculoviruses of marine decapod crustacean: a unique structure, crystal organization, and proposed model. J. Struct. Biol. *120*, 134–145.

Boucias, D.G., Maruniak, J.E., and Pendland, J.C. (1989). Characterization of a non-occluded baculovirus (subgroup C) from the field cricket, *Gryllus rubens*. Arch. Virol. *106*, 93–102.

Burand, J.P., Stiles, B., and Wood, H.A. (1983a). Structural and intracellular proteins of the nonoccluded baculovirus Hz-1. J. Virol. *46*, 137–142.

Burand, J.P., Wood, H.A., and Summers, M.D. (1983b). Defective particles from a persistent baculovirus infection in *Trichoplusia ni* tissue culture cells. J. Gen. Virol. *64*, 391–398.

Burand, J.P. (1991). Molecular Biology of the Hz-1 and *Oryctes* nonoccluded baculoviruses. In: Viruses of Invertebrates (E. Kustak ed.) Marcel Dekker, New York, 111–126.

Burand, J. P., and Lu, H. (1997). Replication of a gonad-specific insect virus in TN-368 cells in culture. J. Invertebr. Pathol. *70*, 88–95.

Burand, J.P. (1998). Nudiviruses. In: The Insect Viruses (Miller L K, Ball L A. ed.) New York: Plenum Press; 69- 90.

Burand, J.P., and Rallis, C.P. (2004). *In vivo* dose–response of insects to Hz-2V infection. Virol. J. *1*, 15.

Burand, J.P., and Tan, W. (2006). Mate preference and mating behavior of male *Helicoverpa zea* (Lepidoptera: Noctuidae) moths infected with the sexually transmitted insect virus Hz-2V. Ann. Entomol. Soc. Am. *99*, 967–973.

Burand, J.P. (2009). The sexually transmitted insect virus, Hz-2V. Virol. Sinica *24*, 428–435.

Chaivisuthangkura, P., Tawilert, C., Tejangkura, T., Rukpratanporn, S., Longyant, S., Sithigorngul, W., and Sithigorngul, P. (2008). Molecular isolation and characterization of a novel occlusion body protein gene from *Penaeus monodon* nucleopolyhedrovirus. Virology *381*, 261–267.

Chao, Y.C., Hamblin, M., and Wood, H.A. (1990). Physical map of Hz-1 baculovirus genome from standard and defective interfering particles. J. Gen. Virol. *71*, 1265–1270.

Chao, Y.C., Wood, H.A., Chang, C.Y., Lee, H.J., Shen, W.C., and Lee, H.T. (1992). Differential expression of Hz-1 baculovirus genes during productive and persistent viral infections. J. Virol. *66*, 1442–1448.

Chao, Y.C., Lee, S.T., Chang, M.C., Chen, H.H., Chen, S.S., Wu, T.Y., Liu, F.H., Hsu, E.L., and Hou, R.F. (1998). A 2.9-kilobase noncoding nuclear RNA functions in the establishment of persistent Hz-1 viral infection. J. Virol. *72*, 2233–2245.

Chen, H.H., Tsai, F.Y., and Chen, C.T. (2001). Negative regulatory regions of the PAT1 promoter of Hz-1 virus contain GATA elements which associate with cellular factors and regulate promoter activity. J. Gen. Virol. *82*, 313–20.

Chen, H.H., Tso, D.J., Yeh, W.B., Cheng, H.J., and Wu, T.F. (2001). The thymidylate synthase gene of Hz-1 virus: a gene captured from its lepidopteran host. Insect Mol. Biol. *10*, 495–503.

Cheng, C.H., Liu, S.M., Chow, T.Y., Hsiao, Y.Y., Wang, D.P., Huang, J.J., and Chen, H.H. (2002). Analysis of the complete genome sequence of the Hz-1 virus suggests that it is related to members of the *Baculoviridae*. J. Virol. *76*, 9024–9034.

Couch, J.A. (1974a). Free and occluded virus, similar to Baculovirus, in hepatopancreas of pink shrimp. Nature *247*, 229–231.

Couch, J.A. (1974b). An enzootic nuclear polyhedrosis virus of pink shrimp: Ultrastructure, prevalence, and enhancement. J. Invertebr. Pathol. *24*, 311–331.

Coulibaly, F., Chiu, E., Ikeda, K., Gutmann, S., Haebel, P.W., Schulze-Briese, C., Mori, H., and Metcalf, P. (2007). The molecular organization of cypovirus polyhedra. Nature. *446*, 97–101.

Crawford, A.M., Ashbridge, K., Sheehan, C., and Faulkner, P. (1985). A physical map of the *Oryctes* baculovirus genome. J. Gen. Virol. *66*, 2649–2658.

Crawford, A.M., and Sheehan, C. (1985). Replication of *Oryctes* baculovirus in cell culture: viral morphogenesis, infectivity and protein synthesis. J. Gen. Virol. *66*, 529–539.

Evans, L.H., and Edgerton, B.F. (2002). Pathogens, parasites and commensals. In: Biology of freshwater crayfish (Holdich D M. ed.). Oxford: Blackwell Science, 377–438.

Federici, B.A. (1986). Ultrastructure of baculoviruses. In: The Biology of Baculovirses. Volume 1 (Granados, R.R., Federici, B.A. Ed.). Boca Raton, FL: CRC Press.

Gorick, B. D. (1980). Release and establishment of the baculovirus disease of *Oryctes rhinoceros* (L.) (Coleoptera: Scarabaeidae) in Papua New Guinea. Bull. Entomol. Res. *70*, 445–453.

Gouranton, J. (1972). Development of an intranuclear non- occluded rod-shaped virus in some midgut cells of an adult insect, *Gyrinus natator* L. (Coleoptera). J. Ultrastruct. Res. *39*, 281–294.

Granados, R.R., Nguyen, T., and Cato, B. (1978). An insect cell line persistently infected with a baculovirus-like particle. Intervirology *10*, 309–317.

Grégoire, C. (1951). Virus-like bodies in the blood of the house cricket. J. Gen. Microbiol. *5*, 121–123.

Hamm, J.J., Carpenter, J.E., and Styer, E.L. (1996). Oviposition day effect on incidence of agonadal progeny of *Helicoverpa zea* (Lepidoptera: Noctuidae) infected with a virus. Ann. Entomol. Soc. Am. *56*, 535–556.

Hamm, J.J., Styer, E.L., and Lewis, W.J. (1988). A baculovirus pathogenic to the parasitoid *Microplitis croceipes* (Hymenoptera: Braconidae). J. Invertebr. Pathol. *52*, 189–191.

Herniou, E.A., Olszewski, J.A., Cory, J.S., and D. R. O'Reilly, D.R. (2003). The genome sequence and evolution of baculoviruses. Annu. Rev. Entomol. *48*, 211–234.

Huang, Y-S., Hedberg, M., and Kawanishi, C.Y. (1982). Characterization of the DNA of a Nonoccluded Baculovirus, Hz-1V. J. Virol. *43*, 174–181.

Huger, AM. (1966). A virus disease of the Indian rhinoceros beetle, *Oryctes rhinoceros* (Linnaeus), caused by a new type of insect virus, Rhabdionvirus oryctes gen. *n.*, sp. *n.* J. Invertebr. Pathol. *8*, 38–51.

Huger, A.M. (1985). A new virus disease of crickets (Orthoptera:Gryllidae) causing macronucleosis of fat body. J. Invertebr. Pathol. *45*, 108–111.

Huger, A.M., and Krieg, A. (1991). Baculoviridae. Nonoccluded Baculoviruses. In: Atlas of Invertebrate Viruses (Adams, J.R., Bonami, J.R. ed.). Boca Raton, CRC Press, Inc., 287–319.

Huger, A.M. (2005). The *Oryctes* virus: its detection, identification, and implementation in biological control of the coconut palm rhinoceros beetle, *Oryctes rhinoceros* (Coleoptera: Scarabaeidae). J. Invertebr. Pathol. *89*, 78–84.

Jacob, T. K. (1996). Introduction and establishment of baculovirus for the control of rhinoceros beetle *Oryctes rhinoceros* (Coleoptera: Scarabaeidae) in the Andaman Islands (India). Bull. Entomol. Res. *86*, 257–262

Jackson, T.A., Crawford, A.M., and Glare, T.R. (2005). *Oryctes* virus – time for a new look at a useful biocontrol agent. J. Invertebr. Pathol. *89*, 91–94.

Jehle, J.A., and Backhaus, H. (1994). The genome organization of the DNA-binding protein gene region of *Cryptophlebia leucotreta* granulosis virus is closely related to that of nuclear polyhedrosis viruses. J. Gen. Virol. *75*, 1815–1820.

Jehle, J.A., Blissard, G.W., Bonning, B.C., Cory, J.S., Herniou, E.A., Rohrmann, G.F., Theilmann, D.A., Thiem, S.M., and Vlak, J.M. (2006). On the classification and nomenclature of baculoviruses: a proposal for revision. Arch. Virol. *151*, 1257–1266.

Johnson, P.T. (1978). Viral disease of the blue crab, *Callinectes sapidus*. Mar. Fish Rev. *40*, 13–15.

Johnson, P.T. (1984). Viral diseases of marine invertebrates. Helgoland Mar. Res. *37*, 65–98.

Johnson, P.T., and Lightner, D.V. (1988) Rod-shaped nuclear viruses of crustaceans: Gut-infecting species. Dis. Aquat. Org. *5*, 123–141.

Kelly, D.C., Lescott, T., Ayres, M.D., Carey, D., Coutts, A., and Harrap, K.A. (1981). Induction of a nonoccluded baculovirus persistently infecting *Heliothis zea* cells by *Heliothis armigera* and *Trichoplusia ni* nuclear polyhedrosis viruses. Virology *112*, 174–189.

Kim, K.S., and Kitajima, E.W. (1984). Nonoccluded baculovirus– and filamentous virus–like particles in the spotted cucumber beetle, *Diabrotica undecimpunctata* (Coleoptera: Chrysomelid). J. Invertebr. Pathol. *43*, 234–241.

Kitajima, E.W., Costa, C.L., and Sá, C.M. (1978). Baculovirus–like particles in two aphid species J. Invertebr. Pathol. *31*, 123–125.

Langridge, W.H. (1981). Biochemical properties of a persistent nonoccluded baculovirus isolated from *Heliothis zea* cells. Virology *30*, 770–774.

Larsson, R. (1984). Baculovirus–like particles in the midgut epithelium of the phantom midge, *Chaoborus crystalllinus* (Diptera, Chaoboridae). J. Invertebr. Pathol. *44*, 178–186.

Lee, J.C., Chen, H.H., Wie, H.L., and Chao, Y.C. (1993). Superinfection-induced apoptosis and its correlation with the reduction of viral progeny in cells persistently infected with Hz-1 baculovirus. J Virol. *67*, 6989–6994.

Lee, J.C., and Chao, Y.C. (1998). Apoptosis resulting from superinfection of *Heliothis zea* virus 1 is inhibited by p35 and is not required for virus interference. J. Gen. Virol. *79*, 2293–300.

Lightner, D.V., and Redman, R.M. (1981). A baculovirus-caused disease of the penaeid shrimp, *Penaeus monodon*. J. Invertebr. Pathol. *38*, 299–302.

Lin, C.L., Lee, J.C., Chen, S.S., Wood, H.A., Li, M.L., Li, C.F., and Chao, Y.C. (1999). Persistent Hz-1 virus infection in insect cells: evidence for insertion of viral DNA into host chromosomes and viral infection in a latent status. J. Virol. *73*, 128–139.

Lu, H., and Burand, J.P. (2001). Replication of the gonad-specific virus Hz-2V in Ld652Y cells mimics replication *in vivo*. J. Invertebr. Pathol. *77*, 44–50.

Marshall, K.J. (1970). Introduction of a new virus disease of the coconut rhinoceros beetle in West Samoa. Nature *225*, 288–289.

McIntosh, A.H., Grasela, J.J., and Ignoffo, C.M. (2007). *In vitro* host range of the Hz-1 nonoccluded virus in insect cell lines. *In Vitro* Cell. Dev. Biol. Anim. *43*, 196–201.

Mayo, M.A. (1995). Unassigned Viruses. In: Virus Taxonomy: The Sixth Report of the International Committee on Taxonomy of Viruses (Murphy, F.A., Fauquet, C.M., Bishop, D.H.L., Ghabrial, S.A., Jarvis, A.W., Martelli, G.P., Mayo, M.A., and Summers, M.D. ed.). Wien: Springer-Verlag, 504–507.

Mohan, K.S., and Gopinathan, K.P. (1991). Physical mapping of the genomic DNA of the *Oryctes rhinoceros* baculovirus, KI. Gene *107*, 343–334.

Monty, J. (1978). The coconut palm rhinoceros beetle, *Oryctes rhinoceros* (L.) (Col., Dynastidae) in Mauritius and its control. Rev. Agricole et Sucrière de l'Ile Maurice *57*, 60–76.

Nadala, E.C., Tapay Jr., L.M., and Loh, P.C. (1998). Characterization of a non-occluded baculovirus-like agent pathogenic to penaeid shrimp. Dis. Aquat. Organ. *33*, 221- 229.

Payne, C.C. (1974). The isolation and characterization of a virus from *Oryctes rhinoceros*. J. Gen. Virol. *25*, 105–116.

Payne, C.C., Compson, D., and De Looze, S.M. (1977). Properties of the nucleocapsids of a virus isolated from *Oryctes rhinoceros*. Virology *77*, 269–280.

Perera, O.P., Valles, S.M., Green, T.B., White, S., Strong, C.A., and Becnel, J.J. (2006). Molecular analysis of an occlusion body protein from *Culex nigripalpus* nucleopolyhedrovirus (CuniNPV). J. Invertebr. Pathol. *91*, 35–42.

Raina, A.K., and Adams, J.R. (1995). Gonad-specific virus of corn earworm. Nature *374*, 770.

Raina, A.K., Adams, J.R., Lupiani, B., Lynn, D.E., Kim, W., Burand, J.P., and Dougherty, E.M. (2000). Further characterization of the gonad-specific virus of corn earworm, *Helicoverpa zea*. J. Invertebr. Pathol. *76*, 6–12.

Raina, A.K., and Lupiani, B. (2006). Acquisition, persistence, and species susceptibility of the Hz-2V virus. J. Invertebr. Pathol. *93*, 71–74.

Rallis, C.P., and Burand, J. P. (2002a). Pathology and ultrastructure of the insect virus, Hz-2V, infecting agonadal male corn earworms, *Helicoverpa zea*. J. Invertebr. Pathol. *80*, 81–89.

Rallis, C.P., and Burand, J.P. (2002b). Pathology and ultrastructure of the insect virus, Hz-2V, infecting agonadal female corn earworms, *Helicoverpa zea*. J. Invertebr. Pathol. *81*, 33–44.

Ralston, A.L., Huang, Y-S., and Kawanishi, C.Y. (1981). Cell culture studies with the IMC-Hz-1 nonoccluded virus. Virology *115*, 33–44.

Ramle, M., Wahid, M.B., Norman, K., Glare, T.R., and Jackson, T.A. (2005). The incidence and use of *Oryctes* virus for control of rhinoceros beetle in oil palm plantations in Malaysia. J. Invertebr. Pathol. *89*, 85–90.

Reed, D.K., and Hall, I.M. (1972). Electron microscopy of a rod-shaped noninclusion virus infecting the citrus red mite. J. Invertebr. Pathol. *20*, 272–278.

Richards, N.K., Glare, T.R., Aloali'I, I., and Jackson, T.A. (1999). Primers for the detection of *Oryctes* virus from Scarabaeidae (Coleoptera). Mol. Ecol. *8*, 1552–1553.

Sano, T., Nishimura, T., Oguma, K., Momoyama, K., and Takeno, N. (1981). Baculovirus infection of cultured Kuruma shrimp, *Penaeus japonicus*, in Japan. Fish Pathol. *15*, 185–191.

Scali V, Montanelli E, Lanfranchi A, and Bedini, C. (1980). Nuclear alterations in a baculovirus-like infection of mid- gut epithelial cells in the stick insect, *Bacillus rossius*. J Invertebr. Pathol. 35, 109–118.

Shyam Prasad, G., Jayakumar, V., Ranganath, H.R., and Bhagwat, V.R. (2008). Bio-suppression of coconut rhinoceros beetle, *Oryctes rhinoceros* L. (Coleoptera: Scarabaeidae) by *Oryctes* baculovirus (Kerala isolate) in South Andaman, India. Crop Prot. *27*, 959–964.

Stentiford, G.D., Bateman, K., and Feist, S.W. (2004). Pathology and ultrastructure of an intranuclear bacilliform virus (IBV) infecting brown shrimp *Crangon crangon* (Decapoda: Crangonidae). Dis. Aquat. Organ. *58*, 89–97.

van Oers, M.M., and Vlak, J.M. (2007). Baculovirus genomics. Curr. Drug. Targets *8*, 1051–1068.

Wang, Y., van Oers, M.M., Crawford, A.M., Vlak, J.M., and Jehle, J.A. (2007a). Genomic analysis of *Oryctes rhinoceros* virus reveals genetic relatedness to *Heliothis zea* virus 1. Arch. Virol. *152*, 519–531.

Wang, Y., Kleespies, R.G., Huger, A.M., and Jehle, J.A. (2007b). The genome of *Gryllus bimaculatus* nudivirus indicates an ancient diversification of baculovirus-related nonoccluded nudiviruses of insects. J. Virol. *81*, 5395–5406.

Wang, Y., Burand, J.P., and Jehle, J.A. (2007c). Nudiviruses genomics: diversity and classification. Virol. Sinica 22,128–136.

Wang, Y., Kleespies, R.G., Ramle, M.B., and Jehle, J.A. (2008). Sequencing of the large dsDNA genome of *Oryctes rhinoceros* nudivirus using multiple displacement amplification of nanogram amounts of virus DNA. J. Virol. Meth. *152*, 106–108.

Wang, Y., and Jehle, J.A. (2009). Nudiviruses and other large, double-stranded circular DNA viruses of invertebrates: new insights on an old topic. J. Invertebr. Pathol. *101*, 187–193.

Wongteerasupaya, C., Vickers, J.E., Sriurairatana, S., Nash, G.L., Akarajamorn, A., Boonsaeng, V., Panyim, S., Tassanakajon, A., Withyachumnarnkul, B., and Flegel, T.W. (1995). A non-occluded, systemic baculovirus that occurs in cells of ectodermal and mesodermal origin and causes high mortality in the black tiger prawn *Penaeus monodon*. Dis. Aquat. Org. *21*, 69–77.

Wu, Y.L., Liu, C.Y., Wu, C.P., Wang, C.H., Lee, S.T., and Chao, Y.C. (2008). Cooperation of ie1 and p35 genes in the activation of baculovirus AcMNPV and HzNV-1 promoters. Virus Res. *135*, 247–254.

Wu, Y.L., Wu, C.P., Lee, S.T., Tang, H., Chang, C.H., Chen, H.H., and Chao, Y.C. (2010). The early gene hhi1 reactivates *Heliothis zea* nudivirus 1 in latently infected cells. J. Virol. *84*,1057–65.

Yuen, L., Dionne, J., Arif, B., and Richardson, C. (1990). Identification and sequencing of the spheroidin gene of *Choristoneura biennis* entomopoxvirus. Virology *175*, 427–433.

Zelazny B., and Alfiler, A.R. (1991). Ecology of baculovirus-infected and healthy adults of *Oryctes rhinoceros* (Coleoptera: Scarabaeidae) on coconut palms in the Philippines. Ecol. Entomol. *16*, 253–259.

Zelazny B., Lolong, A., and Pattang, B. (1992) *Oryctes rhinoceros* (Coleoptera: Scarabaeidae) populations suppressed by baculovirus. J. Invertebr. Pathol. *59*, 61–68.

Polydnaviruses

8

Michael R. Strand

Abstract

Viruses in the family *Polydnaviridae* are symbiotically associated with parasitoid wasps in the families Braconidae and Ichneumonidae. Polydnaviruses (PDVs) exist in two forms. In wasps, they persist and are transmitted to offspring as stably integrated proviruses. Replication is restricted to specialized cells in the ovaries of females. This results in production of the encapsidated form of the virus, which wasps inject into hosts when they oviposit. PDVs do not replicate in the wasp's host but they do infect different tissues and express gene products that facilitate successful development of the wasp's offspring. In this chapter, I first discuss the phylogenetic distribution of PDVs in wasps and key features that distinguish the encapsidated form of these viruses. I then review current understanding of how the proviral form of PDVs is organized in the genome of wasps, how PDVs replicate, and the origins of this virus family. I end the chapter with a discussion of the gene products PDVs produce and the role they play in parasitism.

Introduction

The chapters of this volume collectively indicate that viruses from diverse families infect and cause disease in insects. The focus of this chapter is the *Polydnaviridae*, whose members have evolved a symbiotic rather than strictly parasitic association with insects. Structurally, polydnaviruses (PDVs) are also the only known DNA viruses with segmented genomes. PDVs are only associated with parasitoid wasps (Hymenoptera), whose hosts are primarily larval stage Lepidoptera (moths and butterflies). Each PDV carried by a given wasp species is genetically unique, and all PDVs persist as stably integrated proviruses that are vertically transmitted to offspring. Replication is restricted to cells in the reproductive tracts of adult female wasps. Wasps inject the encapsidated form of PDVs into hosts when laying their eggs. PDVs express multiple gene products in the wasp's host that are essential for survival of its offspring. However, this form of the virus does not replicate (Webb and Strand, 2005; Dupuy *et al.*, 2006; Strand, 2009). Thus, a mutualism exists because PDV transmission as a provirus depends upon survival of the wasp, and survival of the wasp depends upon infection of its host by the encapsidated, non-replicating form of the virus. In recent years, our understanding of PDV function in parasitism has increased greatly. Studies also provide important insights into the organization of the proviral and encapsidated forms of PDV genomes, and the evolutionary origins of PDVs. Here, I summarize this literature by reviewing key findings and synthesizing results from studies conducted in different species of PDV-carrying wasps.

The biology of PDV-carrying parasitoids

Understanding the biology of PDVs first requires some knowledge of the parasitoid wasps that carry them. The Hymenoptera (wasps, bees and ants) is one of the megadiverse orders of holometabolous insects, and its members exhibit a tremendous diversity of biological habits (Fig. 8.1). The

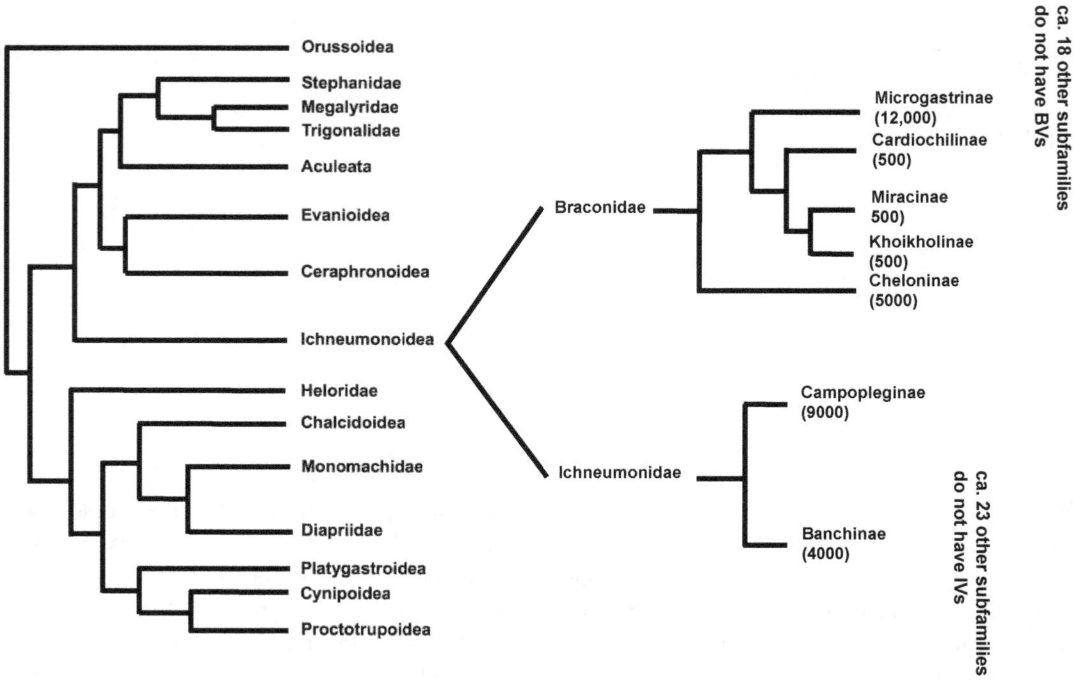

Figure 8.1 Higher order phylogeny of the Hymenoptera and the distribution of polydnaviruses. The Hymenoptera is divided into two suborders: the Symphyta (sawflies) and the Apocrita which contains more than 200,000 species. The left portion of the figure shows the major superfamilies of the Aprocrita. Most of these superfamilies are comprised of species that are parasitoids. The right portion of the figure shows the division of the Ichnemonoidea into the Braconidae and Ichneumonidae, which are further divided into multiple subfamilies. PDVs in the genus *Bracovirus* are associated with all known species in five subfamilies of the Braconidae that form the microgastroid complex. PDVs in the genus *Ichnovirus* are associated with two subfamilies of the Ichneumonidae. The estimated number of species in each PDV-carrying subfamily is indicated in parentheses.

order is divided into the Symphyta (sawflies), which are herbivores, and the Apocrita, which are primarily parasitoids. Parasitoids are defined as insects that are parasites of other arthropods during their immature stages but are free-living as adults (Godfray, 1994; Dowton and Austin, 1994; Whitfield, 1998). Almost all parasitoids develop by laying one or more eggs in or on a single host, which serves as a food source for the wasp's larval-stage offspring. Offspring almost always complete their immature development by feeding on a single host, and the host almost always dies when the parasitoid larva(e) completes its final instar and pupates. Parasitoids that carry PDVs are restricted to a single superfamily of the Hymenoptera called the Ichneumonoidea, which contains an estimated 100,000 species. The Ichneumonoidea in turn is divided into two families, the Braconidae and Ichneumonidae, which are each comprised of many subfamilies.

The *Polydnaviridae* parallels this scheme in that it too is currently divided into two genera: *Bracovirus* (BV), which is associated with an estimated 18,000 species of braconids in five subfamilies, and *Ichnovirus* (IV), which is associated with an estimated 13,000 species of ichneumonids in two subfamilies (Stoltz and Vinson, 1979; Whitfield and Asgari, 2003; Webb and Strand, 2005) (Fig. 8.1). All PDV-carrying parasitoids are endoparasitic, which means the adult female wasp injects her eggs into the body of a host, and the progeny develop within the host's haemocoel. BV-carrying braconids are either solitary, with a female laying one egg and

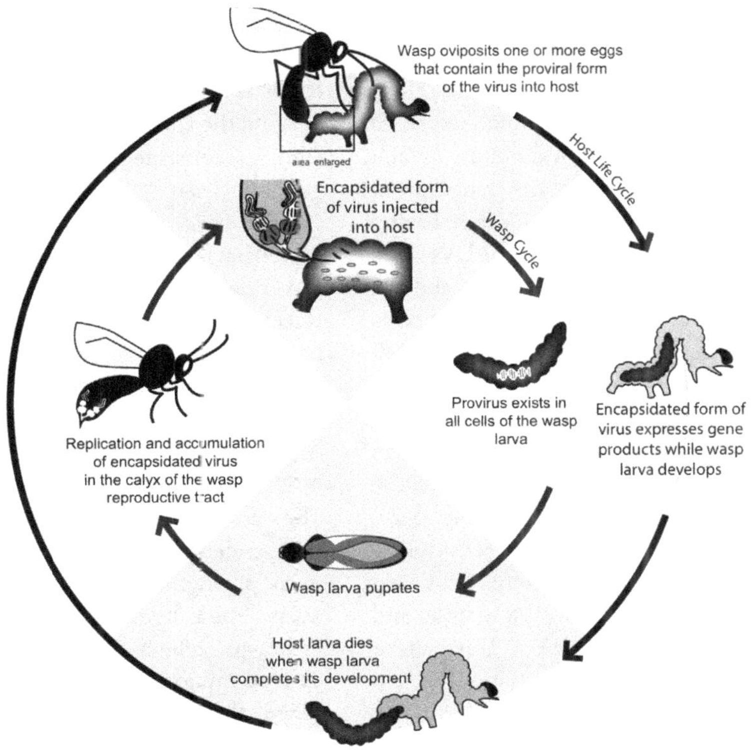

Figure 8.2 The polydnavirus life cycle. Polydnaviruses exist in two forms: a provirus integrated into the genome of wasps, and an encapsidated form that accumulates in the reproductive tract of females. Female wasps inject one or more eggs, containing the provirus, and a quantity of encapsidated virus into hosts. After infection, genes encoded by the encapsidated virus are expressed in the tissues of hosts, which allows the wasp's progeny to develop. The parasitoid's offspring, which contain provirus, complete their larval stage and pupate, while the host dies. Late in the pupal stage replication begins in the reproductive tract of female wasps, which results in production of the encapsidated form of the virus.

a single wasp offspring developing per host, or gregarious with wasps laying multiple eggs and several offspring developing per host. BV-carrying wasp larvae feed primarily on host haemolymph and emerge from the host in their final instar to pupate. Although most BV-carrying species oviposit and emerge from the larval stage of the host, all species of braconids in the subfamily Cheloninae oviposit into the egg stage of the host and the progeny emerge from the host's penultimate larval stadium (Lawrence and Lanzrein, 1993). IV-carrying ichneumonids in contrast are exclusively solitary endoparasitoids, which oviposit into host larvae. Their offspring usually consume the entirety of the host and pupate within its remnant cuticle. Most if not all PDV-carrying parasitoids have highly restricted host ranges that include one or at most a few closely related species (Whitfield, 2002). As previously noted, most of these hosts are in the order Lepidoptera but some PDV-carrying wasps parasitize hosts in other orders including the Hymenoptera and Coleoptera. It is also important to recognize that different host species are commonly parasitized by more than one species of parasitoid (Godfray, 1994; Pennacchio and Strand, 2006).

Classification of the Polydnaviridae

Viruses are usually classified on the basis of several phenotypic characteristics that include nucleic acid type, morphology, replication strategy, and life cycle. The *Polydnaviridae* was originally classified as a virus family on the basis that the encapsidated form of BVs and IVs possess segmented, circular, double-stranded DNA genomes with individual genomic segments being present in non-equimolar amounts (Webb *et al.*, 2000). Genome segments range from 2.0 to more than 30 kb with BVs usually having fewer but larger segments than IVs. PDVs were also recognized as a family because BVs and IVs exhibit similar life cycles (Fig. 8.2). Each persists as a provirus integrated into the genome of all cells in both sexes of the associated wasp species. Replication in contrast only occurs in female wasps in specialized cells that form a region between the ovaries and lateral oviducts of the reproductive tract called the calyx (Fig. 8.2). Replication is first detected in calyx cells during the late pupal stage and usually continues during adulthood (Norton and Vinson, 1983; Stoltz and Vinson, 1979; Volkoff *et al.*, 1995). This pattern of activity suggests that the initiation of BV and IV replication involves both tissue and sex specific factors from the wasp but little is known currently about what these signals are except the observation that replication begins in association with a rise in the wasp's ecdysteroid titre (Webb and Summers, 1992; Webb, 1998; Gruber *et al.*, 1996). Regardless of what stimulates replication, it results in virions accumulating to high densities in the lumen of the calyx to form a suspension of virus called calyx fluid, which wasps inject a quantity of when ovipositing into a host. BV and IV virions infect several different host cell types, followed by expression of multiple viral gene products. These products affect two processes considered essential for survival of the wasp's offspring. First, many PDVs induce alterations that prevent the host's immune system from killing wasp eggs and/or larvae (summarized by Webb and Strand, 2005; Strand, 2009). Second, PDVs often cause alterations in host growth and/or development (Beckage and Gelman, 2004; Pennacchio and Strand, 2006). Neither BVs nor IVs replicate in the wasp's host but both are vertically transmitted to offspring via stable integration of proviral DNA into the genome of eggs and sperm (Fleming and Summers, 1991; Belle *et al.*, 2002; Dejardens *et al.*, 2008) (Fig. 8.2).

The association of BVs and IVs with wasps likely arose independently

Despite the similarities in life cycle, phylogenetic studies combined with morphological and genomic data strongly suggest that the *Polydnaviridae* is not a natural taxon. Instead, current evidence indicates that BVs derive from a common ancestor but are not related to IVs. Recent results further suggest IVs themselves may have more than one origin.

BV-carrying braconids form a monophyletic assemblage separate from IV-carrying ichneumonids

The first line of evidence arguing for the independent origins of BVs and IVs derives from phylogenetic analysis of their associated wasps. These studies indicate that the subfamilies of wasps, which carry BVs form a relatively derived monophyletic lineage in the Braconidae called the microgastroid complex (Fig. 8.1). Fossil calibrations combined with molecular data indicate the microgastroid lineage arose approximately 100 million years ago (MYA) (Whitfield, 2002; Murphy *et al.*, 2008). Since BVs are only transmitted vertically, these data further suggest that the BV–braconid association arose from an interaction established between a single ancestral virus and the common ancestor of the microgastroids. This interaction then diversified into the thousands of BV-carrying braconids that exist today. The phylogenetic relationships among subfamilies in the Ichneumonidae are less clear (Dowton and Austin, 1998; Quicke *et al.*, 2000), but current data indicate that IVs are restricted to the Campopleginae and Banchinae (Stoltz and Whitfield, 1992) (Fig. 8.1). Analyses based on morphological and 28S rDNA data do not resolve the relationship between campoplegines and banchines, although both are derived groups within the Ichneumonidae (Quicke *et al.*, 2000). Taken together, the absence of PDVs from any of the basal lineages of the Ichneumonoidea suggest either a common ancestor of the Braconidae and Ichneumonidae lost its association with PDVs, or the association of BVs with braconids and IVs with ichneumonids arose independently. The absence of PDVs from several subfamilies of ichneumonids more closely related to the

Campopleginae than the Banchinae suggest that the viruses associated with these two subfamilies could also have separate origins.

BV and IV virions differ morphologically

Differences in virion morphology also support an independent origin for BVs and IVs with ichneumonoid wasps. BV virions consist of a single unit membrane enveloping one or more cylindrical nucleocapsids that possess tails (Stoltz *et al.*, 1976; Stoltz and Vinson, 1977; Beck *et al.*, 2007) (Fig. 8.3A and B). Capsids, excluding tails, are uniform in width (30–50 nm) but vary greatly in length (30–150 nm). Early studies further noted that BVs morphologically resemble nudiviruses or baculoviruses (Stoltz and Vinson, 1979) (see below and Chapters 2 and 7). BVs acquire their envelope in the nuclei of calyx cells, presumably by *de novo* membrane synthesis, and are released into the lumen of the calyx by lysis of calyx cells (Stoltz and Vinson, 1977; Stoltz and Vinson, 1979). Thus, production of BV virions requires a continuous differentiation of virus-producing calyx cells from a stem cell precursor (Stoltz and Vinson, 1979). Although cell lysis results in release of other debris, Wyler and Lanzrein (2003) report a structure in the braconid calyx that prevents cellular debris from entering the calyx fluid while allowing entry of virions. Structural analysis indicates that BV virions consist of multiple proteins (Bezier *et al.*, 2009a).

In contrast, IVs from campoplegine wasps have two unit membranes that envelope a single fusiform nucleocapsid with a short tail-like appendage (Stoltz and Vinson, 1979; Volkoff *et al.*, 1995) (Fig. 8.3C). Capsids, minus tail, are approximately 85 × 330 nm, and virions are released into the lumen of the calyx by budding through the plasma membrane of calyx cells. Thus, the inner membrane is acquired *de novo* in the nucleus while the outer membrane is acquired

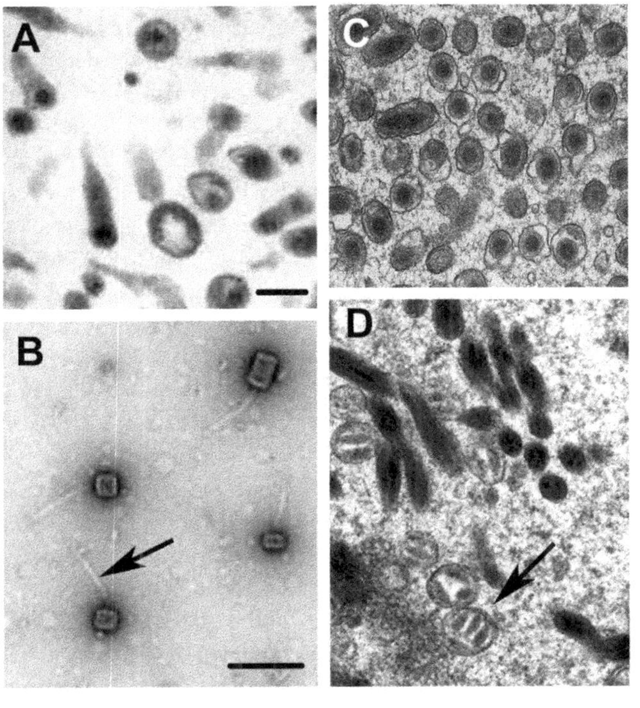

Figure 8.3 Electron micrographs of bracoviruses and ichnoviruses. (A) *Microplitis demolitor* bracovirus (MdBV) virions (scale bar = 1 µm). (B) Negatively stained MdBV nucleocapsids after removal of the enveloping membrane (scale bar = 200 nm). Note the presence of a tail at the base of each barrel-shaped capsid (arrow). C) Virions from the campoplegine icheumonid *Hyposoter fugitivus* (HfIV). D) Virions from the banchine ichneumonid *Glypta fumiferanae* (GfV). Note the multiple virions within the outer envelope (arrow).

during the budding process (Norton and Vinson, 1983; Stoltz and Vinson, 1979; Yin *et al.*, 2003). Several authors have noted that morphologically IVs resemble ascoviruses (Federici and Bigot, 2003; Bigot *et al.*, 2008) (see Chapter 1). Like BVs, IV virions consist of multiple proteins (Webb, 1998; Webb and Strand, 2005). Virions from only one banchine ichneumonid have been examined to date. They too consist of two unit membranes but envelop multiple quasi-cylindrical nucleocapsids distinct from both BVs and campoplegine IVs (Lapointe *et al.*, 2007) (Fig. 8.3D).

The encapsidated forms of BV and IV genomes exhibit limited sequence homology

The encapsidated (episomal) genomes of four BVs (*Cotesia congregata* bracovirus (CcBV) (Espagne *et al.*, 2004), *Microplitis demolitor* bracovirus (MdBV) (Webb *et al.*, 2006), and *Glyptapanteles indiensis* bracovirus (GiBV) plus *Glyptapaneles flavicoxis bracovirus* (GfBV) (Desjardins *et al.*, 2008)), three IVs carried by campoplegine ichneumonids (*Campoletis sonorensis* ichnovirus (CsIV) (Webb *et al.*, 2006), *Hyposoter fugitivus* ichnovirus (HfIV) (Tanaka *et al.* 2007), and *Tranosema rostrale* ichnovirus (TrIV) (Tanaka *et al.*, 2007)), and one IV carried by a banchine ichneumonid (*Glypta fumiferanae* ichnovirus (GfIV); Lapointe *et al.*, 2007) have been fully sequenced. In addition, partial sequence data are available from several other BVs including *Chelonus inanitus* bracovirus (CiBV) (Weber *et al.*, 2007) and *Cotesia plutellae* bracovirus (CpBV) (Choi *et al.*, 2009).

Analysis of these genomes reveals several shared features in addition to being segmented and individual genome segments being non-equimolar in abundance (Kroemer and Webb, 2004a; Webb and Strand, 2005; Strand, 2009) (Fig. 8.4). First, the encapsidated form of BV and IV genomes exhibit low coding densities and strong A + T biases. Second, a majority of predicted genes consist of related variants, which form multimember gene families that are transcribed in parasitized hosts (Fig. 8.4). Third, the encapsidated form of BV and IV genomes largely lack genes with similarity to genes known from other viruses including, importantly, factors required for replication of DNA viruses or virion assembly. Fourth, several but not all genes encoded by BVs and IVs contain introns that are usually 100–200 bp in length.

On the other hand, the encapsidated genomes of BVs and IVs also exhibit several differences in genome organization and gene content.

BV genomes

The number of genomic segments varies from as few as 6 DNAs that range from 6 to 23 kb in the case of *Toxoneuron nigriceps* bracovirus (TnBV) (Varricchio *et al.*, 1999) to 15 segments that range from 5 to greater than 30 kb for *Microplitis demolitor* bracovirus (MdBV) (Webb *et al.*, 2006), and 30 DNAs from 10 to more than 20 kb for *Cotesia congregata* bracovirus (CcBV) (Espagne *et al.*, 2004) (Fig. 8.4). The aggregate size of BV genomes is also variable, ranging from 188 kb for MdBV to more than 600 kb for CcBV, *Glyptapanteles indiensis* bracovirus (GiBV) and *G. flavicoxis* bracovirus (GfBV) (Espagne *et al.*, 2004; Webb *et al.*, 2006; Dejardens *et al.*, 2008). Kleinschmidt preparations indicate that BV nucleocapsids encapsidate only a single genomic DNA segment with capsid size correlating with segment size (Albrecht *et al.*, 1994; Beck *et al.*, 2007). The non-equimolar abundance of different DNAs also correlates with capsid size. For example, the MdBV genome consists of 15 segments, but five of these segments (B, H, J, N, O) packaged into different size classes of nucleocapsids, account for more than 60% of the total viral DNAs in calyx fluid (Beck *et al.*, 2007) (Fig. 8.4).

Studies to date suggest that gene conservation among BVs from wasps in the same or closely related genera is higher than among BVs from more distant taxa. For example, *Cotesia congregata* is in the subfamily Microgastrinae. The encapsidated genome of CcBV contains five main gene families: (1) the protein tyrosine phosphatase (*ptp*) genes (27 genes) that encode predicted proteins related to PTPs of the classical subtype; (2) the ankyrin motif (*ank*) genes (six genes) that are related to inhibitor κB (IκB) proteins which function as negative regulators of NF-κB transcription factors, (3) the *crp* gene family (four genes) that encodes for predicted cysteine-rich proteins, (4) the *cyst* gene family

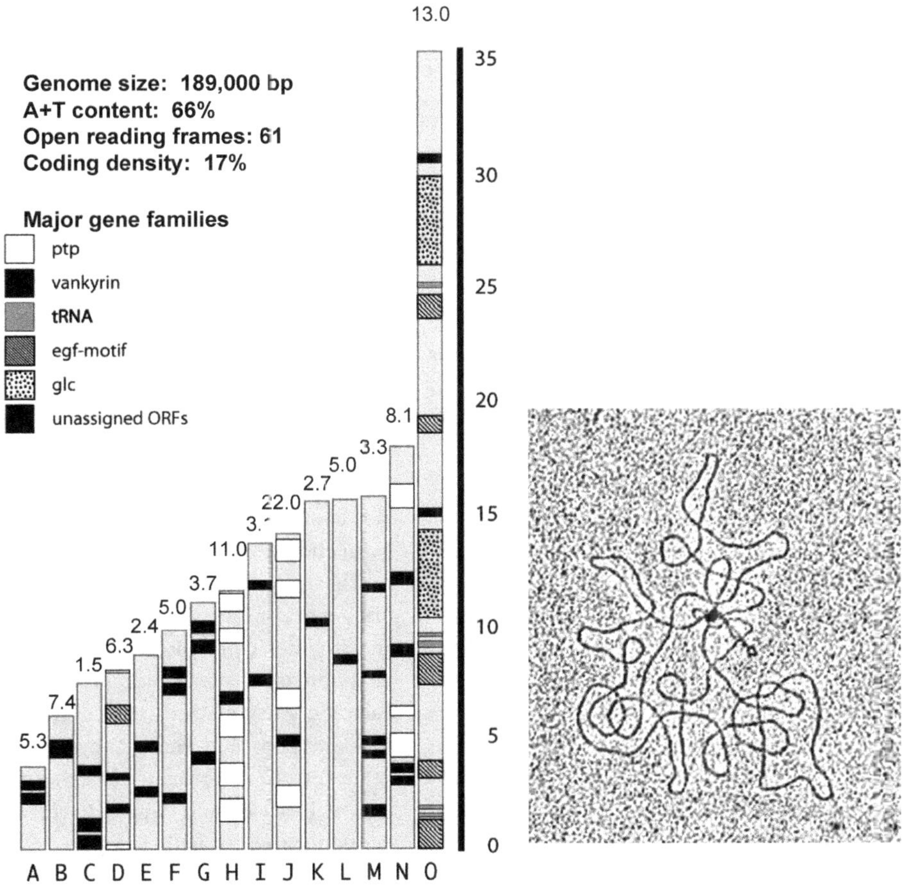

Figure 8.4 Schematic illustrating the segmentation, non-equimolar abundance, and packaging of the encapsidated genomes of PDVs. The left side of the figure shows the segmented genome (linearized) of *Microplitis demolitor* bracovirus (MdBV). Aggregate genome size, A+T content, number of predicted open reading frames (>100 amino acids), and average coding density is indicated. The distribution of the major gene families among the 15 segments of the genome is indicated. Numbers above each segment indicate their relative abundance (%) in calyx fluid. The right side of the figure shows a Kleinschmidt preparation from a single MdBV capsid. The capsid contains only one circular, double-stranded DNA. Adapted from Webb *et al.* (2006) and Beck *et al.* (2007).

(three genes) that encode predicted cysteine proteinases, and (5) the *ep1-like* family (six genes) that encode novel proteins (Espagne *et al.*, 2004). Most other genes in the CcBV genome are single copy. Homologues to these gene families and a number of the single copy genes in the CcBV genome have been identified in BVs from other *Cotesia* species (Serbielle *et al.*, 2009; Choi *et al.*, 2009). BVs associated with wasps in the genus *Glyptapanteles* (GiBV and GfBV, see above), the sister group to *Cotesia* (Murphy *et al.*, 2008), also encode some of the same gene families (*ptps, anks, cyst*) as found in *Cotesia* BVs, although gene family sizes differ (Desjardins *et al.*, 2008). *Glyptapanteles* BVs, however, also encode genes, such as a sugar transporter gene family, that are absent from *Cotesia* BVs (Desjardins *et al.*, 2008). BVs associated with more distantly related taxa share fewer genes still with *Cotesia* BVs. The genus *Microplitis,* for example, is in the Microgastrinae but is distantly related from the genus *Cotesia* (Murphy *et al.*, 2008). Correspondingly, the encapsidated genome of MdBV shares only two gene families (*ptps, anks*) with CcBV (Webb *et al.*, 2006), while *Chelonus inanitus* bracovirus (CiBV), which is in the Cheloninae (Fig. 8.1), appears to share no genes with either CcBV or MdBV (Annaheim and Lanzrein, 2007; Weber *et al.*, 2007). Transcriptional studies of BV gene activity indicate that most genes encoded by the encapsidated genome are expressed in parasitized hosts but not in the adult wasp. In hosts, some BV genes exhibit stage-specific patterns of expression (Asgari *et al.*, 1997; Bonvin *et al.*, 2005; Pfister-Wilhelm and Lanzrein, 2009), while others exhibit tissue-specific patterns of expression (Teramoto and Tanaka, 2003; Pruijssers and Strand, 2007; Beck *et al.*, 2007). Weber *et al.* (2007) also note that genes on a given DNA segment of the CiBV genome exhibit similar stage-specific patterns of expression.

Taken together, these data indicate that each BV from a given wasp species is genetically unique with relatedness among isolates paralleling wasp phylogeny. Comparative sequence analysis within species suggests BV gene families have diversified through a combination of gene duplication events within a given genomic segment and inter-genomic segment recombination events (Espagne *et al.*, 2004; Webb *et al.*, 2006; Friedman and Hughes, 2006). Comparisons among species also provide evidence for acquisition and/or loss of genes at different times in the radiation of BV-associated wasps that have resulted in divergent gene content. Some of this divergence likely reflects wasp ancestry, but evidence of positive selection on certain gene family members also suggests that the gene content of BVs is influenced by the particular host(s) a given wasp species parasitizes (Serbielle *et al.*, 2008; Desjardins *et al.*, 2008; Bezier *et al.*, 2008; Lu *et al.*, 2008). Several genes including *ptps, anks* and *cysts* also exhibit features consistent with having originated from wasps, hosts, or other arthropods (Webb *et al.*, 2006; Pruijssers and Strand, 2007; Desjardins *et al.*, 2008; Bezier *et al.*, 2008). Thus, the genes present in the encapsidated form of BV genomes reflect the influence of both ancestry and adaptations to the different hosts parasitized by wasps.

IV genomes from campoplegine ichneumonids

The encapsidated form of IVs from campoplegine ichneumonids consist of 20–25 non-redundant segments that range in size from 2 to 28 kb (Webb *et al.*, 2006; Tanaka *et al.*, 2007). Aggregate, non-redundant genome sizes are similar for CsIV, HfIV, and TrIV (250 kb) (Webb *et al.*, 2006; Tanaka *et al.*, 2007). Unlike BVs, a subset of the unique genomic segments of campoplegine IVs also give rise to additional segments by intramolecular recombination after excision (Kroemer and Webb, 2004a; Webb and Strand, 2005) (see below). This process is called segment nesting, because a unique 'parental' segment serves as the template for production of several smaller, 'nested' segments (Xu and Stoltz, 1993; Cui and Webb, 1997; Webb and Cui, 1998). In CsIV, for example, segment W (15.8 kb) gives rise to segments R (13.1 kb), M (10.4 kb) and C2 (Cui and Webb, 1997).

IVs from campoplegine ichneumonids share six gene families designated as (1) the ankyrin motif genes (*ank*), (2) the innexin genes (*inex*) that encode predicted proteins involved in forming gap junctions, (3) cysteine-motif genes (*cys-motif*) which share a cysteine-rich domain related to ω-conotoxins, (4) the repeat element genes (*rep*) which share a common 540 bp sequence present in 1–5 copies, (5) the N-family genes (*N-family*), and (6) the polar-residue rich genes (*pol-res*). As with BVs, gene family size varies among species. For example, the *cys-motif* family of CsIV contains 10 members, whereas HfIV has only 5 *cys-motif* genes and TrIV has only 1 (Webb *et al.*, 2006; Tanaka *et al.*, 2007). IVs, like BVs, also encode a number of single copy genes. This level of comparison indicates that campoplegine IVs encode only one gene family that shares any similarity with genes encoded by BVs (*ank* genes). Yet, phylogenetic analysis indicates the IVs *ank* genes are distantly related to BV *ank* genes (Webb *et al.*, 2006). Most genes encoded by the encapsidated form of IV genomes are only transcribed in parasitized hosts. However, a few genes have been identified that are expressed in only the wasp or in both parasitized hosts and wasps. The lone example of the former is a structural protein (p12) associated with CsIV virions (Deng and Webb, 1999), while an example of the latter is members of the *rep* gene family (Theilmann and Summers, 1987; Hilgarth and Webb, 2002; Rasoolizadeh *et al.*, 2009).

IV genomes from banchine ichneumonids

The lone banchine-associated IV near-fully sequenced is *Glypta fumiferanae* ichnovirus (GfIV) (Lapointe *et al.*, 2007). The encapsidated genome of this virus is presumed to excise from unique integration loci like other PDVs and has an aggregate genome size of 290 kb. However, the encapsidated form of the GfV genome differs from BVs and campoplegine IVs in that it is divided into many more segments (105) that on average are much smaller (1.5–5 kb). The combination of predicted genes encoded by the encapsidated form of GfV also differs. For example, GfIV encodes an *ank* (4 genes) and *ptp* family (23 genes), but phylogenetic analysis indicates the *ank* genes are distinct from those of campoplegine IVs, while the *ptp* family members are distinct from BV *ptp* family members. Four putative GfV genes with similarity to a hypothetical gene of *Cotesia plutellae* bracovirus (CpBV) and a single copy gene with similarity to two predicted genes encoded by *Chelonus inanitus* bracovirus (CiBV) were also identified along with a family of 4 genes with similarity to the pox-D5 family of NTPases (*NTPase-like*) whose closest relatives are encoded by viruses in other families (*Mimiviridae, Iridoviridae, Poxviridae,* and *Ascoviridae*) (Lapointe *et al.*, 2007).

Shared organizational features of BV and IV genomes likely reflects convergence

Comparative data on PDVs are still limited with genomic information in particular restricted to only a small number of species in certain subfamilies. At the subfamily level, no data at all exist for the encapsidated genomes of BVs from wasps in the Miracinae and Khoikhoiinae. At the generic level, comparative data are also limited to a few species of the Microgastrinae in three genera (*Cotesia, Glyptapanteles, Microplitis*) and single species in the Cardiochilinae and Cheloninae. Data are similarly limited for IVs. Nonetheless, the collective picture that emerges is that the association of BVs with braconids arose independently from IVs with ichneumonids. Results from GfIV further suggest that IVs associated with campoplegines and banchines may also have independent origins from one another. If so, the traits shared among PDVs, like genome segmentation and the non-equimolar distribution of genomic segments, reflects convergence driven by their analogous roles in parasitism. For example, the non-equimolar production of genomic segments by BVs and IVs is thought to be a shared strategy for altering gene copy number in the wasp's host and in turn the abundance of gene products essential for successful parasitism (Webb and Strand, 2005). Webb and Cui (1998) found a direct correlation between viral gene copy number and the titre of the corresponding protein, while Beck *et al.* (2007) noted that secreted viral gene products whose function depends upon high abundance are encoded upon high copy genomic segments. Segment nesting has also been suggested as a strategy for amplifying the copy number of certain viral genes expressed in hosts (Cui and Webb, 1997). The presence of introns in some genes encoded by BVs and IVs but not others has been suggested as a shared strategy for enhancing expression or mRNA stability, while the evolution of multimember gene families may reflect selection pressure imposed by coevolutionary interactions between the virulence factors PDVs encode to disable host defences (see below) and counterstrategies by hosts to evade parasitoid and virus attack (Webb and Strand, 2005; Dupuy *et al.*, 2006; Bezier *et al.*, 2008).

The proviral form of PDV genomes resides in loci of the wasp genome

The second form in which PDVs exist is as proviruses. In situ hybridization and PCR-based experiments initially suggested the proviral form of BV genomes is clustered in the wasp genome with individual proviral genomic segments organized in tandem array (Savary *et al.*, 1997; Wyder *et al.*, 2002; Belle *et al.*, 2002). These results also led to the suggestion that the entire proviral genome of BVs forms a single macrolocus on one wasp chromosome (Belle *et al.*, 2002) (Fig. 8.5). Analysis of BAC clones from *Glyptapanteles indiensis* and *G. flavicoxis* partially support this idea (Desjardens *et al.*, 2007, 2008). The encapsidated form of the GiBV and GfBV genomes consist of 29 circular dsDNA segments. The corresponding proviral (linear) forms of genomic segments 1–20 (GiBV) and 1–21 (GfBV) exist in tandem array within two partially

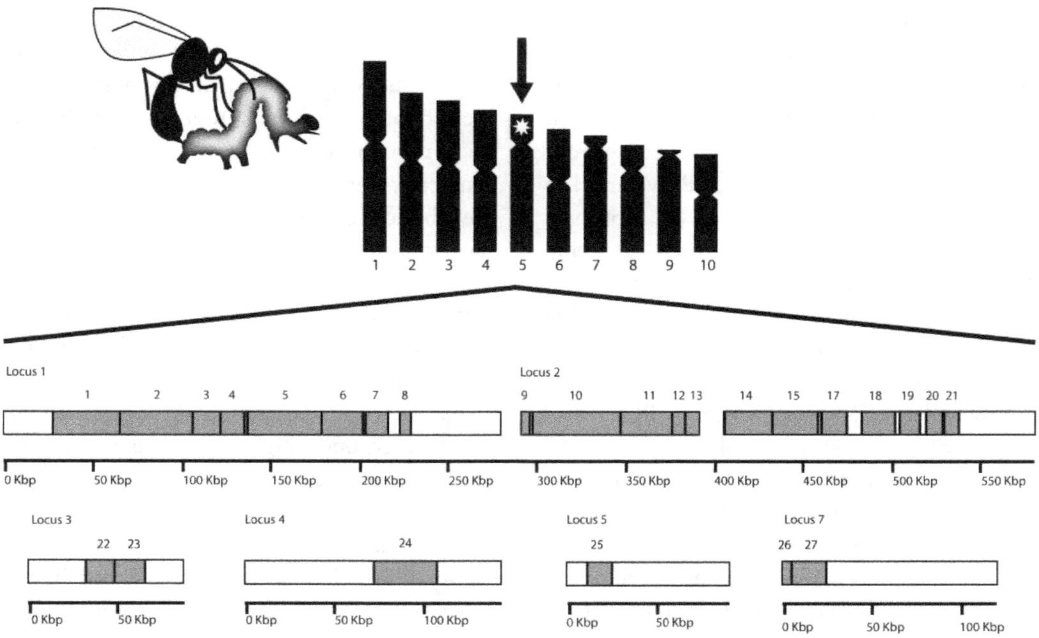

Figure 8.5 Schematic illustrating the distribution of proviral genomic segments from *Glyptapanteles flavicoxis* bracovirus (GfBV) in the genome of *G. flavicoxis*. The upper part of the figure shows a cartoon of wasp chromosomes. The lower part of the figure shows the distribution of proviral segments among seven loci (see text). The genome segments (1–27) packaged into the encapsidated genome are indicated above each proviral locus (grey). Black boxes represent tandem repeats that flank many of the viral genomic segments. Wasp DNA flanking the proviral segments of GfBV is indicated in white. Loci 1 and 2 likely comprise a single macrolocus located on one chromosome of the wasp (arrow on upper figure). The length of each locus is shown in kilobases (kbp). See text for details. Adapted from Desjardens *et al.* (2008).

overlapping genomic loci that together are more than 550 kb in length (Fig. 8.5). These two loci almost certainly represent a single 'macrolocus' within the genomes of *G. indiensis* and *G. flavicoxis*. If so, this macrolocus contains ca. 70% of the genomic segments that are packaged into GiBV and GfBV virions. Desjardens *et al.* (2008), however, also identified five other proviral loci (loci 3–7), which contain either a single proviral segment or two proviral segments in tandem array (Fig. 8.5). It is unclear whether loci 3–7 are in close proximity to loci 1 and 2, and are thus part of the same macrolocus, or reside in other regions of the wasp genome.

Small, intersegmental regions (<1 kb), lacking any predicted open reading frames, separate the tandemly arrayed proviral segments of GiBV and GfBV. However, a number of genes with sequence similarity to insect genes, transposable elements or, interestingly, genes identified in the encapsidated genomes of other BVs reside in regions that flank proviral-containing loci. Notably, all proviral segments also share a common 6 bp motif (AGCTTT) on their 5′ and 3′ ends that identifies the boundary at which each is excised from the wasp genome to produce the circularized (episomal) segments that are packaged into capsids. This excision motif also appears to be conserved among other BVs.

The replication process itself for BVs is only partially understood. Marti *et al.* (2003) used quantitative PCR to show that *Chelonus inanitus* calyx cells become polyploid during the pupal stage of female wasps coincident with the onset of replication by CiBV. Proviral sequences are selectively amplified while still in their proviral form with certain proviral segments being in greater copy number than others (Annaheim and Lanzrein, 2007). Amplification is then followed by excision of the proviral segments, presumably at the sites noted above, followed by circularization via juxtaposition and recombination of the terminal repeats. This results in 'empty' loci in the calyx cell at the site of excision of each proviral segment (Gruber *et al.*, 1996; Desjardins *et al.*, 2008). No further amplification of circularized segments appears to occur before packaging into virions (Marti *et al.*, 2003). The occurrence of similar events during replication of CcBV in *Cotesia congregata* also suggests the replication process among BVs is conserved (Pasquier-Barre *et al.*, 2002; Drezen *et al.*, 2003). Overall then, BV replication begins with differential amplification of proviral segments during the pupal stage of the wasp. Individual proviral segments are excised and circularized, without further replication, followed by encapsidation of one genomic segment per virion (Drezen *et al.*, 2003; Marti *et al.*, 2003; Pasquier-Barre *et al.*, 2002; Beck *et al.*, 2007). Virions are then released from nuclei into the lumen of the calyx by lysis of the calyx cell (see above).

Less is known about the proviral form of IV genomes, although studies with *Campoletis sonorensis* ichnovirus (CsIV) suggest that proviral segments are dispersed within the chromosomal DNA of *C. sonorensis* (Fleming and Summers, 1991; Cui and Webb, 1997). Studies further indicate that some CsIV proviral segments are flanked by short imperfect terminal repeats (example, segment B), while others are flanked by long perfect terminal repeats (segment W) (Cui and Webb, 1997; Fleming and Summers, 1991). Both segment B and W retain a single copy of these repeats after recombination produces the circular forms of these segments that are packaged into virions (Rattanadechakul and Webb, 2003). These observations combined with the phenomenon of segment nesting (see above) has led to the suggestion that replication of campoplegine IVs may involve a rolling-circle type mechanism with excision of proviral segments from the genome of the wasp occurring first followed by amplification of circular episomes, packaging, and exit of virions via budding from calyx cells (Savary *et al.*, 1997; Webb, 1998). The size of the repeat domains in episomes may mediate the copy number of each segment, as well as which segments generate additional nested segments, and thus the molar abundance of each segment in calyx fluid. Sequence data across the junction sites of selected proviral segments indicate that, similar to BVs, proviral sequences of IVs are flanked by wasp genomic DNA (Cui and Webb 1997; Fleming and Summers, 1991; Rattanadechakul and Webb, 2003). In contrast, it is unknown whether the proviral segments of campoplegine IVs reside on one or multiple wasp chromosomes.

BVs evolved from an ancestor in the Nudiviridae

PDV virions clearly look like viruses. However, the near absence of genes in the encapsidated genome with similarity to genes from other known virus families made it unclear whether PDVs originated from a viral ancestor or are a novel wasp secretion that merely look like viruses (Stoltz and Vinson, 1979; Stoltz and Whitfield, 1992; Federici and Bigot, 2003). To address this question, Bezier *et al.* (2009a) sequenced transcripts expressed in wasp ovaries during replication of two BVs: *Cotesia congregata* bracovirus (CcBV) and *Chelonus inanitus* bracovirus (CiBV) whose associated wasps are in different subfamilies of the microgastroid assemblage (see Fig. 8.1). This resulted in identification of 22 genes that share significant similarity with genes encoded by members of the *Nudiviridae*, which is a sister group to the *Baculoviridae* (see Chapters 3 and 7). Importantly, 14 of these genes belong to a core of 20 genes conserved among all nudiviruses and baculoviruses examined to date (Herniou *et al.*, 2003; Wang and Jehle, 2009; Bezier *et al.*, 2009b) (Table 8.1). Although not yet studied in nudiviruses, these core genes are known to play essential roles in replication of baculoviruses. Homologues identified from the ovaries of *C. congregata* and *C. inanitus* include: (1) products required for transcription including multiple subunits of the viral RNA polymerase complex required for expression of 'late' and 'very late' genes, (2) three genes that encode predicted proteins involved in nucleocapsid packaging and assembly, and (3) five genes encoding predicted proteins that are related to components of the envelope that surrounds the occluded form of baculoviruses. Seven other genes were identified that code for protein variants related to ODV-E66; an envelope protein conserved among

Table 8.1 Genes expressed in the ovaries of *Cotesia congregata* and/or *Chelonus inanitus* during replication of BV virions that share similarity with nudivirus/baculovirus core genes

Nudivirus/Baculovirus core genes	
Transcription	RNA polymerase subunit lef-4
	RNA polymerase subunit lef-8
	RNA polymerase subunit p47
	Initiation factor lef-5
Packaging and assembly	38K (Ac98)
	Very late factor-1 (Vlf-1)
	Vp91
Occluded virus envelope components	Per os infectivity factor p74
	Per os infectivity factor pif-1
	Per os infectivity factor pif-2
	Per os infectivity factor pif-3
	odv-e56

Adapted from Bezier (2009a).

baculoviruses and nudiviruses that specifically infect Lepidoptera (Jehle *et al.*, 2006). However, no homologues of baculovirus genes required for DNA replication (*lef1–3, dnapol, helicase, and ie-1*) were identified.

These results lend strong support that BVs originated from an ancestral nudivirus. They also suggest several core genes required for replication of baculoviruses, and presumably nudiviruses, have been retained for replication of BVs in calyx cells. In addition to fitting well with the morphological similarity between BV and nudivirus virions, it is also known that some nudiviruses preferentially infect the reproductive system of insects where they establish persistent infections. Thus, the nudivirus-microgastroid ancestral association may have begun as a persistent infection of the wasp.

In contrast, it remains unclear where the nudivirus-related genes reside in the wasp genome relative to the proviral loci that are amplified, circularized and packaged into virions. There is also no evidence that nudivirus or baculovirus genes are involved in the production of virions by IVs suggesting campoplegine and banchine-associated viruses arose from another ancestor (Bezier *et al.*, 2009a). Bigot *et al.* (2008) report that the *NTPase* genes identified from the encapsidated genome of GfIV (see above) are most closely related to three proteins from an ascovirus and suggest that IVs originated from lateral transfer of ascovirus genes. However, more data than shared homology between one gene family encoded in the encapsidated genome of GfIV is needed before concluding an ancestral relationship between the *Ascoviridae* and IVs. Recently, a more comprehensive analysis of the proviral form of IVs identified multiple genes involved in IV virion formation. These genes cluster within specialized regions of the wasp genome and are not packaged into the encapsidated form of IVs. The genomic structure and the process of virion formation together suggest that these IV structural genes originate from an ancestral virus, but this ancestor appears to not be either an ascovirus or nudivirus. Thus, the progenitor of IVs remains unknown (Volkoff, *et al.*, 2010).

The function of PDVs in parasitism

As previously discussed, PDV-carrying wasps inject a quantity of calyx fluid into hosts together with eggs and other factors like venom when they oviposit into hosts (Webb and Strand, 2005; Strand, 2009). Following infection of haemocytes and other tissues, expression of genes encoded by BVs and IVs have physiological effects in the parasitized host that facilitate survival of the parasitoid's progeny. These include suppression of the host's immune system, changes in host growth and development, and alterations in metabolism (Strand and Pech, 1995; Schmidt *et al.*, 2001; Turnbull and Webb, 2002; Beckage and Gelman, 2003; Pennacchio and Strand, 2006; Strand, 2009).

The lepidopteran immune response involves both humoral and cellular defences

The most important function of PDVs in parasitism appears to be the protection of the wasp's offspring from immune defences of the host. Similar to other insects, the innate immune system of Lepidoptera is comprised of many components, which are broadly divided into humoral and cellular defences. The humoral arm of the immune system refers to pathways that regulate production of soluble effector molecules that are released into the haemocoel. Examples include antimicrobial peptides (AMPs), complement-like proteins, and products generated by the phenoloxidase (PO) cascade (Cerenius and Soderhall, 2004; Theopold *et al.*, 2004; Imler and Bulet, 2005; Kanost and Gorman, 2008). Some of these effector molecules are constitutively produced, while others are inducibly expressed following immune challenge. The cellular arm of the immune system refers to defence responses like phagocytosis, encapsulation, and clotting, which are mediated directly by circulating immune cells (haemocytes) (Strand, 2008a). Lepidopteran haemocytes are usually divided into four main subpopulations called granulocytes (= granular cells), plasmatocytes, spherule cells, and oenocytoids (Lavine and Strand 2002; Strand, 2008a; Nakahara *et al.*, 2009). Granulocytes are generally the most abundant subpopulation whose functions include phagocytosis and initiation of an encapsulation response towards some foreign targets. Plasmatocytes are the second most abundant subpopulation and are the main capsule-forming haemocyte. Granulocytes and plasmatocytes are both strongly adhesive and bind to a variety of foreign surfaces. Oenocytoids are non-adhesive cells that contain PO and other components of the PO cascade, while spherule cells are potential sources of cuticular components. Progenitor cells called prohaemocytes reside in haematopoietic organs and, at low frequency, in circulation (Gardiner and Strand, 2000; Nakahara *et al.*, 2003; Nardi *et al.*, 2003, 2006).

The main defence against parasitoids like PDV-carrying wasps is encapsulation (Strand 2008a; Strand, 2009). An encapsulation response begins when the host recognizes the parasitoid egg or larva as foreign; a response which is mediated by recognition receptors expressed on the surface of haemocytes as well as by humoral pattern recognition receptors which enhance recognition of a target by haemocytes when they bind to its surface (i.e. opsonization). Cell surface receptors with roles in phagocytosis or encapsulation include scavenger receptors (Ramet *et al.*, 2001; Kocks *et al.*, 2005; Philips *et al.*, 2005), Down's syndrome cell adhesion protein (Dscam) (Ramet *et al.*, 2002; Moita *et al.*, 2005; Dong *et al.*, 2006), and integrins (Lavine and Strand, 2003; Irving *et al.*, 2005; Levin *et al.*, 2005; Wertheim *et al.*, 2005; Moita *et al.*, 2005). Humoral pattern recognition receptors include haemolin, lipopolysaccharide (LPS)-binding protein, gram-negative bacteria recognition protein (GNBPs), soluble peptidoglycan recognition proteins (PGRP-SA and PGRP-SD), glucan recognition proteins (GRPs), complement-like thioester-containing proteins (TEPs), and immunolectins (Levashina *et al.*, 2001; Irving *et al.*, 2005; Moita *et al.*, 2005; Dong *et al.*, 2006; Ling and Yu, 2006; Terenius *et al.*, 2007).

Recognition results in the binding of a small number of haemocytes, usually granulocytes, to the surface of the parasitoid (Pech and Strand, 1996; Irving *et al.*, 2005). Granulocytes then release cytokines that recruit additional haemocytes, primarily plasmatocytes, which form a capsule. Tetraspanin proteins have been shown to serve as ligands for integrins in the spingid moth, *Manduca sexta*, which mediate haemocyte binding within a capsule (Zhuang *et al.*, 2007). Granulocyte-stimulated activation and binding of plasmatocytes also involves the immunoglobulin superfamily member neuroglian (Nardi *et al.*, 2006). Once formed, capsules often melanize due to activation of the PO cascade. The parasitoid is then killed by asphyxiation and/or compounds generated as a consequence of producing melanin. Capsule formation in Lepidoptera usually begins 2–6 hours after parasitism and is completed by 48 hours.

PDVs have multiple immunosuppressive effects on hosts

Most studies indicate that PDVs prevent the parasitoid egg and larva from being encapsulated by expressing gene products that disable capsule formation (Strand and Pech, 1995; Schmidt *et al.*, 2001; Webb and Strand, 2005; Strand, 2008b). In some parasitoid-host systems, PDV gene activity appears to only suppress encapsulation of the wasp's eggs or larvae (Doucet and Cusson, 1996), whereas in others PDV genes prevent a variety of foreign targets from being encapsulated and also disable other components of the immune system (summarized by Webb and Strand, 2005; Strand, 2008b, 2009). Some PDVs also have immunoevasive activities that do not involve viral gene expression.

Immunosuppressive effects of BVs

Disruption of encapsulation and other haemocyte-mediated functions differs somewhat among BV species. In the case of *M. demolitor*, for example, MdBV infection of the host, *Pseudoplusia includens* (Lepidoptera: Noctuidae) irreversibly suppresses encapsulation and phagocytosis (Strand and Noda, 1991; Strand and Pech, 1995). Granular cells and plasmatocytes lose the capacity to bind to any foreign surface within 4 hours of infection (Fig. 8.6). Later, granulocytes also undergo a high level of apoptosis. These effects persist for the duration of parasitoid development. Similar alterations in the adhesive properties of host haemocytes have also been described for BVs from other *Microplitis* species (Tanaka, 1987; Kadash *et al.*, 2003).

The effects of BVs associated with wasps in the genus *Cotesia*, in contrast, are more variable. Infection of *Pieris rapae* by *Cotesia rubecula* bracovirus (CrBV) results in transient cytoskeletal alterations in haemocytes (Asgari *et al.*, 1996, 1997). Inhibition of haemocyte binding to foreign surfaces also occurs in hosts infected by *C. melanoscela* bracovirus (CmBV), *C. kariyai* bracovirus (CkBV), and *C. plutellae* bracovirus (CpBV) (Guzo and Stoltz, 1987; Tanaka, 1987; Gad and Kim, 2008). In contrast, inhibition of encapsulation by *C. congregata* bracovirus (CcBV) has been reported to either cause haemocytes of its host, *M. sexta*, to aggregate (Lovallo and Cox-Foster, 1999) or lose the ability to adhere and spread on foreign surfaces (Amaya *et al.*, 2005). Little is known about the role of chelonine BVs in immune suppression, but studies with *Chelonus inanitus* bracovirus (CiBV) indicate that viral infection protects the wasp larva from encapsulation, but does not suppress capsule formation of other foreign targets like latex beads (Stettler *et al.*, 1998).

BVs also disable humoral immune defences. For example, a number of host species exhibit partial or complete suppression of haemolymph melanization after parasitism by BV carrying wasps or infection by BVs alone (Stoltz and

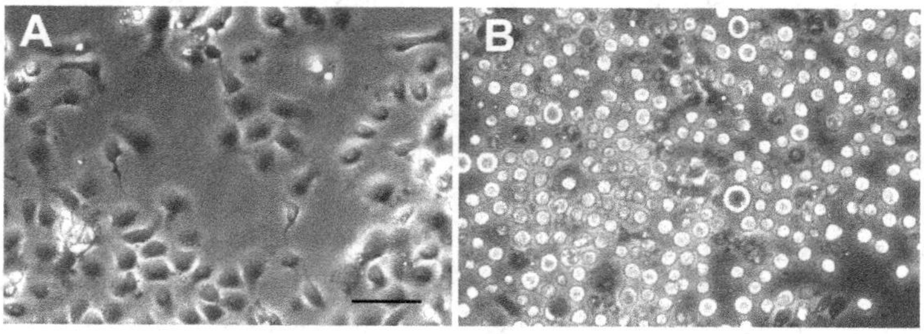

Figure 8.6 Phase contrast micrographs of non-infected haemocytes from *Pseudoplusia includens* (A) and *P. includens* haemocytes 24 hours post infection by *Microplitis demolitor* bracovirus (MdBV). Most haemocytes in (A) are granulocytes and plasmatocytes that have attached and spread on the surface of a culture plate. In contrast, most haemocytes in (B) remain round and are unable to attach or spread on the culture plate.

Cook, 1983; Beckage *et al.*, 1990; Strand and Noda, 1991). Several antimicrobial peptides are up-regulated in the fat body of lepidopteran larvae following immune challenge (Kang *et al.*, 1996; Lundstrum *et al.*, 2002). However, MdBV infection disables inducible expression of some antimicrobial peptides (Thoetkiattikul *et al.*, 2005). Lastly, some BVs have immunoevasive activities that do not involve expression of viral genes but rather protect newly laid parasitoid eggs by coating their surface. The presence of virions, and possibly other factors in calyx fluid on the surface of eggs either prevents host haemocytes from recognizing the wasp egg as foreign or interferes with the ability of haemocytes to bind to the surface of the egg. Eggs of *C. kariyai*, for example, are coated with CkBV virions plus other proteins present in calyx fluid. Two of these proteins, named immunoevasive protein 1 (IEP1) and IEP2, protect *C. kariyai* eggs from encapsulation by the host, *Pseudaletia separata* (Hayakawa and Yazaki, 1997). IEP1 and 2 are detected on the surface of CkBV virions, which suggests they may be components of the virus envelope. IEP-like proteins have also been described to coat the eggs of other parasitoids in the genus *Cotesia* and *Toxoneuron* (Davies and Vinson, 1986; Asgari *et al.*, 1998).

Currently, the mechanisms underlying BV-mediated immunosuppression are best understood for MdBV. Quantitative PCR studies indicate that MdBV primarily infects host haemocytes with lower levels of infection detected in other tissues like the fat body and nervous system (Strand, 1994; Beck *et al.*, 2007). In the case of haemocytes, greater than 99% of circulating cells are infected with each cell being infected by one or more virions containing each segment of the encapsidated genome. Like other BVs from microgastrine braconids, MdBV encodes a *ptp* gene family (13 members) and an *ank* gene family (12 members). It also encodes a family known as the *egf* genes (three members), which share a common trypsin inhibitor-like domain, and a family called *glc* genes (2 members), which encode heavily N-glycosylated cell surface proteins (Webb *et al.*, 2006). Two members of the *ptp* family, PTP-H2 and PTP-H3, and one member of the *glc* family, *glc1.8*, are the most important gene products in disabling encapsulation and phagocytosis (Beck and Strand, 2003, 2005; Pruijssers and Strand, 2007; Suderman *et al.*, 2008). The Toll and immune deficiency (imd) signalling pathways regulate the inducible expression of many antimicrobial peptides as well as several other immune genes. Both pathways involve specific NF-κB transcription factors, which are negatively regulated by endogenous inhibitor κB (IκB) proteins. The *ank* gene family of MdBV shares similarity with the domain of insect IκBs that mediates binding to NF-κBs (Thoetkiattikul *et al.*, 2005). In turn, functional studies indicate two MdBV *ank* family members, H4 and -N5, suppress Toll and Imd signalling by binding insect NF-κBs (Thoetkiattikul *et al.*, 2005). As previously noted, melanin formation is regulated by the PO cascade which consists of multiple serine proteinases that terminate with the zymogen pro-phenoloxidase (Kanost and Gorman, 2008). The serine proteinases that activate pro-PO, called pro-phenoloxidase activating proteinases (PAPs), are themselves activated by multiple upstream proteinases. One member of the MdBV *egf* gene family, Egf1.0, is a dual-activity inhibitor that disables melanization by blocking the processing of pro-PAPs and the enzymatic activity of activated PAPs (Beck and Strand, 2007; Lu *et al.*, 2008).

Most other studies on BV-mediated immunosuppression involve wasps in the genus *Cotesia*. Studies with *C. plutellae* bracovirus (CpBV) implicate a virally encoded PTP, lectin, histone, and an EP1-like gene in disrupting haemocyte adhesion to foreign surfaces and haemocyte proliferation (Gad and Kim, 2008; Ibrahim and Kim, 2008; Kwon and Kim, 2008; Lee *et al.*, 2008). As noted above, infection of *P. rapae* by CrBV results in transient cytoskeletal alterations in haemocytes, which is associated with short-term expression of a gene product called CrV1 (Asgari *et al.*, 1997; Zhang *et al.*, 2004). An orthologue of the CrV1 gene encoded by CcBV is also reported to disable function of the pattern recognition protein hemolin (Labropoulou *et al.*, 2008). Similar to MdBV, parasitism by *Cotesia* sp. inhibits melanization of host haemolymph (Stoltz and Cook, 1983; Beckage *et al.*, 1990), although no *Cotesia* BVs encode homologues of the MdBV *egf* genes. Zhang *et al.* (2004), however, suggests this activity may not be due to a BV-encoded gene but rather a serine proteinase homologue present in the venom of *C. rubecula*. Other BV genes with possible immunosuppressive functions include TnBV1 from *Toxoneuron nigriceps* bracovirus (TnBV), which appears to activate caspases in the absence of apoptosis (Lapointe *et al.*, 2005), and *ank* genes encoded by TnBV that interact with insect NF-κBs (Falabella *et al.*, 2007).

Immunosuppressive effects of IVs

Similar to BVs, IVs infect several host tissues, although haemocytes and the fat body are the principal sites of transcription (Beliveau *et al.*, 2000; Galibert *et al.*, 2003; Yin *et al.*, 2003). IVs also adversely affect host cellular and humoral immune defences. Several IVs have been shown to alter the number of haemocytes in circulation and the ability of granulocytes and plasmatocytes to spread on foreign surfaces (Davies *et al.*, 1987; Guzo and Stoltz, 1987; Doucet and Cusson, 1996; Yin *et al.*, 2003). In the case of *Campoletis sonorensis* ichnovirus (CsIV), host haemocytes are unable to encapsulate parasitoid eggs as well as other foreign targets (Edson *et al.*, 1981; Davies *et al.*, 1987). In contrast, infection of hosts by *Tranosema rostrale* ichnovirus (TrIV) disables encapsulation of *T. rostrale* eggs but has no effect on encapsulation of other targets suggesting suppression of the host encapsulation response is restricted to the parasitoid (Doucet and Cusson, 1996). Like BVs, infection by IVs also suppresses melanization of haemolymph (Stoltz and Cook, 1983; Doucet and Cusson, 1996; Shelby *et al.*, 2000) and expression of at least some antimicrobial peptides (Shelby *et al.*, 1998).

IV genes implicated in causing these alterations include the *cys-motif* proteins VHv1.1 and VHv1.4 from CsIV, which are expressed in the host fat body and secreted into the haemolymph. Both proteins bind to the surface of haemocytes. Expression of VHv1.1 also reduces encapsulation of *C. sonorensis* eggs in hosts infected with a VHv1.1-expressing recombinant baculovirus (Li and Webb 1994; Cui *et al.*, 1997). Other IV genes have been implicated in immunosuppression, because of homology to BV genes or because they are expressed in host immune tissues like haemocytes or the fat body. IV encoded *ank* genes, for example, may function as NF-κB inhibitors, but structural differences with BV *ank* genes as well as recent expression studies suggest that they may have other functions (Kroemer and Webb, 2004b; Webb *et al.*, 2006). For example, stable transfection of the Sf9 cell line with the CsIV *ank1* gene inhibits apoptosis (Fath-Goodin *et al.*, 2009). Products of the CsIV *inex* genes form functional gap junctions, which have diverse functions in cell-cell communication (Turnbull *et al.*, 2005). Given observations that gap junctions form between haemocytes in capsules, it is possible that IV *inex* genes may play a role in altering haemocyte function. The P30 ORF from *Hyposoter didymater* ichnovirus encodes a predicted mucin-like protein with weak similarity to Glc1.8 from MdBV (Galibert *et al.*, 2003). Doucet *et al.* (2008) reports that infection of hosts with TrIV transiently reduced expression of host PO genes, but this effect does not appear to account for the near complete inhibition of melanization activity observed in parasitized hosts.

Like BVs, some IVs from campoplegine ichneumonids also transiently protect wasp eggs from encapsulation by coating their surface. *Venturia canescens* produces virus-like particles (VLPs), which morphologically resemble IVs, but do not encapsidate any viral DNAs. Nonetheless, the pyralid larvae that *V. canescens* parasitizes are unable to encapsulate wasp eggs because VLPs and a haemomucin coat their surface (Theopold *et al.*, 1996). TrIV similarly coats the eggs of *T. rostrale* in a manner that likely confers passive protection from encapsulation (Cusson *et al.*, 1998).

Developmental interactions between PDV-carrying parasitoids and hosts

The second type of alterations associated with PDV infection of the wasp's host is changes in growth, development and/or metabolism. In some cases, these alterations are fully mimicked by infection with PDV alone. In others, the developmental alterations that occur in parasitized hosts involve multiple factors that include PDVs, other components introduced into the host by the adult wasp (venom), and products from the developing wasp embryo or larva.

Regulation of growth and moulting in Lepidoptera and other host insects of wasps

Before considering the effects of PDVs on host growth and development, it is important to summarize how growth and moulting are regulated normally. In brief, all growth in insects occurs during the premetamorphic nymphal or larval stages, and is regulated by the interactions between nutrition, other environmental factors, and hormones (Nijhout, 1994, 2003). Juvenile hormone (JH) is the key endocrine factor involved in maintenance of the larval stage, while juvenile hormone esterase (JHE) is the primary enzyme responsible for JH breakdown. Studies with Lepidoptera indicate that the final size an insect achieves is a function of a species-specific threshold (critical weight) that must be attained during the final instar to moult to a pupa without developmental delay (Nijhout, 1975). Critical weight triggers the release of the neuropeptide prothoracicotropic hormone (PTTH), which stimulates the prothoracic glands (PTGs) to synthesize and release ecdysteroids. This causes the larva to stop feeding and initiate premetamorphic behaviours like wandering and cocoon spinning, while a second release of ecdysteroids causes the larva to pupate (Riddiford, 1976; Gilbert *et al.*, 2002). Far less clear is how insects assess when a size threshold like critical weight has been reached. Positive regulators include nutrient acquisition and metabolic activity mediated through the insulin and TOR signalling pathways (Oldham and Hafem, 2003; Mirth and Riddiford, 2007; Baker and Thummel, 2007), whereas JH, ecdysteroids, and the glucagon-like factor adipokinetic hormone function as negative regulators that can repress growth under conditions of stress (Takaki and Sakurai, 2003; Lee and Park, 2004; Mirth *et al.*, 2005; Colombani *et al.*, 2005; Truman *et al.*, 2006).

Developmental alterations associated with BV infection

The developmental response of hosts after parasitism by BV-carrying braconids differs among taxa and species. Microgastrine braconids like *Microplitis, Cotesia,* and *Glyptapanteles* sp. and cardiochiline braconids like *Toxoneuron* sp. are all larval endoparasitoids that are either solitary (one offspring per host) or gregarious (multiple offspring per host). Most species are able to oviposit into multiple instars of their host but their offspring complete development either in the host's final instar or in an earlier instar (summarized by Pennacchio and Strand, 2006). Hosts parasitized by solitary species usually exhibit severe reductions in weight gain, delays in moulting, and inhibition of metamorphosis. Hosts parasitized by gregarious species are also inhibited from pupating but usually exhibit less severe alterations in weight gain. The exception to these trends is braconids in the subfamily Cheloninae, which are all solitary and oviposit into the egg stage of the host. The host then hatches and exhibits reduced weight gain relative to non-parasitized larvae. Hosts parasitized by chelonine braconids also undergo precocious metamorphosis, which means they initiate developmental processes associated with pupation one instar earlier than normal (Lawrence and Lanzrein, 1993; Lanzrein *et al.*, 2001). The parasitoid completes its larval development and emerges from the prepupal stage host, which then dies before actually pupating.

BV infection contributes to many of these developmental alterations (Beckage and Gelman, 2003; Pennacchio and Strand, 2006). For example, infection of hosts by MdBV alone suppresses weight gain and metamorphosis of the host *P. includens* in a manner indistinguishable from natural parasitism (Pruijssers *et al.*, 2009). MdBV infection causes host JH titres to rise, reduces JHE activity and suppresses any increase in the haemolymph ecdysteroid titre during the host's final instar (Balgopal *et al.*, 1996; Pruijssers *et al.*, 2009). These alterations collectively prevent the host from pupating. *C. congregata*, a gregarious species, also inhibits metamorphosis of the host by elevating haemolymph JH titres and interfering with processes that prevent PTGs from producing or releasing ecdysteroids (Beckage and Riddiford, 1982; Gelman *et al.*, 1998). The rise in JH titres, however, appears to involve both CcBV gene expression, which reduces JHE activity, and developing *C. congregata* larvae, which secrete JH (Cole *et al.*, 2002). Disruption of host metamorphosis by *C. kariyai*, a solitary species, involves inhibition of PTTH release, while TnBV appears to disrupt ecdysteroid biosynthesis by the PTGs (Tanaka *et al.*, 1987; Pennachio *et al.*, 1998). Fewer studies have been conducted with chelonine braconids but experiments with *C. inanitus* indicate that CiBV gene activity developmentally arrests *Spodoptera littoralis* larvae in the prepupal stage by suppressing ecdysteroid synthesis and/or release (Lanzrein *et al.*, 2001). Reed and Brown (1998) also report that *Chelonus* near *curvimaculatus* and *Ascogaster quadridentata* secrete ecdysteroids that may be involved in synchronizing the timing of parasitoid emergence from the host.

Metabolic alterations include changes in the abundance of haemolymph proteins, free amino acids, and carbohydrate levels (Thompson, 1993; Thompson and Dahlman, 1998; Pennacchio and Strand, 2006). For example, infection by CkBV and CcBV down regulates several haemolymph proteins (Nakamatsu *et al.*, 2001; Beckage and Kanost, 1993). MdBV rapidly induces hyperglycaemia in infected hosts and depletes nutrient stores (Pruijssers *et al.*, 2009). Associated with these metabolic changes are reductions in the size of the host fat body and gonads (Yagi and Tanaka, 1992; Reed and Brown, 1998; Doi *et al.*, 2009) suggesting these metabolic alterations enhance nutrient availability to the parasitoid larva at the expense of host tissues.

Developmental alterations associated with IV infection

All IV-carrying parasitoids are solitary, and after parasitism their hosts exhibit large reductions in weight gain, delays in moulting, and are unable to pupate. As with BV-carrying parasitoids, juvenilization of the host and inhibition of pupation are associated with specific alterations in endocrine physiology, which include reductions in JHE activity and a reduced haemolymph ecdysteroid titre (summarized by Pennacchio and Strand, 2006). In the case of *C. sonorensis*, CsIV infection causes PTGs of last instar hosts to degenerate which suppresses pupation (Dover *et al.*, 1995). Injection of TrIV from *Tranosema rostrale* into penultimate instar *Choristoneura fumiferana* reduces JHE activity but does not elevate JH titres (Cusson *et al.*, 2000). TrIV infection also suppresses haemolymph ecdysteroid titres but does not cause PTGs to degenerate (Cusson *et al.*, 2000). Infection of hosts by IVs also causes changes in metabolic physiology. Alterations in the abundance of host haemolymph proteins caused by CsIV infection are selective with the storage protein arylphorin being down-regulated, while other abundant proteins, such as lipophorin and transferrin, remain unchanged relative to non-infected host larvae (Shelby and Webb 1994, 1997). Triglycerides and glycogen deposits in fat body are also greatly reduced in hosts parasitized by *C. sonorensis*, while levels of free sugars in haemolymph increase (Vinson, 1990).

PDV gene products implicated in altering host growth and development

The alterations in endocrine and metabolic physiology described above are likely linked given the role the insect endocrine system plays in regulating nutrient storage and mobilization (Mirth and Riddiford, 2007; Strand and Casas, 2007). Elevating host JH titres and blocking the rise in haemolymph ecdysteroid levels could be partially responsible for altering titres of specific storage proteins, whereas changes in carbohydrate and lipid metabolism could reflect alterations in regulation by the insulin and adipokinetic hormone signalling pathways (Mirth *et al.*, 2005; Colombani *et al.*, 2005; Wu and Brown, 2006). Pennacchio and Strand (2006) emphasized that disabling metabolic sinks in the host, like metamorphosis and/or reproduction, and mobilizing nutrient stores has the net effect of redirecting energetic resources away from host tissues and towards the parasitoid's progeny. Thus, the primary adaptive significance of altering host endocrine physiology and reproduction may be metabolic, while arrested development and inhibition of pupation of the host are indirect consequences of redirecting host nutritional resources.

Less clear is the identity of the PDV gene products responsible for causing these alterations or their molecular targets in hosts. Given that parasitized insects usually exhibit reduced haemolymph ecdysteroid titres and elevated JH titres, most studies have focused on the hypothesis that PDVs produce molecules which interfere with the signalling pathways that regulate biosynthesis or metabolism of these hormones. For example, PTTH signalling and ecdysteroid production by PTGs involves phosphorylation of certain components in the ecdysteroid biosynthetic pathway. Falabella *et al.* (2006) report that expression of some *ptp* gene family members from TnBV in host PTGs correlates with reduced production of ecdysteroids. However, it is unclear whether these viral PTPs are functional enzymes or that a reduction in ecdysteroid production is due to viral PTP activity. Several PTP genes encoded by MdBV are also expressed in the nervous system and PTGs of infected hosts (Pruijssers and Strand, 2007), but it is similarly unknown whether these factors have any functional role in altering PTTH synthesis or release, or in the production of ecdysteroids by PTGs. Two novel genes encoded by CiBV appear involved in causing the developmental arrest of hosts following the onset of precocious metamorphosis, but what they interact with in the host is unknown (Bonvin *et al.*, 2005).

The alternative hypothesis for what causes alterations in host growth and moulting after parasitism is that these changes reflect an indirect response to PDV and/or parasitoid induced changes in nutritional physiology (Thompson, 1993; Pennacchio and Strand, 2006). For example, two members of the *cys* motif gene family, VHv1.1 and VHv1.4, from CsIV reduce protein translation in hosts and lepidopteran-derived cell lines (Shelby and Webb, 1997; Kim, 2005). Injection or feeding of VHv1.1 and VHv1.4 also causes host larvae to gain weight more slowly and to pupate abnormally suggesting altered protein synthesis is responsible for altered growth and metamorphosis (Fath-Goodin *et al.*, 2006). Another factor implicated in growth of lepidopteran larvae is bioactive ENF-peptides (Hayakawa, 1995; Strand *et al.*, 2000). Increases in the titre of ENF-peptides cause larval growth to decline. Concomitantly, CkBV from *C. kariyai* causes the ENF-peptide titre of *Pseudaletia separata* to increase (Ohnishi *et al.*, 1995). Lastly, Pruijssers *et al.* (2009) found that infection by MdBV fully inhibits metamorphosis of the host *P. includens* by preventing larvae from achieving critical weight, which is essential for ecdysteroid biosynthesis by PTGs. In effect, MdBV does not directly target ecdysteroid biosynthesis by PTGs but rather nutrient metabolism and feeding behaviour which prevents the host from attaining the size necessary for high level ecdysteroid biosynthesis by PTGs (Pruijssers *et al.*, 2009).

Concluding remarks

In this chapter, I have reviewed our current understanding of the *Polydnaviridae*. I show that PDVs are exclusively associated with parasitoid wasps in the families Braconidae and Ichneumonidae, and that the association of BVs and IVs with parasitoids likely arose independently. I also describe how recent evidence supports that BVs arose from an ancestor in the *Nudiviridae*, whereas the origin(s) of IVs remains uncertain.

The encapsidated genomes of PDVs are comparatively large but have low coding densities that are dominated by a small number of genes that have diverged into multimember families. A majority of genes that are highly expressed in the wasp's hosts are spliced and located on high copy genomic segments, while low abundance intracellular gene products tend to be encoded by genes that lack introns that are located on low copy segments. These observations suggest that in the absence of replication, BVs and IVs have evolved similar compensatory strategies for regulating the abundance and stability of different viral transcripts. The diversification of virulence genes into families is also likely an adaptation that reflects the role of these viruses in parasitism. The organization of the proviral form of PDV genomes, including where genes encoding the replication machinery are located in the wasp relative to the genomic segments that are packaged into virions, remains incompletely understood.

Nonetheless, the overall picture that currently exists suggests those regions of PDV genomes that are packaged into virions have expanded in size by the acquisition of genes and non-coding sequence from wasps, hosts or other organisms. This expansion likely contributes to the function PDVs play in parasitism. In contrast, those parts of PDV genomes that are required for maintenance of a symbiosis with wasps have been reduced or eliminated entirely from the encapsidated genome (Webb and Strand, 2005; Bezier *et al.*, 2009b). The result is that PDVs maintain their mutualistic association with wasps as proviruses, yet retain the ability to produce replication-defective virions for delivery of virulence genes to the wasp's host.

Despite evidence that at least one genus of PDVs originated from a viral ancestor, the question remains whether the PDVs that exist today should be considered viruses. Some have argued against this proposition, because the encapsidated forms of PDVs encode almost no virus-related genes and are unable to replicate. Organizational features of the encapsidated genome such as large amounts of non-coding sequence, spliced genes, and the existence of gene families are also noted as features that differ from other viruses. Thus, the suggestion is that PDVs should be considered a type of organelle (Whitfield and Asgari, 2003; Federici and Bigot, 2003). Others, including myself, favour that PDVs should continue to be recognized as one or more families of viruses because (1) the proviral form of BV genomes does contain viral genes, (2) PDVs do replicate in wasps, albeit to produce replicatively defective virions, and (3) PDV virions fully resemble other viruses, encapsidate nucleic acid, and are essential for delivery of this nucleic acid to host cells (Stoltz and Vinson, 1979; Webb and Strand, 2005; Dupuy *et al.*, 2006). A number of viruses in addition to PDVs also have life cycles in which transmission depends primarily upon vertical transmission as proviruses (many phages, retroviruses) rather than as infectious particles. Obviously, PDVs are unique given that transmission depends entirely upon the proviral form of the virus and the encapsidated form of the virus is replication defective. However, referring to PDVs as organelles or secretions suggests differences relative to other viruses that are greater than PDVs actually exhibit.

Rather than debating whether PDVs fully satisfy a classical definition of a virus, Webb and Strand (2005) and Bezier *et al.* (2009b) suggest a more constructive perspective would be to shift away from the assumption that all viruses are obligate parasites, and therefore are defined on the basis of traits consistent with only this life history, to one which recognizes that viruses can also have traits consistent with persistence as a symbiont. Intracellular bacteria exhibit numerous changes in genome organization and transcriptome activity, which result in an inability to propagate themselves outside their host and several traits that are more similar to eukaryotes than free-living prokaryotes (Moran, 2007; Guell *et al.*, 2009). Yet, most bacteriologists recognize that these organisms still share enough features to refer to them as bacteria. By analogy, other viruses besides PDVs show evidence of having evolved stable (persisting) rather than parasitic relationships with hosts (Vilarreal, 2007), which has similarly resulted in alterations of some traits exhibited by classical (parasitic) viruses, but also retention of many features associated with viruses more generally. The likelihood that symbiotic associations between viruses and hosts are more prevalent than previously recognized further argues for the need to broaden our thinking about how we define viruses and view their interactions with hosts.

Acknowledgements

I greatly thank Jena Johnson for assistance in preparing this manuscript, and Don Stoltz and Michel Cusson for generously sharing their micrographs of IVs. Some of the work discussed in this chapter was supported by grants from the USDA NRI program and National Science Foundation to MRS.

References

Albrecht, U., Wyler, T., Pfister-Wilhelm, R., Gruber, A., Stettler, P., Heiniger, P., Schumperli, D., and Lanzrein, B. (1994). PDV of the parasitic wasp *Chelonus inanitus* (Braconidae): characterization, genome organization and time point of replication. J. Gen. Virol. *75*, 3353–3363.

Amaya, K.E., Asgari, S., Jung, R., Hongskula, M., and Beckage, N.E. (2005). Parasitization of *Manduca sexta* by the parasitoid *Cotesia congregata* induces an impaired host immune response. J. Insect Physiol. *51*, 505–512.

Annaheim, M., and Lanzrein, B. (2007). Genome organization of the *Chelonus inanitus* polydnavirus: excision sites, spacers, and abundance of proviral and excised segments. J. Gen. Virol. *8*, 450–457.

Asgari, S., Hellers, M., and Schmidt, O. (1996). Host haemocyte inactivation by an insect parasitoid: transient expression of a polydnavirus gene. J. Gen. Virol. *77*, 2653–2662.

Asgari, S., Schmidt, O., and Theopold, U. (1997). A polydnavirus-encoded protein of an endoparasitoid wasp is an immune suppressor. J. Gen. Virol. *78*, 3061–3070.

Asgari, S., Theopold, U., Wellby, C., and Schmidt, O. (1998). A protein with protective properties against the cellular defense reactions in insects. Proc. Natl. Acad. Sci. U.S.A. *95*, 3690–3695.

Balgopal, M.M., Dover, B.A., Goodman, W.G., and Strand, M.R. (1996). Parasitism by *Microplitis demolitor* induces alterations in the juvenile hormone titer of its host, *Pseudoplusia includens*. J. Insect Physiol. *42*, 337–345.

Baker, K.D., and Thummel, C.S. (2007). Diabetic larvae and obese flies-emerging studies of metabolism in *Drosophila*. Cell Metab. *6*, 257–265.

Beck, M., and Strand, M.R. (2003). RNA interference silences *Microplitis demolitor* bracovirus genes and implicates glc1.8 in disruption of adhesion in infected host cells. Virology *314*, 521–535.

Beck, M., and Strand, M.R. (2005). Glc1.8 from *Microplitis demolitor* bracovirus induces a loss of adhesion and phagocytosis in insect high five and S2 cells. J. Virol. *79*, 1861–1870.

Beck, M.H., and Strand, M.R. (2007). A novel polydnavirus protein inhibits the insect prophenoloxidase activation pathway. Proc. Natl. Acad. Sci. U.S.A. *104*, 19267–19272.

Beck, M.H., Inman, R.B., and Strand, M.R. (2007). *Microplitis demolitor* bracovirus genome segments vary in abundance and are individually packaged in virions. Virology *359*, 179–189.

Beckage, N.E., and Riddiford, L.M. (1982). Effects of parasitism by *Apanteles congregatus* on the endocrine physiology of the tobacco hornworm, *Manduca sexta*. Gen. Comp. Endocr. *47*, 308–322.

Beckage, N.E., and Kanost, M.R. (1993). Effects of parasitism by the braconid wasp *Cotesia congregata* on host hemolymph proteins of the tobacco hornworm, *Manduca sexta*. Insect Biochem. Mol. Biol. *23*, 643–653.

Beckage, N.E., and Gelman, D.B. (2004). Wasp parasitoid disruption of host development: implications for new biologically based strategies for insect control. Annu. Rev. Entomol. *49*, 299–330.

Beckage, N.E., Metcalf, J.S., Nesbit, D.J., Schleifer, K.W., Zetlan, S.R., and de Buron, I. (1990). Host hemolymph monophenoloxidase activity in parasitized *Manduca sexta* larvae and evidence for inhibition by wasp polydnavirus. Insect Biochem. *20*, 285–294.

Beliveau, C., Laforge, M., Cusson, M., and Bellemare, G. (2000). Expression of a *Tranosema rostrale* polydnavirus gene in the spruce budworm, *Choristoneura fumiferana*. J. Gen. Virol. *81*, 1871–1880.

Belle, E., Beckage, N.E. Rousselet, J., Poirie, M., Lemeunier, F., and Drezen, J.-M. (2002). Visualization of polydnavirus sequences in a parasitoid wasp chromosome. J. Virol. *76*, 5793–5796.

Bezier, A., Herbiniere, J. Serbielle, C., Lesobre, J., Wincker, P., Huguet, E., and Drezen, J.-M. (2008). Bracovirus gene products are highly divergent from insect proteins. Arch. Insect Biochem. Physiol. *17*, 172–187.

Bezier, A., Annaheim, M. Herbiniere, Wetterwald, G., Gyapay, S., Bernard-Samain, S., Wincker, P., Rodidi, I., Heller, M., Belghaazi, M., Pfister-Wilhem, R., Periquet, G., Dupuy, C., Huguet, E., Volkoff, A.N., Lanzrein, B., and Drezen, J.-M. (2009a). Polydnaviruses of braconid wasps derive from an ancestral nudivirus. Science *323*, 926–930.

Bezier, A., Herbiniere, J., Lanzrein, B., and Drezen, J.-M. (2009b). Polydnavirus hidden face: the genes producing virus particles of parasitic wasps. J. Invert. Pathol. *101*, 194–203.

Bigot, Y., Samain, S., Auge-Gouillou, C., and Federici, B.A. (2008). Molecular evidence for the evolution of ichnoviruses from ascoviruses by symbiogenesis. BMC Evol. Biol. *8*, 253.

Bonvin, M., Marti, D., Wyder, S., Kojic, D., Annaheim, M., and Lanzrein, B. (2005). Cloning, characterization and analysis by RNA interference of various genes of the *Chelonus inanitius* polydnavirus. J. Gen. Virol. *86*, 973–983.

Cerenius, L., and Soderhall, K. (2004). The prophenoloxidse-activating system in invertebrates. Immunol. Rev. *198*, 116–126.

Choi, J.Y., Kwon, S.J., Roh, J. Y., Yang, T. J., Yoon, S. H., Kim, H., Li, M.S., Park, B.S., Woo, S.D., Jin, B.R., Kim, Y., and Je, Y.H. (2009). Sequence and gene organization of 24 circles from the *Cotesia plutellae* bracovirus genome. Arch. Virol. *154*, 1313–1327.

Cole, T.J., Beckage, N.E., Tan, F.F., Srinivasan, A., and Ramaswamy, S.B. (2002). Parasitoid-host endocrine relations: self-reliance or cooptation? Insect Biochem. Mol. Biol. *32*, 1673–1679.

Colombani, J., Bianchini, L., Layelle, S., Pondeville, E., Dauphin-Villemant. C., Antoniewski, C., Carré, C., Noselli, S., and Léopold, P.M. (2005). Antagonistic actions of ecdysone and insulins determine final size in *Drosophila*. Science *310*, 667–670.

Cui, L., and Webb, B.A. (1997). Homologous sequences in the *Campoletis sonorensis* polydnavirus genome are implicated in replication and nesting of the W segment family. J. Virol. *71*, 8504–8513.

Cui, L., Soldevila, A.I., and Webb, B.A. (1997). Expression and hemocyte-targeting of a *Campoletis sonorensis* polydnavirus cysteine-rich gene in *Heliothis virescens* larvae. Arch. Insect Biochem. Physiol. *36*, 251–271.

Cusson, M., Lucarotti, C., Stoltz, D., Krell, P., and Doucet, D. (1998). A polydnavirus from the spruce budworm parasitoid Tranosema rostrale (Ichneumonidae). J. Invertebr. Pathol. *72*, 50–56.

Cusson, M., Laforge, M., Miller, D., Cloutier, C., and Stoltz, D. (2000). Functional significance of parasitism-induced suppression of juvenile hormone esterase activity in developmentally delayed *Choristoneura fumiferana* larvae. Gen. Comp. Endocr. *117*, 343–354.

Davies, H.D., and Vinson, S.B. (1986). Passive evasion by eggs of the braconid parasitoid *Cardiochiles nigriceps* of encapsulation *in vitro* by haemocytes of host *Heliothis virescens*. Possible role for fibrous layer in immunity. J. Insect Physiol. *32*, 1003–1010.

Davies, D.H., Strand, M.R., and Vinson, S.B. (1987). Changes in differential haemocyte count and *in vitro* behaviour of plasmatocytes from the host *Heliothis virescens* caused by *Campoletis sonorensis* polydnavirus. J. Insect Physiol. 33, 143–153.

Deng, L., and Webb, B.A. (1999). Cloning and expression of a gene encoding a *Campoletis sonorensis* polydnavirus structural protein. Arch. Insect Biochem. Physiol. *40*, 30–40.

Desjardins, C.A., Gundersen-Rindal, D.E., Hostetler, J.B., Tallon, L.J., Fuester, R.W., Schatz, M.C., Pedroni, M.J., Fadrosh, D.W., Haas, B.J., Toms, B.S., Chen, D., and Nene, V. (2007). Structure and evolution of a proviral locus of *Glyptapanteles indiensis* bracovirus. BMC Microbiol. *7*, 61.

Desjardins, C.A., Gundersen-Rindal, D.E. Hostetler, J.B., Tallon, L.J., Fadrosh, D.W., Fuester, R.W., Pedroni, M.J., Haas, B.J., Schatz, M.C., Jones, K.M., Crabtree, J., Forgerger, H., and Nene, V. (2008). Comparative genomics of mutualistic viruses of *Glyptapanteles* parasitic wasps. Genome Biol. *9*, R183.

Doi, S.F., Cai, D.Z., Li, X., and Chen, X.X. (2009). Parasitic castration of *Plutella xylostella* larvae induced by polydnavirus and venom of *Cotesia vestalis* and *Diadegma semiclausum*. Arch. Insect Biochem. Physiol. *70*, 30–43.

Dong, Y.M., Taylor, H.E., and Dimopoulos, G. (2006). AgDscam, a hypervariable immunoglobulin domain-containing receptor of the *Anopheles gambiae* immune system. PLoS Biol. *4*, 1137–1146.

Doucet, D., Beliveau, C., Dowling, A., Simard, J., Feng, Q., Krell, P.J., and Cusson, M. (2008). Prophenoloxidase 1 and 2 from the spruce budworm, *Choristoneura fumiferana*: molecular cloning and assessment of transcriptional regulation by a polydnavirus. Arch. Insect Biochem. Physiol. *67*, 188–201.

Doucet, D., and Cusson, M. (1996). Alteration of developmental rate and growth of *Choristoneura fumiferana* parasitized by *Tranosema rostrale* – role of the calyx fluid. Entomol. Exp. et Appl. *81*, 21–30.

Doucet, D., and Cusson, M. (1996). Role of calyx fluid in alterations of immunity in *Choristoneura fumiferana* larvae parasitized by *Tranosema rostrale*. Comp. Biochem. Physiol. *114*, 311–317.

Dover, B.A., Tanaka, T., and Vinson, S.B. (1995). Stadium-specific degeneration of host prothoracic glands by *Campoletis sonorensis* calyx fluid and its association with host ecdysteroid titers. J. Insect Physiol. *41*, 947–955.

Dowton, M., and Austin, A.D. (1994). Molecular phylogeny of the insect order Hymenoptera: apocritain relationships. Proc. Natl. Acad. Sci. U.S.A. *91*, 9911–9915.

Drezen, J-M., Provost, B., Espagne, E., Cattolico, L., Dupuy, C., Poirie, M., Periquet, G., and Huguet, E. (2003). Polydnavirus genome: integrated vs. free virus. J. Insect Physiol. *49*, 407–417.

Dupuy, C., Huguet, E., and Drezen, J.-M. (2006). Unfolding the evolutionary history of polydnaviruses. Virus Res. *117*, 81–89.

Edson, K.M., Vinson, S.B., Stoltz, D.B., and Summers, M.D. (1981). Virus in a parasitoid wasp: suppression of the cellular immune response in the parasitoid's host. Science *211*, 582–583.

Espagne, E., Dupuy, C., Huguet, E., Cattolico, L., Provost, B., Martins, N., Poirie, M., Periquet, G., and Drezen, J.-M. (2004). Genome sequence of a polydnavirus: insights into symbiotic virus evolution. Science *306*, 286–289.

Falabella, P., Cacciaupi, P., Varricchio, P., Malva, C., and Pennacchio, P. (2006). Protein tyrosine phosphatases of *Toxoneuron nigriceps* bracovirus as potential disrupters of host prothoracic gland function. Arch. Insect Biochem. Physiol. *61*, 157–69.

Falabella, P., Varricchio, P., Provost, B., Espagne, E., Ferrarese, R., Grimaldi, A., de Eguileor, M., Fimiani, G., Ursini, M. V., Malva, C., Drezen, J.-M., and Pennacchio, F. (2007). Characterization of the IkappaB-like gene family in polydnaviruses associated with wasps belonging to different Braconid subfamilies. J. Gen. Virol. *88*, 92–104.

Fath-Goodin, A., Gill, T.A., Martin, S.B., and Webb, B.A. (2006). Effect of *Campoletis sonorensis* cys-motif proteins on *Heliothis virescens* development. J. Insect Physiol. *52*, 576–85.

Fath-Goodin, A., Kroemer, J.A., and Webb, B.A. (2009). The *Campoletis sonorensis* ichnovirus vankyrin protein P-vank-1 inhibits apoptosis in insect Sf9 cells. Insect Mol. Biol. *18*, 497–506.

Federici, B.A., and Bigot, Y. (2003). Origin and evolution of polydnaviruses by symbiogenesis of insect DNA viruses in endoparasitic wasps. J. Insect Physiol. *49*, 419–432.

Fleming, J.G., and Summers, M.D. (1991). PDV DNA is integrated in the DNA of its parasitoid wasp host. Proc. Natl. Acad. Sci. U.S.A. *88*, 9770–74.

Friedman, R., and Hughes, A.L. (2006). Pattern of gene duplication in the *Cotesia congregata* bracovirus. Infect. Genet. Evol. *6*, 315–22.

Gad, W., and Kim, Y. (2008). A viral histone H4 encoded by *Cotesia plutellae* bracovirus inhibits haemocyte-spreading behaviour of the diamondback moth, *Plutellae xylostella*. J. Gen. Virol. *89*, 931–938.

Galibert, L., Rocher, J., Ravallec, M., Duonor-Cerutti, M., Webb, B.A., and Volkoff, A.N. (2003). Two *Hyposoter didymator* ichnovirus genes expressed in the lepidopteran host encode secreted or membrane-associated serine and threonine rich proteins in segments that may be nested. J. Insect Physiol. *49*, 441–451.

Gardiner, E.M.M., and Strand, M.R. (2000). Hematopoiesis in larval *Pseudoplusia includens* and *Spodoptera frugiperda*. Arch. Insect Biochem. Physiol. *43*, 147–164.

Gelman, D.B., Reed, D.A., and Beckage, N.E. (1998). Manipulation of fifth-instar host (*Manduca sexta*) ecdysteroid levels by the parasitoid wasp *Cotesia congregata*. J. Insect Physiol. *44*, 833–843.

Godfray, H.C.J. (1994). Insect Parasitoids (Princeton, USA: Princeton University Press).

Gilbert, L.I., Rybczynski, R., and Warren, J.T. (2002). Control and biochemical nature of the ecdysteroidogenic pathway. Annu. Rev. Entomol. *47*, 883–912.

Gruber, A., Stettler, P., Heiniger, P., Schumperli, D., and Lanzrein, B. (1996). Polydnavirus DNA of the braconid wasp *Chelonus inanitus* is integrated in the wasp's genome and excised only in later pupal and adult stages of the female. J. Gen. Virol. *77*, 2873–2879.

Guell, M., van Noort, V., Yus, E., Chen, W.-H., Leigh-Bell, J., Michalodimitrakis, K., Yamada, T., Marumugam, M., Doerks, T., Kuhner, S., Rode, M., Suyama, M. Schmidt, S., Gavin, A.-C., Bork, P., and Serrano. L. (2009). Trascriptome complexity in a genome-reduced bacterium. Science *326*, 1268–1271.

Guzo, D., and Stoltz, D.B. (1987). Observations on cellular immunity and parasitism in the tussock moth. J. Insect Physiol. *33*, 19–31.

Hayakawa, Y. (1995). Growth-blocking peptide: an insect biogenic peptide that prevents the onset of metamorphosis. J. Insect Physiol. *41*, 1–6.

Hayakawa, Y., and Yazaki, K. (1997). Envelope proteins of parasitic wasp symbiont virus, protects the wasp eggs from the cellular immune reactions by the host insect. Eur. J. Biochem. *246*, 820–826.

Herniou, E.A., Olszewski, J.A., Cory, J.S., and O'Reilly, D.R. (2003). The genome sequence and evolution of baculoviruses. Annu. Rev. Entomol. *48*, 211–234.

Hilgarth, R.S., and Webb, B.A. (2002). Characterization of *Campoletis sonorensis* ichnovirus segment I genes as members of the repeat element gene family. J. Gen. Virol. *83*, 2393–2402.

Ibrahim, A.M., and Kim, Y. (2008). Transient expression of protein tyrosine phosphatases encoded by *Cotesia plutellae* bracovirus inhibits insect cellular immune responses. Naturwissenschaften *95*, 25–32.

Imler, J-L., and Bulet, P. (2005). Antimicrobial peptides in *Drosophila*, structures, activities and gene regulation. In Mechanisms of Epithelial Defense, Kabelitz, D., and Schroder, J.M. eds., (Basil, Switzerland: Karger) Vol. 86, pp. 1–21.

Irving, P., Ubeda, J., Doucet, D., Troxler, L., Lagueux, M., Zachary, D., Hoffmann, J., Hetru, C., and Meister, M. (2005). New insights into *Drosophila* larval haemocyte functions through genome-wide analysis. Cell. Microbiol. *7*, 335–350.

Jehle, J.A., Blissard, G.W., Bonning, B.C., Cory, J.S., Herniou, E.A., Rohrmann, G.F., Theilmann, D.A., Theim, S.M., and Vlak, J.M. (2006). On the classification and nomenclature of baculoviruses: a proposal for revision. Arch. Virol. *151*, 1257–1266.

Kadash, K., Harvey, J.A., and Strand, M.R. (2003). Cross-protection experiments with parasitoids in the genus *Microplitis* (Hymenoptera: Braconidae) suggest a high level of specificity in their associated bracoviruses. J. Insect Physiol. *49*, 473–482.

Kanost, M.R., and Gorman, M.J. (2008). Phenoloxidases in insect immunity. In Insect Immunity, Beckage, N.E. ed., (San Diego, USA: Academic Press) pp. 69–96.

Kang D., Liu, G., Gunne, H., and Steiner, H. (1996). PCR differential display of immune gene expression in *Trichoplusia ni*. Insect Biochem. Mol. Biol. *26*, 177–184.

Kim, Y. (2005). Identification of host translation inhibitory factor of *Campoletis sonorensis* ichnovirus on the tobacco budworm, *Heliothis virescens*. Arch. Insect Biochem. Physiol. *59*, 230–244.

Kocks, C., Cho, J-H., Nehme, N., Ulvila, J., Pearson, A.M., Meister, M., Strom, C., Conto, S.L., Hetru, C., Stuart, L.M., Stehle, T., Hoffmann, J.A., Reichhart, J-M., Ferrandon, D., Ramet, M., and Ezekowitz, R.A.B. (2005). Eater, a transmembrane protein mediating pahgocytosis of bacterial pathogens in *Drosophila*. Cell *123*, 335–346.

Kroemer, J.A., and Webb, B.A. (2004a). Polydnavirus genes and genomes: emerging gene families and new insights into polydnavirus replication. Annu. Rev. Entomol. *49*, 431–456.

Kroemer, J.A., and Webb, B.A. (2004b). Divergences in protein activity and cellular localization within the *Campoletis sonorensis* ichnovirus vankyrin family. J. Virol. *80*, 12219–12228.

Kwon, B., and Kim, Y. (2008). Transient expression of an EP1-like gene encoded in *Cotesia plutellae* bracovirus suppresses the hemocyte population in the diamondback moth, *Plutella xylostella*. Dev. Comp. Immunol. *32*, 932–942.

Labropoulou, V., Douis, V., Stefanou, D., Magrioti, C., Swevers, L., and Iatrou, K. (2008). Endoparasitoid wasp bracovirus-mediated inhibition of hemolin function and lepidopteran host immunosuppression. Cell. Microbiol. *10*, 2118–2128.

Lapointe, R., Tanaka, K., Barney, W.E., Whitfield, J.B., Banks, J.C., Beliveau, C., Stoltz, D., Webb, B.A., and Cusson, M. (2007). Genomic and morphological features of a banchine polydnavirus: comparison with bracoviruses and ichnoviruses. J. Virol. *81*, 6491–6501.

Lapointe, R., Wilson, R., Vilaplana, L., O'Reilly, D.R., Falabella, P., Douris, V., Bernier-Cardou, M., Pennacchio, F., Iatrou, K., Malva, C., and Olszewski, J.A. (2005). Expression of a *Toxoneuron nigriceps* polydnavirus (TnBV) encoded protein, causes apoptosis-like programmed cell death in lepidopteran insect cells. J. Gen. Virol. *86*, 963–971.

Lavine, M.D., and Strand, M.R. (2002). Insect hemocytes and their role in immunity. Insect Biochem. Mol. Biol. *32*, 1237–1242.

Lavine, M.D., and Strand, M.R. (2003). Hemocytes from *Pseudoplusia includens* express multiple alpha and beta integrin subunits. Insect Mol. Biol. *12*, 441–452.

Lawrence, P.O., and Lanzrein, B. (1993). Hormonal interactions between insect endoparasites and their host insects. In Parasites and Pathogens of Insects: Parasites Vol. I, Thompson, S.N. Federici, B.A., and Beckage, N.E. eds., (San Diego, USA: Academic Press), pp. 59–86.

Lanzrein, B., Pfister-Wilhelm, R., and von Niederhäusern, F. (2001). Effects of an egg-larval parasitoid and its polydnavirus on development and the endocrine system of the host. In Endocrine Interactions of Insect Parasites and Pathogens, Edwards, J.P., and Weaver, R.J. eds., (Oxford, UK: BIOS Sci. Publishers), pp. 95–109.

Lee, G., and Park, J.H. (2004). Hemolymph sugar homeostatsis and starvation-induced hyperactivity affected by genetic manipulations of the adipokinetic hormone-encoding gene in *Drosophila melanogaster*. Genetics *167*, 311–323.

Lee, S., Nalini, M., and Kim, Y. (2008). A viral lectin encoded in *Cotesia plutellae* bracovirus and its immunosuppressive effect on host hemocytes Comp. Biochem. Physiol. A *149*, 351–361.

Levin, D.M., Breuer, L.N., Zhuang, S.F., Anderson, S.A., Nardi, J.B., and Kanost, M.R. (2005). A hemocyte-specific integrin required for hemocytic encapsulation in the tobacco hornworm, *Manduca sexta*. Insect Biochem. Mol. Biol. *35*, 369–380.

Levashina, E.A., Moita, L.F., Blandin, S., Vriend, G., Lagueux, M., and Kafatos, F.C. (2001). Conserved role of a complement-like protein in phagocytosis revealed by dsRNA knockout in cultured cells of the mosquito, *Anopheles gambiae*. Cell *104*, 709–718.

Li, X., and Webb, B.A. (1994). Apparent functional role for a cysteine-rich polydnavirus protein in suppression of the insect cellular immune response. J. Virol. *68*, 7482–7489.

Ling, E.J., and Yu, X.Q. (2006). Cellular encapsulation and melanization are enhanced by immulectins, pattern recognition receptors from the tobacco hornworm *Manduca sexta*. Dev. Comp. Immunol. *30*, 289–299.

Lovallo, N., and Cox-Foster, D.L. (1999). Alteration in FAD-glucose dehydrogenase activity and hemocyte behavior contribute to initial disruption of *Manduca sexta* immune response to *Cotesia congregata* parasitoids. J. Insect Physiol. *45*, 1037–1048.

Lu, Z., Beck, M.H., Jiang, H., Wang, Y., and Strand, M.R. (2008). The viral protein Egf1.0 is a dual activity inhibitor of prophenoloxidase activating proteinases 1 and 3 from *Manduca sexta*. J. Biol. Chem. *283*, 21325–21333.

Lundstrom, A., Liu, G., Kang, D., Berzins, K., and Steiner, H. (2002). *Trichoplusia ni* gloverin, and inducible immune gene encoding an antibacterial insect protein. Insect Biochem. Mol. Biol. *32*, 795–801.

Marti, D., Grossniklaus-Burgin, C., Wyder, S., Wyler, T., and Lanzrein, B. (2003). Ovary development and polydnavirus morphogenesis in the parasitic wasp *Chelonus inanitus*. I. Ovary morphogenesis, amplification of viral DNA and ecdysteroid titres. J. Gen. Virol. *84*, 1141–1150.

Mirth, C.K., Truman, J.W., and Riddiford, L.M. (2005). The role of the prothoracic gland in determining critical weight for metamorphosis in *Drosophila melanogaster*. Curr. Biol. *15*, 1796–1807.

Mirth, C.K., and Riddiford, L.M. (2007). Size assessment and growth control: how adult size is determined in insects. BioEssays *29*, 344–355.

Moita, L.F., Wang-Sattler, R., Michel, K., Zimmermann, T., Blandin, S., Levashina, E.A., and Kafatos, FC. (2005). *In vivo* identification of novel regulators and conserved pathways of phagocytosis in *A. gambiae*. Immunity *23*, 65–73.

Moran, N.A. (2007). Symbiosis as an adaptive process and source of phenotypic complexity. Proc. Natl. Acad. Sci. U.S.A. *104*, 8627–8633.

Murphy, N., Banks, J.C., Whitfield, J. B., and Austin, A.D. (2008). Phylogeny of the parasitic microgastroid subfamilies (Hymenoptera : Braconidae) based on sequence data from seven genes, with an improved time estimate of the origin of the lineage. Mol. Phylogenet. Evol. *47*, 378–395.

Nakahara, Y., Kanamori, Y., Kiuchi, M., and Kamimura, M. (2003). *In vitro* studies of hematopoiesis in the silkworm, cell proliferation in and hemocyte discharge from the hematopoietic organ. J. Insect Physiol. *49*, 907–916.

Nakahara, Y., Shimura, S., Ueno, C., Kanamori, Y., Mita, K., Kiuchi, M., and Kamimura, M. (2009). Purification and characterization of silkworm hemocytes by flow cytometry. Dev. Comp. Immunol. *33*, 439–448.

Nakamatsu, Y., Gyotoku, Y., and Tanaka, T. (2001). The endoparasitoid *Cotesia kariyai* (Ck) regulates the growth and metabolic efficiency of *Pseudaletia separata* larvae by venom and Ck polydnavirus. J. Insect Physiol. 47, 573–584.

Nardi, J.B., Ujhelyi, E. Pilas, B., Garsha, K., and Kanost, M.R. (2003). Hematopoietic organs of *Manduca sexta* and hemocyte lineages. Dev. Genes Evol. *213*, 477–491.

Nardi, J.B., Pilas, B., Bee, C.M., Zhuang, S., Garsha, K., and Kanost, M.R. (2006). Neuroglian-positive plasmatocytes of *Manduca sexta* and the initiation of hemocyte attachment to foreign surfaces. Dev. Comp. Immunol. *30*, 447–462.

Nijhout, H.F. (1975). A threshold size for metamorphosis in the tobacco hornworm, *Manduca sexta*. Biol. Bull. *149*, 214–225.

Nijhout, H.F. (1994). *Insect Hormones*. (Princeton, USA: Princeton University Press)

Nijhout, H.F. (2003). The control of body size in insects. Dev. Biol. *261*, 1–9.

Norton, W.N., and Vinson, S.B. (1983). Correlating the initiation of virus replication with a specific pupal developmental phase of an ichneumonid parasitoid. Cell Tissue Res. *231*, 387–398.

Oldham, S., and Hafen, E. (2003). Insulin/IGF and target of rapamycin signaling: a TOR de force in growth control. Trends Cell Biol. *13*, 79–85.

Ohnishi, A., Hayakawa, Y., Matsuda, Y., Kwon, K.W., Takahashi, T.A., and Sekiguchi, S. (1995). Growth-blocking peptide titer during larval development of parasitized and cold-stressed armyworm. Insect Biochem. Mol. Biol. *10*, 1121–1127.

Pasquier-Barre, F., Dupuy, C., Huguet, E., Moneiro, F., Moreau, A., Poire, M., and Drezen, J-M. (2002). Polydnavirus replication: The EP1 segment of the parasitoid wasp *Cotesia congregata* is amplified within a larger precursor molecule. J. Gen. Virol. *83*, 2035–2045.

Pech, L.L., and Strand, M.R. (1996). Granular cells are required for encapsulation of foreign targets by insect haemocytes. J. Cell Sci. *109*, 2053–2060.

Pennacchio, F., Falabella, P., and Vinson, S.B. (1998). Regulation of *Heliothis virescens* prothoracic glands by *Cardiochiles nigriceps* polydnavirus (*Cn*PDV). Arch. Insect Biochem. Physiol. *38*, 1–10.

Pennacchio, F., and Strand, M.R. (2006). Evolution of developmental strategies in parasitic Hymenoptera. Annu. Rev. Entomol. *51*, 233–258.

Pfister-Wilhelm, R., and Lanzrein, B. (2009). Stage dependent influences of polydnaviruses and the parasitoid larva on host ecdysteroids. J. Insect Physiol. *55*, 707–715.

Philips, J.A., Rubin, E.J., and Perrimon, N. (2005). *Drosophila* RNAi screen reveals C35 family member required for mycobacterial infection. Science *309*, 1248–1251.

Pruijssers, A.J., and Strand, M.R. (2007). PTP-H2 and PTP-H3 from *Microplitis demolitor* bracovirus localize to focal adhesions and are antiphagocytic in insect immune cells. J. Virol. *81*, 1209–1219.

Pruijssers, A.J., Falabella, P., Eum, J.H., Pennacchio, F., Brown, M.R., and Strand, M.R. (2009). Infection by a symbiotic polydnavirus induces wasting and inhibits metamorphosis of the moth *Pseudoplusia includens*. J. Exp. Biol. *212*, 2998–3006.

Quicke, D.L.J., Fitton, M.G., Notton, D.G., Broad, G.R., and Dolphin, K. (2000). Phylogeny of the subfamilies of Ichneumonidae (Hymenoptera): a simultaneous molecular and morphological analysis. In Hymenoptera: Evolution, Bodiversity and Biological Control, A.D. Austin and M. Dowton, eds. (Collingwood, Australia: CSIRO Publishing), pp 74–83.

Ramet, M., Pearson, A., Manfruelli, P., Li, X., Koziel, H., Gobel, V., Chung, E., Krieger, M., and Ezekowitz, R.A.B. (2001). *Drosophila* scavenger receptor CI is a pattern recognition receptor for bacteria. Immunity *15*, 1027–1038.

Ramet, M., Manfruelli, P., Pearson, A., Mathey-Prevot, B., and Ezekowitz, R.A.B. (2002). Functional genomic analysis and identification of a *Drosophila* receptor for *E. coli*. Nature *416*, 644–648.

Rattanadechakul, W., and Webb, B.A. (2003). Characterization of *Campoletis sonorensis* ichnovirus unique segment B and excision locus structure. J. Insect Physiol. *49*, 523–532.

Rasoolizadeh, A., Beliveau, C., Steward, D., Cloutier, C., and Cusson, M. (2009). *Tranosema rostrale* ichnovirus repeat element genes display distinct transcriptional patterns in caterpillar and wasp hosts. J. Gen. Virol. *90*, 1505–1514.

Reed, D.A., and Brown, J.J. (1998). Host/parasitoid interactions: critical timing of parasitoid-derived products. J. Insect Physiol. *44*, 721–732.

Riddiford, L.M. (1976). Hormonal control of insect epidermal cell commitment *in vitro*. Nature *259*, 115–117.

Savary, S., Beckage, N., Tan, F., Periquet, G., and Drezen, J-M. (1997). Excision of the polydnavirus chromosomal integrated EP1 sequence of the parasitoid wasp *Cotesia congregata* (Braconidae, Microgastrinae) at potential recombinase binding sites. J. Gen. Virol. *78*, 3125–3134.

Serbielle, C., Moreau, S., Veillard, F., Voldoire, E., Bezier, A., Mannucci, M.A., Volkoff, A.N., Drezen, J.-M., Lalmanach, G., and Huguet, E. (2009). Identification of parasite-responsive cysteine proteases in *Manduca sexta*. Biol. Chem. *390*, 493–502.

Shelby, K.S., and Webb, B.A. (1994). Polydnavirus infection inhibits synthesis of an insect plasma protein, arylphorin. J. Gen. Virol. *75*, 2285–2292.

Shelby, K.S., and Webb, B.A. (1997). Polydnavirus infection inhibits translation of specific growth-associated host proteins. Insect Biochem. Mol. Biol. *27*, 263–270.

Shelby, K.S., Cui, L., and Webb, B.A. (1998). Polydnavirus-mediated inhibition of lysozyme gene expression and the antibacterial response. Insect Mol. Biol. *7*, 265–272.

Shelby, K.S., Adeyeye, O.A., Okot-Kotber, B.M., and Webb, B.A. (2000). Parasitism-linked block of host plasma melanization. J. Invertebr. Pathol. *75*, 218–225.

Schmidt, O., Theopold, U., and Strand, M. (2001) Innate immunity and its evasion and suppression by hymenopteran endoparasitoids. Bioessays *23*, 344–51.

Stettler, P., Trenczek, T., Wyler, T., Pfister-Wilhelm, R., and Lanzrein, B. (1998). Overview of parasitism associated effects on host haemocytes in larval parasitoids and comparison with effects of the egg-larval parasitoid *Chelonus inanitus* on its host *Spodoptera littoralis*. J. Insect Physiol. *44*, 817–831.

Stoltz, D.B., and Vinson, S.B. (1977). Baculovirus-like particles in the reproductive tracts of female parasitoid wasps. II. The genus *Apanteles*. Can. J. Microbiol. *23*, 28–37.

Stoltz, D.B., and Vinson, S.B. (1979). Viruses and parasitism in insects. Adv. Virus Res. *24*, 125–171.

Stoltz, D.B., and Cook, D.I. (1983). Inhibition of host phenoloxidase activity by parasitoid Hymenoptera. Experientia *39*, 1022–1024.

Stoltz, D.B., Vinson, S.B., and MacKinnon, E.A. (1976). Baculovirus-like particles in the reproductive tracts of female parasitoid wasps. Can. J. Microbiol. *22*, 1013–1023.

Stoltz, D.B., and Whitfield, J.B. (1992). Viruses and virus-like entities in the parasitic Hymenoptera. J. Hymen. Res. *1*, 125–139.

Strand, M.R. (2008a). Insect hemocytes and their role in immunity. In Insect Immunity, Beckage, N.E. ed., (San Diego, USA: Academic Press) pp. 25–47.

Strand, M.R. (2008b). Polydnavirus abrogation of the insect immune system. In Encyclopedia of Virology, 3rd Edition, Mahy, ed B.W. J., and van Regenmortel, M.H.V. eds., (London: Elsevier), Vol. 4 pp. 250–256.

Strand, M.R. (2009). The interactions between polydnavirus-carrying parasitoids and their lepidopteran hosts. In Molecular Biology and Genetics of the Lepidoptera, M. R. Goldsmith, and F. Marec, eds. (Boca Raton, USA: CRC Press), pp. 321–336.

Strand, M.R., and Casas, J. (2007). Parasitoid and host nutritional physiology in behavioral ecology. In Parasitoid Behavioral Ecology, Wajnberg, E., van Alphen, J., and Bernstein, C. eds., (Oxford, UK: Blackwell Press), pp. 113–28.

Strand, M.R., and Noda, T. (1991). Alterations in the haemocytes of *Pseudoplusia includens* after parasitism by *Microplitis demolitor*. J. Insect. Physiol. *37*, 839–850.

Strand, M.R., and Pech, L.L. (1995). Immunological basis for compatibility in parasitoid–host relationships. Annu. Rev. Entomol. *40*, 31–56.

Strand, M.R., Hayakawa, Y., and Clark, K.D (2000). Plasmatocyte spreading peptide (PSP1) and growth blocking peptide (GBP) are multifunctional homologs. J. Insect Physiol. *46*, 817–824.

Suderman, R.J., Pruijssers, A.J., and Strand, M.R. (2008). Protein tyrosine phosphatase-H2 from a polydnavirus induces apoptosis of insect cells. J. Gen. Virol. *89*, 1411–1420.

Takaki, K., and Sakurai, S. (2003). Regulation of prothoracic gland ecdysteroidogenic activity leading to pupal metamorphosis. Insect Biochem. Mol. Biol. *33*, 1189–1199.

Tanaka, K., Lapointe, R., Barney, W.E., Makkay, A.M., Stoltz, D., Cusson, M., and Webb, B.A. (2007). Shared and species-specific features among ichnovirus genomes. Virology *363*, 26–35.

Tanaka, T. (1987). Morphological changes in haemocytes of the host *Pseudaletia separata* parasitized by *Microplitis mediator* or *Apanteles kariyae*. Dev. Comp. Immunol. *1*, 57–67.

Tanaka, T., Agui, N., and Hiruma, K. (1987). The parasitoid *Apanteles kariyai* inhibits pupation of its host, *Pseudaletia separata*, via disruption of prothoracicotropic hormone release. Gen. Comp. Endocrinol. *67*, 364–374.

Teramato, T., and Tanaka, T. (2003). Similar polydnavirus genes of two parasitoids, *Cotesia kariyai* and *Cotesia ruficrus*, of the host *Pseudaletia separata*. J. Insect Physiol. *49*, 463–471.

Terenius, O., Bettencourt, R., Lee, S.Y., Li, W., Soderhall, K., and Faye, I. (2007). RNA interference of Hemolin causes depletion of phenoloxidase activity in *Hyalophora cecropia*. Dev. Comp. Immunol. *31*, 571–575.

Theopold, U., Samakovlis, C., Erdjumentbromage, H., Dillon, N., Axelsson, B., Schmidt, O., Tempst, P., and Hultmark, D. (1996). *Helix pomatia* lectin, an inducer of *Drosophila* immune response, binds to hemomucin, a novel surface mucin. J. Biol. Chem. *271*, 12708–12715.

Theopold, U., Schmidt, O., Soderhall, K., and Dushay, M.S. (2004). Coagulation in arthropods, defence, wound closure and healing. Trends Immunol. *25*, 289–294.

Theilmann, D.A., and Summers, M.D. (1987). Physical analysis of the *Campoletis sonorensis* virus multipartite genome and identification of a family of tandemly repeated elements. J. Virol. *61*, 2589–2598.

Thoetkiattikul, H., Beck, M.H., and Strand, M.R. (2005). Inhibitor kappaB-like proteins from a polydnavirus inhibit NF-kappaB activation and suppress the insect immune response. Proc. Natl. Acad. Sci. U.S.A. *102*, 11426–11431.

Thompson, S.N. (1993). Redirection of host metabolism and effects on parasite nutrition. In Parasites and Pathogens of Insects Vol 1, Beckage, N.E., Thompson, S.N., and Federici, B.A. eds., (New York: Academic Press), pp. 125–144.

Thompson, S.N., and Dahlman, D.L. (1998). Aberrant nutritional regulation of carbohydrate synthesis by parasitized *Manduca sexta*. J. Insect Physiol. *44*, 745–754.

Truman, J.W., Hiruma, K., Allee, J.P., Macwhinnie, S.G., Champlin, D.T., and Riddiford, L.M. (2006). Juvenile hormone is required to couple imaginal disc formation with nutrition in insects. Science *310*, 1385–1388.

Turnbull, M.W., and Webb, B.A. (2002). Perspectives on polydnavirus origin and evolution. Adv. Virus Res. *58*, 203–254.

Turnbull, M.W., Volkoff, A.N., Webb, B.A., and Phelan, P. (2005). Functional gap junction genes are encoded by insect viruses. Curr. Biol. *15*, R491–492.

Varricchio, P., Falabella, P., Sordetti, R., Graziani, F., Malva, C., and Pennacchio, F. (1999). *Cardiochiles nigriceps* polydnavirus: molecular characterization and gene expression in parasitized *Heliothis virescens* larvae. Insect Biochem. and Mol. Biol. *29*, 1087–1096.

Villarreal, L.P. (2007). Virus–host symbiosis mediated by persistence. Symbiosis *44*, 1–9.

Vinson, S.B. (1990). Physiological interactions between the host genus *Heliothis* and its guild of parasitoids. Arch. Insect Biochem. Physiol. *13*, 63–81.

Volkoff, A-N., Ravallec, M., Bossy, J., Cerutti, P., Rocher, J., Cerutti, M., and Devauchelle, P. (1995). The replication of *Hyposoter didymator* PDV: cytopathology of the calyx cells in the parasitoid. Biol. Cell *83*, 1–13.

Volkoff, N.A., Jouan, V., Urbach, S., Samain, S., Bergoin, M., Wincker, P., Demettre, E., Cousserans, F., Provost, B., and Coulibaly, F., *et al.* (2010). Analysis of virion structural components reveals vestiges of the ancestral inchnovirus genome. Plos Pathogens *6*, e1000923.

Wang, Y., and Jehle, J. (2009). Nudiviruses and other large, double-stranded circular DNA viruses of invertebrates: new insights on an old topic. J. Invert. Pathol. *101*, 187–193.

Webb, B.A. (1998). Polydnavirus biology, genome structure, and evolution. In The Insect Viruses, Miller, L.K., and Balls, L.A. eds, (New York: Plenum Press), pp. 105–139.

Webb, B.A., and Cui, L. (1998). Relationships between polydnavirus genomes and viral gene expression. J. Insect Physiol. *44*, 785–793.

Webb, B.A., and Summers, M.D. (1992). Stimulation of PDV replication by 20-hydroxyecdysone. Experientia *48*, 1018–1022.

Webb, B.A., Beckage, N.E., Hayakawa, Y., Krell, P.J., Lanzrein, B., Stoltz, D.B., Strand, M.R., and Summers, M.D. (2000). Polydnaviridae. In Virus Taxonomy: The Classification and Nomenclature of Viruses. The seventh report of the International Committee on Taxonomy of Viruses, M.H.V. van Regenmortel, C.M. Fauquet, D.H. L. Bishop, E.B. Carstens, M.K. Estes, S.M. Lemon, J. Maniloff, M.A. Mayo, D.J. McGeoch, C.R. Pringle and R.B. Wickner, eds. (San Diego, USA: Academic Pres), pp. 1167.

Webb, B.A., and Strand, M.R. (2005). The biology and genomics of polydnaviruses. In Comprehensive Molecular Insect Science, Vol. 6, Gilbert, L.I., Iatrou, K., and Gill, S.S. eds., (San Diego, USA: Elsevier), pp. 323–60.

Webb, B.A., Strand, M.R., Deborde, S.E., Beck, M., Hilgarth, R.S., Kadash, K., Kroemer, J.A., Lindstrom, K.G., Rattanadechakul, W., Shelby, K.S., Thoetkiattikul, L., Turnbull, M.W., Barney, W.E., and Witherell, R.A. (2006). Polydnavirus genomes reflect their dual roles as mutualists and pathogens. Virology *347*, 160–174.

Weber, B., Annaheim, M., and Lanzrein, B., (2007). Transcriptional analysis of polydnavirus gene in the course of parasitization reveals segment-specific patterns. Arch. Insect Biochem. Physiol. *66*, 9–22.

Wertheim, B., Kraaijeveld, A.R. Schuster, E., Blanc, E., Hopkins, M., Pletcher, S.D., Strand, M.R., Godfray, H.C.J., and Partridge, L. (2005). Genome wide expression in response to parasitoid attack in *Drosophila*. Genome Biol. *6*, R94.

Whitfield, J.B. (1998). Phylogeny and evolution of host–parasitoid interactions in Hymenoptera. Annu. Rev. Entomol. *43*, 129–151.

Whitfield, J.B. (2002). Estimating the age of the polydnavirus/braconid wasp symbiosis. Proc. Natl. Acad. Sci. U.S.A. *99*, 7508–7513.

Whitfield, J.B, and Asgari, S. (2003). Virus or not? Phylogenetics of polydnaviruses and their wasp carriers. J. Insect Physiol. *49*, 397–405.

Wu, Q., and Brown, M.R. (2006). Signaling and function of insulin-like peptides in insects. Annu. Rev. Entomol. *51*, 1–24.

Wyder, S., Tschannen, A., Hochuli, A., Gruber, A., Saladin, V., Zumbach, S., and Lanzrein, B. (2002). Characterization of *Chelonus inanitus* polydnavirus segments: sequences and analysis, excision site and demonstration of clustering. J. Gen. Virol. *83*, 247–256.

Wyler, T., and Lanzrein, B. (2003). Ovary development and polydnavirus morphogenesis in the parasitic wasp *Chelonus inanitus*. II. Ultrastructural analysis of calyx cell development, virion formation and release. J. Gen. Virol. *84*, 1151–1163.

Xu, D., and Stoltz, D. (1993). Polydnavirus genome segment families in the ichneumonid parasitoid *Hyposoter fugitivus*. J. Virol. *67*, 1340–1349.

Yagi, S., and Tanaka, T. (1992). Retardation of testis development in the armyworm, *Pseudaletia separata*, parasitized by the braconid wasp, *Cotesia kariyai*. Int. J. Inver. Rep. Dev. *22*, 151–157.

Yin, L., Zhang, C., Qin, J., and Wang, C. (2003). Polydnavirus of *Campoletis chlorideae*: characterization and temporal effect on host *Helicoverpa armigera* cellular immune response. Arch. Insect Biochem Physiol. *52*, 104–113.

Zhang, G., Lu, Z.-Q., Jiang, H., and Asgari, S. (2004). Negative regulation of prophenoloxidase (proPO) activation by a clip-domain serine proteinase homolog (SPH) from endoparasitoid venom. Insect Biochem. Mol. Biol. *34*, 477–483.

Zhuang, S., Kelo, L., Nardi, J.B., and Kanost, M.R. (2007). An integrin–tetraspanin interaction required for cellular innate immune responses of an insect, *Manduca sexta*. J. Biol. Chem. *282*, 22563–22572.

Part II

RNA Viruses

Dicistroviruses

9

Bryony C. Bonning and Karyn N. Johnson

Abstract

The *Dicistroviridae* family is comprised of viruses that infect arthropods including beneficial and pest species as well as the model insect *Drosophila melanogaster*. The 30 nm virions have pseudo $T = 3$ icosahedral symmetry and are non-enveloped. The defining character of this group of viruses is that the two open reading frames encoded by the positive-sense RNA genome are both translated directly from the genomic RNA. This is achieved via two internal ribosome entry sites that independently guide the initiation of translation of the structural and non-structural viral polyproteins; one by elegant molecular mimicry of the initiator tRNA. This chapter describes our current understanding of the molecular biology of these viruses, their interactions with their hosts and potential future impacts of dicistrovirus research.

Introduction

The *Dicistroviridae* is a recently defined family of positive-sense RNA viruses of arthropods with non-enveloped $T = 3$ icosahedral particles (Christian *et al.*, 2005). While we have a comprehensive understanding of the particle structure of these viruses and of the mechanisms of internal ribosome entry in viral RNA, knowledge of other fundamental aspects of dicistrovirus biology is lacking (Bonning and Miller, 2010). The construction of the first infectious clones of dicistrovirus genomes may facilitate future research to fill these gaps in knowledge (Boyapalle *et al.*, 2008; Pal *et al.*, 2007). Dicistroviruses have been identified in metagenomic analyses of such disparate environments as fresh water (Djikeng *et al.*, 2009), the ocean (Culley *et al.*, 2006), and the human intestinal tract (Victoria *et al.*, 2009). Hence, these viruses which profoundly impact arthropod populations are abundant in the environment.

Taxonomy

Similarity with other groups (Picornavirales)

The dicistroviruses share many features with a number of other taxonomic groups of small, non-enveloped icosahedral viruses with positive-sense RNA genomes. Dicistroviruses were historically referred to as 'picorna-like viruses' as the *Picornaviridae* is one of the largest and best-known taxonomic groups that have these features. Early analysis suggested that *Drosophila C virus* (DCV) and *Cricket paralysis virus* (CrPV) might have belonged to the *Picornaviridae* family. However, bioinformatics analysis of limited sequence from the 3′ end of the CrPV genome (King *et al.*, 1987) suggested that CrPV, in contrast to picornaviruses, may encode the capsid proteins at the 3′ end of the genome (Koonin and Gorbalenya, 1992). In 1998 the genomes of three dicistroviruses (DCV, *Rhopalosiphum padi virus* – RhPV and *Plautia stali intestine virus* – PSIV) were sequenced and together with direct sequencing of the capsid proteins, this confirmed the 3′ location of the structural genes. The genome organization of these three viruses clearly established that they were distinct from the picornaviruses and all other known virus families (Johnson and

Christian, 1998; Moon *et al.*, 1998; Sasaki *et al.*, 1998). Since that time the genome sequences of many dicistroviruses have been determined (Table 9.1). Phylogenetic analysis based on the conserved sequence of the RNA-dependent RNA polymerase (RdRp) and the 3C-like protease from the virus families commonly considered as 'picorna-like' indicate that the dicistroviruses are most similar to a group of viruses that infect marine algae belonging to the family *Marnaviridae* (Koonin *et al.*, 2008; Le Gall *et al.*, 2008), and analysis based on the RNA helicase sequence indicates that the dicistroviruses form an independent clade (Koonin *et al.*, 2008).

The *Dicistroviridae* family belongs to the order *Picornavirales* (Le Gall *et al.*, 2008). This order of viruses includes the families: *Dicistroviridae, Marnaviridae, Iflaviridae, Picornaviridae, Sequiviridae, Comoviridae* and the unassigned genera *Cheravirus* and *Sadwavirus*. These viruses infect a wide range of hosts including insects and other invertebrates, vertebrates, plants and algae. The order *Picornavirales* is defined by a group of characteristics including: non-enveloped icosahedral capsid with pseudo-$T = 3$ symmetry which is about 30 nm in diameter (Fig. 9.1A); a positive-sense RNA genome that encodes a type III helicase, 3C-like protease (usually cysteine nucleophile) followed by the RdRp; conserved order of structural proteins in the genome; presence of a small (< 5 kDa) virus protein (VPg) linked to the 5′ end of the genome; and synthesis of one or more polyproteins directly from the genomic RNA with no synthesis of a subgenomic RNA. Additional features that are common between the dicistroviruses and the majority of the groups in the *Picornavirales* order are that the genome is monopartite and polyadenylated at the 3′ end. The unique feature of the dicistroviruses is the presence of two open reading frames (ORFs, or cistrons) within the genomic RNA (Fig. 9.1B). It is this feature that represents the derivation of the name *Dicistroviridae*.

Similarities within the group

The family *Dicistroviridae* contains more than a dozen virus species (Table 9.1). The type species of the *Cripavirus* genus is *Cricket paralysis virus*. The majority of other species also belong to this genus. A second genus *Aparavirus*, has been proposed to the International Committee for the Taxonomy of Viruses. The two genera differ in characteristics of the intergenic region (IGR) and the internal ribosome entry site (IRES), which is supported by phylogenetic analysis of both the capsid protein and RdRp regions (Nakashima and Uchiumi, 2009).

Virion properties and structure

The virion of dicistroviruses has a diameter of approximately 30 nm and no envelope. An electron micrograph of CrPV particles is shown in Fig. 9.1. The virion is comprised of the positive-sense ssRNA genome and 60 protomers.

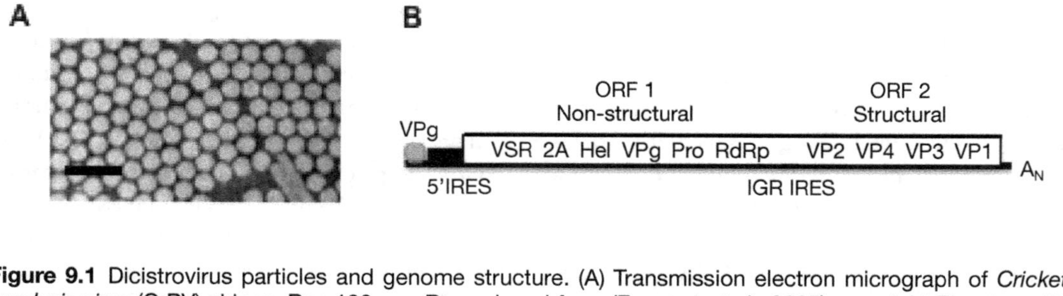

Figure 9.1 Dicistrovirus particles and genome structure. (A) Transmission electron micrograph of *Cricket paralysis virus* (CrPV) virions. Bar, 100 nm. Reproduced from (Fauquet *et al.*, 2005) copyright Elsevier, with permission. (B) Schematic diagram showing common features of the dicistrovirus genome. The RNA is represented by the thick black line and includes a polyadenylated tail (A_N). The 5′ and intergenic region (IGR) internal ribosome entry sites (IRES) are represented by black rectangles. The two open reading frames (ORF) of the virus are indicated by white rectangles. The coding domains of ORF1 are designated: viral suppressor of silencing (VSR), 2A-like motif (2A), helicase (Hel), the virus genome linked protein (VPg, which is often present in multiple copies: see Table 9.1), protese (Pro) and the RNA dependent RNA polymerase (RdRp). The structural proteins encoded by ORF2 are indicated in order (VP1-VP4).

Each protomer includes one copy of each of the large structural proteins VP1, VP2, VP3 and the small protein VP4. The cleavage of VP3 and VP4 from the capsid protein precursor VP0 likely occurs as a maturational autocatalytic cleavage, similar to the well studied cleavage of the capsid proteins that occurs in the virions of insect nodaviruses and tetraviruses (Liljas *et al.*, 2002 and see Chapter 14). This cleavage has yet to be investigated experimentally for dicistroviruses.

The X-ray crystal structure of CrPV has been solved to 2.4 Å resolution (Liljas *et al.*, 2002; Tate *et al.*, 1999) and the low resolution structure of *Triatoma virus* (TrV) is consistent with the major features identified in the CrPV structure (Estrozi *et al.*, 2008). The CrPV structure shows that the three large capsid proteins of the dicistroviruses form a pseudo-$T = 3$ icosahedral capsid. VP1, VP2 and VP3 each have a core β-barrel (jelly roll) structure, with VP1 positioned around the 5-fold axis and VP2 and VP3 alternating around the 3-fold axis. The small capsid protein VP4 lies under the 5-fold axis of symmetry at the interface between the capsid and the RNA (Tate *et al.*, 1999). The overall capsid structure is similar between the dicistroviruses and other members of the *Picornavirales*. Further detail on the structure of CrPV is discussed in Chapter 14.

Genome coding strategy and translation

Dicistroviruses have a single genomic RNA comprised of 9000–11,000 nt (Table 9.1, Fig. 9.1). There is a small virus protein (VPg) covalently linked to the 5′ end of the genome and the 3′ end is polyadenylated (Eaton and Steacie, 1980; King and Moore, 1988; Nakashima and Shibuya, 2006). The 400–1,000 nt 5′ UTR is followed by a large ORF (ORF1) that encodes the non-structural proteins necessary for virus replication. A non-translated intergenic region (IGR), usually of about 200 nt, separates ORF1 from ORF2. ORF2 encodes the structural proteins and this is followed by a 150–500 nt 3′ UTR. The genome is unique in that both ORF1 and ORF2 are independently translated directly from the genomic RNA without synthesis of a subgenomic RNA (Johnson and Christian, 1998; Nakashima *et al.*, 1998; Sasaki *et al.*, 1998). Independent regulation of protein synthesis from the two ORFs allows the capsid proteins to be synthesized in supramolar excess compared to the replication proteins (Moore *et al*, 1980, 1981). Each of the ORFs encodes a polyprotein that is processed into the mature non-structural (ORF1) and structural (ORF2) proteins.

ORFs

The large precursor polyprotein(s) encoded by ORF1 is processed into a yet to be specified number of mature proteins (Moore *et al.*, 1980; Moore *et al.*, 1981; Nakashima and Nakamura, 2008). To date, five common functional domains have been identified (Fig. 9.1). The functions of some regions of ORF1 have been inferred from sequence homology, whereas others have been experimentally demonstrated. It is likely that there are domains of the ORF1 protein(s) that are utilized for as yet unidentified functions. The identified functional domains from N-terminus to C-terminus are: viral suppressor of RNA interference (VSR), helicase, VPg, 3C-like protease and RdRp. Additional motifs including the 2A-like NPGP sequence and BIR have been identified in some dicistroviruses (Table 9.1).

The structural protein ORF2 encodes a single polyprotein that is post-translationally cleaved. The order of the capsid proteins from the N-terminus is VP2, VP4, VP3 then VP1. Cleavage of the precursor into VP2, VP0 and VP1 occurs via the viral 3C-like protease. The amino acid preceding these cleavage sites is a conserved glutamine in all dicistroviruses except TSV where cleavage is predicted to occur following a histidine (Nakashima and Nakamura, 2008). It is likely that capsids are assembled from VP2, VP3 and a precursor protein VP0 and that VP0 is then proteolytically cleaved to liberate the mature capsid proteins VP3 and VP4. This model is supported by the detection of low levels of uncleaved VP0 in dicistrovirus particles and the close proximity of the C-terminus of VP4 to the N-terminus of VP3 within mature virus particles (Liljas *et al.*, 2002). In contrast to the initial ORF2 proteolytic processing, the amino acid preceding the VP3 cleavage site is commonly a phenylalanine.

Table 9.1 Isolation, genomic sequencing and biophysical properties of the dicistroviruses[a]

Virus	Acronym	Original description	Genome sequence	Genome characteristics[b]	M_r capsid proteins[c]	Buoyant density in CsCl (g/ml)	Diameter (nm)
Cripavirus							
Cricket paralysis virus	CrPV	Reinganum *et al.* (1970)	Wilson *et al.*, (2000b)	9185 nt, VPg4, 2Am	35, 34, 30, 6 (43)	1.34	27
Aphid lethal paralysis virus	ALPV	Williamson *et al.* (1988)	van Munster *et al.* (2002)	9812 nt, VPg1	34, 32, 31 (41)	1.34	26–28
Black queen cell virus	BQCV	Bailey and Woods (1977)	Leat *et al.* (2000)	8550 nt, VPg4	34, 32, 29,6	1.34	30
Drosophila C virus	DCV	Jousset *et al.* (1972)	Johnson and Christian (1998)	9264 nt, VPg4, 2Am	31, 30, 28, 9 (37)	1.34	27
Himetobi P virus	HiPV	Toriyama *et al.* (1992)	Nakashima *et al.* (1999)	9257 nt, VPg4	36, 33, 28 (41)	1.35	29
Homalodisca coagulata virus-1	HoCV-1	Hunnicutt *et al.* (2006)	Hunnicutt *et al.* (2006)	9345 nt	32, 31, 30, 5.6	–	–
Plautia stali intestine virus	PSIV	Nakashima *et al.* (1998)	Sasaki *et al.*, (998)	8797 nt, VPg3	33, 30, 26, 4.5 (35)	–	30
Rhopalosiphum padi virus	RhPV	D'Arcy *et al.* (1981)	Moon *et al.* (1998)	10,011 nt, VPg1	31, 30, 28 (41)	1.37	27
Triatoma virus	TrV	Muscio *et al.* (1987)	Czibener *et al.* (2000)	9010 nt, VPg2	39, 37, 33	1.39	30
Aparavirus							
Acute bee paralysis virus	ABPV	Bailey *et al.*, (963)	Govan *et al.* (2000)	9491 nt, VPg3, 2Am	35, 33, 24, 9	1.34	30
Israeli acute paralysis virus	IAPV	Maori *et al.* (2007a)	Maori *et al.* (2007b)	9487 nt, 2Am	35, 33, 26, 17 (19, 30, 46)	–	27
Kashmir bee virus	KBV	Bailey and Woods (1977)	de Miranda *et al.* (2004)	9524 nt, VPg4, 2Am	41, 37, 25, 6	1.37	30
Solenopsis invicta virus-1	SINV-1	Valles *et al.* (2004)	Valles *et al.* (2004)	8026 nt, VPg6			31
Taura syndrome virus	TSV	Hasson *et al.* (1995)	Mari *et al.* (2002)	10,205 nt, VPg4	55, 40, 24 (58)	1.33	31

[a]After Christian and Scotti (2006).

[b]Characteristics of genomic RNA: number of nt; VPg, genome-linked protein; VPg#, number of VPg-like sequences present where known (Nakashima and Shibuya, 2006); 2Am, 2A-like NPGP motif present in ORF1; IAP, inhibitor of apoptosis repeat domain.

[c]Minor proteins identified in virion preparations are shown in parentheses.

IRES elements and translational control

Viruses utilize host protein translational machinery for synthesis of the viral proteins. To ensure efficient viral protein synthesis many viruses, including the dicistroviruses, down-regulate host protein synthesis. For this strategy, the translation of the viral RNA must be controlled differently from that of the host. Host mRNA translation involves recognition of the cap structure (m^7GpppN) on the 5′ end of the mRNA by ribosomes and translational initiation factors. A common strategy for viruses with non-capped positive-sense RNA genomes is to interfere with cap-dependent translational initiation, which will decrease host translation without interfering with cap-independent translation of the viral RNA. To achieve cap-independent initiation many viruses (and some host mRNAs) have IRES elements. This enables host ribosomes to interact with the viral RNA and initiate translation in the absence of a cap.

Canonical eukaryotic translation initiation involves a number of steps. The translation initiation factor complex eIF4F recognizes the 5′ capped mRNA. The eIF4F complex comprises the cap-binding protein (eIF4E), an RNA helicase (eIF4A) and a protein (eIF4G) that acts as a scaffold between the mRNA and the small ribosomal subunit. The small ribosomal subunit has two tRNA binding sites and one of these (the ribosomal P site) is charged with the initiator methionine tRNA (Fig. 9.2). Once the small ribosomal subunit interacts with the eIF4F complex and eIF3, it scans along the mRNA until it reaches the first AUG start codon, at which time the large ribosomal subunit is recruited to the complex. Elongation starts when a second tRNA (specified by the second codon of the ORF) interacts with the ribosomal A site and a peptide bond is formed between the methionine and amino acid on the tRNA that interacts with the ribosomal A site. Viruses that do not have capped RNAs can usually initiate protein synthesis in the absence of the cap-binding protein. Some viruses, for example picornaviruses, actively disrupt cap-dependent translation initiation via protease cleavage of the eIF4G by a viral protease (Roberts *et al.*, 1998).

The dicistrovirus genome includes two different types of IRES elements; one in the 5′ UTR which controls translation of the non-structural protein precursor from ORF1, and the second which lies in the IGR between the two ORFs and controls the synthesis of the structural polyprotein from ORF2. Both the 5′ and IGR IRES elements are atypical as compared to IRES elements from other virus systems. One unusual feature is that, unlike most well characterized IRES elements, both the 5′ IRES and IGR IRES are not restricted in function to their natural host system. While IRES function is clearly a requirement for infectivity, there are further host/virus factors that are involved in determining the permissiveness of a particular host to virus infection (Masoumi *et al.*, 2003). Dicistrovirus IRES function has been demonstrated in a range of different systems including mammalian, plant and insect cells as well as *in vitro* in *Drosophila*, mammalian and wheat germ extracts (Table 9.2). This feature has focused research on the utility of dicistrovirus IRES elements for non-host expression systems such as baculovirus expression vectors (Chen *et al.*, 2005; Pijlman *et al.*, 2006).

The mechanism of function of the two dicistrovirus IRES elements is very different, which at least in part, can account for the differential regulation of translation between the two ORFs. Translation from each of these IRES elements has been shown to be sensitive to ribosomal machinery depletion under conditions where non-IRES mRNA translation is relatively unaffected (Cherry *et al.*, 2005). There is some evidence that a subset of ribosomal proteins is required for IRES initiated translation that is not essential for cap-dependent translation. For example, Rps25 has been identified as essential for CrPV IGR IRES function in yeast and mammalian cells (Landry *et al.*, 2009); interestingly this contrasts with the lack of effect of RNAi knockdown of the homologous protein in DCV infected *Drosophila* cells (Cherry *et al.*, 2005). This inconsistency may indicate a difference between invertebrate cells and other cell types, or may reflect the difference in IRES initiated translation in a reporter gene system as opposed to a viral infection model.

Table 9.2 Use of IRES elements for expression

		Functionality		
Virus	IRES	*In vivo*	*In vitro*	References
CrPV	5′	*Drosophila melanogaster, Aedes albopictus, Ae. aegypti, Anticarcia gemmatalis, Plutella xylostella, Trichoplusia ni*	RRL[2]	Masoumi *et al.* (2003), Wilson *et al.* (2000b)
	IGR	*D. melanogaster, Ae. albopictus, Ae. aegypti, T. ni, Spodoptera frugiperda, Bombyx mori,* yeast[1]	RRL, WGE[2]	Carter *et al.* (2008), Masoumi *et al.* (2003), Reavy and Moore (1981), Thompson *et al.* (2001), Wilson *et al.* (2000b)
DCV	5′	*D. melanogaster*		Cherry *et al.* (2005)
	IGR	*Ae. aegypti, D. melanogaster, S. frugiperda, B. mori*	RRL	Carter *et al.* (2008), Cherry *et al.* (2005), Reavy *et al.* (1983), Schubot *et al.* (2005)
RhPV	5′	*S. frugiperda*	*Drosophila*, *Spodoptera*, RRL, WGE	Domier and McCoppin (2003), Royall *et al.* (2004), Woolaway *et al.* (2001)

Table 9.2 continued

Virus	IRES	Functionality: *In vivo*	Functionality: *In vitro*	References
	IGR	*S. frugiperda*	RRL	Domier and McCoppin (2003), Domier *et al.* (2000)
PSIV	5′		*S. frugiperda*	Shibuya and Nakashima (2006)
	IGR		RRL, WGE	Sasaki and Nakashima (1999), Shibuya *et al.* (2003)
TrV	5′	*Xenopus* oocytes, BHK cells, *Ae. albopictus*		Czibener *et al.* (2005)
	IGR	*Xenopus* oocytes, BHK cells, *Ae. albopictus*		Czibener *et al.* (2005)
ALPV	5′			
	IGR	*Ae. aegypti, D. melanogaster, S. frugiperda, B. mori*		Carter *et al.* (2008)
BQCV	5′			
	IGR	*Ae. aegypti, D. melanogaster, S. frugiperda, B. mori*		Carter *et al.* (2008)
KBV				
	IGR	*B. mori*		Carter *et al.* (2008)
SINV				
	IGR	*B. mori*		Carter *et al.* (2008)
TSV				
	IGR	*B. mori*	RRL, WGE	Carter *et al.* (2008), Cevallos and Sarnow (2005) Hatakeyama *et al.* (2004)

[1]Note only functions efficiently in yeast with mutations that lower eIF2-GTP/initiator $tRNA^{met}$ complex levels.

[2]RRL, rabbit reticulolysate; WGE, wheat germ extract.

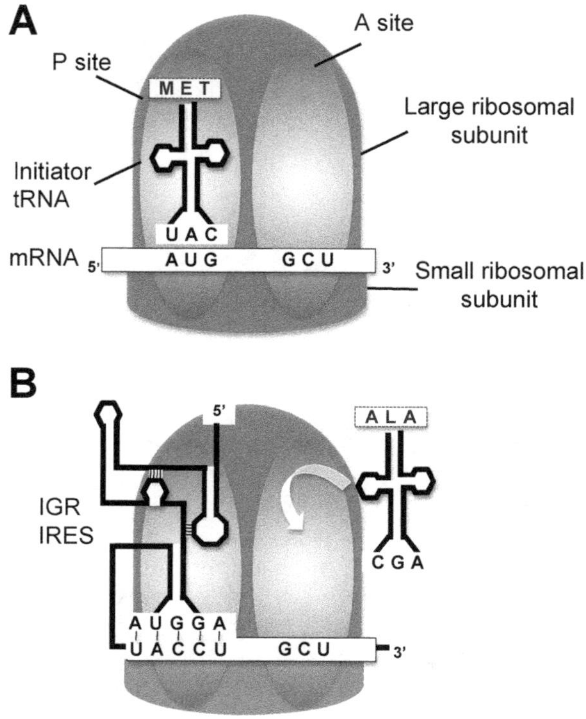

Figure 9.2 Comparison of canonical cap-dependent and IGR IRES translation initiation. A. Standard initiation of eukaryotic translation involves interaction of a methionine charged tRNA encoding the anticodon to the initiation sequence AUG. This interaction occurs in the P site of the ribosome. A second tRNA then interacts with the second codon of the ORF which sits in the A site of the ribosome. A peptide bond is formed between the methionine and the second amino acid, the ribosome translocates to the next codon and translation continues. B. To initiate translation of the capsid protein ORF, the dicistrovirus IGR IRES interacts with the P site of the ribosome. A pseudoknot is formed between the IRES and five nucleotides immediately preceding the translation initiation site. The 3′ most three of these 5 base pairs are recognized by the ribosome as a mimic for the codon–anticodon interaction that occurs in canonical translation. A tRNA charged with an amino acid (alanine for CrPV as shown in diagram) can then enter the A site and pair with the first codon of the capsid ORF. In this way, translation is initiated in a cap and methionine independent manner.

5′ IRES

It was predicted that the 5′ UTR of the dicistroviruses encodes an IRES element as the RNA is not capped and the long (400–1000 nt) sequence before initiation of translation of ORF1 at an AUG codon argues against the canonical scanning mechanism described for standard eukaryotic protein translation. The 5′ IRES of several dicistroviruses has been shown to function using reporter gene assays. Using this approach, the 5′ IRES facilitates translation of the second cistron in a bicistronic message in *in vitro* and *in vivo* assays in systems from multiple kingdoms (Table 9.2). Interestingly, not all 5′ IRES elements function in all systems (Carter *et al.*, 2008; Masoumi *et al.*, 2003; Sasaki *et al.*, 1998; Shibuya and Nakashima, 2006; Wilson *et al.*, 2000b) and there is little nucleotide sequence conservation in the 5′ UTR among the dicistroviruses.

Current data indicate that there are some specificity and functional differences between the 5′ IRES elements found in dicistroviruses and future research is required to determine if there are indeed different classes of IRES within this group. The RhPV 5′ IRES has little secondary structure and does not require any specific sequences for function (Terenin *et al.*, 2005; Woolaway *et al.*, 2001). Deletion of large regions of the 5′ UTR does not abrogate

IRES function (Groppelli *et al.*, 2007) and two domains (nt 300–429 and 426–579 in RhPV) can independently act as an IRES in experimental systems. This indicates that there are multiple and redundant domains involved in recruitment of the translational machinery. The host translational initiation factors shown to be required for RhPV 5′ IRES function are eIF1–3, whereas the eIF4 factors are not required (although eIF4G and eIF4A stimulate translation) (Terenin *et al.*, 2005). The requirement of the scanning factor eIF1 suggests that in contrast to other viral IRES elements, some scanning is involved in the process of translation initiation from the RhPV 5′ UTR. Functionally, the RhPV IRES is similar to some cellular IRES elements and it has been proposed by Roberts and Groppelli (2009) that initiation of translation may be dependent on binding of the poly(A) binding protein to the A-rich tracts in this region.

In addition to RhPV, the functionality of the 5′ IRES elements has been confirmed for four other viruses TrV, DCV, CrPV and PSIV (Table 9.2). In contrast to RhPV, the PSIV 5′ UTR is predicted to contain stem–loop structures and a possible pseudo-knot that have an impact on the functionality of the IRES element (Shibuya and Nakashima, 2006). The 5′ 225 nt of the PSIV 5′ UTR is dispensable for IRES function, but other regions of the 5′ UTR were shown by deletion mapping to be required for IRES function (Shibuya and Nakashima, 2006). In addition, there is variation between the different IRES elements as to which *in vitro* translation systems support initiation of translation (Shibuya and Nakashima, 2006; Wilson *et al.*, 2000b; Woolaway *et al.*, 2001) (Table 9.2). Hence, it seems likely that there is not high conservation in the functionality of the dicistrovirus 5′ IRES element.

IGR IRES

The first sequences of full-length dicistrovirus genomes showed some unexpected features. These virus genomes had two ORFs separated by an IGR of approximately 200 nt. There was no subgenomic RNA and the N-terminus of the capsid protein, as determined by N-terminal sequencing, was 5′ of the first in frame AUG codon of the second ORF. These features led to the suggestion that the IGR may represent the first naturally occurring IRES that lay between two cistrons in a eukaryotic message, and further, that this putative IRES likely facilitated initiation of the capsid proteins in a methionine independent manner (Johnson and Christian, 1998; Moon *et al.*, 1998; Sasaki *et al.*, 1998). Since that time the dicistrovirus IGR IRES has become intensively studied and this hypothesis has been confirmed (Domier *et al.*, 2000; Sasaki and Nakashima, 1999, 2000; Wilson *et al.*, 2000a,b).

The dicistrovirus IGR IRES is an elegant example of a virus adaptation to divert host cell machinery to promote virus production (Jan, 2006; Kieft, 2009; Nakashima and Uchiumi, 2009). The IGR IRES initiates translation of the capsid proteins without any requirement for either canonical initiation factors or methionine tRNA (Jan *et al.*, 2003; Pestova and Hellen, 2003, 2005). This allows synthesis of the capsid proteins in conditions where cellular translation is suboptimal. The IGR IRES forms a complex secondary structure including a series of stem–loop structures and the formation of three pseudo-knots (PKI, PKII and PKIII) (Kanamori and Nakashima, 2001). The structure of the IGR IRES has been determined both by chemical and enzymatic analysis and also the structure has been solved by X-ray crystallography (Costantino *et al.*, 2008; Jan and Sarnow, 2002; Nishiyama *et al.*, 2003; Pfingsten *et al.*, 2006). There is little nucleotide sequence conservation outside the loop regions, although the overall structure of the IGR IRES is conserved among the dicistroviruses suggesting that the structure is important for function. This RNA structure includes independent domains that recruit the ribosomal subunits and a smaller domain that functionally mimics a tRNA and docks in the P-site of the ribosome (Costantino and Kieft, 2005; Jan *et al.*, 2003; Wilson *et al.*, 2000a). The pseudo-knot of the IRES mimics the codon-anticodon binding of the tRNA and mRNA (Fig. 9.2) (Costantino *et al.*, 2008). The interaction of the IRES with the P-site of the ribosome positions the first codon of the ORF and is sufficient to initiate protein synthesis from the A-site in an RNA sequence specific manner. Once initiated, translation proceeds as normal. Dicistrovirus IGR IRES elements fall into two classes based on the detail of some structural elements (reviewed in Nakashima and Uchiumi, 2009); however, the functional significance of the difference between the classes is not known.

Protein processing and products

The ORF1 protein precursor is processed to produce the mature non-structural proteins. Some dicistroviruses (DCV, CrPV, *Israeli acute paralysis virus* - IAPV, *Kashmir bee virus* – KBV and *Acute bee paralysis virus* – ABPV) encode a 2A-like NPGP motif, which if functional, would indicate that the ORF is translated into two long polyproteins. In the picornaviruses from the *Enterovirus* and *Rhinovirus* genera, 2A protease cleaves between the structural and non-structural regions of the polyprotein. In other genera within the family *Picornaviridae*, the 2A protein guides a co-translational processing event that separates the 2A and 2B regions of the polyprotein via a ribosomal 'skipping' process (Donnelly *et al.*, 2001). The conserved 2A-like motif DxExNPGP is found in many dicistroviruses towards the N-terminus of the non-structural polyprotein and these domains have been shown to be functional in plasmid based translational assays (Donnelly *et al.*, 2001; Luke *et al.*, 2008). Interestingly, 2A-like motifs are common in several insect virus groups including the *Tetraviridae* and the *Iflaviridae* (Luke *et al.*, 2008). Further processing of the ORF1 polyprotein precursor involves the virus encoded cysteine protease, which in PSIV has been shown to cleave at the C-terminus of the helicase and at the junction of the protease and the RdRp (Nakashima and Nakamura, 2008).

For DCV and CrPV, the N-terminal region encoded by ORF1 functions as a suppressor of RNAi (See below and Chapter 14) (van Rij *et al.*, 2006; Wang *et al.*, 2006). The RNAi suppressor in DCV (1A) is found in the N-terminal 99 amino acids of ORF1 and has sequence similarity with dsRNA binding motifs. DCV-1A has been shown to bind with high affinity to long dsRNA, thus preventing the enzyme Dicer from cleaving the RNA. The CrPV RNAi suppressor is also N-terminal and lies within the first 150 amino acids. However, there is little sequence homology and the CrPV-1A protein does not bind dsRNA. The divergence in sequence is particularly interesting as these two viruses otherwise share high homology in their genomes.

The RdRp domain is found in the C-terminal region of ORF1 and includes the conserved motifs that are commonly found in Supergroup I replicases (Koonin and Dolja, 1993). The helicase domain lies approximately 400 amino acids from the N-terminus of ORF1 and was identified by homology with three conserved motifs found in supergroup type 3 helicases (Koonin and Dolja, 1993). Functional analysis has not been undertaken on either of these proteins in the dicistroviruses.

A small virus protein VPg linked to the genomic RNA has been identified for several dicistroviruses. Analysis of radioactive labelling of proteins associated with the genome of DCV and CrPV confirmed that each of these viruses has protein covalently linked to the virion RNA, although the size and sequence of these proteins was not determined (King and Moore, 1988). Direct protein sequencing of the PSIV VPg indicated that for this virus there are three VPg proteins that are each comprised of 14 amino acids. The PSIV VPg proteins are encoded in tandem in the genome between the helicase and 3C-like protease, are highly related in sequence and are cleaved following a C-terminal glutamic acid (Nakashima and Shibuya, 2006). The sequence of the multiple copies of VPg from the majority of the dicistroviruses have been deduced based on genomic position, repeated nature of sequences and glutamic acid cleavage (Nakashima and Shibuya, 2006) (Table 9.1). Interestingly, RhPV and ALPV do not appear to encode multiple VPg sequences.

The VP4 protein has been identified as present in the virion of several members of the *Dicistroviridae*. This small protein can be difficult to detect by electrophoresis and its presence has not been confirmed in a number of viruses. However, the use of alternative techniques to analyse the polypeptides present in the virion is necessary before the presence of VP4 is ruled out. For example, the presence of VP4 in the CrPV virion was not originally identified by gel electrophoresis of capsid protein components, but was confirmed both by mass spectrometry analysis of whole virus particles (Despeyroux *et al.*, 1996) and by structural analysis of particles (Tate *et al.*, 1999).

Infection cycle and replication

Comparatively little is known about the cellular infection and replication cycle of dicistroviruses, with most information coming from studies of DCV. Replication of DCV first involves virus particles being taken up by the host cell, which occurs via endocytosis of DCV by clathrin-coated vesicles (Cherry and Perrimon, 2004) (Fig. 9.3). Cell attachment probably involves virus interaction with a cell surface receptor, although this is yet to be demonstrated experimentally. Entry of dicistroviruses is predicted to be distinct from the well-studied picornaviruses. The CrPV capsid lacks the canyon identified in the polio virus structure that is important for cell receptor interaction (Tate *et al.*, 1999), suggesting a distinct cell surface interaction. Dicistroviruses also lack the flexible 'pocket factor' (likely fatty acid) found in some picornaviruses and fixed amino acid side chains replace this. It has been suggested that this lack of flexibility would likely impact the mechanism of receptor-mediated destabilization of the capsid and subsequent release of the genomic RNA (Tate *et al.*, 1999).

Once entry to the cell has occurred, the viral genome is translated and replicated. The exit of virus from the endocytic vesicle and the release of the genome from the capsid occur by unknown mechanisms. The viral genome is then translated by host cell machinery with viral protein detected by 7 hours post infection. Following translation, cytoplasmic vesicular compartments derived from the Golgi apparatus are induced within 10 hours of infection. Colocalization of the viral helicase protein with the compartments suggest that this is the site of viral RNA replication (Cherry *et al.*, 2006) and infection with DCV induces a 2-fold increase in total cell membranes (Cherry *et al.*, 2006). Depletion of the master regulator of the fatty acid biosynthesis pathway leads to a decrease in accumulation of DCV proteins and RNA, suggesting that suppression of fatty acid metabolism interferes with DCV replication (Cherry *et al.*, 2006). The replication of genomic RNA involves generation of a dsRNA intermediate (Reavy *et al.*, 1983).

Following RNA replication, virus particles assemble from the DCV genome and capsid proteins, but the steps involved in dicistrovirus assembly have not been experimentally addressed. Virus particles can exit host cells early in infection and the cell is lysed late in infection

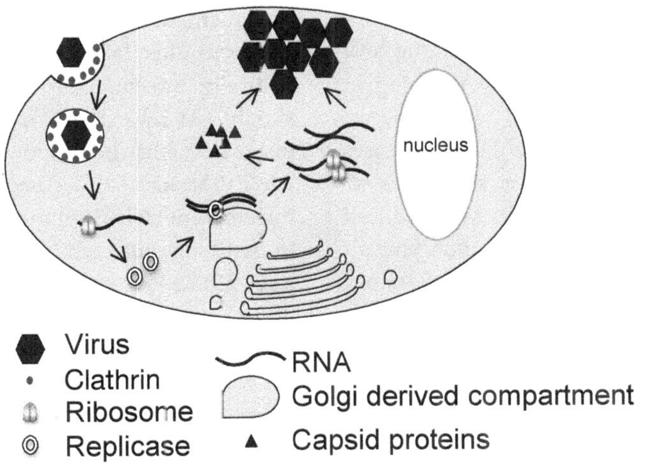

Figure 9.3 Dicistrovirus infection cycle. The schematic diagram represents a model of the dicistrovirus infection process in *Drosophila* cells. Virus particles enter the cell via endocytosis in a clathrin-coated vesicle. Following release, the genomic RNA is translated. The replicase proteins associate with vesicular compartments that are derived from the Golgi and this is the proposed site of RNA replication via a double-stranded RNA intermediate. The capsid proteins are translated from the accumulating genomic RNA. New virus particles are assembled from the 60 copies each of the large capsid proteins and one copy of the genomic RNA. Assembled particles accumulate in paracrystalline arrays in infected cells.

such that remaining virus is released (Cherry and Perrimon, 2004).

Mixed infections

There is considerable variation between bee virus isolates, for KBV and IAPV in particular, with pronounced variation at the 5′ end of the genome (de Miranda *et al.*, 2004; Palacios *et al.*, 2008). This variation in the 5′ UTR may allow for identification of the correct genomic RNA in mixed infections, which are common in bees (de Miranda *et al.*, 2004). Variation resulting from insertions and deletions into the viral genome could be random, or result from recombination between viruses, or evolutionary convergence. RNA viruses are highly variable and evolve rapidly and recombination among viruses such as the ABPV–KBV-IAPV complex of bee viruses is common. Based on the 3C protease cleavage sites, it is hypothesized that each of the three bee dicistroviruses is able to cleave the ORF2 polyprotein of the other two viruses. This strategy could provide a competitive advantage by disrupting the proteolytic processing of the structural proteins of competing viruses (de Miranda *et al.*, 2010; Ryan and Flint, 1997).

Pathology and host defence

Pathology

The impact of dicistrovirus infection on the host arthropod may be acute, typically resulting in paralysis (Fig. 9.4 and Table 9.3), but commonly, infection is chronic and covert. Dicistroviruses tend to persist within the host at low titres. Chronic infection often results in reduced longevity and fecundity but does not prevent maintenance of infected arthropods in laboratory cultures. Chronic infections may become acute in response to stress induced by overcrowding, limited resources, or secondary infection for example, leading to epizootics particularly in lab cultures. Dicistroviruses may also persist in a host without an increase in titre. As for other insect viruses, young larvae may be more susceptible to infection than older larvae (Jousset, 1975; Reinganum *et al.*, 1970). There is a relative absence of dicistroviruses in the larval stages of the social insects (de Miranda *et al.*, 2010).

Virus titres can be determined in cell culture (plaque assay, $TCID_{50}$) with indirect detection (immunoblot) for viruses that do not cause cell lysis (Hedges and Johnson, 2008; Plus, 1978; Scotti, 1977; Scotti *et al.*, 1981). Bioassays of dicistroviruses are typically conducted by injecting virus into the host organism. With the exception of parasite-mediated virus transmission (bee viruses transmitted by the varroa mite for example), injection of virus represents an unnatural route for infection. As the gut is a key site for antiviral defence (see below), injection likely underestimates the actual dose of virus that would be required for per os infection.

The time to death is highly variable according to the virus – host combination and environmental conditions. When infected by injection, mortality within 4–5 days is typical for DCV and within 6 days for CrPV (Plus *et al.*, 1978). The virulence of different isolates of DCV following injection is highly variable with some resulting in mortality within 3 days and yielding high viral titres (10^8 to 10^{10} infectious particles per fly), and others resulting in mortality in 10 days with low viral yields (Plus *et al.*, 1978). Until just prior to death, DCV-injected adult flies often do not exhibit strong symptoms of infection.

The pathogenic effects after injection of DCV into the haemocoel of *D. melanogaster* are well documented. The reports of pathogenic effects observed after larval feeding experiments or in naturally infected laboratory stocks are more variable. At one extreme, ingestion of DCV by first instar larvae has been shown to lead to high egg-adult mortality (40% compared to 10% in mock infected controls) and greatly reduced longevity in naturally infected stocks (Golubovsky and Plus, 1982; Jousset and Plus, 1975). In contrast, under certain circumstances, DCV infection (natural or by feeding) is reported to be beneficial to *D. melanogaster* by allowing them to develop more quickly, produce more ovarioles and show increased fecundity (Gravot *et al.*, 2000; Thomas-Orillard, 1984). It is notable, however, that larval-to-adult mortality was shown to be increased by DCV infection using this same experimental system (Thomas-Orillard, 1984, 1988) and mortality increased as the concentration of virus fed to larvae was increased (Thomas-Orillard,

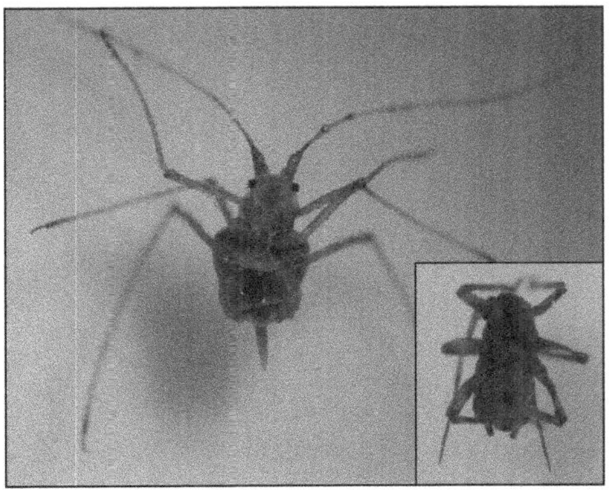

Figure 9.4 A pea aphid, *Acyrthosiphon pisum*, killed by *Aphid lethal paralysis virus* (ALPV)-induced paralysis, with abnormal distension of the legs. Aphids paralysed by ALPV infection die within 2 days of infection and fall from the host plant. Inset: A pea aphid that died of natural causes with characteristically infolded legs. (Inset image from Bonning and Miller, 2010.)

1988). The pathogenicity of the virus isolate or the susceptibility of the host population may contribute to variation in the impact of DCV infection on *D. melanogaster* (Plus, 1978; Plus and Golubovsky, 1980; Scotti *et al.*, 1981). An alternative explanation for these seemingly conflicting reports is that the subset of the fly population that survive virus infection includes the individuals that develop more quickly and are more fecund. The potential role of the bacterium *Wolbachia pipientis* on this phenotype remains to be examined (see *Host defence* below).

Paralysis is associated with infection by ALPV, ABPV, or IAPV and has also been reported in some insects infected with CrPV or TrV (Table 9.3). In Hymenoptera, paralysis of individual insects does not translate to widespread paralysis of infected insects within the colony but the progression of paralysis may be too rapid to be observed. TrV and ALPV have been detected in neural tissue (Hatfill *et al.*, 1990; Muscio *et al.*, 1987) leading to the hypothesis that severe pathology caused by dicistrovirus infection results from neurotropism. CrPV infection of field crickets appears to result in paralysis of some hosts and not others (Wigley and Scotti, 1983), suggesting that the virus was maintained as a persistent, asymptomatic infection in this species. A similar observation was made for CrPV in honey bees (Anderson and Gibbs, 1988). Different pathologies may reflect key genetic differences in different virus isolates, or physiological differences between host insect populations or species that prevent paralysis.

Transmission

The mode of transmission of dicistroviruses relates in part to the severity of disease (Table 9.3). The highly virulent and paralytic viruses are transmitted primarily by horizontal per os transmission, with virus shed in the faeces of infected hosts serving as the source of infection (Chen *et al.*, 2006a). In the case of viruses of Hemiptera, horizontal transmission may also occur via the phloem of the plant on which the host insect feeds (Gildow and D'Arcy, 1988), or via virus-contaminated faeces (Nakashima *et al.*, 1998). Transmission from females to males has also been noted for DCV (Gomirez-Zilber and Thomas-Orillard, 1993). In at least two cases, hemipteran dicistroviruses have been isolated during the purification of a plant pathogenic virus from infected plants: HiPV during purification of Rice stripe tenuivirus (Toriyama *et al.*, 1992), and RhPV during purification of Barley yellow dwarf virus (Boyapalle *et al.*, 2008). Viruses that have more subtle effects on the host, such as reduced fecundity and longevity, are also transmitted vertically (Clayton and Tompkins, 1994). Transmission of RhPV, PSIV and DCV occurs both horizontally and vertically (Table 9.3). There may also be variation in the mode of transmission between different virus–host

Table 9.3 Biological characteristics of dicistroviruses

Virus	Orders of natural hosts (number of species per order)	Symptoms	Tissue tropism, where known	Transmission, where known
CrPV	Diptera (4), Hemiptera (2), Hymenoptera (2), Lepidoptera (12), Orthoptera (4)	Paralysis[a]	Gut, fat body, trachea, muscle, brain	H, V[b]
ALPV	Hemiptera (3)	Paralysis[a]	Gut, brain	H (V)*
BQCV	Hymenoptera (1)			H
DCV	Diptera (2)	Reduced longevity, L-P mortality	Multiple tissues, varies according to host	H, V transovum
HiPV	Hemiptera (3)	Reduced longevity	Gut	
HoCV-1	Hemiptera (1)	Reduced longevity	Gut, eggs	(V)
PSIV	Hemiptera (1)	Reduced longevity	Gut	H, V transovum
RhPV	Hemiptera (7)	Reduced longevity, fecundity	Gut	H (via plant), V
TrV	Hemiptera (2)	Paralysis[a]	Nerves, gut	H, V
ABPV	Hymenoptera (6)	Paralysis[a]	Brain, hypopharyngeal glands	H
IAPV	Hymenoptera (1)	Paralysis[a]	Gut	H
KBV	Hymenoptera (4)	Reduced longevity	Gut	H, V
SINV-1	Hymenoptera (1)	None observed	Gut	H
TSV	Decapoda (3)	Reduced longevity	Cuticular epithelium, haemolymph	H

[a] Individuals that are not paralysed exhibit reduced longevity and fecundity.

[b]H, horizontal transmission; V, vertical transmission. Laboratory studies have shown that transovum transmission (on the surface of the egg) is possible for CrPV, DCV.

*ALPV detected in embryos of infected *R. padi* (Hatfill *et al.*, 1990).

combinations. For example, RhPV is transmitted from the leaf surface to *R. padi* and *Schizaphis graminum*, but not to other aphid species (*Metopolophium dirhodum*, *R. maidis*, *Sitobion avenae*) that are natural hosts of RhPV (Gildow and D'Arcy, 1990).

The dense populations within hives provide ideal conditions for horizontal transmission of bee viruses and mixed virus infections in bees are common (Chen *et al.*, 2006b). Honey bee viruses are transmitted in saliva, via pollen (ABPV, BQCV, KBV), in honey (BQCV), in royal jelly, faeces, and during mating. KBV is also transmitted vertically (Chen *et al.*, 2006a). The varroa mite, *Varroa destructor* transmits bee viruses (ABPV, KBV, BQCV) directly into the haemocoel of honey bee pupae with some viruses also replicating in the mite (Chen *et al.*, 2006b; Rosenkranz *et al.*, 2010; Shen *et al.*, 2005b). Lateral transmission of dicistroviruses between host species has also been proposed via consumption of virus-infected cadavers by other arthropods (Reinganum, 1975). CrPV-killed cadavers of the lepidopteran *Antheraea eucalypti* fed to crickets, resulted in infection of the crickets (Reinganum, 1975).

Host defence

Gut barrier

Given that the primary route for dicistrovirus infection is per os, it is logical that key antiviral defence mechanisms are activated in the gut of the host organism. As for many other insect–virus systems (Sparks *et al.*, 2008), bypassing the natural route of infection by injecting infectious virus directly into the haemocoel of an insect can result in (a) infection in host species that are not normally susceptible to a particular virus by normal modes of transmission, and (b) extreme virulence in species that show only a mild response following per os infection. The gut-based anti-dicistrovirus response is exemplified by the differential response of *D. melanogaster* to infection with DCV, and dicistrovirus infection of bees per os versus by injection: While injection of 100 virus particles results in death of a honey bee within days, it takes 10^{11} virus particles to achieve the same effect when virus is delivered per os (de Miranda *et al.*, 2010). Hence the antiviral responses recorded following virus injection may not be representative of the authentic antiviral response, if the natural route of infection is per os. Little is known of how dicistroviruses spread within the host insect, or indeed, whether CrPV or DCV enter the haemocoel following per os infection of the host insect (see Basal lamina below).

While it is known that DCV and CrPV virions are stable at acid pH as low as 2.5–3 (Scotti *et al.*, 1981), the susceptibility of virus particles to the highly variable gut pH and proteolytic enzymes in the gut of the various host arthropods has not been addressed. At approximately 30 nm in diameter, dicistrovirus particles may be sufficiently small to readily penetrate the peritrophic matrix when present, which may have pore diameters of up to 36 nm (Lehane, 1997).

Basal lamina

The basal lamina or basement membrane that overlies the gut epithelium presents a barrier to dissemination of virus from the gut epithelium to other tissues. In some insects, the size exclusion limit of the basal lamina is 15 nm (Reddy and Locke, 1990). Some viruses that have more extensive tissue tropism are clearly able to overcome this barrier by an unknown mechanism. While 'On a few occasions, RhPV-like virions were observed free in the haemocoel of infected aphids' (Gildow and D'Arcy, 1990), conclusive evidence for movement of dicistroviruses from the gut epithelium, through the basal lamina and into the haemocoel is lacking.

RNA interference

Following receptor-mediated endocytosis into the *Drosophila* cell (Cherry and Perrimon, 2004), production of dsRNA on replication of the virus in the cytoplasm would elicit the RNAi response (Cherry *et al.*, 2006; Galiana-Arnoux *et al.*, 2006; van Rij *et al.*, 2006; Wang *et al.*, 2006; see also Chapter 16). The enzyme Dicer, recognizes double-stranded RNA and cleaves it into small interfering RNAs (siRNAs) of about 20–22 nt in length (Meister and Tuschl, 2004). The RNA-induced silencing complex (RISC) binds the antisense strand of the siRNA and degrades viral RNA complementary to the siRNA (Wang *et al.*, 2008). The nuclease Argonaute 2, which is a component of the RISC, has been shown to be essential for antiviral defence against DCV and CrPV (van Rij *et al.*, 2006). There is experimental evidence that antiviral RNAi response spreads systemically throughout the fly via a dsRNA uptake pathway (Saleh *et al.*, 2009).

Viral suppressors of RNAi (VSR) have been identified in both CrPV and DCV at the amino terminus of ORF 1 (Cherry *et al.*, 2006; van Rij *et al.*, 2006; Wang *et al.*, 2006). *In vitro* expression of the VSRs in *D. melanogaster* cells resulted in increased virus production, blocked dsRNA-induced silencing and degradation of assay reporter genes, and reduced production of siRNAs (Wang *et al.*, 2006).

The VSR encoded by DCV (1A) shares sequence similarity with a dsRNA binding motif (van Rij *et al.*, 2006), preferentially binds long (>200 nt) over short (<30 nt) dsRNA, and blocks the processing of long dsRNA into siRNAs by Dicer. The high affinity for long dsRNAs may protect from interference with host microRNA processing which is essential for insect development (van Rij *et al.*, 2006). However, the VSR of DCV also suppresses the piwiRNA-mediated silencing of retrotransposons by endogenous siRNAs (Berry *et al.*, 2009). In contrast and somewhat intriguingly, the CrPV suppressor lacks a recognizable dsRNA binding motif, does not inhibit dsRNA processing, and lacks homology to any other dicistrovirus protein (van Rij *et al.*, 2006).

Apoptosis

In addition to the RNAi pathway, a second cellular response, apoptosis, has evolved to restrict virus spread (see Chapter 17). Apoptosis is not known to be involved in defence against dicistroviruses. However, the caspase *Damm*, was induced by infection of *Drosophila* with DCV (Dostert *et al.*, 2005), and an inhibitor of apoptosis protein (IAP) repeat domain has been identified at the N-terminus of the TSV ORF1 polyprotein (Mari *et al.*, 2002). A similar domain is also present in the kelp fly virus, an unclassified picorna-like virus (Hartley *et al.*, 2005). IAPs can also function in cell proliferation, differentiation and cell signalling (Dubrez-Daloz *et al.*, 2008), and it remains to be determined whether the IAP domains in TSV and kelp fly virus function to block apoptosis of virus-infected cells.

Humoral immunity

The host response at the subcellular level to dicistrovirus infection has been studied most extensively in *D. melanogaster* infected with DCV or CrPV. Dicistrovirus infection induces a response regulated in part by the kinase Jak and transcription factor STAT (Beutler *et al.*, 2007; Cherry *et al.*, 2006; Dostert *et al.*, 2005; Galiana-Arnoux *et al.*, 2006; Huszar and Imler, 2008; Lemaitre and Hoffmann, 2007; van Rij *et al.*, 2006; Wang *et al.*, 2006). However, many of the studies on antiviral immunity in *Drosophila* deliver virus directly into the haemocoel by injection rather than per os, the more natural route of infection for these viruses. Hence the relationship of these findings to the authentic immune response elicited by these viruses during the natural infection process remains to be explored.

In contrast to per os infection of *D. melanogaster* with DCV, which may result in few apparent symptoms (Roxstrom-Lindquist *et al.*, 2004), injected DCV is highly pathogenic indicating that injection allows DCV to bypass key antiviral defences in the gut (Sparks *et al.*, 2008). Infection of *D. melanogaster* with DCV per os results in up-regulation of 11 genes including those that encode the antimicrobial peptides, Attacin A, Cecropins A1 and A2, and Drosomycin (Roxstrom-Lindquist *et al.*, 2004). The 140 genes up-regulated by injection of *D. melanogaster* with DCV (Dostert *et al.*, 2005) did not include genes regulated by the NF-κB signalling pathways Toll and immune deficiency (Imd), which are induced by infection with Drosophila X virus, a dsRNA virus (*Birnaviridae*) (Zambon *et al.*, 2005). Some of the genes induced by DCV injection were regulated by the Jak-STAT pathway, including the virus specific gene *vir-1*. A role for this pathway in DCV infection was confirmed by demonstrating that mutation of the gene *hopscotch*, which encodes the Jak kinase of *Drosophila* (Agaisse and Perrimon, 2004), resulted in a moderate increase in DCV accumulation (about 5-fold) and increased DCV-induced mortality (Cherry *et al.*, 2006; Dostert *et al.*, 2005). Up-regulation of *vir-1* and two other genes putatively regulated by Jak-STAT was also observed by infection with three different DCV isolates and the nodavirus Flock House virus (FHV) (Dostert *et al.*, 2005; Hedges and Johnson, 2008), indicating that up-regulation is likely a generalized virus induced process.

The elicitor mechanisms for different RNA viruses, which may result from different aspects of the virus life cycle, remain to be determined. The Jak-STAT regulated antiviral response was not induced by non-infectious DCV particles or dsRNA, indicating that virus replication and/or non-structural proteins may serve as elicitors of the immune response (Hedges and Johnson, 2008). Given that only a subset of induced genes was regulated by Jak-STAT, and that the Toll and Imd pathways did not appear to be involved in the antiviral response to DCV, it was postulated that other, as yet undefined regulatory pathways may function in the antiviral response following injection of DCV (Cherry and Silverman, 2006; Dostert *et al.*, 2005). Further analysis of one of the DCV-induced genes, *vago*, showed that this gene is involved in regulating DCV viral load in the fat body of flies injected with virus (Deddouche *et al.*, 2008). *Vago* appears to be induced directly by the presence of DCV. In addition, the DExD/H-box helicase domain of Dicer-2 was necessary for *vago* induction. These data suggest that an inducible antiviral response exists in *Drosophila*, which is independent of the Toll, Imd and Jak-STAT pathways, and regulates innate immunity. Dicer-2, in addition to functioning in RNAi, regulates induction of Vago that functions to suppress viral load in some tissues (Deddouche *et al.*, 2008) implicating a connection between RNAi and an additional antiviral response.

In contrast to the lack of evidence from microarray analysis for a role of the Imd pathway in antiviral response following injection of DCV, using a different experimental approach the Imd pathway was implicated in the antiviral response to injected CrPV in *Drosophila* (Costa *et al.*, 2009). In this study, increased sensitivity to infection and increased viral load in flies lacking some of the functional Imd pathway genes was observed. Testing of several Imd pathway mutant lines suggested that the *imd* gene itself was not necessary for the anti-CrPV response. The Imd pathway functions via two branches, one of which (dTAK1-kenny-ird5) results in phosphorylation of Rel and the other (dFADD-Dredd) in cleavage of phosphorylated Rel. Two factors suggested that these branches of the Imd pathway function differently in response to virus infection. First, transcription of antimicrobial peptides (AMP) is directed by the Toll and/or Imd pathways. AMP transcript levels measured by RT-qPCR were not induced following injection of CrPV. Second, flies homozygous for loss of function mutations in *imd* or *dFADD* did not show increased sensitivity to CrPV infection, in contrast to genes that function upstream of Imd (*PGRP-LC*), or in the other Imd pathway branch (*Tak1*, *ird5*, *key*, and *rel*). The dFADD-Dredd branch is also implicated in regulation of apoptosis and hence apoptosis would have been blocked in the *dFADD* mutant flies.

Phagocytosis by haemocytes was also implicated in the anti-CrPV response: Phagocytosis was blocked by injection of flies with polystyrene beads, and these flies were more susceptible to the impact of CrPV infection than flies in control treatments. CrPV infection also resulted in depletion of haemocytes, possibly as a result of CrPV replication within the haemocytes (Costa *et al.*, 2009).

Wolbachia pipientis

The presence of the maternally transmitted, gram-negative, obligate intracellular bacterium *Wolbachia pipientis*, has been shown to confer anti-virus protection to *D. melanogaster* and *D. simulans* by an unknown mechanism (Hedges *et al.*, 2008; Osborne *et al.*, 2009; Teixeira *et al.*, 2008). Infection by *Wolbachia* reduced replication of DCV and CrPV and associated host mortality, and conferred protection against two other RNA viruses (Nora virus and FHV), but not a DNA virus (*Insect iridescent virus 6*) (Teixeira *et al.*, 2008). The response of *Wolbachia*-infected *D. melanogaster* was slightly different for DCV and FHV. While the rate of DCV accumulation was reduced, which may be indicative of resistance to virus accumulation, tolerance of virus infection, or both (Hedges *et al.*, 2008; Teixeira *et al.*, 2008). For FHV, no reduction in virus accumulation was observed, suggesting the flies may be tolerant to infection (Teixeira *et al.*, 2008). In *D. simulans* lines infected with different strains of *Wolbachia*, variation was observed both in protection from virus-induced mortality and DCV accumulation (Osborne *et al.*, 2009). This antiviral defence may provide a selective advantage to *Wolbachia*-infected flies that may account in part for the wide distribution of *Wolbachia* in *Drosophila* populations. Indeed, *Wolbachia* is estimated to infect more than 20% of insect species (Hilgenboecker *et al.*, 2008; Jeyaprakash and Hoy, 2000; Werren and Windsor, 2000). *Wolbachia*-mediated protection of *D. melanogaster* against RNA viruses provides one of an increasing number of examples of symbiont-mediated protection of insects (Brownlie and Johnson, 2009). Interestingly, early studies noted that different strains of *D. melanogaster* exhibited varying levels of susceptibility to DCV infection, and following genetic crosses, Plus and Golubovsky suggested in 1980 that maternal effects play an important part in the transmission of phenotypic resistance from parent to progeny (Plus and Golubovsky, 1980; Scotti *et al.*, 1981). The presence of *Wolbachia* in *D. melanogaster* may also account for observations of increased reproduction following infection with DCV (Gravot *et al.*, 2000; Thomas-Orillard, 1984). DCV infected hosts develop more rapidly than uninfected flies, and females produce greater numbers of progeny.

Behavioural resistance

As described above, the varroa mite plays an important role in the vertical transmission of the dicistroviruses of honey bees. Bees exhibit a number of behaviours that reduce disease within the hive, including grooming, hygienic behaviour and undertaking (Evans and Spivak, 2010). Grooming in particular can result in removal of mites including the varroa mite, and if the mite is dislodged and damaged by the bee, varroa-mediated virus transmission may be reduced. The propensity for grooming behaviour varies with the species of bee and is not as pronounced in *Apis mellifera* as in the original host of the varroa mite, *A. cerana* (Evans and Spivak, 2010).

Ecology and distribution

Virus identification

Preliminary analyses of dicistrovirus distribution relied on immunological assays and nucleic acid hybridization to distinguish between different viruses (Christian and Scotti, 1994; Scotti *et al.*, 1981), although some such as ABPV and KBV are serologically related. RT-PCR followed by restriction enzyme profile analysis of 12 DCV isolates from distinct geographical regions gave estimates of nucleotide divergence of 0 to 10% between isolates (Johnson and Christian, 1999). A lower level of similarity was established for DCV isolates by using a cDNA hybridization approach (King *et al.*, 1984). A biogeographical split has been suggested between the Australasian isolates of CrPV, and those from North America, with 7–8% variation at the nucleotide level between the two regions, and less than 3% variation within each group (Johnson and Christian, 1996). Indeed, CrPV isolates from these two regions represent distinct serotypes (Plus and Scotti, 1984).

Similarly, bee dicistroviruses have been identified by serology and more recently by using molecular techniques. Cell lines are not available for *in vitro* production of bee viruses, and contamination of bee virus preparations used for production of antiserum was a common problem (de Miranda *et al.*, 2010). Many IAPV isolates have been misclassified as KBV in the literature and online databases as the original primers designed for KBV would also amplify IAPV and APBV sequences.

Ecology

Dicistrovirus infection can have diverse impacts on the ecology of the host insect. In chronic infections, population dynamics are altered through reduction of longevity and fecundity. Paralysis of dicistrovirus-infected insects, which may cause the insect to fall to the ground, may change the ecological dynamics in relation to predatory arthropods, and the routes of virus transmission. Scavenging beetles on the ground, for example, may have greater access to virus-laden cadavers. Within a bee hive, ABPV and KBV can result in sharp declines in the adult bee population which leads to increased disease in larvae and pupae in the absence of adult attendance of the brood, and removal of diseased individuals.

Few studies have been done on the broader impacts of dicistrovirus infection that impact the ecology of the host organism, with the exception of RhPV infection of *R. padi*. Aphids infected with RhPV have reduced sensitivity to chemical cues (Ban *et al.*, 2008). Healthy aphids are attracted to other healthy aphids via an aggregation pheromone, but RhPV-infected aphids did not respond to the presence of other aphids, or to the plant attractant methyl salicylate. Infected aphids were also more sensitive to alarm pheromone and more vulnerable to a predatory ladybird, *Coccinella septempunctata*, and the parasitoid *Aphidius ervi* (Ban *et al.*, 2008).

Distribution

Many of the dicistroviruses of insects are known to be widespread (Hunnicutt *et al.*, 2008; Scotti *et al.*, 1981). CrPV and DCV were first isolated in the 1970s (Jousset *et al.*, 1972; Reinganum *et al.*, 1970), and there are large numbers of isolates for both viruses (Plus *et al.*, 1975; Plus and Scotti, 1984). It is estimated that one third of both laboratory cultures and natural populations of *D. melanogaster* are infected with DCV (Plus *et al.*, 1975). In contrast, a low detection frequency of only 0.27% was found by dot–blot hybridization for DCV in field populations in Australia (Christian and Scotti, 1998). Natural infections may be more prevalent between the tropics and 30°S latitude, the southernmost limit for *Drosophila*. Similarly, HiPV has been isolated from laboratory cultures of host Hemiptera, but is present at low incidence in the field (Guy *et al.*, 1992; Toriyama *et al.*, 1992).

CrPV has a wide host range, and is common in nature. Analysis of the field cricket, *Teleogryllus commodus* in Victoria, Australia showed that CrPV was present in 43% of the populations sampled (Reinganum *et al.*, 1981). Infection of the New Zealand cricket, *Pteronemobius nigrovus* varied between 10% and 43% (Wigley and Scotti, 1983). Despite the high prevalence of CrPV in the field, no epizootics have been reported in the wild for CrPV or DCV (Scotti *et al.*, 1981). DCV infection of laboratory populations results in decreased longevity and can lead to loss of fly populations.

Typical of the bee viruses, KBV is widely distributed and commonly occurs as an inapparent infection. KBV causes mortality only in the presence of an additional stress such as other pathogens or the varroa mite. Given the extensive transportation of hives (for almond pollination in California, USA for example), and the ready supply of bees and bee products from different countries, it is no surprise that the dicistroviruses of bees are widespread across the globe. There is ample opportunity for virus and parasite spread between bees within and between hives working the same crop, and via bee products. BQCV, ABPV, KBV and IAPV are highly prevalent in apiaries (Cox-Foster *et al.*, 2007; Johnson *et al.*, 2009; van Engelsdorp *et al.*, 2009). The prevalence of ABPV and KBV within a hive (and also of SINV-1 within an ant colony), increases as the season progresses and it is hypothesized that the viral titre is related to the growth rate of the colony.

Potential applications of dicistrovirus research

In addition to the use of IRES elements for heterologous gene expression detailed above (Table 9.2), several potential applications have arisen from dicistrovirus research.

Infectious clone of *Rhopalosiphum padi* virus

The first dicistrovirus infectious clone was constructed for RhPV and launched into cells by two different means (Boyapalle *et al.*, 2008; Pal *et al.*, 2007). The infectious clone was initially expressed using the baculovirus expression system for production in lepidopteran cells that do not support replication of RhPV (Boyapalle, 2005; Boyapalle *et al.*, 2007; Pal *et al.*, 2007). Interestingly, the RhPV infectious clone appeared to replicate in the baculovirus infected cells, based on detection of negative-sense RhPV RNA (Pal *et al.*, 2007). Presumably, the baculovirus suppressed an antiviral response that normally blocks replication of RhPV, a phenomenon that has also been noted for two insect nodaviruses (Krishna *et al.*, 2003; K.N. Johnson, unpublished data). Baculovirus expressed RhPV was infectious to aphids following feeding. Non-viral bases (totalling more than 100) derived from the expression vector at each end of the RhPV transcript were retained in the genome following 32 days of replication in aphids. This finding is similar to other viruses that tolerate the presence of non-viral bases at their termini (Agrawal and Johnson, 1995; Boyer and Haenni, 1994; Krishna *et al.*, 2003). Baculovirus expression of infectious RhPV provides an important precedent for *in vitro* production of small RNA viruses for which permissive cell lines are not available. The lack of bee-derived cell lines has been a particular challenge for the study of the many RNA viruses of bees.

The second approach for production of infectious RhPV was to transfect glassy-winged sharpshooter, *Homalodisca vitripennis,* cells with *in vitro* transcripts of the infectious RhPV RNA genome. These transcripts included 15 and five non-viral bases at the 5′ and 3′ ends, respectively, and lacked a 3′ poly(A) tail. Infection resulted in cytopathic effects, production of negative-sense RNA, and production of virions that were infectious to aphids (Boyapalle *et al.*, 2008). The RhPV infectious clone will allow for dissection of the mechanisms of dicistrovirus replication and for reverse genetics studies. The infectious clone may also provide a tool for biopesticide production (see below).

Biological control

Although the potential use of small RNA viruses for management of insect pests has long been recognized (Christian *et al.*, 1993; Hanzlik *et al.*, 1999; Scotti *et al.*, 1981), this application has yet to come to fruition. Eight dicistroviruses are pathogenic to insect pests of agricultural importance (CrPV, ALPV, RhPV, PSIV, HiPV, HoCV-1, SINV-1) and two are pathogenic to insects of medical importance (TrV, SINV-1). The hosts of RhPV, ALPV, HiPV, and PSIV are phytophagous Hemiptera that cause both direct damage by weakening the host plant, and indirect damage through the transmission of plant viruses. SINV-1 infects the red imported fire ant, *Solenopsis invicta,* which is an invasive species in the United States, Australia and parts of Asia. The fire ant incurs more than US$5 billion annually to cover costs associated with medical treatment of victims, crop and livestock losses, and control measures. HoCV-1 infects the polyphagous glassy-winged sharpshooter, *H. vitripennis,* which is also an invasive species in the United States (Hunnicutt *et al.*, 2006). In addition to its voracious appetite, this pest transmits several plant pathogens including *Xylella fastidiosa,* the causative agent of Pierce's disease of grapes (Bi *et al.*, 2007; Chatterjee *et al.*, 2008), and poses a significant threat to agricultural and horticultural operations. TrV infects a haematophagous reduviid bug that vectors the protozoan parasite *Trypanosoma cruzi. T. cruzi* infection results in Chagas disease in an estimated 8 to 11 million people in Latin America (Muscio *et al.*, 1987).

For effective biological control, the virus would ideally kill the host insect relatively rapidly to minimize damage, and show some persistence in the environment for sustained control. In this respect, a virus such as ALPV that results in rapid paralysis that may persist for some time in the plant vector, or a virus that is vertically transmitted would seem like an ideal candidate. However, large scale production of small RNA viruses for use as biological control agents is currently limiting (Christian and Scotti, 1996). *In vitro* production requires identification or establishment of a suitably permissive cell line, and in this respect CrPV is the best studied. CrPV infects no fewer than six host orders, and indeed has the broadest host range of any invertebrate small RNA virus. Consequently, CrPV infects several insect cell lines including the dipteran cell line DL2 derived from *D. melanogaster* and lepidopteran cell lines derived from *Trichoplusia ni* (Christian and Scotti, 1996). Methods for titration and production of CrPV in suspension culture have been developed (Scotti, 1977; Scotti and Dearing, 1996; Scotti *et al.*, 1996). However, given that dicistroviruses can be produced using the baculovirus expression system (Pal *et al.*, 2007), there is also the potential to take advantage of the extensive research conducted for large scale bioreactor production of baculovirus insecticides (Agathos, 1996; Black *et al.*, 1997; Ikonomou *et al.*, 2003) for the production of dicistrovirus insecticides.

An alternative strategy for use of dicistroviruses as biopesticides that is particularly attractive for dicistroviruses that use plants as vectors (Gildow and D'Arcy, 1988), is to construct transgenic plants that express these viruses (Bonning and Miller, 2010). A precedent for this approach using *Helicoverpa armigera stunt virus* (HaSV; *Tetraviridae*) expressed in tobacco was thwarted by silencing of the HaSV RNA 1 (Gordon and Waterhouse, 2006). Whether or not a dicistrovirus can be produced *in planta* will depend on both the ability of the virus to replicate in the plant milieu and on RNA secondary structure (i.e. production of dsRNA), either of which could elicit RNAi. Although it would present similar hurdles with respect to RNAi induction, dicistrovirus virus-like particles could also be employed to deliver dsRNAs that target essential genes in the host insect (Baum *et al.*, 2007; Gordon and Waterhouse, 2006). Clearly there are several avenues that could be pursued for exploitation of dicistroviruses as biopesticides for management of insect pests.

Protection of beneficial arthropods

Shrimp

The shrimp industry is worth more than US$10 billion annually with increasing demand for shrimp for human consumption. Taura syndrome virus (TSV) is one of five viruses that negatively impact the global shrimp industry (Lightner, 1996; Lightner *et al.*, 1997). TSV infects penaeid shrimp species, and has caused significant losses in Latin America, the United States (Brock, 1997) and Taiwan (Yu and Song, 2000). Efforts to manage virus infection have involved both traditional and modern, cutting edge approaches: Classical breeding for TSV resistance has met with some success (Srisuvan *et al.*, 2006) and investigation of the potential for vaccination of shrimp against viral disease is ongoing (Johnson *et al.*, 2008). Partial resistance to TSV was demonstrated for transgenic shrimp engineered to express an antisense transcript from the TSV VP gene (Lu and Sun, 2005; Robalino *et al.*, 2005). In addition, the administration of non-specific dsRNA conferred resistance to TSV in addition to white spot syndrome virus, a DNA virus (*Nimaviridae*) (Robalino *et al.*, 2007; Robalino *et al.*, 2004), suggesting that non-specific stimulation of antiviral immunity could be used for protection of shrimp against diverse viruses.

Honey bees

Of the more than 18 viruses that infect bees (Allen and Ball, 1996; Ball and Bailey, 1991; Chen and Siede, 2007), five are dicistroviruses: ABPV, KBV, BQCV, IAPV and CrPV (Christian and Scotti, 1998). Bee viruses typically persist as chronic infections until activated by stress. For example, BQCV infection is associated with the presence of the microsporidian *Nosema apis*, which increases susceptibility of the honey bees to BQCV (Chen and Siede, 2007). Infestation of hives with the varroa mite has increased incidence and awareness among bee keepers of the honey bee viruses and their detrimental impact on hives. The varroa mite suppresses the immune system of the honey bee, thereby facilitating virus replication (de Miranda *et al.*, 2010; Rosenkranz *et al.*, 2010; Shen *et al.*, 2005a; Shen *et al.*, 2005b; Yang and Cox-Foster, 2007; Yang and Cox-Foster, 2005). Varroa mites also transmit bee viruses including ABPV and KBV directly into the haemolymph (Shen *et al.*, 2005b); the mites transmit ABPV with 50 to 80% efficiency.

Based on the similarity of IAPV to KBV, it is likely that some early reports of KBV infection in honey bees were misreported IAPV infections. IAPV was isolated following severe honey bee mortality in Israel (Maori *et al.*, 2007a), and is widespread (Blanchard *et al.*, 2008; Cox-Foster *et al.*, 2007; Maori *et al.*, 2007b; Palacios *et al.*, 2008). IAPV infection results in shivering wings, followed by paralysis and death. Interestingly, KBV infected bees do not display these symptoms (de Miranda *et al.*, 2010).

Although integration of RNA virus sequences into a host genome is rare, fragments of the IAPV genome were reported to be found in the honey bee genome in 30% of the bees tested (Maori *et al.*, 2007b). Further, the authors suggest that the presence of these IAPV sequences in the honey bee genome conferred resistance to IAPV (Maori *et al.*, 2007b), likely mediated by an RNAi-based resistance mechanism. Feeding bees with sugar solution containing IAPV dsRNA effectively protected bees from subsequent IAPV infection (Maori *et al.*, 2009). The use of RNAi for virus control relies, however, on near perfect sequence complementarity between the siRNA and the targeted transcript. Given the propensity of RNA virus genomes to mutate, feeding dsRNA to bees may rapidly select for different IAPV genomic variants. While effective for silencing viruses in the gut epithelium, oral administration of dsRNA may not be effective against viruses in more distal tissues.

The role of viruses in colony collapse disorder

'Colony collapse disorder' (CCD) is typified by loss of the adult worker population from the bee hive. In contrast to other types of colony collapse, the capped brood and food reserves in hives affected by CCD are left untouched by insects that typically scavenge from deserted hives, and dead bees are not found proximal to the hive. CCD has jeopardized apiaries in Europe and the United States, and resulted in colony losses of up to 50% in some operations (van Engelsdorp *et al.*, 2008). Because of the economic importance of crops that depend primarily on bees for pollination (fruit, vegetables and nuts), there is great incentive to elucidate the causative agents of CCD and to develop technologies for hive protection. Crops pollinated by bees have an estimated value of US$215 billion worldwide (Gallai *et al.*, 2008).

A metagenomic analysis of healthy and CCD honey bee colonies revealed a significant correlation between CCD and the presence of IAPV (Cox-Foster *et al.*, 2007). However, a more recent analysis of 61 quantifiable variables that could contribute to CCD did not confirm a link between CCD and IAPV. The authors suggested that the combined impact of pathogens and other stress factors likely accounts for CCD (van Engelsdorp *et al.*, 2009). As exemplified by the finding that *Wolbachia* protects insects from a range of RNA viruses, a complex interaction within the microbiome of bees may determine the development of CCD within a colony.

Whole genome microarray analysis of gut samples from CCD and non-CCD colonies revealed 65 transcripts as potential markers for CCD. Notably, these markers did not include genes involved in response to pesticide exposure, or immunity genes (Johnson *et al.*, 2009). The authors did not address however, the potential role of the varroa mite in CCD or presence or absence of the mite in their source colonies. Atypical ribosomal fragments were observed, which were more abundant in the guts of CCD bees. Although not addressed experimentally, and by analogy with the impact of picornaviruses on mammalian hosts, the authors speculated that these fragments could result from picornavirus-like infection and relate to arrested translation and subsequent ribosome degradation (Johnson *et al.*, 2009). This hypothesis clearly warrants further investigation to establish whether there is a link between CCD and viruses, including the dicistroviruses. There was no clear correlation between CCD and the presence or abundance of any single virus of those examined in this study (ABPV, KBV, IAPV, deformed wing virus, Sacbrood virus). However, more than 55% of CCD colonies examined in a separate study were infected with three or more viruses (van Engelsdorp *et al.*, 2009), increasing the likelihood for a role of RNA viruses in CCD.

Concluding remarks

In the last comprehensive historical review of this virus group, Christian and Scotti stated that 'until some great commercial or scientific advantage for these viruses is perceived' this area of virology will remain understudied (Christian and Scotti, 1998). Over the last twelve years, tremendous strides have been made in our understanding of dicistroviruses for both commercial and scientific reasons: the potential role of dicistroviruses in CCD of honey bees is of enormous economic importance for agriculture, and the unique nature of the dicistrovirus IGR IRES has made this one of the best studied of any RNA structures. With a few notable exceptions, progress has lagged in analysis of the ecological role played by dicistroviruses and in assessment of the field distribution and diversity of dicistroviruses. The generation of dicistrovirus infectious clones, and production of infectious dicistrovirus using the baculovirus expression system which bypasses the need for susceptible host cell lines, provide the tools for the next decade of dicistrovirus research.

Acknowledgements

This journal paper of the Iowa Agriculture and Home Economics Experiment Station,

Ames, Iowa, Project No. 6673 and 3608 was supported by the Iowa State University Plant Sciences Institute and the Consortium for Plant Biotechnology Research. KNJ is supported by funding from the Australian Research Council

References

Agaisse, H., and Perrimon, N. (2004). The roles of JAK/STAT signaling in *Drosophila* immune responses. Immunol. Rev. *198*, 72–82.

Agathos, S.N. (1996). Insect cell bioreactors. Cytotechnology *20*, 173–189.

Agrawal, D.K., and Johnson, J.E. (1995). Assembly of the $T = 4$ *Nudaurelia capensis ω virus* capsid protein, post-translational cleavage, and specific encapsidation of its mRNA in a baculovirus expression system. Virology *207*, 89–97.

Allen, M., and Ball, B. (1996). The incidence and world distribution of honey bee viruses. Bee World *77*, 141–162.

Anderson, D.L., and Gibbs, A.J. (1988). Inapparent infections and their interactions in pupae of the honeybee (*Apis mellifera* Linnaeus) in Australia. J. Gen. Virol. *69*, 1619.

Bailey, L., Gibbs, A.J., and Woods, R.D. (1963). Two viruses from adult honey bees (*Apis mellifera* Linneaus). Virology *21*, 390–395.

Bailey, L., and Woods, R.D. (1977). Two small RNA viruses from honey bees and further observations on sacbrood and acute bee-paralysis. J. Gen Virol. *37*, 175–182.

Ball, B., and Bailey, L. (1991). Viruses of honey bees. In Atlas of Invertebrate Viruses, J.R. Adams, and J.R. Bonami, eds. (Boca Raton, FL, CRC Press), pp. 525–551.

Ban, L., Ahmed, E., Ninkovic, V., Delp, G., and Glinwood, R. (2008). Infection with an insect virus affects olfactory behaviour and interactions with host plant and natural enemies in an aphid. Entomol. Exp. Appl. *127*, 108–117.

Baum, J.A., Bogaert, T., Clinton, W., Heck, G.R., Feldmann, P., Ilagan, O., Johnson, S., Plaetinck, G., Munyikwa, T., Pleau, M., *et al.* (2007). Control of coleopteran insect pests through RNA interference. Nat. Biotechnol. *25*, 1322–1326.

Berry, B., Deddouche, S., Kirschner, D., Imler, J.L., and Antoniewski, C. (2009). Viral suppressors of RNA silencing hinder exogenous and endogenous small RNA pathways in *Drosophila*. PLoS ONE *4*, e5866.

Beutler, B., Eidenschenk, C., Crozat, K., Imler, J.L., Takeuchi, O., Hoffmann, J.A., and Akira, S. (2007). Genetic analysis of resistance to viral infection. Nat. Rev. Immunol. *7*, 753–766.

Bi, J.L., Dumenyo, C.K., Hernandez-Martinez, R., Cooksey, D.A., and Toscano, N.C. (2007). Effect of host plant xylem fluid on growth, aggregation, and attachment of *Xylella fastidiosa*. J. Chem. Ecol. *33*, 493–500.

Black, B.C., Brennan, L.A., Dierks, P.M., and Gard, I.E. (1997). Commercialization of baculoviral insecticides. In The Baculoviruses, L.K. Miller, ed. (New York, Plenum Press), pp. 341–388.

Blanchard, P., Schurr, F., Celle, O., Cougoule, N., Drajnudel, P., Thiery, R., Faucon, J.P., and Ribiere, M. (2008). First detection of Israeli acute paralysis virus (IAPV) in France, a dicistrovirus affecting honeybees (*Apis mellifera*). J. Invertebr. Pathol. *99*, 348–350.

Bonning, B.C., and Miller, W.A. (2010). Dicistroviruses. Annu. Rev. Entomol. *55*, 129–150.

Boyapalle, S. (2005). A cell culture system and production of an infectious clone of *Rhopalosiphum padi* virus (Dicistroviridae). In Department of Entomology (Ames, Iowa State University), pp. 115.

Boyapalle, S., Beckett, R.J., Pal, N., Miller, W.A., and Bonning, B.C. (2008). Infectious genomic RNA of *Rhopalosiphum padi* virus transcribed *in vitro* from a full-length cDNA clone. Virology *375*, 401–411.

Boyapalle, S., Pal, N., Miller, W.A., and Bonning, B.C. (2007). A glassy-winged sharpshooter cell line supports replication of *Rhopalosiphum padi* virus (*Dicistroviridae*). J. Invertebr. Pathol. *94*, 130–139.

Boyer, J.C., and Haenni, A.L. (1994). Infectious transcripts and cDNA clones of RNA viruses. Virology *198*, 415–426.

Brock, J.A. (1997). Taura syndrome, a disease imported to shrimp farms in the Americas. World J. Micro. Biotech. *13*, 415–418.

Brownlie, J.C., and Johnson, K.N. (2009). Symbiont-mediated protection in insect hosts. Trends Microbiol. *17*, 348–354.

Carter, J.R., Fraser, T.S., and Fraser, M.J., Jr. (2008). Examining the relative activity of several dicistrovirus intergenic internal ribosome entry site elements in uninfected insect and mammalian cell lines. J. Gen. Virol. *89*, 3150–3155.

Cevallos, R.C., and Sarnow, P. (2005). Factor-independent assembly of elongation-competent ribosomes by an internal ribosome entry site located in an RNA virus that infects penaeid shrimp. J. Virol. *79*, 677–683.

Chatterjee, S., Almeida, R.P., and Lindow, S. (2008). Living in two worlds: the plant and insect lifestyles of *Xylella fastidiosa*. Annu. Rev. Phytopathol. *46*, 243–271.

Chen, Y., Evans, J., and Feldlaufer, M. (2006a). Horizontal and vertical transmission of viruses in the honey bee, *Apis mellifera*. J. Invertebr. Pathol. 92, 152–159.

Chen, Y.J., Chen, W.S., and Wu, T.Y. (2005). Development of a bi-cistronic baculovirus expression vector by the *Rhopalosiphum padi* virus 5′ internal ribosome entry site. Biochem. Biophys. Res. Commun. 335, 616–623.

Chen, Y.P., Pettis, J.S., Collins, A., and Feldlaufer, M.F. (2006b). Prevalence and transmission of honeybee viruses. Appl. Environ. Microbiol. 72, 606–611.

Chen, Y.P., and Siede, R. (2007). Honey bee viruses. Adv. Virus Res. *70*, 33–80.

Cherry, S., Doukas, T., Armknecht, S., Whelan, S., Wang, H., Sarnow, P., and Perrimon, N. (2005). Genome-wide RNAi screen reveals a specific sensitivity of IRES-containing RNA viruses to host translation inhibition. Genes Dev. *19*, 445–452.

Cherry, S., Kunte, A., Wang, H., Coyne, C., Rawson, R.B., and Perrimon, N. (2006). COPI activity coupled with fatty acid biosynthesis is required for viral replication. PLoS Pathog. *2*, e102.

Cherry, S., and Perrimon, N. (2004). Entry is a rate-limiting step for viral infection in a *Drosophila melanogaster* model of pathogenesis. Nat. Immunol. *5*, 81–87.

Cherry, S., and Silverman, N. (2006). Host–pathogen interactions in drosophila: new tricks from an old friend. Nat. Immunol. *7*, 911–917.

Christian, P., Carstens, E., Domier, L., Johnson, J., Johnson, K., Nakashima, N., Scotti, P., and van Der Wilk, F. (2005). Family *Dicistroviridae*. In Virus Taxonomy Eighth Report of the International Committee on Taxonomy of Viruses, C.M. Fauquet, M.A. Mayo, J. Maniloff, U. Desselberger, and L.A. Ball, eds. (San Diego, Elsevier), pp. 783–788.

Christian, P.D., Hanzlik, T.N., Dall, D.J., and Gordon, J.H. (1993). Insect viruses: new strategies for pest control. In Molecular Approaches to Fundamental and Applied Entomology, J. Oakeshott, and M.J. Whitten, eds. (New York, Springer-Verlag), pp. 128–163.

Christian, P.D., and Scotti, P. (1994). A suggested taxonomy and nomenclature for the cricket paralysis and *Drosophila* C virus complex. J. Invertebr. Pathol. *63*, 157–162.

Christian, P.D., and Scotti, P.D. (1996). Biopesticides from small RNA viruses of insects: aspects of their *in vitro* production. In Invertebrate Cell Culture: Looking Toward the Twenty First Century. Proceedings of the IX International Conference on Invertebrate Cell Culture (San Francisco, Society for *In Vitro* Biology), *p*. 141.

Christian, P.D., and Scotti, P.D. (1998). Picornalike viruses of insects. In The Insect Viruses, L.K. Miller, and L.A. Ball, eds. (New York, Plenum Press), pp. 301–336.

Christian, P.D., and Scotti, P.D. (2006). The invertebrate small RNA viruses; changing perceptions and 40 years of evolving taxonomy. In Bee Research and Virus in Europe (BRAVE) Symposium Proceedings (Sophia-Antipolis, France), pp. 11–43.

Clayton, D.H., and Tompkins, D.M. (1994). Ectoparasite virulence is linked to mode of transmission. Proc. Biol. Sci. *256*, 211–217.

Costa, A., Jan, E., Sarnow, P., and Schneider, D. (2009). The Imd pathway is involved in antiviral immune responses in *Drosophila*. PLoS ONE *4*, e7436.

Costantino, D., and Kieft, J.S. (2005). A preformed compact ribosome-binding domain in the cricket paralysis-like virus IRES RNAs. RNA *11*, 332–343.

Costantino, D.A., Pfingsten, J.S., Rambo, R.P., and Kieft, J.S. (2008). tRNA-mRNA mimicry drives translation initiation from a viral IRES. Nat. Struct. Mol. Biol. *15*, 57–64.

Cox-Foster, D.L., Conlan, S., Holmes, E.C., Palacios, G., Evans, J.D., Moran, N.A., Quan, P.L., Briese, T., Hornig, M., Geiser, D.M., *et al.* (2007). A metagenomic survey of microbes in honey bee colony collapse disorder. Science *318*, 283–287.

Czibener, C., Alvarez, D., Scodeller, E., and Gamarnik, A.V. (2005). Characterization of internal ribosomal entry sites of Triatoma virus. J. Gen. Virol. *86*, 2275–2280.

Czibener, C., La Torre, J.L., Muscio, O.A., Ugalde, R.A., and Scodeller, E.A. (2000). Nucleotide sequence analysis of Triatoma virus shows that it is a member of a novel group of insect RNA viruses. J. Gen. Virol. *81*, 1149–1154.

Culley, A.I., Lang, A.S., and Suttle, C.A. (2006). Metagenomic analysis of coastal RNA virus communities. Science *312*, 1795–1798.

D'Arcy, C.J., Burnett, P.A., Hewings, A.D., and Goodman, R.M. (1981). Purification and characterization of a virus from the aphid *Rhopalosiphum padi*. Virology *112*, 346–349.

de Miranda, J.R., Cordoni, G., and Budge, G. (2010). The Acute bee paralysis virus-Kashmir bee virus-Israeli acute paralysis virus complex. J. Invertebr. Pathol. *103*.

de Miranda, J.R., Drebot, M., Tyler, S., Shen, M., Cameron, C.E., Stoltz, D.B., and Camazine, S.M. (2004). Complete nucleotide sequence of Kashmir bee virus and comparison with Acute bee paralysis virus. J. Gen. Virol. *85*, 2263–2270.

Deddouche, S., Matt, N., Budd, A., Mueller, S., Kemp, C., Galiana-Arnoux, D., Dostert, C., Antoniewski, C., Hoffmann, J.A., and Imler, J.L. (2008). The DExD/H-box helicase Dicer-2 mediates the induction of antiviral activity in drosophila. Nat. Immunol. *9*, 1425–1432.

Despeyroux, D., Phillpotts, R., and Watts, P. (1996). Electrospray mass spectrometry for detection and characterisation of purified Cricket paralysis virus (CrPV). Rapid Commun. Mass Sp. *10*, 937–941.

Djikeng, A., Kuzmickas, R., Anderson, N.G., and Spiro, D.J. (2009). Metagenomic analysis of RNA viruses in a fresh water lake. PLoS ONE *4*, e7264.

Domier, L.L., and McCoppin, N.K. (2003). *In vivo* activity of *Rhopalosiphum padi virus* internal ribosome entry sites. J. Gen. Virol. *84*, 415–419.

Domier, L.L., McCoppin, N.K., and D'Arcy, C.J. (2000). Sequence requirements for translation initiation of *Rhopalosiphum padi* virus ORF2. Virology *268*, 264–271.

Donnelly, M.L., Luke, G., Mehrotra, A., Li, X., Hughes, L.E., Gani, D., and Ryan, M.D. (2001). Analysis of the aphthovirus 2A/2B polyprotein 'cleavage' mechanism indicates not a proteolytic reaction, but a novel translational effect: a putative ribosomal 'skip'. J. Gen. Virol. *82*, 1013–1025.

Dostert, C., Jouanguy, E., Irving, P., Troxler, L., Galiana-Arnoux, D., Hetru, C., Hoffmann, J.A., and Imler, J.L. (2005). The Jak-STAT signaling pathway is required but not sufficient for the antiviral response of drosophila. Nat. Immunol. *6*, 946–953.

Dubrez-Daloz, L., Dupoux, A., and Cartier, J. (2008). IAPs: more than just inhibitors of apoptosis proteins. Cell Cycle *7*, 1036–1046.

Eaton, B.T., and Steacie, A.D. (1980). Cricket paralysis virus RNA has a 3′ terminal poly (A). J. Gen. Virol. *50*, 167–171.

Estrozi, L.F., Neumann, E., Squires, G., Rozas-Dennis, G., Costabel, M., Rey, F.A., Guerin, D.M., and Navaza, J. (2008). Phasing of the Triatoma virus diffraction data using a cryo-electron microscopy reconstruction. Virology *375*, 85–93.

Evans, J.D., and Spivak, M. (2010). Socialized medicine: Individual and communal disease barriers in honey bees. J. Invertebr. Pathol. *103*, S62-S72.

Fauquet, C.M., Mayo, M.A., Maniloff, J., Desselberger, U., and Ball, L.A. (2005). 8th Report of the International Committee on Taxonomy of Viruses (Academic Press; Elsevier).

Galiana-Arnoux, D., Dostert, C., Schneemann, A., Hoffmann, J.A., and Imler, J.L. (2006). Essential function *in vivo* for Dicer-2 in host defense against RNA viruses in drosophila. Nat. Immunol. *7*, 590–597.

Gallai, N., Salles, J.M., Settele, J., and Vaissiere, B.E. (2008). Economic valuation of the vulnerability of world agriculture confronted with pollinator decline. Ecol. Econ. *68*, 810–821.

Gildow, F.E., and D'Arcy, C.J. (1988). Barley and oats as reservoirs for an aphid virus and the influence on barley yellow dwarf virus transmission. Phytopathology *78*, 811–816.

Gildow, F.E., and D'Arcy, C.J. (1990). Cytopathology and experimental host range of *Rhopalosiphum padi* virus, a small isometric RNA virus infecting cereal grain aphids. J. Invertebr. Pathol. *55*, 245–257.

Golubovsky, M.D., and Plus, N. (1982). Mutability studies in two *Drosophila melanogaster* isogenic stocks, endemic for C picornavirus and virus-free. Mutat. Res. *103*, 29–32.

Gomirez-Zilber, E., and Thomas-Orillard, M. (1993). *Drosophila* C virus and *Drosophila* hosts: A good association in various environments. J. Evol. Biol. *6*, 677.

Gordon, K.H.J., and Waterhouse, P.M. (2006). Small RNA viruses of insects: Expression in plants and RNA silencing. In Insect Viruses: Biotechnological Applications, B.C. Bonning, ed. (San Diego, Elsevier), pp. 459–502.

Govan, V.A., Leat, N., Allsopp, M., and Davison, S. (2000). Analysis of the complete genome sequence of acute bee paralysis virus shows that it belongs to the novel group of insect-infecting RNA viruses. Virology *277*, 457–463.

Gravot, E., Thomas-Orillard, M., and Jeune, B. (2000). Virulence variability of the Drosophila C virus and effects of the microparasite on demographic parameters of the host (*Drosophila melanogaster*). J. Invertebr. Pathol. *75*, 144–151.

Groppelli, E., Belsham, G.J., and Roberts, L.O. (2007). Identification of minimal sequences of the *Rhopalosiphum padi* virus 5′ untranslated region required for internal initiation of protein synthesis in mammalian, plant and insect translation systems. J. Gen. Virol. *88*, 1583–1588.

Guy, P.L., Toriyama, S., and Fuji, S. (1992). Occurrence of a picorna-like virus in planthopper species and its transmission in *Laodelphax striatellus*. J. Invertebr. Pathol. *59*, 1021–1023.

Hanzlik, T.N., Zeddam, J.L., Gordon, K.H.J., and Christian, P.D. (1999). A new view of small RNA viruses of insects. Pestic. Outlook *10*, 22–26.

Hartley, C.J., Greenwood, D.R., Gilbert, R.J., Masoumi, A., Gordon, K.H., Hanzlik, T.N., Fry, E.E., Stuart, D.I., and Scotti, P.D. (2005). Kelp fly virus: a novel group of insect picorna-like viruses as defined by genome sequence analysis and a distinctive virion structure. J. Virol. *79*, 13385–13398.

Hasson, K.W., Lighner, D.V., Poulos, B.T., Redman, R.M., White, B.L., Brock, J.A., and Bonami, J.R (1995). Taura syndrome in *Penaeus vannamei*: demonstration of viral etiology. Dis. Aquat. Organ. *23*, 115–126.

Hatakeyama, Y., Shibuya, N., Nishiyama, T., and Nakashima, N. (2004). Structural variant of the intergenic internal ribosome entry site elements in dicistroviruses and computational search for their counterparts. RNA *10*, 779–786.

Hatfill, S.J., Williamson, C., Kirby, R., and Wechmar, M.B.V. (1990). Identification and localization of Aphid lethal paralysis virus particles in thin tissue sections of the *Rhopalosiphum padi* aphid by in situ nucleic acid hybridization. J. Invertebr. Pathol. *55*, 265–271.

Hedges, L.M., Brownlie, J.C., O'Neill, S.L., and Johnson, K.N. (2008). *Wolbachia* and virus protection in insects. Science *322*, 702.

Hedges, L.M., and Johnson, K.N. (2008). Induction of host defence responses by *Drosophila* C virus. J. Gen. Virol. *89*, 1497–1501.

Hilgenboecker, K., Hammerstein, P., Schlattmann, P., Telschow, A., and Werren, J.H. (2008). How many species are infected with *Wolbachia*? – a statistical analysis of current data. FEMS Microbiol. Lett. *281*, 215–220.

Hunnicutt, L.E., Hunter, W.B., Cave, R.D., Powell, C.A., and Mozoruk, J.J. (2006). Genome sequence and molecular characterization of *Homalodisca coagulata* virus-1, a novel virus discovered in the glassy-winged sharpshooter (Hemiptera: Cicadellidae). Virology *350*, 67–78.

Hunnicutt, L.E., Mozoruk, J., Hunter, W.B., Crosslin, J.M., Cave, R.D., and Powell, C.A. (2008). Prevalence and natural host range of *Homalodisca coagulata* virus-1 (HoCV-1). Arch. Virol. *153*, 61–67.

Huszar, T., and Imler, J.L. (2008). *Drosophila* viruses and the study of antiviral host-defense. Adv. Virus Res. *72*, 227–265.

Ikonomou, L., Schneider, Y.J., and Agathos, S.N. (2003). Insect cell culture for industrial production of recombinant proteins. Appl. Microbiol. Biotechnol. *62*, 1–20.

Jan, E. (2006). Divergent IRES elements in invertebrates. Virus Res. *119*, 16–28.

Jan, E., Kinzy, T.G., and Sarnow, P. (2003). Divergent tRNA-like element supports initiation, elongation, and termination of protein biosynthesis. Proc. Natl. Acad. Sci. U.S.A. *100*, 15410–15415.

Jan, E., and Sarnow, P. (2002). Factorless ribosome assembly on the internal ribosome entry site of Cricket paralysis virus. J. Mol. Biol. *324*, 889–902.

Jeyaprakash, A., and Hoy, M.A. (2000). Long PCR improves *Wolbachia* DNA amplification: wsp sequences found in 76% of sixty-three arthropod species. Insect Mol. Biol. *9*, 393–405.

Johnson, K., and Christian, P. (1998). The novel genome organization of the insect picorna-like virus *Drosophila* C virus suggests this virus belongs to a previously undescribed virus family. J. Gen. Virol. *79*, 191–203.

Johnson, K.N., and Christian, P.D. (1996). A molecular taxonomy for Cricket paralysis virus including two new isolates from Australian populations of *Drosophila* (Diptera: *Drosophilidae*). Arch. Virol. *141*, 1509–1522.

Johnson, K.N., and Christian, P.D. (1999). Molecular characterization of *Drosophila* C virus isolates. J. Invertebr. Pathol. 73, 248–254.

Johnson, K.N., van Hulten, M.C., and Barnes, A.C. (2008). 'Vaccination' of shrimp against viral pathogens: phenomenology and underlying mechanisms. Vaccine *26*, 4885–4892.

Johnson, R.M., Evans, J.D., Robinson, G.E., and Berenbaum, M.R. (2009). Changes in transcript abundance relating to colony collapse disorder in honey bees (*Apis mellifera*). Proc. Natl. Acad. Sci. U.S.A. *106*, 14790–14795.

Jousset, F.-X., and Plus, N. (1975). Etude de la transmission verticale des picornaviruses de *Drosophila melanogaster* et de *Drosophila immigrans*. Ann. Microbiol. (Paris) *126B*, 231- 249.

Jousset, F.-X., and Plus, N., Croizier, G., and Thomas, M. (1972). Existence chez *Drosophila* de deux groupes de picornaviruaea de propietes serologiques et biologiques differentes. C. R. Acad. Sci. (Paris) *275*, 3043–3046.

Kanamori, Y., and Nakashima, N. (2001). A tertiary structure model of the internal ribosome entry site (IRES) for methionine-independent initiation of translation. RNA *7*, 266–274.

Kieft, J.S. (2009). Comparing the three-dimensional structures of Dicistroviridae IGR IRES RNAs with other viral RNA structures. Virus Res. *139*, 148–156.

King, L.A., Massalski, P.R., Cooper, J.I., and Moore, N.F. (1984). Comparison of the genome RNA sequence homology between Cricket paralysis virus and strains of *Drosophila* C virus by complementary DNA hybridisation analysis. J. Gen Virol. *65*, 1193–1196.

King, L.A., and Moore, N.F. (1988). Evidence for the presence of a genome-linked protein in two insect picornaviruses, cricket paralysis and *Drosophila* C viruses. FEMS Microbiol. Lett. *50*, 41–44.

King, L.A., Pullin, J.S.K., Stanway, G., Almond, J.W., and Moore, N.F. (1987). Cloning of the genome of Cricket paralysis virus: sequence of the 3′ end. Virus Res. *6*, 331–344.

Koonin, E.V., and Dolja, V.V. (1993). Evolution and taxonomy of positive-strand RNA viruses: implications of comparative analysis of amino acid sequences. Crit. Rev. Biochem. Mol. *28*, 375–430.

Koonin, E.V., and Gorbalenya, A.E. (1992). An insect picornavirus may have genome organization similar to that of caliciviruses. FEBS Lett. *297*, 81–86.

Koonin, E.V., Wolf, Y.I., Nagasaki, K., and Dolja, V.V. (2008). The Big Bang of picorna-like virus evolution antedates the radiation of eukaryotic supergroups. Nat. Rev. Microbiol. *6*, 925–939.

Krishna, N.K., Marshall, D., and Schneemann, A. (2003). Analysis of RNA packaging in wild-type and mosaic protein capsids of Flock House virus using recombinant baculovirus vectors. Virology *305*, 10–24.

Landry, D.M., Hertz, M.I., and Thompson, S.R. (2009). RPS25 is essential for translation initiation by the *Dicistroviridae* and hepatitis C viral IRESs. Genes Dev. *23*, 2753–2764.

Le Gall, O., Christian, P., Fauquet, C.M., King, A.M., Knowles, N.J., Nakashima, N., Stanway, G., and Gorbalenya, A.E. (2008). *Picornavirales*, a proposed order of positive-sense single-stranded RNA viruses with a pseudo-T = 3 virion architecture. Arch. Virol. *153*, 715–727.

Leat, N., Ball, B., Govan, V., and Davison, S. (2000). Analysis of the complete genome sequence of Black queen cell virus, a picorna-like virus of honey bees. J. Gen. Virol. *81*, 2111–2119.

Lehane, M.J. (1997). Peritrophic matrix structure and function. Annu. Rev. Entomol. *42*, 525–550.

Lemaitre, B., and Hoffmann, J. (2007). The host defense of *Drosophila melanogaster*. Annu. Rev. Immunol. *25*, 697–743.

Lightner, D.V. (1996). Epizootiology, distribution and the impact on international trade of two penaeid shrimp viruses in the Americas. Rev. Sci. Tech. *15*, 579–601.

Lightner, D.V., Redman, R.M., Poulos, B.T., Nunan, L.M., Mari, J.L., and Hasson, K.W. (1997). Risk of spread of penaeid shrimp viruses in the Americas by the international movement of live and frozen shrimp. Rev. Sci. Tech. *16*, 146–160.

Liljas, L., Tate, J., Lin, T., Christian, P., and Johnson, J.E. (2002). Evolutionary and taxonomic implications of conserved structural motifs between picornaviruses and insect picorna-like viruses. Arch. Virol. *147*, 59–84.

Lu, Y., and Sun, P.S. (2005). Viral resistance in shrimp that express an antisense Taura syndrome virus coat protein gene. Antivir. Res. *67*, 141–146.

Luke, G.A., de Felipe, P., Lukashev, A., Kallioinen, S.E., Bruno, E.A., and Ryan, M.D. (2008). Occurrence, function and evolutionary origins of '2A-like' sequences in virus genomes. J. Gen. Virol. *89*, 1036–1042.

Maori, E., Lavi, S., Mozes-Koch, R., Gantman, Y., Peretz, Y., Edelbaum, O., Tanne, E., and Sela, I. (2007a). Isolation and characterization of Israeli acute paralysis virus, a dicistrovirus affecting honeybees in Israel: evidence for diversity due to intra- and inter-species recombination. J. Gen. Virol. *88*, 3428–3438.

Maori, E., Paldi, N., Shafir, S., Kalev, H., Tsur, E., Glick, E., and Sela, I. (2009). IAPV, a bee-affecting virus associated with Colony Collapse Disorder can be silenced by dsRNA ingestion. Insect. Mol. Biol. *18*, 55–60.

Maori, E., Tanne, E., and Sela, I. (2007b). Reciprocal sequence exchange between non-retro viruses and hosts leading to the appearance of new host phenotypes. Virology *362*, 342–349.

Mari, J., Poulos, B.T., Lightner, D.V., and Bonami, J.-R. (2002). Shrimp Taura syndrome virus: genomic characterization and similarity with members of the genus Cricket paralysis-like viruses. J. Gen. Virol. *83*, 915–926.

Masoumi, A., Hanzlik, T.N., and Christian, P.D. (2003). Functionality of the 5′ - and intergenic IRES elements of Cricket paralysis virus in a range of insect cell lines, and its relationship with viral activities. Virus Res. *94*, 113–120.

Meister, G., and Tuschl, T. (2004). Mechanisms of gene silencing by double-stranded RNA. Nature *431*, 343–349.

Moon, J.S., Domier, L.L., McCoppin, N.K., D'Arcy, C.J., and Jin, H. (1998). Nucleotide sequence analysis shows that *Rhopalosiphum padi* virus is a member of a novel group of insect-infecting RNA viruses. Virology *243*, 54–65.

Moore, N.F., Kearns, A., and Pullin, J.S. (1980). Characterization of Cricket paralysis virus-induced polypeptides in *Drosophila* cells. J.Virol. 33, 1–9.

Moore, N.F., Reavy, B., Pullin, J.S., and Plus, N. (1981). The polypeptides induced in *Drosophila* cells by *Drosophila* C virus (strain Ouarzazate). Virology *112*, 411–416.

Muscio, O.A., LaTorre, J.L., and Scodeller, E.A. (1987). Small nonoccluded viruses from triatomine bug *Triatoma infestans* (Hemiptera: *Reduviidae*). J. Invertebr. Pathol. *49*, 218–220.

Nakashima, N., and Nakamura, Y. (2008). Cleavage sites of the 'P3 region' in the nonstructural polyprotein precursor of a dicistrovirus. Arch. Virol. *153*, 1955–1960.

Nakashima, N., Sasaki, J., and Toriyama, S. (1999). Determining the nucleotide sequence and capsid-coding region of Himetobi P virus: a member of a novel group of RNA viruses that infect insects. Arch. Virol. *144*, 2051–2058.

Nakashima, N., Sasaki, J., Tsuda, K., Yasunaga, C., and Noda, H. (1998). Properties of a new picorna-like virus of the brown-winged green bug, *Plautia stali*. J. Invertebr. Pathol. *71*, 151–158.

Nakashima, N., and Shibuya, N. (2006). Multiple coding sequences for the genome-linked virus protein (VPg) in dicistroviruses. J. Invertebr. Pathol. *92*, 100–104.

Nakashima, N., and Uchiumi, T. (2009). Functional analysis of structural motifs in dicistroviruses. Virus Res. *139*, 137–147.

Nishiyama, T., Yamamoto, H., Shibuya, N., Hatakeyama, Y., Hachimori, A., Uchiumi, T., and Nakashima, N. (2003). Structural elements in the internal ribosome entry site of *Plautia stali intestine virus* responsible for binding with ribosomes. Nucleic Acids Res. *31*, 2434–2442.

Osborne, S.E., Leong, Y.S., O'Neill, S.L., and Johnson, K.N. (2009). Variation in antiviral protection mediated by different *Wolbachia* strains in *Drosophila simulans*. PLoS Pathog. *5*, e1000656.

Pal, N., Boyapalle, S., Beckett, R., Miller, W.A., and Bonning, B.C. (2007). A baculovirus-expressed dicistrovirus that is infectious to aphids. J. Virol. *81*, 9339–9345.

Palacios, G., Hui, J., Quan, P.L., Kalkstein, A., Honkavuori, K.S., Bussetti, A.V., Conlan, S., Evans, J., Chen, Y.F., van Engelsdorp, D., *et al.* (2008). Genetic analysis of Israel acute paralysis virus: distinct clusters are circulating in the United States. J. Virol. *82*, 6209–6217.

Pestova, T.V., and Hellen, C.U. (2003). Translation elongation after assembly of ribosomes on the Cricket paralysis virus internal ribosomal entry site without initiation factors or initiator tRNA. Genes Dev. *17*, 181–186.

Pestova, T.V., and Hellen, C.U. (2005). Reconstitution of eukaryotic translation elongation *in vitro* following initiation by internal ribosomal entry. Methods *36*, 261–269.

Pfingsten, J.S., Costantino, D.A., and Kieft, J.S. (2006). Structural basis for ribosome recruitment and manipulation by a viral IRES RNA. Science *314*, 1450–1454.

Pijlman, G.P., Roode, E.C., Fan, X., Roberts, L.O., Belsham, G.J., Vlak, J.M., and van Oers, M.M. (2006). Stabilized baculovirus vector expressing a heterologous gene and GP64 from a single bicistronic transcript. J. Biotechnol. *123*, 13–21.

Plus, N. (1978). Endogenous viruses of *Drosophila melanogaster* cell lines: their frequency, identification and origin. *In Vitro 14*, 1015–1021.

Plus, N., Croizier, G., Reinganum, C., and Scott, P.D. (1978). Cricket paralysis virus and *Drosophila* C virus: serological analysis and comparison of capsid polypeptides and host range. J. Invertebr. Pathol. *31*, 296–302.

Plus, N., Croizier, G., Jousset, F-X., and David, J. (1975). Picornaviruses of laboratory and wild *Drosophila melanogaster*: Geographical distribution and serotypic composition. Ann. Microbiol. (Inst. Pasteur) *126A*, 107–117.

Plus, N., and Golubovsky, M.D. (1980). Resistance to *Drosophila* C virus of fifteen 1 (2) gl/Cy stocks carrying 1(2) gl lethals from different geographical origins. Genetika *12*, 227–231.

Plus, N., and Scotti, P.D. (1984). The biological properties of eight different isolates of Cricket paralysis virus. Ann. Virol. (Inst. Pasteur) *135E*, 257–268.

Reavy, B., Crump, W.A.L., and Moore, N.F. (1983). Characterization of cricket paralysis virus and *Drosophila* C virus-induced RNA species synthesized in infected *Drosophila melanogaster* cells. J. Invertebr. Pathol. *41*, 397–400.

Reavy, B., and Moore, N.F. (1981). *In vitro* translation of Cricket paralysis virus RNA. Arch. Virol. *67*, 175–180.

Reddy, J.T., and Locke, M. (1990). The size limited penetration of gold particles through insect basal laminae. J. Insect Physiol. *36*, 397–407.

Reinganum, C. (1975). The isolation of cricket paralysis virus from the emperor gum moth, *Antheraea eucalypti* Scott, and its infectivity towards a range of insect species. Intervirology *5*, 97–102.

Reinganum, C., Gagen, S.J., Sexton, S.B., and Vellacott, H.P. (1981). A survey of pathogens of the black field cricket, *Teleogryllus commodus*, in the Western District of Victoria, Australia. J. Invertebr. Pathol. *38*, 153.

Reinganum, C., O'Loughlin, G.T., and Hogan, T.W. (1970). A non-occluded virus of the field crickets *Teleogryllus aceanicus* and *T. commodus* (Orthoptera: Gryllidae). J. Invertebr. Pathol. *16*, 220–314.

Robalino, J., Bartlett, T., Shepard, E., Prior, S., Jaramillo, G., Scura, E., Chapman, R.W., Gross, P.S., Browdy, C.L., and Warr, G.W. (2005). Double-stranded RNA induces sequence-specific antiviral silencing in addition to nonspecific immunity in a marine shrimp: convergence of RNA interference and innate immunity in the invertebrate antiviral response? J. Virol. *79*, 13561–13571.

Robalino, J., Bartlett, T.C., Chapman, R.W., Gross, P.S., Browdy, C.L., and Warr, G.W. (2007). Double-stranded RNA and antiviral immunity in marine shrimp: inducible host mechanisms and evidence for the evolution of viral counter-responses. Dev. Comp. Immunol. *31*, 539–547.

Robalino, J., Browdy, C.L., Prior, S., Metz, A., Parnell, P., Gross, P., and Warr, G. (2004). Induction of antiviral immunity by double-stranded RNA in a marine invertebrate. J. Virol. *78*, 10442–10448.

Roberts, L.O., and Groppelli, E. (2009). An atypical IRES within the 5′ UTR of a dicistrovirus genome. Virus Res. *139*, 157–165.

Roberts, L.O., Seamons, R.A., and Belsham, G.J. (1998). Recognition of picornavirus internal ribosome entry sites within cells; influence of cellular and viral proteins. RNA *4*, 520–529.

Rosenkranz, P., Aumeier, P., and Ziegelmann, B. (2010). Biology and control of *Varroa destructor*. J. Invertebr. Pathol. *103*, S96-S119.

Roxstrom-Lindquist, K., Terenius, O., and Faye, I. (2004). Parasite-specific immune response in adult *Drosophila melanogaster*: a genomic study. EMBO Rep. *5*, 207–212.

Royall, E., Woolaway, K.E., Schacherl, J., Kubick, S., Belsham, G.J., and Roberts, L.O. (2004). The *Rhopalosiphum padi virus* 5′ internal ribosome entry site is functional in *Spodoptera frugiperda* 21 cells and in their cell-free lysates: implications for the baculovirus expression system. J. Gen. Virol. *85*, 1565–1569.

Ryan, M.D., and Flint, M. (1997). Virus-encoded proteinases of the picornavirus super-group. J. Gen. Virol. *78*, 699–723.

Saleh, M.C., Tassetto, M., van Rij, R.P., Goic, B., Gausson, V., Berry, B., Jacquier, C., Antoniewski, C., and Andino, R. (2009). Antiviral immunity in *Drosophila* requires systemic RNA interference spread. Nature *458*, 346–350.

Sasaki, J., and Nakashima, N. (1999). Translation initiation at the CUU codon is mediated by the internal ribosome entry site of an insect picorna-like virus *in vitro*. J. Virol. 73, 1219–1226.

Sasaki, J., and Nakashima, N. (2000). Methionine-independent initiation of translation in the capsid protein of an insect RNA virus. Proc. Natl. Acad. Sci. U.S.A. *97*, 1512–1515.

Sasaki, J., Nakashima, N., Saito, H., and Noda, H. (1998). An insect picorna-like virus, Plautia stali intestine virus, has genes of capsid proteins in the 3′ part of the genome. Virology *244*, 50–58.

Schubot, F.D., Jackson, M.W., Penrose, K.J., Cherry, S., Tropea, J.E., Plano, G.V., and Waugh, D.S. (2005). Three-dimensional structure of a macromolecular assembly that regulates type III secretion in *Yersinia pestis*. J. Mol. Biol. *346*, 1147–1161.

Scotti, P.D. (1977). End-point dilution and plaque assay methods for titration of Cricket paralysis virus in cultured *Drosophila* cells. Arch. Virol. *35*, 393–396.

Scotti, P.D., and Dearing, S.C. (1996). An agarose cell suspension plaque assay for titration of cricket paralysis virus. J. Invertebr. Pathol. *67*, 190–191.

Scotti, P.D., Hoefakker, P., and Dearing, S. (1996). The production of Cricket paralysis virus in suspension cultures of insect cell lines. J. Invertebr. Pathol. *68*, 109–112.

Scotti, P.D., Longworth, J.F., Plus, N., Croizier, G., and Reinganum, C. (1981). The biology and ecology of strains of an insect small RNA virus complex. Adv. Virus Res. *26*, 117–143.

Shen, M., Cui, L., Ostiguy, N., and Cox-Foster, D. (2005a). Intricate transmission routes and interactions between picorna-like viruses (Kashmir bee virus and sacbrood virus) with the honeybee host and the parasitic varroa mite. J. Gen. Virol. *86*, 2281–2289.

Shen, M., Yang, X., Cox-Foster, D., and Cui, L. (2005b). The role of varroa mites in infections of Kashmir bee virus (KBV) and Deformed wing virus (DWV) in honey bees. Virology *342*, 141–149.

Shibuya, N., and Nakashima, N. (2006). Characterization of the 5′ internal ribosome entry site of *Plautia stali intestine virus*. J. Gen. Virol. *87*, 3679–3686.

Shibuya, N., Nishiyama, T., Kanamori, Y., Saito, H., and Nakashima, N. (2003). Conditional rather than absolute requirements of the capsid coding sequence for initiation of methionine-independent translation in *Plautia stali* intestine virus. J. Virol. *77*, 12002–12010.

Sparks, W.O., Bartholomay, L., and Bonning, B.C. (2008). Insect immunity to viruses. In Insect Immunology, N.E. Beckage, ed. (Academic Press), pp. 209–242.

Srisuvan, T., Noble, B.L., Schofield, P.J., and Lightner, D.V. (2006). Comparison of four Taura syndrome virus (TSV) isolates in oral challenge studies with *Litopenaeus vannamei* unselected or selected for resistance to TSV. Dis. Aquat. Organ. *71*, 1–10.

Tate, J., Liljas, L., Scotti, P., Christian, P., Lin, T., and Johnson, J.E. (1999). The crystal structure of cricket paralysis virus: the first view of a new virus family. Nat. Struct. Biol. *6*, 765–774.

Teixeira, L., Ferreira, A., and Ashburner, M. (2008). The bacterial symbiont *Wolbachia* induces resistance to RNA viral infections in *Drosophila melanogaster*. PLoS Biol. *6*, e2.

Terenin, I.M., Dmitriev, S.E., Andreev, D.E., Royall, E., Belsham, G.J., Roberts, L.O., and Shatsky, I.N. (2005). A cross-kingdom internal ribosome entry site reveals a simplified mode of internal ribosome entry. Mol. Cell. Biol. *25*, 7879–7888.

Thomas-Orillard, M. (1984). Modifications of mean ovariole number, fresh weight of adult females and developmental time in *Drosophila melanogaster* induced by *Drosophila* C virus. Genetics *107*, 635–644.

Thomas-Orillard, M. (1988). Interaction between a picornavirus and a wild population of *Drosophila melanogaster*. Oecologia (Berlin) *75*, 516–520.

Thompson, S.R., Gulyas, K.D., and Sarnow, P. (2001). Internal initiation in *Saccharomyces cerevisiae* mediated by an initiator tRNA/eIF2-independent internal ribosome entry site element. Proc. Natl. Acad. Sci. U.S.A. *98*, 12972–12977.

Toriyama, S., Guy, P.L., Fuji, S., and Takahashi, M. (1992). Characterization of a new picorna-like virus, himetobi P virus, in planthoppers. J. Gen. Virol. *73*, 1021–1023.

Valles, S.M., Strong, C.A., Dang, P.M., Hunter, W.B., Pereira, R.M., Oi, D.H., Shapiro, A.M., and Williams, D.F. (2004). A picorna-like virus from the red imported fire ant, *Solenopsis invicta*: initial discovery, genome sequence, and characterization. Virology *328*, 151–157.

van Engelsdorp, D., Evans, J.D., Saegerman, C., Mullin, C., Haubruge, E., Nguyen, B.K., Frazier, M., Frazier, J., Cox-Foster, D., Chen, Y., *et al.* (2009). Colony collapse disorder: a descriptive study. PLoS ONE *4*, e6481

van Engelsdorp, D., Hayes, J., Jr., Underwood, R.M., and Pettis, J. (2008). A survey of honey bee colony losses in the U.S., fall 2007 to spring 2008. PLoS ONE 3, e4071.

van Munster, M., Dullemans, A.M., Verbeek, M., van Den Heuvel, J.F., Clerivet, A., and van Der Wilk, F. (2002). Sequence analysis and genomic organization of Aphid lethal paralysis virus: a new member of the family *Dicistroviridae*. J. Gen. Virol. *83*, 3131–3138.

van Rij, R P., Saleh, M.C., Berry, B., Foo, C., Houk, A., Antoniewski, C., and Andino, R. (2006). The RNA silencing endonuclease Argonaute 2 mediates specific antiviral immunity in *Drosophila melanogaster*. Genes Dev. *20*, 2985–2995.

Victoria, J.G., Kapoor, A., Li, L., Blinkova, O., Slikas, B., Wang, C., Naeem, A., Zaidi, S., and Delwart, E. (2009). Metagenomic analyses of viruses in the stool of children with acute flaccid paralysis. J. Virol. *83*, 4642–4651.

Wang, X.H., Aliyari, R., Li, W.X., Li, H.W., Kim, K., Carthew, R., Atkinson, P., and Ding, S.W. (2006). RNA interference directs innate immunity against viruses in adult *Drosophila*. Science *312*, 452–454.

Wang, Y., Sheng, G., Juranek, S., Tuschl, T., and Patel, D.J. (2008). Structure of the guide-strand-containing argonaute silencing complex. Nature *456*, 209–213.

Werren, J.H., and Windsor, D.M. (2000). *Wolbachia* infection frequencies in insects: evidence of a global equilibrium? Proc. Biol. Sci. *267*, 1277–1285.

Wigley, P.J., and Scotti, P.D. (1983). The seasonal incidence of cricket paralysis virus in a population of the New Zealand small field cricket, *Pteronemobius nigrovus* (Orthoptera: Gryllidae). J. Invertebr. Pathol. *41*, 378–380.

Williamson, C., Rybicki, E.P., Kasdorf, G.G.F., and von Wechmar, M.B. (1988). Characterization of a new picorna-like virus isolated from aphids. J. Gen. Virol. *69*, 787–795.

Wilson, J.E., Pestova, T.V., Hellen, C.U., and Sarnow, P. (2000a). Initiation of protein synthesis from the A site of the ribosome. Cell *102*, 511–520.

Wilson, J.E., Powell, M.J., Hoover, S.E., and Sarnow, P. (2000b). Naturally occurring dicistronic cricket paralysis virus RNA is regulated by two internal ribosome entry sites. Mol. Cell. Biol. *20*, 4990–4999.

Woolaway, K.E., Lazaridis, K., Belsham, G.J., Carter, M.J., and Roberts, L.O. (2001). The 5′ untranslated region of *Rhopalosiphum padi* virus contains an internal ribosome entry site which functions efficiently in mammalian, plant, and insect translation systems. J. Virol. *75*, 10244–10249.

Yang, X., and Cox-Foster, D. (2007). Effects of parasitization by *Varroa destructor* on survivorship and physiological traits of *Apis mellifera* in correlation with viral incidence and microbial challenge. Parasitology *134*, 405–412.

Yang, X., and Cox-Foster, D.L. (2005). Impact of an ectoparasite on the immunity and pathology of an invertebrate: evidence for host immunosuppression and viral amplification. Proc. Natl. Acad. Sci. U.S.A. *102*, 7470–7475.

Yu, C.I., and Song, Y.L. (2000). Outbreaks of Taura syndrome in pacific white shrimp *Penaeus vannamei* cultured in Taiwan. Fish Pathol. *35*, 21–24.

Zambon, R.A., Nandakumar, M., Vakharia, V.N., and Wu, L.P. (2005). The Toll pathway is important for an antiviral response in *Drosophila*. Proc. Natl. Acad. Sci. U.S.A. *102*, 7257–7262.

Genomics and Biology of Iflaviruses

10

Monique M. van Oers

Abstract

Iflaviruses are insect viruses that form non-enveloped, icosahedral particles approximately 30 mm in diameter. The particles contain one copy of the single-stranded RNA genome, which has a positive polarity. The genome encodes one large polyprotein, which is post-translationally processed. The viral coat proteins are located in an N-terminal domain of the polyprotein, while the non-structural proteins, involved in replication and polyprotein processing, are present in the C-terminal region. The order of the individual proteins in the polyprotein is strictly conserved. Polyprotein translation may initiate at an internal ribosome entry site (IRES) in the 5′ untranslated region (UTR) of the RNA. The 3′ UTR is followed by a poly(A) tail. Based on genome organization, virion structure and phylogeny, iflaviruses form a distinct group in the order *Picornavirales,* justifying their classification in a virus family separate from for instance *Dicistroviridae* and *Picornaviridae*. The type species of the family *Iflaviridae*, genus *Iflavirus*, is *Infectious flacherie virus* (IFV) of silkworms.

Introduction

Iflaviruses have been described so far in insects belonging to the orders Lepidoptera, Hymenoptera and Hemiptera, and in bee parasitic mites (class Arachnida, order Acarina). Iflaviruses form icosahedral particles and have a single-stranded RNA genome of positive polarity. They were initially discovered in insects of economic importance such as honey bees and silkworms and may cause diarrhoea, developmental malformations and death of the infected insects (Fig. 10.1). Examples of symptomless infections have also been reported.

Sacbrood disease is an example of a fatal iflavirus infection, causing a failure of honey bee larvae (*Apis mellifera*) to pupate. Sacbrood was first described in bee colonies in the USA (White, 1913). Bee larvae with sacbrood cannot shed the last larval skin. Fluid accumulates under this skin and in this way larvae become trapped in a fluid-filled sac, hence the name sacbrood disease. After bee larvae succumb to the disease they turn dark brown and finally dry out and flatten to form gondola-shaped scales (Fig. 10.1A) (Bailey and Ball, 1991; Gochnauer, 1990). In 1917 the causative agent of sacbrood was the first filterable pathogen reported in an insect (White, 1917). Sacbrood virus (SBV) was morphologically characterized in 1964 showing isometric, 28-nm particles in electron microscopy (Bailey *et al.*, 1964). These days, Sacbrood virus is known to be widely spread around the world (Bradbear, 1988).

The aetiological agent of infectious flacherie, a type of infectious diarrhoea in the silkworm *Bombyx mori* and a major cause of cocoon loss, was shown to be a virus in 1964 (Aizawa *et al.*, 1964). *Infectious flacherie virus* (IFV) was the first iflavirus, for which the complete genome sequence was published (Isawa *et al.*, 1998), followed by SBV a year later (Ghosh *et al.*, 1999).

Deformed wing virus (DWV) is an example of an iflavirus that induces morphological abnormalities in emerging honey bees, resulting in crippled bees (Fig. 10.1B). DWV can also

Figure 10.1 Iflavirus-infected insects. (A) Cadavers of honey bee larvae that succumbed to Sacbrood virus. (B) Honey bee with crippled wings as a consequence of Deformed wing virus infection. Note the varroa mite sitting on the back of the bee (arrow). Pictures were a courtesy of E. Genersch (A) and M. Traynor (B)

cause covert, symptomless infections (Yue and Genersch, 2005). The genetically very closely related Kakugo virus (KV) was isolated from brains of aggressive worker bees, that showed an increased attack rate for their natural enemies, such as hornets ('*kakugo*' is Japanese for 'ready to attack') (Fujiyuki *et al.*, 2004). A causative correlation between aggressiveness and Kakugo virus infections has not yet been confirmed.

Viruses infecting bees have gained increased scientific attention in recent years in relation to colony collapse disorder (CCD), a phenomenon observed in the USA since 2006, in which complete bee colonies suddenly disappear (van Engelsdorp *et al.*, 2009). The biological and economical impact of this phenomenon is immense (van Engelsdorp *et al.*, 2008; Stankus, 2008). One hypothesis is that CCD is caused by a combination of pathogen manifestation and stress factors, such a pesticides (van Engelsdorp *et al.*, 2009). A recent comparison of transcript abundance in guts between CCD and healthy bees, however, did not show an up-regulation of pesticide-reactive or immune response genes (Johnson *et al.*, 2009). In this study viral RNAs, including those of DWV and Israeli acute paralysis virus (IAPV) were present in the CCD bees. DWV has been implicated as an important pathogen in the parasitic mite syndrome and DWV can cause collapse of bee hives in autumn, when *Varroa destructor* mite densities increase (Nordström *et al.*, 1999). However, DWV is not the causative agent of CCD. Currently, IAPV is thought to be the most important viral factor in CCD (Maori *et al.*, 2007, 2009; Blanchard *et al.*, 2008). Since IAPV belongs to the *Dicistroviridae* (Christian *et al.*, 2005), CCD is further addressed in Chapter 9. For a recent review on honey bee virus literature see (Chen and Siede, 2007).

This chapter focuses on the molecular virology and biology of iflaviruses. As this chapter is written, ten iflavirus genomes have been completely sequenced and the similarities and differences in genome organization with the dicistroviruses and the mammalian picornaviruses have become visible.

Taxonomy

As of 2008, iflaviruses are taxonomically classified in the family *Iflaviridae* with a single genus called *Iflavirus* (ICTV, 2009). These two taxonomic names are derived from the type species *Infectious flacherie virus* (IFV), that infects silkworms. Five other iflavirus species have an official status as of January 2010 (Table 10.1), including the above mentioned bee viruses SBV and DWV. *Varroa destructor virus-1* (VDV-1, first isolated from bee parasitic mites but also present in bees) (Ongus *et al.*, 2004), and two other lepidopteran insect-infecting viruses *Ectropis obliqua virus (EoV)* and *Perina nuda virus* (PnV) (Wu *et al.*, 2002a; Wang *et al.*, 2004). Tentative species may include Slow bee paralysis virus (SBPV), originally called Honey bee slow paralysis virus (HBSPV) (de Miranda *et al.*, 2007), *Venturia canescens* small RNA-containing virus (VcSRV), *Nasonia*

Table 10.1 Overview of iflavirus and dicistrovirus names used in this chapter*

Abbrevation*	Virus name	Host
Iflaviridae		
BpPV	*Brevicoryne brassicae* (picorna-like) virus	Cabbage aphid
CSBV	Chinese sacbrood virus	Honey bee
DWV	*Deformed wing virus*	Honey bee
EoV	*Ectropis obliqua virus*	Tee looper
GlmoV	*Glossinia morsitans* virus	Tsetse fly (EST)
IFV	*Infectious flacherie virus*	Silkworm *Bombyx mori*
KV	Kakugo virus	Honey bee
MudoV	*Musca domestica* virus	Housefly (EST)
NvV	*Nasonia vitripens* virus	Parasitic wasp
PnV	*Perina nuda virus*	Ficus transparent wing moth
SBPV	Slow bee paralysis virus	Honey bee
SBV	*Sacbrood virus*	Honey bee
VcSVR	*Venturia canescens* small RNA virus	Parasitic wasp
VDV-1	*Varrao destructor* virus-1	Honey bee[1]
Dicistroviridae		
CrPV	*Cricket paralysis virus*	Cricket
IAPV	Israeli acute paralysis virus	Honey bee
RhPV	*Rhopalosiphum padi virus*	Oat aphid

*Abbreviation of viruses with an ICTV status in bold.

[1]VDV-1 was originally identified in mites, but based on homology to DWV, honey bees are probably the primary hosts.

vitripens virus (NvV), the last two from parasitic wasps (Reineke and Asgari, 2005; Oliveira *et al.*, 2009), and *Brevicoryne brassicae* (picorna-like) virus (BbPV) from cabbage aphids (Ryabov, 2007). Whether Chinese sacbrood virus (CSBV) (Zhang *et al.*, 2001) should have its own species status or should be regarded as a strain of SBV awaits further sequence information. KV of bees (Fujiyuki *et al.*, 2004) is best regarded as a strain of DWV given the high similarity in sequence (only 28 nt are different).

In addition, a cDNA sequence recently derived from salivary glands of the tsetse fly *Glossina morsitans morsitans* most likely has an iflavirus origin (Ribeiro *et al.*, 2009) and the same holds true for a housefly (*Musca domestica*) RNA sequence with homology to Sacbrood virus (D. Boucias, personal communication). This would extend the insect orders in which iflaviruses have been found to the Diptera.

The family *Iflaviridae* is classified under the order *Picornavirales* (Le Gall *et al.*, 2008). This order harbours five other, related virus families, *Comoviridae*, *Dicistroviridae*, *Marnaviridae*, *Picornaviridae* and *Sequiviridae*, and two genera not assigned to a family yet, *Cheravirus* and *Sadwavirus*. The viruses classified under the order *Picornavirales* share a number of important characteristics (Le Gall *et al.*, 2008): (1) icosahedral particles with a pseudo $T = 3$ symmetry, approximately 30 nm in diameter, that are non-enveloped and made up of 60 protomers;

(2) a single-stranded, linear RNA genome of positive polarity with usually a small protein (< 5 kDa) covalently linked to the viral genome (VPg) at the 5′ end; (3) the genomic RNA translates without subgenomic mRNAs directly into post-translationally cleaved polyprotein(s); (4) The coat proteins are characterized by a 'jelly-roll' domain (an eight-stranded β-barrel) (Chandrasekar and Johnson, 1998); (5) The polyprotein has a three-domain unit encoding in a conserved order: a superfamily III helicase, a (cysteine) proteinase with a chymotrypsin-like fold and an RNA-dependent RNA polymerase (RdRp); (6) The RNA contains a 3′ polyA tail, except in members of the *Sequiviridae*. Other virus families previously also referred to as 'picorna-like' viruses, e.g. *Caliciviridae*, *Potyviridae* and *Hypoviridae*, do not share all the features mentioned above and are therefore placed outside the order *Picornavirales*.

Iflaviruses have been isolated from insects and insect parasitic mites, where as dicistroviruses have been found in both insects and crustaceans. Picornaviruses cause disease in vertebrate hosts, for instance the common cold, hepatitis A and poliomyelitis in humans, and foot and mouse disease in cattle. The plant-infecting viruses in the order *Picornavirales* are divided over the sequiviruses with mono-partite genomes, and the como-, chera- and sadwaviruses, where the genome consist of two RNA segments. Marnaviruses occur in algae (Lang *et al.*, 2004) and so far only one species has been assigned.

Iflavirus nomenclature

Several insect-infecting viruses with single-stranded, linear RNA of positive polarity that share important features with vertebrate picornaviruses and which are now classified either in the family *Iflaviridae* or *Dicistroviridae*, have been described as 'insect picorna-like' viruses in the past. Since the name 'picorna-like viruses (PV)' does not refer unambiguously to a definite taxonomic unit, it should not be used in names of (tentative) virus species. In line with this, the old name *Ectropis obliqua* picorna-like virus (EoPV) (Wang *et al.*, 2004) has recently been changed into *Ectropis obliqua virus (EoV)*, when this species received an official status (ICTV, 2009) similar to the renaming of *Perina nuda* picorna-like virus (PnPV) (Wang *et al.*, 1999) to PnV. In line with this, the name BbPV should be changed in the near future to *Brevicoryne brassicae virus-1* (BbV-1) in case this becomes an official species.

To prevent confusion with other viruses isolated from the same insect species, the addition of the genus name iflavirus in the species name might be considered to allow easy discrimination between distinct viruses isolated from the same host, i.e. *Perina nuda iflavirus or* PnIV versus *Perina nuda* nuclear polyhedrovirus (PnNPV) (*Baculoviridae*).

Iflavirus genome organization

The viral capsid harbours an RNA genome as was first shown for SBV (Lee and Furgala, 1965) and later for IFV (Himeno *et al.*, 1974). The single-stranded, linear RNA molecule is non-segmented and of positive polarity. Hence, the genomic RNA can directly be translated into protein as was shown *in vitro* in rabbit reticulocyte and wheat germ lysates (Hashimoto *et al.*, 1984). The 3′-terminus of the RNA has a poly(A) tail (Isawa *et al.*, 1998; Ongus *et al.*, 2004). As of October 2009, complete genome sequences of 10 iflavirus genomes (two IFV strains, SBV, PnV, EoV, DWV, KV, VDV-1, SBPV and BbPV) are available in the public domain. The main characteristics of each individual genome are summarized in Table 10.2. The genome size of these viruses varies between 8,577 and 10,116 nucleotides, without counting the nucleotides in the poly(A) tail at the 3′ end.

Iflavirus RNA genomes have one open reading frame (ORF), which is translated into one large polyprotein, as has been shown by *in vitro* translation experiments of viral RNA (Hashimoto *et al.*, 1984). In line with this translation mechanism, subgenomic mRNAs are not observed (Wu *et al.*, 2002a). Polyprotein sizes vary between 2858 (SBV) and 3085 amino acids (BbPV) (Table 10.2). The large polyprotein (200 kDa in IFV) is cleaved post-translationally into functional viral proteins (Fig. 10.2). An intermediate processing product (130 kDa in IFV) encompassing the four capsid proteins (VP1-VP4) has also been found (Hashimoto *et al.*, 1984). The genetic information for the four viral capsid proteins is located in the 5′ part of the genome and is followed downstream by a

Table 10.2 Iflavirus genomes*

Host	Virus	Abbr.	Total RNA	ORF	Polyprotein	5′ UTR	3′ UTR	Accession number	Reference
Lepidoptera	*Ectropis obliqua* virus	EoV	9394	8964	2987	390	40	AY365064	Wang *et al.* (2004)
	Perina nuda virus	PnV	9476	8961	2986	473	42	AF323747	Wu *et al.* (2002a)
	Infectious flacherie virus, Kinshu–Syowa strain	IFV	9650	9258	3085	156	236	AB000906	Isawa *et al.* (1998)
	Infectious flacherie virus, ZheJiang01/Ch strain	IFV	9655	9258	3085	155	242	EU868609	Li *et al.* (2009)
Hymenoptera	*Sacbrood virus*	SBV	8832	8577	2858	178	77	AF092924	Ghosh *et al.* (1999)
	Slow bee paralysis virus	SBPV	9505	8895	2964	316	294	EU035616	de Miranda *et al.* (2007)
	Varro destructor virus-1[1]	VDV-1	10,112	8682	2893	1117	313	AY251269	Ongus *et al.* (2004)
	Deformed wing virus	DWV	10,140	8682	2893	1144	314	AJ489744	Lanzi *et al* (2006)
	Kakugo virus	KV	10,152	8682	2893	1156	314	AB070959	Fujiyuki *et al.* (2004)
Hemiptera	*Brevicoryne brassicae* (picorna-like) virus	BpPV	10,161	8952	2983	792	417	EF517277	Ryabov (2007)

*Genome size without poly(A) tail. The stop codon is counted in the ORF, all sizes in nucleotides, except polyprotein (amino acids).

[1] VDV-1, originally isolated from mites, was placed with the bee viruses based on homology and replication in bees as well as mites.

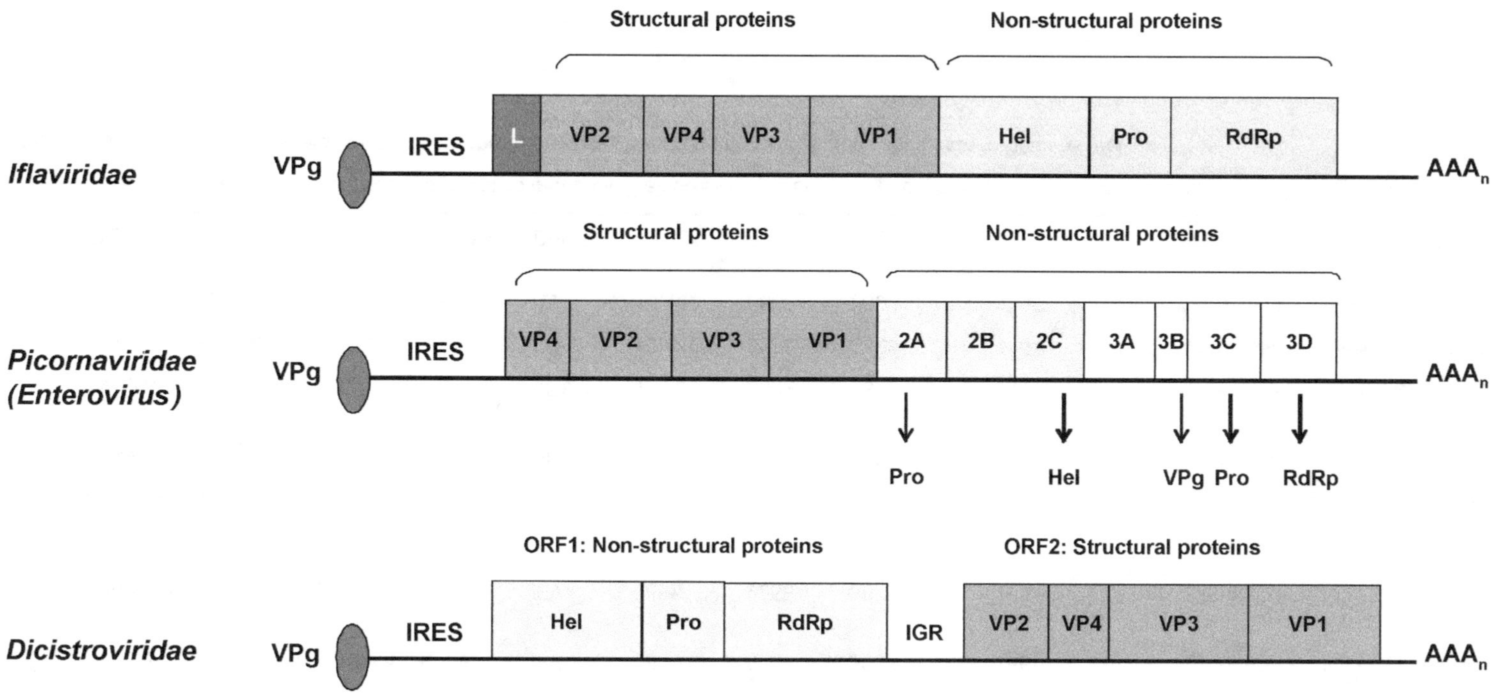

Figure 10.2 Genome organization of viruses in the family *Iflaviridae* compared to *Picornaviridae* and *Dicistroviridae*. Iflavirus genomes encode a single polyprotein with the structural proteins VP1-VP4 located in the N-terminal region and the non-structural proteins, helicase, protease and RNA-dependent RNA polymerase (RdRp) in the C-terminal region. The N-terminal L polypeptide is not conserved in all iflavirus genomes. The presence of an IRES element in the 5′ UTR has been confirmed for several iflaviruses. Dicistroviruses have a bicistronic genome, encoding two polyproteins separated by an intergenic region (IGR) that serves as an IRES for the second ORF. ORF1 encodes the non-structural proteins, ORF 2 the structural proteins.

conserved cluster encoding the non-structural proteins (Fig. 10.2). Since the names used in various publications to address the structural proteins of iflaviruses is sometimes confusing, in this review the convention is followed as given in the Eighth Report of the International Committee on Taxonomy of Viruses (Fauquet *et al.*, 2005). This means that VP1 is the largest and VP4 the smallest capsid protein. The capsid proteins appear in the order VP2, VP4, VP3 and VP1 from N- to C-terminus in the iflavirus polyprotein. The C-terminal region of the polyprotein contains a region for the protease, the helicase and the RdRp, which are found in this order from N- to C-terminus in all members of the *Picornavirales*. This region may also contain other not yet identified or not yet mapped proteins, such as the VPg. The non-structural proteins play crucial roles in replication and polyprotein processing.

Comparison to genome organization of other viruses

The genome organization of iflaviruses most strongly resembles that of the vertebrate-infecting picornaviruses, explaining the previously used name 'insect picorna-like viruses'. Viruses classified in the families *Iflaviridae* and *Picornaviridae* produce one single polyprotein in which the structural proteins are located N-terminally to the non-structural ones (Fig. 10.2A and B). The order of the capsid proteins in the polyprotein is different between these two virus families. The smallest capsid protein, VP4, is located at the N-terminus of the capsid protein module in picornaviruses, whereas in iflavirus genomes it is positioned between VP2 and VP3.

In the iflavirus polyprotein, the structural proteins may be preceded by a leader peptide (L) of unknown function as predicted for VDV-1 (Ongus *et al.*, 2004) (Fig. 10.2A). In IFV the N-terminus of VP4 corresponds to the N-terminus of the polyprotein, indicating that IFV lacks an L protein (Choi *et al.*, 1992). This iflavirus L protein is not homologous to the leader proteins present in the family *Picornaviridae* (Ryan and Flint, 1997), such as the L protein of aphthoviruses, a protease that cleaves the translation initiation factor eIF4G, (Hinton *et al.*, 2002) or the cardiovirus L protein, which is a manipulator of nucleocytoplasmic trafficking and an inhibitor of interferon production (Porter and Palmenberg, 2009; Ricour *et al.*, 2009). Enteroviruses (*Picornaviridae*) encode a 2A protease (Seipelt *et al.*, 1999), for which homologues are absent from iflavirus genomes.

The order of the capsid proteins in dicistroviruses is the same as in iflaviruses (Fig. 10.2C). The major difference in genome organization between ifla- and dicistroviruses is that iflaviruses encode one large polyprotein, that is cleaved post-translationally into both structural (capsid) and non-structural proteins, whereas the genome of dicistroviruses, such as *Cricket paralysis virus* (CrPV), carries two ORFs or cistrons (e.g. Wilson *et al.*, 2000). This bicistronic genome organization provided the family name *Dicistroviridae* (Christian *et al.*, 2005), see Chapter 9. The 5′ ORF encodes the non-structural proteins, while the structural proteins are encoded by the second ORF. In iflavirus genomes, on the other hand, the genetic information for the structural proteins precedes the coding region for the non-structural ones. Within the sets for structural and non-structural proteins, the order of the encoded proteins is the same in ifla- and dicistroviruses.

The viral capsid

Iflavirus particles have an icosahedral appearance (Fig. 10.3A) and are often found densely packed in cytoplasmic areas in infected cells (Fig. 10.3B). The typical size of an iflavirus particle is approximately 30 nm. The buoyant density in CsCl of SBV particles is 1.35 g/ml, and 1.37 g/ml for DWV, and sedimentation coefficients are 160 S for SBV and 165 S for DWV (Bailey, 1976; Bailey and Ball, 1991). For IFV two types of virus particles were observed. They vary in their sedimentation coefficient in sucrose gradients, one of 180S and one of 134S, corresponding to diameters of 27 and 22 nm, and buoyant densities of 1.29 and 1.32 g/ml, respectively (Himeno *et al.*, 1974). Both these particle types were infectious. Mixtures of virus particles varying in size have been reported by several authors and the various forms may represent different stages in virus capsid maturation. In some cases, however, it can not be excluded that the variation in particle

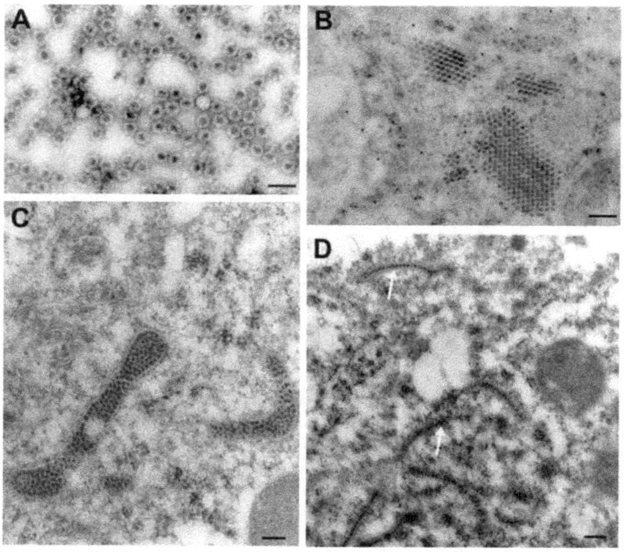

Figure 10.3 Transmission electron microscope images of *Varroa* destructor virus-1. (A) Purified VDV-1 particles, (B) a thin section of mite tissue with densely packed VDV-1 particles, (C) vesicular structures filled with virus particles and (D) membranes with associated virus particles (arrows). The size bars in the photographs represent 50 nm (A) and 180 nm (B, C and D). Images for this figure were supplied by D. Peters.

sizes results from mixed infections with different viruses. Empty capsids without RNA are also a regularly observed phenomenon.

Structural analyses have been performed for virus particles of CSBV, showing that 60 heterotrimers formed of the three large viral capsid proteins (VP1-VP3) serve as the building blocks (protomers or capsomers) for 12 pentamers, that finally form the viral capsid. The exact position of VP4 in the capsid is not resolved. The capsid has a $T = 1$ or a pseudo $T = 3$ structure ($P = 3$) (Zhang *et al.*, 2001). This basic architecture is similar to that of mammalian picornaviruses and CrPV (*Dicistroviridae*) (Rossmann and Tao, 1999; Tate *et al.*, 1999). A comparable packaging is also seen in comovirus capsids, although two VP proteins form the protomers in that case (Chen *et al.*, 1989). The CSBV structure did not reveal the 'Rossmann canyon' surrounding the 5-fold axis, a region that contains the receptor binding site in enteroviruses (*Picornaviridae*) in a location shielded from the immune system. CrPV also does not have a canyon (Tate *et al.*, 1999). This points towards a difference in the receptor binding site in insect-infecting members of the *Picornavirales* compared to mammalian enteroviruses, which may be related to differences in the immune systems of their hosts. In CSBV particles, holes were observed in the middle of each unit of five protomers. A hole with a diameter of 2.0 nm reached into the inside of the capsid (Zhang *et al.*, 2001).

Each protomer contains three distantly related 8-stranded β-barrel domains described as coat protein modules (Le Gall *et al.*, 2008). Each mature capsid protein (VP1-VP3), derived by proteolytic processing of the polyprotein, has one of these domains. The hypothesis is that the three large capsid protein domains originated from a triplicated ancestral capsid protein gene. The individual domains evolved further and proteolytic cleavage sites evolved in the interdomain regions, as was proposed for *Tobacco ringspot virus* (*Comoviridae,* genus *Nepovirus*) (Chandrasekar and Johnson, 1998). Due to this triplication capsid protein sequences are less suitable for phylogenetic analysis.

For PnV and VDV-1 nucleotide sequence information was combined with N-terminal protein sequencing data for capsid proteins, allowing for demarcation of the protease cleavage sites and therefore the borders of the individual capsid proteins (Wu *et al.*, 2002a; Lanzi *et al.*, 2006; Ongus, 2006).

The sizes of the capsid proteins vary among iflaviruses. The three major capsid proteins of SBV are 31, 28 and 26 kDa in size (Bailey, 1976). For IFV the capsid proteins have sizes of 35, 33 and 31 kDa (Hashimoto and Kawase, 1983; Choi *et al.*, 1992). In VDV-1, VP1 is 46 kDa and VP2 is 32 kDa. Three bands ranging form 17–25 kDa are also detected with an anti-capsid antibody (Ongus *et al.*, 2004). The capsid proteins in the related DWV are 44, 32 and 28 kDa in size (Lanzi *et al.*, 2006). The smaller less prominent structural protein described as VP4 is unrelated to VP1-VP3 and has a size of 11.6 kDa in IFV (Hashimoto and Kawase, 1983). VP4 is probably located on the inside of the capsid (Zhang *et al.*, 2001). The exact position of the VP4 protein in the iflavirus capsids cannot be extrapolated from related viruses due to the variation in VP4 conformations observed in the various virion structures and awaits further structural analysis. The exact order of polyprotein processing has not been determined for iflaviruses, but the final step is most likely the cleavage between VP3 and VP4. In picornaviruses, where a different order of VP proteins is found in the polyprotein (VP4, VP2, VP3 and VP1 from N- to C- terminus) (Fig. 10.2), the final maturation cleavage occurs between VP4 and VP2 (Hogle, 2002).

The non-structural proteins

The non-structural proteins appear in the distal part of the iflavirus polyprotein. They appear in the order helicase, protease and RdRp (Fig. 10.2), as is the case in all members of the order *Picornavirales* (Le Gall *et al.*, 2008).

The helicase domain corresponds to the picornavirus 2C domain. It belongs to the helicase superfamily III (Gorbalenya *et al.*, 1990; Kadare and Haenni, 1997; Le Gall *et al.*, 2008), which is characterized by three sequence motifs A, B and C. Motif A in iflavirus helicases has the canonical sequence GxxGxGK(S/T), motif B is characterized by $(Y/F/W)_2X_5QX_5(Q/D)D$. Motif C is less well conserved. It is anticipated that domain A and B are involved in NTP-binding, as in the membrane-associated helicase of poliovirus (official name *Human enterovirus C*, HEV-C) (Mirzayan and Wimmer, 1992, 1994).

The iflavirus protease resembles the picornavirus 3C protease. 3C proteases have a chymotrypsin-like fold, but with a cysteine instead of a serine in the catalytic site (Allaire *et al.*, 1994; Matthews *et al.*, 1994). In the iflavirus pro-domain the catalytic cysteine is absolutely conserved and is followed 10–15 amino acids downstream by a conserved GXH motif. For enteroviruses it was shown that the histidine in this motif is crucial for enzymatic activity (Cheah *et al.*, 1990; Miyashita *et al.*, 1995). It is likely to be involved in substrate recognition as shown for hepatitis A virus (Allaire *et al.*, 1994). Iflavirus 3C needs to be active within the polyprotein to release itself by intermolecular (*trans*) cleavage, as in picornaviruses (Allaire *et al.*, 1994; Matthews *et al.*, 1994; Schultheiss *et al.*, 1995) in order to facilitate cleavage of the rest of the polyprotein. In DWV the VP3 cleavage site ($AIPE_{901}GEES$) follows the classic pattern for viral 3C proteases that cut after either glutamine (Q) or glutamic acid (E) (Lanzi *et al.*, 2006). The cleavage also occurs downstream of a Q residue in IFV and often between Q and Glycine (G) (Isawa *et al.*, 1998). More variation in cleavage sites between individual capsid proteins is reported for PnV and VDV-1 (Wu *et al.*, 2002a; Ongus *et al.*, 2004).

The RdRp domain belongs to RNA polymerase superfamily I (Kamer and Argos, 1984), which is characterized by three motifs: motif I (KxE) (KD(E/T) in iflaviruses), motif V (SGxxxTxxxN(S/T)), and motif VI (YGDD) (MGDD in IFV), which is located in an hydrophobic context. The catalytic centre probably involves motifs V and VI, as in the virus causing poliomyelitis (HEV-C) (Hansen *et al.*, 1997).

The GDD motif together with the poly(A) tail is a useful target for primers in RT-PCR, especially since the RdRp sequence is located proximal to the 3′ end of the virus. This knowledge can be applied for instance to obtain the first piece of sequence information for a new iflavirus (Ongus *et al.*, 2004).

Phylogenetic relationships among iflaviruses

Among the viruses in the order *Picornavirales* the sequence similarities are the highest in the polyprotein regions encoding the non-structural proteins, which have key roles in replication and processing. Therefore, these regions are particular useful for phylogenetic analysis. The RdRp sequences are most often used for this purpose. Capsid proteins are less suitable due to the putative amplification of an ancestor capsid protein unit in the RNA genome followed by subsequent modifications (See above). Since new sequences are published regularly, a new tree was constructed with all sequenced iflavirus RdRps (Fig. 10.4). Sequences with similarity to iflaviruses derived from a transcript sequencing project of a laboratory colony of the housefly *Musca domestica* (D. Boucias, personal communication) and from expressed sequence tag (EST) sequences obtained from salivary glands of the tsetse fly *Glossina morsitans morsitans* (Ribeiro *et al.*, 2009) were also included in this analysis. CSBV was not included in this phylogeny, since only the capsid protein region has been sequenced (Zhang *et al.*, 2001). Based on capsid homology CSBV most likely groups with SBV, a Japanese isolate.

Clearly visible is the segregation of ifla- and dicistroviruses, which together with their distinct genome structures, justifies their taxonomical classification into separate virus families. Iflaviruses do not group with vertebrate picornaviruses either, despite similarities in genome organization. From the analysis we can see that iflaviruses infecting insects in the same order, for instance IFV, PnV and EoV, all infecting lepidopteran insects, do not form a single clade.

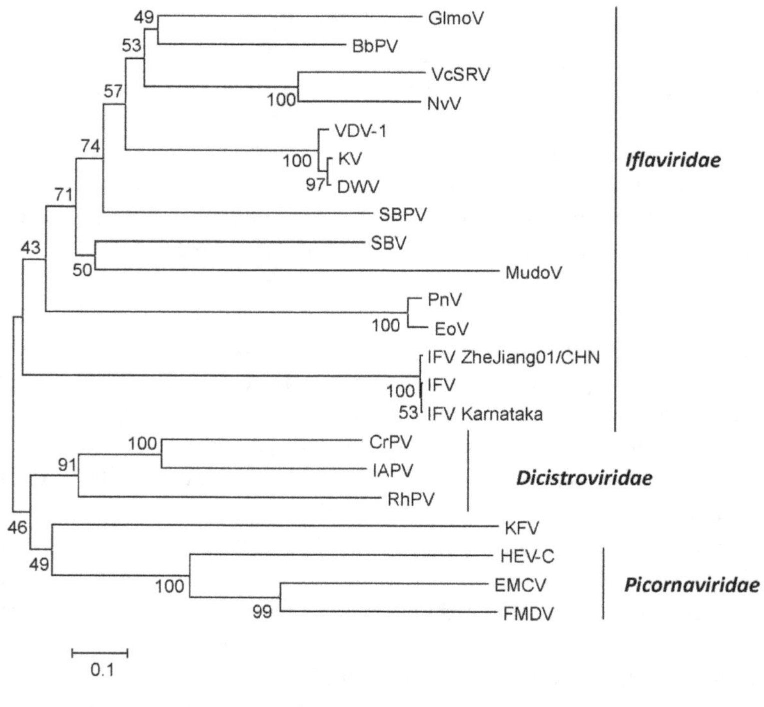

Figure 10.4 Phylogenetic relationships between iflaviruses, dicistroviruses, and picornaviruses. Phylogenetic analyses were conducted using MEGA version 3.1. Amino acid sequences were aligned for the RdRp region using the ClustalW module and phylogenetic relationships and evolutionary distances were calculated using the neighbour-joining method with pairwise deletion of gaps and uniform mutation rates among sites. Bootstrap values are indicated and based on 2000 replicates. The distance marker indicates 0.1 substitutions per amino acid. RdRp sequences extracted from GenBank are: BbPV [ABP57198], CrPV [NP_647481], DWV [AAP49283], EoV [NP_919029], SBPV [ABS84820], IAPV [YP001040002], IFV [NC_003781], IFV-Karnataka [EF422866], IFV-ZheJiang/CHN [ACH57393], KV [YP_015696], KFV [YP_415507], NvV [ACN94442], PnV [NP_277061], RhPV [NP_046155], SBV [AAL79021], VcSRV [AY534885], VDV-1 [AY251269], *Encephalomyocarditis virus* (EMCV [ABC25550]), *Human enterovirus C* (HEP-C [CAA24465]), and *Foot and mouth disease virus* (FMDV [AY304994]). For full names of other viruses see Table 10.1.

In a similar way, the genetically related DWV/VDV-1/KV group is separated from SBV, and also from SBPV, although all these viruses are found in honey bees. The two iflavirus sequences found in dipteran insects, tentatively addressed as GlmoV and MudoV, are found in distant positions as well. These examples suggest that iflaviruses did not follow the same evolutionary path as their insect hosts. This is in contrast to for instance baculoviruses, where a much more evident co-evolution between host and virus is observed (Herniou and Jehle, 2007).

To be able to say whether iflaviruses and their host insects followed a different evolutionary path we need to know more about the host range of iflaviruses and whether they may have switched hosts at some point in time. Host range studies have hardly been performed for iflaviruses, which makes it difficult to say whether the organism from which the virus was purified is the original host in which it evolved. DWV has also been reported in crippled bumble bees (Genersch *et al.*, 2006). VDV-1 was first isolated from mites, but shows close similarity to DWV of bees and replicates in mites and bees (Ongus *et al.*, 2004; Ongus, 2006). A similar 'Which organism is the original host for the virus?' issue might apply to VcSRC and NdV (Reineke and Asgari, 2005; Oliveira *et al.*, 2009), isolated from wasps parasitizing on moths and flies, respectively. The fact though that these viruses group together in the phylogenetic analysis (Fig. 10.4) may point to the parasitic wasp as the primary host. It is not known whether these iflaviruses can replicate in the parasitized insects. It may as well be that within the *Iflaviridae* family, there are in fact separate evolutionary lineages and not just one genus, and that within those lineages they do follow the evolution of their host more strictly. This can not be resolved without the availability of sequence information of many more iflavirus genomes.

RdRp sequences are available for IFV strains from three different geographical locations and these form a single clade in the phylogenetic tree (Fig. 10.4). Their RdRps have a variation of 1% in nucleotide sequence, but an almost identical amino acid composition. Between the two completely sequenced IFV strains, 46 polymorphisms are found genome wide. DWV, VDV-1 and KV also group together on a branch. Surveys of variation in genomes have been performed for DWV and VDV-1 in mite samples collected from beehives in distinct European locations. This study showed similarities of 96–99% between the various isolates in selected domains in the 5′ UTR (Ongus, 2006), one of the most variable parts of the genome (Lanzi *et al.*, 2006). Even between DWV samples isolated from individual brood cells within Chinese hives polymorphisms were observed (Forsgren *et al.*, 2009). The discovery of a virus in aggressive bees with a sequence similarity closer to DWV than to KV (Terio *et al.*, 2008) is another illustration of the genome flexibility of iflaviruses. The IFV and DWV examples clearly show that the quasi species concept (Domingo, 1992) is applicable to iflaviruses. The iflavirus genome should therefore not be seen as a stable entity and the 'species' as a collection of variants. VDV-1, although more distantly related, clearly belongs to the DWV/KV group as well.

Kelp fly virus (KFV), isolated from *Chaetocoelopa sydneyensis*, remains to be a non-classified virus. The KFV genome encodes one large polyprotein with an inhibitor of apoptosis (IAP) preceding the capsid and non-structural protein segments (Hartley *et al.*, 2005). The encoded IAP protein has homology with baculovirus IAP proteins (Clem, 2007). KFV is probably not a recent recombinant virus originating from an iflavirus and an IAP sequence. In the phylogeny KFV is positioned outside both the *Iflaviridae* and *Dicistroviridae* families (Fig. 10.4).

Virus replication

Iflavirus replication and accumulation occurs in the cytoplasm, as has been concluded from the incorporation of RNA precursors in the form of [^{3}H]uridine in the cytoplasm of IFV-infected cells (Watanabe 1991), from immune fluorescence studies (Inoue and Ayuzawa, 1972), and from the densely packed aggregates of progeny virus observed in the cytoplasm (Fig. 10.3B) of for instance IFV (Watanabe, 1991) or VDV-1 infected cells (Zhang *et al.*, 2007). When SBV infected honey bee pupae were examined by electron microscopy, an association between virus capsids and virus-induced vesicular structures was observed (Dall, 1987). These structures were located in the cytoplasm in proximity of the nucleus. Empty virus-like particles were also observed in the nucleus. In VDV-1 infected cells virus particles were also seen in association with membranes or within vesicular structures (Fig. 10.3C and D), indicating that membranes play an important role in virus replication. It is known that cellular membrane structures also play a crucial role in picornaviruses RNA synthesis (Schlegel *et al.*, 1996).

When RNA was purified from IFV particles and labelled with N-succinimidyl 3- (4-hydroxy, 5-[^{125}I]iodophenyl) propionate, an 11.5 kDa protein was detected by SDS-PAGE. This protein co-sedimented with the viral RNA in sucrose gradients, and could not be removed by treatment with heat, SDS or DMSO (Hashimoto *et al.*, 1986). Proteinase K treatment removed the labelling from the RNA, but this did not affect *in vitro* translation of the RNA in rabbit reticulocyte lysates. The 11.5 kDa protein did not react with antiserum against the viral capsid proteins. Together these results strongly point towards the presence of a VPg protein covalently linked to the viral genome. This would imply that the iflavirus VPg, at least for IFV, is larger than in other members of the *Picornavirales*, where it is smaller than 5 kDa (Le Gall *et al.*, 2008). The 5′ end of the negative-sense strand, which serves as replication intermediate, most likely also has this covalently attached terminal VPg protein. The purpose of this protein is most likely to serve as a primer for RNA synthesis as in picornaviruses. The location of the VPg module in the polyprotein is unresolved, but with current mass-spectrometry based proteomic analysis methods, this should not be too difficult to tackle. In picornaviruses, the VPg sequence is located within the non-structural protein area as region 3B between the helicase and protease domains (Fig. 10.2).

The presence of negative strand iflavirus RNA in cells is a hallmark of replication of these positive-strand RNA viruses. For DWV and VDV-1 for example, negative strand RNA has been identified in honey bees, but also in bee-parasitic varroa mites by negative strand-specific RT-PCR reactions (Ongus *et al.*, 2004). The presence of negative strand RNA in mites (Ongus *et al.*, 2004; Yue and Genersch, 2005) clearly points towards replication of DWV and VDV-1 in mites, as do the high levels of virus (Bowen-Walker *et al.*, 1999; Gisder *et al.*, 2009) and the densely packaged virions observed in mite tissues (Zhang *et al.*, 2007) (Fig. 10.3B). However, in another study, which used immunohistochemistry to detect DWV in mites, the virus was only detected in the gut lumen (Santillan-Galicia *et al.*, 2008). The reason for this discrepancy is not entirely clear, but may result from differences in sensitivity between the methods used (RT-PCR and electron microscopy versus immunohistochemistry), the number of mites analysed, or the time post infection at which the analysis was performed.

Studies of iflavirus replication have not been performed extensively, probably due to a limited number of cell lines that support iflavirus replication. For *Perinu nuda* the cell line NTU-PN-HH is available (Wang *et al.*, 1999). The cell line NTU-Pn-HF was shown to be persistently infected with PnPV (Wu *et al.*, 2002b). *Bombyx mori* cell lines could be a useful tool to study IFV replication, but whether these cells support replication has not been reported.

Translation mechanism

Mammalian cells, including rabbit reticulocyte lysates, contain a mechanism that removes VPg prior to translation (Ambros and Baltimore, 1980; Ambros *et al.*, 1978). IFV-RNA also looses the covalently attached protein in rabbit reticulocyte lysates (Hashimoto *et al.*, 1986). Whether insect cells also contain such a mechanism to remove VPg prior to translation has not been analysed.

The sequenced iflavirus genomes can be grouped based on the length of their 5′ UTRs (Table 10.2). SBV and IFV have very short 5′ UTRs (156–178 nt) with hardly any predicted secondary structure. PnV and EoV have 5′ UTRs ranging from 390 to 473 nt, which are homologous to each other and for which a mostly non-branched secondary structure was predicted (Lu *et al.*, 2006; Ongus *et al.*, 2006). PnV and EoV also group together in phylogenetic trees (Fig. 10.4). The bee viruses DWV, KV and VDV-1 have much longer 5′ UTRs, which are also homologous to each other. For DWV and VDV-1 complex-branched secondary structures have been predicted for this region (Ongus *et al.*,

2006). The existence of co-variation in the stem structures predicted for the 5′ UTRs of DWV and VDV-1 points to an important biological function of these stem–loop structures (Ongus *et al.*, 2006). Co-variation means that a point mutation observed at one site is accompanied by a second mutation on the other site of the stem that restores basepairing and in that way preserves secondary structure. Co-variation is generally excepted as a measure for the likelihood of these predicted secondary structures (Witwer *et al.*, 2001). The predicted three-branched or clover-leaf structure in DWV/VDV-1/KV (Ongus *et al.*, 2006) is also seen in members of the *Enterovirus* genus within the mammalian picornaviruses.

The presence of a functional internal ribosomal entry site (IRES) in the 5′ UTR, which would allow the translation of the polyprotein in a cap-independent manner has been shown for both VDV-1 and EoV, using bicistronic plasmid constructs (Lu *et al.*, 2006; Ongus *et al.*, 2006; Lu *et al.*, 2007). For EoV mutational analysis allowed the identification of functional domains in the IRES element and the functional core was located in the third and fourth stem loop (out of six) (Lu *et al.*, 2006). The EoV IRES was not only active in insect cells but also showed activity in mammalian cells. In plant cells it was not functional (Lu *et al.*, 2007). Due to the high degree of similarity in sequence and predicted secondary structure of the 5′ UTR for EoV and PnV on one hand and for VDV-1, DWV and KV on the other hand, functional IRES elements are also predicted for the 5′ UTRs of these related viruses. The presence of a functional IRES element in the very short, unstructured IFV and SBV 5′ UTR sequences seems unlikely, but awaits further analysis. The interaction of host translation initiation factors with these IRES elements has not been studied.

Since iflaviruses make only one polyprotein, they cannot regulate protein subsets directly at the translational level. The bicistronic genome organization of dicistroviruses, on the other hand, facilitates separate regulation of the genes for the structural and non-structural protein subsets at the translational level. In order to translate these two ORFs into polyproteins dicistroviruses have both a 5′ IRES and an intergenic IRES element (Fig. 10.2). These two IRES elements vary from each other in sequence, structure and mechanism of action, hence allowing individual regulation of expression levels for structural and non-structural proteins (see Chapter 9). In transient assays though, the *Rhopalosiphum padi* virus (RhPV) 5′ and intergenic IRES elements were equally efficient (Domier and McCoppin, 2003).

For Foot and mouth disease virus (FMDV), genus *Aphthovirus*, it has been reported that both the L protein and the 3C protease cleave translation initiation factor 4G (eIF4G) (Hinton *et al.*, 2002; Strong and Belsham, 2004). The 2A proteases of enteroviruses, also cleave eIF4G (Kraussiich *et al.*, 1987; Kerekatte *et al.*, 1999). This cleavage prevents initiation of cap-dependent, host mRNA translation and as such induces shut off of host protein synthesis. It would be interesting to determine whether host shut off also occurs in iflavirus infections. The presence of an IRES element, as shown for VDV-1 and EoV (Ongus *et al.*, 2006; Lu *et al.*, 2007), may suggest so. Important for the success of such studies is the availability of permissive cell lines. If iflaviruses induce host translational arrest, the next step is to determine whether eIF4G cleavage occurs and whether the viral 3C-like protease is responsible for this cleavage. Whether the L protein found in some iflaviruses could mediate eIF4G cleavage is not clear, since the L protein of iflaviruses is not homologous to that of aphthoviruses.

The codon usage of six honey bee-infecting viruses, including the iflaviruses VDV-1, SBV and KV as well as three dicistroviruses, was compared with that of their host *A. mellifera*, and three other insects with completely sequenced genomes, the vinegar fly (*Drosophila melanogaster*), the malaria mosquito (*Anopheles gambiae*), and the red flour beetle (*Tribolium castaneum*) (Chantawannakul and Cutler, 2008). The codon usage among the six bee viruses under study was remarkably similar. A clear relationship was also seen with the codon usage of bees, which was biased in the same directions. The red flour beetle was more distantly related in terms of codon usage, while the dipteran insects clearly showed a different pattern. The dicistroviruses followed the codon abundance of the honey bee more closely than the iflaviruses, and it was speculated that this may be one of the reasons for the high virulence of bee dicistroviruses. The idea is that abundant codons are more easily translated. As a consequence, a virus that follows the codon abundance patterns of its host would have a replication advantage (Chantawannakul and Cutler, 2008). The similarity in codon usage observed here cannot be explained by relatedness-by-descent, since in phylogenetic trees these viruses do not group together, but is a sign of convergent evolution of these small RNA viruses towards their host, the honey bee. The fact that VDV-1, which was originally isolated from varroa mites (Ongus *et al.*, 2004), follows the codon usage of the bee, may be an extra indication that this virus was originally a bee virus and a variant of DWV.

Pathology

The symptoms of sacbrood disease in honey bees (see Introduction) lead to a failure to pupate (Gochnauer, 1990; Bailey and Ball, 1991). IFV primarily infects goblet cells of the midgut epithelium of *B. mori* larvae as was shown with immunofluorescent labelling (Inoue and Ayuzawa, 1972). Electron micrographs (Watanabe 1991) revealed that in the early stages of infection membrane-bound vesicles, 100–300 nm in diameter, filled with filamentous material and surrounded by virus particles, were formed in the goblet cells. Later in infection the cells became densely packed with virus particles and vesicles, organelles disappeared and goblet cells were finally discharged into the gut lumen. Infection of neighbouring columnar cells appeared to occur through phagocytosis, explaining the immunofluorescence observed at later stages in these cells.

DWV causes wing deformities, shortened abdomens and miscolouring of adult bees, as was demonstrated by DWV injection into young pupae (Bailey and Ball, 1991). DWV infection spreads in the whole body of the bee, including queen ovaries, queen fat body and drone seminal vesicles (Fievet *et al.*, 2006). However, the spatial distribution in bees with asymptomatic DWV-infections and DWV-infected crippled bees (Fig. 10.1) is different. Head, thorax and abdomen were all strongly positive in RT-PCR reactions for DWV RNA in crippled bees, but DWV RNA was rarely found in the head of asymptomatic bees (Yue and Genersch, 2005). This result seems to contrast with VDV-1 infections in Dutch bee hives, where symptoms were occasionally observed, but where both positive and negative strand RNA of both VDV-1 and DWV was regularly detected in head, thorax and abdomen of adult bees (Ongus, 2006). KV, which varies only slightly from DWV in sequence, was found in brains of aggressive bees (Fujiyuki *et al.*, 2004; Fujiyuki *et al.*, 2005).

Iflaviruses may reside at low levels in bees without showing clear signs of infection. An example follows from a survey in Devon, UK, using RT-PCR techniques to detect various bee viruses despite the absence of clear symptoms (Baker and Schroeder, 2008). In nearly all apiaries DWV was detected, often in combination with other ifla- or dicistroviruses. Mixed virus infections are often seen in bee hives and even individual bees can have mixed infections. DWV and VDV-1 have been found within single bees in Dutch bee hives (Ongus, 2006). Individual bees can also have mixed infections of ifla- and dicistroviruses (Chen *et al.*, 2004) and in the study referred to 93% of the queens carried multiple viruses (Chen *et al.*, 2005).

The pathogenicity of DWV is related to the presence of *V. destructor* mites during the pupal stages (Bailey and Ball, 1991; Bowen-Walker *et al.*, 1999). In the absence of *V. destructor*, DWV infection gives covert infections without visible symptoms. Whether crippled bees develop or an asymptomatic infection occurs is correlated not only with the presence of *V. destructor* mites in the pupal stage of the bees, but also with the ability of the virus to replicate in the mites (Yue and Genersch, 2005; Gisder *et al.*, 2009). The DWV titre in the parasitizing mites is crucial and 10^{10} to 10^{12} genome copies are required per mite for overt infections in bees. Mites with less than 10^{8} viral genome equivalents induce asymptomatic infections (Gisder *et al.*, 2009). An additional effect of these parasitic mites is that the constant feeding of the nymphs and adult female mites makes the bee larvae and pupae more susceptible to pathogens in general, due to exhaustion and suppression of the insects immune system (Gregory *et al.*, 2005; Yang and Cox-Foster, 2005).

For several iflaviruses the pathology is not clear as these viruses have been isolated from dead insects that were simultaneously infected with for instance a baculovirus. This holds true for EoV isolated from tea loopers that probably succumbed to a granulovirus (Wang *et al.*, 2004). PnV was isolated together with PnNPV from ficus transparent wing moths showing flacherie (Wang *et al.*, 1999), a symptom said to be caused by the simultaneous infection with these two viruses. IFV is often observed together with a densovirus (Tanada and Kaya, 1993).

Transmission

Covert DWV infections can be vertically transmitted from queens to both worker and drone offspring, as was shown for virgin queens that laid DWV-positive unfertilized eggs, giving rise to DWV-positive drones and for DWV-negative queens inseminated with DWV-positive semen, resulting in DWV-positive workers (Yue *et al.*, 2007). Viral sequences in offspring can therefore originate from either infected drones or infected queens. Further proof for vertical transmission was the visualization by *in situ* hybridization of DWV RNA in the ovary of a DWV-negative queen inseminated with semen from a DWV-positive drone. Detection of DWV in the ovaries of artificially inseminated virgin queens followed by subsequent vertical transmission was also demonstrated (de Miranda and Fries, 2008). The ability to cause symptomless infections and the possibility for vertical transmission allows the virus to persist in the population.

The transmission of DWV by varroa mites and the relation with pathology has been described above. Recently, DWV was also found in bee parasitic mites belonging to the genus *Tropilaelaps*. Although originally known from the Asian giant honey bee *Apis dorsata*, these mites now also use brood of the Western honey bee *A. mellifera*, introduced into Southeast Asia, to feed on. DWV was observed with quantitative RT-PCR in *Tropilaelaps mercedesae* mites and the infested honey bee brood and a quantitative correlation between mite-infested pupae and DWV infections was observed (Forsgren *et al.*, 2009), suggesting that these mites also serve as vectors for the transmission of iflaviruses.

Infectivity tests showed that 10^7–10^8 SBV particles are needed to cause disease in 2-day-old bee larvae, while the amount of SBV particles released per larva ranges between 10^{11} and 10^{13} (Bailey, 1969). This corresponds to approximately 1 mg of progeny virus per larva, said to be enough to spread disease in 1000 bee colonies. However, the infectivity of the virus reduces quickly and the dried scales hardly contain any infectious virus (Hitchcock, 1966). Young workers with the task of removing sacbrood diseased/killed larvae are thought to play a major role in spreading the disease by ingesting fluid from these larvae and transmitting the virus during nursing to other larvae. This is in line with the accumulation of SBV in the hypopharyngeal glands of these worker bees (Bailey and Ball, 1991). Horizontal transmission is also very likely among adult bees and from adult workers to larvae through contaminated food resources (brood food, honey, pollen and royal jelly) (Shen *et al.*, 2005). Adult bees can also artificially be infected by injection (Lee and Furgala, 1967). Transmission from year to year is mostly caused by adult bees in which the virus multiplies without showing obvious symptoms (Bailey and Ball, 1991). Larvae can become infected until they are about 4 days old (Hitchcock, 1966), a time point after which non-infected bee larvae pupate. When larvae become infected at an age older than 2 days, however, they may survive the infection and become symptomless infected adults (Anderson and Gibbs, 1989), which transmit the disease to the next year's generation. SBV may also use a transovarial transmission route as SBV RNA has been detected in queens and eggs (Shen *et al.*, 2005). Transmission may also be achieved by varroa mites, since SBV has been detected in mites and mite saliva (Shen *et al.*, 2005).

As the goblet cells of the midgut epithelium are the primary targets for IFV, virus uptake most likely occurs via food contamination and this may also be the case for PnV and EoV, which also infect lepidopteran insects.

Concluding remarks

The properties of iflaviruses position them firmly within the order *Picornavirales*. More and more iflavirus sequences are being detected, but molecular, virological and biological studies are restricted to IFV, SBV and DWV, VDV, and PnV, with varying level of detail. The main questions remaining relate to the host range of iflaviruses, the nature of the VPg protein, the effect of virus infection on the translation machinery of the host, the interaction of host factors with the IRES and the role of the leader peptide. So far we have seen iflaviruses infecting insects of the orders Hymenoptera, Hemiptera and Lepidoptera. RNA-derived sequences from two fly species clearly point to their presence in dipteran insects as well. With the ongoing effort to discover viruses in species belonging to other invertebrate orders and in other ecological niches, the known spectrum of iflavirus hosts is likely to expand. Especially whole insect-based EST sequencing projects and bulk sequencing of nucleotides in soil, sea or fresh water samples may prove worthwhile in this aspect (Culley *et al.*, 2003) and are likely to produce a collection of new (ifla-) virus sequences. It is my guess that in the future when more sequence data become available we will be able to distinguish subgroups of iflaviruses allowing classification into different genera. From the phylogenetic studies performed so far, the borders of these iflavirus genera probably will not closely follow the taxonomy of their hosts. It will be interesting to see if we can finally come to an iflavirus taxonomy that follows the host at a higher order level (classes), e.g. Insecta-infecting iflaviruses separated from crustacean-infecting iflaviruses. However, viruses replicating in very distinct organisms (such as bees and mites) may complicate future taxonomic grouping, and analysis of codon usage may be needed to understand which organism is the primary host.

Acknowledgments

Elke Genersch and Michael Traynor are acknowledged for the photographs of a DWV infected crippled bee and SBV-infected bee larvae (Fig. 10.1). Dick Peters is acknowledged for supplying the photographs for Fig. 10.3. I am grateful to Frank van Kuppeveld, Elke Genersch and Dick Peters for critical reading of this manuscript prior to submission. Sarap Aksoy and Drion Boucias are acknowledged for sharing the tsetse fly and housefly RNA sequences prior to publication, respectively.

References

Aizawa, K., Furuta, Y., Kurata, K., and Sato, F. (1964). On the etiologic agent of the infectious flacherie of the silkworm, *Bombyx mori* (Linnaeus). Bull. Sericult. Exp. Sta. *19*, 223–240.

Allaire, M., Chernala, M.M., Malcolm, B.A., and James, M.N.G. (1994). Picornaviral 3C cysteine proteinases have a fold similar to chymotrypsin-like serine proteinases. Nature *369*, 72–76.

Ambros, V., and Baltimore, D. (1980). Purification and properties of a HeLa cell enzyme able to remove the 5′-terminal protein from poliovirus RNA. J. Biol. Chem. *255*, 6739–6744.

Ambros, V., Pettersson, R.F., and Baltimore, D. (1978). An enzymatic activity in uninfected cells that cleaves the linkage between poliovirion RNA and the 5′ terminal protein. Cell *15*, 1439–1446.

Anderson, D., and Gibbs, A. (1989). Transpuparial transmission of Kashmir bee virus and Sacbrood virus in the honey bee (*Apis melifera*). Ann. Appl. Biol. *114*, 1–7.

Bailey, L. (1969). The multiplication and spread of Sacbrood virus of bees. Ann. Appl. Biol. *63*, 483–491.

Bailey, L. (1976). Viruses attacking the honey bee. Adv. Virus Res. *20*, 271–304.

Bailey, L., and Ball, B. (1991). Honey Bee Pathology., 2nd edn (London: Academic Press).

Bailey, L., Gibbs, A.J., and Woods, R.D. (1964). Sacbrood virus of the larval honey bee (*Apis mellifera* Linnaeus). Virology 23, 425–429.

Baker, A., and Schroeder, D. (2008). Occurrence and genetic analysis of picorna-like viruses infecting worker bees of *Apis mellifera* L. populations in Devon, South West England. J. Invertebr. Pathol. *98*, 239–242.

Blanchard, P., Schurr, F., Celle, O., Cougoule, N., Drajnudel, P., Thiery, R., Faucon, J.P., and Ribiere, M. (2008). First detection of Israeli acute paralysis virus (IAPV) in France, a dicistrovirus affecting honeybees (*Apis mellifera*). J. Invertebr. Pathol. *99*, 348–350.

Bowen-Walker, P.L., Martin, S.J., and Gunn, A. (1999). The transmission of Deformed wing virus between honeybees (*Apis mellifera* L.) by the ectoparasitic mite *Varroa jacobsoni* Oud. J. Invertebr. Pathol. *73*, 101–106.

Bradbear, N. (1988). World distribution of major honey bee diseases and pests. Bee World *69*, 15–39.

Chandrasekar, V., and Johnson, J.E. (1998). The structure of Tobacco ringspot virus: A link in the evolution of icosahedral capsids in the picornavirus superfamily. Structure *6*, 157–171.

Chantawannakul, P., and Cutler, R.W. (2008). Convergent host-parasite codon usage between honeybee and bee associated viral genomes. J. Invertebr. Pathol. *98*, 206–210.

Cheah, K.C., Leong, L.E.C., and Porter, A.G. (1990). Site-directed mutagenesis suggests close functional relationship between a human rhinovirus 3C cysteine protease and cellular trypsin-like serine proteases. J. Biol. Chem. *265*, 7180–7187.

Chen, Y., Zhao, Y., Hammond, J., Hsu, H.T., Evans, J., and Feldlaufer, M. (2004). Multiple virus infections in the honey bee and genome divergence of honey bee viruses. J. Invertebr. Pathol. *87*, 84–93.

Chen, Y., Pettis, J.S., and Feldlaufer, M.F. (2005). Detection of multiple viruses in queens of the honey bee *Apis mellifera* L. J. Invertebr. Pathol. *90*, 118–121.

Chen, Y.P., and Siede, R. (2007). Honey bee viruses. Adv. Virus. Res. *70*, 33–80.

Chen, Z., Stauffacher, C., Li, Y., Schmidt, T., Bomu, W., Kamer, G., Shanks, M., Lomonossoff, G., and Johnson, J.E. (1989). Protein–RNA interactions in an icosahedral virus at 3.0 A resolution. Science *245*, 154–159.

Choi, H., Sasaki, T., Tomita, T., Kobayashi, M., and Kawase, S. (1992). Processing of structural polypeptides of Infectious flacherie virus of the silkworm, *Bombyx mori*: VP1 and VP4 are derived from VP0. J. Invertebr. Pathol. *60*, 113–116.

Christian, P., Carstens, E., Domier, L., Johnson, J., Johnson, K.., Nakashima, N., Scotti, P., and van der Wilk, F. (2005). *Dicistroviridae*, In Virus Taxonomy: VIIIth Report of the International Committee on Taxonomy of Viruses, C.M. Fauquet, M.A. Mayo, J. Maniloff, U. Desselberger, and L.A. Ball eds. (London: Elsevier Academic Press), pp. 783–788.

Clem, R.J. (2007). Baculoviruses and apoptosis: a diversity of genes and responses. Curr. Drug Targets *8*, 1069–1074.

Culley, A.I., Lang, A.S., and Suttle, C.A (2003). High diversity of unknown picorna-like viruses in the sea. Nature *424*, 1054–1057.

Dall, D.J. (1987). Intracellular structures associated with the multiplication of Sacbrood virus. J. Invertebr. Pathol. *50*, 261–268.

de Miranda, J.R., and Fries, I. (2008). Venereal and vertical transmission of deformed wing virus in honeybees (*Apis mellifera* L.). J. Invertebr. Pathol. *98*, 184–189.

de Miranda, J.R., Ball, B.V., Stoltz, D.B., and Paxton, R.A. (2007). Honey bee slow paralysis virus polyprotein gene, complete cds. (GenBank Accesion [EU035616]).

Domier, L.L., and McCoppin, N.K. (2003). *In vivo* activity of *Rhopalosiphum padi* virus internal ribosome entry sites. J. Gen. Virol. *84*, 415–419.

Domingo, E. (1992). Genetic variation and quasi-species. Curr. Opin. Genet. Dev. *2*, 61–63.

Fauquet, C.M., Mayo, M.A., Maniloff, J., Desselberger, U., Ball, L.A., and eds. (2005). Virus Taxonomy: VIIIth Report of the International Committee on Taxonomy of Viruses, (London: Elsevier Academic Press).

Fievet, J., Tentcheva, D., Gauthier, L., De Miranda, J., Cousserans, F., Colin, M.E., and Bergoin, M. (2006). Localization of Deformed wing virus infection in queen and drone *Apis mellifera* L. Virol. J. *3:16*.

Forsgren, E., De Miranda, J.R., Isaksson, M., Wei, S., and Fries, I. (2009). Deformed wing virus associated with *Tropilaelaps mercedesae* infesting European honey bees (*Apis mellifera*). Exp. Appl. Acarol. *47*, 87–97.

Fujiyuki, T., Takeuchi, H., Ono, M., Ohka, S., Sasaki, T., Nomoto, A., and Kubo, T. (2004). Novel insect picorna-like virus identified in the brains of aggressive worker honeybees. J. Virol. *78*, 1093–1100.

Fujiyuki, T., Takeuchi, H., Ono, M., Ohka, S., Sasaki, T., Nomoto, A., and Kubo, T. (2005). Kakugo virus from brains of aggressive worker honeybees. Adv. Virus Res. *65*, 1–27.

Genersch, E., Yue, C., Fries, I., and de Miranda, J.R. (2006). Detection of Deformed wing virus, a honey bee viral pathogen, in bumble bees (*Bombus terrestris* and *Bombus pascuorum*) with wing deformities. J. Invertebr. Pathol. *91*, 61–63.

Ghosh, R.C., Ball, B.V., Willcocks, M.M., and Carter, M.J. (1999). The nucleotide sequence of Sacbrood virus of the honey bee: An insect picorna-like virus. J. Gen. Virol. *80*, 1541–1549.

Gisder, S., Aumeier, P., and Genersch, E. (2009). Deformed wing virus: Replication and viral load in mites (*Varroa destructor*). J. Gen. Virol. *90*, 463–467.

Gochnauer, T. (1990). Viruses, In Honey Bee Pests, Predators, and Diseases, R. Morse, and R. Nowogrodski, eds. (Ithaca: Cornell University Press), pp. 12–26.

Gorbalenya, A.E., Koonin, E.V., and Wolf, Y.I. (1990). A new superfamily of putative NTP-binding domains encoded by genomes of small DNA and RNA viruses. FEBS Lett. *262*, 145–148.

Gregory, P.G., Evans, J.D., Rinderer, T., and De Guzman, L. (2005). Conditional immune-gene suppression of honeybees parasitized by *Varroa* mites. J. Insect Sci. *5*, 1–5.

Hansen, J.L., Long, A.M., and Schultz, S.C. (1997). Structure of the RNA-dependent RNA polymerase of poliovirus. Structure *5*, 1109–1122.

Hartley, C.J., Greenwood, D.R., Gilbert, R.J.C., Masoumi, A., Gordon, K.H.J., Hanzlik, T.N., Fry, E.E., Stuart, D.I., and Scotti, P.D. (2005). Kelp fly virus: A novel group of insect picorna-like viruses as defined by genome sequence analysis and a distinctive virion structure. J. Virol. *79*, 13385–13398.

Hashimoto, Y., and Kawase, S. (1983). Characteristics of structural proteins of Infectious flacherie virus from the silkworm, *Bombyx mori*. J. Invertebr. Pathol. *41*, 68–76.

Hashimoto, Y., Watanabe, A., and Kawase, S. (1984). *In vitro* translation of Infectious flacherie virus RNA in a wheat germ and a rabbit reticulocyte system. Biochim. Biophys. Acta *781*, 76–80.

Hashimoto, Y., Watanabe, A., and Kawase, S. (1986). Evidence for the presence of a genome-linked protein in Infectious flacherie virus. Arch. Virol. *90*, 301–312.

Herniou, E.A., and Jehle, J.A. (2007). Baculovirus phylogeny and evolution. Curr. Drug Targets *8*, 1043–1050.

Himeno, M., Onodera, K., and Tanami, Y. (1974). Properties of flacherie virus of the silkworm, *Bombyx mori*. J. Invertebr. Pathol. 23, 164–171.

Hinton, T.M., Ross-Smith, N., Warner, S., Belsham, G.J., and Crabb, B.S. (2002). Conservation of L and 3C proteinase activities across distantly related aphthoviruses. J. Gen. Virol. *83*, 3111–3121

Hitchcock, J. D. (1966). Transmission of sacbrood disease to individual honey bee larvae. *J. Econ. Ent. 59*, 1154–1156.

Hogle, J.M. (2002). Poliovirus cell entry: Common structural themes in viral cell entry pathways. Ann. Rev. Microbiol. *56*, 677–702.

ICTV (2009). Virus Taxonomy 2009 (International Committee on Taxonomy of Viruses). web site:

http://www.ictvonline.org/virusTaxonomy.asp?version = 2009&bhcp = 1

Inoue, H., and Ayuzawa, C. (1972). Fluorescent antibody study of an infectious flacherie of the silkworm, *Bombyx mori*. J. Invertebr. Pathol. *19*, 227–230.

Isawa, H., Asano, S., Sahara, K., Iizuka, T., and Bando, H. (1998). Analysis of genetic information of an insect picorna-like virus, Infectious flacherie virus of silkworm: Evidence for evolutionary relationships among insect, mammalian and plant picorna(-like) viruses. Arch. Virol. *143*, 127–143.

Johnson, R.M., Evans, J.D., Robinson, G.E., and Berenbaum, M.R. (2009). Changes in transcript abundance relating to colony collapse disorder in honey bees (*Apis mellifera*). Proc. Natl. Acad. Sci. U.S.A. *106*, 14790–14795.

Kadare, G., and Haenni, A.L. (1997). Virus-encoded RNA helicases. J. Virol. *71*, 2583–2590.

Kamer, G., and Argos, P. (1984). Primary structural comparison of RNA-dependent polymerases from plant, animal and bacterial viruses. Nucl. Acids Res. *12*, 7269–7282.

Kerekatte, V., Keiper, B.D., Badorff, C., Cat, A., Knowlton, K.U., and Rhoads, R.E. (1999). Cleavage of poly(A)-binding protein by Coxsackievirus 2A protease *in vitro* and *in vivo*: Another mechanism for host protein synthesis shutoff? J. Virol. 73, 709–717.

Kraussl ich, H.G., Nicklin, M.J.H., Toyoda, H., Etchison, D., and Wimmer, E. (1987). Poliovirus proteinase 2A induces cleavage of eucaryotic initiation factor 4F polypeptide p220. J. Virol. *61*, 2711–2718.

Lang, A.S., Culley, A.I., and Suttle, C.A. (2004). Genome sequence and characterization of a virus (HaRNAV) related to picorna-like viruses that infects the marine toxic bloom-forming alga *Heterosigma akashiwo*. Virology *320*, 206–217.

Lanzi, G., de Miranda, J.R., Boniotti, M.B., Cameron, C.E., Lavazza, A., Capucci, L., Camazine, S.M., and Rossi, C. (2006). Molecular and biological characterization of Deformed wing virus of honeybees (*Apis mellifera* L.). J. Virol. *80*, 4998–5009.

Le Gall, O., Christian, P., Fauquet, C.M., King, A.M., Knowles, N.J., Nakashima, N., Stanway, G., and Gorbalenya, A.E. (2008). Picornavirales, a proposed order of positive-sense single-stranded RNA viruses with a pseudo-T = 3 virion architecture. Arch. Virol. *153*, 715–727.

Lee, P.E., and Furgala, B. (1965). Sacbrood virus: Some morphological features and nucleic acid type. J. Invertebr. Pathol. 7, 502–505.

Lee, P.E., and Furgala, B. (1967). Virus-like particles in adult honey bees (*Apis mellifera* Linnaeus) following injection with sacbrood virus. Virology 32, 11–17.

Li, M.Q., Lu, Q.N., Wu, X.F., Man, N.N., Chen, X.X., Jin, W., and Lu, X.M. (2009). Analysis of RNA-dependent RNA polymerase sequence of Infectious flacherie virus isolated in China and its expression in BmN Cells. Agricult. Sci. China 8, 872–879.

Lu, J., Zhang, J., Wang, X., Jiang, H., Liu, C., and Hu, Y. (2006). *In vitro* and *in vivo* identification of structural and sequence elements in the 5′ untranslated region of *Ectropis obliqua* picorna-like virus required for internal initiation. J. Gen. Virol. *87*, 3667–3677.

Lu, J., Hu, Y., Hu, L., Zong, S., Cai, D., Wang, J., Yu, H., and Zhang, J. (2007). *Ectropis obliqua* picorna-like virus IRES-driven internal initiation of translation in cell systems derived from different origins. J. Gen. Virol. *88*, 2834–2838.

Maori, E., Lavi, S., Mozes-Koch, R., Gantman, Y., Peretz, Y., Edelbaum, O., Tanne, E., and Sela, I. (2007). Isolation and characterization of Israeli acute paralysis virus, a dicistrovirus affecting honeybees in Israel: Evidence for diversity due to intra- and inter-species recombination. J. Gen. Virol. *88*, 3428–3438.

Maori, E., Paldi, N., Shafir, S., Kalev, H., Tsur, E., Glick, E., and Sela, I. (2009). IAPV, a bee-affecting virus associated with colony collapse disorder can be silenced by dsRNA ingestion. Insect Mol. Biol. *18*, 55–60.

Matthews, D.A., Smith, W.W., Ferre, R.A., Condon, B., Budahazi, G., Sisson, W., Villafranca, J.E., Janson, C.A., McElroy, H.E., Gribskov, C.L., and Worland, S. (1994). Structure of Human rhinovirus 3C protease reveals a trypsin-like polypeptide fold, RNA-binding site, and means for cleaving precursor polyprotein. Cell 77, 761–771.

Mirzayan, C., and Wimmer, E. (1992). Genetic analysis of an NTP-binding motif in poliovirus polypeptide 2C. Virology *189*, 547–555.

Mirzayan, C., and Wimmer, E. (1994). Biochemical studies on poliovirus polypeptide 2C: Evidence for ATPase activity. Virology *199*, 176–187.

Miyashita, K., Utsumi, R., Utsumi, T., Komano, T., and Satoh, N. (1995). Mutational analysis of the putative substrate-binding site of 3C proteinase of coxsackievirus B3. Biosci. Biotechnol. Biochem. *59*, 121–122.

Nordström, S., Fries, I., Aarhus, A., Hansen, H., and Korpela, S. (1999). Virus infections in Nordic honey bee colonies with no, low or severe *Varroa jacobsoni* infestations. Apidologie *30*, 475–484.

Oliveira, D.C.S.G., Hunter, W.M., Ng, J., Desjardins, C.A., Dang, P.M., and Werren, J.H. (2009). Data mining cDNAs reveals 3 new ssRNA viruses in *Nasonia* (*Hymenoptera*: *Pteromalidae*). GenBank Accession [ACN94442].

Ongus, J. (2006) *Varroa destructor* virus 1: a new picorna-like virus in *Varroa* mites as well as honey bees, PhD thesis, Wageningen University.

Ongus, J.R., Peters, D., Bonmatin, J.M., Bengsch, E., Vlak, J.M., and van Oers, M.M. (2004). Complete sequence of a picorna-like virus of the genus *Iflavirus* replicating in the mite *Varroa destructor*. J. Gen. Virol. *85*, 3747–3755.

Ongus, J.R., Roode, E.C., Pleij, C.W., Vlak, J.M., and van Oers, M.M. (2006). The 5′ non-translated region of *Varroa destructor* virus 1 (genus *Iflavirus*): structure prediction and IRES activity in *Lymantria dispar* cells. J. Gen. Virol. *87*, 3397–3407.

Porter, F.W., and Palmenberg, A.C. (2009). Leader-induced phosphorylation of nucleoporins correlates with nuclear trafficking inhibition by cardioviruses. J. Virol. *83*, 1941–1951.

Reineke, A., and Asgari, S. (2005). Presence of a novel small RNA-containing virus in a laboratory culture of the endoparasitic wasp *Venturia canescens* (*Hymenoptera: Ichneumonidae*). J. Insect Physiol. *51*, 127–135.

Ribeiro, J.M.C., Lehane, M.J., Berriman, M., van Oers, M.M., and Aksoy, S. (2009). *Glossina morsitans morsitans* (tsetse fly) salivary gland EST sequence with homology to iflavirus protease and RNA-dependent-RNA-polymerase. GenBank Accession Number [EZ407281]

Ricour, C., Delhaye, S., Hato, S.V., Olenyik, T.D., Michel, B., van Kuppeveld, F.J.M., Gustin, K.E., and Michiels, T. (2009). Inhibition of mRNA export and dimerization of interferon regulatory factor 3 by Theiler's virus leader protein. J. Gen. Virol. *90*, 177–186.

Rossmann, M.G., and Tao, Y. (1999). Structural insight into insect viruses. Nature Struct. Biol. *6*, 717–719.

Ryabov, E.V. (2007). A novel virus isolated from the aphid *Brevicoryne brassicae* with similarity to *Hymenoptera* picorna-like viruses. J. Gen. Virol. *88*, 2590–2595.

Ryan, M.D., and Flint, M. (1997). Virus-encoded proteinases of the picornavirus super-group. J. Gen. Virol. *78*, 699–723.

Santillan-Galicia, M.T., Carzaniga, R., Ball, B.V., and Alderson, P.G. (2008). Immunolocalization of Deformed wing virus particles within the mite *Varroa destructor*. J. Gen. Virol. *89*, 1685–1689.

Schlegel, A., Giddings Jr, T.H., Ladinsky, M.S., and Kirkegaard, K. (1996). Cellular origin and ultrastructure of membranes induced during Poliovirus infection. J. Virol. *70*, 6576–6588.

Schultheiss, T., Sommergruber, W., Kusov, Y., and Gauss-Muller, V. (1995). Cleavage specificity of purified recombinant Hepatitis A virus 3C proteinase on natural substrates. J. Virol. *69*, 1727–1733.

Seipelt, J., Guarné, A., Bergmann, E., James, M., Sommergruber, W., Fita, I., and Skern, T (1999). The structures of picornaviral proteinases. Virus Res. *62*, 159–168.

Shen, M., Cui, L., Ostiguy, N., and Cox-Foster, D. (2005). Intricate transmission routes and interactions between picorna-like viruses (Kashmir bee virus and Sacbrood virus) with the honeybee host and the parasitic Varroa mite. J. Gen. Virol. *86*, 2281–2289.

Stankus, T. (2008). A review and bibliography of the literature of honey bee colony collapse disorder: A poorly understood epidemic that clearly threatens the successful pollination of billions of dollars of crops in America. J.Agri. Food Inf. *9*, 115–143.

Strong, R., and Belsham, G.J. (2004). Sequential modification of translation initiation factor eIF4GI by two different Foot-and-mouth disease virus proteases within infected baby hamster kidney cells: Identification of the 3Cpro cleavage site. J. Gen. Virol. *85*, 2953–2962.

Tanada, Y., and Kaya, H. (1993). Other RNA-viral infections, In Insect Pathology (San Diego: Academic Press), pp. 296–310.

Tate, J., Liljas, L., Scotti, P., Christian, P., Lin, T., and Johnson, J.E. (1999). The crystal structure of Cricket paralysis virus: the first view of a new virus family. Nature Struct. Biol. *6*, 765–774.

Terio, V., Martella, V., Camero, M., Decaro, N., Testini, G., Bonerba, E., Tantillo, G., and Buonavoglia, C. (2008). Detection of a honeybee iflavirus with intermediate characteristics between Kakugo virus and Deformed wing virus. New Microbiol. *31*, 439–444.

van Engelsdorp, D., Hayes Jr, J., Underwood, R.M., and Pettis, J. (2008). A survey of honey bee colony losses in the U.S., Fall 2007 to Spring 2008. PLoS ONE 3, art no. e4071.

van Engelsdorp, D., Evans, J.D., Saegerman, C., Mullin, C., Haubruge, E., Nguyen, B.K., Frazier, M.A., Frazier, J., Cox-Foster, D., Chen, Y., *et al.* (2009). Colony collapse disorder: A descriptive study. PLoS ONE *4*, art no. e6481.

Wang, C.H., Wu, C.Y., and Lo, C.F. (1999). A new picorna-like virus, PnPV, isolated from ficus transparent wing moth, *Perina nuda* (Fabricius). J. Invertebr. Pathol. *74*, 62–68.

Wang, X., Zhang, J., Lu, J., Yi, F., Liu, C., and Hu, Y. (2004). Sequence analysis and genomic organization of a new insect picorna-like virus, *Ectropis obliqua* picorna-like virus, isolated from *Ectropis obliqua*. J. Gen. Virol. *85*, 1145–1151.

Watanabe, H. (1991). Infectious flacherie virus, In Atlas of Invertebrate Viruses, J. Adams, and J. Bonami, eds. (Boca Raton: CRC Press), pp. 515–523.

White, G. (1913). Sacbrood, a disease of bees. US Dept. Agr. Bur. Entomol. Circ. *169*.

White, G. (1917). Sacbrood. US Dept. Agr.Bull. *437*, 1–55.

Wilson, J.E., Powell, M.J., Hoover, S.E., and Sarnow, P. (2000). Naturally occurring dicistronic Cricket paralysis virus RNA is regulated by two internal ribosome entry sites. Mol. Cell Biol. *20*, 4990–4999.

Witwer, C., Rauscher, S., Hofacker, I.L., and Stadler, P.F. (2001). Conserved RNA secondary structures in *Picornaviridae* genomes. Nucl. Acids Res. *29*, 5079–5089.

Wu, C.Y., Lo, C.F., Huang, C.J., Yu, H.T., and Wang, C.H. (2002a). The complete genome sequence of *Perina nuda* picorna-like virus, an insect-infecting RNA virus with a genome organization similar to that of the mammalian picornaviruses. Virology *294*, 312–323.

Wu, C.Y., Yang, H.N., Lo, C.F., and Wang, C.H. (2002b). A *Perina nuda* cell line (NTU-Pn-HF) from pupal ovary that is persistently infected with a picorna-like virus (PnPV). Appl. Entomol. Zool. *37*, 171–179.

Yang, X., and Cox-Foster, D.L. (2005). Impact of an ectoparasite on the immunity and pathology of an invertebrate: Evidence for host immunosuppression and viral amplification. Proc. Natl. Acad. Sci. U.S.A. *102*, 7470–7475.

Yue, C., and Genersch, E. (2005). RT-PCR analysis of Deformed wing virus in honeybees (*Apis mellifera*) and mites (*Varroa destructor*). J. Gen. Virol. *86*, 3419–3424.

Yue, C., Schroder, M., Gisder, S., and Genersch, E. (2007). Vertical-transmission routes for Deformed wing virus of honeybees (*Apis mellifera*). J. Gen. Virol. *88*, 2329–2336.

Zhang, J., Feng, J., Liang, Y., Chen, D., Zhou, Z.H., Zhang, Q., and Lu, X. (2001). Three-dimensional structure of the Chinese Sacbrood bee virus. Sci. China, C, Life Sci. *44*, 443–448.

Zhang, Q., Ongus, J.R., Boot, W.J., Calis, J., Bonmatin, J.M., Bengsch, E., and Peters, D. (2007). Detection and localisation of picorna-like virus particles in tissues of *Varroa destructor*, an ectoparasite of the honey bee, *Apis mellifera*. J. Invertebr. Pathol. *96*, 97–105.

Insect Nodaviruses

11

P. Arno Venter, Juan Jovel and Anette Schneemann

Abstract

Nodaviruses represent a family of small, RNA-containing viruses that naturally infect insects and fish. This review describes recent advances in our understanding of the biology of insect nodaviruses, all of which are currently classified in the genus *Alphanodavirus*. The small genome size of nodaviruses combined with their abundant proliferation in a wide variety of cells, has allowed their use as model systems for studies on a variety of topics relevant to RNA virology. Many of these studies have been performed with Flock House virus, the most thoroughly characterized insect nodavirus. Here we summarize how research on Flock House virus and other members of this family has shed light on entry, replication and assembly of alphanodaviruses, as well as on mechanisms employed by insect viruses to counter the antiviral immune responses of their hosts. In addition, we review potential applications of nodaviruses in nanotechnology and nanomedicine.

Introduction

The nodaviruses are a family of non-enveloped, icosahedral viruses that have a bipartite, positive-strand RNA genome. The type member of this virus family, *Nodamura virus* (NoV), was discovered in 1956 in mosquitoes trapped near the village of Nodamura in Japan during an outbreak of Japanese encephalitis (Scherer and Hurlbut, 1967). It was initially thought to represent a novel type of ether- and chloroform-resistant arbovirus but subsequent characterization revealed that NoV had features unlike those of any other virus known at the time and it thus became the first member of a new virus family. Additional members of the *Nodaviridae* family have since been isolated from a variety of insects, including black beetle virus (BBV) (Longworth and Carey, 1976), Flock House virus (FHV) (Scotti *et al.*, 1983), Boolara virus (BoV) (Reinganum *et al.*, 1985), Pariacoto virus (PaV) (Johnson *et al.*, 2000) and most recently Wuhan nodavirus (WhNV) (Liu *et al.*, 2006b) (Table 11.1). A thorough description of individual insect nodaviruses, their host range and pathology has been presented previously (Ball and Johnson, 1998) and is not included in this chapter. It should be noted, however, that NoV is unique in its ability to infect mammals, specifically suckling mice and suckling hamsters, in which it causes hind limb paralysis and death (Garzon and Charpentier, 1991; Scherer and Hurlbut, 1967; Scherer *et al.*, 1968). Moreover, in addition to their natural hosts, nodaviruses can infect other insect species. For example, FHV which was originally isolated from third instar larvae of the grass grub *Costelytra zealandica* (Scotti *et al.*, 1983), also infects adult *Drosophila melanogaster* flies (Galiana-Arnoux *et al.*, 2006; Wang *et al.*, 2006), several species of mosquitoes (Dasgupta *et al.*, 2003), the tsetse fly (Dasgupta *et al.*, 2007), a vector for transmission of sleeping sickness, and the reduviid bug, which transmits Chagas disease (Dasgupta *et al.*, 2007). Additionally, nodaviruses can be propagated in plants (Selling *et al.*, 1990), yeast (Price *et al.*, 1996), *Caenorhabditis elegans* (Lu *et al.*, 2005) and mammalian cells (Ball *et al.*, 1992), although this requires transfection of the

Table 11.1 Natural hosts and geographic origin of insect viruses belonging to the family *Nodaviridae*

	Acronym	Host	Geographic origin
Assigned species in the genus *Alphanodavirus**			
Black beetle virus	BBV	Scarab beetles (*Heteronychus arator*)	New Zealand
Boolarra virus	BoV	Underground grass grub (*Oncopera intricoides*)	Australia
Flock House virus	FHV	Grass grub (*Costelytra zealandica*)	New Zealand
Nodamura virus	NoV	Mosquitoes (*Culex tritaeniorhynchus*)	Japan
Pariacoto virus	PaV	Southern armyworm (*Spodoptera eridania*)	Peru
Tentative species in the genus *Alphanodavirus**			
Gypsy moth virus	GMV	Gypsy moth (*Lymantria ninayi*)	Papua New Guinea
Manawatu virus	MwV	Grass grub (*Costelytra zealandica*)	New Zealand
New Zealand virus	NZV	Unknown	Unknown
Putative *Alphanodavirus***			
Wuhan nodavirus	WhNV	European cabbage butterfly (*Pieris rapae*)	China

*As designated by the International Committee on Taxonomy of Viruses (ICTV) in the committee's most recent report (Schneemann *et al.*, 2005).

**Yet to be considered by the ICTV for inclusion into this genus.

viral RNAs due to lack of suitable receptors in these hosts.

In 1992, the first nodavirus infecting fish was identified in farmed striped jack (Mori *et al.*, 1992). This virus causes nervous necrosis, abnormal swimming behaviour and high mortality in larvae and juveniles. Numerous additional fish nodaviruses were subsequently identified, all of which cause mass mortality in a wide range of fish species including sea bass (Comps *et al.*, 1994), halibut (Grotmol *et al.*, 1997), barramundi (Comps *et al.*, 1994), flounder (Nguyen *et al.*, 1994) and others. Fish nodaviruses represent a significant threat to the aquaculture industry and efforts are under way to contain the global spread of these viruses to commercial fisheries. Fish nodaviruses will not be further discussed in this chapter, except for a brief description of nodaviral taxonomy in the next section. For a recent review on fish nodaviruses the reader is directed to the reference by Munday *et al.* (2002).

In contrast to fish nodaviruses, insect nodaviruses have neither medical nor economic significance. Nonetheless, they have served as exceptionally rich model systems for understanding fundamental aspects in virology that go far beyond this particular family of viruses. Here we summarize recent results that have shed new light on nodaviral cell entry, RNA replication, particle assembly, genome packaging and inhibition of RNA interference. Potential applications of nodaviruses in nanotechnology and biotechnology are also reviewed.

Nodaviral taxonomy

Nodaviruses are taxonomically divided into two genera: *Alphanodavirus* and *Betanodavirus* (Schneemann *et al.*, 2005). Currently, the genus *Alphanodavirus* comprises insect-pathogenic viruses whereas the genus *Betanodavirus* comprises fish pathogenic viruses. Assignment to the *Alpha-* or *Betanodavirus* genus, however, is not based on the host that a given virus infects but rather on the sequence of the coat protein. Within the *Alphanodavirus* genus, coat protein sequence identity ranges from 39% to 87%, and in the *Betanodavirus* genus from 67% to 99%. Sequence identity between the two genera does not exceed 17%. The marked difference in coat protein sequence between viruses in the two genera gives rise to particles with distinct features when viewed by electron microscopy (EM). While negatively stained preparations of alphanodavirus particles have relatively smooth surface features, betanodavirus particles exhibit distinct surface protrusions. These differences are a manifestation of the single versus two domain structure of the alpha- and betanodavirus coat proteins, respectively (Tang *et al.*, 2002).

A recently isolated insect nodavirus, WhNV (Liu *et al.*, 2006a; Liu *et al.*, 2006b), appears to represent a point of departure from the simple segregation of insect viruses with the *Alphanodavirus* genus and fish viruses with the *Betanodavirus* genus. WhNV is unusual in that its coat protein sequence shows very little identity with alphanodaviruses (14–16%) and a somewhat higher level of identity with betanodaviruses (17–24%) (Liu *et al.*, 2006a). Protein A, the viral replicase of WhNV, shares highest sequence identity with that of the insect virus PaV (Liu *et al.*, 2006b). WhNV has not yet been assigned to either nodavirus genus pending further characterization. On the basis of current data, it could well form the basis for a third genus in the family *Nodaviridae*.

Nodaviral genome organization

Viral RNAs

Fig. 11.1 illustrates the general organization of the nodaviral genome. Its defining feature is the division into two segments of positive-sense, single-strand RNAs ((+)ssRNAs), designated RNA1 (3.0–3.2 kb) and RNA2 (1.3–1.4 kb). A subgenomic RNA3, corresponding to the 3′ end of RNA1, is generated during genome replication but not packaged into virus particles (Friesen and Rueckert, 1982).

The lengths of RNA1, RNA2 and RNA3 are known for several alphanodaviruses and listed in Table 11.2. All RNAs are capped at their 5′ ends but lack a poly-A tail at the 3′ end (Dasgupta *et al.*, 1984; Newman and Brown, 1976). The cap structure of BBV RNA1 and RNA2 has been identified as 7-methylguanosine (Dasgupta *et al.*, 1984) and is likely to be identical in other alphanodaviral RNAs. The 3′ ends of alphanodaviral RNAs are unreactive to RNA ligase and RNA polymerase, indicating the presence of some sort of covalent block by an unidentified moiety or a strong secondary structure (Dasmahapatra *et al.*, 1985; Guarino *et al.*, 1984; Kaesberg, 1987). In favour of the latter possibility is the observation that the 3′ termini of the viral RNAs are rendered accessible to RNA modifying enzymes in their double-stranded form in which intramolecular RNA folding is expected to be at a minimum (Dasmahapatra *et al.*, 1985; Guarino *et al.*, 1984).

Like all (+)ssRNA viruses, nodaviruses amplify their genomic RNA through negative sense RNA intermediates. Such intermediates are readily detectable for RNAs 1, 2 and 3 in infected cells, albeit at quantities approximating 1–2% of the level of their positive-sense counterparts (Ball, 1994; Kopek *et al.*, 2007). Double-stranded (ds) versions of all three nodaviral RNA species have also been purified from infected cells (Guarino *et al.*, 1984). Other minor forms of viral RNA that have been shown to result from nodaviral replication include RNA1- and RNA2-derived defective interfering RNAs, as well as RNA dimers formed by head-to-tail junction of RNAs 1, 2 and 3 (Albarino *et al.*, 2001; Ball, 1992, 1994; Dasgupta *et al.*, 1994).

Coding potential

The genomes of alphanodaviruses have an exceedingly limited coding potential, containing a mere three genes to establish productive infections within their insect hosts (Friesen and Rueckert, 1981; Gallagher *et al.*, 1983; Guarino *et al.*, 1984; Harper, 1994). These genes encompass those for protein A, the viral RNA-dependent RNA polymerase (RdRp), protein B2, a suppressor of RNA silencing, and protein α, the precursor of the viral coat protein (Fig. 11.1) (Ball, 1995; Friesen and Rueckert, 1981;

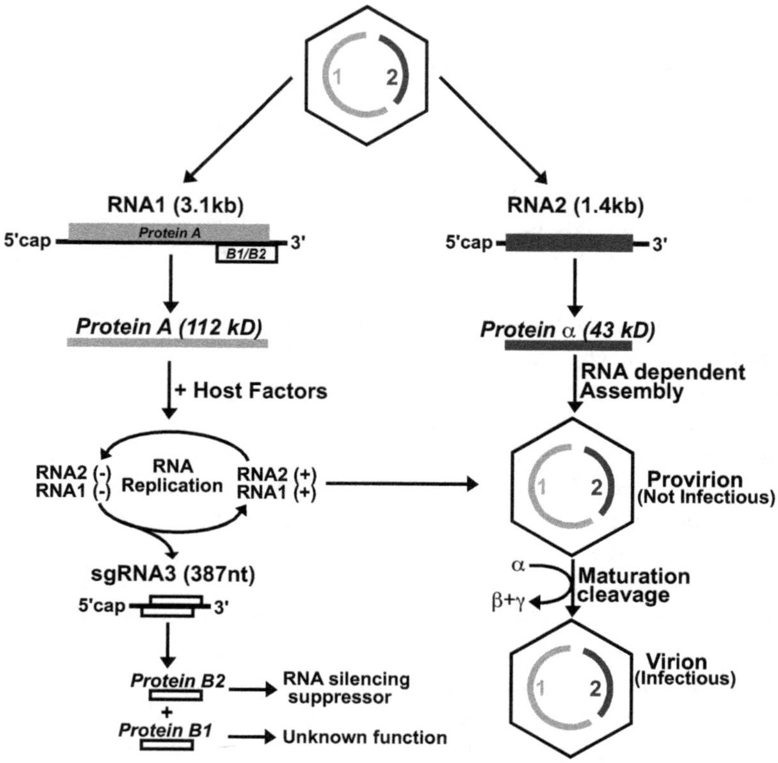

Figure 11.1 Genome organization and replication strategy of *Flock House virus*. Reproduced from Venter and Schneemann (2008) with permission from Birkhäuser Verlag.

Gallagher *et al.*, 1983; Li *et al.*, 2002). Proteins A and B2 are expressed from RNAs 1 and 3, respectively, while the open reading frame (ORF) for protein α is harboured on RNA2. An ORF for a fourth protein called B1 is also present for all the species within the *Alphanodavirus* genus, with the notable exception of BoV (Harper, 1994; Johnson *et al.*, 2001). No function has yet been ascribed to this protein and it has been shown to be non-essential for virus replication in cultured cells (Harper, 1994).

The ORF of protein A spans almost the entire length of RNA1, ranging from 973 to 1042 codons (Table 11.2). In the case of FHV, this protein has a molecular weight of 112 kDa and is sufficient for all of the processes associated with viral RNA synthesis. Protein B2 is encoded in the +1 reading frame relative to protein A and has a molecular weight of 11.6 kDa in FHV. The expression of this protein is contingent on RNA3 synthesis due to the 3′-proximal location of its ORF on RNA1 (Friesen and Rueckert, 1982; Guarino *et al.*, 1984). Mutations that block RNA3 synthesis are therefore also expected to render the virus defenceless to RNA silencing and unable to replicate in its insect host. Where present, the reading frame for protein B1 is identical to that of protein A and overlaps with that of B2. In BBV and FHV, the B1 ORF begins downstream of the B2 ORF, whereas in NoV and PaV this arrangement is reversed (Johnson *et al.*, 2001).

Nodaviral coat precursor protein α (approx. 43 kDa) is translated from a single ORF in RNA2 ranging from 399 to 407 codons (Table 11.2). After assembly into progeny virus particles, protein α autocatalytically cleaves into proteins β and γ with approximate sizes of 40 kDa and 4 kDa, respectively (Fig. 11.1) (Gallagher and Rueckert, 1988). This cleavage event, which is required for the maturation of nodaviral particles into their infectious forms (Schneemann *et al.*, 1992), occurs near the C-terminus of protein α between an Asn and Ala. An exception is PaV, where the scissile bond lies between Asn and Ser residues

Table 11.2 Genomic and subgenomic RNAs of insect nodaviruses and lengths of the ORFs present on these RNAs.

Virus	RNA1 size (nt)/ Accession no.	RNA2 size (nt)/ accession no.	RNA3 size (nt)	Protein A ORF (aa)	Protein α ORF (aa)	Protein B1 ORF (aa)	Protein B2 ORF (aa)
BBV	3106[a] X02396	1399[b] X00956	387	998	407	102	106
BoV	3096[c] AF329080	1305[d] X15960	~390*	998	403	No ORF	106
FHV	3107 X77156	1400[e] X15959	387	998	407	102	106
NoV	3204[f] AF174533	1335[g] X15961	471	1042	399	131	137
PaV	3011[h] AF171942	1311[i] AF171943	414	973	401	94	90

[a]Dasmahapatra *et al.* (1985)
[b]Dasgupta *et al.* (1984)
[c]Johnson *et al.* (2001)
[d]Dasgupta and Sgro (1989)
[e]Dasgupta and Sgro (1989)
[f]Johnson *et al.* (2001)
[g]Dasgupta and Sgro (1989)
[h]Johnson *et al.* (2000)
[i]Johnson *et al.* (2000)
*Precise termini not confirmed

(Johnson *et al.*, 2000). PaV is also unusual in that translation of coat protein occasionally occurs from a second, downstream AUG which gives rise to a slightly smaller version of precursor α. The smaller precursor protein assembles into particles that package the viral genome and then mature normally by cleavage of α into a smaller form of β and standard size γ (Johnson and Ball, 2003).

Alphanodavirus life cycle

Cell entry

Formation of eluted particles

The molecular events associated with nodaviral cell entry are still incompletely understood. It is assumed that nodaviruses bind to a specific, proteinaceous receptor on susceptible cells but no such receptor has yet been unambiguously identified for any member of the *Nodaviridae.* The most detailed information regarding cell entry has come from studies of FHV in cultured *Drosophila* cells (Schneider's line 1). FHV binds to *Drosophila* cells in a saturable manner with approximately 50,000 particle binding sites per cell (Walukiewicz *et al.*, 2006). Interestingly, experiments using radiolabelled virions revealed that although all input virus binds to the cells at 4°C, about one third of the particles elute from the cell surface upon warming to physiological temperature (Walukiewicz *et al.*, 2006). The particles recovered in the medium are no longer able to re-associate with the cells and are thus non-infectious by the normal cell entry route. Based on analogous observations that had been made previously for picornaviruses (Abraham and Colonno, 1988; Fenwick and Cooper, 1962; Joklik and Darnell, 1961; Lonberg-Holm and Korant, 1972), these particles are referred to as eluted virus particles. Eluted FHV particles differ from naïve virus particles, i.e. particles that have never been incubated with *Drosophila* cells, in several respects. First, they are sensitive to low pH as revealed by their tendency to aggregate under these conditions. Second, they show a distinct stain-permeable phenotype under the electron microscope and third, they have lost about a quarter of the 180 γ-peptides normally resident inside the mature virion (Walukiewicz *et al.*, 2006).

It has been proposed that eluted particles represent an aborted infection intermediate. Interaction of the particles with the cell surface receptor is thought to induce structural changes that lead to exposure of previously buried regions of the capsid protein as well as release of internal components, such as the γ peptides. Why a considerable fraction of virions elutes from the receptor at this point is unclear but could be a result of reduced binding affinity as the particle initiates its conformational transitions. Alternatively, the possibility that eluted particles are a result of internalization and recycling to the cell surface cannot be formally excluded at this point. Whatever the mechanism, the formation of eluted particles partly explains the high particle:pfu ratio (300:1) that has been observed for FHV.

The acid-sensitive phenotype of eluted particles suggests that particles, which are internalized by receptor-mediated endocytosis, undergo additional transitions under the low-pH conditions of the endosome. These secondary transitions may further promote release of γ peptides, which are known to have membrane-permeabilizing capacity (see below), ultimately leading to release of the encapsidated viral RNA into the cytosol. The programmed formation of defined structural intermediates presumably allows the virus to control at what point during entry the genome is delivered thereby ensuring maximum efficiency of the infection process.

Cell entry requires low pH

Additional experiments provided support for a model in which low pH plays a role in FHV cell entry. Specifically, it was shown that the lysosomotropic agent NH_4Cl and the vacuolar H^+-ATPase inhibitor bafilomycin A1 significantly decrease FHV infectivity in a dose-dependent manner (Odegard *et al.*, 2009). These results suggest that FHV does not release its genome at the cell membrane but requires an endocytic pathway and that low endosomal pH is an important trigger for viral infectivity. Preliminary, as yet unpublished experiments furthermore suggest that endocytosis is clathrin-mediated based on the observation that chlorpromazine also leads to a significant decrease in viral infectivity (H. Walukiewicz, personal communication).

The requirement for low pH during viral entry was interpreted to indicate that the associated conformational changes in the particle are required to initiate virus–membrane interactions. Indeed, it was shown that naïve FHV particles can disrupt liposomes composed of phosphatidylcholine and that this disruption is more efficient at low pH (Odegard *et al.*, 2009). Furthermore, non-infectious, mutant virus particles in which the maturation cleavage was inhibited, i.e. particles in which the γ peptide had not been released from the precursor protein, cannot mediate liposome disruption (Odegard *et al.*, 2009). These results strongly suggest that the γ-peptides play a critical role during viral entry and that they likely function in disruption of endosomal membranes.

The notion that γ peptides have membrane permeabilization function is supported by studies conducted *in vitro*. Specifically, a synthetic version of the γ peptide, containing the first 21 residues of the full-length 44 amino acid peptide chain, was shown to cause leakage of fluorescent dyes from large unilamellar vesicles and to induce ion conductance across planar lipid bilayers (Bong *et al.*, 1999). Moreover, various biophysical investigations, including CD spectroscopy, NMR spectroscopy and ATR-FT-IR measurements, revealed that the 21 amino acid peptide forms an amphipathic alpha helix and spontaneously partitions into lipid bilayers (Bong *et al.*, 2000; Bong *et al.*, 1999; Janshoff *et al.*, 1999; Maia *et al.*, 2006). However, it does not appear to form bona fide transmembrane channels nor does it cause complete disruption of the lipid bilayer. Rather the peptide induces local membrane permeability to hydrophilic solutes (Bong *et al.*, 1999).

γ peptide functions in trans *to rescue infectivity of maturation-defective FHV particles*

The nodaviral maturation cleavage, which gives rise to the γ-peptide after assembly of virus particles, is a critical prerequisite for virion infectivity (Schneemann *et al.*, 1992). It was therefore surprising to learn that maturation-defective particles can in fact initiate viral infection as long as the γ-peptide is provided *in trans* (Walukiewicz *et al.*, 2008). This was shown in a clever experimental approach that involved two types of FHV particles: the first was a particle that contained two mutations in capsid protein α (N363T/D75N) which inhibited cleavage into β and γ. These particles contained the viral genome, RNA1 and RNA2. The second type of particle was a virus-like particle (VLP) generated with a recombinant baculovirus in Sf21 cells. The VLPs were assembled from wt capsid protein α that efficiently cleaved into β and γ. However, these VLPs did not contain the viral genome, instead they contained random cellular RNA. Both types of particles on their own are non-infectious: the mutant particles because they do not contain the γ-peptide and the VLPs because they do not contain the viral RNAs. Nonetheless, they both efficiently bind to and are internalized by *Drosophila* cells (Walukiewicz *et al.*, 2006). Surprisingly, when these particles were mixed and used to 'co-infect' cells, viable progeny emerged that had the phenotype of the cleavage-defective mutant (Walukiewicz *et al.*, 2008). Apparently, γ-peptides associated with the VLPs could act *in trans* to permeabilize endosomal membranes and allow the genome packaged in the cleavage-defective mutant to be released into the cytosol to initiate infection. This result implies that γ-peptides do not have to be associated with the particle that delivers the genome, but clearly the two types of particles have to be present in the same endosome. Additionally, the results indicate that the structural transitions required for genome release from the particle are independent of the presence or absence of γ peptide.

The development of this experimental system had important practical consequences as it allowed investigations of the role of full-length γ in viral entry, independent of its role in specific genome packaging (see below). Mutagenesis experiments, in which VLPs were used that contained deletions in the C-terminal portion of γ established that the first 21 amino acids of the γ-peptide are not sufficient to rescue infectivity *in trans* but that the entire γ-peptide is required (Banerjee *et al.*, 2009). This was unexpected as *in vitro* experiments had shown that the N-terminal 21 residues alone are highly membrane-active (Bong *et al.*, 1999, 2000; Janshoff *et al.*, 1999; Maia *et al.*, 2006). Three phenylalanine residues near the C-terminal end of γ, which had already been shown to play a role in specific encapsidation of the viral genome, were also found to be required for efficient viral entry (Banerjee *et al.*, 2009). Structural analysis of VLPs containing only the first 21 residues of gamma showed that the γ-peptides are less well icosahedrally ordered inside the particles (Banerjee *et al.*, 2009). Whether this structural change has any consequence for viral entry, however, is not clear.

Temporal pattern of gene expression

Following the delivery of viral genomic RNA into the cytosol, alphanodaviruses face two immediate tasks in setting up a productive infection. First, the virus needs to establish functional replication complexes to ensure a continued supply of viral RNAs for all of the downstream processes mediating viral proliferation. Second, rapid protection must be attained for the viral RNAs against targeting and degradation by the RNA interference (RNAi) response of the host. Corresponding to these requirements, proteins A and B2 are expressed from the earliest stages of infection, resulting in accumulation of detectable levels of intracellular progeny RNAs as soon as 5 hours post infection in cell culture (Friesen and Rueckert, 1981; Go *et al.*, 2006). Conversely, the synthesis of coat protein α trails behind that of proteins A and B2, in line with its role in the encapsidation of viral genomic RNA. As described below, RNA3 plays a pivotal role in facilitating this temporal pattern of viral gene expression and is thus central to the highly productive life cycles of nodaviruses in cell culture, with yields reaching 1–2 mg of virus particles from 10^8 *Drosophila* cells (Friesen *et al.*, 1980).

Synthesis of viral RNA.

From the onset of infection, the rates of RNA1 and RNA2 synthesis continuously increase until about 16 hours post infection, when they become constant for at least another 12 hours (Gallagher and Rueckert, 1988). RNA1 and RNA2 accumulate at a roughly equimolar ratio throughout this period to levels approximating those of the most abundant ribosomal RNA species in the cell (Ball *et al.*, 1992; Friesen and Rueckert, 1984; Kopek *et al.*, 2007). Subgenomic RNA3 coordinates the synthesis of genomic RNA1 and RNA2 in two ways:

1. RNA3 acts as a transactivator of RNA2 replication, an activity that most probably favours the replication of RNA1 over RNA2 during initial stages of infection when the translation of protein A from RNA1 is critically required for the establishment of RNA replication complexes (Eckerle and Ball, 2002; Lindenbach *et al.*, 2002; Price *et al.*, 2005). It still needs to be established how RNA3 exerts this effect on RNA2. Possible mechanisms include a potential role for RNA3 in the recruitment of a host factor required for RNA2 replication, or an ability of RNA3 to extend the template specificity of protein A (Eckerle and Ball, 2002). Another intriguing possibility is that RNA2/RNA3 head-to-tail dimers serve as templates for the initiation of RNA2 replication, given the ability of protein A to resolve these dimers into their monomeric components (Albarino *et al.*, 2001; Eckerle and Ball, 2002).
2. RNA2 acts as a feedback inhibitor of RNA3 replication, evident in the dramatic down-regulation of RNA3 by increasing levels of RNA2 at about 10 hours post infection (Zhong and Rueckert, 1993). This type of regulation presumably prevents RNA3 from promoting disproportionately high RNA2 to RNA1 ratios in infected cells (Eckerle and Ball, 2002). It has been proposed that RNA3 is down-regulated by RNA2 via an enhanced ability of RNA2 to serve as a template for RNA replication (Zhong and Rueckert, 1993). A competitive advantage for RNA2 in this regard is evident in the continuous and abundant replication of RNA3 when RNA1 replicates in the absence of RNA2 (Gallagher *et al.*, 1983).

Synthesis of viral protein

Proteins A and B2 attain peak rates of synthesis at 5 hours and 6–10 hours post infection, respectively (Friesen and Rueckert, 1981). Subsequent to this, the rates of synthesis of both proteins decrease rapidly. The RNA2-mediated down-regulation of RNA3 is at least in part responsible for the decreased rate of B2 production, given the role of RNA3 as a template for the translation of this protein. Another possible explanation for the reduced synthesis of B2 that could also be pertinent to protein A is that RNA2 outcompetes RNA1 and/or RNA3 for binding to a limiting translation factor (Friesen and Rueckert, 1984). Indeed, the AUG codon for protein α in FHV RNA2 is in a more favourable context for translation initiation than those for proteins A and B2 (Dasmahapatra *et al.*, 1985). Moreover, the synthesis of proteins A and B2 is not suppressed when RNA1 is replicated in the absence of RNA2. However, protein A still accumulates to lower than expected levels under these circumstances, suggesting that the synthesis of this protein is subject to additional mechanisms of regulation.

Despite the early decline in their synthesis both protein A and B2 are detectable in cell lysates as late as 36 hours post infection and thus appear to be very stable (Friesen and Rueckert, 1981). Protein B2 accumulates to a much higher level than protein A, comprising up to 5% of the total cellular protein (Harper, 1994). However, this value is minor in comparison with that for coat protein α, whose levels significantly exceed those of any other protein in infected cells (Friesen and Rueckert, 1981). Pulse-chase experiments in BBV-infected cells showed that protein α is synthesized about 2 to 3 hours later than proteins A and B2, presumably due to the RNA3-mediated delay in RNA2 replication relative to RNA1 replication (Friesen and Rueckert, 1981). A similar lag between the expression of non-structural and structural genes of nodaviruses has recently been observed in FHV-infected cells via a quantitative mass spectrometry technique (Go *et al.*, 2006). Subsequent to its initial detection in infected cells, protein α is translated at an unprecedented rate that peaks at about 16 hours post infection (Gallagher and Rueckert, 1988). The protein is then rapidly incorporated into virus particles, followed by autocatalytic cleavage into β and γ with a half-life of about 4 hours at 26°C (Gallagher and Rueckert, 1988). Both β and γ remain associated with mature virions and are readily detectable in lysates from infected cells at later stages of infection.

Genome replication

General features of alphanodaviral genome replication

All (+)ssRNA viruses replicate their genomes in association with intracellular membranes. The *Nodaviridae* establish replication complexes in the outer membranes of mitochondria (Miller *et al.*, 2001). Replication is thought to occur within small invaginations, or spherules, that are open to the cytoplasm (Fig. 11.2a) (Kopek *et al.*, 2007). In these spherules (−)-sense RNA intermediates are thought to serve as templates for the transcription of (+)-sense genomic and subgenomic RNAs, which are released to the cytoplasm through the membranous necks of the spherules. In the case of nodaviruses, protein A suffices for the subcellular targeting and establishment of RNA replication complexes, as well as for all steps of viral RNA synthesis including the addition of cap structures to the 5′ ends of progeny RNA (Dye *et al.*, 2005; Johnson *et al.*, 2001; Miller and Ahlquist, 2002; Van Wynsberghe *et al.*, 2007). This is evident in the ability of RNA1 to replicate autonomously when it is transfected into susceptible cells (Gallagher *et al.*, 1983). Autonomous replication of RNA1 in *Drosophila* cells does, however, require protein B2 to counteract the RNAi response, but is dispensable for replication in mammalian cells which do not appear to use RNAi as an antiviral defence (Ball, 1995; Johnson *et al.*, 2004a). Interestingly, autonomous replication of RNA1 in BHK21 cells is known to cease after several days despite the presence of replication-competent RNA (Johnson and Ball, 1999). Subsequent to this shutoff, replication can be reinitiated by the introduction of fresh RNA1 transcripts into the cells, suggesting that cessation of replication is not due to a cellular environment that is unfavourable for viral RNA synthesis. While the underlying mechanism for this phenomenon remains unclear, the results are suggestive of a natural endpoint for autonomous RNA1 replication (Johnson and Ball, 1999).

In keeping with its role in RNA replication, canonical RNA polymerase motifs are contained within the protein A sequences from all five of the currently established viral species in the genus

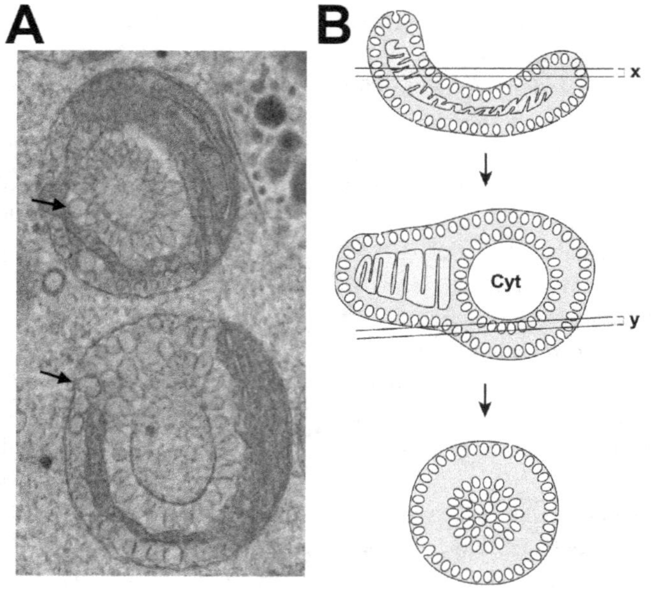

Figure 11.2 FHV-induced mitochondrial modifications. (A) Electron micrograph of mitochondria in FHV infected *Drosophila* cells. Arrows point to virus-induced spherule-like structures within the outer membranes of mitochondria where viral RNA synthesis occurs. (B) Schematic illustration of how different planes of sectioning through the same mitochondrion gives rise to three distinct morphologies commonly observed in EM thin-sections of FHV-infected cells: elongated organelles with compressed matrices (top), mitochondrial ring-like structures with occluded cytoplasmic pockets (middle) and cytoplasmic packets of vesicles (bottom). Reproduced from Kopek *et al.* (2007).

alphanodavirus (Johnson *et al.*, 2001). In FHV, these motifs are located within a region spanning amino acids 513–752 that is likely to be central to the RdRp activities of protein A (Johnson *et al.*, 2001). One of these motifs corresponds to a conserved glycine-aspartate-aspartate (GDD) box, which is required for metal coordination in the active sites of all known polymerases (Argos, 1988; Benzaghou *et al.*, 2004; Kamer and Argos, 1984; Koonin, 1991). Mutation of this motif severely inhibits the activity of protein A, analogous to what has been shown for the RdRps of several other viruses (Benzaghou *et al.*, 2004; Jablonski and Morrow, 1995; Price *et al.*, 2000). A putative guanylylation site has also been identified in all five of the currently available alphanodaviral protein A sequences, in support of a role for this protein in RNA capping (Johnson *et al.*, 2001). Structural information for protein A is currently limited to a 3D homology model of the putative polymerase domain of PaV protein A, which was generated using the poliovirus RdRp 3D^{pol} as a template (Johnson *et al.*, 2001).

Unless noted otherwise, information in the remainder of this section focuses on FHV, whose RNA replication complexes have been studied in much greater structural detail than those of other nodaviruses (Kopek *et al.*, 2007). In addition, critical insights into establishment and functional characteristics of these complexes have been obtained through the development of effective experimental RNA replication systems for this virus. Plasmid-based systems allowing uncoupling of the replication and translation functions of RNA1 have been invaluable for the identification of *cis* acting signals required for viral RNA synthesis, as well as FHV-encoded requirements for replication complex establishment (Albarino *et al.*, 2001; Price *et al.*, 2000). In these systems, protein A is translated from replication-incompetent mRNAs and RNA transcripts are provided *in trans* to serve as replication templates. In addition, the ability of FHV to replicate in the yeast *Saccharomyces cerevisiae* has aided studies aimed at the identification of important host factors for RNA replication (Price *et al.*, 1996). Research in this area benefited from the vast collection of available mutant yeast strains and the highly tractable genetic tools for this organism.

Assembly of mitochondrial replication complexes

FHV protein A is directed to mitochondria via its N-terminal 46 residues, which comprise a self-sufficient mitochondrial targeting signal that bears resemblance to the targeting motifs found in other mitochondrial outer membrane proteins (Kanaji *et al.*, 2000; Kuroda *et al.*, 1998; Miller *et al.*, 2001). A transmembrane domain spanning residues 15–36 mediates the integration of protein A into the mitochondrial outer membrane, where it resides with its N-terminus in the intermembrane space and remainder exposed to the cytoplasm (Miller *et al.*, 2001). Interactions with the outer mitochondrial membrane also appear to be mediated by other regions within protein A. This is evident in the propensity of certain N-terminal truncation mutants to maintain membrane association during subcellular fractionation (Miller and Ahlquist, 2002; Van Wynsberghe and Ahlquist, 2009).

The mechanism underlying the integration of protein A into mitochondrial membranes remains unclear. It has, however, been established that this process is independent of the mitochondrial outer membrane (TOM) complex, which is utilized for the mitochondrial import of most proteins. In a recent study, mitochondrial association of protein A was analysed subsequent to *in vitro* translation of this protein with mitochondria isolated from different yeast strains, including strains with deletions or mutations in TOM components Tom20 and Tom70 (Stapleford *et al.*, 2009). The binding of protein A to mitochondria was affected neither by the absence of the two TOM receptors nor by the removal of protease sensitive outer mitochondrial membrane proteins from the purified organelles via proteinase K digestion. In addition, association with mitochondria was found to be independent of heat shock protein 90 (Hsp90), a chaperone implicated in the mitochondrial trafficking and import of several cellular proteins (Castorena *et al.*, 2007; Kampmueller and Miller, 2005). Interestingly, in *Drosophila* cells Hsp90 plays a role in efficient translation for protein A (Castorena *et al.*, 2007). This role does not extend to lower eukaryotes, such as *S. cerevisiae*, in which RNA replication/translation appears to be independent of functional Hsp90 complexes but dependent on Hsp70 chaperones and co-chaperones (Weeks and Miller, 2008; Weeks *et al.*, 2010). This discrepancy is most probably attributed to known differences in the specific functions of Hsp90 chaperone complexes between yeast and insects. Interestingly, FHV RNA replication can either be up- or down-regulated by chaperones in *S. cerevisiae*. It was recently shown that RNA replication is enhanced by Hsp70 members belonging to the SSA chaperone family, but suppressed by those belonging to the SSB family (Weeks *et al.*, 2010). Contrasting effects on FHV RNA replication were also observed for Hsp40 chaperones that serve as co-chaperones for Hsp70 chaperones. For example, chaperones *YDJ1* and *APJ1* were shown to have positive regulatory effects on RNA replication, while *ZUO1*, *JJJ1* and *JJJ2* significantly down-regulated the cellular accumulation of FHV RNA (Weeks and Miller, 2008; Weeks *et al.*, 2010).

Lipid composition has been implicated as an important determinant for the specific manner in which protein A is associated with the outer membrane of mitochondria. This is evident in the selectivity with which protein A binds to certain anionic phospholipids (e.g. cardiolipin but not phosphocholine), as well as its inefficient interactions with neutral phospholipids (Stapleford *et al.*, 2009) The high affinity of protein A for cardiolipin is of particular interest, given its almost exclusive location in mitochondrial membranes. It is furthermore possible that the unique conical structure of this lipid is required to impart curvature to the membranes of spherules during the assembly of nodaviral replication complexes (Stapleford *et al.*, 2009).

Despite the selectivity of protein A for specific lipids during membrane integration, protein A does not appear to require a particular membrane environment for RNA synthesis. Indeed, retargeting of protein A to an alternative intracellular membrane has been shown to result in increased replication activity (Miller *et al.*, 2003). In the relevant study, yeast-expressed protein A was relocated to the ER through the replacement of its mitochondrial targeting signal with known ER-targeting signals, including those of hepatitis C-virus (HCV) NS5B polymerase, NADP cytochrome P450 and the yeast t-SNARE Ufe1p. The retargeted protein amplified FHV RNA to significantly higher levels relative to wt protein A in yeast cells and in cell-free extracts. Interestingly, retargeting coincided with extensive remodelling of the ER into stacked membrane layers that were exclusively located within perinuclear regions of the cells. Spherule-like structures such as those typically observed for wt protein A were absent from the layered stacks, but high-magnification EM images revealed that the surfaces of individual tubules were nonetheless irregular.

RNA replication templates are essential components of functional RNA replication complexes and protein A plays a key role in the recruitment of such templates (Van Wynsberghe and Ahlquist, 2009; Van Wynsberghe *et al.*, 2007). First, RNA1 was shown to be much more stable in the presence than in the absence of a replication deficient protein A mutant called protein A_{D692} in *S. cerevisiae* (Van Wynsberghe *et al.*, 2007). The increased stability correlated with the association of RNA1 with intracellular membranes, in agreement with the notion that RNA1 was sequestered in spherules where it was protected from degradation by cytosolic ribonucleases. The recruitment of FHV RNA to mitochondria by protein A_{D692} was directly visualized in cells by a fluorescence microscopic technique in which selected portions of RNA1 were linked with tandem copies of the binding site of bacteriophage MS2 coat protein and detected using an MS2 coat protein–GFP fusion protein (Van Wynsberghe and Ahlquist, 2009). This technique was used in combination with other molecular biological approaches to map the regions within FHV RNA1 required for protein-A-mediated association with mitochondria. Two predicted stem–loop structures located between nt 68–205 and centred at nt 90 and nt 170 were found to be essential for this recruitment. The stem–loop structures were probed with various RNases and shown to contain loops that are 6 nt in length with nearly identical sequences. Interestingly, these structures appear to be conserved among insect nodaviruses, because similar structures with a consensus sequence of AAUYGB are predicted for the 5′-proximal regions of RNA1 in BBV, BoV and NoV, where Y = U or C and B = C, G or U (Van Wynsberghe and Ahlquist, 2009). In protein A several domains were shown to be required for the recruitment of RNA1, including the N-terminal mitochondrial targeting domain and a region that both surrounds and includes the RdRp domain (Van Wynsberghe *et al.*, 2007). It remains to be determined whether direct interactions between any of these domains and the recruitment element on RNA1 are responsible for the association of RNA1 with mitochondria.

Spherule topology, organization and functional significance

The replication of (+)ssRNA viruses is generally linked to the formation of cytoplasmic vesicles that are frequently associated with common cellular organelles. As a result, these organelles may bear little resemblance to those observed in healthy cells (Novoa *et al.*, 2005; Salonen *et al.*, 2005). Research on nodaviruses, particularly FHV, has dramatically influenced our understanding of the structural organization and functional significance of these cytoplasmic vesicles and their correlation with organellar deformation.

An early connection between mitochondrial modification and vesicle formation was made in an EM study of BoV-infected *Drosophila* cells, where a coincidence between the appearance of cytoplasmic vesicles and disappearance of mitochondria was noted (Bashiruddin and Martin, 1987). A subsequent study reported the observation of vesicles as attachments to outer mitochondrial membranes in NoV-infected *G. mellonella* muscle tissue (Garzon *et al.*, 1990). More than a decade passed until the role of mitochondrial spherules as sites of nodaviral RNA replication was proposed. Initial insight into the possible role of these vesicles was obtained by Miller *et al.* (2001) who showed that viral RNA synthesis occurs on mitochondria and that protein A is an integral outer mitochondrial membrane protein. This study also highlighted several virus-induced mitochondrial modifications, including the significant expansion of intermembrane spaces due to the presence of numerous 40–60 nm spherules between inner and outer mitochondrial membranes, and the dissolution of mitochondrial cristae. A more detailed depiction of these modifications was provided by Kopek *et al.* (2007), who used EM tomography to reconstruct three-dimensional maps of the deformed organelles and their associated spherules. These maps provided an explanation for three mitochondrial morphologies commonly observed in thin sections of FHV-infected cells: elongated organelles with compressed matrices, ring-like structures with occluded cytoplasmic pockets and 'packets' of cytoplasmic vesicles. It was shown that these seemingly distinct morphologies represent similar structures viewed from different perspectives. Fig. 11.2b shows how different planes of sectioning give rise to such divergent views of the same mitochondrion. Three-dimensional analyses of individual spherules in turn demonstrated that the membranes of these vesicles are continuous with outer mitochondrial membranes and that a channel links the contents of each spherule to the cytoplasm. A diameter of approximately 10 nm for these channels makes them wide enough to serve as conduits for the entry of ribonucleotides and exit of progeny RNAs during RNA replication. Importantly, Kopek and colleagues also provided strong support for the proposed role of spherules in nodaviral RNA replication via immunoelectron microscopy studies, which pinpointed both protein A and FHV-specific RNA synthesis to the cavities within spherules (Kopek *et al.*, 2007).

Comparative biochemical and tomographic analysis of FHV spherules offered valuable insights into the structural organization of nodaviral RNA replication complexes (Kopek *et al.*, 2007). Biochemical measurements suggest that each spherule contains approximately 100 molecules of protein A, one (–)-sense copy of RNA1 and two (–)-sense copies of RNA2. These measurements correlate well with the dimensions computed for an average spherule, which has sufficient interior volume for all three RNA replication intermediates and interior surface area to accommodate approximately 150 imbedded molecules of protein A. Such a close packing arrangement for protein A agrees with the tendency of protein A to oligomerize *in vivo* as gauged by fluorescence resonance energy transfer (FRET) (Dye *et al.*, 2005). In that study, regions in protein A that are capable of self-interaction were identified and it was demonstrated that protein A mutants exhibiting reduced FRET also exhibit decreased replicase activity. Such a correlation between oligomerization and RNA replication for protein A is reminiscent of recognized roles for RdRp oligomerization in cooperative RNA synthesis for poliovirus and HCV (Lyle *et al.*, 2002; Wang *et al.*, 2002).

Determinants of RNA synthesis

Replication of nodaviral RNAs is highly sensitive to nucleotide changes at the 5′ and 3′ ends with modifications at the 3′ ends being particularly deleterious (Ball, 1994, 1995; Ball and Li, 1993). For this reason, efficient replication of RNAs generated from cDNA clones depends on termini that closely mimic authentic viral RNAs (Ball and Johnson, 1998, 1999). The inhibitory effect of heterologous terminal extensions does, however, not extend to some of the minor, dimeric forms of the viral RNAs observed in infected cells. This is evident in the ability of protein A to resolve and replicate RNA2 from an RNA2–RNA3 heterodimer and from complete or partial RNA2 homodimers (Albarino *et al.*, 2001). Replication signals in the terminal regions of RNA1 and RNA2 therefore appear to be sensitive to a particular sequence context, as they are functional in the context of certain FHV-derived sequences but compromised by protruding heterologous sequences (Albarino *et al.*, 2001).

RNA sequence elements important for the replication of FHV genomic RNAs have been mapped out in considerable detail. Sequence requirements at the 5′ ends of RNAs 1 and 2 are minimal, with a prerequisite for a mere 3–14 terminal nucleotides, while more sizeable *cis*-acting elements appear to be located both internal to and at the 3′ ends of the RNAs (Albarino *et al.* 2003; Ball, 1994; Ball and Li, 1993; Lindenbach *et al.* 2002). The internally located *cis*-acting element of RNA2, encompassing nt 520–720, is not strictly necessary for FHV RNA2 replication but has been shown to significantly enhance

RNA2 synthesis (Albarino *et al.*, 2001, 2003; Ball and Li, 1993; Li and Ball, 1993). The situation is less clear for the interior element in FHV RNA1 (nt 2322–2501) given two conflicting reports regarding its requirement (Albarino *et al.*, 2003; Lindenbach *et al.*, 2002). The discrepancy is most probably the result of different hosts and/or experimental strategies used in the two studies (Albarino *et al.*, 2003). Variations such as these may also account for the two different boundaries reported for the 3′-terminal element of RNA1, which were set between nt 2735 and 3011 by Lindenbach *et al.* (2002) or within 48–108 nucleotides from the 3′ end by Albariño *et al.* (2003). For FHV RNA2, a 3′-terminal *cis*-acting element has been demarcated to the last 50 nucleotides, which contain a signal for the RNA3-mediated transactivation of RNA2 replication (Albarino *et al.*, 2003). This region overlaps with a motif consisting of two stem loops and a preceding C-rich stretch that may act as a conserved 3′-terminal *cis*-acting element, because similar motifs are discernible for BBV, BoV and NoV RNA2 (Kaesberg *et al.*, 1990). In contrast to RNA2, conserved sequences or predicted structures are not obvious in the 3′ -proximal region of RNA1 (Johnson *et al.*, 2001).

Unlike the RdRps of most (+)-strand RNA viruses, protein A capably initiates RNA replication from (–)-sense FHV RNA1, RNA2 and RNA3 (Ball, 1994). However, analyses of the transcription products from (–)-sense RNA3 templates suggest that, unlike RNA1 and RNA2, RNA3 does not undergo full replication but hemi-replication, i.e. it can be copied from (–) to (+) but (+)-strand is a dead-end product. While the underlying mechanism for such a block in (–)-strand synthesis remains unclear, it raises the intriguing possibility that nascent RNA3 chains may be transported away from replication complexes to a cellular region that prohibits access to protein A for (–)-strand synthesis (Eckerle *et al.*, 2003).

Two models have been proposed for the synthesis of RNA3 from RNA1 (Zhong and Rueckert, 1993). In one model, RNA3 is transcribed from an internal promoter on (–)-sense strands of RNA1. In the other, the synthesis of (–)-sense RNA1 from its (+)-sense template terminates prematurely to generate (–)-sense RNA3, which then serves as a template for the synthesis of (+)-sense RNA3. The ability of (–)-sense RNA3 to serve as a template for (+)-strand synthesis is at odds with a requirement for an internal promoter on (–)-sense RNA1 and thus favours the possibility that RNA3 could be synthesized via the premature termination mode (Eckerle *et al.*, 2003). Additional support for this model was obtained through the use of an RNA1 derivative with a GFP insertion in RNA3 that supports replication-dependent GFP expression from RNA3. It was shown that a G-to-U mutation at the start of the RNA3 sequence in this RNA derivative inhibited the accumulation of (+)-sense RNA3 but not (–) sense RNA3 (Price *et al.*, 2000). This result suggests that (–)-sense RNA3 synthesis precedes (+)-sense RNA3 synthesis, in agreement with the premature termination model.

Lindenbach *et al.* (2002) discovered a secondary structure in RNA1 that might serve to promote premature transcription termination and thus (–)-sense RNA3 synthesis. The formation of this structure is dependent on a long-distance basepairing interaction between two *cis*-acting elements on RNA1: a proximal element located close to the RNA3 start site (nt 2282–2777) and a short distal element located 1.5 kb upstream of the start site (nt 1229–1239). Importantly, RNA3 synthesis was shown to be highly dependent on the formation of two 6-bp helices via an interaction between the distal element and two discrete regions in the proximal element. The positioning of these helices adjacent to the RNA3 start site raises the possibility that the resulting secondary structure could act as a terminator to produce RNA3 during (–)-strand RNA1 synthesis. It is important to note, however, that this structural motif by no means represents conclusive evidence in favour of the premature termination model, because an analogous structure in (–)-sense RNA1 could be equally suited as an internal promoter for (+)-sense RNA3 synthesis (Lindenbach *et al.*, 2002).

RNA recombination

Nodaviral RNA replication often results in the appearance of minor forms of viral RNA, which include defective interfering RNAs (DI-RNAs) and RNA dimers, and studies on these RNA species have shed light on the recombination activities of protein A. DI-RNAs represent shortened RNA1 and RNA2 molecules with internal deletions and, in some cases, rearranged sequence elements, while RNA dimers represent homo- and heterotypic covalent fusions between RNAs 1, 2 or 3 (Albarino *et al.*, 2001; Dasgupta *et al.*, 1994; Li and Ball, 1993). The formation of these atypical RNAs has been attributed to an inherent trait of RdRps to switch templates during RNA replication. RNA dimers are most probably generated by template switching when RNA synthesis is resumed without prior release of the previously synthesized strand of RNA (Albarino *et al.*, 2001). Albarino *et al.* (2001) proposed two mechanisms whereby protein A could resolve RNA dimers into monomeric RNA replication products. In one, protein A initiates or terminates the transcription of the monomer in response to a *cis*-acting element that is located internal to the dimeric template. In the other, protein A switches from the 3′ component to a homologous position in the 5′ component via homologous recombination.

The generation of DI-RNAs, on the other hand, involves intramolecular recombination events at specific crossover points that preserve functionally important sequences within the viral RNA (Li and Ball, 1993). This enables DI-RNAs to reduce viral titres by out-competing full-length RNA1 and RNA2 as templates for replication, translation and encapsidation. Interestingly, the appearance of DI-RNAs in cell culture correlates with the establishment of persistent nodaviral infections (Dasgupta *et al.*, 1994) (see below).

The bipartite nature of nodaviral genomes suggests that reassortment between RNA1 and RNA2 could play an important role in the evolution of nodaviruses. Indeed, phylogenetic analysis of fish nodaviruses has already provided strong evidence for the natural reassortment of the genome segments of these viruses (Toffolo *et al.*, 2007). Analogous support for such an evolutionary role for intergenic recombination in insect nodaviruses is currently lacking. However, Gallagher (1987) demonstrated that reassortment of genome segments among different alphanodaviruses is at least possible in the laboratory, evident in the ability of the RNA1 and RNA2 molecules of FHV, BBV and BoV to reassort into all six of the possible reassortant viruses. Of note is that NoV failed to exchange its RNAs with any of these viruses, a result that is at least partly attributable to the distinct template specificities of NoV RdRp (Gallagher, 1987; Price *et al.*, 2005).

Alphanodavirus assembly

Kinetics of particle assembly

Subsequent to their synthesis within mitochondrial spherules, the two genomic RNAs of nodaviruses are selectively incorporated into provirions by protein α, the only structural protein that is required for nodavirus assembly. The RNA does not appear to be encapsidated during its egress from spherules, as had been proposed for other (+)ssRNA viruses (Larson and McPherson, 2001; Nugent *et al.*, 1999). Instead, it appears as if nodaviral RNA accumulates into substantial cytoplasmic pools prior to its encapsidation, evident in the slow rate of incorporation of this RNA into provirions in pulse-chase experiments (Gallagher and Rueckert, 1988). Protein α, on the other hand, is rapidly detected in provirions, with as much as 85% of this protein being incorporated into particles within 10 minutes after its synthesis. Provirion assembly is therefore a very quick process, in contrast to the next step of the assembly pathway, when protein α is autocatalytically cleaved with a slow overall half-life of about 4 hours into proteins β and γ to produce mature virions (Gallagher and Rueckert, 1988). The maturation cleavage confers both physicochemical stability and infectivity to nodavirus particles, despite the lack of any obvious morphological differences between provirions and mature virions.

Structure of the protein capsid

Detailed structural information on nodavirus particles is provided in Chapter 14. Here we only supply a brief overview to highlight structural features with known or predicted function in specific genome recognition and virus assembly. Nodavirus capsids are approximately 32–33 nm in diameter and have a $T = 3$ icosahedral shell composed of 60 triangular asymmetric units (Fisher and Johnson, 1993; Hosur *et al.*, 1987; Tang *et al.*, 2001). Each of these units consists of three protein α conformers, A, B and C, related by quasi-3-fold icosahedral symmetry (Fig. 11.3a). The structural differences between A, B and C subunits are slight and dictated by the location of each subunit within the asymmetric unit. The spherical geometry of nodavirus particles arises from an arrangement in which the triangular asymmetric units form flat contacts with their 2-fold-related neighbours and bent contacts with their quasi-2-fold-related neighbours (Fig. 11.3a). Importantly, the crystal structure of FHV revealed that an ordered RNA duplex of 20 nucleotides serves as a molecular switch by forcing flat contacts between asymmetric units at the 2-fold axes of symmetry (Fisher and Johnson, 1993). Conversely, these duplexes are absent at quasi-2-fold axes, allowing the formation of bent contacts. The structure of encapsidated RNA, described in more detail below, thus plays an important role in nodaviral architecture.

At the tertiary level, protein α contains a central β-barrel motif that forms the closed shell of the capsid. Scissile bonds for the maturation cleavage are buried deep within the capsid and inaccessible to external proteases, in keeping with the autoproteolytic mode of provirion maturation. A number of prominent loops between the strands of the β-barrel give rise to characteristic pyramid-shaped protrusions on the exterior surface of nodavirus particles (Fig. 11.3b). The N- and C-termini of protein α are located proximal to the packaged RNA within the capsid interior.

Subsequent to the specific selection of viral RNAs 1 and 2 for encapsidation, assembly generally involves two processes: in one, the viral RNA is extensively condensed and folded into a compact structure, while in the other specific protein-protein and RNA–protein interactions mediate the formation of the T = 3

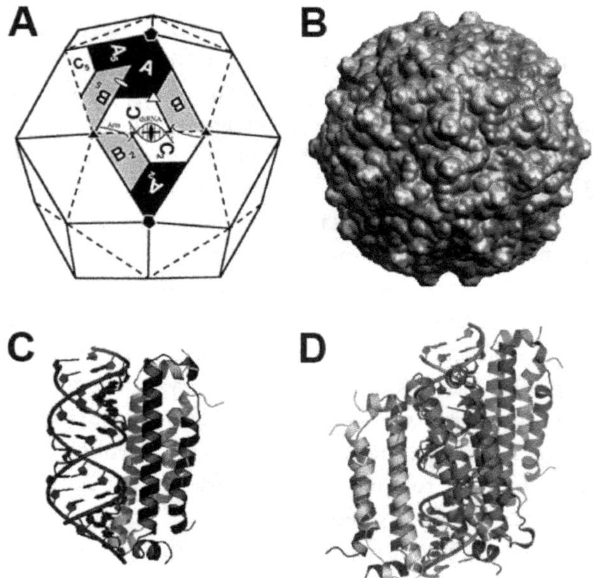

Figure 11.3 Structure of FHV particle and FHV protein B2 in complex with dsRNA. (A) Schematic representation of FHV T = 3 icosahedral lattice. The 2-fold, 3-fold and 5-fold axes of symmetry are indicated by solid eclipses, triangles and pentagons, respectively, while empty eclipses point to quasi-2-fold axes and empty triangles to quasi-3-fold axes. Also indicated is an ordered segment of duplex RNA that forces flat contacts between 2-fold-related triangular asymmetric units and a peptide arm comprising residues 20–31 of protein α that aids the duplex RNA in its role in T = 3 particle formation. (B) Image of an FHV particle reconstructed from the X-ray crystallographic coordinates of this virus (available at viperdb.scripps.edu). (C) Structure of FHV protein B2 dimer in complex with an 18 bp dsRNA duplex. (D) Model of three protein B2 dimers coating an RNA duplex containing 24 bps.

icosahedrally closed shell. A recently described 'all atom' computational assembly model for PaV predicts that RNA condensation results from the non-specific binding of the positively charged N-terminal tails of PaV coat protein to the negatively charged phosphodiester backbone of genomic RNA (Devkota *et al.*, 2009). According to this model, these interactions result in the collapse of the RNA once a sufficient quantity of the negative charge associated with RNA has been neutralized. The globular protein domains are tethered to the collapsed structural intermediate, enabling the second step of the assembly pathway, when specific subunit–subunit interactions are formed and a closed capsid shell is assembled.

RNA is indispensable for nodavirus assembly as evident in its structural requirement for the formation of $T = 3$ particles but also in the inability of coat proteins to assemble into empty particles *in vivo* and *in vitro* (Fisher and Johnson, 1993; Schneemann *et al.*, 1994). Viral RNA, however, can be replaced by heterologous RNAs, which are readily packaged into VLPs by protein α when it is expressed in the absence of RNAs 1 and 2 from recombinant baculoviruses (Schneemann *et al.*, 1993). While the VLPs are morphologically indistinguishable from authentic virions, their RNA contents are made up of small amounts of coat protein mRNA and an assortment of cellular RNAs. However, despite this apparent high affinity for heterologous RNAs, nodaviral coat proteins exhibit exceptional RNA packaging specificity under native conditions. Effective safety mechanisms must therefore be in place to ensure accurate formation of infectious virion progeny.

Structure of encapsidated RNA

The crystal structure of the nodaviruses FHV and PaV has provided unique insights into the organization of the viral genome in the assembled $T = 3$ particle (Fisher and Johnson, 1993; Tang *et al.*, 2001). For most icosahedral viruses, information on the encapsidated nucleic acid is not available at high resolution partly because structure determination involves icosahedral averaging of the electron density and the packaged genome usually does not have that symmetry. In FHV and PaV, however, portions of the viral RNA interact with the N and C-termini of the coat protein subunits tightly enough that they conform to the symmetry of the protein capsid and thus are visible in the final structure.

For FHV, electron density is detectable at the icosahedral 2-fold axes and can be modelled as a piece of double-stranded RNA containing 10 basepairs and approximating an A-type helix (Fisher and Johnson, 1993). Since the RNA density represents an average of the 30 icosahedrally related locations, the sequence of the bases cannot be identified. It is clear, however, that perfect RNA duplexes cannot exist at all symmetrically related positions because the primary sequence of the FHV genome is not compatible with such a high degree of Watson-Crick basepairing. Therefore, most of the duplexes are likely to contain non-Watson-Crick basepairs and/or unpaired nucleotides. Combined, the 20 nucleotides at each of the 30 icosahedral 2-fold contacts account for approximately 13% of the FHV genome. Considering the single-stranded nature of the nodavirus genome, the presence of dsRNA in the particle suggests that the packaged RNA contains extensive secondary structural elements most likely as a consequence of RNA condensation during encapsidation into virus particles.

The RNA duplex interacts with the N and the C-termini of the capsid protein (Fisher and Johnson, 1993). However, the available information is limited by the fact that most of the N and C-termini are disordered in the crystal structure. Nonetheless, the interactions that can be observed are electrostatic in nature and involve the negatively charged phosphate backbone of the RNA and the positively charged side chains of residues Lys 68, 371 and 375 of the C subunits. It is likely that other portions of the N and C-termini are contacting the packaged RNA. The N-terminus in particular is very basic containing 17 arginines within the first 50 residues. It has been proposed that these termini penetrate deep into the viral interior to neutralize the negatively charged RNA core (Devkota *et al.*, 2009).

In PaV a much longer stretch of dsRNA is visible at the 2-fold contacts (Tang *et al.*, 2001). Specifically, a 25-basepair A-type RNA duplex could be modelled with high fidelity into the corresponding density overall representing approximately 35% of the PaV genome. The dsRNA spans the entire length of the 2-fold contact with the ends extending towards the 3-fold related duplexes in a head-to-tail manner. This arrangement gives the impression of a dodecahedral cage of dsRNA located directly beneath the interior surface of the viral capsid. The density of the RNA is strong near the centre of the 2-fold contact but becomes significantly weaker at the ends near the 3-fold axes. This suggests that the RNA is not continuous at all of the symmetrically related 3-fold junctions but takes excursions towards the interior of the particle or other areas beneath the capsid shell.

As in FHV, the interactions between the RNA and the coat protein are primarily electrostatic in nature. In PaV, however, almost the entire N-terminus was observed in the A subunit revealing additional interactions not detected in FHV (Tang *et al.*, 2001). The N-terminus is highly basic and contains a cluster of 12 arginines and lysines between amino acids 11–40. This region runs above the RNA duplex and interacts extensively with the phosphate backbone. Near the 3-fold axis, the N-terminus takes a reverse turn and the visible end ultimately inserts into the major groove of the RNA. The 12 basic residues in the 180 N-termini of the capsid produce a positively charged inner surface that partly neutralizes the negative charge of the packaged RNA. In addition, Lys61 of the C subunits and Arg370 of the B and C subunits engage in electrostatic interactions with the duplex RNA.

The observation that a considerable portion of the nodaviral genome is involved in secondary structure interactions and is highly ordered in the particle raises the question whether these properties are dictated by the primary sequence of the genomic RNAs or whether the fold and position of the RNA is controlled by other factors. This issue was investigated by heterologous expression of the coat protein in Sf21 cells using recombinant baculovirus vectors (Johnson *et al.*, 2004b; Tihova *et al.*, 2004). In this system, the coat protein spontaneously assembles into VLPs, which randomly package cellular RNA. Electron cryomicroscopy (cryoEM) reconstruction of FHV and PaV VLPs showed that in both cases the encapsidated RNA forms the same dodecahedral cage already observed for PaV at high resolution (Johnson *et al.*, 2004b; Tihova *et al.*, 2004). Moreover, density ascribed to RNA was observed below the 5-fold axes, in the centre of the rings that form the dodecahedral cage, but it is not clear what conformation the RNA assumes at these positions. The presence of duplex RNA at the 2-fold contacts regardless of the sequence and length of the encapsidated RNA implies that the observed secondary and tertiary structures are not a result of spontaneous folding of the RNA but that this organization is imposed during the assembly process, presumably by the coat protein. More specifically, the coat protein captures appropriate elements of secondary structures from a range of accessible structural elements and enforces the dodecahedral cage as the subunits come together to form a spherically closed shell. Although heterologous RNAs can conform to these constraints, it is likely that the viral genomic segments have adapted particularly well to these restrictions.

The location of duplex RNA at the icosahedral 2-fold contacts provided a starting point for generating a structural model that incorporates the non-visible remainder of the packaged RNA (Tihova *et al.*, 2004). In this model, the symmetry of the dodecahedral cage had to be broken because it requires 12 interlinked closed rings, an arrangement that is incompatible with the linear structure of the viral genomes. In addition, points of connection linking the RNA in the cage with that elsewhere in the particle had to be incorporated. These considerations were satisfied by postulating that select vertices of the dodecahedral cage represent 'stalactites' where the RNA deviates from the cage and either penetrates the interior or follows a path towards the 5-fold axes, then returns to the same stalactite. These stalactites can be distributed in such a way that allows the RNA to cover each edge of the dodecahedron twice to give antiparallel helices. Vertices that do not have stalactites have four-way or three-way RNA junctions. Using these parameters a path around the edges of the entire dodecahedral cage could be modelled for the full-length PaV genome (Tihova *et al.*, 2004).

Molecular parameters of specific genome recognition

Studies on FHV assembly in *Drosophila* cells have implicated a putative RNA packaging signal as well as several coat protein regions in the specific packaging of the viral genome during assembly (Marshall and Schneemann, 2001; Schneemann and Marshall, 1998; Zhong *et al.*, 1992). Evidence for a role of a particular RNA sequence or structure in genome encapsidation is currently limited to a study by Zhong *et al.* (1992), who showed that nucleotides 186–217 of RNA2 are required for the packaging of an RNA2-derived DI-RNA. This region is thought to fold into a bulged stem–loop that is also predicted at similar locations in BBV, BoV and NoV RNA2. However, a role for this structure as a packaging signal within the context of full-length RNA2 remains to be established.

Coat protein domains implicated in genome recognition map to the N- and C-termini of protein α. Schneemann *et al.* (1998) showed that the C-terminal six residues are required for packaging of both RNA1 and RNA2. Three phenylalanines in this stretch of amino acids are particularly important, suggesting that their aromatic side chains may interact with signals in the RNA via base stacking interactions. Structural evidence for such interactions is, however, still unavailable because a sizeable portion of the C-terminus is not visible in the crystal structure of FHV. Specific genome recognition is also dependent on the N-terminus of FHV coat protein. Marshall and Schneemann (2001) showed that the first 31 residues are required for RNA2 packaging, evident in the predominate encapsidation of RNA1 by a mutant that lacks these residues. Interestingly, this N-terminal region also fulfils an important structural role during assembly. Specifically, residues 20–30 form a peptide arm that aids duplex RNA to form flat contacts at the 2-fold-related subunit contacts within the capsid (Fig. 11.3a) (Fisher and Johnson, 1993). Consequently, heterologous expression of the N-terminally deleted coat protein in the absence of RNA1 and RNA2 results in the formation of a diverse mixture of particles with different shapes and sizes, including 'egg-shaped' particles whose architecture encompasses a combination of $T = 1$ and $T = 3$ symmetrical interfaces (Dong *et al.*, 1998).

Cellular parameters of specific genome recognition

Nodaviruses synthesize their coat proteins on ribosomes and RNA on mitochondrial membranes raising the question of how these two viral components are brought together in infected cells to enable genome encapsidation and virion assembly. Valuable insight into this problem was obtained using a baculovirus expression system that supports the assembly of infectious FHV particles (Krishna *et al.*, 2003). In this system, Sf21 cells are co-infected with two specialized baculovirus vectors that direct the transcription of RNA1 and RNA2 molecules that closely match authentic viral RNA. Translation of protein A from the RNA1 transcripts then launches robust replication of RNA1 and RNA2, which in turn are efficiently encapsidated into infectious FHV virions. Using this system, Venter *et al.* (2005) uncoupled the replication and translation functions of RNA2 to determine if coat proteins synthesized from non-replicating RNA are able to *trans*-encapsidate replicating RNA1 and RNA2 molecules. Interestingly, it was observed that coat protein translated in this manner was unable to encapsidate the genomic RNAs and instead packaged random cellular RNA. This suggested that specific genome packaging requires coat protein translation from RNA molecules derived by RdRp-dependent replication. The proposed linkage between RNA replication and encapsidation led to the hypothesis that coat protein translation from RNAs synthesized by RdRp by default occurs proximal to sites of viral RNA replication at mitochondria, and that this facilitates retrieval of viral RNA by coat protein during assembly (Fig. 11.4). In other words, genome encapsidation could be dependent on the coordinated synthesis of viral components within specialized compartments, or viral factories, in infected cells. If so, it would explain why neither RNA1 nor RNA2 are efficiently packaged by coat proteins translated from non-replicating mRNA templates. These coat protein subunits are translated distally from the compartments where RNA1 and RNA2 are synthesized and therefore

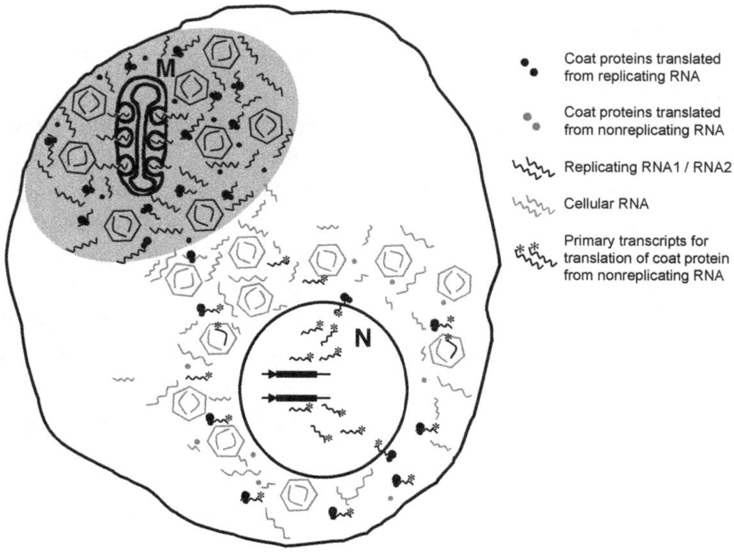

Figure 11.4 Schematic illustration of a model for RNA encapsidation when FHV coat protein is either translated from replicating or non-replicating RNA templates. The top left quarter (shaded grey) of this stylized cell shows replication of RNA1 and RNA2 on the outer surface of mitochondria (M). RNA2 transcripts synthesized in this manner are translated in close proximity to the mitochondria leading to the production of coat protein in the vicinity of the viral RNAs. This, in turn, facilitates the retrieval of viral RNA for encapsidation and the production of infectious virion progeny. As shown at the right near the bottom, primary transcripts for the translation of coat protein from non-replicating RNA are generated in the nucleus (N). These transcripts are then translated shortly after exiting the nucleus in subcellular sites that are distal to mitochondria. The absence of significant levels of RNA1 or RNA2 at these sites leads to the preferential packaging of cellular RNA over viral RNA. Adapted from Venter *et al.* (2005) with permission from the American Society for Microbiology.

package cellular RNA instead of viral RNA (Fig. 11.4).

The genome encapsidation model in Fig. 11.4 has been experimentally evaluated by two studies. In one study, it was shown that, as predicted by the model, coat protein translation from replicating and non-replicating RNA templates in the same cell gives rise to two independent populations of virus particles: a population of infectious virions derived from the synthesis of coat protein from replicating RNA, and a population of VLPs containing random RNA derived from the synthesis of coat protein from non-replicating RNA (Venter and Schneemann, 2007). In the other study, confocal fluorescence microscopy was used to determine if, in agreement with the model, coat protein is localized proximal to mitochondria in FHV infected *Drosophila* cells (Venter *et al.*, 2009). Indeed, significant amounts of coat protein were detected close to mitochondrial RNA replication complexes. However, considerable amounts were also found along the cellular periphery, where it followed reticular distribution patterns that matched those of the ER. While the underlying reason for the association of coat protein with the ER remains unclear, it probably does not represent unassembled subunits. Such a scenario implies transient association of coat protein with the ER prior to assembly, a possibility that is incompatible with the fast rates with which these proteins are incorporated into virions subsequent to their synthesis (Gallagher and Rueckert, 1988).

The two genomic RNAs of nodaviruses are known to be packaged into a single particle. The most direct evidence comes from the observation that a stable hetero-complex is formed between RNAs1 and 2 when FHV particles are heated to 65°C (Krishna and Schneemann, 1999). It is still not understood how nodaviruses package multiple genome segments into the same virion.

A mechanism whereby RNA1 and RNA2 are encapsidated as preformed complexes is unlikely, given that these RNAs can be packaged in an independent manner (Marshall and Schneemann, 2001; Venter *et al.*, 2005). Instead, it is currently thought that the genome segments are recognized for encapsidation in a sequential manner, starting with RNA2. This notion is based on data from a study by Venter *et al.* (2009) who tested if an arginine-rich motif (ARM), comprising residues 32–50 of FHV coat protein, is involved in viral genome recognition. Site-directed mutagenesis indicated that the ARM is required for the recognition of RNA1, because a mutant in which all arginines in this motif were substituted with alanine exclusively packages RNA2. Interestingly, the positively charged nature of the ARM was shown to be all that is required for RNA1 encapsidation, arguing against a role for the ARM in the specific recognition of a packaging signal on this genomic segment. Instead, confocal microscopy analysis pointed to an alternate role for the ARM as a signal required for the trafficking of coat proteins to mitochondria. Given the established role of mitochondria in viral RNA synthesis, these results led to the proposal that RNA2 and RNA1 are recognized by coat protein for assembly sequentially and in distinct cellular microenvironments. Specific recognition of RNA2 may occur co-translationally at sites proximal to mitochondria. The resultant RNA–protein complexes are then redirected to mitochondrial replication complexes in an ARM-dependent manner, where RNA1 is incorporated into the assembling virion. While this proposed mechanism is still speculative, it serves as a testable model for future research on multipartite genome encapsidation.

Particle release by cellular apoptosis

Apoptosis is a type of programmed cell death activated by many viruses. Although it has been regarded as a defence response that restricts virus accumulation and spread, its induction also represents an adaptive strategy of viruses to release infectious particles and to disseminate them to neighbouring cells (Best, 2008). In *Drosophila*, apoptosis requires the main effector caspase DrICE and its upstream initiator DRONC (Cooper *et al.*, 2009). DrICE and DRONC are constitutively repressed by the *Drosophila* inhibitor of apoptosis 1 (DIAP1) protein, whose depletion results in initiation of apoptosis and subsequent cell death.

Settles and Friesen demonstrated that *Drosophila* DL1 cells infected with FHV undergo caspase-dependent apoptosis (Settles and Friesen, 2008). Specifically, FHV was found to promote cleavage of the DrICE zymogen via activation of DRONC. Knockdown of *drice*, *dronc* or *dark*, a DRONC cofactor, with dsRNA prevented FHV-induced cell death. Interestingly, apoptosis did not impair FHV replication, as prolonged survival of infected cells in the presence of the caspase inhibitor z-VAD-fmk did not increase particle yield. Additional investigations revealed that activation of DrICE occurs as a result of DIAP1 depletion upon FHV infection. Accordingly, *diap1* knockdown was sufficient to induce apoptosis in naïve DL1 cells whereas overexpression of DIAP1 prevented apoptosis in FHV-infected cells. It was proposed that the general shutoff of host protein synthesis in FHV-infected cells contributes to reduced levels of DIAP1, given the relatively short (30–45 min) half-life of this protein (Settles and Friesen, 2008). It is not yet known which of the FHV proteins, if any, elicits the apoptotic response or whether it is a combination of RNA replication, protein synthesis and viral assembly. Additionally, it has not yet been investigated what role apoptosis plays in viral pathology and spread in infected insect hosts.

Alphanodaviral interaction with the host defence response

Induction and suppression of RNA interference by alphanodaviruses

Replication of (+)ssRNA viruses invariably gives rise to dsRNAs, primarily in the form of replicative intermediates but also as a result of secondary structure formation in single-stranded viral RNA progeny. The presence of these dsRNAs activates the RNA interference response, a major defence pathway used by insects to protect themselves from viral infection. As part of this response, dsRNAs are sensed and cleaved by the RNaseIII-type endonuclease Dicer into viral small interfering RNAs (vsiRNA) that are 21–24 basepairs (bp) in length. vsiRNAs are loaded into the RNA-induced silencing complex, RISC, which contains an RNaseH enzyme of the family Argonaute (Ago). Ago selectively cleaves one of the vsiRNA strands and preserves the so-called guide strand, which is methylated at its 3′ end by the methyltransferase Hen1. The vsiRNA guide strand directs RISC to specifically degrade viral RNA molecules that harbour complementary sequences (Ding and Voinnet, 2007). Many insect viruses have developed strategies to counter this defence by encoding proteins that interfere with various steps in the RNAi pathway. FHV was the first insect virus shown to induce and then suppress the RNAi response in infected hosts.

The ability of FHV to induce and suppress RNAi was first reported by Li *et al.*, (2002). At that time, it was already known that (+) ssRNA plant viruses are triggers, targets and suppressors of a homologous mechanism called RNA silencing (Voinnet, 2001). The similarities in genome location and expression mechanism of FHV protein B2 and the plant cucumoviral protein 2b, a known viral suppressor of RNA silencing (VSR), prompted Li and co-workers to investigate the potential suppression of silencing activity of FHV B2 in *Nicotiana benthamiana* plants harbouring a green fluorescent protein (GFP) transgene. Surprisingly, in co-infiltration experiments FHV B2 was able to suppress GFP silencing with potency comparable to that of the plant virus protein 2b. The presence of a VSR in the FHV genome suggested the existence of an antiviral RNA-degradation mechanism in insects. In cultured *Drosophila* S2 cells, a series of experiments demonstrated that this was indeed the case (Li *et al.*, 2002). First, vsiRNAs of both polarities were detected during the course of infection, suggesting that FHV triggered a dicing mechanism, presumably activated by virus-derived dsRNAs. Second, knockout of the B2 ORF in a FHV RNA1 replicon, R1ΔB2, prevented RNA1 and RNA3 accumulation. This could be partially rescued by down-regulating Ago2 with dsRNA or by expressing protein B2 protein *in trans*. Third, co-transfection of FHV RNA1 with dsRNA encompassing the sequence of the last 500 nts of RNA1 abolished viral RNA accumulation (Li *et al.*, 2002).

That insects use the RNAi pathway as a defence against virus infection and that Dcr-2 plays an essential role was subsequently shown in *Drosophila* embryos and adult flies. Specifically, it was demonstrated that both embryos and adult flies support FHV replication and that this critically depends on the presence of protein B2 (Galiana-Arnoux *et al.*, 2006; Wang *et al.*, 2006). Analogous to the observations in *Drosophila* cell culture, RNA replication was completely inhibited when the replicon R1ΔB2 was employed. However, RNA replication of R1ΔB2 was fully restored in embryos and flies carrying homozygous *dcr-2* null mutations (Galiana-Arnoux *et al.*, 2006; Wang *et al.*, 2006). This observation and the fact that vsiRNAs were not detected in *dcr-2* mutants established a mechanistic link between Dcr-2 and specific degradation of viral RNA. Interestingly, Dcr-2 mutant flies showed enhanced susceptibility to FHV and harboured higher levels of viral RNA compared to wild type flies (Galiana-Arnoux *et al.*, 2006; Wang *et al.*, 2006). This suggests that the steps associated with RNAi slow the course of FHV infection and in doing so lead to prolonged survival of the host.

Characterization of viral siRNAs

Analysis of FHV-derived vsiRNAs isolated from persistently and lytically infected *Drosophila* S2 cells revealed that they map to distinct regions of full-length RNA1 and 2. Two approaches were used to characterize vsiRNAs in persistently infected cells. In the first approach, vsiRNAs associated with Ago-2, i.e. siRNAs that had been loaded into RISC, were selectively isolated, sequenced and mapped to the FHV genome (Czech *et al.*, 2008; van Rij and Berezikov, 2009). Two findings were of particular interest. First, vsiRNAs mapped in roughly equal proportion to (+) and (−)sense RNA1 and RNA2. This implies that the primary substrates of Dcr-2 are dsRNA replication intermediates and not secondary structural motifs in the (+)sense progeny RNAs, because FHV (+)sense RNA progeny exceeds (−)sense RNA by about 50–100 fold (Ball, 1994; Kopek *et al.*, 2007). Second, although vsiRNAs could be mapped to the entire FHV genome, the majority originated from defined regions on RNA1 and RNA2. Specifically, vsiRNAs derived from RNA1 formed a cluster at the 5′ end (nts 1–250) and the 3′ end (nts 2500–3100). vsiRNAs derived from RNA2 formed three clusters roughly mapping to nts 1–250, 500–750 and 1250–1400. Intriguingly, the regions on RNA2 that are densely populated with vsiRNAs are known to constitute a natural RNA2-derived defective interfering RNA that was isolated from persistently infected *Drosophila* cells (Zhong *et al.*, 1992). The second approach, employed by Flynt *et al.* (2009), involved isolation and sequence analysis of bulk vsiRNAs from persistently infected cells. The results were essentially identical except for the additional identification of a minor cluster of vsiRNAs that mapped to nts 1000–1300 of RNA1.

Obtaining comparable data on vsiRNAs in lytically infected *Drosophila* cells is technically more challenging because vsiRNA abundance is far lower than in persistently infected cells. To circumvent this problem, Aliyari and colleagues (2008) first depleted *Drosophila* cells of Ago2 and then infected them with FHV R1ΔB2 particles thereby achieving replication of the viral RNAs in the absence of B2 (Aliyari *et al.*, 2008). Cloning and sequence analysis of the resulting vsiRNAs revealed that they, too, mapped in roughly equal proportion to (+) and (−)sense RNA1 and RNA2. However, they corresponded to different regions of the FHV genome than vsiRNAs isolated from persistently infected cells. Specifically, vsiRNAs derived from RNA1 mapped primarily to the 400 nucleotides at the 5′ end, whereas vsiRNAs derived from RNA2 were more evenly distributed along the entire segment with a somewhat higher abundance in the 200 nucleotides at the 5′ end. Interestingly, *Drosophila* cells infected with wt FHV gave a different result. In this case, because of the low abundance of vsiRNAs, a Northern blotting procedure was used in which bulk siRNAs were hybridized to seven 445 nt segments spanning the entire RNA1. This analysis revealed that most RNA1-derived siRNAs mapped to the 3′ end of RNA1. Based on these analyses the authors proposed that the newly synthesized double-stranded end of replicative intermediates, which emerges after the polymerase has begun transcription of (−)strand templates into progeny (+)strand templates, is a particularly accessible target for Dcr-2 and that B2 normally binds to this end to protect it from degradation (Aliyari *et al.*, 2008).

Biochemical and structural properties of B2 protein

The B2 proteins of FHV and NoV have been characterized in considerable detail and shown to represent dsRNA binding proteins (Chao *et al.*, 2005; Lu *et al.*, 2005; Sullivan and Ganem, 2005). *In vitro* analyses demonstrated that both proteins bind tightly to long dsRNAs and siRNAs in a sequence-independent manner. Binding is specific for dsRNA and no interaction is detected with ssRNA, ss and dsDNA and RNA/DNA hybrids (Chao *et al.*, 2005). Binding to siRNAs occurs irrespective of chemical modifications at the ends of these molecules and the dissociation constant is in the nanomolar range (Chao *et al.*, 2005; Lu *et al.*, 2005). Binding of B2 to long dsRNAs prevents cleavage of these RNAs by Dicer (Aliyari *et al.*, 2008; Chao *et al.*, 2005; Körber *et al.*, 2009; Lu *et al.*, 2005; Sullivan and Ganem, 2005).

B2 also binds to long and short dsRNAs *in vivo*. Alphanodavirus-derived siRNAs as well as heterologous siRNAs are co-immunoprecipitated with antibodies to NoV B2 and FHV B2 from *Drosophila* cell lysates infected with these viruses or under conditions where B2 is ectopically expressed (Aliyari *et al.*, 2008). In addition, immunoprecipitation of FHV B2 yields FHV RNA 1, 2 and 3 of positive and negative polarity from lytically and persistently infected cells (Aliyari *et al.*, 2008; Flynt *et al.*, 2009). The latter result was interpreted to indicate that B2 binds to dsRNA replicative intermediates during infection. Co-immunoprecipitation experiments also revealed that FHV B2 forms a complex with protein A and that this complex is independent of the presence of RNA, indicative of a direct interaction between the two proteins (Aliyari *et al.*, 2008). All combined, these observations led to the proposal that B2 associates with RdRp during RNA replication and that it functions at least in part to protect nascent viral dsRNA replicative intermediates from degradation by Dicer (Aliyari *et al.*, 2008). Interestingly, an interaction of FHV B2 with the PAZ domain of *Drosophila* Dcr-2 was detected by yeast-two-hybrid analysis and appears to involve the C-terminus of B2 (Singh et al., 2009). Whether this interaction occurs in FHV-infected *Drosophila* cells and whether it plays a role in suppression of RNAi has not yet been investigated.

The high resolution crystal structure of FHV B2 in complex with dsRNA revealed how the protein recognizes nucleic acid and suggested two modes by which it suppresses RNAi (Chao *et al.*, 2005). Prior to structure determination it was determined that the RNA binding domain of FHV B2 resides in the N-terminal 73 residues. This domain was crystallized in complex with a palindromic 18 bp piece of dsRNA. $B2_{1-73}$ forms a homodimer by antiparallel association of two monomers (Fig. 11.3c). Each monomer contains three alpha-helices, the first two of which fold into a helix–turn–helix motif. In the dimer, the helix–turn–helix motif gives rise to a four helix bundle. Only one face of the dimer is involved in interactions with the dsRNA substrate. The dimer recognizes two adjacent minor grooves and the intervening major groove on one side of the RNA, which maintains A-form geometry although the major groove is somewhat wider and the minor groove somewhat narrower than in standard dsRNA. This mode of recognition rationalizes the specificity of B2 for dsRNA as it is dependent on the specific geometry of an A-type RNA duplex. B2 dimer formation and the existence of the four helix bundle are independent of the presence of substrate as shown in an NMR structure of FHV B2 determined in the absence of RNA (Lingel *et al.*, 2005).

The interactions between B2 and dsRNA are solely based on contacts with the sugar-phosphate backbone of the RNA explaining the sequence-independent manner of B2 binding to dsRNA. The interactions are based on electrostatic contacts, hydrogen bonds and van der Waals contacts. Of particular importance are the side chains of Lys47 and Arg54, which make direct contact with the phosphate backbone in the minor groove. Consistent with this is the report that a mutation of Arg54 to Glu reduces B2 binding to dsRNA 20-fold (Lu *et al.*, 2005). Another important, and relatively unusual, interaction in protein–RNA complexes is the van der Waals contact made by Cys44, which is located in the centre of the minor groove recognition site. Mutation of Cys44 to Ala or Ser reduce dsRNA binding affinity by more than 100-fold (Chao *et al.*, 2005).

The mode of B2 binding to dsRNA suggests that multiple copies of B2 can bind to substrates longer than 18 bp in length and protectively coat long dsRNAs (Fig. 11.3d). Indeed, results from binding stoichiometry and ultracentrifugation experiments are consistent with this hypothesis (Chao *et al.*, 2005). Based on these considerations, it was proposed that B2 performs dual roles in suppression of RNAi by inhibiting cleavage of long dsRNAs by Dicer and the incorporation of cleaved vsiRNAs into RISC.

Although FHV B2 and NoV B2 share only marginal sequence similarity, the residues that contact RNA in FHV B2 are conserved in NoV B2. Indeed, a crystal structure of the RNA binding domain of NoV B2 (residues 1–79), determined in the absence of RNA, showed that it adopts the same fold and likely recognizes dsRNA in an analogous manner (Körber *et al.*, 2009).

Alphanodaviral persistence

Alphanodavirus infections are generally lytic and result in death of the host. In light of the efficiency with which nodaviruses shut down RNAi, the major host defence pathway, it is surprising that persistently infected (PI) cells frequently emerge from infected cell cultures. The first indication that alphanodaviruses establish persistent infections came from the observation that a subline of *Drosophila* DL-1 cells was refractory to infection with BBV and contained endogenous particles that were highly similar to native BBV (Friesen *et al.*, 1980). Subsequently, Selling reported a detailed analysis of *Drosophila* DL-1 cells experimentally persistently infected with FHV (Selling, 1986). Typically, infection of DL-1 cells with FHV causes extensive lysis, but about 1% of the cells survive and emerge in a persistently infected state. This percentage is independent of the multiplicity of infection used. The PI cells show no cytopathology, can be passaged for years in the laboratory and produce progeny virus in yields of only a few percent compared to a normal lytic infection. Interestingly, PI cells are resistant to superinfection with genetically related viruses, e.g. the nodaviruses BBV and BoV, but remain fully susceptible to infection with unrelated viruses such as *Cricket paralysis virus*, a dicistrovirus, and *Drosophila X virus*, a birnavirus. Infection with these unrelated viruses is lytic, indicating that persistent infection is not based on a general inhibition of cellular apoptosis. Furthermore, Selling showed that the resistance to superinfection is based on inability of the cells to support replication of the incoming viral RNAs and not on inability of the particles to enter the cell (Selling, 1986).

A follow-up analysis of 20 independent *Drosophila* cell lines persistently infected with FHV demonstrated that establishment of persistence does not require a mutation in the viral genome but that such mutations begin to accumulate as PI cells are continuously passaged (Dasgupta *et al.*, 1994). Mutations seem to occur preferentially in RNA2 which manifest themselves in altered coat proteins. The resulting progeny virus particles commonly form small plaques most likely because mutations on the surface of the virions interfere with efficient binding to the cellular receptor and/or inefficient cell entry and uncoating (Dasgupta *et al.*, 1994). One mutation in particular, Ala265Val, was linked to the small plaque phenotype and could indicate that this amino acid is located in a region on the virion that is involved in receptor binding (Dasgupta *et al.*, 1994).

The observation that PI cells are refractory to superinfection with genetically related viruses but fully susceptible to infection with unrelated viruses is consistent with the hypothesis that persistently infected cells contain FHV-specific RISC complexes with complementarity to the genomes of related nodaviruses such that these genomes are inactivated before their replication can be launched. Indeed, several studies confirmed the presence of vsiRNAs associated with RISC in such cell lines. Moreover, knockdown of genes involved in the RNAi pathway, such as Ago2 and Loqs, led to an increase in FHV RNA replication in PI cells (Czech *et al.*, 2008; Flynt *et al.*, 2009). Thus, alphanodaviral persistence may at least in part be a result of an equilibrium established between RNAi by the host and its suppression by the virus. Another factor likely to play a role in alphanodaviral persistence is the generation and replication of defective interfering RNAs which are commonly associated with PI cells. Defective interfering RNAs derived from both RNA1 and RNA2 have been reported in such cell lines (Dasgupta *et al.*, 1994; Zhong *et al.*, 1992) and sequences contained within these RNAs contribute substantially to the formation of vsiRNAs (Czech *et al.*, 2008; Flynt *et al.*, 2009). Whether defective interfering RNAs are a cause or an effect of persistently infected cells, however, is not yet clear.

Applications in biotechnology

Genetic display of peptides and proteins on coat protein subunits and particles

Monomeric FHV coat protein subunits and fully assembled particles can be used as carriers to display foreign peptides and proteins. Specifically, insertions or amino acid replacements have been introduced into surface-exposed loops that connect the strands of the central beta barrel motif of the FHV coat protein subunit. Loops that have been targeted include those at amino acid positions 131, 206, 264 and 305.

Peptides derived from HIV proteins gp120 and gp41, HCV E1 glycoprotein, HBV surface antigen and others (Buratti *et al.*, 1996, 1997; Chen *et al.*, 2006; Lorenzi and Burrone, 1999; Peng *et al.*, 2005; Tisminetzky *et al.*, 1994; Xiong *et al.*, 2005) have been displayed on single coat protein subunits. The chimeric proteins were expressed in *Escherichia coli* in the form of inclusion bodies. These inclusion bodies were isolated, protein denatured and further purified. Because generally no attempts were made to renature the purified protein, it is not known whether the final product had any structural resemblance to native FHV coat protein and in what conformation the foreign peptide was displayed. Nonetheless, such recombinant proteins proved useful as diagnostic tools, particularly as antigens in ELISA assays to test for the presence of certain antibodies in patient sera. The value of such chimeric proteins as vaccines to induce neutralizing antibodies is limited, however, in part because the foreign peptide is not necessarily presented in the same conformation that it has in its native context.

The effectiveness of FHV as a platform for vaccine design became apparent when fully assembled VLPs were employed to present peptides and proteins to the immune system.

Given the $T = 3$ icosahedral nature of FHV VLPs, any surface-exposed antigen is displayed in a multivalent manner, thereby allowing efficient cross-linking of B cell receptors. Moreover, the particulate nature of FHV VLPs likely enhances their uptake by dendritic cells. This can be expected to prompt a strong T helper cell response, which is ultimately required for a robust antibody response.

The first example of FHV VLPs as a presentation platform for foreign antigens involved the insertion of seven amino acids from the HIV-1 V3 loop at FHV coat protein positions 131 and 305 (Scodeller *et al.*, 1995). The chimeric proteins were expressed in Sf21 cells using recombinant baculovirus vectors, a strategy that leads to formation of VLPs in case of the wt FHV coat protein (Schneemann *et al.*, 1993). Similarly, chimeric protein 131, carrying the HIV-1 epitope at position 131, also appeared to assemble into higher order structures as suggested by their sedimentation pattern on a sucrose gradient (Scodeller *et al.*, 1995). No attempts were made, however, to purify VLPs and confirm their presence by EM. Whether chimeric protein 305 assembled into VLPs was not tested. Immunization of guinea pigs with clarified cell lysates containing the putative VLPs gave rise to antibodies, but only those induced by the construct in which the insertion was placed at position 131 neutralized certain HIV-1 strains. There are several explanations for this observation including the possibility that chimeric protein 305 did not assemble into particles and/or that the peptide at that position did not have the conformation necessary for induction of antibodies that recognize the same epitope in its native context (Scodeller *et al.*, 1995).

A more systematic approach involving the display of entire proteins at positions 206 and 264 was described more recently (Destito *et al.*, 2009; Manayani *et al.*, 2007). This approach involved molecular modelling to determine whether proteins inserted at these locations would be likely to retain their native fold and to ensure that their presence would not interfere with the ability of the chimeric proteins to assemble into particles. Obviously, such an approach requires that a high resolution structure of the target protein is available. In one example, the 137 amino acid Ig-like domain from the coat protein of the insect tetravirus Nudaurelia ω capensis virus was inserted at position 206 (Destito *et al.*, 2009). The chimeric protein formed large amounts of VLPs that could be purified to homogeneity. Three dimensional image reconstruction of the particles by cryoEM revealed how the 180 Ig-like domains are oriented on the surface of the VLP relative to the underlying FHV protein shell (Destito *et al.*, 2009). The structure suggested that the Ig-like domain was folded correctly because its high resolution X-ray structure could be fitted easily into the cryoEM density. Furthermore the chimeric particles could be precipitated with protein A and G sepharose (Destito *et al.*, 2009).

In another example the 181 amino acid extracellular domain of the anthrax toxin receptor 2 (ANTXR2) was inserted at position 206 and 264 (Manayani *et al.*, 2007). ANTXR2 is the receptor for anthrax protective antigen (PA), a component of the anthrax toxins. Both chimeric proteins formed a homogeneous population of VLPs that were purified and structurally characterized by cryoEM. Pseudo-atomic models based on the cryoEM reconstructions revealed that depending on the insertion site, the ANTXR2 domains are arranged differently on the surface of the particles (Manayani *et al.*, 2007). Insertion at position 206 yields three closely juxtaposed domains, which form an offset cluster of six with their 2-fold related counterparts. Insertion of ANTXR2 at position 264, on the other hand, provides more space between the individual domains and leads to a more even distribution on the surface of the particles. Both types of VLP protected cultured cells from intoxication with anthrax lethal toxin suggesting that they could ligate PA and thereby inhibit entry of the toxins into susceptible cells (Manayani *et al.*, 2007).

Alphanodavirus particles as tools in structure-based vaccine design

Chimeric FHV VLPs displaying the extracellular domain of ANTXR2 were not only used as anthrax antitoxins but also as effective vaccines against anthrax (Manayani *et al.*, 2007). Both types of particles were shown to bind multiple copies of PA on their surface (Manayani *et al.*,

2007). The binding affinity was sufficiently strong to allow formation of stable, non-covalent complexes. Because of the different arrangement of ANTXR2 on VLP206 and VLP264, the number of PA molecules that could be docked onto the surface of the particles differed, with VLP206 binding approximately 90 copies whereas VLP264 was estimated to bind 130 copies of PA. Neither particle was able to accommodate PA on all 180 displayed receptor units as a result of steric hindrance.

Complexes of VLP264 decorated with PA were next used to immunize rats and the resulting antibody response to PA, ANTXR2 and FHV coat protein was determined by ELISA (Manayani *et al.*, 2007). Adjuvants were not included in these experiments. The particles induced an extremely robust antibody response to PA and FHV coat protein but not to ANTXR2. The antibody response to PA was strong enough to protect the animals from challenge with anthrax lethal toxin four weeks after a single injection. This represents a dramatic improvement over the current anthrax vaccine, which requires six injections of adjuvanted, monomeric PA over the course of 18 months (Centers for Disease Control and Prevention, 2000), and is a testament to the power of FHV VLPs as a vaccine platform.

Organic and inorganic alphanodavirus hybrid structures

FHV particles can also be modified using conjugation chemistry as an alternate method for surface modification. Conjugation chemistry permits addition of natural and synthetic compounds and offers a much higher degree of flexibility than display of peptides and proteins by genetic insertion into the FHV coat protein. Most conjugation strategies employ free amino groups on lysine residues or sulphydryl groups on cysteine residues to covalently attach the moiety of interest. FHV does not have any lysines or cysteines on its surface but these can be engineered into the coat protein at defined positions using straightforward mutagenesis. Using an engineered mutant of FHV, in which alanine at position 205 was replaced with a lysine, various compounds could be coupled to the surface of the particle. These included simple fluorescent dyes such as fluorescein using fluorescein isothiocyanate or fluorescein-N-hydroxyl-succinimidoester (Portney *et al.*, 2005). More advanced applications involved the conjugation of multivalent quantum dots or carbon nanotubes, which resulted in the formation of hybrid stuctures that produced micron-scale networks (Portney *et al.*, 2005). Such composite materials have potential in fabrication of novel biofilms and nanoelectronic devices.

Concluding remarks

It should be evident from the work summarized in this chapter that alphanodavirus research is vigorous and has provided valuable contributions to our understanding of (+)ssRNA virus biology. There can be no doubt that future studies of these viruses will continue to play a role in creating an integrated view of (+)ssRNA virology. Areas in nodavirus research that are of immediate interest include identification of cellular receptors, determination of the high resolution structure of the RNA polymerase and investigations of the formation and organization of RNA replication complexes in the outer mitochondrial membrane. Elucidation of the molecular mechanism by which the bipartite nodaviral genome is packaged into a single virion is also a high priority. This mechanism has remained mysterious not only for nodaviruses but all multipartite RNA viruses, including influenza viruses, reoviruses, bunyaviruses and others. A new area of research to which nodaviruses have already made seminal contributions are studies of host–pathogen interactions, particularly investigations of RNA interference and its suppression by microbes. Future studies using nodaviruses and *Drosophila* as a model system can be expected to reveal additional insights into antiviral immunity in invertebrates. Lastly, potential applications of nodaviruses in nanotechnology and nanomedicine provide further motivation for continued research with these viruses and may increase their visibility as an important tool in virology.

Acknowledgements

Work in the authors' laboratory was funded by grants from the National Institutes of Health.

References

Abraham, G., and Colonno, R.J. (1988). Many rhinovirus serotypes share the same cellular receptor. J. Virol. *51*, 340–345.

Albarino, C.G., Eckerle, L.D., and Ball, L.A. (2003). The cis-acting replication signal at the 3′ end of *Flock House virus* RNA2 is RNA3-dependent. Virology *311*, 181–191.

Albarino, C.G., Price, B.D., Eckerle, L.D., and Ball, L.A. (2001). Characterization and template properties of RNA dimers generated during *Flock House virus* RNA replication. Virology *289*, 269–282.

Aliyari, R., Wu, Q., Li, H.W., Wang, X.H., Li, F., Green, L.D., Han, C.S., Li, W.X., and Ding, S.W. (2008). Mechanism of induction and suppression of antiviral immunity directed by virus-derived small RNAs in *Drosophila*. Cell Host Microbe *4*, 387–397.

Argos, P. (1988). A sequence motif in many polymerases. Nucleic Acids Res. *16*, 9909–9916.

Ball, L.A. (1992). Cellular expression of a functional nodavirus RNA replicon from vaccinia virus vectors. J. Virol. *66*, 2335–2345.

Ball, L.A. (1994). Replication of the genomic RNA of a positive-strand RNA animal virus from negative-sense transcripts. Proc. Natl. Acad. Sci. U.S.A. *91*, 12443–12447.

Ball, L.A. (1995). Requirements for the self-directed replication of *Flock House virus* RNA 1. J. Virol. *69*, 720–727.

Ball, L.A., Amann, J.M., and Garrett, B.K. (1992). Replication of *Nodamura virus* after transfection of viral RNA into mammalian cells in culture. J. Virol. *66*, 2326–2334.

Ball, L.A., and Johnson, K.L. (1998). Nodaviruses of insects. In The insect viruses, L.K. Miller, and L.A. Ball, eds. (New York, Plenum Press), pp. 225–267.

Ball, L.A., and Johnson, K.L. (1999). Reverse genetics of nodaviruses. Adv. Virus Res. *53*, 229–244.

Ball, L.A., and Li, Y. (1993). cis-acting requirements for the replication of *Flock House virus* RNA 2. J. Virol. *67*, 3544–3551.

Banerjee, M., Khayat, R., Walukiewicz, H.E., Odegard, A.L., Schneemann, A., and Johnson, J.E. (2009). Dissecting the functional domains of a nonenveloped virus membrane penetration peptide. J. Virol. *83*, 6929–6933.

Bashiruddin, J.B., and Martin, J.L. (1987). Proteins induced by Boolarra virus in *Drosophila melanogaster* line 1 cells. Intervirology *28*, 122–124.

Benzaghou, I., Bougie, I., and Bisaillon, M. (2004). Effect of metal ion binding on the structural stability of the hepatitis C virus RNA polymerase. J. Biol. Chem. *279*, 49755–49761.

Best, S.M. (2008). Viral subversion of apoptotic enzymes: escape from death row. Annu. Rev. Microbiol. *62*, 171–192.

Bong, D.T., Janshoff, A., Steinem, C., and Ghadiri, M.R. (2000). Membrane partitioning of the cleavage peptide in Flock House virus. Biophys. J. *78*, 839–845.

Bong, D.T., Steinem, C., Janshoff, A., Johnson, J.E., and Ghadiri, M.R. (1999). A highly membrane-active peptide in Flock House virus: implications for the mechanism of nodavirus infection. Chem. Biol. *6*, 473–481.

Buratti, E., Di Michele, M., Song, P., Monti-Bragadin, C., Scodeller, E.A., Baralle, F.E., and Tisminetzky, S.G. (1997). Improved reactivity of hepatitis C virus core protein epitopes in a conformational antigen-presenting system. Clin. Diagn. Lab. Immunol. *4*, 117–121.

Buratti, E., Tisminetzky, S.G., Scodeller, E.S., and Baralle, F.E. (1996). Conformational display of two neutralizing epitopes of HIV-1 gp41 on the Flock House virus capsid protein. J. Immunol. Methods *197*, 7–18.

Castorena, K.M., Weeks, S.A., Stapleford, K.A., Cadwallader, A.M., and Miller, D.J. (2007). A functional heat shock protein 90 chaperone is essential for efficient flock house virus RNA polymerase synthesis in *Drosophila* cells. J. Virol. *81*, 8412–8420.

Centers for Disease Control and Prevention. (2000). Use of anthrax vaccine in the United States: recommendations of the advisory committee on immunization practices (ACIP). MMWR *49 (RR-15)*, 1–20.

Chao, J.A., Lee, J.H., Chapados, B.R., Debler, E.W., Schneemann, A., and Williamson, J.R. (2005). Dual modes of RNA-silencing suppression by *Flock House virus* protein B2. Nat. Struct. Mol. Biol. *12*, 952–957.

Chen, Y., Xiong, X., Liu, X., Li, J., Wen, Y., Dai, Q., Cao, Z., and Yu, W. (2006). Immunoreactivity of HCV/HBV epitopes displayed in an epitope-presenting system. Mol. Immunol. *43*, 436–442.

Comps, M., Pepin, J.F., and Bonami, J.R. (1994). Purification and characterization of two fish encephalitis viruses (FEV) infecting *Lates calcarifer* and *Dicentrarchus labrax*. Aquaculture *123*, 1–10.

Cooper, D.M., Granville, D.J., and Lowenberger, C. (2009). The insect caspases. Apoptosis *14*, 247–256.

Czech, B., Malone, C.D., Zhou, R., Stark, A., Schlingeheyde, C., Dus, M., Perrimon, N., Kellis, M., Wohlschlegel, J.A., Sachidanandam, R., *et al.* (2008). An endogenous small interfering RNA pathway in Drosophila. Nature *453*, 798–802.

Dasgupta, R., Cheng, L.L., Bartholomay, L.C., and Christensen, B.M. (2003). Flock House virus replicates and expresses green fluorescent protein in mosquitoes. J. Gen. Virol. *84*, 1789–1797.

Dasgupta, R., Free, H.M., Zietlow, S.L., Paskewitz, S.M., Aksoy, S., Shi, L., Fuchs, J., Hu, C., and Christensen, B.M. (2007). Replication of *Flock House virus* in three genera of medically important insects. J. Med. Entomol. *44*, 102–110.

Dasgupta, R., Ghosh, A., Dasmahapatra, B., Guarino, L.A., and Kaesberg, P. (1984). Primary and secondary structure of Black beetle virus RNA2, the genomic messenger for BBV coat protein precursor. Nucleic Acids Res. *12*, 7215–7223.

Dasgupta, R., Selling, B., and Rueckert, R. (1994). Flock House virus: a simple model for studying persistent infection in cultured Drosophila cells. Arch. Virol. Suppl. *9*, 121–132.

Dasgupta, R., and Sgro, J.-Y. (1989). Nucleotide sequences of three nodaviral RNA2's: The messengers of their coat protein precursors. Nucleic Acids Res. *27*, 7525–7526.

Dasmahapatra, B., Dasgupta, R., Ghosh, A., and Kaesberg, P. (1985). Structure of the Black beetle virus genome and its functional implications. J. Mol. Biol. *182*, 183–189.

Destito, G., Schneemann, A., and Manchester, M. (2009). Biomedical nanotechnology using virus-based nanoparticles. Curr. Top. Microbiol. Immunol. *327*, 95–122.

Devkota, B., Petrov, A.S., Lemieux, S., Boz, M.B., Tang, L., Schneemann, A., Johnson, J.E., and Harvey, S.C. (2009). Structural and electrostatic characterization of *Pariacoto virus*: implications for viral assembly. Biopolymers *91*, 530–538.

Ding, S.W., and Voinnet, O. (2007). Antiviral immunity directed by small RNAs. Cell *130*, 413–426.

Dong, X.F., Natarajan, P., Tihova, M., Johnson, J.E., and Schneemann, A. (1998). Particle polymorphism caused by deletion of a peptide molecular switch in a quasiequivalent icosahedral virus. J. Virol. *72*, 6024–6033.

Dye, B.T., Miller, D.J., and Ahlquist, P. (2005). *In vivo* self-interaction of nodavirus RNA replicase protein a revealed by fluorescence resonance energy transfer. J. Virol. *79*, 8909–8919.

Eckerle, L.D., Albarino, C.G., and Ball, L.A. (2003). *Flock House virus* subgenomic RNA3 is replicated and its replication correlates with transactivation of RNA2. Virology *317*, 95–108.

Eckerle, L.D., and Ball, L.A. (2002). Replication of the RNA segments of a bipartite viral genome is coordinated by a transactivating subgenomic RNA. Virology *296*, 165–176.

Fenwick, M.L., and Cooper, P.D. (1962). Early interactions between poliovirus and ERK cells: some observations on the nature and significance of the rejected particles. Virology *18*, 212–223.

Fisher, A.J., and Johnson, J.E. (1993). Ordered duplex RNA controls capsid architecture in an icosahedral animal virus. Nature *361*, 176–179.

Flynt, A., Liu, N., Martin, R., and Lai, E.C. (2009). Dicing of viral replication intermediates during silencing of latent Drosophila viruses. Proc. Natl. Acad. Sci. U.S.A. *106*, 5270–5275.

Friesen, P., Scotti, P., Longworth, J., and Rueckert, R. (1980). Black beetle virus: Propagation in *Drosophila* Line 1 cells and an infection-resistant subline carrying endogenous Black beetle virus-related particles. J. Virol. 35, 741–747.

Friesen, P.D., and Rueckert, R.R. (1981). Synthesis of Black beetle virus proteins in cultured *Drosophila* cells: Differential expression of RNAs 1 and 2. J. Virol. *37*, 876–886.

Friesen, P.D., and Rueckert, R.R. (1982). Black beetle virus: Messenger for protein B is a subgenomic viral RNA. J. Virol. *42*, 986–995.

Friesen, P.D., and Rueckert, R.R. (1984). Early and late functions in a bipartite RNA virus: evidence for translational control by competition between viral mRNAs. J. Virol. *49*, 116–124.

Galiana-Arnoux, D., Dostert, C., Schneemann, A., Hoffmann, J.A., and Imler, J.L. (2006). Essential function *in vivo* for Dicer-2 in host defense against RNA viruses in drosophila. Nat. Immunol. *7*, 590–597.

Gallagher, T.M. (1987). Synthesis and assembly of nodaviruses (PhD thesis, University of Wisconsin-Madison).

Gallagher, T.M., Friesen, P.D., and Rueckert, R.R. (1983). Autonomous replication and expression of RNA 1 from Black beetle virus. J. Virol. *46*, 481–489.

Gallagher, T.M., and Rueckert, R.R. (1988). Assembly-dependent maturation cleavage in provirions of a small icosahedral insect ribovirus. J. Virol. *62*, 3399–3406.

Garzon, S., Strykowski, H., and Charpentier, G. (1990). Implication of mitochondria in the replication of Nodamura virus in larvae of the Lepidoptera, *Galleria mellonella* (L.) and in suckling mice. Arch. Virol. *113*, 165–176.

Garzon, S., and Charpentier, G. (1991). *Nodaviridae*. In Atlas of Invertebrate Viruses, J.R. Adams and J.R. Bonami, eds. (Boca Raton, CRC Press), pp. 351–370.

Go, E.P., Wikoff, W.R., Shen, Z., O'Maille, G., Morita, H., Conrads, T.P., Nordstrom, A., Trauger, S.A., Uritboonthai, W., Lucas, D.A., *et al.* (2006). Mass spectrometry reveals specific and global molecular transformations during viral infection. J. Proteome Res. *5*, 2405–2416.

Grotmol, S., Totland, G.K., Thorud, K., and Hjeltnes, B.K. (1997). Vacuolating encephalopathy and retinopathy with a nodavirus-like agent: a probable cause of mass mortality of cultured larval and juvenile Atlantic halibut *Hippoglossus hippoglossus*. Dis. Aquat. Organ. *29*, 85–97.

Guarino, L.A., Ghosh, A., Dasmahapatra, B., Dasgupta, R., and Kaesberg, P. (1984). Sequence of the Black beetle virus subgenomic RNA and its location in the viral genome. Virology *139*, 199–203.

Harper, T.A. (1994). Characterization of the proteins encoded from the nodaviral subgenomic RNA. (PhD Thesis, University of Wisconsin-Madison).

Hosur, M.V., Schmidt, T., Tucker, R.C., Johnson, J.E., Gallagher, T.M., Selling, B.H., and Rueckert, R.R. (1987). Structure of an insect virus at 3.0 Å resolution. Proteins 2, 167–176.

Jablonski, S.A., and Morrow, C.D. (1995). Mutation of the aspartic acid residues of the GDD sequence motif of poliovirus RNA-dependent RNA polymerase results in enzymes with altered metal ion requirements for activity. J. Virol. *69*, 1532–1539.

Janshoff, A., Bong, D., Steinem, C., Johnson, J.E., and Ghadiri, M.R. (1999). An animal virus derived peptide switches membrane morphology: possible relevance to nodaviral transfection processes. Biochemistry, *38*, 5328–5336.

Johnson, K.L., and Ball, L.A. (1999). Induction and maintenance of autonomous *Flock House virus* RNA1 replication. J. Virol. 73, 7933–7942.

Johnson, K.L., Price, B.D., Eckerle, L.D., and Ball, L.A. (2004a). *Nodamura virus* nonstructural protein B2 can enhance viral RNA accumulation in both mammalian and insect cells. J. Virol. *78*, 6698–6704.

Johnson, K.N., and Ball, L.A. (2003). Virions of *Pariacoto virus* contain a minor protein translated from the second AUG codon of the capsid protein open reading frame. J. Gen. Virol. *84*, 2847–2852.

Johnson, K.N., Johnson, K.L., Dasgupta, R., Gratsch, T., and Ball, L.A. (2001). Comparisons among the larger genome segments of six nodaviruses and their encoded RNA replicases. J. Gen. Virol. *82*, 1855–1866.

Johnson, K.N., Tang, L., Johnson, J.E., and Ball, L.A. (2004b). Heterologous RNA encapsidated in Pariacoto virus-like particles forms a dodecahedral cage similar to genomic RNA in wild-type virions. J. Virol. *20*, 11371–11378.

Johnson, K.N., Zeddam, J.L., and Ball, L.A. (2000). Characterization and construction of functional cDNA clones of *Pariacoto virus*, the first Alphanodavirus isolated outside Australasia. J. Virol. *74*, 5123–5132.

Joklik, W.K., and Darnell, J.E. (1961). The adsorption and early fate of purified poliovirus in HeLa cells. Virology *13*, 439–447.

Kaesberg, P. (1987). Organization of bipartite insect virus genomes: the genome of black beetle virus. In The molecular biology of the positive strand RNA viruses D.J. Rowlands, M.A. Mayo, and B.W.J. Mahy, eds. (London, Academic Press), pp. 207–218.

Kaesberg, P., Dasgupta, R., Sgro, J.Y., Wery, J.P., Selling, B.H., Hosur, M.V., and Johnson, J.E. (1990). Structural homology among four nodaviruses as deduced by sequencing and X-ray crystallography. J. Mol. Biol. *214*, 423–435.

Kamer, G., and Argos, P. (1984). Primary structural comparison of RNA-dependent polymerases from plant, animal and bacterial viruses. Nucleic Acids Res. *12*, 7269–7282.

Kampmueller, K.M., and Miller, D.J. (2005). The cellular chaperone heat shock protein 90 facilitates *Flock House virus* RNA replication in *Drosophila* cells. J. Virol. *79*, 6827–6837.

Kanaji, S., Iwahashi, J., Kida, Y., Sakaguchi, M., and Mihara, K. (2000). Characterization of the signal that directs Tom20 to the mitochondrial outer membrane. J. Cell Biol. *151*, 277–288.

Koonin, E.V. (1991). The phylogeny of RNA-dependent RNA polymerases of positive-strand RNA viruses. J. Gen. Virol. *72*, 2197–2206.

Kopek, B.G., Perkins, G., Miller, D.J., Ellisman, M.H., and Ahlquist, P. (2007). Three-dimensional analysis of a viral RNA replication complex reveals a virus-induced mini-organelle. PLoS Biol. *5*, e220.

Körber, S., Shaik Syed Ali, P., and Chen, J.C. (2009). Structure of the RNA-Binding domain of Nodamura virus protein B2, a suppressor of RNA interference. Biochemistry *48*, 2307–2309.

Krishna, N.K., Marshall, D., and Schneemann, A. (2003). Analysis of RNA packaging in wild-type and mosaic protein capsids of *Flock House virus* using recombinant baculovirus vectors. Virology *305*, 10–24.

Krishna, N.K., and Schneemann, A. (1999). Formation of an RNA heterodimer upon heating of nodavirus particles. J. Virol. *73*, 1699–1703.

Kuroda, R., Ikenoue, T., Honsho, M., Tsujimoto, S., Mitoma, J.Y., and Ito, A. (1998). Charged amino acids at the carboxyl-terminal portions determine the intracellular locations of two isoforms of cytochrome b5. J. Biol. Chem. *273*, 31097–31102.

Larson, S.B., and McPherson, A. (2001). Satellite tobacco mosaic virus RNA: structure and implications for assembly. Curr. Opin. Struct. Biol. *11*, 59–65.

Li, H., Li, W.X., and Ding, S.W. (2002). Induction and suppression of RNA silencing by an animal virus. Science *296*, 1319–1321.

Li, Y., and Ball, L.A. (1993). Nonhomologous RNA recombination during negative-strand synthesis of Flock House virus RNA. J. Virol. *67*, 3854–3860.

Lindenbach, B.D., Sgro, J.Y., and Ahlquist, P. (2002). Long-distance basepairing in Flock House virus RNA1 regulates subgenomic RNA3 synthesis and RNA2 replication. J. Virol. *76*, 3905–3919.

Lingel, A., Simon, B., Izaurralde, E., and Sattler, M. (2005). The structure of the Flock House virus B2 protein, a viral suppressor of RNA interference, shows a novel mode of double-stranded RNA recognition. EMBO Rep. *6*, 1149–1155.

Liu, C., Zhang, J., Wang, J., Lu, J., Chen, W., Cai, D., and Hu, Y. (2006a). Sequence analysis of coat protein gene of Wuhan nodavirus isolated from insect. Virus Res. *121*, 17–22.

Liu, C., Zhang, J., Yi, F., Wang, J., Wang, X., Jiang, H., Xu, J., and Hu, Y. (2006b). Isolation and RNA1 nucleotide sequence determination of a new insect nodavirus from *Pieris rapae* larvae in Wuhan city, China. Virus Res. *120*, 28–35.

Lonberg-Holm, K., and Korant, B.D. (1972). Early interaction of rhinoviruses with host cells. J. Virol. *9*, 29–40.

Longworth, J.F., and Carey, G.P. (1976). A small RNA virus with a divided genome from *Heteronychus arator* (F.) [Coleoperai Scarabaeidae]. J. Gen. Virol. *33*, 31–40.

Lorenzi, R., and Burrone, O.R. (1999). Sequence-specific antibodies against human IgE isoforms induced by an epitope display system. Immunotechnology *4*, 267–272.

Lu, R., Maduro, M., Li, F., Li, H.W., Broitman-Maduro, G., Li, W.X., and Ding, S.W. (2005). Animal virus replication and RNAi-mediated antiviral silencing in *Caenorhabditis elegans*. Nature *436*, 1040–1043.

Lyle, J.M., Bullitt, E., Bienz, K., and Kirkegaard, K. (2002). Visualization and functional analysis of RNA-dependent RNA polymerase lattices. Science *296*, 2218–2222.

Maia, L.F., Soares, M.R., Valente, A.P., Almeida, F.C., Oliveira, A.C., Gomes, A.M., Freitas, M.S., Schneemann, A., Johnson, J.E., and Silva, J.L. (2006). Structure of a membrane-binding domain from a non-enveloped animal virus: insights into the mechanism of membrane permeability and cellular entry. J. Biol. Chem. *281*, 29278–29286.

Manayani, D.J., Thomas, D., Dryden, K.A., Reddy, V., Siladi, M.E., Marlett, J.M., Rainey, G.J., Pique, M.E., Scobie, H.M., Yeager, M., *et al.* (2007). A viral nanoparticle with dual function as an anthrax antitoxin and vaccine. PLoS Pathog. *3*, 1422–1431.

Marshall, D., and Schneemann, A. (2001). Specific packaging of nodaviral RNA2 requires the N-terminus of the capsid protein. Virology *285*, 165–175.

Miller, D.J., and Ahlquist, P. (2002). *Flock House virus* RNA polymerase is a transmembrane protein with amino-terminal sequences sufficient for mitochondrial localization and membrane insertion. J. Virol. *76*, 9856–9867

Miller, D.J., Schwartz, M.D., and Ahlquist, P. (2001). *Flock House virus* RNA replicates on outer mitochondrial membranes in *Drosophila* cells. J. Virol. *75*, 11664–11676.

Miller, D.J., Schwartz, M.D., Dye, B.T., and Ahlquist, P. (2003). Engineered retargeting of viral RNA replication complexes to an alternative intracellular membrane. J. Virol. *77*, 12193–12202.

Mori, K., Nakai, T., Muroga, K., Arimoto, M., Mushiake, K., and Furusawa, I. (1992). Properties of a new virus belonging to *Nodaviridae* found in larval striped jack (*Pseudocaranx dentex*) with nervous necrosis. Virology *187*, 368–371.

Munday, B.L., Kwang, J., and Moody, N. (2002). Betanodavirus infections of teleost fish: a review. J. Fish Dis. *25*, 127–142.

Newman, J.F., and Brown, F. (1976). Absence of poly (A) from the infective RNA of Nodamura virus. J. Gen. Virol. *30*, 137–140.

Nguyen, H.D., Mekuchi, T., Imura, K., Nakai, T., Nishizawa, T., and Muroga, K. (1994). Occurrence of viral nervous necrosis (VNN) in hatchery-reared juvenile japanese flounder *Paralichthys olivaceus*. Fisheries Sci. *60*, 551–554.

Novoa, R.R., Calderita, G., Arranz, R., Fontana, J., Granzow, H., and Risco, C. (2005). Virus factories: associations of cell organelles for viral replication and morphogenesis. Biol. Cell. *97*, 147–172.

Nugent, C.I., Johnson, K.L., Sarnow, P., and Kirkegaard, K. (1999). Functional coupling between replication and packaging of poliovirus replicon RNA. J. Virol *73*, 427–435.

Odegard, A.L., Kwan, M.H., Walukiewicz, H.E., Banerjee, M., Schneemann, A., and Johnson, J.E. (2009). Low endocytic pH and capsid protein autocleavage are critical components of *Flock House virus* cell entry. J. Virol. *83*, 8628–8637.

Peng, M., Dai, C.B., and Chen, Y.D. (2005). Expression and immunoreactivity of an epitope of HCV in a foreign epitope presenting system. World J. Gastroenterol. *11*, 3363–3367.

Portney, N.G., Singh, K., Chaudhary, S., Destito, G., Schneemann, A., Manchester, M., and Ozkan, M. (2005). Organic and inorganic nanoparticle hybrids. Langmuir *21*, 2098–2103.

Price, B.D., Eckerle, L.D., Ball, L.A., and Johnson, K.L. (2005). *Nodamura virus* RNA replication in *Saccharomyces cerevisiae*: heterologous gene expression allows replication-dependent colony formation. J. Virol. *79*, 495–502.

Price, B.D., Roeder, M., and Ahlquist, P. (2000). DNA-Directed expression of functional *Flock House virus* RNA1 derivatives in *Saccharomyces cerevisiae*, heterologous gene expression, and selective effects on subgenomic mRNA synthesis. J. Virol. *74*, 11724–11733.

Price, B.D., Rueckert, R.R., and Ahlquist, P. (1996). Complete replication of an animal virus and maintenance of expression vectors derived from it in *Saccharomyces cerevisiae*. Proc. Natl. Acad. Sci. U.S.A. *93*, 9465–9470.

Reinganum, C., Bashiruddin, J.B., and Cross, G.F. (1985). Boolarra virus: a member of the *Nodaviridae* isolated from *Oncopera intricoides* (Lepidoptera: Hepialidae). Intervirology *24*, 10–17.

Salonen, A., Ahola, T., and Kaariainen, L. (2005). Viral RNA replication in association with cellular membranes. Curr. Top. Microbiol. Immunol. *285*, 139–173.

Scherer, W.F., and Hurlbut, H.S. (1967). Nodamura virus from Japan: a new and unusual arbovirus resistant to diethyl ether and chloroform. Am. J. Epidemiol. *86*, 271–285.

Scherer, W.F., Verna, J.E., and Richter, W (1968). Nodamura virus, an ether- and chloroform-resistant arbovirus from Japan: physical and biological properties, with ecologic observations. Am. J. Trop. Med. Hyg. *17*, 120–128.

Schneemann, A., Ball, L.A., Delsert, C., Johnson, J.E., and Nishizawa, T. (2005). Family *Nodaviridae*. In Virus Taxonomy: Eighth Report of the International Committee on Taxonomy of Viruses, C.M. Fauquet, M.A. Mayo, J. Maniloff, U. Desselberger, and L.A. Ball, eds. (New York, Elsevier), pp. 865–872.

Schneemann, A., Dasgupta, R., Johnson, J.E., and Rueckert, R.R. (1993). Use of recombinant baculoviruses in synthesis of morphologically distinct viruslike particles of Flock House virus, a nodavirus. J. Virol. *67*, 2756–2763.

Schneemann, A., Gallagher, T.M., and Rueckert, R.R. (1994). Reconstitution of Flock House provirions: a model system for studying structure and assembly. J. Virol. *68*, 4547–4556.

Schneemann, A., and Marshall, D. (1998). Specific encapsidation of nodavirus RNAs is mediated through the C-terminus of capsid precursor protein alpha. J. Virol. *72*, 8738–8746.

Schneemann, A., Zhong, W., Gallagher, T.M., and Rueckert, R.R. (1992). Maturation cleavage required for infectivity of a nodavirus. J. Virol. *66*, 6728–6734.

Scodeller, E.A., Tisminetzky, S.G., Porro, F., Schiappacassi, M., De Rossi, A., Chiecco-Bianchi, L., and Baralle, F.E. (1995). A new epitope presenting system displays a HIV-1 V3 loop sequence and induces neutralizing antibodies. Vaccine *13*, 1233–1239.

Scotti, P.D., Dearing, S., and Mossop, D.W. (1983). Flock House virus: a nodavirus isolated from *Costelytra zealandica* (White) (Coleoptera: Scarabaeidae). Arch. Virol. *75*, 181–189.

Selling, B.H. (1986). Infectivity of black beetle virus in cultured *Drosophila* cells. (PhD Thesis, University of Wisconsin, Madison).

Selling, B.H., Allison, R.F., and Kaesberg, P. (1990). Genomic RNA of an insect virus directs synthesis of infectious virions in plants. Proc. Natl. Acad Sci. U.S.A. *87*, 434–438.

Settles, E.W., and Friesen, P.D. (2008). Flock House virus induces apoptosis by depletion of *Drosophila* inhibitor-of-apoptosis protein DIAP1. J. Virol. *82*, 1378–1388.

Singh, G., Popli, S., Hari, Y., Malhotra, P., Mukherjee, S., and Bhatnagar, R.K. (2009). Suppression of RNA silencing by Flock House virus B2 protein is mediated through its interaction with the PAZ domain of Dicer. FASEB J. *23*, 1845–1857.

Stapleford, K.A., Rapaport, D., and Miller, D.J. (2009). Mitochondrion-enriched anionic phospholipids facilitate *Flock House virus* RNA polymerase membrane association. J. Virol. *83*, 4498–4507.

Sullivan, C.S., and Ganem, D. (2005). A virus-encoded inhibitor that blocks RNA interference in mammalian cells. J. Virol. *79*, 7371–7379.

Tang, L., Johnson, K.N., Ball, L.A., Lin, T., Yeager, M., and Johnson, J.E. (2001). The structure of *Pariacoto virus* reveals a dodecahedral cage of duplex RNA. Nat. Struct. Biol. *8*, 77–83.

Tang, L., Lin, C.S., Krishna, N.K., Yeager, M., Schneemann, A., and Johnson, J.E. (2002). Virus-like particles of a fish nodavirus display a capsid subunit domain organization different from that of insect nodaviruses. J. Virol. *76*, 6370–6375.

Tihova, M., Dryden, K.A., Le, T.V., Harvey, S.C., Johnson, J.E., Yeager, M., and Schneemann, A. (2004). Nodavirus coat protein imposes dodecahedral RNA structure independent of nucleotide sequence and length. J. Virol. *78*, 2897–2905.

Tisminetzky, S.G., Scodeller, E.A., Evangelisti, P., Chen, Y., Schiappacassi, M., Porro, F., Bizik, F., Zacchi, T., Lunazzi, G., Miertus, S., *et al.* (1994). Immunoreactivity of chimeric proteins carrying the HIV-1 epitope IGPGRAF. Correlation between predicted conformation and antigenicity. FEBS Lett. *353*, 1–4.

Toffolo, V., Negrisolo, E., Maltese, C., Bovo, G., Belvedere, P., Colombo, L., and Valle, L.D. (2007). Phylogeny of betanodaviruses and molecular evolution of their RNA polymerase and coat proteins. Mol. Phylogenet. Evol. *43*, 298–308.

van Rij, R.P., and Berezikov, E. (2009). Small RNAs and the control of transposons and viruses in Drosophila. Trends Microbiol. *17*, 163–171.

Van Wynsberghe, P.M., and Ahlquist, P. (2009). 5′ cis elements direct nodavirus RNA1 recruitment to mitochondrial sites of replication complex formation. J. Virol. *83*, 2976–2988.

Van Wynsberghe, P.M., Chen, H.R., and Ahlquist, P. (2007). Nodavirus RNA replication protein a induces membrane association of genomic RNA. J. Virol. *81*, 4633–4644.

Venter, P.A., Krishna, N.K., and Schneemann, A. (2005). Capsid protein synthesis from replicating RNA directs specific packaging of the genome of a multipartite, positive-strand RNA virus. J. Virol. *79*, 6239–6248.

Venter, P.A., Marshall, D., and Schneemann, A. (2009). Dual roles for an arginine-rich motif in specific genome recognition and localization of viral coat protein to RNA replication sites in Flock House virus-infected cells. J. Virol. *83*, 2872–2882.

Venter, P.A., and Schneemann, A. (2007). Assembly of two independent populations of Flock House virus particles with distinct RNA packaging characteristics in the same cell. J. Virol. *81*, 613–619.

Venter, P.A., and Schneemann, A. (2008). Recent insights into the biology and biomedical applications of Flock House virus. Cell Mol. Life Sci. *65*, 2675–2687.

Voinnet, O. (2001). RNA silencing as a plant immune system against viruses. Trends Genet. *17*, 449–459.

Walukiewicz, H.E., Banerjee, M., Schneemann, A., and Johnson, J.E. (2008). Rescue of maturation-defective *Flock House virus* infectivity with noninfectious, mature, viruslike particles. J. Virol. *82*, 2025–2027.

Walukiewicz, H.E., Johnson, J.E., and Schneemann, A. (2006). Morphological changes in the $T = 3$ capsid of *Flock House virus* during cell entry. J. Virol. *80*, 615–622.

Wang, Q.M., Hockman, M.A., Staschke, K., Johnson, R.B., Case, K.A., Lu, J., Parsons, S., Zhang, F., Rathnachalam, R., Kirkegaard, K., *et al.* (2002). Oligomerization and cooperative RNA synthesis activity of hepatitis C virus RNA-dependent RNA polymerase. J. Virol. *76*, 3865–3872.

Wang, X.H., Aliyari, R., Li, W.X., Li, H.W., Kim, K., Carthew, R., Atkinson, P., and Ding, S.W. (2006). RNA interference directs innate immunity against viruses in adult *Drosophila*. Science *312*, 452–454.

Weeks, S.A., and Miller, D.J. (2008). The heat shock protein 70 cochaperone YDJ1 is required for efficient membrane-specific *Flock House virus* RNA replication complex assembly and function in *Saccharomyces cerevisiae*. J. Virol. *82*, 2004–2012.

Weeks, S.A., Shield, W.P., Sahi, C., Craig, E.A., Rospert, S., and Miller, D.J. (2010). A targeted analysis of cellular chaperones reveals contrasting roles for heat shock protein 70 in *Flock House virus* RNA replication. J. Virol. *84*, 330–339.

Xiong, X.Y., Liu, X., and Chen, Y.D. (2005). Expression and immunoreactivity of HCV/HBV epitopes. World J. Gastroenterol. *11*, 6440–6444.

Zhong, W., Dasgupta, R., and Rueckert, R. (1992). Evidence that the packaging signal for nodaviral RNA2 is a bulged stem-loop. Proc. Natl. Acad. Sci. U.S.A. *89*, 11146–11150.

Zhong, W., and Rueckert, R.R. (1993). Flock House virus: down-regulation of subgenomic RNA3 synthesis does not involve coat protein and is targeted to synthesis of its positive strand. J. Virol. *67*, 2716–2722.

Tetraviruses

12

Rosemary A. Dorrington and James R. Short

Abstract

The *Tetraviridae* are a relatively unknown family of small RNA viruses that exclusively infect the larvae of lepidopteron insect species, many of which are important agricultural pests. While the structure of tetraviral capsids has been well studied, little is known about their biology and the interactions between virus and host that determine their unusually narrow host range and tissue tropism. This is because of the problems associated with developing suitable experimental systems for studying their biology *in vivo*. This chapter will review recent advances in the understanding of the biology of tetraviruses, highlighting new information on the function and expression of viral translation products and the development of biological systems for elucidating the molecular mechanisms of infection and viral replication.

Introduction

The *Tetraviridae* are unique in that they are the only known family of RNA viruses that exclusively infects insects. The host range of tetraviruses is limited to the order Lepidoptera (butterflies and moths) with most viruses isolated from Saturniid, Limacodid and Noctuid moths, many of which are important agricultural pests. Tetravirus particles consist of one or two single-stranded (ss), positive-sense (+ve) genomic RNAs encapsidated in a non-enveloped icosahedral capsid with unique $T = 4$ icosahedral symmetry. The viral genome may encode two or three genes that are translated directly into proteins, which may be processed further into smaller, functionally distinct peptides (Fauquet *et al.*, 2005).

The first described tetravirus was isolated from diseased larvae of the pine emperor moth (*Nudaurelia cytherea capensis*) in South Africa (Hendry *et al.*, 1968). These larvae were found to be infected with at least five structurally distinct small RNA viruses, designated *Nudaurelia capensis α, β, γ, δ,* and *ε virus* according to their electrophoretic migration in a sucrose gradient (Juckes, 1970, 1974). *Nudaurelia capensis β virus* (NβV), the most plentiful and extensively characterized of these viruses, was thought to have a single coat protein of between 60 and 62 kDa and a (+ve) ssRNA genome (Struthers and Hendry, 1974; Juckes, 1979). Analysis of the structure of the virus particles subsequently revealed a new type of virus capsid with unusual $T = 4$ icosahedral symmetry (Finch *et al.*, 1974) and NβV became the type-member of a new group of viruses named the *Nudaurelia* β-like viruses (Matthews, 1982). In an earlier report a small virus from the Australian gum emperor moth, *Antheraea eucalypti* (Grace and Mercer, 1965), was shown to be serologically related to NβV (Grace and Mercer 1976), and *Antheraea eucalypti* virus (AeV) became the second member of the new group of viruses. The *Nudaurelia* β-like virus group was renamed in the fifth report of the International Committee on Taxonomy of Viruses (ICTV) as the *Tetraviridae* family (Francki *et al.*, 1991) due to the unique $T = 4$ quasi-symmetry of their capsids (the name derived from the Greek *tettares* which means 'four').

In 1985 Hendry *et al.* isolated another small virus from diseased *N. capensis* larvae, which they named *Nudaurelia capensis* ω virus (NωV). Although it displayed the same $T = 4$ icosahedral surface symmetry as NβV, NωV was serologically distinct from NβV and contained a bipartite genome, as opposed to the monopartite genomes of NβV and the seven other members of the *Tetraviridae* (Hendry *et al.*, 1985). Based on this, NωV became the type member of a second genus (*Nudaurelia* ω-like tetraviruses) within the family, with the β-like tetraviruses being the other genus (Murphy *et al.*, 1995). As with other families of insect viruses that are likely to act as natural biocontrol agents for wild populations of their hosts, there was considerable interest in using tetraviruses such as *Darna trima* virus (DtV), as pest control agents (Moore *et al.*, 1985). The discovery of a second *Nudaurelia* ω-like tetravirus, *Helicoverpa armigera stunt* virus (HaSV), in a collapsed laboratory colony of *Helicoverpa armigera* (Hanzlik *et al.*, 1993), focused renewed attention on the *Tetraviridae* and their potential applications as biocontrol agents for agricultural pests (Christian *et al.*, 1993, 2005; Hanzlik *et al.*, 1999).

Over the past thirty-five years, research on the *Tetraviridae* has focused on the structure and assembly of their unique capsids (discussed in detail in Chapter 14 of this volume), the molecular characterization of their genomes, and developing their potential as biocontrol agents for important agricultural pests. On the other hand, our understanding of their pathogenicity and replication biology has lagged behind. This has been primarily due to a lack of tissue culture cell lines susceptible to infection and difficulties in establishing laboratory colonies of most of the host species. The discovery in 2003 of *Providence* virus (PrV), which replicates in *Helicoverpa zea* and *Spodoptera exigua* cell lines (Pringle *et al.*, 2003), has provided exciting new opportunities for studying tetravirus replication and virus–host interactions *in vivo*. This chapter will review recent advances in the understanding of the biology of tetraviruses, highlighting new information on the expression and function of viral translation products and the development of biological systems for elucidating the molecular mechanisms of infection and viral replication.

Taxonomy, host range and tissue culture

Taxonomy

While the *Tetraviridae* are not the only RNA virus family that infects insects, they are distinguished by the characteristic $T = 4$ icosahedral symmetry of their capsids and their unusually narrow host range. Tetraviruses are generally identified by the size of their capsids (approximately 40 nm in diameter) and the presence of major and minor capsid proteins of 58–64 kDa and 6–8 kDa in size, respectively. Members of this family are currently classified into two genera according to their genome organization, namely the betatetraviruses with monopartite genomes and the omegatetraviruses, which have bipartite genomes (Table 12.1). There are eleven assigned members of the *Tetraviridae* family with a further 14 unassigned viruses (Fauquet *et al.*, 2005; Walter *et al.*, 2008). Of these, nine species are classified within the genus *Betatetravirus* and two within the genus *Omegatetravirus* (Fauquet *et al.*, 2005). In the absence of cDNA sequence data, many of the tetraviruses were originally classified according to their serological relationship, with no cross-reaction occurring between members of the two genera (Fauquet *et al.*, 2005). Most of the betatetraviruses, with the exception of *Pseudoplusia includens* virus (PiV), are serologically related to some degree (Grace and Mercer, 1976; Reinganum *et al.*, 1978; Greenwood and Moore, 1981; Chao *et al.*, 1983; Hendry and Agrawal, 1993). The betatetraviruses, AeV and NβV, which infect species belonging to the subfamily *Saturniinae*, are serologically indistinguishable (Grace and Mercer, 1976). However, it is unlikely that they are isolates of the same species since their capsid sizes (32 nm and 39.7 nm in diameter for AeV and NβV, respectively) are significantly different (Grace and Mercer, 1965; Struthers and Hendry, 1974; Juckes, 1979). Not all omegatetraviruses are serologically related (Hanzlik *et al.*, 1993; Yi *et al.*, 2005), despite significant sequence conservation between their capsid proteins (Hanzlik *et al.*, 1995; Yi *et al.*, 2005). Thus the use of serological criteria may not be the most reliable method for distinguishing between tetravirus species.

Table 12.1 Members of the family *Tetraviridae* (adapted from Fauquet *et al., 2005)*

Virus species	Acronym	Host family	Geographical location	Particle diameter (nm)	GENBANK Accession No	References[5]
Genus: *Betatetravirus*						
Darna trima virus	DtV	*Lymacodidae*	Malaysia	38	NA	1–3
Dasychira pudibunda virus	DpV	*Lymantriidae*	UK	38	NA	2–3
Calliteara pudibunda virus[1]	CpV					
Euprosterna elaeasa virus	EeV	*Zygaenidae*	Peru	40	[AF461742]	4, 5
Nudaurelia capensis β *virus*[2]	NβV	*Saturniidae*	South Africa	38	[AF102884]	1, 6–8
Antheraea eucalypti virus[3]	AeV		Australia	32		
Philosamia cynthia x ricini virus	PxV	*Saturniidae*	UK	35	NA	2
Pseudoplusia includens virus	PiV	*Noctuidae*	USA	40	NA	9
Providence virus	PrV	*Noctuidae*	USA	40	[GU991616]	10
Thosea asigna virus	TaV	*Lymacodidae*	Malaysia, USA	38	[AF282930], [AF062037]	2, 11
Setothosea asigna virus[4]	SaV				NA	
Trichoplusia ni virus	TnV	*Noctuidae*	USA	38	NA	12
Genus: *Omegatetravirus*						
Helicoverpa armigera stunt virus	HaSV	*Noctuidae*	Australia	38	RNA1 [U18246], [EU345431], RNA2 [L37299]	3
Nudaurelia capensis ω *virus*[2]	NωV	*Saturnidae*	South Africa	41	RNA2 [DQ054400]	14–15
Dendrolimus punctatus tetravirus	DpTV	*Lasiocampidae*	China	40	RNA1 [AY594352] RNA2 [AY594353]	16

[1]Host renamed *Calliteara pudibunda.*

[2]Type strain for the genus.

[3]Serologically indistinguishable from NβV (Grace and Mercer, 1976).

[4]Host renamed *Thosea asigna*.

[5]References: 1, Juckes *et al.* (1973); 2, Reinganum *et al.* (1978); 3, King and Moore (1984); 4, Gorbalenya *et al.* (2002); 5, Zeddam *et al.* (2009); 6. Struthers and Hendry (1974); 7, Gordon *et al.* (1999); 8, Grace and Mercer (1965); 9, Chao *et al.* (1983); 10, Pringle *et al.* (2003); 11, Pringle *et al.* (1999); 12, Morris *et al.* (1979); 13, Hanzlik *et al.* (1993); 14, Hendry *et al.* (1985); 15, Agrawal and Johnson (1992); 16, Yi *et al.* (2005).

Host range and tissue culture

The majority of tetraviruses have been isolated from Saturniid, Lymacodid and Noctuid moths (Table 12.1). In general, the most striking feature of the *Tetraviridae* is their narrow host rage, often confined to a single species or selected species within one family (Fauquet *et al.*, 2005). An exception to this rule is a report by Greenwood and Moore (1984) that *Trichoplusia ni* virus (TnV) and *Dasychira pudibunda* virus (DpV) infect species in families other than that of their normal host. The host range of the *Omegatetravirus* HaSV is restricted to Noctuid species, including *H. armigera, H. zea, Heliothis virescens, Heliocheilus pallida* (subfamily *Heliothenae*) as well as *Earias huegeliana* and *Chrysodiexis argentifera* (Christian *et al.*, 2001). PrV replicates in *H. zea* and *S. exigua* tissue culture cell lines, but not in *Spodoptera frugiperda* cells (Pringle *et al.*, 2003). The Saturniid, *N. capensis* is host to at least two (NβV and NωV) and probably three other tetraviruses including NεV, which was isolated along with NβV (Juckes 1979). Recently, Walter *et al.* (2008) surveyed a wild population of *N. capensis* over a three year period. They isolated three distinct small RNA viruses during this time, including NβV, which had not been found in wild South African *Nudaurelia* populations for almost 30 years. The other two viruses, designated *Nudaurelia capensis* ζ virus (NζV) and *Nudaurelia capensis* ψ virus (NψV), were serologically related to NβV and NωV, respectively, but were distinct in terms of virion diameter, surface morphology and the sizes of their major capsid proteins. These data suggested that tetraviruses might be ubiquitous in the wild and that the majority of species remain, as yet, undiscovered.

The major problem with studying the biology of tetraviruses is the lack of tissue culture cell lines permissive to tetravirus replication. Tripconey (1970) reported detection of NβV replication in primary cultures of ovarian cells of *Bombyx mori* and *N. capensis*. However, several researchers have been unable to repeat these experiments using continuous culture cells (Moore, 1991). Bawden *et al.* (1999) attempted to infect 13 insect tissue culture cell lines and one mammalian cell line with HaSV particles, or transfection with HaSV genomic RNAs. Included amongst these were several pupal ovarian cell lines derived from heliothine species, the larvae of which were later shown to be susceptible to HaSV infection by Christian *et al.* (2001). However, Northern analysis failed to detect replication in any of the tissue culture cell lines (Bawden *et al.*, 1999). The negative results were attributed to the absence of host factors required for replication of HaSV and the lack of suitable receptors to mediate virus binding and cell entry in the tissue culture cell lines used, because analysis of infected *H. armigera* larvae showed that replication was confined to midgut cells.

The discovery of PrV in 2003, which is the only tetravirus to replicate in cell culture, significantly broadened the scope of research into tetravirus biology. PrV was isolated as a persistent infection in a *H. zea* midgut cell line (MG8) and shown to infect *H. zea* fat body (FB33) cells and replicate in an embryonic *S. exigua* (Se-1) cell line (Pringle *et al.*, 2003). A multiplicity of infection (MOI) of 10^3 particles per cell was required for a productive infection of the FB33 cell line while an MOI of 10^5 per cell was needed to infect the Se-1 cell line, leading to the conclusion that PrV is probably not very infectious. Since early passages of the MG8 cell line (passage 8) were also infected with PrV, Pringle *et al.* (2003) concluded that the laboratory colony of *H. zea* from which the cell line was derived (Goodman and McIntosh, 1995), was likely to have been infected with PrV, suggesting the virus may be present in field populations of *H. zea*. To date, our attempts to establish PrV infection of *H. armigera* neonates have been inconclusive (Short, J.R. and Dorrington, R.A., unpublished results).

Pathology

While there is general agreement that tetraviruses target the midgut tissues of the larvae of host insects, the severity of symptoms observed in infected animals varies from one virus species to another and appears also to be dependent upon the developmental phase of the larva. There are several reports that tetraviruses are highly pathogenic and that infection results in significant mortality (Grace and Mercer, 1965; Tripconey, 1970; Greenwood and Moore, 1984; Hanzlik *et al.*, 1993). In contrast, Vail *et al.* (1983) observed that exposure to high doses of TnV resulted in growth retardation but low mortality in *T. ni* larvae, while Greenwood and Moore (1984) observed only slightly retarded growth of TnV-infected larvae. Virus biofeed assays on *N. capensis* larvae hatched from eggs collected from the wild, showed that infection with NβV or the unassigned NβV-like virus NζV, were discoloured, lethargic and flaccid and most died 16 days post-feeding. *N. capensis* neonate larvae infected with NωV or the unassigned NωV-like virus, NψV, had a 100% mortality rate after 16 days with infected animals showing symptoms of regurgitation and diarrhoea (Walter *et al.*, 2008). These symptoms were consistent with earlier observations of larvae infected with NβV and NωV (Juckes, 1970; Hendry *et al.*, 1985).

The apparently conflicting results obtained by different groups working on the same viruses may be the result of bioassays having been conducted on larvae in different developmental stages or differences between virus preparations used in the different studies. Support for developmental resistance to tetravirus infection is provided in a study on the pathology of HaSV by Christian *et al.* (2001). Surface contamination bioassays showed that HaSV was highly pathogenic to early instar larvae of *H. armigera*, with an average EC_{50} (50% effective concentration of virus) below 1300 virus particles per mm^2 and a high mortality rate. These results were in agreement with similar studies on TnV pathogenicity, although TnV infection resulted in significantly reduced mortality (Vail *et al.*, 1983; Moore, 1991). Infection of later instar larvae with similar doses of HaSV resulted in asymptomatic infection and the virus was carried through by larvae that went on to pupate (Christian *et al.*, 2001), which is probably the route by which tetraviruses are transmitted from one generation to the next in the wild. An important observation was the rapid onset of symptoms following exposure to HaSV in first instar larvae, growth retardation occurred within 24 hours of exposure to the virus. Furthermore, a significant reduction in feeding was observed in larger larvae after 24 hours (Christian *et al.*, 2001).

Tetraviruses target the midgut tissues of their hosts (Moore *et al.*, 1985; Bawden *et al.*, 1999; Brooks *et al.*, 2002). TnV particles were detected by enzyme-linked immunosorbent assay (ELISA) in midgut tissues of *T. ni* larvae and later in the infection cycle, also associated with immature sex organs and fat bodies (Greenwood and Moore, 1984). Northern analysis of HaSV-infected larvae showed that this omegatetravirus also replicates in midgut tissues (Bawden *et al.*, 1999). Transmission electron microscopy (TEM) analysis of tissues isolated from larvae infected with TnV and DpV demonstrated the presence of virus particles organized in crystalline arrays in the cytoplasm, within what appeared to be cytoplasmic vesicles (Greenwood and Moore, 1984; Moore, 1991). Pringle *et al.* (2003) conducted a TEM study of persistently-infected MG8 and PrV-infected FB33 cells and found that virus particles were also confined to the cytoplasm and, in some cases, within a structure that could be membrane-bound. Very few PrV particles were observed in *S. exsigua* Se-1 cells, but they were also associated with a cytoplasmic structure.

Brooks *et al.* (2002) conducted a detailed study on the pathology of HaSV infection in *H. armigera* larvae. Virus particles were detected in the cytoplasm of three of the four midgut cell types, namely columnar, goblet and regenerative cells but not in endocrine cells. When first instar larvae were infected, HaSV particles were detected in basal regenerative cells and 5 days post infection approximately 75% of the midgut cells were infected. When older larvae were infected, there appeared to be smaller foci of infection within the midgut and in the later stages (7 days post infection), no virus particles could be detected in the midgut cells. HaSV-infection resulted in an accumulation of lysosomes and increased cytoplasmic vacuolation and endoplasmic vesiculation in infected cells. There was an increase in shedding (up to 45%) of cells into the midgut lumen. Finally, TdT-mediated dUTP nick end-labelling (TUNEL) tests demonstrated a correlation between HaSV infection and apoptosis, which would account for the observed increase in cell shedding (Brooks *et al.*, 2002).

It is likely that the primary site of HaSV infection is the basal regenerative cells from where the infection spreads as these cells differentiate into columnar and goblet cells (for a recent review on insect midgut growth and development, see Hakim *et al.*, 2009). This hypothesis would account for the spread of HaSV infection in neonate larvae to almost all the midgut cells because the major increase in midgut cell number occurs during the first and second instars (Brooks *et al.*, 2002). Exposure of older larvae to the virus would result in smaller foci of infection because in later stages of larval development the number of basal regenerative cells relative to differentiated midgut cells decreases, although they are present throughout all stages of larval development preceding pupation (Hakim *et al.*, 2009). Since stem cells are inherently more resistant to apoptosis, infection of the basal regenerative cells of larvae in late stages of development could result in a persistent infection that could carry over into pupation. Support for this hypothesis is provided by the isolation of PrV as a persistent infection in the *H. zea* MG8 cell line. Interestingly, unlike other insect tissue culture cell lines, the MG8 cells do not appear to be a primary cell culture. These cells do not require foetal calf serum in their culture medium and can be cultured to high passage number (Goodman and McIntosh, 1995), suggesting that this cell line might be derived from basal regenerative cells. If this hypothesis holds, then it is possible that other tetraviruses may also establish persistent infections in the midgut stem cells of their hosts.

Virus composition and properties

The tetravirus capsid

Tetravirus virions have non-enveloped, spherical particles that exhibit $T = 4$ icosahedral shell quasi-symmetry, ranging in diameter from 32 to 41 nm (Table 12.1). Betatetravirus virions have an average M_R of 18×10^6, a sedimentation coefficient of 194–217S and a buoyant density in CsCl of between 1.28 and 1.33 g/cm^3. Omegatetravirus particles have a lower M_R (16×10^6), similar sedimentation coefficients to those of betatetraviruses (194–217S) and their buoyant densities in CsCl vary from 1.28 to 1.30 g/cm^3 depending upon the species (Fauquet *et al.*, 2005). NβV and NωV have been found to be stable under acidic conditions (Hendry *et al.*, 1985), while HaSV virions are stable between pH 2.8 and 10.0 and at temperatures up to 55°C (Christian *et al.*, 2001). In addition, HaSV virions are resistant to treatment with chloroform, chymotrypsin, trypsin, or RNase A (Christian *et al.*, 2001).

Tetravirus particles assemble as procapsids composed of 240 identical capsid protein precursor subunits. During maturation, autoproteolytic cleavage of the precursor protein at its C-terminus produces the β and γ subunits (Agrawal and Johnson, 1995; Hanzlik *et al.*, 1995; Gordon *et al.*, 1999; Pringle *et al.*, 1999, 2003). The cleavage of the γ-peptide results in the stabilization of the mature virus particle, which is thought to be important for infectivity (Munshi *et al.*, 1996; Taylor *et al.*, 2002; Helgstrand *et al.*, 2004). In the betatetraviruses *Thosea asigna* virus (TaV), *Euprosterna elaeasa* virus (EeV) and PrV, the capsid protein precursor undergoes additional co-translational processing at the amino terminus prior to assembly via the activity of a 2A-like site (Pringle *et al.*, 1999, 2003; Gorbalenya *et al.*, 2002; Zeddam *et al.*, 2010). Whilst TEM analysis shows that only mature tetravirus particles are isolated from their natural hosts, small amounts of the capsid protein precursor can be present in these virus preparations (Fauquet *et al.*, 2005) suggesting that not all the capsid protein subunits in mature particles have undergone maturation-dependent autoproteolytic cleavage. This phenomenon has also been observed in capsids of the structurally related nodaviruses (Gallagher and Rueckert, 1988), where it seems that incomplete cleavage of the capsid subunits does not necessarily affect infectivity.

Baculovirus expression of omegatetravirus capsid coding sequences in *S. frugiperda* cells results in the assembly of virus-like particles (VLPs) (Hanzlik *et al.*, 1999; Canady *et al.*, 2000; Taylor *et al.*, 2002). These VLPs can be isolated either as procapsids or mature capsids, depending on the purification protocol (Agrawal and Johnson, 1995; Canady *et al.*, 2000; Pringle *et al.*, 2001). Cryo-EM analysis of NωV VLPs has shown that procapsids and capsids are structurally distinct with the procapsid being rounded and the mature capsid polyhedral in shape (Canady *et al.*, 2000). The molecular mechanism of transition from procapsid to capsid in NωV VLPs has been well characterized *in vitro* (discussed in detail in Chapter 14). Incubation of NωV procapsids at or below pH 5.0 results in their maturation into VLPs that are structurally indistinguishable from the wild type virus capsids (Canady *et al.*, 2000, 2001). It has been proposed that lowering the pH results in protonation of glutamate and aspartate side chains that facilitate a conformational change and shrinking of the capsid shell enabling autoproteolytic cleavage of the γ-peptide. The conformational change is reversible, provided autoproteolytic cleavage of the γ-peptide is inhibited (Taylor *et al.*, 2002). How the conformational changes observed *in vitro* relate to the infectious life cycle of tetraviruses in their host cells is not clear, but data reported by Tomasicchio *et al.* (2007) demonstrate that HaSV VLPs undergo maturation following induction of apoptosis in yeast cells. Programmed cell death is known to be associated with lysosomal and mitochondrial destabilization that leads to cytosolic acidification (Matsuyama and Reed, 2000; Nilsson *et al.*, 2006). Tomasicchio *et al.* (2007) hypothesize that viral replication may promote apoptosis in infected cells, which results in the cytosolic acidification required for the maturation of virus particles. An alternative hypothesis is that the virus may exploit the native defence system of the insect host, which is to shed virus-infected cells via apoptosis, as a trigger for maturation of virus particles.

Viral RNAs

Omegatetravirus particles encapsidate two (+ve) ssRNA molecules of approximately 5.3 kb and 2.5 kb, that are not polyadenylated at their 3′ ends. Omegatetraviral genomic RNAs are capped at their 5′ ends and, as is the case for NβV, terminate at their 3′ ends with tRNA-like structures (Gordon *et al.*, 1995; Hanzlik *et al.*, 1995; Du Plessis *et al.*, 2005; Yi *et al.*, 2005). The HaSV tRNA-like structures can be charged with valine *in vitro* (unpublished data cited in Gordon *et al.*, 1999). tRNA-like structures that can be aminoacylated with valine, histidine or lysine, are also found at the 3′ ends of (+ve) ssRNA plant virus genomic RNAs. In plant viruses, these structures play a role in promoting replication of the minus strand RNAs, encapsidation of viral RNAs and maintaining the integrity of the 3′ end of the viral RNA (reviewed in Dreher, 2009). Betatetravirus virions contain a single (+ve) ssRNA segment of between 5.8 and 6.6 kb representing approximately 10% of the particle mass (Fauquet *et al.*, 2005). In addition, betatetravirus particles may also encapsidate a subgenomic message for the capsid protein precursor, derived from the 3′ end of the genomic RNA (Gordon *et al.*, 1999; Pringle *et al.*, 1999, 2003). The genomic RNA is not polyadenylated at its 3′ end, nor is it blocked at the 5′ end (King and Moore, 1984; Gordon *et al.*, 1999; Pringle *et al.*, 2003). Bioinformatic analysis suggests the presence of a pseudoknot at the 3′ ends of the EeV and TaV genomic RNAs, which may be involved in the regulation of genome replication in these two betatetraviruses or play a role in preventing degradation of the genomic RNA by nucleases (Zeddam *et al.*, 2010). The genomic RNA of NβV was found to terminate in a tRNA-like structure similar to that found in the omegatetraviruses (Gordon *et al.*, 1999) leading to the suggestion that these tRNA-like structures might fulfil similar functions in the life cycle of HaSV and NβV. Other predicted secondary structures have also been identified on the HaSV and NωV genomic RNAs, most notably a stem and loop at the 5′ end of both HaSV RNAs and NωV RNA2, where the sequence 'GGUAAA' is conserved in the loop region (Gordon *et al.*, 1995). However, this hairpin loop is not present in DpTV (Yi *et al.*, 2005).

Genome organization and viral translation products

Genome organization of the betatetraviruses

Currently, the complete genome sequences of four betatetraviruses, namely NβV, TaV, EeV and PrV, have been determined (Gordon *et al.*, 1999; Pringle *et al.*, 1999; Gorbalenya *et al.*, 2002; Zeddam *et al.*, 2010; Walter *et al.*, 2010). The genomes vary in size from 5698 nts (EeV) and 5714 (TaV) to 6625 nts (NβV) in length and encode two overlapping open reading frames (ORFs), with the viral replicase located at the 5′ end and the capsid protein precursor (VCAP) at the 3′ end (Gordon *et al.*, 1999; Gorbalenya *et al.*, 2002; Zeddam *et al.*, 2010). The replicase ORF of EeV and TaV overlaps *VCAP* by 536 and 529 nts, respectively, while the NβV replicase ORF is significantly larger and almost entirely overlaps *VCAP* (Fig. 12.1). Accordingly, the replicases of both TaV and EeV comprise 1257 amino acids, with a M_R of approximately 140 kDa (Zeddam *et al.*, 2010). This corresponds to the larger of two translation products obtained from *in vitro* translation experiments using genomic RNAs extracted from TaV particles (King and Moore, 1984). In NβV and TaV, the VCAP is translated from a subgenomic RNA originating just upstream of the *VCAP* coding sequence (Gordon *et al.*, 1999; Pringle *et al.*, 1999) and it is likely that the same is true for EeV. The TaV and EeV VCAP ORFs encode proteins with a predicted M_R of 82.5 kDa (Pringle *et al.*, 1999; Zeddam *et al.*, 2010), while the protein encoded by the NβV VCAP ORF has a predicted M_R of 66 kDa (Gordon *et al.*, 1999). There is a 2A-like processing site located near the N-terminus of the TaV VCAP, the activity of which results in the production of a 17 kDa amino-terminal peptide (p17) as well as the capsid protein precursor of 65 kDa. The capsid protein precursor is then autoproteolytically cleaved at its C-terminus during particle maturation to produce the 58 kDa and 7 kDa proteins found in mature capsids (Pringle *et al.*, 1999). This 2A-like site is conserved in the same position in the EeV capsid protein precursor ORF (Zeddam *et al.*, 2010). The activity of these 2A-like sites and the

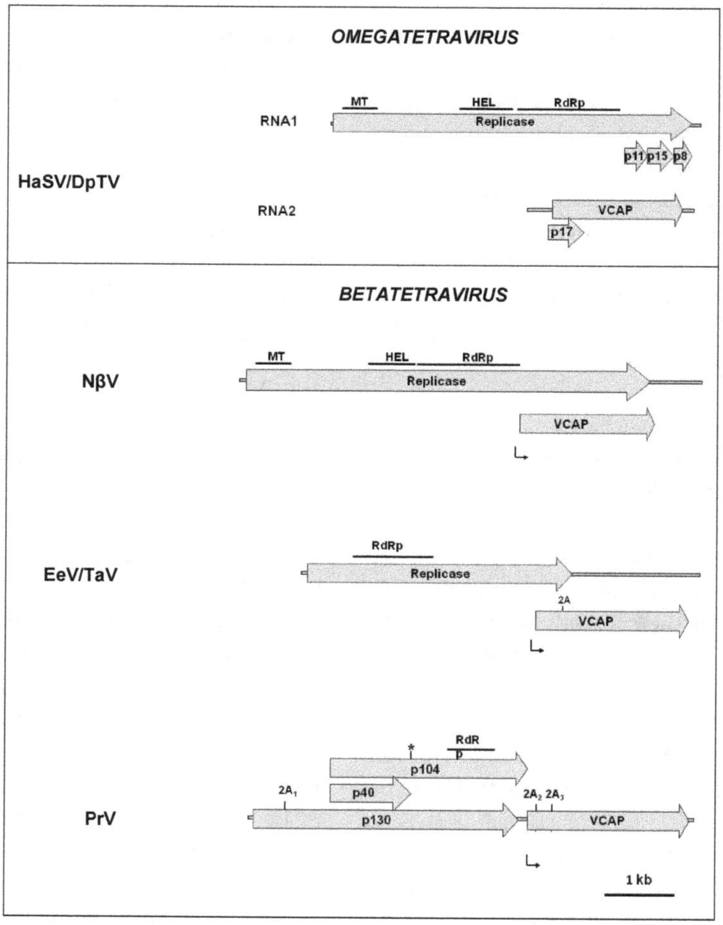

Figure 12.1 Genome organization of the *Tetraviridae*. The positions of the betatetraviral 2A-like processing sequences (2A) and read-through stop codon (*) are indicated as are the functional domains of the viral replicases (MT, methyltransferase; HEL, Helicase; RdRp, RNA-dependent RNA polymerase). The bent arrow shows the start of the betatetraviral subgenomic RNA. Full names of the viruses and their GenBank Accession numbers for NβV, TaV, EeV, HaSV, DpTV and PrV are listed in Table 12.1.

potential function of the translation products produced will be discussed later in this chapter.

Analysis of the full-length PrV cDNA sequence has revealed an atypical genome organization raising the possibility that PrV might be quite different to other *Tetraviridae* (Walter *et al.*, 2010). Northern analysis demonstrated that PrV has a monopartite genome, resulting in the classification of this virus within the genus *Betatetravirus*. When Pringle *et al.* (2003) analysed the sequence of two PrV cDNA clones which included the ORF encoding a capsid protein precursor with a predicted M_R of 81 kDa, two putative 2A-like processing sites (PrV-2A_2 and PrV-2A_3) were discovered near the amino terminus of the VCAP coding sequence. The activity of these 2A-like sites would result in the production of two small peptides of 7 and 8 kDa along with the capsid protein precursor of 68 kDa. As in other betatetraviruses, the ORF encoding VCAP is located at the 3′ end of the viral genome (Fig. 12.1). Upstream of *VCAP* is an ORF encoding a putative replicase (p104), which ends with a stop codon (UGA) four nucleotides upstream of the AUG start of *VCAP*. The putative PrV replicase ORF is distinguished from other tetravirus replicases in that it encodes a read-through stop codon that results in the translation of a smaller product (p40) in addition to p104. The genome of PrV encodes a unique third, large ORF (p130) at its 5′ end and upstream of the replicase ORF. A third 2A-like processing site is present in the N-terminus of p130, the activity of which is predicted to result in the translation of two products of 17 kDa and 113 kDa, respectively. The functions of p130 or the products of 2A-like processing are at present still unknown (Walter *et al.*, 2010).

Genome organization of omegatetraviruses

Currently, the full cDNA sequence of two omegatetraviruses, HaSV and *Dendrolimus punctatus* tetravirus (DpTV), are known (Gordon *et al.*, 1995; Hanzlik *et al.*, 1995; Yi *et al.*, 2005). In addition, the cDNA sequence of NωV RNA2 has also been determined (Agrawal and Johnson, 1992; Du Plessis *et al.*, 2005). As stated before, omegatetraviruses have a bipartite genome with two capped, non-polyadenylated strands of RNA (Fauquet *et al.*, 2005). HaSV RNA1 (5312 nt) and DpTV RNA1 (5492 nts) encode the viral replicase with predicted M_R of 187 and 180 kDa, respectively (Gordon *et al.*, 1995; Yi *et al.*, 2005). RNA1 also encodes three additional small ORFs (p11, p15 and p8), all three of which overlap with the 3′ end of the replicase ORF in HaSV (Fig. 12.1), while only p11 and p15 overlap with the replicase ORF in DpTV. In both HaSV and DpTV the p11 stop codon is immediately upstream of and in-frame with the AUG start of p15. In HaSV the p8 AUG is three nucleotides downstream of the p15 stop codon, while the DpTV p8 is in a different reading frame and is 91 nts downstream of the stop codon of p15 (Yi *et al.*, 2005; J.R. Short, and R.A. Dorrington, unpublished data). A survey of wild-type HaSV cDNA sequences in this region confirms the conservation of the positions of the three in-frame ORFs as well as their amino acid sequence, suggesting that these ORFs may play a role in the life cycle of the virus (V.A. Bailey, J.R. Short and R.A. Dorrington, unpublished data). Interestingly, we have identified three similar small ORFs in the unusually long 3′ untranslated region (UTR), downstream of the NβV replicase ORF. Whether or not these ORFs are translated into functional proteins is not known. The RNA2 strands of NωV, HaSV and DpTV are 2445, 2478 and 2490 nts in length, respectively (Du Plessis *et al.*, 2005; Hanzlik *et al.*, 1995; Yi *et al.*, 2005). The RNA2 5′ UTR is followed by two ORFs, the first encoding a protein with a predicted M_R of 17 kDa (p17) and the second encoding the capsid protein precursor, ranging between 70 and 71 kDa in size (Fig. 12.1).

The capsid protein

A phylogenetic analysis of available tetravirus capsid sequences along with a member of the *Dicistroviridae* family, *Cricket paralysis virus* (CrPV), as a phylogenetic outlier, shows that the three omegatetraviruses are closely related, with an average of 66% sequence identity between their capsid proteins (Fig. 12.2). The phylogram also implies a close relationship between the capsid proteins of the two betatetraviruses, TaV and EeV, which share 77% sequence identity with each other and approximately 27% identity with the NβV capsid, which shares a common

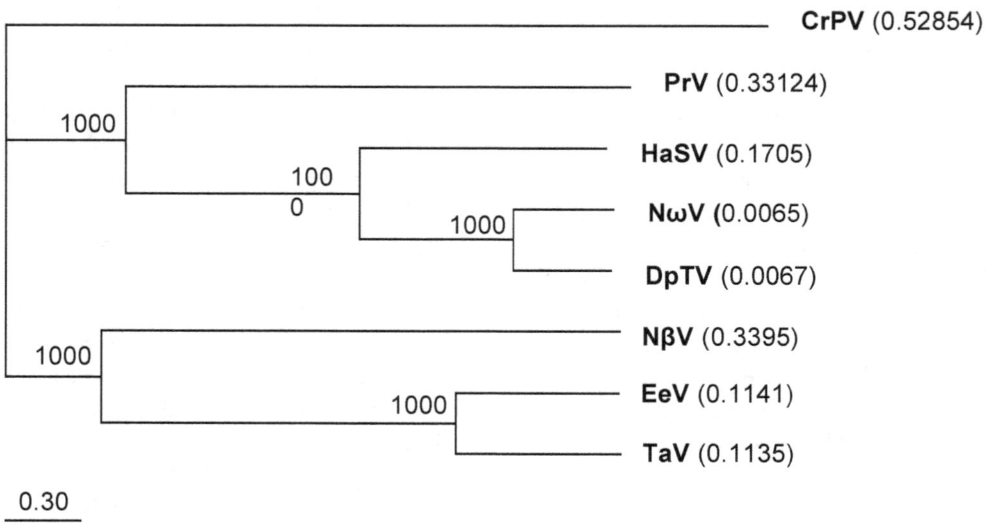

Figure 12.2 Phylogram of aligned tetravirus capsid precursor sequences. Capsid amino acid sequences were aligned using ClustalW and TreeView was used to generate the phylogram (gap extension penalty = 2 and gap creation penalty = 8). Bootstrap values are indicated at the node of each branch. The phylogenetic distance between each sequence is represented next to the sequence name in brackets and the distance scale is shown in the bottom left-hand corner of the phylogram. Accession numbers for tetravirus capsid precursor amino acid sequences are listed in Table 12.1 and the structural protein of the genetic outlier, *Cricket paralysis virus* (CrPV), was derived from information deposited on GenBank Accession Number NP647482.

ancestor with these two viruses. The data suggests that the two betatetraviruses, EeV and TaV, co-evolved from a common unidentified ancestor, and that the same is true for NωV and DpTV. The capsid protein of PrV clusters together with those of the three omegatetraviruses and not the betateraviruses (Walter *et al.*, 2010).

The crystal structure of the omegatetravirus, NωV, has been solved at 2.8 Å resolution using wild type virus prepared from infected insects collected in the field (Munshi *et al.*, 1996; Helgestrand *et al.*, 2004). As stated earlier, tetravirus capsids comprise 240 subunits of a single capsid protein precursor that undergoes C-terminal autoproteolytic cleavage during particle maturation. Each subunit consists of three regions: (1) a positively charged internal helical domain that interacts with genomic RNA; (2) a beta-barrel (jelly roll domain) that comprises the capsid shell and (3) an external immunoglobulin-like (Ig-like) domain, proposed to function in receptor binding during infection of host cells (Munshi *et al.*, 1996; Helgestrand *et al.*, 2004). There is significant sequence conservation between the coat proteins of the three omegatetraviruses (67% identity between HaSV and NωV, 68% between DpTV and NωV and 66% between DpTV and HaSV). The X-ray crystallographic structures of the NωV and HaSV capsid proteins can be almost entirely superimposed, suggesting that these proteins are functionally equivalent (Taylor, 2003). However, there are significant differences between these two omegatetraviral capsid proteins, within the approximately 165 amino acids comprising the Ig-like domain (37% identity) as well as a 49 amino acid region located at the N-terminus corresponding to the internal helical domain (40% identify). The same applies when one compares the capsid sequences of TaV and EeV, where the jelly roll domain and C-terminal γ-peptide are highly conserved while the N-terminal region and Ig-like domain are not conserved. This is not unexpected since the Ig-like domains would be responsible for binding host cell specific receptors and the internal helical domain may interact in some specific way with viral genomic RNAs during particle assembly.

Interestingly, there is little sequence conservation between the Ig-like domains of NωV and NβV, which infect the same host (Gordon *et al.*, 1999), suggesting that these two viruses may bind different receptors on the host cell surface.

As mentioned above, expression of NωV and HaSV capsid proteins in insect cells and yeast results in the efficient assembly of VLPs that are morphologically indistinguishable from the wild type virus particles (Agrawal and Johnson, 1995; Canady *et al.*, 2000; Tomasicchio *et al.*, 2007). NβV VLPs also readily assemble in *S. frugiperda* cells expressing the NβV capsid protein precursor (Walter, 2005). Pringle *et al.* (1999) observed the assembly of TaV VLPs from insect cells infected with recombinant baculovirus expressing the TaV capsid coding sequence. VLP preparations appeared to contain a significant number of partially assembled particles or particles displaying large holes, suggesting that assembly of TaV VLPs was not as efficient as observed with the omegatetraviruses. Pringle *et al.* (1999) also found a decrease in capsid assembly in cells expressing the capsid coding sequence with the region upstream of the 2A-like processing site deleted, leading to the suggestion that the peptide released by 2A-like processing, p17, was involved in stabilising the capsids. Taylor *et al.* (2005) were unable to detect the assembly of PrV VLPs in insect cells expressing either the entire capsid coding sequence (p81) or the coding sequence downstream of the 2A-processing sites (p68). Together, these data suggest that there might be different assembly mechanisms for TaV and PrV particles, with a more stringent requirement for the presence of viral RNA sequences (Taylor *et al.*, 2005).

Omegatetravirus P17

The p17 ORF is found upstream of and overlapping with the VCAP ORF in all three omegatetraviruses. In contrast to VCAP, there is less sequence conservation between the omegatetravirus p17 ORFs with 38.4% amino acid identity between NωV and HaSV p17 and 67.7% and 44.0% identity between DpTV and NωV and DpTV and HaSV, respectively (Hanzlik *et al.*, 1995; Du Plessis *et al.*, 2005; Yi *et al.*, 2005). All three predicted proteins contain an unusually high proportion of proline, glutamate, serine and threonine (PEST) amino acids (Hanzlik *et al.*, 1995; du Plessis *et al.*, 2005; Yi *et al.*, 2005), a characteristic of transcription factors and cell cycle regulators that are subject to high turnover rates (Salaman *et al.*, 1994; Rechsteiner and Rogers, 1996). This led Hanzlik *et al.* (1995) to suggest that p17 might have a regulatory function in the virus life cycle. They concluded that HaSV p17 was not a structural protein because they were unable to detect it in virus particles or infected insect tissues using rabbit polyclonal antibodies raised against recombinant His-tagged p17. Expression of HaSV p17 in *Escherichia coli* resulted in the production of hexagonal tube-like structures leading Hanzlik *et al.* (1995) to propose a role as a viral movement protein. Recently it has been shown that recombinant DpTV p17 can be cross-linked to RNA corresponding to the 177 nt stretch of the DpTV RNA2 3′ UTR. Recombinant DpTV p17 also binds to the RNA2 5′ UTR (373 nt) as well as single-stranded mRNA and double-stranded RNA transcribed from the enhanced green fluorescent protein ORF, but has no affinity for DNA or the 3′ UTR (506 nts) of RNA1 (Zhou *et al.*, 2008). Thus, while the authors demonstrated that p13 was able to interact with RNA, they were unable to show that these interactions were specific. These results led Zhou *et al.* (2008) to speculate that p17 is likely to be involved in viral RNA replication. We have detected p17 expression in HaSV-infected midgut cells at levels several orders of magnitude lower than that of VCAP and have also been able to show that p17 co-purifies with wild type HaSV particles at a ratio of approximately 2–4 molecules per virus particle (Vlok, M., Tomasicchio, M. and Dorrington, R.A., unpublished data). These results lead us to propose that omegatetravirus p17 might be involved in the assembly-dependent packaging of viral RNAs (discussed in more detail below under 'Regulation of viral gene expression').

The viral replicase

While their capsid structure is conserved, tetravirus replicases are widely divergent with respect to their physical characteristics (M_R), amino acid sequence conservation and functional organization. A phylogenetic analysis of the known tetravirus RdRp sequences reveals that they are clustered into three distinct groups (Fig. 12.3), each belonging to a different superfamily of viral RNA replicases (Koonin, 2010; Walter *et al.*, 2010). Firstly, as shown by Gorbalenya *et al.* (2002) and Zeddam *et al.* (2009), TaV and EeV form a group distantly related to the picorna-like viruses along with the birnaviruses, Infectious pancreatic necrosis virus (IPNV) and Infectious bursal disease virus (IBDV), while the alpha-like RdRps of NβV and HaSV form a second, distinct cluster. Unexpectedly, PrV is isolated in this phylogram, being more closely related to the tombusvirus, Pelargonium chlorotic ring pattern virus (PCRPV), and other carmo-like viruses than to any of the tetraviruses (Walter *et al.*, 2010). These phylogenetic relationships are supported by the observation that the seemingly diverse replicase molecular masses observed across the family are surprisingly similar within these groupings. The TaV and EeV replicases both have predicted molecular masses of 140 kDa while NβV and HaSV encode far larger proteins (predicted molecular masses of 215 and 187 kDa respectively) and PrV makes use of a read-through stop to process two putative replication proteins of 40 and 104 kDa (Gordon *et al.*, 1995; Gordon *et al.*, 1999; Gorbalenya *et al.*, 2002; Zeddam *et al.*, 2009; Walter *et al.*, 2010).

Sequence analysis of the NβV, HaSV and DpTV replicases (the DpTV RdRp domain is 67% identical to HaSV) has led to the proposal that these viruses be classified within the alpha-like virus superfamily (Gordon *et al.*, 1995; Gordon *et al.*, 1999; Yi *et al.* 2005). Each replicase encodes the conserved methytransferase (MT), helicase (HEL) and RdRp domains in an N- to C-terminal modular arrangement that is characteristic of all alpha-like replicases (reviewed in Van der Heijden and Bol, 2002). The three replicases display similarity in terms of predicted

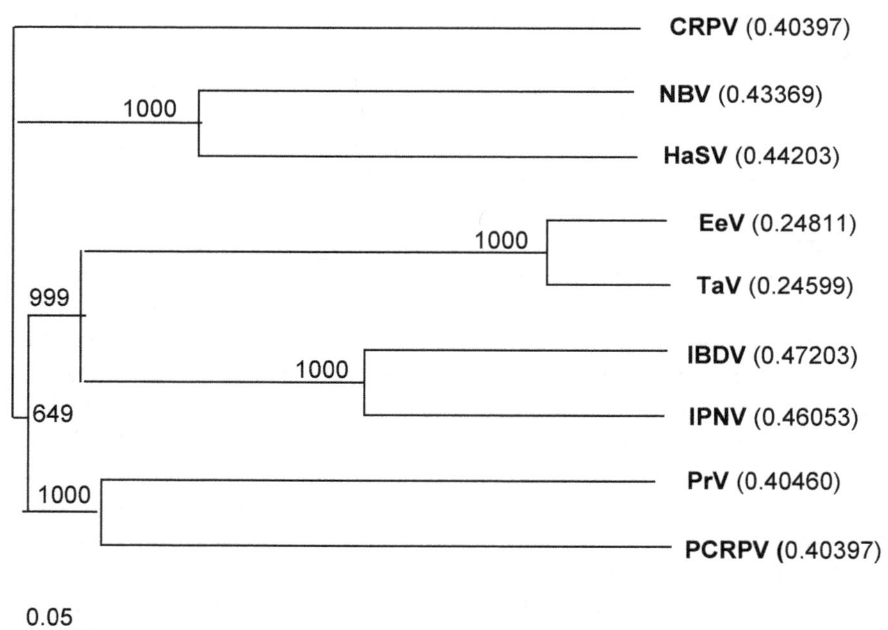

Figure 12.3 Phylogram of aligned tetravirus RdRp sequences, two birnaviruses related to TaV and EeV, PCRPV related to PrV and the genetic outlier CrPV. ClustalW was used to align the RdRp amino acid sequences of HaSV, NβV, PrV, TaV, EeV, Infectious pancreatic necrosis virus (IPNV), Infectious bursal disease virus (IBDV), Pelargonium chlorotic ring pattern virus (PCRPV) and CrPV and a phylogam was generated using TreeView. For full names of the viruses and accession numbers, refer to Table 12.1. The phylogenetic distance between each sequence is represented next to the sequence name and the distance scale is shown in the bottom left-hand corner of the phylogram. Bootsrap values are indicated at the each branch's node. Accession numbers of the tetravirus sequences are listed in Table 12.1 and the sequences for CrPV and PCRPV were derived from GenBank (Accession numbers NP647482 and YP052925.1 respectively).

size (215, 187 and 180 kDa, respectively) and amino acid sequence with DpTV and HaSV being more closely related to each other than to NβV. DpTV and HaSV display 67%, 63% and 76% sequence identity over the MT, HEL and RdRp domains, respectively. In contrast to that observed between DpTV and NβV (38% for the MT domain and 36% for the RdRp domain) and 45%, 42% and 33% identity between the MT, HEL and RdRp domains of HaSV and NβV, respectively (Gordon *et al.*, 1995; Gordon *et al.*, 1999; Yi *et al.*, 2005). The RdRp domain of the DpTV replicase has been expressed in *E. coli* and shown to initiate primer-independent synthesis of viral RNA *in vitro* (Zhou *et al.*, 2006). It is worth noting that although there are a number of insect-borne animal (alphaviruses) and plant viruses (marafiviruses) within the alpha-like viruses, these tetraviruses represent the only insect-specific family within the group (Van der Heijden and Bol, 2002). Another feature of these three tetraviruses that is unusual among the alpha-like viruses, is an apparent lack of co- or post-translational processing, a characteristic shared with a small group of plant viruses that includes the *Foveavirus* and *Potexvirus* genera (Van der Heijden and Bol, 2002).

In contrast, TaV and EeV form part of a separate group that, along with two members of the dsRNA birnaviruses and the *Drosophila* A virus, is distantly related to members of the picorna-like virus superfamily (Gorbalenya *et al.*, 2002; Ambrose, *et al.*, 2009; Zeddam *et al.*, 2010). This distinct cluster carries an internally permuted C-A-B arrangement of the canonical A, B and C motifs found in palm subdomain of all template-dependent polynucleotide polymerases (Poch *et al.*, 1989; Gorbalenya *et al.*, 2002; Ambrose *et al.*, 2009). Gorbalenya *et al.* (2002) were able to modify the sequence of the Poliovirus RdRp and re-model the loop regions such that it was able to accommodate the non-canonical arrangement showing that the permutation was compatible with the architecture of the canonical palm subdomain. A putative nucleotidylation (VPg) signal is present upstream of the TaV and EeV RdRp domains, a feature shared with the *Birnaviridae* and apparently absent from other tetravirus replicases (Zeddam *et al.*, 2010). The distinct grouping of TaV and EeV is further supported by the fact that their replicases share 68% identity over the full length of the protein (1257 aa for both) and both seem to lack discernible MT and HEL domains (Gorbalenya *et al.*, 2002), which is unusual for picorna-like replicases that usually encode a HEL domain (Van der Heijden and Bol, 2002). This data has led Zeddam *et al.* (2010) to propose that EeV and TaV be reclassified as a new family, the *Permutoviridae*, which is distinct from other members of the *Tetraviridae* family.

Finally, the PrV replicase is again distinct from the above two groups. The only similarity with the replicases of other betatetraviruses is the apparent absence of MT and HEL domains, as is the case with TaV and EeV. This along with the sequence similarity to replicases of the tombus- and umbraviruses (both (+ve) ssRNA plant viruses) places PrV within the carmo-like supergroup (Walter *et al.*, 2010). The PrV RdRp domain carries yet another unusual arrangement of the canonical A-B-C palm subdomain, with an additional A and (poorly conserved) B motif spatially arranged in a similar manner to the permuted RdRps of TaV and EeV. A further distinction between PrV and the other tetraviruses replicases is the presence of a read-through stop resulting in the production of two translation products from the replicase (Walter *et al.*, 2010). This feature is shared with the tombusviruses, where a truncated, accessory protein is produced in addition to the full length replicase, both of which are required to establish viral replication in infected plant cells. Taken together, these observations establish a compelling argument for the reclassification of the *Tetraviridae* and the establishment of two new virus families to accommodate TaV and EeV as one family and PrV as the first member of the other.

Regulation of viral gene expression

It is becoming increasingly evident that tetraviruses employ diverse systems to regulate expression of their genes. In the omegatetraviruses, and HaSV in particular, these include temporal regulation of the replication of the two viral genomic RNAs (discussed below under 'Virus replication'), down regulation of the initiation of translation of p17 versus the capsid protein precursor, p71, and increased p17 turnover. In contrast to Hanzlik *et al.* (1995), we have been able to detect expression of p17 in HaSV-infected *H. armigera* larval midgut tissues at levels several orders of magnitude lower than that of the viral capsid protein precursor protein. In addition, p17 was found to co-purify with wild type virus particles, but at low levels relative to the capsid protein (M. Vlok, M. Tomasicchio and R.A. Dorrington, unpublished data). The most likely explanation for low levels of expression of p17 is the poor context of the translational start (AUUUAUGA) as compared with the consensus for invertebrates (A/CAAA/CAUGG) (Cavener, 1987). This poor translational context for p17 is conserved in DpTV (UUAGAAUGG) and NωV (UAGAAUGG). In contrast, there is a better translational context for the VCAP ORF, downstream of the p17 AUG in all three omegatetraviruses. Our attempts to express p17 in yeast cells were unsuccessful, even with an optimized yeast Kozak and a high-level promoter. However, deletion of the 5′ UTR together with replacement of the native p17 Kozak with the invertebrate consensus Kozak sequence results in an increase in p17 expression, but still does not account for the low levels of expression relative to p71 in *S. frugiperda* cells (M. Vlok, M. Tomasicchio, and R.A. Dorrington, unpublished data; unpublished data cited in Gordon and Waterhouse, 2005). The high level of PEST amino acids predicted to be present in all three omegatetravirus *p17* translation products might account for these low levels of expression. It has been shown in yeast that proteins rich in PEST amino acids are targeted for degradation and the addition of a PEST sequence results in rapid turnover of a stable protein (Salaman *et al.*, 1994). While we currently have no direct evidence to support this hypothesis, we believe the same is true for *S. frugiperda* and other insect cells and that the presence of these PEST amino acids ensures that p17 is present at low levels relative to p71 (M. Vlok, M.Tomasicchio, and R.A. Dorrington, unpublished data).

The presence of active 2A-like processing sites at the amino-termini of the VCAP ORFs of TaV (TrV-2A), EeV (EeV-2A) and PrV (PrV-$2A_2$, PrV-$2A_3$) as well as in the PrV p130 ORF (PrV-$2A_1$), suggests that the expression of these betatetraviral genes is co-translationally regulated (Pringle *et al.*, 2001; Luke *et al.*, 2008; Walter *et al.*, 2010). In contrast to PrV-$2A_1$ and PrV-$2A_3$, processing at the PrV-$2A_2$ site is less efficient *in vitro* (Luke *et al.*, 2008). These data is consistent with our *in vitro* translation experiments that show translation products of approximately 15 kDa and 68 kDa, corresponding to a peptide with a predicted M_R of 13.2 kDa (p13) resulting from processing only at PrV-$2A_3$ and the capsid protein precursor, p68 (Walter *et al.*, 2010). The 15 kDa peptide does not cross-react with antibodies raised against PrV virus particles nor does it co-purify with wild type PrV virus particles. This has been observed with the N-terminal 2A-like processing product of TaV, p17 (Pringle *et al.*, 1999, 2001, 2003; Walter *et al.*, 2010). 2A sequences were first identified in the *Picornaviridae*, but they are present in other mammalian viruses (e.g. human type C rotaviruses) as well as a wide range of insect viruses (Luke *et al.*, 2008). The mechanism of 2A processing has been well characterized *in vitro* and a *Saccharomyces cerevisiae* experimental system has been developed to identify host-cell (ribosomal translation) factors that interact with these 2A sequences (Doronina *et al.*, 2008a,b). Importantly, attention is now being focused on the function of the products of 2A-like processing in the virus life cycle. An analysis of the PRV p13 amino acid sequence reveals an unusually high concentration of PEST residues, similar to the omegatetraviral p17. This, together with the position of the p13 coding sequence immediately upstream of the capsid protein precursor ORF, suggests p13 may fulfil a similar function in PrV as p17 in the omegatetraviruses (discussed in more detail below under 'Packaging of viral RNAs').

The PrV replicase ORF (p104) contains an in-frame read-through stop codon following amino acid 359, which facilitates in leaky read-through by the ribosome. This results in the production of a shorter translation product (p40) in addition to the full length replicase (p104) at a ratio of approximately 10:1 (Walter *et al.*, 2010). This translation signal enables the virus to produce two translation products from the same ORF as well as regulating the ratio of these two proteins relative to each other. Interestingly, the read-through stop appears to be suppressed in mammalian systems because we have observed that only p104 is produced in a rabbit reticulocyte coupled *in vitro* transcription-translation system. In contrast, as observed in PrV-infected cells, p40 is the major translation product with ten-fold less p104 in a wheat germ coupled *in vitro* transcription-translation system (Walter *et al.*, 2010). Read-through stop signals, which have been best studied in plant viruses, are also found in some animal viruses including reoviruses and retroviruses, where they are involved in regulation of *gag-pol* expression (Beier and Grimm, 2001; Goff, 2004). The PrV stop signal shares a high sequence similarity with the Group 1 set of read-through stop signals, all of which have thus far been found only in plant viruses. In contrast, the PrV signal shares little homology with read-through stop signals found in insect and animal viruses, raising interesting questions as to how it functions in insect cells (Walter *et al.*, 2010).

Virus replication

The absence of tissue culture cell lines permissive for tetravirus replication (until the discovery of PrV by Pringle *et al.* in 2003) and the difficulties associated with studies on whole animals (Bawden *et al.*, 1999; Brooks *et al.*, 2002) has meant that most research on the tetraviruses has focused on the structure, assembly and maturation of the viral capsid, molecular characterization of their genomes and potential uses as biocontrol agents. As a result very little is known about how tetraviruses replicate in their host cells although the stringent tissue tropism of these viruses suggests unique interactions required for tetravirus replication in host cells.

In the first study on tetravirus replication, Bawden *et al.* (1999) analysed *H. armigera* for the presence of viral RNAs and detected RNA1 three days after infection with HaSV, while small amounts of RNA2 could be detected after a further two days. This data suggests that RNA1, encoding the replicase, is replicated during the early stages of infection and that replication of RNA2 occurs during the later stages, effectively ensuring that expression of the capsid protein precursor (p71) and p17 would occur later in the virus life cycle. This phenomenon has been well characterized in the replication of the nodavirus, Flock House virus. Nodaviruses also have a bipartite genome with RNA1 encoding the viral replicase and two small ORFs which overlap with the 3′ end of the replicase ORF. RNA2 encodes the virus capsid protein precursor. Nodaviruses produce a subgenomic RNA (RNA3) originating from the 3′ end of RNA1. RNA3 encodes two small ORFs, B1 and B2. B2 has been shown to inhibit Dicer function in infected cells (Galiana-Arnoux *et al.*, 2006). RNA3 is also required for promoting replication of RNA2 during the later stages of infection (Eckerle *et al.*, 2003) Bawden *et al.* (1999) did not detect the presence of any subgenomic RNAs in HaSV-infected larvae, which suggests that the mechanism for regulating viral genome replication in HaSV might be different. However, it is worth noting that omegatetravirus RNA1 encodes three small ORFs that overlap with the 3′ end of the replicase ORF and there are also three small ORFs at the 3′ end of the NβV replicase ORF. The translation products of these small ORFs might fulfil a similar function to those found in the nodaviruses. Pringle *et al.* (2003) detected the synthesis of negative sense copies of the PrV subgenomic RNA indicating that subgenomic RNA may be replicated by this tetravirus, rather than simply being transcribed off the negative sense genomic RNA. Whether or not this is true for other betatetraviruses, remains to be determined.

Recent work in our laboratory has focused on developing experimental systems for the study of tetravirus replication *in vivo* and the results have begun to shed some light on this area, particularly with regards to HaSV and PrV. The observations of a correlation between the onset of apoptosis and HaSV infection in larval midgut cells (Brooks *et al.*, 2002) together with maturation of VLPs (Tomasicchio *et al.*, 2007) led to the proposal that virus replication might induce apoptosis to promote particle maturation during its life cycle in infected cells (Tomasicchio *et al.*, 2007). Since all RNA viruses are known to replicate in association with host cell membranes, it is plausible that these membranes might be derived from organelles that are involved in the onset of programmed cell death and acidification of the cytoplasm (Matsuyama and Reed, 2000; Nilsson *et al.*, 2006). We investigated the potential site of HaSV replication using transient expression of HaSV replicase-enhanced green fluorescent protein (EGFP) fusions in *S. frugiperda* Sf9 and mammalian (HeLa and BHK-21) cells and viewed the subcellular localization of fluorescence by confocal microscopy (Short *et al.*, 2010). Expression of the HaSV replicase with EGFP fused to its C-terminus results in a punctate distribution of fluorescence throughout the cytoplasm in live and fixed cells (Fig. 12.4A and B). However this punctate pattern is abolished when EGFP is fused to the N-terminus of the replicase, implying a function for the N-terminus in subcellular localization of the replicase (Fig. 12.4C). We have identified two domains, located at the N-terminus and within the C-terminal half of the HaSV replicase that are both required for targeting of the replicase–EGFP fusion to punctate structures within the cytoplasm (J. R. Short, and R.A. Dorrington, unpublished data).

Live cell imaging has shown that the EGFP fluorescence associated with the punctate structures is distinct from the mitochondria and lysosomes in both insect and mammalian cells (Short *et al.*, 2010). The absence of suitable antibodies against other organelle markers precluded immunofluorescence co-localization studies in the insect cells so we focused on mammalian (HeLa) cells for which such marker antibodies are available. These experiments showed that the punctate structures were not associated with the Golgi apparatus, the endoplasmic reticulum and early endosomes, but there was partial overlap of the EGFP signal with the late endosome marker protein, CD63, 16 and 18 hours post-transfection but not after 20 or 24 hours (Short *et al.*, 2010). Additional evidence for the endosomal origin of the punctate structures was obtained though time-lapse live cell imaging of transfected cells. The punctate structures appeared to undergo fusion, fission and 'kiss-and-run' events as well as what appeared to be cellular transport along defined pathways (Short *et al.*, 2010), which have been described as characteristics of the behaviour of late endosomes (Bright *et al.*, 2005). To further investigate the origin of the punctate structures, cells expressing the replicase–EGFP fusion were incubated with Cascade Blue®-conjugated dextran, which is incorporated into the endocytic pathway via endocytosis. We were unable to detect any overlap between the fluorescent punctate structures and endocytosed dextran, suggesting that the punctate structures may represent non-functional endocytic vesicles, possibly resulting from the exclusion of host cell proteins and/or destabilization of the membranes (Short *et al.*, 2010). These data lead us to propose that the HaSV replicase may recruit and remodel membranes from the endocytic pathway as the first step towards establishing viral replication complexes. The site of tetravirus replication is the larval midgut cells, which would have a well-developed endocytic pathway given their function in the absorption of nutrients. Thus it is not entirely surprising that replication might occur in association with membranes derived from the endocytic pathway (Short *et al.*, 2010).

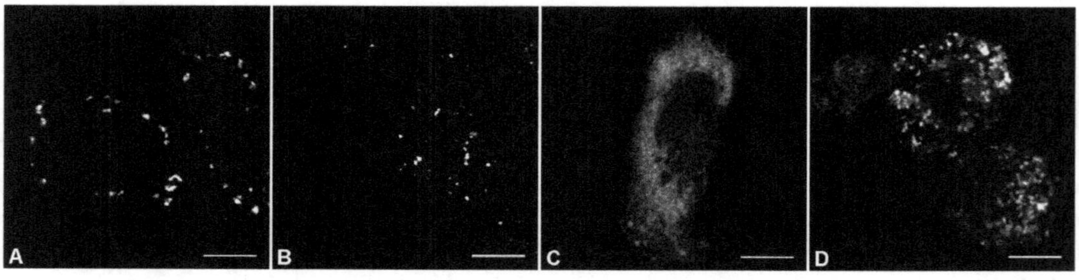

Figure 12.4 Distribution of tetravirus replicase proteins in tissue culture cells. Fluorescence microscopy images of HeLa cells expressing the full length HaSV replicase fused at its C-terminus to EGFP as live (A) or fixed cells (B). A cell expressing an HaSV replicase with EGFP fused to its N-terminus (C) is shown alongside the PrV replicase detected in persistently infected *H. zea* MG8 cells by immunofluorescence using antibodies directed against PrV p40 (D). The scale bar represents 10 μm

We have observed a similar, punctate distribution of the PrV replicase protein, p40, in the persistently infected *H. zea* MG8 cell line using immunofluorescence microscopy with anti-p40 antibodies (Fig. 12.4D). Subcellular fractionation experiments indicate that p40 is tightly membrane-associated (Walter, 2008), and is resistant to solubilization in high salt (1 M NaCl), high pH (pH 11.1) or in 1% Triton X-100 (J.R. Short and R.A. Dorrington, unpublished data). The HaSV replicase, expressed in *S. frugiperda* Sf9 cells using a recombinant baculovirus, is also resistant to solubilization in the same conditions, suggesting that the PrV and HaSV replicases may be using a similar mechanism for subcellular localization. (J.R. Short and R.A. Dorrington, unpublished data).

The data from our subcellular localization studies on the HaSV and PrV replicases (Short *et al.*, 2010), together with observations by Brooks *et al.* (2002) and Tomasicchio *et al.* (2007) lead us to propose the following speculative model for tetravirus replication in their host cells. First, the tetraviruses target the basal regenerative cells in the insect midgut and establish an infection that persists throughout the life of the larva, through pupation and possibly even associated with the reproductive organs of the adult moths. The viral replicase is trafficked from the site of translation to the endocytic vesicles where it is able to sequester membranes from late endosomes and exclude host marker proteins by an unknown mechanism. The accumulation of the replicase (and viral replication) would then result in membrane rearrangement and destabilization to the extent that luminal contents are released into the cytosol of the host cell, causing localized cytosolic acidification resulting in the induction of apoptosis and maturation of assembled tetravirus particles. We are now engaged in developing new experimental systems that will enable us to test this hypothesis *in vivo*.

Packaging of tetraviral RNAs

There have been several apparently conflicting reports as to whether or not tetravirus VLPs produced in the absence of viral replication, package viral RNAs. Agrawal and Johnson (1995) demonstrated the assembly of VLPs in insect cells with evidence that these VLPs packaged VCAP mRNA. Subsequently however, the observation has been that omegatetravirus VLPs produced by recombinant baculovirus expression of VCAP in insect tissue culture cells do not specifically encapsidate viral RNAs (Taylor, 2003; unpublished data, cited in Gordon and Waterhouse, 2005) as is also the case with VLPs produced in yeast (Tomasicchio, 2007). These VLPs package mostly cellular RNAs as well as some mRNA expressing the capsid protein precursor. Gordon *et al.* (2001) reported the replication independent assembly of infectious HaSV particles in tobacco protoplasts. Their experimental system was based on the co-expression of HaSV RNA1 and RNA2 cDNAs fused at their 5′ ends to the (−1) nucleotide of the Cauliflower mosaic virus promoter and flanked at their 3′ ends by a *cis*-acting ribozyme, producing precise copies of viral RNA. In addition, the plant protoplasts were co-transfected with a third plasmid expressing HaSV VCAP message after which they were fed to neonate *H. armigera* larvae. Evidence for assembly of infectious particles was the recovery of infectious HaSV from these larvae and detection of virus-like particles by immunoelectron microscopy, but the yields of virus particles were too low to confirm successful encapsidation of viral RNAs (Gordon *et al.*, 2001). In subsequent experiments described in Gordon and Waterhouse (2005), the same experimental system was used to demonstrate delivery of non-viral RNA encoding EGFP to larval midgut cells by recombinant HaSV VLPs assembled in tobacco protoplasts. This was achieved by transfecting protoplasts with plasmids expressing RNA1, VCAP and a third with RNA2 in which VCAP had been replaced with EGFP just downstream of the p17 stop codon. The presence of EGFP fluorescence in larval midgut cells up to 10 days after ingestion suggested replication of recombinant RNA2-EGFP transcripts in the midgut cells. However, the low level of VLP assembly in the system precluded analysis to show specific encapsidation of the viral RNAs (Gordon and Waterhouse, 2005).

Our observation that p17 is expressed in infected insect cells and that the protein is present at low concentrations in wild type virus particles purified from HaSV-infected insects, led us to propose that p17 is involved in packaging of viral RNA during capsid assembly (M. Vlok, M. Tomasicchio and R.A. Dorrington, unpublished data). To test this hypothesis, *p17* was co-expressed with *VCAP* in *S. frugiperda* cells, the resulting VLPs purified and analysed for the presence of viral RNA2. VLPs assembled in cells co-expressing VCAP and p17 contained viral mRNA, but when p17 was absent, no viral RNA could be detected in purified capsids. Deletion analysis showed that the 5′ and 3′ UTR sequences of HaSV RNA2 were not required for packaging. This data, together with that of Gordon and Waterhouse (2005) suggest the presence of a specific sequence within the p17 ORF or immediately downstream of the p17 ORF, that may be required for packaging of HaSV RNA2. The observation that p17 co-purifies with wild type virus particles suggests that omegatetravirus p17 functions in facilitating the packaging of viral RNAs during capsid assembly. Interestingly, an analysis of the amino acid sequence of p13, produced by 2A-like processing of the PrV VCAP, reveals a high incidence of PEST amino acid residues, raising the possibility that this peptide might have a similar function in facilitating packaging of viral RNAs by PrV particles. Similarly, a peptide of approximately 17 kDa is also produced by 2A-like processing of the TaV (and probably EeV) capsid protein precursor. The observation by Pringle *et al.* (1999) that the presence of p17 appears to stabilize TaV VLPs suggests that this peptide may serve a similar function as the omegatetraviral p17 in these two betatetraviruses. There is no evidence of the presence of a p17-like ORF on the NβV genome sequence, which may either indicate an alternative packaging mechanism in this virus, or mutations in cDNA fragments used to assemble the NβV genome sequence.

Biotechnology applications

Tetraviruses as biopesticides

Since the discovery of the first tetraviruses, there has been much interest in the potential of applying tetraviruses as biopesticides because of their unusually narrow host range and pathogenicity. DtV, which infects *D. trima*, an important pest of oil palms in Malaysia, was the first tetravirus to be used as a pest control agent (Moore *et al.*, 1985). However, there was some resistance to applying tetraviruses as biopesticides following reports of the presence of DtV neutralising antibodies in some people in Malaysia (MacCallum *et al.*, 1979; Moore *et al.*, 1981). There were also similar observations relating to *Nudaurelia capensis* virus antibodies in human sera in South Africa (D.A. Hendry, personal communication). The reason for this was thought to be due either to replication of a serologically related virus or the insect virus itself in mammals (to date there have been no reports of either) or, more likely, given the heavy load of virus in a severely affected insect population, an immunological response caused by ingestion or inhalation of the insect virus (Moore *et al.*, 1985). HaSV has been the focus of most of the attention because of the economic importance of its host, *H. armigera* in Africa, Asia and Australasia. In addition, there are other equally important related heliothine species that are also susceptible to HaSV infection, including *H. zea* and *H. virescens* (New World) and *Helicoverpa punctigera*, prevalent in Australasia (Christian *et al.*, 2001). Field trials on sorghum have demonstrated that HaSV applied at 4×10^{15} and 4×10^{14} virus particles/ha is an efficient control agent for *H. armigera*, comparable with a commercially available nucleopolyhedrovirus virus (Christian *et al.*, 2005).

Despite the potential of tetraviruses as biocontrol agents for agricultural pets, the problem of production of sufficient quantities of virus for commercial applications is still a major obstacle. This has prompted research into the development of non-host production systems that rely on replication-independent assembly of infectious virus particles. Gordon *et al.* (2001) used a plant protoplast system in which they were able to co-express authentic copies of HaSV RNA1 and RNA2 together with the capsid coding sequence (*VCAP*). The system was unable to produce sufficient quantities of virus particles to confirm packaging of viral RNAs. However, the assembly of infectious particles was indirectly demonstrated by feeding the protoplasts to neonate larvae and recovering infectious virus (Gordon *et al.*, 2001). This plant protoplast system was also used to show that virus particles could package a chimeric RNA2 in which the capsid coding sequence had been replaced with EGFP. EGFP fluorescence was detected in the midgut cells of larvae for up to 10 days post feeding, leading to the conclusion that virus particles were able to deliver non-viral sequences and that these sequences were being replicated by the viral replicase (Gordon and Waterhouse, 2005). Further to this line of research were attempts to generate transgenic tobacco plants co-expressing HaSV RNA1, RNA2 and *VCAP* mRNA. Initial bioassays suggested that infectious virus particles were assembling in the transgenic plants, because stunting symptoms were observed in *H. virescens* larvae that had been fed on plant material and viral RNA was detected in these animals (unpublished data cited in Gordon and Waterhouse, 2005). Analysis of the progeny of these transgenic plants, however, showed that the HaSV genes were unstable and there was evidence of silencing in transgenic lines where all three genes were still present. While some VLP assembly was observed in plants expressing VCAP, it was evident that the levels of expression were very low and that the transgene was unstable and probably subjected to silencing (Gordon and Waterhouse, 2005).

The potential of using the yeast, *S. cerevisiae* as an alternative non-host expression system for assembling infectious HaSV particles, has been assessed. As with the system developed by Gordon *et al.* (2001), authentic copies of RNA1 and RNA2 were co-expressed with VCAP, resulting in the assembly of VLPs. As observed in the plant protoplast system, VLP assembly was inefficient, but biofeed assays did result in stunting of insects and viral RNA was detected in the midgut tissues of these animals (Venter, 2001). Further optimization resulted in a significant increase in VCAP expression and VLP assembly, but no packaging of viral RNAs (Tomasicchio, 2007). The lack of viral RNA encapsidation is likely to be due to the fact that the yeast expression system did not include co-expression of HaSV p17, which has been shown to be required for packaging of viral RNAs in insect tissue culture cells (M. Vlok, M. Tomasicchio and R.A. Dorrington, unpublished data).

Retargeting of tetravirus capsids

Modification of the Ig-like domain of the capsid protein precursor of the omegatetravirus, NωV, has resulted in the ability to redirect the tropism of VLPs, prompting interest in the application of tetravirus capsids for targeted drug and gene delivery termed 'Capsivector Technology' by Hanzlik and Gordon (1995). Furthermore, insertion of a 6xHis tag into the GH loop of the Ig-like domain results in the assembly of NωV VLPs that are morphologically similar to the wild type capsids, displaying the 6xHis tag on the virus particle surface (Maree *et al.*, 2006). This data suggests that tetraviral capsids have potential applications as chemical platforms for other applications such as targeted drug and gene delivery systems (Speir and Johnson, 2008).

Concluding remarks

The *Tetraviridae* are still a relatively unknown family of small RNA viruses because of the problems associated with developing suitable experimental systems for studying their biology *in vivo*. The discovery of PrV and recent advances in understanding key aspects of their life cycle have provided the tools to now make rapid advances towards finding the answers to several fundamental questions relating to the interaction between virus and host. First, despite having a detailed understanding of the structure of the omegatetravirus capsid, nothing as yet is known about the factors on the host cell surface that serve as receptors for virus binding and entry into the cells. Second, while preliminary results suggest that replication may occur in association with membranes derived from the endocytic pathway, nothing is known about the host factors involved and the assembly of virus replication complexes. The answers to these two questions will shed light on the molecular mechanisms that determine the unusually narrow host range and tissue tropism of the *Tetraviridae*. Another important question is whether the three distinct types of replicases reflect fundamental mechanistic differences in the replication biology of the three groups of tetraviruses. Third, there is mounting evidence that tetraviruses employ several mechanisms, including ribosome recoding and protein turnover, to regulate their gene expression, which are also used by many important RNA viruses.

Studies on viral ribosome recoding have greatly increased our knowledge of the host–pathogen interactions essential for the viral life cycle. More importantly, they have also made groundbreaking contributions to the understanding of general translational control systems in their host cells, providing novel opportunities for the development of antiviral drugs and gene therapies. Finally, what is the molecular basis underlying the ability of tetraviruses to establish persistent infections in their hosts? This question is not only relevant to the *Tetraviridae*, but also to many other insect-borne RNA viruses that are maintained as persistent infections in their insect host. The tetraviruses may provide a useful model system for studying persistent viral infections in insects and the development of new technologies for combating insect-borne diseases.

Acknowledgements

Unpublished research was funded by Grants from the National Research Foundation of South Africa (GUN: 207986 and GUN: 65512). James Short was supported by fellowships from the Andrew Mellon Trust, the Ernst and Ethel Eriksen Trust and the German Academic Exchange Service (DAAD). Opinions expressed and conclusions arrived at, are those of the authors and are not necessarily to be attributed to any of the above-mentioned donors. The authors wish to thank Val Hodgson and Don Hendry for their help in the preparation of this manuscript.

References

Agrawal, D.K., and Johnson, J.E. (1992). Sequence and analysis of the capsid protein of *Nudaurelia capensis* omega virus, an insect virus with $T = 4$ icosahedral symmetry. Virology *190*, 806–814.

Agrawal, D.K., and Johnson, J.E. (1995). Assembly of the $T = 4$ *Nudaurelia capensis* omega virus capsid protein, post-translational cleavage, and specific encapsidation of its mRNA in a baculovirus expression system. Virology *207*, 89–97.

Ambrose, R.L., Lander, G.C., Maaty, W.S., Bothner, B., Johnson, J.E., and Johnson, K.N. (2009). Drosophila A virus is an unusual RNA virus with a $T = 3$ icosahedral core and permuted RNA-dependant RNA polymerase. J. Gen. Virol. *90*, 2191–2200.

Bawden, A.L., Gordon, K.H.J., and Hanzlik, T.N. (1999). The specificity of *Helicoverpa armigera* stunt virus infectivity. J. Invertebr. Pathol. *74*, 156–163.

Beier, H., and Grimm, M. (2001). Misreading of termination codons in eukaryotes by natural nonsense suppressor tRNAs. Nucleic Acids Res. *29*, 4767–4782.

Bright, N.A., Gratian, M.J., and Luzio, J.P. (2005). Endocytic delivery to lysosomes mediated by concurrent fusion and kissing events in living cells. Curr. Biol. *15*, 360–365.

Brooks, E.M., Gordon, K.H.J., Dorrian, S.J., Hines, E.R., and Hanzlik, T.N. (2002). Infection of its lepidopteran host by the *Helicoverpa armigera* stunt virus (*Tetraviridae*). J. Invertebr. Pathol. *80*, 97–111.

Canady, M.A., Tihava, M., Hanzlik, T.N., Johnson, J.E., and Yeager, M. (2000). Large conformational changes in the maturation of a simple RNA virus, *Nudaurelia capensis* ω virus (NωV). J. Mol. Biol. *299*, 573–584.

Canady, M.A., Tsuruta, H., and Johnson, J.E. (2001). Analysis of Rapid, Large-scale Protein Quaternary Structural Changes: Time-resolved X-ray solution scattering of *Nudaurelia capensis* ω virus (NωV) maturation. J. Mol. Biol. *311*, 803–814.

Cavener, D.R. (1987). Comparison of the consensus sequence flanking translational start sites in Drosophila and vertebrates. Nucleic Acids Res. *15*, 1353–1361.

Christian, P.D., Dorian, S.J., Gordon, K.H.J., and Hanzlik, T.N. (2001). Pathology and properties of the tetravirus *Helicoverpa armigera* stunt virus. Biol. Control *20*, 65–75.

Christian, P.D., Hanzlik, T.N., Dall, D.J., and Gordon, K.H.J. (1993). Insect viruses: new strategies for pest control. In Molecular approaches to fundamental and applied entomology. J.G. Oakeshott and M.J. Whitten, eds. (Springer-Verlag, New York). pp. 128–163.

Christian, P.D., Murray, D., Powell, R., Hopkinson, J., Gibb, N.N., and Hanzlik, T.N. (2005). Effective control of a field population of *Helicoverpa armigera* by using the small RNA virus *Helicoverpa armigera* stunt virus (*Tetraviridae: Omegatetravirus*). J. Econ. Entomol. *98*, 1839–1847.

Chao, Y.C., Scott, H.A., and Young, S.Y. (1983). An icosahedral RNA virus of the soybean looper (*Pseudoplusia includens*). J. Gen. Virol. *64*, 1835–1838.

Doronina, V.A., de Felipe, P., Wu, C., Sharma, P., Sachs, M.S., Ryan, M.D., and Brown, J.D. (2008a). Dissection of a co-translational nascent chain separation event. Biochem. Soc. Trans. *36*, 712–716.

Doronina, V.A., Wu, C., de Felipe, P., Sachs, M.S., Ryan, M.D., and Brown, J.D. (2008b). Site-specific release of nascent chains from ribosomes at a sense codon. Mol. Cell Biol., *28*, 4227–4239.

Dreher, T.W. (2009). Role of tRNA-like structures in controlling plant virus replication. Virus Res. *139*, 217–29.

du Plessis, L., Hendry, D.A., Dorrington, R.A., Hanzlik, T.N., Johnson, J.E., and Appel, M. (2005). Revised RNA2 sequence of the tetravirus, *Nudaurelia capensis* ω virus (NωV). Arch. Virol. *150*, 2397–2402.

Eckerle, L.D., Albariño, C.G., and Ball, L.A. (2003). Flock House virus subgenomic RNA3 is replicated and its replication correlates with transactivation of RNA2. Virology *317*, 95–108.

Fauquet, C.M., Mayo, M.A., Maniloff, J., Desselberger, U., and Ball, L.A. (2005). Virus Taxonomy: Eighth Report of the International Committee on Taxonomy of Viruses. (Academic Press).

Finch, J.T., Crowther, R.A., Hendry, D.A., and Struthers, J.K. (1974). The structure of *Nudaurelia capensis* β virus: the first example of a capsid with icosahedral surface symmetry $T = 4$. J. Gen. Virol. *24*, 191–200.

Francki, R.I.B., Fauquet, C.M., Knudson, D.L., and Brown, F. (1991). Classification and nomenclature of viruses: Fifth Report of the International Committee on Taxonomy of Viruses. (Springer-Verlag, Vienna).

Galiana-Arnoux D., Dostert, C., Schneemann, A., Hoffmann, J.A., and Imler, J.L. (2006). Essential function *in vivo* for Dicer-2 in host defence against RNA viruses in Drosophila. Nat. Immunol. 7, 590–597.

Gallagher T.M., and Rueckert, R.R. (1988). Assembly-dependent maturation cleavage in provirions of a small icosahedral insect ribovirus. J. Virol. *62*, 3399–3406.

Goff, S.P. (2004). Genetic reprogramming by retroviruses: Enhanced suppression of translational termination. Cell Cycle 3, 123–125.

Goodman, C.L., and McIntosh, A.H. (1995). Characteristics of midgut-derived insect cell lines. *In Vitro* Cell. Dev. Biol. *31*, 85A.

Gorbalenya, A.E., Pringle, F.M., Zeddam, Z-L., Luck, B.T., Cameron, C.E., Kalmakoff, J., Hanzlik, T.N., Gordon, K.H., and Ward, V.K. (2002). The palm subdomain-based active site is internally permuted in viral RNA-dependent RNA polymerases of an ancient lineage. J. Mol.Biol. *324*, 47–62.

Gordon, K.H.J., Johnson, K.N., and Hanzlik, T.N. (1995). The larger genomic RNA of *Helicoverpa armigera* stunt virus encodes the viral RNA polymerase and has a novel 3′-terminal tRNA-like structure. Virology *208*, 84–98.

Gordon, K.H.J., and Waterhouse, P.M. (2005). Small RNA viruses of insects: expression in plants and RNA silencing. Adv. Virus Res. *68*, 459–562.

Gordon, K.H.J., Williams, M.R., Baker, J.S., Gibson, J.N., Bawden, A.L., Millgate, A.G., Larkin, J., and Hanzlik, T.N. (2001). Replication-independent assembly of an insect virus (*Tetraviridae*) in plant cells. Virology *288*, 36–50.

Gordon, K.H.J., Williams, M.R., Hendry, D.A., and Hanzlik, T.N. (1999). Sequence of the genomic RNA of *Nudaurelia* β virus (*Tetraviridae*) defines a novel genome organization. Virology 258, 42–53.

Grace, T.D.C., and Mercer, E.H. (1965). A new virus of the Saturniid *Antherea eucalypti* Scott. J. Invertebr. Pathol 7, 241–244.

Grace, T.D.C., and Mercer, E.H. (1976). Serological relations between twelve small RNA viruses of insects. J. Gen. Virol. *31*, 131–134.

Greenwood, L.K., and Moore, N.F. (1981). The *Nudaurelia* β group of small RNA viruses of insects: serological comparison of four members. Microbiologica *4*, 271–280.

Greenwood, L.K., and Moore, N.F. (1984). Determination of the location of an infection in *Trichoplusia ni* larvae by a small RNA-containing virus using enzyme-linked immunosorbent assay and electron microscopy. Microbiologica 7, 97–102.

Hakim, R.S., Baldwin, K., and Smagghe, G. (2009). Regulation of midgut growth, development, and metamorphosis. Annu. Rev. Entomol. *55*, 593–608.

Hanzlik, T.N., Dorrian, S.J., Gordon, K.H.J., and Christian, P.D. (1993). A novel small RNA virus isolated from the cotton bollworm, *Helicoverpa armigera*. J. Gen. Virol. *74*, 1805–1810.

Hanzlik, T.N., Dorrian, S.J., Johnson, K.N., Brooks, E.M., and Gordon, K.H.J. (1995). Sequence of RNA 2 of the *Helicoverpa armigera* stunt virus (*Tetraviridae*) and bacterial expression of its genes. J. Gen. Virol. *76*, 799–811.

Hanzlik, T.N., Zeddam, J.-L., Gordon, K.H.J., and Christian, P.D. (1999). A new view of small RNA viruses of insects. Pestic. Outlook *10*, 22–26.

Helgstrand, C., Munshi, S., Johnson, J.E., and Liljas, L. (2004). The refined structure of *Nudaurelia capensis* ω virus reveals control elements for a $T = 4$ capsid maturation. Virology *318*, 192–203.

Hendry, D.A., and Agrawal, D.K. (1993). Tetraviruses. In Encyclopaedia of Virology. A. Granoff and R.G. Webster eds. (Academic Press, New York). pp. 1416–1422.

Hendry, D.A., Becker, M.F., and van Regenmortel, M.V.H. (1968). A non inclusion virus of the pine emperor moth *Nudaurelia cytherea capensis* Stoll. S. Afr. Med. J. *42*, 117.

Hendry, D.A., Hodgson, V., Clark, R., and Newman, J. (1985). Small RNA viruses co-infecting the pine emperor moth (*Nudaurelia cytherea capensis*). J. Gen. Virol. *66*, 627–632.

Juckes, I.R.M. (1970). Viruses of the pine emperor moth. Bull. S. Afr. Soc. Plant. Pathol. Microbiol. *4*, 18.

Juckes, I.R.M., Longworth, J.F., and Reinganum, C. (1973). A serological comparison of some non-occluded insect viruses. J. Invertebr. Pathol. *21*, 19–125.

Juckes, I.R.M. (1974). Purification and characterization of viruses of the pine emperor moth, *Nudaurelia cytherea capensis* (Fabr.). PhD thesis. University of Cape Town.

Juckes, I.R.M. (1979). Comparison of some biophysical properties of the *Nudaurelia* β and ε viruses. J. Gen. Virol. *42*, 89–94.

King L.A., and Moore, N.F. (1984). The RNAs of two viruses of the *Nudaurelia* β family share little homology and have no terminal poly(A) tracts. FEMS Microbiol. Lett. *26*, 41–43.

Koonin, E.V. (1991). The phylogeny of RNA-dependent RNA polymerases of positive strand RNA viruses. J. Gen. Virol. 72, 2197–2206.

Luke, G.A., de Felipe, P., Lukashev, A., Kallioinen, S.E., Bruno, E.A., and Ryan, M.D. (2008). Occurrence, function and evolutionary origins of '2A-like' sequences in virus genomes. J. Gen. Virol. *89*, 1036–1042.

MacCallum, F., Brown, G., and Tinsley, T (1979). Antibodies in human sera reacting with an insect pathogenic virus. Intervirology *11*, 234–237.

Maree, H.J., Van der Walt, E., Tiedt, F.A.C., Hanzlik, T.N., and Appel, M. (2006). Surface display of an internal His-tag on virus-like particles of *Nudaurelia capensis* ω virus (NωV) produced in a baculovirus expression system. J. Virol. Methods *136*, 283–288.

Matsuyama, S., and Reed, J.C. (2000). Mitochondria-dependant apoptosis and cellular pH regulation. Cell Death Differ. 7,1155–1165.

Matthews, R.E.F. (1982). Classification and nomenclature of viruses. Fourth report of the international committee on taxonomy of viruses. Intervirology *17*, 1–160.

Moore, N.F. (1991). The *Nudaurelia* β family of insect viruses. In Viruses of Invertebrates. E. Kursak, ed. (Marcel Dekker, New York). pp 277–285.

Moore, N.F., Greenwood, L.K., and Rixon, K.R. (1981). Studies on the capsid protein of several members of the *Nudaurelia* β group of small RNA viruses of insects. Microbiologica *4*, 59–71.

Moore, N.F., Reavy, B., and King, L.A. (1985). General characteristics, gene organization and expression of small RNA viruses of insects. J. Gen. Virol. *66*, 647–559.

Morris, R.J., Hess, R.T., and Pinnock, O.E. (1979). Physicochemical characterization of a small RNA virus associated with baculovirus infection in *Trichoplusia ni*. Intervirology *11*, 238–247.

Munshi, S., Liljas, L., Cavarelli, J., Bomu, W., McKinney, B., Reddy, V., and Johnson, J.E. (1996). The 2.8 Å structure of a $T = 4$ animal virus and its implications for membrane translocation of RNA. J. Mol. Biol. *261*, 1–10.

Murphy, F.A., Fauquet, C.M., Bishop, D.H.L., Ghabrial, S.A., Jarvis, A.W., Martelli, G.P., Mayo, M.A., and Summers, M.D. (1995). Virus Taxonomy: Sixth Report of the International Committee on Taxonomy of Viruses (Springer-Verlag, Vienna).

Nilsson, C., Johansson, U., Johansson, A-C., Kågedal, K., and Öllinger, K. (2006). Cytosolic acidification and lysosomal alkalinization during TNF-α induced apoptosis in U937 cells. Apoptosis *11*, 1149–1159.

Poch, O., Sauvaget, I., Delarue, M., and Tordo, N. (1989). Identification of four conserved motifs among the RNA-dependant polymerase encoding elements. EMBO J. *8*, 3867–3874.

Pringle, F.M., Gordon, K.H.J., Hanzlik, T.N., Kalmakoff, J., Scottii, P.D., and Ward, V. (1999). A novel capsid expression strategy for *Thosea asigna* virus (*Tetraviridae*). J. Gen. Virol. *80*, 1855–1863.

Pringle, F.M., Kalmakoff, J., and Ward, V.K. (2001). Analysis of the capsid processing strategy of *Thosea asigna* virus using baculovirus expression of virus-like particles. J. Gen. Virol. *82*, 259–266.

Pringle, F.M., Johnson, K.N., Goodman, C.L., McIntosh, A.H., and Ball, L.A. (2003). *Providence virus*: a new member of the *Tetraviridae* that infects cultured insect cells. Virology *306*, 359–370.

Rechsteiner, M., and Rogers, S.W. (1996). PEST sequences and regulation by proteolysis. Trends Biochem. Sci. *21*, 267–271.

Reinganum, C., Robertson, J.S., and Tinsley, T.W. (1978). A new group of RNA viruses from insects. J. Gen. Virol. *40*, 195–202.

Salaman, S.R., Hendricks, K.B., and Thorner, J. (1994). G1 cyclin degradation: the PEST motif of yeast Cln2 is necessary, but not sufficient, for rapid protein turnover. Mol. Cell. Biol. *14*, 7953–7966.

Short, J.R., Knox, C., and Dorrington, R.A. (2010). Subcellular localization and live cell imaging of the *Helicoverpa armigera* stunt virus replicase in mammalian and *Spodpoptera frugiperda* cells. J. Gen. Virol. *91*, 1514–1523.

Speir, J.A., and Johnson, J.E. (2008). Tetraviruses. In Encyclopaedia of Virology, 3rd edition, B.W.J. Mahy and M.H.V. van Regenmortel, eds. (Oxford, Elsevier), pp. 27–37.

Struthers, J.K., and Hendry, D.A. (1974). Studies of the protein and nucleic acid components of *Nudaurelia capensis* β virus. J. Gen. Virol. *22*, 355–362.

Taylor, T.J. (2003). Molecular and biophysical analysis of a conformational change in a nonenveloped RNA virus. PhD thesis, University of California, San Diego, CA, USA.

Taylor, D.J., Krishna, N.K., Canady, M.A., Schneemann, A., and Johnson, J.E. (2002). Large-scale, pH-dependant, quaternary structure changes in an RNA virus capsid are reversible in the absence of subunit autoproteolysis. J. Virol. *76*, 9972–9980.

Taylor, D.J., Speir, J.A., Reddy, V., Cingolani, G., Pringle, F.M., Ball, L.A., and Johnson, J.E. (2005). Preliminary X-ray characterization of authentic Providence virus and attempts to express its coat protein gene in recombinant baculovirus. Arch. Virol. *151*, 155–165.

Tomasicchio, M. (2007). Assembly of *Omegatetravirus* virus-like particles in the yeast *Saccharomyces cerevisiae*. PhD Thesis, Rhodes University, Grahamstown, South Africa.

Tomasicchio, M., Venter, P.A., Gordon, K.H.J., Hanzlik, T.N., and Dorrington, R.A. (2007). Induction of apoptosis in *Saccharomyces cerevisiae* results in the spontaneous maturation of tetravirus procapsids *in vivo*. J. Gen. Virol. *88*, 1576–1582.

Tripconey, D. (1970). Studies on a nonoccluded virus of the pine tree emperor moth. J. Invertebr. Pathol. *I5*, 268–275.

Vail, P. V., Morris, T. J., and Collier, S. S. (1983). An RNA virus in *Autographa californica* nuclear polyhedrosis virus preparations: Gross pathology and infectivity. J. Invertebr. Pathol. *41*, 179–183.

Van der Heijden M.W., and Bol, J.F. (2002). Composition of alphavirus-like replication complexes: involvement of virus and host encoded proteins. Arch. Virol. *147*, 875–898.

Venter, P.A. (2001). Heterologous expression of the *Helicoverpa armigera* stunt virus in *Saccharomyces cerevisiae*. PhD Thesis. Rhodes University, Grahamstown, South Africa.

Walter, C.T. (2005). Development of experimental systems for studying the biology of *Nudaurelia capensis* β virus. MSc Thesis, Rhodes University, Grahamstown, South Africa.

Walter, C.T. (2008). Establishing experimental systems for studying the replication biology of Providence virus. PhD Thesis, Rhodes University, Grahamstown, South Africa.

Walter, C.T., Tomasicchio, M., Hodgson, V., Hendry, D.A., Hill, M.P., and Dorrington, R.A. (2008).

Characterization of a succession of small insect viruses in a wild South African population of *Nudaurelia cytherea capensis*. S. Afr. J. Sci. *104*, 147–152.

Walter, C.T., Pringle, F.M., Nakayinga, R., de Felipe, P., Ryan, M.D., Ball, L.A., and Dorrington, R.A. (2010). Genome organization and translation products of Providence virus: insight into a unique tetravirus. J. Gen. Virol. In press.

Yi, F., Zhang, J., Yu, H., Liu, C., Wang, J., and Hu, Y. (2005). Isolation and identification of a new tetravirus from *Dendrolimus punctatus* larvae collected from Yunnan province, China. J. Gen. Virol. *86*, 789–796.

Zeddam, J.-L., Gordon, K.H.J., Lauber, C., Felipe Alves, C.A., Luke, B.T., Hanzlik, T.N., Ward, V.K., and Gorbalenya, A.E. (2010). *Euprosterna elaeasa* virus genome sequence and evolution of the *Tetraviridae* family: Emergence of bipartite genomes and conservation of the VPg signal with the dsRNA *Birnaviridae* family. Virology 397,145–154.

Zhou, L., Zhang, J., Wang, X., Jiang, H., Yi, F. and Hu, Y. (2006). Expression and characterisation of RNA-dependent polymerase of *Dendromilus punctatus* tetravirus. J. Biochem. Mol. Biol. *39*, 571–577.

Zhou, L., Zheng, Y., Jiang, H., Zhou, W., Lin, M., Han, Y., Cao, X., Zhang, J., and Hu, Y. (2008). RNA-binding properties of *Dendrolimus punctatus* tetravirus p17 protein. Virus Res. *138*, 1–6.

Cypoviruses

13

Hajime Mori and Peter Metcalf

Abstract

Cypoviruses are members of the *Reoviridae* family of segmented dsRNA genome viruses which are unique amongst *Reoviridae* because the infectious forms of the virus are polyhedra, crystalline occlusion bodies containing thousands of virus particles that form in the cytoplasm of infected cells. Polyhedra are micron-size infectious protein crystals which function to protect the virus particles within the crystals from hostile environmental conditions. Polyhedra are stable, only dissolving at pH > 10.5 in the larval midgut, allowing virus particles released from ingested polyhedra to infect the epithelia. Here we briefly review current knowledge of cypovirus and summarize structures of both the virus particle and of polyhedra. The two structures provide a framework for understanding both the formation and stability of polyhedra, and the molecular events that cause virus particles to be incorporated into polyhedra in the cytoplasm of infected cells. This information has been exploited to develop a general method to stabilize proteins by incorporating them into modified polyhedra. We describe these modified polyhedra or 'nano-containers', which have a variety of applications, including the development of stabilized growth factors for cell-culture. Finally we discuss evidence for the possible evolution of cypovirus from members of plant *Reoviridae*.

Introduction

Cytoplasmic polyhedrosis viruses (CPVs) belong to the genus *Cypovirus* of the family *Reoviridae* (Mertens *et al.*, 2004a,b). Viruses belonging to the family *Reoviridae* are characterized by the presence of capsids made up of concentric icosahedrally symmetric layers of proteins organized in one, two or three shells containing 10 to 12 segments of linear, double-stranded RNA (dsRNA). Members of the *Reoviridae* are grouped into 12 distinct genera: *Orthoreovirus, Orbivirus, Rotavirus, Coltivirus, Aquareovirus, Cypovirus, Fijivirus, Mycoreovirus, Phytoreovirus, Oryzavirus, Seadornavirus* and *Idnoreovirus* (Payne and Mertens, 1983; Belloncik, 1989; Belloncik and Mori, 1998). The natural hosts of the *Reoviridae* include vertebrates, invertebrates, and plants. The virulence they exhibit toward these hosts differs widely. Insect larvae infected by CPV are characterized by the production of massive amounts of polyhedrin protein, which crystallizes in the cell cytoplasm forming occlusion bodies named polyhedra that each incorporate numerous virus particles (Fig. 13.1A). CPVs are unique amongst the *Reoviridae* because of this unusual encapsulation of virus within protein micro-crystals and also because the virus particles have only a single capsid shell.

CPVs have been isolated from more than 250 species of insects (Hukuhara and Bohami, 1991). Different cypovirus 'types' were initially identified on the basis of differences in the migration patterns of the genome segments during polyacrylamide gel electrophoresis, i.e. electropherotypes (Payne and Merten, 1983; Belloncik, 1989; Belloncik and Mori, 1998; Shapiro *et al.*, 2005; Li *et al.*, 2006, 2007; Tan *et al.*, 2008). Subsequent studies showed a good correlation between electropherotype and the

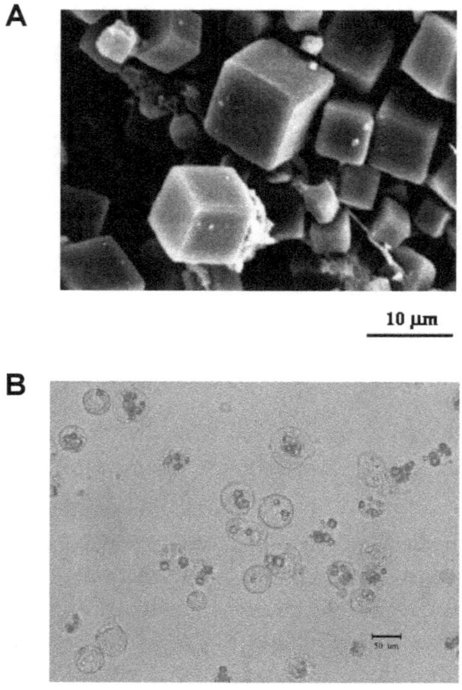

Figure 13.1 BmCPV polyhedra. (A) Scanning electron micrograph of BmCPV polyhedra purified from the midgut of the silkworm. (B) Polyhedra produced by a recombinant baculovirus (AcCP-H) expressing the BmCPV polyhedrin.

grouping of strains on the basis of serological properties or RNA cross-hybridization. More recently, nucleotide sequences, for example those of the genome segment encoding polyhedrin, have been used to distinguish different cypovirus types. Twenty cypovirus types are currently recognized (Mertens *et al.*, 2004a,b; Graham *et al.*, 2006) (Table 13.1). The complete nucleotide sequence of *Bombyx mori* CPV (BmCPV) from type 1 was the first to be determined (Mori *et al.*, 1987; Arella *et al.*, 1988; Mori *et al.*, 1989; Hagiwara *et al.*, 1998a,b, 2001, 2002; Hagiwara and Matsumoto, 2000; Ikeda, 2001; Rao *et al.*, 2003) and now genome sequences for types 5, 14, and 15 are available (Li *et al.*, 2006, 2007; Tan *et al.*, 2008).

The importance of CPV infections was first recognized in the silkworm industry, where these viruses caused economically significant losses of cocoon production (Aruga, 1963, 1971; Hukuhara, 1971; Hukuhara and Midorikawa, 1983; Hukuhara, 1985). For this reason BmCPV has been extensively studied in sericultural institutes. In *B. mori* larvae, CPV primarily infects epithelial midgut cells. Late in infection, the gut becomes enlarged and milky-white in colour, a result of the formation of masses of polyhedra. Ruptured cells release polyhedra into the gut lumen and these together with desquamated epithelial cells are eliminated in the faeces and reach the environment. These polyhedra contaminate leaves or artificial diets and serve as a stable source of cypovirus that infect feeding larvae. Ingested polyhedra dissolve in the alkaline midgut releasing virus particles that initially infect the midgut epithelia. The *Cypovirus* genera is a unique group within the family *Reoviridae,* as they lack the outer protective shells that exist in viruses from all other genera. Yet CPVs exhibit unusual stability and the integrity of the capsids was unaffected under high-pH conditions (Zhang *et al.*, 2002). *In vivo* entry of BmCPV into the silkworm midgut was studied by electron microscopy of ultrathin sections of midguts from silkworm larvae (Tan *et al.*, 2003). Virions were seen adhering to the plasma membrane

Table 13.1 Cypovirus types

CPV-1	BmCPV-1	*Bombyx mori*	CPV-7	MbCPV-7	*Mamestra brassicae*
	DsCPV-1	*Dendrolimus spectabilis*		NpCPV-7	*Noctua pronuba*
	LdCPV-1	*Lymantria dispar*		TpCPV-7	*Triphena pronuba*
CPV-2	AuCPV-2	*Aglais urticae*	CPV-8	AgCPV-8	*Abraxas grossulariata*
	AvaCPV-2	*Agraulis vanillae*		HaCPV-8	*Helicoverpa armigera*
	AcCPV-2	*Arctia caja*		MdCPV-8	*Malacosoma disstria*
	AviCPV-2	*Arctia villica*		NcCPV-8	*Nudaurelia cytherea*
	BdCPV-2	*Boloria dia*		PmCPV-8	*Phlogophora meticulosa*
	CdCPV-2	*Clossiana dia*		SexmCPV-8	*Spodoptera exempta*
	DpCPV-2	*Dasychira pudibunda*	CPV-9	AsCPV-9	*Agrotis segetum*
	ElCPV-2	*Eriogaster lanestris*	CPV-10	AlCPV-10	*Aporophyla lutulenta*
	HpCPV-2	*Hyloicus pinastri*	CPV-11	HaCPV-11	*Helicoverpa armigera*
	IiCPV-2	*Inachis io*		HzCPV-11	*Helicoverpa zea*
	LoCPV-2	*Lacanobia oleracea*		LdCPV-11	*Lymantria dispar*
	MnCPV-2	*Malacosoma neustria*		MbCPV-11	*Mamestra brassicae*
	MbCPV-2	*Mamestra brassicae*		PgCPV-11	*Pectinophora gossypiella*
	ObCPV-2	*Operophtera brumata*		PuCPV-11	*Pseudaletia unipuncta*
	PmCPV-2	*Papilio machaon*		SexmCPV-11	*Spodoptera exempta*
	PbCPV-2	*Phalera bucephala*		SexgCPV-11	*Spodoptera exigua*
	PrCPV-2	*Pieris rapae*	CPV-12	AgCPV-12	*Autographa gamma*
CPV-3	ApCPV-3	*Anaitis plagiata*		MbCPV-12	*Mamestra brassicae*
	AcCPV-3	*Arctia caja*		PrCPV-12	*Pieris rapae*
	AvCPV-3	*Arctia villica*		SexmCPV-12	*Spodoptera exempta*
	AiCPV-3	*Automeris io*	CPV-13	PhCPV-13	*Polistes hebraeus*
	DpCPV-3	*Danaus plexippus*	CPV-14	HaCPV-14	*Helicoverpa armigera*
	GrCPV-3	*Gonometa rufibrunnea*		HasCPV-14	*Helicoverpa assulta*
	MnCPV-3	*Malacosoma neustria*		LdCPV-14	*Lymatria dispar*
	ObCPV-3	*Operophtera brumata*	CPV-15	TnCPV-15	*Trichoplusia ni*
	PmCPV-3	*Phlogophera meticulosa*		UsCPV-15	*Uranotaenia sapphirina*
	PrCPV-3	*Pieris rapae*	CPV-16	CfCPV-16	*Choristoneura fumiferana*
	SexmCPV-3	*Spodoptera exempta*	CPV-17	CrCPV-17	*Culex restuans*
CPV-4	AsCPV-4	*Actias selene*		UsCPV-17	*Uranotaenia sapphirina*
	AaCPV-4	*Antheraea assamensis*	CPV-18	ObCPV-18	*Operophtera brumata*
	AmCPV-4	*Antheraea mylitta*	CPV-19	ObCPV-19	*Operophtera brumata*
	ApCPV-4	*Antheraea pernyi*	CPV-20	SuCPV-20	*Simulium ubiquitum*
	ApCPV-4	*Antheraea proylei*			

Table 13.1 continued

CPV-5	EsCPV-5	*Euxoa scandens*
	HaCPV-5	*Helicoverpa armigera*
	HasCPV-5	*Heliothis assulta*
	OpCPV-5	*Orgyia pseudosugata*
	SexmCPV-5	*Spodoptera exempta*
	TnCPV-5	*Trichoplusia ni*
CPV-6	AuCPV-6	*Aglais urticae*
	AhCPV-6	*Agrochola helvolva*
	AlCPV-6	*Agrochola lychnidis*
	ApCPV-6	*Anaitis plagiata*
	AhCPV-6	*Anchoscelis helvola*
	AxCPV-6	*Antitype xanthomista*
	BbCPV-6	*Biston betularia*
	ElCPV-6	*Eriogaster lanestris*
	LqCPV-6	*Lasiocampa quercus*
	PbCPV-6	*Phalera bucephala*

of microvilli, embedded in the membrane and settling themselves intact inside the microvilli of the columnar cells. These results suggested that intact virions entered columnar cells by means of direct penetration through the cell membrane.

The virus particle

Polyhedra are easily obtained from infected larvae and virus particles purified from alkaline treated polyhedra have been investigated using a variety of analytical, enzymatic and biophysical methods. Negative-stained electron microscopy revealed that cypovirus particles are approximately 60-nm-diameter spheres with 12 prominent turrets which extend approximately 4 nm outwards from the surface at icosahedral 5-fold symmetry axes (Miura *et al.* 1969). The size and shape of the cypovirus particles are similar to the core particles of orthoreovirus and Blue tongue virus, members of the *Reoviridae* for which atomic structures have been determined by protein X-ray crystallography (Grimes *et al.*, 1998; Reinisch *et al.*, 2000). SDS–PAGE analysis revealed that BmCPV virus particles contain five structural proteins VP1 (148 kDa), VP2 (136 kDa), VP3 (130 kDa), VP4 (64 kDa), and VP7 (LPP, 50 kDa) which are encoded by dsRNA segments 1, 2, 4, 6 and 7, respectively (Arella *et al.*, 1988; Aruga, 1963; Hagiwara *et al.*, 1998a,b, 2001, 2002; Hagiwara and Matsumoto, 2000; Ikeda *et al.*, 2001; Mertens *et al.*, 2004a,b; Mori *et al.*, 1987, 1989; Rao *et al.*, 2003; Zhou *et al.*, 2003). VP1 and VP3 are major components of the capsid while VP2 is a minor capsid protein (Lewandowski and Traynor, 1972) and *in vitro* labelling of BmCPV with ^{125}I indicated that both VP1 and VP3 were outer components of the virus particle (Lewandowski and Traynor, 1972; McCrae and Mertens, 1983; Payne and Mertens, 1983). Purified CPV particles have both mRNA capping enzyme and RNA-dependent RNA polymerase (RdRp) enzymatic activities (Furuichi, 1974, 1978; Furuichi and Miura, 1975). The segmented dsRNA genomes are transcribed and replicated within viral capsids by the RdRp. VP2 and VP3 are considered to be the RdRp and guanylyl transferase capping enzyme, respectively (Mertens *et al.*, 2004a,b). Cypovirus proteins, protein function and corresponding dsRNA segments are tabulated at the *Reoviridae*

Web Database (http://www.reoviridae.org/dsRNA_virus_proteins/).

Two groups reported detailed structures of *Orgyia pseudosugata* CPV (CPV-5) and *Dendrolimus spectabilis* CPV (CPV-1) (Hill *et al.*, 1999; Zhou *et al.*, 2003), respectively. A high-resolution electron density map of the BmCPV (CPV-1) particle has been recently obtained (Fig. 13.2A) from 12,814 cryo-EM images of individual particles using icosahedral symmetry to combine these multiple images into a single three-dimensional icosahedrally averaged electron density map (Yu *et al.*, 2008; Zhou, 2008a,b). The 3.88 Å resolution obtained was sufficient to allow a tentative alignment of some protein amino-acid sequences into the electron density map. In particular, the positions of the C-alpha atoms for 70% of VP1 and for 33% of

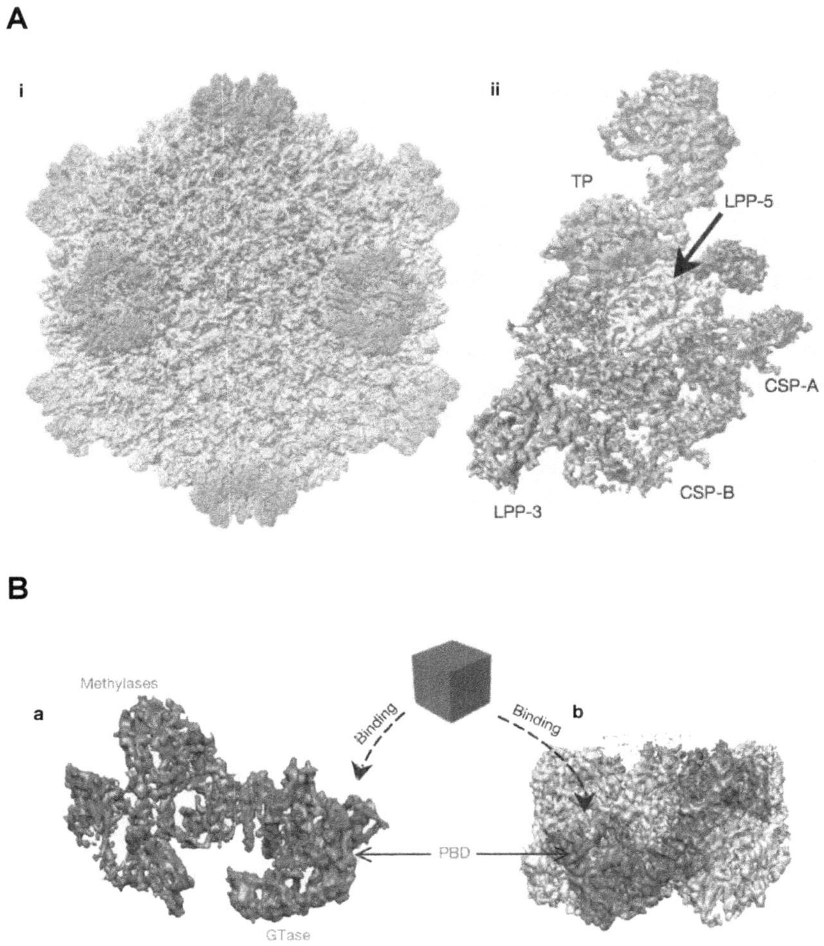

Figure 13.2 BmCPV 3.88 Å resolution electron density map derived from 12,814 cryo-electron microscopy single particle images using icosahedral averaging. (A) (i) The whole particle viewed down a two-fold symmetry axis. Eight of the twelve turret projects are visible in this orientation. The colours correspond to radial distance. (ii) The icosahedral asymmetric unit, which is repeated 60 times to generate the whole particle. The asymmetric unit contains one TP molecule, two CSP molecules (CSP-A and CSP-B) and 2 LPP molecules (LPP-3 and LPP-5). (B) Turret protein (TP, VP3) electron density. (i) A single TP protein, showing the GTase and methylase domains. (ii) Turret (like one of the eight turrets visible in A) containing five TP proteins. The location of the N-terminal 79 residue polyhedrin binding domain (PBD) is indicated. A colour version of this figure is located in the plate section at the back of the book.

VP3 have been deposited at the protein data bank (code 3CNF). The icosahedral subunit consists of two VP1 molecules with different conformations (CSP-A and CSP-B), two VP7 molecules with different conformations (LPP-3 and LPP-5) and a single VP3 molecule (the turret protein). The virus particle consists of 60 of these icosahedral subunits together with components that do not have icosahedral symmetry – e.g. the dsRNA molecules and the 'A spikes', fibrous molecules extending from the centre of each of the 12 turrets. Protein density corresponding to the CSP-A, CSP-B, the two LPP conformers and two N-terminal domains of the turret protein could be identified in the map, which also revealed averaged out density corresponding to the A-spike and outer parts of the dsRNA genome.

Consistent with the overall structural similarity of cypovirus with the orthoreovirus core, VP1 has a similar fold and secondary structure elements as the corresponding reovirus protein lambda 1, with the two molecules with different conformations CSP-A and CSP-B together forming a connected thin plate shell for the particle. An exception to this overall structural homology is the small protrusion domain (SPP), unique to VP1. Two non-equivalent VP7 molecules (reovirus sigma 2 homologues) close the VP1 shell forming clamps near the 3-fold (LPP-3) and 5-fold (LPP-5) symmetry axes. Each turret is formed from five equivalent VP3 molecules. A unique feature of cypovirus is a cleft at the base of the turret, formed by apical domains of two CSP-As and one CSP-B and proposed to be the exit channel for nascent mRNAs because it is adjacent to the guanyltransferase active site of VP3. The turret protein VP3 has a similar overall domain architecture to the orthoreovirus equivalent lambda 2, but lacks the immunoglobulin fold flap domain, and instead forms a hollow pentameric structure with the central chamber plugged by the A-spike protein. The N-terminal 79 amino acids of VP3 form a separate domain, also unique to cypovirus, which extends outwards from the base of the turret (Fig. 13.2B). This largely beta-strand domain corresponds to the minimal N-terminal region of VP3 used to target foreign proteins into recombinant cypovirus polyhedra (Ikeda *et al.*, 2006) and for that reason has been termed the polyhedrin binding domain (Yu *et al.*, 2008).

Polyhedra

The matrix protein of BmCPV polyhedra is called polyhedrin and is encoded by the smallest RNA segment 10 (Mori *et al.*, 1987, 1989, 1993; Arella *et al.*, 1988; Nakazawa *et al.*, 1996; Ikeda *et al.*, 1998). The BmCPV polyhedrin gene consists of 944 basepairs encoding a polypeptide of 248 amino acids with predicted molecular mass of 29 kDa. Many BmCPV strains have been isolated according to the shape and localization of the polyhedra (Hukuhara, 1971, 1985; Hukuhara and Midorikawa, 1983). The polyhedrin gene from wild-type isolate (strain H) was introduced into a baculovirus expression vector and the insect cell line *Spodoptera frugiperda* IPLB-Sf21AE (Sf21 cells) was infected with the recombinant baculovirus designated AcCP-H. The polyhedrin of strain H was massively produced and many cubic polyhedra were observed (Mori *et al.*, 1993) (Fig. 13.1B). This result demonstrated that other cypovirus-encoded proteins were unnecessary for the crystallization of BmCPV polyhedrin.

Polyhedra function to facilitate the survival of the virus in the environment, stabilizing the virus particles and allowing them to remain viable for long periods (Payne and Mertens, 1983; Belloncik, 1989; Hukuhara and Bohami, 1991; Belloncik and Mori, 1998). Polyhedra are remarkably stable under physiological conditions, but do dissolve at pH > 10.5. Infection occurs by the release of the virus particles in the alkaline larval midgut after ingestion. Only three virus families, all from insects, share this mode of propagation within infectious protein crystals. Besides cypovirus, DNA viruses in the baculovirus group (nucleopolyhedrovirus and granulovirus) and the entomopoxvirus group (Rohrmann, 1986) also produce polyhedra. Baculovirus polyhedrin has a molecular weight similar to CPV polyhedrin; however, there is no homology between the amino acid sequences of the two types of polyhedrin. Baculovirus polyhedra incorporate fewer virus particles than cypovirus polyhedra, which each occlude thousands of virus particles as they grow in the cytoplasm of infected cells (Payne and Mertens, 1983; Belloncik, 1989; Belloncik and Mori, 1998). The maintenance of occlusion body proteins in such a diverse group of viruses indicates the importance of polyhedrin in the life cycles of these insect viruses.

The atomic structure of BmCPV polyhedra from a wild type strain has been determined to a resolution of 2Å (Coulibaly *et al.*, 2007). The crystals have body centred cubic symmetry with a 103 Å unit cell, the same symmetry and unit cell as baculovirus polyhedra, but a different underlying protein structure (Coulibaly *et al.*, 2009; Ji *et al.*, 2010). Each unit cell contains 24 identical polyhedrin molecules in different positions and orientations related by I23 space group symmetry. A typical 2-micron crystal contains about 200 million polyhedrin molecules, 8 million cubic unit cells and thousands of virus particles. The virus particles within polyhedra each displace approximately 200 unit cells from the crystalline polyhedrin lattice. The proportion of the volume occupied by virus particles in a 2-micron crystal containing 1000 particles is 2.5%.

The atomic structure revealed that cypovirus polyhedra are extraordinarily intricate and densely packed, unlike laboratory-crystallized proteins where only a tiny fraction of the surface area of each molecule in the crystal interacts with other molecules. Analysis of all the contacts to other molecules in polyhedra revealed that the strongest crystalline interactions formed trimers of molecules within the crystal. For this reason, the structure of polyhedra is conveniently described as being based on trimeric building blocks (although all the molecules in the crystal are identical). Individual polyhedrin molecules have a distinctive extended N-terminal α helix (the H1 helix) which projects from the main part of the molecule, a compact β-sandwich domain with three layered β-sheets and peripheral small α helices (Fig. 13.3A). The trimeric building blocks are linked to each other in polyhedra by multiple interactions, primarily involving these extended H1 N-terminal arms. The trimers themselves are formed by extensive hydrophobic interactions between monomers, particularly between the H4 helices which surround the 3-fold axis (Fig. 13.3B and C). The trimeric building blocks may represent an assembly intermediate for polyhedra, but this is yet to be determined.

Polyhedra are stable in reagents that would normally be expected to denature proteins (concentrated urea, SDS, acid etc.). The structure is consistent with this, with numerous protein–protein contacts shielding over 70% of the monomer surface from solvent. Polyhedra do however dissolve readily above pH 10.5 and the deprotonation of a buried cluster of tyrosines (pK_a ~10.1) may provide a clue for the disassembly mechanism. Larval alkaline proteinases degrade polyhedrin further contributing to the efficient release of virus from polyhedra in the midgut (Coulibaly *et al.*, 2007).

The nucleotides ATP, GTP and CTP were identified in cypovirus polyhedra by mass-spectrometry and the structure revealed a single molecule of each nucleotide bound to each polyhedrin molecule. The ATP and GTP molecules are bound at the interface between four trimers and the CTP molecule is located near the central cavity (Coulibaly *et al.*, 2007). The function of the bound nucleotides is unknown. No other adducts, other than virus particles, have been discovered within polyhedra.

Incorporation of cypovirus particles into polyhedra

In the case of baculovirus, it has been hypothesized that virus particle occlusion and polyhedral growth are initiated by specific interactions between polyhedrin molecules and the virus particle envelope (Eason *et al.*, 1998). For cypovirus, similar specific interactions between CPV polyhedrin and a viral capsid protein may lead to occlusion of virus particles into polyhedra. Iodination of BmCPV virus particle and analysis of the labelled polypeptides by SDS-PAGE first indicated that VP1 and VP3 are outer components of the BmCPV particle (Lewandowski and Traynor, 1972; McCrae and Mertens, 1983). VP3 was first selected to investigate interaction with BmCPV polyhedrin in the occlusion of virus particles into the polyhedra (Ikeda *et al.*, 2001). Green fluorescent protein (GFP) was fused to the C-terminus of BmCPV VP3 and introduced into the baculovirus expression vector, resulting in AcVP3/GFP (Ikeda *et al.*, 2001). The fusion protein of VP3 and GFP was co-expressed with AcCP-H polyhedrin in Sf21 cells. The polyhedra were purified and green fluorescence observed under UV irradiation. The fluorescence indicated that the VP3-GFP chimera was incorporated into

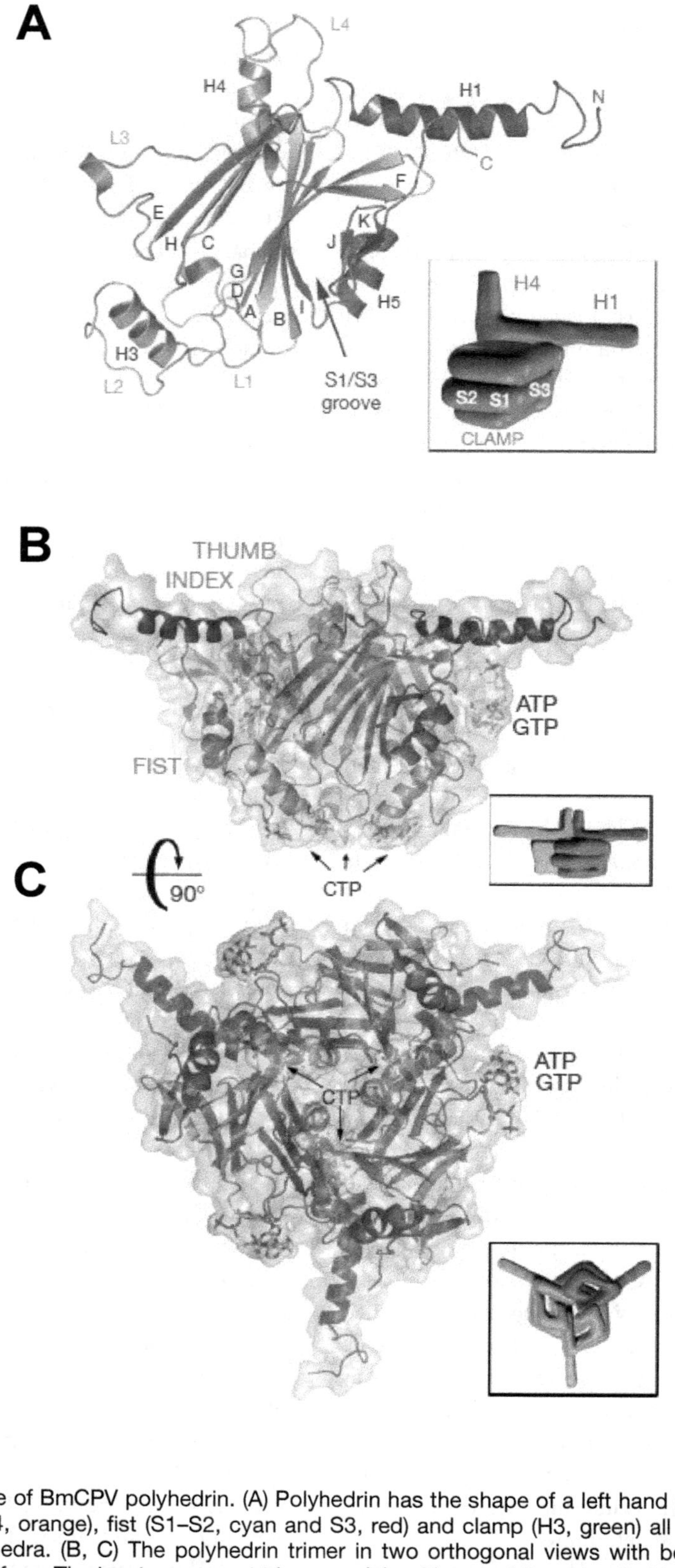

Figure 13.3 Structure of BmCPV polyhedrin. (A) Polyhedrin has the shape of a left hand (inset). The finger (H1, blue), thumb (H4, orange), fist (S1–S2, cyan and S3, red) and clamp (H3, green) all contribute to the organization of polyhedra. (B, C) The polyhedrin trimer in two orthogonal views with bound nucleotides shown in a yellow surface. The insets represent trimer models with one of the subunits highlighted in red. A colour version of this figure is located in the plate section at the back of the book.

these polyhedra (Fig. 13.4A). The fluorescence disappeared when the polyhedra were suspended in acetate buffer at pH 4 and reappeared after the pH was increased (Fig. 13.4B). There was no change in the appearance of BmCPV polyhedra produced by AcCP-H below pH 10, however, when the pH increased above 10, the polyhedra dissolved rapidly and the green fluorescence was intensified (Fig. 13.4B) (Ikeda *et al.*, 2001). This effect could be due to the release of the VP3-GFP chimera from the dissolving polyhedra. No fluorescence was detected in polyhedra obtained from cells infected with both AcGFP and AcCP-H, indicating that the chimeric protein occlusion was initiated by specific interactions between BmCPV polyhedrin and VP3 (Ikeda *et al.*, 2001).

In order to map an immobilization signal to a fragment of the BmCPV VP3 protein, either the N-terminal or C-terminal half of VP3 was fused to the N-terminus of EGFP and recombinant baculoviruses encoding these two chimeric proteins were constructed (Ikeda *et al.*, 2006). One of these viruses, designated AcVP3(N)/GFP, encoded amino acids 1–448 of VP3 fused to EGFP and the other, designated AcGFP/VP3(C), encoded EGFP fused to amino acids 428–1057 of VP3. Each recombinant virus was

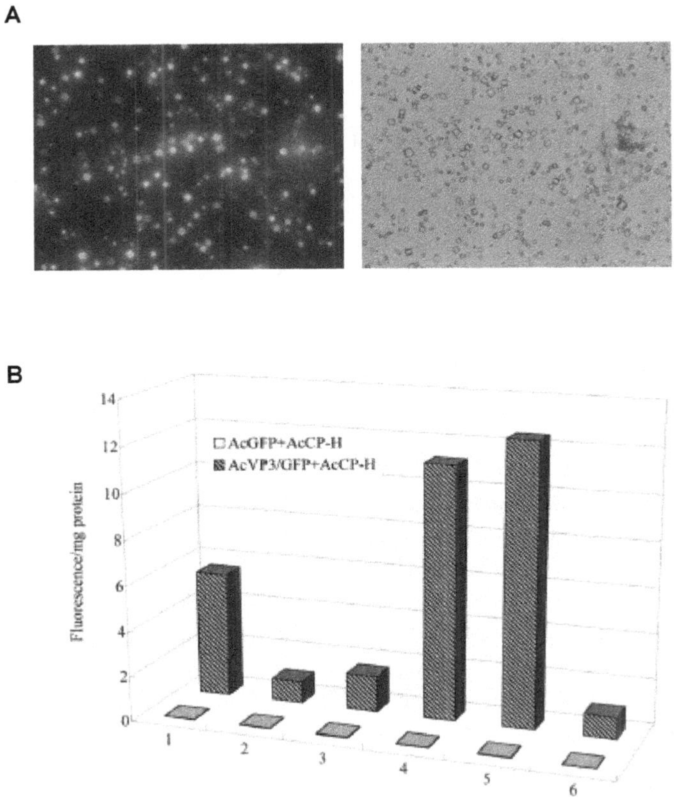

Figure 13.4 Immobilization of EGFP into the BmCPV polyhedra. (A) The green fluorescence from the polyhedra by double infection with AcVP3/GFP and AcCP-H. The green fluorescence under the irradiation of UV (left) and micrograph of BmCPV polyhedra (right). (B) Measurement of green fluorescence of the VP3-GFP chimera in BmCPV polyhedra. Polyhedra obtained by double infection with AcGFP and AcCP-H or with AcVP3/GFP and AcCP-H were resuspended in distilled water, acetate buffer (pH 4.0), and carbonate buffer (pH 11.0). After resuspension in acetate buffer, the pH was adjusted to 6.0, 10.0, and 12.5 by the addition of 5 N NaOH. Fluorescence intensity was represented as fluorescence emission at 510 nm per milligram of protein which was illuminated with 475 nm light. Column; 1, distilled water; column 2, pH 4; column 3, pH 6; column 4, pH 10; column 5, pH 11; column 6, pH 12.5. A colour version of this figure is located in the plate section at the back of the book.

used for double infections together with AcCP-H, which is a third recombinant that produces a cubic form of BmCPV polyhedra. Green fluorescence was observed, predominantly on the surface, with polyhedra isolated from cells that had been co-infected with AcVP3 (N)/GFP and AcCP-H (Ikeda *et al.*, 2006). In contrast, no green fluorescence was observed with polyhedra isolated from cells that had been co-infected with AcGFP/VP3(C) and AcCP-H.

A large number of recombinant baculoviruses encoding different VP3–GFP fusion proteins were subsequently constructed to identify the minimal region of VP3 required to target the GFP fusion construct to polyhedra. Sf21 cells were co-infected with these recombinant baculoviruses and AcCP-H, polyhedra were purified from each doubly-infected culture, and then the presence of each fusion protein in the polyhedra was determined by confocal microscopy. Based upon the results of these analyses, it was concluded that the occlusion of GFP into polyhedra required the region of VP3 between amino acids 1 and 79 and this amino acid sequence also was able to direct the immobilization of other foreign proteins into polyhedra (Ikeda *et al.*, 2006). The N-terminal 79 residues of VP3 form a separate domain on the outside of the turret (Yu *et al.*, 2008). The N-terminal 79 residues of VP3 were named the immobilization signal (Ikeda *et al.*, 2006) or polyhedrin-binding domain (PBD) (Yu *et al.*, 2008), since these residues form a separate domain at the base of the turret, as seen in the 3.88 Å resolution cryo-EM map (Yu *et al.*, 2008).

Polyhedra as protein nanocontainers

The atomic structure of BmCPV polyhedra revealed trimeric building blocks connected by extensive interactions, many involving the polyhedrin H1 helix (Coulibaly *et al.*, 2007). Each building block consists of a cluster of three identical polyhedrin molecules and the three corresponding N-terminal H1 helices project outwards from the centre of the trimer. The building blocks are interlocked with other identical building blocks to form a tight scaffold largely generated by H1-helix (Coulibaly *et al.*, 2007). Because the H1 helix is at the N-terminus of the molecule and projects outwards from the structure it may form into a helix independently as the molecule folds. This possibility, together with its role in cross-linking polyhedra, led to the suggestion that the polyhedrin H1 helix might also prove to be a useful tag for incorporating foreign proteins into polyhedra, like the VP3 residue 1–79 immobilization signal.

The H1-helix sequence was added to either the N-terminus or the C-terminus of enhanced green fluorescent protein (EGFP) (Ijiri *et al.*, 2009). Each recombinant EGFP (H1/EGFP and EGFP/H1) was co-expressed with BmCPV polyhedrin using the baculovirus expression vector system. Immobilization of EGFP fused with H1 signal was compared with VP3 signal. The emission of green fluorescence from H1/EGFP polyhedra displayed a greater intensity compared with either EGFP/H1 or EGFP/VP3 polyhedra. The VP3 signal and the H1signal were used together to produce polyhedra containing both EGFP and DsRed. EGFP and DsRed were fused with VP3 signal (EGFP/VP3) and H1 signal (H1/DsRed), respectively, and the reverse conformations (H1/EGFP and DsRed/VP3) were also constructed. The double-labelled polyhedra with EGFP and DsRed were isolated and imaged using dual wavelength confocal fluorescence microscopy, showing that a multiple proteins can be incorporated into single polyhedra using both VP3 and H1 immobilization signals (Ijiri *et al.*, 2009).

Fibroblast growth factor-2 (FGF-2) is a well-characterized heparin-binding growth factor that regulates cell proliferation, differentiation, or migration (Gospodarowicz and Cheng, 1986; Bailly *et al.*, 2000; Thisse and Thisse, 2005). When FGF-2 was immobilized using the VP3 signal into polyhedra, the polyhedra were functional and possessed biological activities (Mori *et al.*, 2007). FGF-2 was also fused with the H1 signal at the N-terminal (H1/FGF-2) and the biological activity of H1/FGF-2 polyhedra was assayed using mouse fibroblast NIH3T3 cells (Mori *et al.*, 2007; Ijiri *et al.*, 2009). After 4 days in culture, cell proliferation was measured by intracellular dehydrogenase activity. In the presence of either FGF-2/VP3 polyhedra or H1/FGF-2 polyhedra, NIH3T3 cells proliferated in a dose-dependent manner to an extent that was comparable with the addition of recombinant human FGF-2 (rhFGF-2) (Fig. 13.5A). When

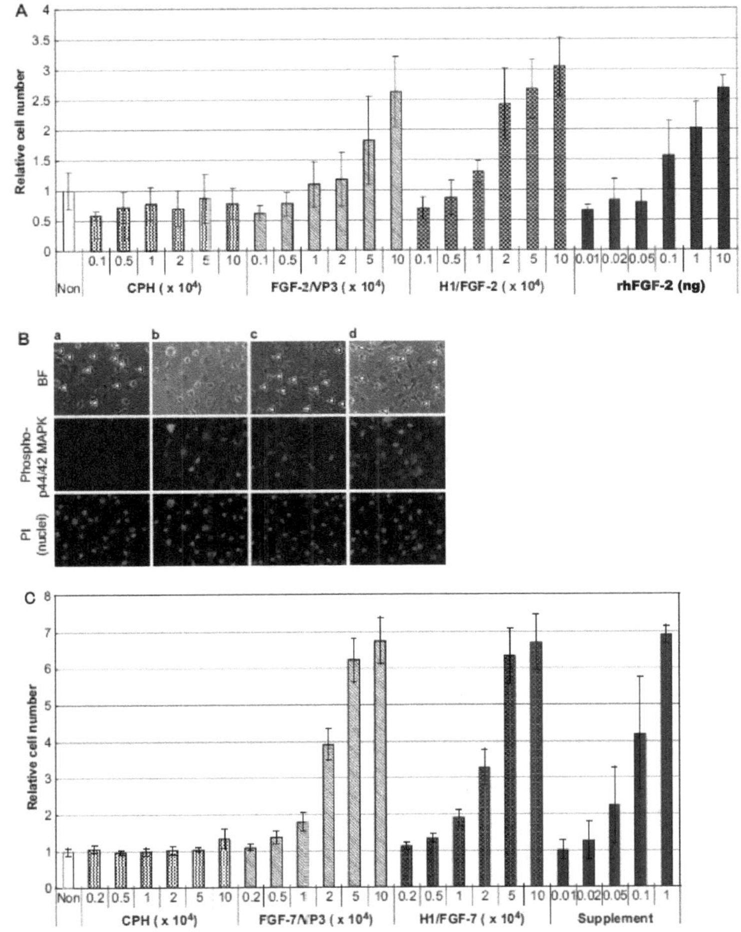

Figure 13.5 Effects of FGF-2 polyhedra on the proliferation of NIH/3T3 cells in culture. (A) Proliferation of NIH/3T3 cells by FGF-2 polyhedra. NIH3T3 cells were seeded onto a 96-well cell culture plate at a density of 4,000 cells per well and cultured overnight in DMEM medium containing 10% FCS followed by 6 hours culture in α-MEM medium without FCS to starve the cells. Cells were stimulated with CP-H polyhedra and FGF-2 polyhedra (FGF-2/VP3 and H1/FGF-2) or recombinant human FGF-2 (rhFGF-2). Number of polyhedra or the dose of rhFGF-2 is indicated on the horizontal axis. After 5 days, the growth rate of respective cell group was determined by CCK assay using a microplate reader. Cell proliferation is reported relative to the negative control arbitrarily set to 1. The mean value of triplicate samples and its standard deviation (SD) are shown as a bar and an error bar, respectively. (B) Immunocytochemistry of phosphorylated MAPK in NIH3T3 cells. The serum-starved NIH3T3 cells on 96 well culture plate were stimulated with CP-H polyhedra (a), 5 ng of rhFGF-2 (b), FGF-2/VP3 polyhedra (c) or H1/FGF-2 polyhedra (d) (each 2.5×10^4 polyhedra) for 15 min. The cells were then fixed and nuclei were stained with propidium iodide (PI). Furthermore, the intracellular phosphorylated MAP kinase was detected by immunocytochemistry using a primary anti-phospho p44/42 MAP kinase antibody and a secondary antibody conjugated with Alexa 488 fluorescent dye. The phosphorylated cells were detected under a fluorescent microscope. Each image indicates phosphorylated p44/42 MAP kinase (green) and nuclei (red) and, in BF, polyhedra are identified with white triangles. (C) Culture of keratinocytes with FGF-7 polyhedra. NHEK at passage 3 with DK-SFM containing the growth supplement were seeded at a density of 900 cells/well and cultured for 5 days at 37°C under 5% CO_2 in air. NHEK cells were grown with or without a commercial growth supplement, and with CP-H polyhedra or FGF-7 polyhedra (FGF-7/VP3 and H1/FGF-7) at concentrations of 0.2 to 10 x 10^4 polyhedra per well. The rate of cell proliferation was detected by CCK assay for 4 hours at 37°C under 5% CO_2 in air and measure of absorbance at 450 nm. Cell proliferation is reported relative to the negative control arbitrarily set to 1. The mean value of triplicate samples and its standard deviation (SD) are shown as a bar and an error bar, respectively. A colour version of this figure is located in the plate section at the back of the book.

2×10^4 H1/FGF-2 polyhedra were added to the culture, the NIH3T3 cell proliferation was twice that of non-stimulated cells. In contrast, 5×10^4 cubes of FGF-2/VP3 polyhedra were required for a similar two-fold proliferation. In the control experiment, empty polyhedra did not induce NIH3T3 cell proliferation. The number of FGF-2 polyhedra required to obtain an equivalent proliferative response to 1 ng of rhFGF-2 was 1.5×10^4 for H1/FGF-2 polyhedra and 7.5×10^4 for FGF-2/VP3 polyhedra.

It is well known that intracellular signalling events such as phosphorylation of p44/42 mitogen-activated protein kinase (MAPK) are observed after FGF-2 stimulation (Pages *et al.*, 1993; Kovalenko *et al.*, 2003; Mackeigan *et al.*, 2005). FGF-2 modified polyhedra also have this activity, as demonstrated by the addition of FGF-2/VP3 to NIH3T3 cells which induced the phosphorylation of p44/42 MAPK (Mori *et al.*, 2007). To address the biological characteristics of H1/FGF-2 polyhedra, a similar experiment was attempted by immunocytochemistry with anti-phospho-p42/44 MAPK-specific antibodies (Fig. 13.5B). It was found that 2.5×10^4 cubes of H1/FGF-2 and FGF-2/VP3 polyhedra induced phosphorylation of p44/42 similarly to the phosphorylation induced by addition of rhFGF-2. Moreover, the brighter fluorescence observed for H1/FGF-2 suggests that the dose of the biologically active form of FGF-2 that a single cube of polyhedra can release is higher in H1/FGF-2 polyhedra than in FGF-2/VP3 polyhedra (Ijiri *et al.*, 2009).

The epithelium of human skin is mainly composed of keratinocytes (approximately 95%) together with melanocytes and fibroblasts (Metcalfe and Ferguson, 2007). The proliferation and migration of keratinocytes are pivotal events for re-epithelialization of skin during wound healing (Haase *et al.*, 2003; Werner and Grose, 2003; Beer *et al.*, 2005; Eming *et al.*, 2007; Pastore *et al.*, 2008). Keratinocytes are known to proliferate in the presence of FGF-7 (also called keratinocyte growth factor) (Igarashi *et al.*, 1998; Werner and Grose, 2003). Hence, we attempted to generate polyhedra containing FGF-7 and examined its proliferative effect on human epidermal keratinocytes (NHEK). FGF-7 polyhedra were generated by fusing VP3 immobilization signal to FGF-7 C-terminal and H1 signal to FGF-7 N-terminal, respectively (Ijiri *et al.*, 2009). FGF-7 polyhedra were deposited on the bottom of culture well and air dried. NHEK were cultured for 5 days in serum-free medium on the FGF-7 polyhedra-coated wells (Fig. 13.5C). The positive control was cells cultured in medium with a commercial growth supplement specific for cultivation for keratinocyte. This supplement contains FGF-7, EGF and other undisclosed cytokines. The exact quantity of theses cytokines in the growth supplement is unknown. The proliferation of keratinocyte was confirmed in a dose-dependent manner either by addition of H1/FGF-7 polyhedra or FGF-7/VP3 polyhedra. Unlike immobilization of EGFP or that of FGF-2 into polyhedra, there was no significant difference between FGF7/VP3 and H1/FGF7 polyhedra in terms of cell proliferation of NHEK.

These studies suggest that modified cypovirus polyhedra can function as nanocontainers stabilizing incorporated proteins in a functional form. One application is a mimic of the extracellular matrix (ECM) where cytokines including growth factors are normally bound to structural scaffolds such as collagen, fibronectin and heparan sulphate (Taipale and Keski-Oja, 1997; Kitajima *et al.*, 2007; Macri *et al.*, 2007). In the ECM, enhanced stability and activity of the embedded cytokines results in cell proliferation and differentiation regulated in time and space (Hubbell, 2006; Sato *et al.*, 2007; Stefonek and Masters, 2007). A cell chip where polyhedra containing various combinations of cytokines are arranged in a specific design might achieve the same function and regulate cellular signalling, proliferation and differentiation at the single cell level (Andreadis and Geer, 2006). Such perspectives will certainly drive active research on the application of polyhedra in stem cell research, tissue engineering and regenerative medicine (Boontheekul and Mooney, 2003; Werner and Grose, 2003; Finch and Rubin, 2004; Spielberger *et al.*, 2004; Choi *et al.*, 2008). Another application is as a stable delivery vehicle for cargos such as vaccine components, fluorescent probes or drugs (MacBeath and Schreiber, 2000; Zhu *et al.*, 2001; Rusmini *et al.*, 2007).

Origin of CPV

Fig. 13.6 shows a radial neighbour-joining tree derived from a multiple sequence alignment of viral RdRps (Attoui *et al.*, 2006). Viruses from the genus *Cypovirus* are related to those from the genera *Oryzavirus* and *Fijivirus*, which both include turreted viruses. Comparison of the VP3 turret protein amino acid sequences encoded by BmCPV S4 with GenBank and EMBL databases using BLAST or FASTA programs revealed partial homology with plant reoviruses. Fig. 13.7 shows locations of the homologous regions in the amino acid sequences between BmCPV S4, rice ragged stunt virus (RRSV) S2, *Nilaparvata lugens* reovirus (NLRV) S4, and *Fiji disease virus* (FDV) S4. There is 22% identity and 39% sequence similarity with an S2-encoded P2 protein of RRSV, a member of the genus *Oryzavirus* (Upadhyaya *et al.*, 1995; Miyazaki *et al.*, 2008), 17% identity and 33% sequence similarity with an S4-encoded 130-kDa protein of NLRV, a putative member of the genus *Fijivirus* (Nakashima *et al.*, 1996; Soo *et al.*, 1998), and 17% identity and 36% sequence similarity with an S4-encoded

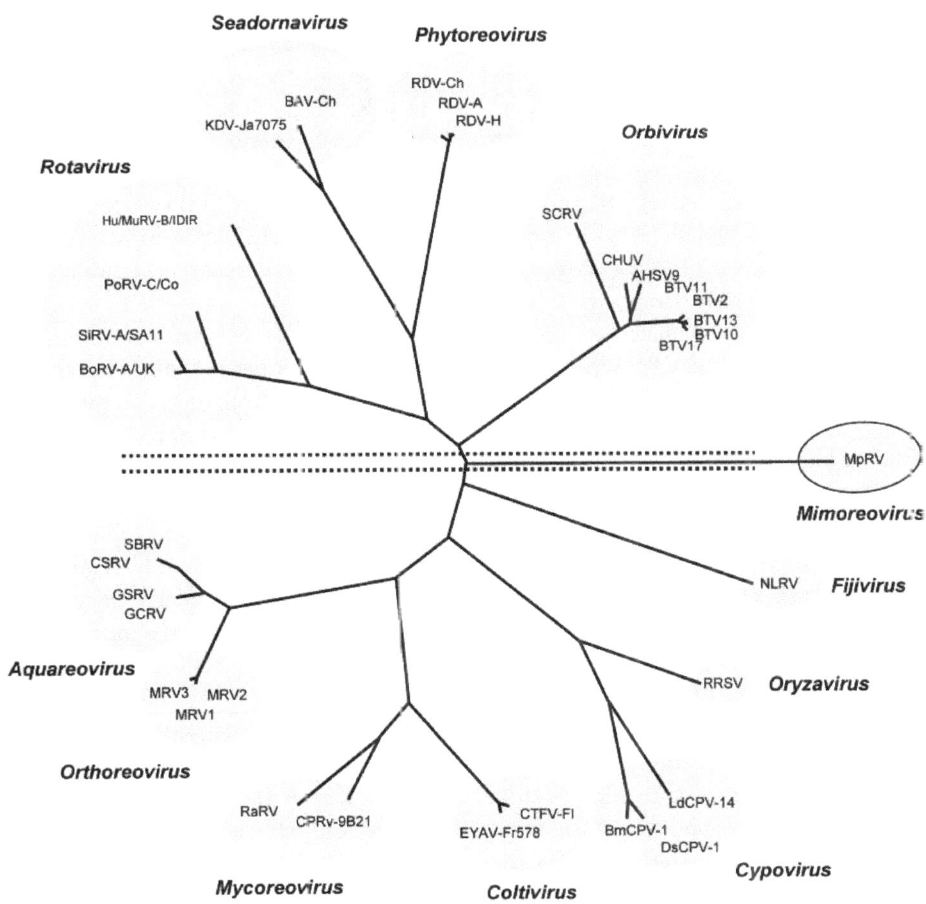

Figure 13.6 Phylogenetic relationships of members of the family *Reoviridae* based on the sequence of the putative RNA-dependent RNA polymerase. Neighbour-joining phylogenetic tree built with available polymerase sequences (using the Poisson-correction or gamma-distribution algorithms) for representative members of 11 genera of the family *Reoviridae*.

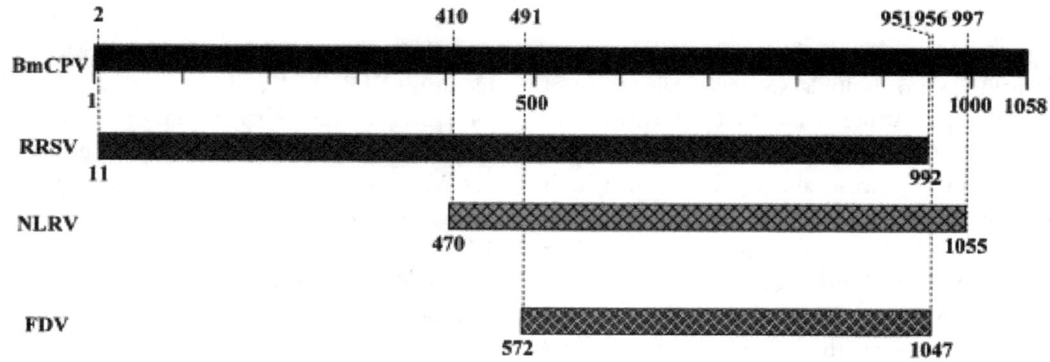

Figure 13.7 Locations of the homologous regions in the amino acid sequences between BmCPV S4, RRSV S2, NLRV S4, and FDV S4.

protein of (FDV), a member of the genus *Fijivirus* (Nakashima *et al.*, 1996) to the BmCPV VP3. For RRSV, the homologous region covered almost the full length of the amino acid sequence (residues 2 to 951 in BmCPV and residues 11 to 992 in RRSV). Other groups have also reported that the capsid proteins encoded by S1 and S3 of BmCPV showed about 38% and 47% homology with P3 and P1 protein of RRSV (Hagiwara *et al.*, 2002). It has been recently reported that the RRSV has a structure intermediate between that of the Orthoreovirus and the CPV virus particle (Li *et al.*, 2006,2007). The homology of BmCPV to NLRV and FDV was less evident and observed only in the carboxy-terminal half (residues 470 to 1055 for NLRV and residues 572 to 1047 for FDV) (Fig. 13.7). These results might indicate that BmCPV is more closely related to RRSV than to the other two viruses and that these two viruses split more recently. According to the estimation of the rate of nucleic acid substitution (2.2×10^{-3} substitutions/site/year) in the case of the genus *Orbivirus* in the family *Reoviridae* (Kowalik and Li, 1991), the divergence of BmCPV and RRSV can be estimated to have occurred as recently as about 180 years ago. However, CPVs are extremely stable in the environment and can survive for long periods without replication in insects, and it is supposed that they spent many years without changing significantly in terms of their RNA sequences. Therefore, some caution must be exercised in interpretation of the data and in any speculation of a date for divergence of two genera within the *Reoviridae*. Comparison of the nucleotide sequences encoding the highly conserved amino acid sequences shown in Fig. 13.7 (BmCPV S4, RRSV S2, NLRV S4, and FDV S4) yielded the following results. Between BmCPV S4 and RRSV S2, the levels of identity of the first and the second nucleotides in the codons were 50% and 76%, respectively. The third nucleotide, having a higher degree of freedom, showed a much lower level of homology, only 25%. In the case of NLRV S4 and FDV S4, comparison of the first and the second nucleotides in the codons showed similar homologies of about 55% and the identity of the third nucleotide was about 35%. The non-random distribution of nucleotide exchanges was shown in the codons of ORFs of BmCPV S4 and RRSV S4. In conjunction with the amino acid sequence data, these findings might be a good example of the continuing divergence in viral evolution, which takes place under the functional constraint of maintaining the necessary capsid properties.

Concluding remarks

Structural analyses of CPV polyhedra and virus particles open the door for many exciting new applications of insect virus. These studies suggest that CPV polyhedra can be used as protein nanocontainers. One application is a mimic of the ECM where cytokines including growth factors are normally bound to structural scaffolds such as collagen, fibronectin and heparan sulphate. In the ECM, an enhanced stability and activity of the embedded cytokines results in cell proliferation and differentiation regulated in time and space.

A cell chip where polyhedra containing various combinations of cytokines are arranged in a specific design might achieve the same function and regulate cellular signalling, proliferation and differentiation at the single-cell level. Such perspectives will certainly drive active research on the application of polyhedra in stem cell research, tissue engineering and regenerative medicine. Another application is to deliver valuable cargos such as vaccine components, fluorescent probes or drugs.

References

Andreadis, S.T., and Geer, D.J. (2006). Biomimetic approaches to protein and gene delivery for tissue regeneration. Trends Biotechnol. *24*, 331–337.

Arella, M., Lavallée, C., Belloncik, S., and Furuichi, Y. (1988). Molecular cloning and characterization of cytoplasmic polyhedrosis virus polyhedrin and a viable deletion mutant gene. J. Virol. *62*, 211–217.

Aruga, H. (1963). Induction of virus infection. In An Advence Treatise, E. A. Steinhaus, ed. (New York, Academic Press), pp 499–530.

Aruga, H. (1971). Cytoplasmic polyhedrosis of the silkworm⊠Historical, economical and epizootiological aspects. In The Cytoplasmic Polyhedrosis Virus of the Silkworm, H. Aruga and Y. Tanada, eds. (Tokyo,University of Tokyo Press), pp 3–57.

Attoui, H., Jaafar, F.M., Belhouchet, M., de Micco, P., de Lamballerie, X., and Brussaard, C.P.D. (2006). Micromonas pusilla reovirus: a new member of the family Reoviridae assigned to a novel proposed genus (Mimoreovirus). J. Gen.Virol *87*, 1375–1383.

Bailly, K., Soulet, F., Leroy, D., Amalric, F., and Bouche, G. (2000). Uncoupling of cell proliferation and differentiation activities of basic fibroblast growth factor. FASEB J *14*, 333–344.

Beer, H.D., Bittner, M., Niklaus, G., Munding, C., Max, N., Goppelt, A., and Werner, S. (2005). The fibroblast growth factor binding protein is a novel interaction partner of FGF-7, FGF-10 and FGF-22 and regulates FGF activity: implications for epithelial repair. Oncogene *24*, 5269–5277.

Belloncik, S. (1989). Cytoplasmic polyhedrosis viruses-Reoviridae. Adv. Virus Res. *37*, 173–209.

Belloncik, S., and Mori, H. (1998). Cypoviruses. In The Insect Viruses, L. K. Miller and L. A. Ball, eds. (New York, Plenum), pp. 337–369.

Boontheekul, T., and Mooney, D.J. (2003). Protein-based signaling systems in tissue engineering. Curr. Opin. Biotechnol. *14*, 559–565.

Choi, J.S., Leong, K.W., and Yoo, H.S. (2008). *In vivo* wound healing of diabetic ulcers using electrospun nanofibers immobilized with human epidermal growth factor (EGF). Biomaterials *29*, 587–596.

Coulibaly, F., Chiu, E., Ikeda, K., Gutmann, S., Haebel, P.W., Schulze-Briese, C., Mori, H., and Metcalf, P. (2007). The molecular organization of cypovirus polyhedra. Nature *446*, 97–101.

Coulibaly, F., Chiua, E., Gutmann, S., Rajendran, C., Haebel, P.W., Ikeda, K., Mori, H., Ward, V.K., Schulze-Briese, C., and Metcalf, P. (2009). The atomic structure of baculovirus polyhedra reveals the independent emergence of infectious crystals in DNA and RNA viruses. Proc. Nat. Acad. Sci. USA, *106*, 22205–22210

Eason, J.E., Hice, R.H., Johnson, J.J., and Federici, B.A. (1998). Effects of substituting granulin or a granulin-polyhedrin chimera for polyhedrin on virion occlusion and polyhedral morphology in *Autographa californica* multinucleocapsid nuclear polyhedrosis virus. *J. Virol.* *72*, 6237–6243.

Eming, S.A., Krieg, T., and Davidson, J.M. (2007). Inflammation in wound repair: molecular and cellular mechanisms. J. Invest. Dermatol. *12*, 514–525.

Finch, P.W., and Rubin, J.S. (2004). Keratinocyte growth factor/fibroblast growth factor 7, a homeostatic factor with therapeutic potential for epithelial protection and repair. Adv. Cancer Res. *91*, 69–136.

Furuichi, Y. (1974). 'Methylation-coupled' transcription by virus associated transcriptase of cytoplasmic polyhedrosis virus containing double-stranded RNA, Nucleic Acids Res. *1*, 809–822.

Furuichi, Y. (1978). 'Pretranscriptional capping' in the biosynthesis of cytoplasmic polyhedrosis virus mRNA. Proc. Natl. Acad. Sci. U.S.A. *75*, 1086–1090.

Furuichi, Y., and Miura, K. (1975). A blocked structure at the 5′ terminus of mRNA from cytoplasmic polyhedrosis virus. Nature *253*, 374–375.

Gospodarowicz, D., and Cheng, J. (1986). Heparin protects basic and acidic FGF from inactivation. J. Cell Physiol. *128*, 475–484.

Graham, R.I., Rao, S.J., Possee, R.D., Sait, S.M., Mertens, P.P.C., and Hails, R.S. (2006). Detection and characterisation of three novel species of reovirus (Reoviridae), isolated from geographically separate populations of the winter moth *Operophtera brumata* (Lepidoptera: *Geometridae*) on Orkney. J Invertebr. Pathol. *91*, 79–87.

Grimes, J.M., Burroughs, J.N., Gouet, P., Diprose, J.M., Malby, R., Zientara, S., Mertens, P.P., and Stuart, D.I. (1998). The atomic structure of the bluetongue virus core. Nature *395*, 470–478

Haase, I., Evans, R., Pofahl, R., and Watt, F.M. (2003). Regulation of keratinocyte shape, migration and wound epithelialization by IGF-1- and EGF-dependent signalling pathways. J. Cell Sci. *116*, 3227–3238.

Hagiwara, K., and Matsumoto, T. (2000). Nucleotide sequences of genome segments 6 and 7 of *Bombyx mori* cypovirus 1, encoding the viral structural proteins V4 and V5, respectively. *J. Gen. Virol.* *81*, 1143–1147.

Hagiwara, K., Kobayashi, J., Tomita, M., and Yoshimura, T. (2001). Nucleotide sequence of genome segment 5 from *Bombyx mori* cypovirus 1. *Arch. Virol.* *146*, 181–187.

Hagiwara, K., Rao, S., Scott, S.W., and Carner, G.R. (2002). Nucleotide sequences of segments 1, 3 and 4 of the genome of *Bombyx mori* cypovirus 1 encoding putative capsid proteins VP1, VP3 and VP4, respectively. *J. Gen. Virol*, *83*, 1477–1482.

Hagiwara, K., Tomita, M., Kobayashi, J., Miyajima, S., and Yoshimura, T. (1998a). Nucleotide sequence of

Bombyx mori cytoplasmic polyhedrosis virus segment 8. Biochem. Biophys. Res. Commun. *247*, 549–553.

Hagiwara, K., Tomita, M., Nakai, K., Kobayashi, J., Miyajima, S., and Yoshimura, T. (1998b). Determination of the nucleotide sequence of *Bombyx mori* cytoplasmic polyhedrosis virus segment 9 and its expression in BmN4 cells. J. Virol. *72*, 5762–5768.

Hill, C.L., Booth, T.F., Prasad, B.V., Grimes, J.M., Mertens, P.P., Sutton, G.C., and Stuart, D.I. (1999). The structure of a cypovirus and the functional organization of dsRNA viruses. Nat. Struct. Biol. *6*, 565–568.

Hubbell, J. (2006). Matrix-bound growth factors in tissue repair. Swiss Med. Wkly. *136*, 387–391.

Hukuhara, T. (1971). Variations in cytoplasmic-polyhedrosis virus. In The Cytoplasmic Polyhedrosis of the Silkworm, H. Aruga and Y. Tanada, eds. (Tokyo, University of Tokyo Press), pp. 79–100.

Hukuhara, T. (1985). Pathology associated with cytoplasmic polyhedrosis viruses. In Viral Insecticides for Biological Control, H. Aruga and Y. Tanada, eds. (New York, Academic Press), pp. 121–161.

Hukuhara, T., and Bohami, J.R. (1991). Reoviridae. In Atlas of Invertebrate Viruses, J. R. Adams, and J. R. Bohami (Boca Raton, Fla, CRC Press, Inc.), pp. 393–434.

Hukuhara, T., and Midorikawa, M. (1983). Pathogenesis of cytoplasmic polyhedrosis in the silkworm. In Double-Stranded RNA Viruses, R. W. Compans and D. H. L. Bishop, eds. (New York, Elsevier Biomedical), pp. 405–414.

Igarashi, M., and Finch, P.W. And Aaronson S.A. (1998). Characterization of recombinant human fibroblast growth factor (FGF)-10 reveals functional similarities with keratinocyte growth factor (FGF-7). J. Biol. Chem. *273*, 13230–13235.

Ijiri, H., Coulibaly, F., Nishimura, G., Nakai, D., Chiu, E., Takenaka, C., Ikeda, K., Nakazawa, H., Hamada, N., Kotani, E., *et al.* (2009). Structure-based targeting of bioactive proteins into cypovirus polyhedra and application to immobilized-cytokines for mammalian cell culture. Biomaterials *30*, 4297–4308.

Ikeda, K., Nagaoka, S., Winkler, S., Kotani, K., Yagi, H., Nakanishi, K., Miyajima, S., Kobayashi, J., and Mori, H. (2001). Molecular characterization of *Bombyx mori* cytoplasmic polyhedrosis virus genome segment 4. *J. Virol., 75*, 988–995.

Ikeda, K., Nakazawa, H., Alain, R., Belloncik, S., and Mori, H. (1998). Characterizations of natural and induced polyhedrin gene mutants of *Bombyx mori* cytoplasmic polyhedrosis viruses. Arch. Virol. *143*, 241–248.

Ikeda, K., Nakazawa, H., Shimo-Oka, A., Ishio, K., Miyata, S., Hosokawa, Y., Matsumura, S., Masuhara, H., Belloncik, S., *et al.* (2006). Immobilization of diverse foreign proteins in viral polyhedra and potential application for protein microarrays. Proteomics. *6*, 54–66.

Ji, X., Sutton, G., Evans, G., Axford, D., Owen, R., and Stuart, D.I. (2010). How baculovirus polyhedra fit square pegs into round holes to robustly package viruses. EMBO J. *29*, 505–514

Kitajima, T., Terai, H., and Ito, Y. (2007). A fusion protein of hepatocyte growth factor for immobilization to collagen. Biomaterials *28*, 1989–1997.

Kovalenko, D., Yang, X., Nadeau, R.J., Harkins, L.K., and Friesel, R. (2003). Sef inhibits fibroblast growth factor signaling by inhibiting FGFR1 tyrosine phosphorylation and subsequent ERK activation. J. Biol. Chem. *278*, 14087–14091.

Kowalik, T.F., and Li, J.K. (1991). Bluetongue virus evolution: sequence analyses of the genomic S1 segments and major core protein VP7. Virology *181*, 749–755.

Lewandowski, L.J., and Traynor, B.L. (1972). Comparison of the structure and polypeptide composition of three double-stranded ribonucleic acid-containing viruses (diplornaviruses): Cytoplasmic polyhedrosis virus, wound tumor virus and reovirus, J. Virol. *10*, 1053–1070.

Li, Y., Tan, L., Li, Y., Chen, W., Zhang, J., and Hu, Y. (2006). Identification and genome characterization of *Heliothis armigera* cypovirus types 5 and 14 and *Heliothis assulta* cypovirus type 14. J. Gen. Virol. *87*, 387–394.

Li, Y., Zhang, J., Li, Y., Tan, L., Chen, W., Luo, H., and Hu, Y. (2007). Phylogenetic analysis of *Heliothis armigera* cytoplasmic polyhedrosis virus type 14 and a series of dwarf segments found in the genome. J. Gen. Virol. *88*, 991–997.

MacBeath, G., and Schreiber, S.L. (2000). Printing proteins as microarrays for high-throughput function determination. Science *289*, 1760–1763.

Mackeigan, J.P., Murphy, L.O., Dimitri, C.A., and Blenis, J. (2005). Graded mitogen-activated protein kinase activity precedes switch-like c-Fos induction in mammalian cells. Mol. Cell Biol. *25*, 4676–4682.

Macri, L., Silverstein, D., and Clark, R.A. (2007). Growth factor binding to the pericellular matrix and its importance in tissue engineering. Adv. Drug Deliv. Rev. *59*, 1366–1381.

McCrae, M.A., and Mertens, P.P.C. (1983). *In vitro* translation studies on and RNA coding assignments for cytoplasmic polyhedrosis viruses. In Double-Stranded RNA Viruses, R. W. Compans and D. H. L. Bishop, eds. (New York, Elsevier Biomedical), pp.35–41.

Mertens, P.P.C., Rao, S., and Zhou, H. (2004a). *Cypovirus*, In Virus taxonomy: eighth report of the International Committee on Taxonomy of Viruses, C. M. Fauquet, M. A. Mayo, J. Maniloff, U. Desselberger, and L. A. Ball, eds. (London England, Elsevier Academic Press), pp. 522–533.

Mertens, P.P.C., Duncan, R., Attoui, H., and Dermody, T.S. (2004b). *Reoviridae*, In Virus taxonomy: eighth report of the International Committee on Taxonomy of Viruses, C.M. Fauquet, M.A. Mayo, J. Maniloff, U. Desselberger, and L.A. Ball, eds. (London, England, Elsevier Academic Press), pp. 447–454.

Metcalfe, A.D., and Ferguson, M.W. (2007). Tissue engineering of replacement skin: the crossroads of biomaterials, wound healing, embryonic development, stem cells and regeneration. J. R. Soc. Interface *4*, 413–437.

Miura, K., Fujii-Kawata, I., Iwata, H., and Kawase, S. (1969). Electron-microscopic observation of a cytoplasmic-polyhedrosis virus from the silkworm. J. Invertbr. Pathol., *14*, 262–265.

Miyazaki, N., Uehara-Ichiki, T., Xing, L., Bergman, L., Higashiura, A., Nakagawa, A., Omura, T., and Cheng,

R.H. (2008). Structural evolution of *Reoviridae* revealed by *Oryzavirus* in acquiring the second capsid shell. J. Virol. *82*, 11344–11353.

Mori, H., Minobe, Y., Sasaki, T., and Kawase, S. (1989). Nucleotide sequence of the polyhedrin gene of *Bombyx mori* cytoplasmic polyhedrosis virus A strain with nuclear localization of polyhedra. J. Gen. Virol. *70*, 1885–1888.

Mori, H., Sasaki, T., Minobe, Y., Miyajima, S., and Kawase, S. (1987). Difference of proteins from inclusion bodies formed in the nucleus and cytoplasm of the cytoplasmic polyhedrosis virus infected midgut in the silkworm, *Bombyx mori*. J. Invertebr. Pathol. *50*, 26–32.

Mori, H., Ito, R., Nakazawa, H., Sumida, M., Matsubara, F., and Minobe, Y. (1993). Expression of *Bombyx mori* cytoplasmic polyhedrosis virus polyhedrin in insect cells by using a baculovirus expression vector, and its assembly into polyhedra. J. Gen. Virol. *74*, 99–102.

Mori, H., Shukunami, C., Furuyama, A., Notsu, H., Nishizaki, Y., and Hiraki, Y. (2007). Immobilization of bioactive FGF-2 into cubic proteinous micro-crystals (*Bombyx mori* cypovirus polyhedra) that are insoluble in a physiological cellular environment. J. Biol. Chem. *282*, 17289–17296.

Nakashima, N., Koizumi, M., Watanabe, H., and Noda, H. (1996). Complete nucleotide sequence of the Nilaparvata lugens reovirus: a putative member of the genus Fijivirus. J. Gen. Virol. 77, 139–146.

Nakazawa, H., Kendirgi, F., Belloncik, S., Ito, R., Takagi, S., Minobe, Y., Higo, K., Sumida, M., Matsubara, F., and Mori, H. (1996). Effect of mutations on the intracellular localization of *Bombyx mori* cytoplasmic polyhedrosis virus polyhedron. J. Gen. Virol. 77, 147–153.

Pages, G., Lenormand, P., L'Allemain, G., Chambard, J.C., Meloche, S., and Pouyssegur, J. (1993). Mitogen-activated protein kinases p42mapk and p44mapk are required for fibroblast proliferation. Proc. Natl. Acad. Sci. U.S.A. *90*, 8319–8323.

Pastore, S., Mascia, F., Mariani, V., and Girolomoni, G. (2008). The epidermal growth factor receptor system in skin repair and inflammation. J. Invest. Dermatol. *128*, 1365–1374.

Payne, C.C., and Mertens, P.P.C. (1983). Cytoplasmic polyhedrosis viruses. In The *Reoviridae*, W. K. Joklik ed. (New York, Plenum), pp. 425–504.

Rao, S., Carner, G.R., Scott, S.W., Omura, T., and Hagiwara, K. (2003). Comparison of the amino acid sequences of RNA-dependent RNA polymerases of cypoviruses in the family *Reoviridae. Arch. Virol. 148*, 209–219.

Reinisch, K.M., Nibert, M.L., and Harrison, S.C. (2000). Structure of the reovirus core at 3.6 A resolution. Nature *404*, 960–967

Rohrmann, G.F. (1986). Polyhedrin structure. J. Gen. Virol. *67*, 1499–1513.

Rusmini, F., Zhong, Z., and Feijen, J. (2007). Protein immobilization strategies for protein biochips. Biomacromolecules *8*, 1775–1789.

Sato, M., Ishihara, M., Kaneshiro, N., Mitani, G., Nagai, T., Kutsuna, T., Asazuma, T., Kikuchi, M., and Mochida, J. (2007). Effects of growth factors on heparin-carrying polystyrene-coated atelocollagen scaffold for articular cartilage tissue engineering. J. Biomed. Mater. Res. B Appl. Biomater. *83*, 181–188.

Shapiro, A., Green, T., Rao, S., White, S., Carner, G., Mertens, P.P.C., and Becnel, J.J. (2005). Morphological and molecular characterization of a cypovirus (*Reoviridae*) from the mosquito *Uranotaenia sapphirina* (Diptera: Culicidae) J. Virol. *79*, 9430–9438.

Soo, H.M., Handley, J.A., Maugeri, M.M., Burns, P., Smith, G.R., Dale, J.L., and Harding, R.M. (1998). Molecular characterization of Fiji disease fijivirus genome segment 9. J. Gen. Virol. *79*, 3155–3161.

Spielberger, R., Stiff, P., Bensinger, W., Gentile, T., Weisdorf, D., Kewalramani, T., Shea, T., Yanovich, S., Hansen, K., *et al.* (2004). Palifermin for oral mucositis after intensive therapy for hematologic cancers. N Engl. J. Med. *351*, 2590–2598.

Stefonek, T.J., and Masters, K.S. (2007). Immobilized gradients of epidermal growth factor promote accelerated and directed keratinocyte migration. Wound Repair Regen. *15*, 847–855.

Taipale, J., and Keski-Oja, J. (2007). Growth factors in the extracellular matrix. FASEB J. *11*, 51–59.

Tan, Y.R., Sun, J.C., Lu, X.Y., Su, D.M., and Zhang, J.Q. (2003). Entry of *Bombyx mori* cypovirus 1 into midgut cells *in vivo*. J Electron Microsc (Tokyo) *52*, 485–489.

Tan, L., Zhang, J., Li, Y., Li, Y Jiang, H., Cao, X., and Hu, Y. (2008). The complete nucleotide sequence of the type 5 *Helicoverpa armigera* cytoplasmic polyhedrosis virus genome. Virus Genes *36*, 587–593.

Thisse, B., and Thisse, C. (2005). Functions and regulations of fibroblast growth factor signaling during embryonic development. Dev. Biol. *287*, 390–402.

Upadhyaya, N.M., Yang, M., Kositratana, W., Ghosh, A., and Waterhouse, P.M. (1995). Molecular analysis of rice ragged stunt oryzavirus segment 9 and sequence conservation among isolates from Thailand and India. Arch. Virol. *140*, 1945–1956.

Werner, S., and Grose, R. (2003). Regulation of wound healing by growth factors and cytokines. Physiol. Rev. *83*, 835–870.

Yu, X., Jin, L., and Zhou, Z. H. (2008). 3.88 Å structure of cytoplasmic polyhedrosis virus by cryo-electron microscopy. Nature *453*, 415–419.

Zhang, H., Yu, X.K., Lu, X.Y., Zhang, J.Q., and Zhou, Z.H. (2002). Molecular interactions and viral stability revealed by structural analyses of chemically treated cypovirus capsids. Virology *298*, 45–52.

Zhou, Z.H., Zhang, H., Jakana, J., Lu, X.Y., and Zhang, J.Q. (2003). Cytoplasmic polyhedrosis virus structure at 8 A by electron cryomicroscopy: structural basis of capsid stability and mRNA processing regulation. Structure *11*, 651–663.

Zhou, Z.H. (2008a). Towards atomic resolution structural determination by single-particle cryo-electron microscopy. Curr. Opin. Struct. Biol. *18*, 218–28.

Zhou, Z.H. (2008b). Cypovirus, in segmented double-stranded RNA viruses. In Structure and Molecular Biology, J. T. Patton ed. (Norfolk, UK, Caister Academic Press), pp 27–43.

Zhu, H., Bilgin, M., Bangham, R., Hall, D., Casamayor, A., Bertone, P., Lan, N., Jansen, R., Bidlingmaier, S., *et al.* (2001). Global analysis of protein activities using proteome chips. Science *293*, 2101–2105.

Part III

Current Topics in Insect Virology

Structural Comparison of Insect RNA Viruses

14

Manidipa Banerjee, Jeffrey A. Speir and John E. Johnson

Abstract

In recent years, small insect RNA viruses with non-enveloped capsids, such as members of the *Nodaviridae, Tetraviridae, Dicistroviridae* and *Cypoviridae* families, have been characterized using X-ray crystallography or cryoelectron microscopy. The capsids of these viruses are icosahedral, with diameters of 30 – 70 nm, and triangulation numbers ranging from 1 to 4 (T = 1 to T = 4). Viruses packaging single-stranded RNA such as nodaviruses, tetraviruses and dicistroviruses contain capsid proteins with closely similar cores comprising an antiparallel β-barrel, jellyroll fold. These capsid proteins undergo analogous post-assembly autocatalytic maturation cleavage. Tetraviruses are unique among this group in containing an immunoglobulin-like domain inserted in their jellyroll capsid proteins, and in undergoing large conformational changes during maturation. Cypoviruses, which package double-stranded RNA, have T = 2 capsids with structural similarities to the inner shell of reovirus particles. Structural studies of insect RNA viruses have provided valuable information regarding the principles of icosahedral capsid formation, genomic RNA organization, RNA–protein interaction, the mechanism of maturation and associated conformational changes, and the positioning of virus-encoded membrane lytic peptides in capsids. This has resulted in compelling hypotheses regarding capsid disassembly and RNA translocation during infection, some of which have been validated using biochemical and biological studies.

Introduction

Owing to their small size and high degree of stability and symmetry, the icosahedral capsids of small insect RNA viruses have proved amenable to a variety of biophysical and structural studies. Attention has focused on specific families, particularly *Nodaviridae, Tetraviridae, Dicistroviridae* and *Cypoviridae,* due to the availability of robust expression systems for preparing purified particles at high concentrations. Medium to high resolution structures are available for several members of these families. Among *Nodaviridae,* X-ray crystal structures in the 2–3 Å resolution range are available for *Flock House virus* (FHV), *black beetle virus* (BBV), *Nodamura virus* (NoV) and *Pariacoto virus* (PaV) (Fisher and Johnson, 1993; Hosur *et al.*, 1987; Tang *et al.*, 2001; Wery *et al.*, 1994; Zlotnick *et al.*, 1993). In addition, X-ray and cryo-electron microscopy (cryoEM) structures are available for several FHV variants with mutations in the capsid (Banerjee *et al.*, 2009; Fisher *et al.*, 1993; Z. Chen and J.E. Johnson, unpublished observation). Among tetraviruses, X-ray crystal structures are available for *Providence virus* (PrV), a betatetravirus (Speir *et al.*, 2010); and *Nudaurelia capensis omega virus* (NωV), an omegatetravirus (Helgstrand *et al.*, 2004; Munshi *et al.*, 1996). In addition, *Nudaurelia capensis beta virus* (NβV) as well as mutants of NωV, have been studied at subnanometer resolution using cryoEM and image reconstruction (Olson *et al.*, 1990; Tang *et al.*, 2009; T. Matsui, G. Lander, R. Khayat, and J.E. Johnson, unpublished data).

Of the small, single-stranded RNA (ssRNA) containing picorna-like viruses, medium to high resolution structures are available for Drosophila A virus (DAV) (Ambrose *et al.*, 2009), and members of the *Dicistroviridae* family including *Cricket paralysis virus* (CrPV) (Tate *et al.*, 1999) and *Triatoma virus* (TrV) (Estrozi *et al.*, 2008). Several members of the *Iflaviridae*, another family of picorna-like viruses with insect hosts, have also been structurally studied. Low resolution cryoEM structures are available for *infectious flacherie virus* (IFV) (Xie *et al.*, 2009) and Chinese sacbrood bee virus (CSBV) (Zhang *et al.*, 2001). The double-stranded RNA (dsRNA) containing insect viruses are represented by members of the *Cypoviridae* family, of which Cytoplasmic polyhedrosis virus (otherwise known as cypovirus, CPV) is the best studied structurally. The capsid structure of *Cypovirus-5* (CPV-5) has been determined at different resolutions using cryoEM and image reconstruction (Hill *et al.*, 1999; Yu *et al.*, 2008; Zhou *et al.*, 2003). Structural information is not available currently for some other well-studied insect RNA viruses, like *Drosophila X virus* or birnaviruses with insect hosts. A description of the host specificity, infectivity and biochemical studies for individual members of these families is available in other chapters.

We begin by providing a detailed description of the size and geometry of the virus capsids being studied, followed by a close look at the similarities and differences in capsid organization, protein folds, RNA organization and RNA–protein interaction. We then discuss the capsid structural changes associated with virus maturation, host cell entry and disassembly, and the organization of biologically important dynamic polypeptides in the capsid. The applicability of these studies towards understanding the principles behind assembly, stability and infectivity of more complex viruses affecting higher order organisms is also discussed. We finish with a description of how this wealth of structural information can be utilized for designing biopesticides or nanobiomedical tools.

Capsid architecture

Capsid size, geometry and surface features

The capsid organization of the insect RNA viruses is summarized in Table 14.1. The ssRNA insect viruses display icosahedral symmetry, with triangulation numbers ranging from $T = 1$ to $T = 4$. The $T = 3$ capsids of nodaviruses are approximately 30 nm in diameter, and constructed from 180 copies of the capsid protein alpha, organized into alternating pentamers and hexamers on the surface (Fisher and Johnson, 1993; Kaesberg *et al.*, 1990) (Fig. 14.1). Each icosahedral asymmetric unit (iASU) in the capsid is composed of three copies of alpha, which are designated subunits A, B and C (Kaesberg *et al.*, 1990) (Fig. 14.1). The capsid contains 12 5-fold, 20 3-fold and 30 2-fold axes of symmetry. Five A subunits form each icosahedral 5-fold axis of symmetry, while alternating B and C subunits (three of each) are grouped at the icosahedral 3-fold or quasi 6-fold axes of symmetry.

Tetraviruses are the only insect virus family with $T = 4$ icosahedral symmetry. The mature capsids of tetraviruses NβV, NωV and PrV are approximately 40 nm in diameter (Speir and Johnson, 2008). The capsid shell consists of 240 copies of capsid protein alpha, with each iASU being composed of four quasiequivalent subunits A, B, C and D (Speir and Johnson, 2008) (Fig. 14.1). The classification of tetraviruses into omega- and beta-subgroups is based partly on the appearance of their capsid surfaces (Fauquet *et al.*, 2005). While betatetraviruses contain pitted surfaces with prominent protrusions, omegatetraviruses display a relatively smooth surface. PrV, in spite of having a higher sequence similarity with omegatetraviruses, has nonetheless been classified as a betatetravirus because of its characteristic pitted surface, as well as its monopartite genome. A recent crystal structure of PrV demonstrated that the differences in the capsid surface arise because of different orientations of the immunoglobulin-like (Ig-like) domains in the capsid proteins (Speir *et al.*, 2010).

Table 14.1 Capsid organization of insect RNA viruses

Family	Viruses	Mode of structure determination	Diameter	Geometry	Capsid structural proteins	Maturation cleavage	Genome type
Nodaviridae	Flock House virus (FHV)	X-ray crystallography	30 nm	T = 3 icosahedral	α	$\alpha \rightarrow \beta + \gamma$	ssRNA
	Black beetle virus (BBV)						
	Nodamura virus (NoV)						
	Pariacoto virus (PaV)						
Tetraviridae	*Nudaurelia capensis* beta virus (NβV)	Cryoelectron microscopy	Provirion ~48 nm	T = 4 icosahedral	α	$\alpha \rightarrow \beta + \gamma$	ssRNA
	Nudaurelia capensis omega virus (NωV)	X-ray crystallography	Mature virus ~41 nm				
	Providence virus (PrV)	X-ray crystallography					
Dicistroviridae	Cricket paralysis virus (CrPV)	X-ray crystallography	30 nm	T = 1 (P = 3) icosahedral	VP1, VP2, VP3, VP4	VP0 $\rightarrow$ VP4 + VP3	ssRNA
	Drosophila A virus (DAV)	Cryoelectron microscopy	30 nm	T = 3 icosahedral (partially)	50 kDa protein	Ambiguous	
Cypoviridae	Cytoplasmic polyhedrosis virus (CPV)	Cryoelectron microscopy	71 nm	T = 2 icosahedral	CSP, LPP, TP	Not known	dsRNA

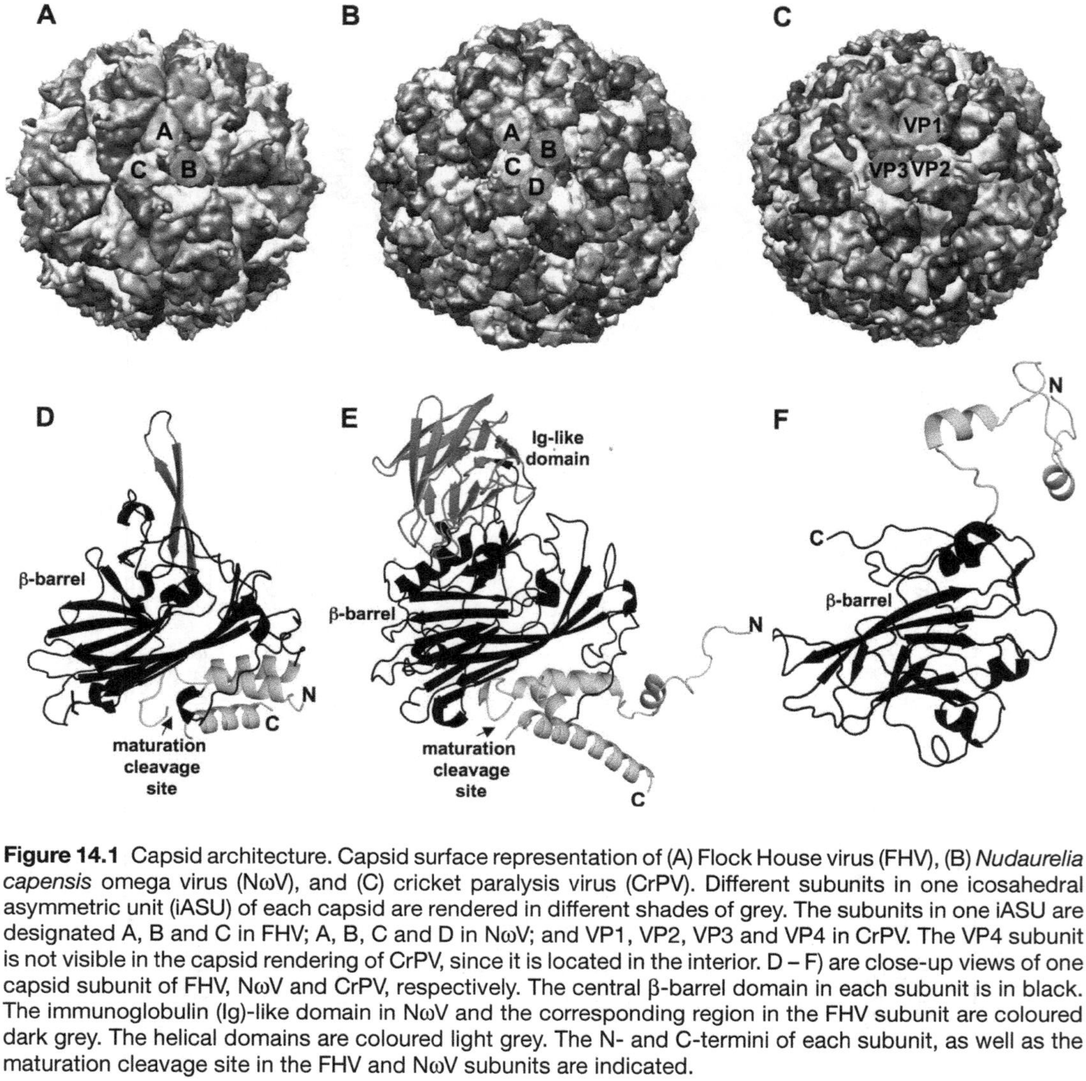

Figure 14.1 Capsid architecture. Capsid surface representation of (A) Flock House virus (FHV), (B) *Nudaurelia capensis* omega virus (NωV), and (C) cricket paralysis virus (CrPV). Different subunits in one icosahedral asymmetric unit (iASU) of each capsid are rendered in different shades of grey. The subunits in one iASU are designated A, B and C in FHV; A, B, C and D in NωV; and VP1, VP2, VP3 and VP4 in CrPV. The VP4 subunit is not visible in the capsid rendering of CrPV, since it is located in the interior. D – F) are close-up views of one capsid subunit of FHV, NωV and CrPV, respectively. The central β-barrel domain in each subunit is in black. The immunoglobulin (Ig)-like domain in NωV and the corresponding region in the FHV subunit are coloured dark grey. The helical domains are coloured light grey. The N- and C-termini of each subunit, as well as the maturation cleavage site in the FHV and NωV subunits are indicated.

The dicistrovirus, CrPV, forms icosahedral particles about 30 nm in diameter. CrPV is designated a 'picorna-like' virus due to similarities in capsid architecture (P = 3 or $T = 1$) and genome organization. The P = 3 (pseudo $T = 3$) capsid of CrPV is organized similarly to that of nodaviruses, except the subunits in the iASU are sequentially distinct polypeptide chains (VP1, VP2 and VP3). Sixty copies of a small polypeptide VP4 (57 residues) are also associated with the CrPV capsid (Tate *et al.*, 1999). Five VP1 moieties form the 5-fold axis of symmetry, while three subunits each of VP2 and VP3 alternate around the 3-fold axis of symmetry (Tate *et al.*, 1999) (Fig. 14.1). There are some prominent structural differences between the CrPV capsid and those of true picornaviruses. One of these is the lack of a 'canyon' surrounding the 5-fold axes of symmetry, which is considered the site of receptor binding in animal picornaviruses (Belnap *et al.*, 2000b; He *et al.*, 2000). CrPV contains deletions and conformational changes in the loop regions of capsid proteins VP1 and VP2 that constitute the canyons (Tate *et al.*, 1999). Other insect picorna-like viruses, such as TrV and IFV also appear to lack these canyons (Estrozi *et al.*, 2008; Xie *et al.*, 2009). Another difference is the lack of a pocket factor bound inside the β-barrel region of VP1. The pocket factor is a fatty acid molecule present in most picornaviruses (Mosser and Rueckert, 1993), and has also been detected in FHV (Cheng *et al.*, 1994). Also, the small polypeptide VP4 in CrPV is in a more compact conformation compared to its positioning in animal picornaviruses (Tate *et al.*, 1999), where it is extended along the inner surface of the capsids (Hogle *et al.*, 1985). In true picornaviruses like poliovirus, VP4 is myristoylated and plays an essential role during cellular entry (Danthi *et al.*, 2003; Moscufo *et al.*, 1993). Another member of the *Dicistroviridae* family, *Drosophila C virus* (DCV), is similar to CrPV in its genome organization, and the CrPV and DCV capsid structures are also expected to be closely similar (Johnson and Christian, 1998; Moore *et al.*, 1980).

A recent cryoEM structure of DAV demonstrates that its capsid is about 30 nm in diameter and contains distinct inner and outer domains (Ambrose *et al.*, 2009), similar to the fish nodavirus Malabaricus grouper nervous necrosis virus (MGNNV) (Tang *et al.*, 2002). While the inner domain of the DAV capsid is organized in a $T = 3$ shell as seen in nodaviruses, the outer domain appears to lack icosahedral symmetry. The loose, cage-like density that makes up the outer domain probably indicates that it is dynamic, in contrast to the rigid inner shell. This domain is spatially separated from the inner shell by a distance of approximately 10 Å (Ambrose *et al.*, 2009). DAV contains only one major capsid protein with a 42 kDa molecular weight, and presumably this protein constitutes both the inner and outer domains of the capsid. Small quantities of two separate proteins of molecular weight 50 kDa and 36 kDa were also detected using the Western blot technique, however, their presence in the capsid has not been confirmed (Ambrose *et al.*, 2009).

Among dsRNA insect viruses, the cypovirus CPV-5 has an icosahedral particle, of approximately 58 nm in diameter, containing 120 copies of the capsid shell protein (CSP) (Hill *et al.*, 1999). The particles also contain turret proteins (TP, 60 copies per particle), which form turrets on the capsid surface, as well as large protrusion proteins (LPP, 120 copies per particles), which serve as molecular clamps to secure the interaction between the CSP proteins. The diameter of the particles including turrets is 71 nm. Each iASU is composed of one TP molecule, and two copies each of CSP and LPP. One LPP molecule (LPP-3) stabilizes the interaction between two CSP molecules (CSP-A and CSP-B), while the other LPP protein (LPP-5) helps in the interaction of CSP-B with two other CSP molecules from neighbouring iASUs (Zhou *et al.*, 2003). At the 5-fold axes of symmetry of the particle, the interaction between five CSP-A molecules is enhanced by the presence of five TP moieties, and this pentamer of TP constitutes the prominent turrets (Zhou *et al.*, 2003). The cypovirus capsid is very similar to other members of the *Reoviridae* family, although it lacks the protective outer layer that the latter contain (Yu *et al.*, 2008; Zhou *et al.*, 2003). In contrast to the ssRNA virus capsids, the *Cypoviridae* capsids package copies of the viral RNA-dependent RNA polymerase for transcription of packaged RNA. Thus transcription of genome occurs inside the capsid without its complete dissociation or exposure of the genome to the cellular environment. The nascent mRNA generated inside the capsid is released in the cytosol. Cypoviruses are very often occluded in polyhedra. These are unique 'biological crystals', and have been characterized using X-ray crystallography (Coulibaly *et al.*, 2007). Each polyhedron contains thousands of viral capsids organized in a lattice.

Structure of capsid proteins

Among nodaviruses, FHV and BBV share 87% similarity in the sequence of protein alpha (Kaesberg *et al.*, 1990). FHV shares substantially lower sequence similarity with NoV (about 50%) and PaV (37%). There is, however, a high degree of homology in the capsid protein structures. The backbone atoms in one iASU of PaV superposes with analogous regions in BBV with a root mean square deviation of 0.94 Å, and in FHV with a root mean square deviation of 1.20 Å (Tang *et al.*, 2001). The core of all four capsid proteins is composed of an eight-stranded antiparallel β-barrel jellyroll, which forms the exterior shell of the capsids (Fisher and Johnson, 1993; Kaesberg *et al.*, 1990; Tang *et al.*, 2001; Wery *et al.*, 1994) (Fig. 14.1D). The interior surfaces contain the termini of the capsid proteins, which are either helical, lack secondary structure or are disordered.

The capsid protein is initially produced as a 43 kDa precursor protein alpha. After capsid assembly, protein alpha cleaves into beta (38 kDa) and a small peptide, designated gamma (4 kDa), in an autoproteolytic process (Schneemann *et al.*, 1992) that will be detailed in a later section. The N-termini of beta are mostly disordered in the nodavirus crystal structures, with some exceptions. Twelve amino acids (residues 20–31) from the N-terminus are ordered in subunit A of FHV (Fisher and Johnson, 1993), while all but the initial six residues are ordered in the A subunits of PaV. The C-terminus of the gamma peptide is also typically disordered in nodaviruses, except in PaV. Eight amino acids from this region are ordered in the A subunits of PaV, and extend into the neighbouring iASU (Tang *et al.*, 2001). Thus, of all nodaviruses, PaV displays the most order in the N- and C-termini of capsid proteins.

The capsid protein of tetraviruses is also generated as a precursor protein alpha (71 kDa), which undergoes post-assembly maturation cleavage into a beta protein (62 kDa) and a gamma peptide (approximately 8 kDa) (Speir and Johnson, 2008). The beta protein consists mostly of a β-barrel region similar to that seen in nodaviruses (Johnson *et al.*, 1994), which forms the surface of the capsid (Fig. 14.1E). The β-barrel cores in nodaviruses and tetraviruses overlap well, and the additional regions in the larger tetravirus capsid proteins form loops and other insertions between the β-strands (Johnson *et al.*, 1994). Both NωV and PrV contain an Ig-like domain inserted between the E and F strands of the β-barrel, which are on the outer surface of the capsid, and occupy different rotational conformations relative to the β-barrel in the beta- and omega-tetravirus subgroups (Johnson *et al.*, 1994; Speir *et al.*, 2010). As in nodaviruses, the N- and C-termini of capsid proteins form the inner surface of the capsid shell. The gamma peptides in tetravirus capsids display different degrees of order depending on their quasi-equivalent environment (Munshi *et al.*, 1996; Helgstrand *et al.*, 2004; Speir *et al.*, 2010). This type of switching between order and disorder is necessary for maintaining capsid architecture.

The capsid proteins VP1, VP2 and VP3 of CrPV also contain cores with antiparallel β-barrel or 'jellyroll' folds similar to that seen in the capsid proteins of nodaviruses and tetraviruses (Tate *et al.*, 1999) (Fig. 14.1F). The inner, icosahedral shell of DAV has a β-barrel topology similar to those of the plant viruses of the *Tombusviridae* family, while the structural details of the outer shell are not detectable because of its dynamic nature and associated lack of symmetry in the reconstruction (Ambrose *et al.*, 2009).

The capsid proteins of cypoviruses do not contain β-barrels, and display a combination of β-sheets and α-helices. An 8 Å cryoEM structure, combined with structural homology to other members of the *Reoviridae* family, has allowed the assignment of density for individual proteins in the capsid (Yu *et al.*, 2008; Zhou *et al.*, 2003). CSP contains 4 prominent domains, of which the N-terminal Apical domain and the small Protrusion domain mainly consist of β-sheets, while the Carpase and Dimerization domains are mainly helical. LPP contains a central helical region, with two β-sheets on either end. TP contains two domains with the conserved methylase fold of a β-sheet sandwiched by helical pairs on either side. A third, mainly helical domain, maintains contact with the capsid shell (Zhou *et al.*, 2003).

Molecular switches

Noda- and tetraviruses utilize a molecular switch in order to control quasi-symmetry in their capsids (Johnson *et al.*, 1994) (Fig. 14.2). In these capsids, sequentially identical subunits are positioned in structurally different environments making different types of contacts with their neighbours. Distinct 'bent' and 'flat' contacts are generated at subunit interfaces by using a variety of capsid elements in a wedge-like fashion (Fig. 14.2A). This produces the molecular switch essential for generating a capsid surface with the appropriate geometry. The capsid elements involved in forming distinct types of interface contacts were first identified from the 2.7 Å crystal structure of FHV (Fisher and Johnson, 1993). In FHV, the molecular switch is produced by ordered duplex RNA and N-termini of the A subunits (residues 20–31), designated the 'N-arms'. These are ordered at the icosahedral 2-fold axes, thus preventing the formation of a bent inter subunit contact, and resulting in a flat contact surface instead (Fig. 14.2A). These

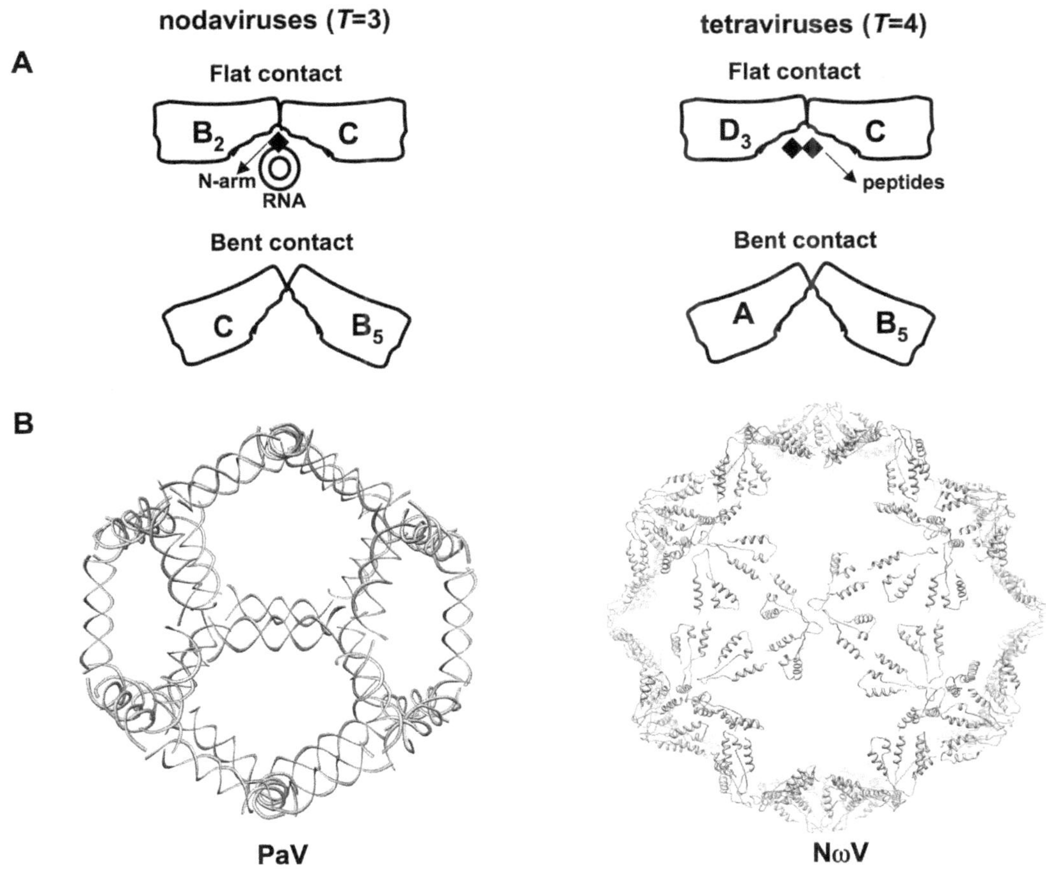

Figure 14.2 Molecular switch. (A) Schematic of the flat and bent contacts between quasi-equivalent subunits in the capsid of nodaviruses (left) and tetraviruses (right). The elements involved in molecular switching are indicated. While the N-arm of the A subunits and ordered RNA stabilize the flat contacts in nodaviruses, in tetraviruses, this function is carried out by either the gamma peptides or the N-termini. (B) Ordered RNA in the capsid of Pariacoto virus (PaV) (left), and ordered gamma peptides from the C and D subunits of the *Nudaurelia capensis* omega virus (NωV) capsid (right). These elements are responsible for the formation of flat contacts in the capsids. Approximately 35% of the total RNA in PaV is ordered in the X-ray crystal structure.

elements are not ordered at the quasi 2-fold axes of symmetry, thus allowing the formation of a bent inter subunit contact, with a dihedral angle of 144°. RNA is an integral part of the nodavirus capsids and incorporation of RNA is essential to form an icosahedral shell in this case (Fig. 14.2B). Empty capsids of nodaviruses have never been detected in either *in vitro* or *in vivo* expression systems.

In contrast to the arrangement seen in nodaviruses, ordered RNA has no role to play in the formation of molecular switches in tetraviruses (Fig. 14.2A). In the omegatetravirus NωV, the molecular switch consists of the gamma peptides contributed from subunits C and D (Helgstrand *et al.*, 2004; Munshi *et al.*, 1996) (Fig. 14.2B). The C-terminal region of the gamma peptides is ordered in these subunits, and the helical regions (residues 624–641) from both C and D subunit gamma peptides in one iASU are wedged between the morphological units ABC and DDD, resulting in the formation of a flat contact at this location. Interaction with other regions of the gamma peptide, as well as with the N-termini (residues 42–57) of β, holds the C-termini of gamma in this groove. The interface between two ABC units related through 5-fold symmetry, forms a bent contact, since the corresponding region of gamma is not ordered in the A or B subunits. The molecular switch in the betatetravirus PrV (J.A. Speir and J.E. Johnson, unpublished data) is more similar in nature to nodaviruses than to the omegatetraviruses. In PrV, the C-termini of gamma are not ordered in the C and D subunits. The groove between the ABC and DDD units is filled instead by the well-ordered N-termini of the C and D subunits (residues 40–78), forming a flat contact. The partially ordered gamma peptides from two surrounding subunits support the N-termini (Fig. 14.2A). Thus, the roles of the subunit termini have been reversed while conserving the molecular switch.

Capsid stability

Apart from inter subunit and protein-RNA contacts, heterologous elements like pocket factors and divalent metal ions may play essential roles in stabilizing insect virus capsids. Ordered density corresponding to calcium ions was detected in several members of the *Nodaviridae* family, including FHV, BBV and PaV (Fisher and Johnson, 1993; Tang *et al.*, 2001; Wery *et al.*, 1994). The identity of the metal ions in FHV was confirmed by soaking FHV crystals in the calcium-chelating agent EGTA, and calculating a difference map with untreated crystals (Fisher and Johnson, 1993). FHV contains four calcium ions in each iASU, which are chelated by the acidic side chains of glutamate or aspartate residues. One calcium ion at the quasi 3-fold axis of symmetry between subunits A, B and C is coordinated by asp249 and glu251 from each subunit. Three calcium ions are present at the interface between subunits where asp161, asp221, and glu257 from quasi-equivalent subunits contribute to their coordination. FHV also contains sulfate ions at the quasi-3-fold axes and subunit interfaces at each iASU. The residues coordinating calcium ions at the quasi-3-fold axes are conserved in other members of the family like BBV and PaV. We have recently shown that the removal of calcium ions from FHV capsids by mutation of the acidic residues or by treatment with metal chelators drastically decreases the stability of the capsid, but does not lead to disassembly (Banerjee *et al.*, 2010). Metal ions are also bound to the capsid of CrPV (Tate *et al.*, 1999). Density corresponding to a calcium ion is coordinated by the carboxyl oxygens of threonine174 contributed from five VP1 molecules at each 5-fold axis of symmetry of the capsid. This calcium-binding site is situated in a depression approximately 15 Å below the surface of the capsid, and metal ion binding to this site is expected to enhance the stability of the capsid, as in FHV.

Several animal picornaviruses contain a small hole or 'pocket' in the β-barrel region of VP1 molecules, and a small but prominent section of density has been consistently discovered nestled there. This density, referred to as a 'pocket factor', is considered to represent a fatty acid molecule. It has been suggested that the release of the pocket factor during receptor binding decreases stability of picornavirus capsids and leads to disassembly (Belnap *et al.*, 2000a; Mosser and Rueckert, 1993). Small but continuous sections of density were also detected at the 5-fold axes of symmetry of FHV capsids, and this density was conspicuously absent from the cryoEM reconstructions of FHV particles which contained deletions in the gamma peptide region and were defective in cellular entry (Banerjee *et al.*, 2009). The identity of the pocket factors in nodaviruses, as well as their function in the virus life cycle, remains unknown. Interestingly, the picorna-like CrPV does not contain any pocket factors, and the pockets in the β-barrels of VP1 moieties are filled instead by hydrophobic side chains (Tate *et al.*, 1999). This indicates that the stability of CrPV capsids is not modulated by the release of the pocket factors.

Although the CPV capsid is very similar to other members of the *Reoviridae* family, CPV lacks the outer, protective protein layers present in the latter. In spite of having just one shell, the CPV capsid is resistant to a variety of chemicals, protease or nuclease treatment and large temperature fluctuations (Zhou *et al.*, 2003). This stability is attributed to the presence of the LPP proteins in the capsid, which serves as a 'molecular bolt' across the CSP molecules within the iASU or between neighbouring iASUs. The turret proteins further enhance capsid stability by interacting around the 5-fold axes of symmetry. The small protrusion domains of CSP participate in intermolecular contact between CSPs and also contribute to capsid stability (Zhou *et al.*, 2003).

Organization of RNA in capsids

Dodecahedral cage of RNA

The X-ray crystal structures of nodaviruses demonstrate the presence of organized genomic RNA in the capsids (Fig. 14.2B). The crystal structure of FHV first displayed the specific organization of genomic RNA in a dodecahedral cage-like structure (Fisher and Johnson, 1993). Density corresponding to approximately 10 basepairs of duplex RNA was detected at the icosahedral 2-fold axes of symmetry. The generation of duplex RNA is probably due to formation of secondary structures through association of ssRNA strands packaged in the genome. This RNA is also instrumental in the formation of flat inter subunit contacts as described above. The crystal structure of the nodavirus PaV contains ordered density for about 35% of the total genome (approximately 1500 of 4300 nucleotides) (Tang *et al.*, 2001). As in FHV, the RNA is organized in the form of a dodecahedral cage, with the duplex RNA strands contacting one another at the icosahedral 3-fold axes of symmetry (Fig. 14.2B). The rest of the RNA is thought to be disordered. In all of these cases, the density of the RNA represents an average structure of the RNA strands present inside the virion. The capsid of the tetravirus PrV contains some ordered density corresponding to packaged RNA, although in a smaller amount compared to the nodaviruses (Speir *et al.*, 2010).

Interaction with capsid protein

In nodaviruses, the genomic RNA interacts with specific regions of the capsid protein through electrostatic and van der waals contacts. The N-terminal 50 amino acids of nodaviruses contain a very high concentration of basic lysine and arginine residues. In FHV, this region is disordered in subunits B and C, and a partially ordered section from subunit A forms the flat contacts along with ordered RNA (Fisher and Johnson, 1993). This region has been shown to be essential for RNA packaging in FHV (Marshall and Schneemann, 2001), and possibly makes contacts with bulk RNA inside the particle. In the crystal structure of PaV, this basic N-termini of the A subunit inserts into the major groove of the RNA duplex at the icosahedral 2-folds and forms extensive electrostatic interactions with the sugar-phosphate backbone of RNA (Tang *et al.*, 2001). In FHV, the gamma peptide region is also involved in contacting genomic RNA. The gamma peptide contains two distinct domains – the N-terminus is amphipathic (Bong *et al.*, 1999), while the C-terminus, which is predominantly hydrophobic, is required for specific recognition and packaging of genomic RNA (Schneemann and Marshall, 1998). Mutation of any one of the three phenylalanine residues present at the extreme C-terminal end of gamma results in a particle that packages non-genomic cellular RNA (Schneemann and Marshall, 1998). The C-terminal end of gamma is disordered in the crystal structure of FHV, and thus its interactions with RNA, specific or otherwise, has not been visualized in the particle (Fisher and Johnson, 1993). However, two lysine residues from the N-terminal amphipathic region of gamma, K371 and K375, form salt bridges with the sugar-phosphate backbone of RNA. These lysines are conserved in all members of the *Nodaviridae* family, which underscores their importance in capsid formation (Kaesberg *et al.*, 1990).

Ordered duplex RNA is also present at the icosahedral 2-folds in the tetravirus PrV. In this case, basic residues contributed from the gamma peptides form a concave binding groove for RNA (Speir *et al.*, 2010).

Transcription complex in dsRNA viruses

Ordered dsRNA was detected in the cryoEM reconstruction of the RNA-containing capsids of CPV (Zhang *et al.*, 1999). Two distinct shells of RNA, with a gap of 25 Å between the shells, were visible in the reconstruction. While the outermost RNA shell displays a dodecahedral shape, the inner shells do not display regular symmetry and appear featureless. In order to detect the transcription enzyme complex (TEC) inside the particles, cryoEM reconstructions of the RNA-containing and empty capsids of CPV were compared (Zhang *et al.*, 1999). By analogy with reoviruses, TEC in CPV capsids is made of a 140 kDa protein designated V2. TECs were detected as a flower-shaped mass bound to the base of each 5-fold axis of symmetry, and attached to the inner surface of the capsid through a stalk-like density (Zhang *et al.*, 1999). However, the existing reconstructions of CPV do not provide a clear path for the release of nascent RNA from inside the CPV capsid. The obvious paths of release are through the channels of the turrets at the 5-fold axes of symmetry, however, these channels are blocked at several points by various capsid protein elements. It has been suggested that conformational adjustments in the capsid, culminating in the removal of these blockages, occurs during the transcription and release of mRNA from capsid interior (Zhou *et al.*, 2003).

Viruses and virus-like particles (VLPs)

Advantages of producing VLPs

Among the high-resolution structures reported for insect RNA viruses, several are of virus-like particles (VLPs) instead of authentic virions. VLPs are typically generated by expressing the capsid protein(s) from a baculovirus vector (Schneemann *et al.*, 1993). The expressed proteins spontaneously assemble into particles that encapsidate non-genomic heterologous cellular RNA present in the expression system. There are several advantages to these particles:

1. VLPs are useful for structural studies of those viruses for which there are no established host cell lines or reverse genetics system for production of authentic particles *in vitro* and characterization of mutations in the capsid. For example, mutants of the tetravirus NωV defective in maturation have been designed and expressed as VLPs (Taylor *et al.*, 2002).
2. VLPs can be generated and purified in large quantities (grams) relatively easily. This is particularly useful for structural studies of non-infectious mutant viruses, which are produced in very small quantities as authentic virions, even in well-established cell lines (Fisher *et al.*, 1993). Maturation defective and other entry defective mutants of FHV have been structurally studied as VLPs (Banerjee *et al.*, 2009; Fisher *et al.*, 1993). There are exceptions like PrV, which cannot be generated as VLPs. The extensive, specific interaction of the viral RNA with capsid protein in this particle (Speir *et al.*, 2010) indicates that genomic RNA might for required for particle assembly (Taylor *et al.*, 2006). Expression of the virus capsid protein (VCAP) of another betatetravirus, *Thosea asigna virus* (TaV) also produces low capsid yields (Pringle *et al.*, 1999).
3. VLPs are structurally similar, but non-infectious, mimics of authentic virions. Thus, VLPs are ideal for studying the biochemical properties of virions *in vitro* (Banerjee *et al.*, 2009).

Comparison with infectious particles

Comparison of infectious, authentic virions with non-infectious VLPs is possible only for a few systems, where both forms have been structurally characterized. Virions and VLPs of FHV and PaV have both been studied at high resolution. A comparison of the crystal structures of FHV virions and VLPs shows that the capsids are indistinguishable at the surface, and that the capsid proteins assume identical folds and are organized identically in the shell. Interestingly, the non-genomic RNA in FHV and PaV capsids assumes a dodecahedral shape similar to that detected in authentic virions (Johnson *et al.*, 2004; Tihova *et al.*, 2004). This observation indicates that the organization of RNA inside capsids is imposed by the capsid protein, and does not depend on the sequence or size of RNA. In the crystal structure of FHV, the interaction of protein subunits is mainly with the phosphate backbone of RNA, and is therefore equally applicable to genomic or non-genomic RNA. However, VLPs of FHV are digested seven times faster compared to authentic virus in a partial proteolysis and mass spectrometry analysis (Bothner *et al.*, 1998; Bothner *et al.*, 1999), indicating that interaction of subunits with authentic, genomic RNA imparts additional stability to the capsid shell.

Maturation cleavage

Trigger for cleavage

After assembly, the capsid proteins of noda- and tetraviruses undergo an autocatalytic cleavage (Fig. 14.3). This event is referred to as particle maturation and has been shown to be required for the infectivity of the nodavirus FHV (Schneemann *et al.*, 1992). During this event, all copies of the capsid protein (alpha) cleave into a beta fragment, which forms the capsid shell, and a small peptide, designated gamma, which is present in the interior of the capsid and interacts with RNA. In nodaviruses, the cleavage occurs after assembly, and does not appear to depend on any other triggers. In tetraviruses, however, the trigger for maturation is provided by low pH conditions. VLPs of NωV are purified as large procapsids at pH 7.6. When the pH is lowered to 5.0, particles undergo a large conformational change and subsequently, maturation cleavage (Canady *et al.*, 2000; Matsui *et al.*, 2009; Taylor *et al.*, 2002).

In CrPV, capsid proteins VP3 and VP4 are generated through autocatalytic cleavage of precursor protein VP0 (Tate *et al.*, 1999). This is different from true picornaviruses, where the precursor autocatalytically cleaves into VP2 and VP4 (Basavappa *et al.*, 1994). The trigger for maturation cleavage in CrPV is not known. Recent studies on DAV have postulated the generation of the 42 kDa major capsid protein and an approximately 6 kDa peptide from an initial 50 kDa precursor (Ambrose *et al.*, 2009). The details of this process have not been elucidated yet. No specific cleavage of the capsid proteins has been detected in the dsRNA virus CPV.

Mechanism

The mechanism of capsid autoproteolytic cleavage is closely similar in noda- and tetraviruses, and the sites of cleavage in members of these families are superimposable (Fig. 14.3). The site of cleavage

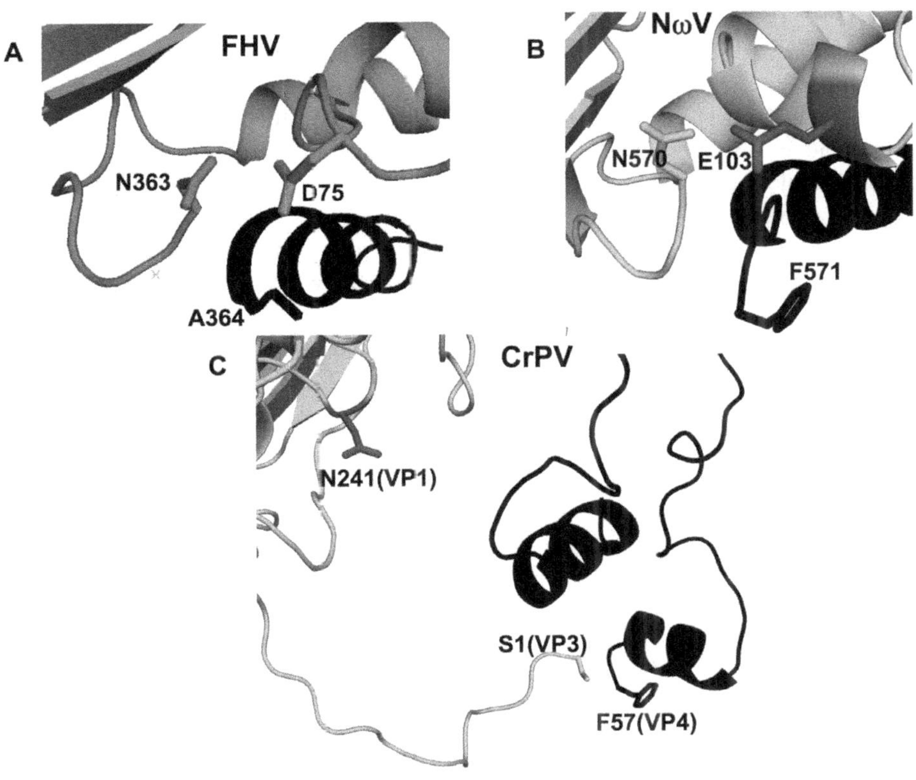

Figure 14.3 Autocatalytic maturation cleavage. Close-up view of the maturation cleavage site in (A) FHV, (B) NωV and (C) CrPV. In each case, ribbon diagrams of the capsid proteins are shown in grey, while the cleaved polypeptide (gamma in FHV and NωV, VP4 in CrPV) is highlighted in black. The respective N- and C-terminal residues at the cleavage site, as well as the catalytic residues (D75 of protein alpha in FHV, E103 of protein alpha in NωV and N241 of VP1 in CrPV) are labelled.

in FHV and BBV is between asparagine363 (N363) and glycine364 (G364) (Schneemann *et al.*, 1992). An aspartic acid at position 75 (D75) is part of a loop region close to the cleavage site, and is thought to catalyse the process (Fig. 14.3). Combined mutational, biochemical and structural studies has resulted in the proposal of an acid-catalysed mode of cleavage in FHV. D75 is present in a hydrophobic environment and is therefore protonated even at neutral pH conditions. It has been postulated that D75 forms a hydrogen bond with the carbonyl oxygen at the N363-G364 site, followed by a nucleophilic attack at that site by a water molecule, and the disruption of the peptide backbone (Zlotnick *et al.*, 1994). Mutating either the catalytic residue (D75N), or asparagine 363 (N363T, N363A or N363D), inhibits cleavage, which supports this mechanism (Schneemann *et al.*, 1992).

In NωV, the site of cleavage is between residues asparagine570 (N570) and phenylalanine571 (F571) (Helgstrand *et al.*, 2004; Munshi *et al.*, 1996) (Fig. 14.3). The cleavage is probably catalysed by a glutamic acid at position 103 (E103), which resides in a hydrophobic environment and therefore probably has an elevated pKa. Mutation of N570 (to T570) prevents cleavage (Taylor *et al.*, 2002), as in nodaviruses, and also makes the large conformational change reversible under low and neutral pH conditions.

The autocatalytic cleavage of CrPV precursor capsid protein VP0 into the VP3 and VP4 moieties occurs at the site GF/SKP (Tate *et al.*, 1999). This region of sequence is identical in several other members of the *Dicistroviridae*. The cleavage is probably catalysed by an aspartate residue of VP1, N241 (Fig. 14.3). The predicted acid catalysed mechanism of cleavage is similar to that seen in noda- and tetraviruses (Tate *et al.*, 1999), and differs from true picornaviruses like poliovirus, where the cleavage is catalysed by a histidine residue (Hindiyeh *et al.*, 1999).

Cleavage associated structural changes

The size and overall conformation of nodaviruses does not undergo any major change upon maturation cleavage. Crystal structures of maturation defective mutants of FHV (D75N FHV, N363T FHV) are closely similar to that of wild type FHV. There are slight differences at the cleavage site, where the density between N363 and G364 shows a clear break of about 8 Å in wild type FHV (Z. Chen and J.E. Johnson, unpublished observation). The conformation of a loop, which contains the catalytic D75, is also subtly altered in the mutated particles compared to wild type. Biochemical studies have demonstrated that cleavage increases the stability of the particle as well as the dynamicity of the gamma peptides (Bothner *et al.*, 1999), which is probably essential during infection.

In contrast to nodaviruses, tetravirus capsids undergo a major contraction upon maturation (Speir and Johnson, 2008). The structural changes occurring in VLPs of NωV upon maturation has been studied extensively. The procapsids purified from baculovirus-infected cells at neutral pH conditions are porous spheres of approximately 480 Å diameter. There are large gaps at the icosahedral 5-fold and 2-fold axes of these particles, making them permeable to stain (Tang *et al.*, 2009; Taylor *et al.*, 2002). Upon reducing the pH to 5.0, the procapsid is converted into a stable, angular capsid, approximately 410 Å in diameter (Taylor *et al.*, 2002), which contains more ordered C-terminal helices and a surface morphology different from procapsids. Solution X-ray scattering (SAXS) studies have shown that the time required for the conformational change to occur is about 100ms, and it is reversible until about 10% of the subunits have undergone cleavage (Lee *et al.*, 2004). Conformational change can occur at pH values higher than 5.0, and the particle size is similar to mature capsids at pH 5.8. However, the speed of maturation cleavage is optimal at pH 5.0, with a half-life of approximately 0.5 hours. At pH values higher or lower than 5.0, the speed of maturation cleavage is significantly reduced. Interestingly, if partially cleaved particles at pH 5.0 are transferred to pH 7.0, the cleavage reaction will continue, albeit at a slower rate. Taken together, the data indicates that the decrease in pH is not directly responsible for cleavage, however, it indirectly brings about particle maturation by allowing the conformational change resulting in the formation of a more compact capsid (Bothner *et al.*, 2005; Matsui *et al.*, 2009; Taylor *et al.*, 2002). This reduction in capsid size probably brings the active site together, and the cleavage can subsequently proceed even at non-optimal pH conditions. The maturation defective N570T mutant of NωV displays continuously reversible conformational change at neutral and low pH conditions. Since the particle does not cleave, it cannot get 'locked' into the mature conformation (Taylor *et al.*, 2002).

Although an uncleaved form of the CrPV capsid is not available for structural study, it is unlikely that this virus undergoes major conformational changes during maturation cleavage, by homology with picornaviruses (Basavappa *et al.*, 1994). The termini of cleaved VP3 and VP4, at the cleavage site, are situated close to each other in the crystal structure (Tate *et al.*, 1999), indicating that only small conformational adjustments have taken place after autoproteolytic cleavage.

Order of cleavage

A recent study from the author's laboratory has investigated the order in which the quasi-equivalent subunits in the NωV iASU undergo maturation cleavage. Since the subunits A, B, C and D are in different chemical environments, the change in their conformation might proceed at different rates upon lowering of pH. This is indeed the case, and the order of maturation of the quasi-equivalent subunits has now been established by time-resolved cryoEM and image reconstruction of the particles at different pH conditions (T. Matsui, G. Lander, R. Khayat, and J.E. Johnson, unpublished data). It is possible, although not proven, that maturation cleavage also proceeds at different rates in the quasi-equivalent subunits in the *Nodaviridae* family.

Dynamic capsid elements and their biological role

Virus-encoded lytic polypeptides in capsids

A subset of virus-encoded polypeptides that are more dynamic than the rest of the capsid, are thought to be essential for biological membrane disruption and translocation of RNA in several insect virus families (Banerjee and Johnson, 2008; Tsai, 2007). The role of the gamma peptide in the life cycle of FHV has been studied extensively. During maturation of FHV, this 44 residue peptide is cleaved from capsid protein alpha (Fig. 14.3), and remains non-covalently attached to the interior of the particle. Partial proteolysis and mass spectrometry demonstrated that maturation cleavage increases the exposure of the gamma peptide without making any major changes to the capsid shell (Bothner *et al.*, 1999). The increased dynamic behaviour of the gamma peptide is essential for the biological function of the virus, since the maturation defective mutants are non-infectious (Schneemann *et al.*, 1992).

The gamma peptide of FHV has two distinct domains. The N-terminal 21 amino acids (residue 364–384 of alpha, designated the γ1 peptide) constitute an amphipathic helix. A variety of biophysical experiments have indicated that the γ1 peptide disrupts cellular membranes during FHV entry (Bong *et al.*, 1999, 2000; Janshoff *et al.*, 1999; Maia *et al.*, 2006). A synthetic version of the γ1 peptide partitions into lipid membranes and increases the permeability of liposomes. This peptide is a random coil in solution, but assumes an α-helical conformation upon interaction with artificial membranes. It has been suggested that this peptide causes localized disruption of cellular membranes, since it packs parallel to artificial membranes *in vitro* (Bong *et al.*, 1999, 2000; Janshoff *et al.*, 1999). The C-terminal region of gamma is predominantly hydrophobic, disordered in the crystal structure, and is expected to contact bulk RNA packaged inside capsids (Schneemann and Marshall, 1998). Three phenylalanines at the C-terminal region have been deemed responsible for specific recognition of FHV genomic RNA. Although the gamma peptide contains distinct functional domains, it was recently demonstrated that in context of the virus, the γ1 peptide is unable to disrupt membranes on its own and requires assistance from the C-terminus (Banerjee *et al.*, 2009).

The gamma peptides generated during the maturation cleavage of tetraviruses are also hydrophobic in nature and expected to be similarly involved in membrane penetration during host cell entry. The 8-kDa gamma peptide of NωV has been shown to be membrane active *in vitro* (A.L. Odegard and J.E. Johnson, unpublished observation). Capsid protein VP4 of CrPV is expected to have a role in cellular membrane penetration by analogy with homologues in the *Picornaviridae* family (Danthi *et al.*, 2003; Moscufo *et al.*, 1993), although this has not been directly tested.

Structural organization of lytic polypeptides and putative role in RNA translocation

At neutral pH conditions, the gamma peptide of FHV is accessible to proteases and small molecules, in spite of being secluded in the interior of the capsid shell (Bothner *et al.*, 1998). This indicates that gamma is transiently exposed to the environment. Gamma becomes fully exposed at acidic pH, corresponding to the environment inside early endosomes (Odegard *et al.*, 2009). This event probably signals the initiation of host membrane penetration during FHV entry. Interestingly, a comparative study of FHV and NωV indicates that the gamma peptides in the latter are much less accessible to proteases and specific labelling by small molecules (Bothner *et al.*, 2005). The difference in accessibility of gamma peptides in the two viruses is readily explained by the difference in their structural organization in the capsid.

In FHV, the amphipathic helices of gamma from the A subunits are organized in a pentameric bundle at the 5-fold axes of symmetry of the capsid (Cheng *et al.*, 1994) (Fig. 14.4). Gamma peptides from the B and C subunits are organized as dimers around the icosahedral 3-fold axes, and form extensive contacts with bulk RNA. This striking difference in organization of the peptide on the basis of quasi-equivalence lead to the hypothesis that differently organized gamma peptides might perform separate functions (Cheng *et al.*, 1994). It has been suggested that the bundle-like amphipathic helices at the 5-fold axes (Fig. 14.4) are primarily involved in membrane penetration during virus entry; and form a pore or channel in the endosomal membranes. This is partially validated by recent cryoEM studies of particles with mutations in the gamma peptide, where disorder in the A and B subunit gamma peptides, but not in the C subunits, correlated with decrease in membrane disruption *in vitro* and cellular entry defects *in vivo* (Banerjee *et al.*, 2009). The gamma peptides could also have a role in RNA translocation, since the C-terminal regions of gamma peptides, which are disordered in the crystal structure, make specific contacts with genomic RNA (Schneemann and Marshall, 1998). It is possible that the N-terminal amphipathic helices of gamma peptides disrupt membranes, while the C-terminal region pulls the genomic RNA through the membrane.

The gamma peptides in the tetraviruses NωV and PrV gather in 10-helix bundles at the 5-fold axes of symmetry (Helgstrand *et al.*, 2004; Munshi *et al.*, 1996; Speir *et al.*, 2010), in contrast to the 5-helix bundles seen in nodaviruses (Fig. 14.4). In the crystal structure of NωV, this 10-helix bundle is oriented towards the interior of the capsid and consists of the gamma peptides from the A subunits, and the N-termini of capsid protein beta contributed by the B-subunits (Fig. 14.4). The resultant bundle has a diameter of 20 Å, large enough to allow passage to ssRNA. However, unlike the gamma helices in nodaviruses, these 10-helix bundles are unlikely to be released from the capsid, since the N-termini of the beta proteins are still covalently attached to the capsid shell (Helgstrand *et al.*, 2004). Partial proteolysis experiments demonstrated that both gamma and the N-termini of the beta proteins are transiently exposed to the exterior of the capsid (Bothner *et al.*, 2005), therefore it is possible that conformational changes in the capsid might allow the mixed 10-helix bundle to integrate into membranes.

The peptide bundles in PrV are homogeneous in nature, being made entirely of covalently separate gamma peptides contributed from the A subunits (Speir *et al.*, 2010). These gamma peptides contain two anti-parallel helices formed by residues 560–574 and 579–588, which extend vertically along the 5-fold axes, and culminate in a reverse turn. The reverse turn contacts the first gamma peptide helix from the next symmetry related A subunit, to form a ten-helix bundle from five gamma peptides. The height of the bundle is less than that in NωV, however, the diameter is slightly larger at 25 Å. The 10-helix bundle of PrV (Fig. 14.4) could potentially detach from the capsid and integrate into membranes during entry.

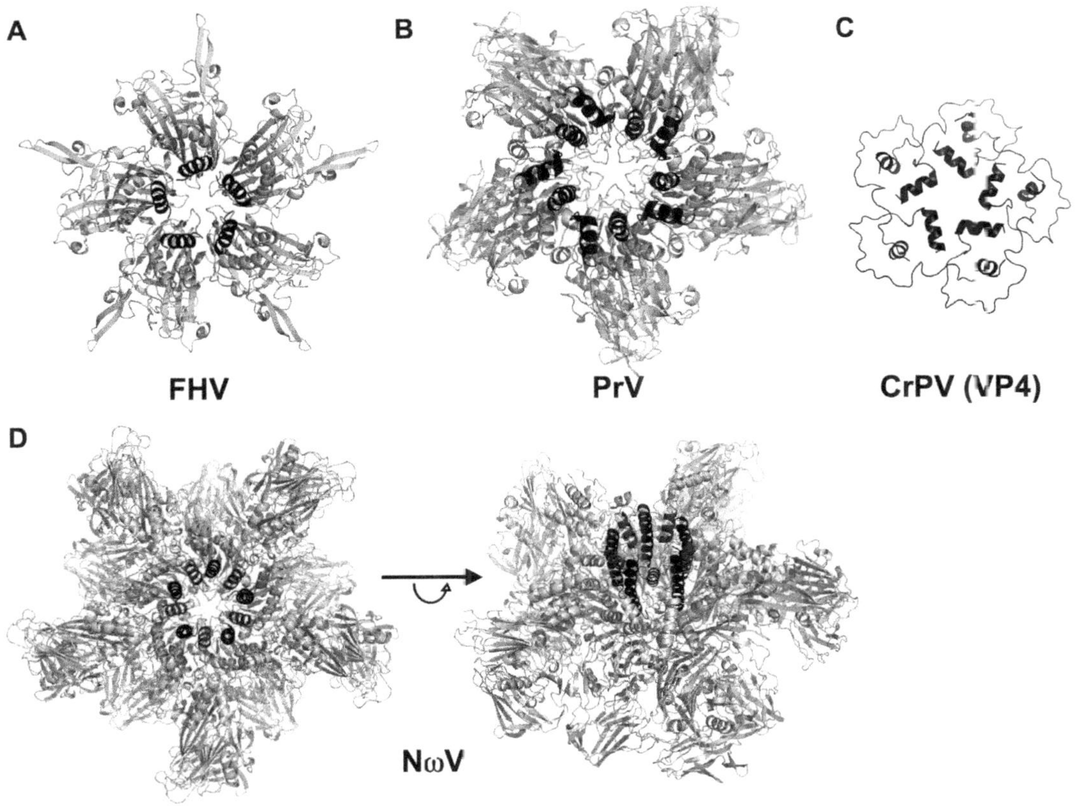

Figure 14.4 Organization of putative membrane penetrating polypeptides in capsids. (A) View of a 5-fold axis of symmetry of the FHV capsid, from the capsid interior. The subunits are shown in grey, while the gamma amphipathic helices (from A subunits) are highlighted in black. B) View, from capsid interior, of a 5-fold axis of symmetry of Providence virus (PrV). The capsid proteins (A subunits) are shown in grey, and the gamma peptides are in black. The gamma peptides in PrV form a 10-helix bundle, instead of the 5-helix bundle seen in nodaviruses. C) Ribbon diagram of five VP4 polypeptides of CrPV, organized at a 5-fold axis of symmetry of the capsid. D) A 5-fold axis of symmetry of NωV, viewed from capsid interior. A and B subunits are in grey, whereas the gamma peptides from A are in black and the N-termini of B are shown in dark grey. Together, the helical regions form a 10-helix bundle. The figure on the right is a side-view, showing that the N-termini of B subunits are shorter than the gamma peptides. The images in A–D are not drawn to scale.

The analogous polypeptides in dicistrovirus capsids have not been studied biochemically. By comparison with true picornaviruses, the 57 amino acid polypeptide VP4 could be involved in membrane disruption in CrPV. The crystal structure of CrPV shows that helical regions at the N- and C-termini of VP4 are organized around the 5-fold axes of symmetry (Fig. 14.4), under the capsid surface (Tate *et al.*, 1999). This organization is reminiscent of the helical bundles in noda- and tetravirus capsids, and could be involved in forming channels across cellular membranes.

Structural changes during cellular entry and disassembly

The stable, icosahedral capsids of insect RNA viruses must disassemble to release packaged genomic RNA, but the conformational changes induced in capsids during cellular entry are largely unknown. Intermediates formed during entry of FHV have been isolated from infected cells or prepared *in vitro* by mimicking conditions likely to exist inside cellular compartments.

Eluted particles

During a standard infection of *Drosophila* DL-1 cells with FHV, approximately 30% of particles are 'eluted' from cells 1 hour post infection (Walukiewicz *et al.*, 2006). These particles are thought to have gone through an abortive infection process, and therefore display some structural changes associated with entry. Briefly, they are sensitive to low pH, non-infectious, and have lost 25% of their gamma peptides. This indicates that cellular membrane penetration is probably carried out by gamma peptides released from FHV capsids. Low-resolution electron microscopy of the eluted particles showed that they are stain permeable and asymmetric in their staining pattern. A high-resolution reconstruction of these particles will provide valuable information regarding the identity of the capsid subunits that have lost gamma peptides, and the structural rearrangements that have occurred in the shell. The eluted particles are expected to represent a step in disassembly prior to genome release, since they retain the packaged RNA (Walukiewicz *et al.*, 2006).

Other putative entry intermediates

Efforts are ongoing in the author's laboratory to generate and structurally characterize disassembly intermediates of FHV. FHV particles become more hydrophobic under low pH conditions; however, their overall structure appears to remain similar to wild type particles upon negative staining and electron microscopy (Odegard *et al.*, 2009). FHV mutants that have released calcium ions appear to be structurally unstable and contain gamma in a more exposed conformation (Banerjee *et al.*, 2010). Calcium deficient mutants of FHV could correspond to an entry intermediate, since protonation of metal-chelating acidic residues in the endosome could lead to loss of calcium ions from the capsid. Heating authentic FHV particles at 65°C for 10 minutes produced another possible candidate, the 'puffed particles' (Cheng *et al.*, 1994). These particles appear to be releasing RNA from one specific region of the capsid. It is possible that heating the mature, metastable particles provides sufficient energy for conformational changes, similar to that provided by receptor binding during entry. CryoEM and asymmetric reconstruction of puffed particles may be able to pinpoint the mechanism of genome release from particles. Additional intermediates can also be generated by treating FHV particles with soluble receptor molecules *in vitro*. Structural changes in capsids are expected to occur in a stepwise manner during entry, and there are possibilities for the existence of several distinct entry intermediates leading to partial or complete disassembly.

Benefits to structural characterization of insect viruses

Understanding principles of icosahedral capsid organization and maturation

The high-resolution crystal structures of noda- and tetraviruses were instrumental in understanding the principles of icosahedral capsid architecture. The crystal structure of FHV provided early information about the molecular switches required to generate distinct interactions between quasi-equivalent subunits, the dodecahedral organization assumed by packaged RNA in particles, and protein–RNA interaction in capsids (Fisher and Johnson, 1993). The PaV crystal structure contains the highest proportion of organized genomic RNA (35%) detected in virus capsids to date (Tang *et al.*, 2001). A close-up view of the autoproteolytic cleavage site in FHV was also instrumental in identifying the catalytic residue (D75) and elucidating the mechanism of cleavage (Zlotnick *et al.*, 1994). Similar mechanisms of autoproteolytic cleavage were later identified in more complex non-enveloped viruses infecting higher organisms like reovirus. Tetraviruses contain the only non-enveloped $T = 4$ capsids discovered so far, and structural studies of this family provided information regarding molecular switching, organization of distinct domains such as the Ig-like domains and large conformational changes accompanying maturation.

Nano-biotechnology applications

FHV is currently being used as a platform for generating more effective vaccines as well as antitoxins against a variety of bacterial and viral pathogens including *B. anthracis* and the influenza virus (Manayani *et al.*, 2007; J.A. Speir and A. Schneemann, unpublished observation). This involves molecular engineering of the particle for accurate placement of antigenic sites from these pathogens, which requires an extensive knowledge of the capsid structure. Efforts are also ongoing to utilize tetraviruses as biopesticides for commercially important crops or as self-replicating platforms for catalysis (Speir and Johnson, 2008). The high-resolution structures of these viruses are essential for designing such potential nano-devices.

Concluding remarks

A variety of biophysical techniques including X-ray crystallography, cryoEM, SAXS and mass spectrometry has delivered comprehensive structural information for several insect RNA viruses. Attention has focused on the *Nodaviridae* and *Tetraviridae* families, and structural studies of their members have provided valuable information regarding the architecture of icosahedral capsids, principles of quasi-equivalence, mechanisms of autoproteolytic maturation cleavage and packaging of genome inside capsids. Indeed, these studies have established that in spite of sequential dissimilarities, the mechanistic basis for most of these processes remain closely similar in non-enveloped icosahedral capsids, and are also applicable to viruses infecting higher organisms.

The high-resolution structures of several members of the *Dicistroviridae* and *Cypoviridae* families are yet to be determined. The inability to purify infectious virus from the insect hosts, in amounts sufficient for structural studies, might constitute the rate-determining step for such studies. In this scenario, virus-like particles produced from capsid proteins could provide a practical alternative for obtaining high resolution structural information. Previous studies have indicated that VLPs and authentic virions of the same virus, where available, contain identical surface organizations. Of particular importance is an understanding of the roles of specific capsid regions in the life cycle of viruses. Hypotheses on the role of capsid elements, derived from structural information, need to be validated through mutagenesis and biological studies. So far, FHV and NωV have provided ample opportunities for such studies, due to the existence of established cell culture system to produce altered particles as authentic virus or VLPs or both.

The well-ordered capsids of insect RNA viruses do not remain in the static state during their life cycles. Viruses undergo significant conformational changes during maturation and cellular entry and disassembly. Understanding the pathway of disassembly is particularly essential, since this constitutes one of the initial and key steps in virus infection. In this context, insect RNA viruses like members of the *Nodaviridae* and *Tetraviridae* families, could be extremely important. The easy production and genetic tractability, coupled with the enormous amount of available structural and biochemical information, makes these viruses useful model systems for understanding conformational changes in the capsid. We, and others are currently involved in determining maturation and entry related structural changes in the capsids of some of these viruses. Thus, a combination of novel structure determination and analysis of dynamic changes occurring in capsids using available structures, is essential for a more complete understanding of the assembly, infection and pathogenesis of insect RNA viruses and their eukaryotic counterparts.

References

Ambrose, R.L., Lander, G.C., Maaty, W.S., Bothner, B., Johnson, J.E., and Johnson, K.N. (2009). Drosophila A virus is an unusual RNA virus with a T = 3 icosahedral core and permuted RNA-dependent RNA polymerase. J. Gen. Virol. *90*, 2191–2200.

Banerjee, M., and Johnson, J.E. (2008). Activation, exposure and penetration of virally encoded, membrane-active polypeptides during non-enveloped virus entry. Curr. Prot. Pept. Sci. *9*, 16–27.

Banerjee, M., Khayat, R., Walukiewicz, H.E., Odegard, A.L., Schneemann, A., and Johnson, J.E. (2009). Dissecting the functional domains of a nonenveloped virus membrane penetration peptide. J. Virol. *83*, 6929–6933.

Banerjee, M., Speir, J.A., Kwan, M.H., Huan, R., Aryanpur, P.P., Bothner, B., and Johnson, J.E. (2010). Structure and function of a genetically engineered mimic of a nonenveloped virus entry intermediate. J. Virol. *84*, 4737–4746.

Basavappa, R., Syed, R., Flore, O., Icenogle, J.P., Filman, D.J., and Hogle, J.M. (1994). Role and mechanism of the maturation cleavage of VP0 in poliovirus assembly: structure of the empty capsid assembly intermediate at 2.9 Å resolution. Protein Sci. 3, 1651–1669.

Belnap, D.M., Filman, D.J., Trus, B.L., Cheng, N., Booy, F.P., Conway, J.F., Curry, S., Hiremath, C.N., Tsang, S.K., Steven, A.C., and Hogle, J.M. (2000a). Molecular tectonic model of virus structural transitions: the putative cell entry states of poliovirus. J. Virol. *74*, 1342–1354.

Belnap, D.M., McDermott, B.M., Jr., Filman, D.J., Cheng, N., Trus, B.L., Zuccola, H.J., Racaniello, V.R., Hogle, J.M., and Steven, A.C. (2000b). Three-dimensional structure of poliovirus receptor bound to poliovirus. PNAS *97*, 73–78.

Bong, D.T., Janshoff, A., Steinem, C., and Ghadiri, M.R. (2000). Membrane partitioning of the cleavage peptide in Flock House virus. Biophys. J. *78*, 839–845.

Bong, D.T., Steinem, C., Janshoff, A., Johnson, J.E., and Ghadiri, M.R. (1999). A highly membrane-active peptide in Flock House virus: implications for the mechanism of nodavirus infection. Chem. Biol. *6*, 473–481.

Bothner, B., Dong, X.F., Bibbs, L., Johnson, J.E., and Siuzdak, G. (1998). Evidence of viral capsid dynamics using limited proteolysis and mass spectrometry. J. Biol. Chem. *273*, 673–676.

Bothner, B., Schneemann, A., Marshall, D., Reddy, V., Johnson, J.E., and Siuzdak, G. (1999). Crystallographically identical virus capsids display different properties in solution. Nat. Struct. Biol. *6*, 114–116.

Bothner, B., Taylor, D., Jun, B., Lee, K.K., Siuzdak, G., Schultz, C.P., and Johnson, J.E. (2005). Maturation of a tetravirus capsid alters the dynamic properties and creates a metastable complex. Virology *334*, 17–27.

Canady, M.A., Tihova, M., Hanzlik, T.N., Johnson, J.E., and Yeager, M. (2000). Large conformational changes in the maturation of a simple RNA virus, *Nudaurelia capensis* omega virus (NomegaV). J. Mol. Biol. *299*, 573–584.

Cheng, R.H., Reddy, V.S., Olson, N.H., Fisher, A.J., Baker, T.S., and Johnson, J.E. (1994). Functional implications of quasi-equivalence in a T = 3 icosahedral animal virus established by cryo-electron microscopy and X-ray crystallography. Structure 2, 271–282.

Coulibaly, F., Chiu, E., Ikeda, K., Gutmann, S., Haebel, P.W., Schulze-Briese, C., Mori, H., and Metcalf, P. (2007). The molecular organization of cypovirus polyhedra. Nature *446*, 97–101.

Danthi, P., Tosteson, M., Li, Q.H., and Chow, M. (2003). Genome delivery and ion channel properties are altered in VP4 mutants of poliovirus. J. Virol. *77*, 5266–5274.

Estrozi, L.F., Neumann, E., Squires, G., Rozas-Dennis, G., Costabel, M., Rey, F.A., Guerin, D.M., and Navaza, J. (2008). Phasing of the Triatoma virus diffraction data using a cryo-electron microscopy reconstruction. Virology 375, 85–93.

Fauquet, C.M., Mayo, M.A., Maniloff, J., Desselberger, U., and Ball, L.A. (2005). Virus Taxonomy: Eighth Report of the International Committee on Taxonomy of Viruses. (Academic Press).

Fisher, A.J., and Johnson, J.E. (1993). Ordered duplex RNA controls capsid architecture in an icosahedral animal virus. Nature *361*, 176–179.

Fisher, A.J., McKinney, B.R., Schneemann, A., Rueckert, R.R., and Johnson, J.E. (1993). Crystallization of viruslike particles assembled from Flock House virus coat protein expressed in a baculovirus system. J. Virol. *67*, 2950–2953.

He, Y., Bowman, V.D., Mueller, S., Bator, C.M., Bella, J., Peng, X., Baker, T.S., Wimmer, E., Kuhn, R.J., and Rossmann, M.G. (2000). Interaction of the poliovirus receptor with poliovirus. Proc. Natl. Acad. Sci. U.S.A. *97*, 79–84.

Helgstrand, C., Munshi, S., Johnson, J.E., and Liljas, L. (2004). The refined structure of *Nudaurelia capensis* omega virus reveals control elements for a T = 4 capsid maturation. Virology *318*, 192–203.

Hill, C.L., Booth, T.F., Prasad, B.V., Grimes, J.M., Mertens, P.P., Sutton, G.C., and Stuart, D.I. (1999). The structure of a cypovirus and the functional organization of dsRNA viruses. Nat. Struct. Biol. *6*, 565–568.

Hindiyeh, M., Li, Q.H., Basavappa, R., Hogle, J.M., and Chow, M. (1999). Poliovirus mutants at histidine 195 of VP2 do not cleave VP0 into VP2 and VP4. J.Virol. *73*, 9072–9079.

Hogle, J.M., Chow, M., and Filman, D.J. (1985). Three-dimensional structure of poliovirus at 2.9 Å resolution. Science *229*, 1358–1365.

Hosur, M.V., Schmidt, T., Tucker, R.C., Johnson, J.E., Gallagher, T.M., Selling, B.H., and Rueckert, R.R. (1987). Structure of an insect virus at 3.0 Å resolution. Proteins 2, 167–176.

Janshoff, A., Bong, D.T., Steinem, C., Johnson, J.E., and Ghadiri, M.R. (1999). An animal virus-derived peptide switches membrane morphology: possible relevance to nodaviral transfection processes. Biochemistry *38*, 5328–5336.

Johnson, J.E., Munshi, S., Liljas, L., Agrawal, D., Olson, N.H., Reddy, V., Fisher, A., McKinney, B., Schmidt, T., and Baker, T.S. (1994). Comparative studies of T = 3 and T = 4 icosahedral RNA insect viruses. Arch. Virol. *9*, 497–512.

Johnson, K.N., and Christian, P.D. (1998). The novel genome organization of the insect picorna-like virus *Drosophila C virus* suggests this virus belongs to a previously undescribed virus family. J. Gen. Virol. *79*, 191–203.

Johnson, K.N., Tang, L., Johnson, J.E., and Ball, L.A. (2004). Heterologous RNA encapsidated in Pariacoto virus-like particles forms a dodecahedral cage similar to genomic RNA in wild-type virions. J. Virol. *78*, 11371–11378.

Kaesberg, P., Dasgupta, R., Sgro, J.Y., Wery, J.P., Selling, B.H., Hosur, M.V., and Johnson, J.E. (1990). Structural homology among four nodaviruses as deduced by sequencing and X-ray crystallography. J. Mol. Biol. *214*, 423–435.

Lee, K.K., Tang, J., Taylor, D., Bothner, B., and Johnson, J.E. (2004). Small compounds targeted to subunit interfaces arrest maturation in a nonenveloped, icosahedral animal virus. J. Virol. *78*, 7208–7216.

Maia, L.F., Soares, M.R., Valente, A.P., Almeida, F.C., Oliveira, A.C., Gomes, A.M., Freitas, M.S., Schneemann, A., Johnson, J.E., and Silva, J.L. (2006). Structure of a membrane-binding domain from a non-enveloped animal virus: insights into the mechanism of membrane permeability and cellular entry. J. Biol. Chem. *281*, 29278–29286.

Manayani, D.J., Thomas, D., Dryden, K.A., Reddy, V., Siladi, M.E., Marlett, J.M., Rainey, G.J., Pique, M.E., Scobie, H.M., Yeager, M., *et al.* (2007). A viral nanoparticle with dual function as an anthrax antitoxin and vaccine. PLoS Path. 3, 1422–1431.

Marshall, D., and Schneemann, A. (2001). Specific packaging of nodaviral RNA2 requires the N-terminus of the capsid protein. Virology *285*, 165–175.

Matsui, T., Lander, G., and Johnson, J.E. (2009). Characterization of large conformational changes and autoproteolysis in the maturation of a T = 4 virus capsid. J. Virol. *83*, 1126–1134.

Moore, N.F., Kearns, A., and Pullin, J.S. (1980). Characterization of Cricket paralysis virus-induced polypeptides in *Drosophila* cells. J. Virol. *33*, 1–9.

Moscufo, N., Yafal, A.G., Rogove, A., Hogle, J., and Chow, M. (1993). A mutation in VP4 defines a new step in the late stages of cell entry by poliovirus. J. Virol. *67*, 5075–5078.

Mosser, A.G., and Rueckert, R.R. (1993). WIN 51711-dependent mutants of poliovirus type 3: evidence that virions decay after release from cells unless drug is present. J. Virol. *67*, 1246–1254.

Munshi, S., Liljas, L., Cavarelli, J., Bomu, W., McKinney, B., Reddy, V., and Johnson, J.E. (1996). The 2.8 Å structure of a T = 4 animal virus and its implications for membrane translocation of RNA J. Mol. Biol. *261*, 1–10.

Odegard, A.L., Kwan, M.H., Walukiewicz, H.E., Banerjee, M., Schneemann, A., and Johnson, J.E. (2009). Low endocytic pH and capsid protein autocleavage are critical components of Flock House virus cell entry. J. Virol. *83*, 8628–37.

Olson, N.H., Baker, T.S., Johnson, J.E., and Hendry, D.A. (1990). The three-dimensional structure of frozen-hydrated *Nudaurelia capensis* beta virus, a T = 4 insect virus. J. Struct. Biol. *105*, 111–122.

Pringle, F.M., Gordon, K.H., Hanzlik, T.N., Kalmakoff, J., Scotti, P.D., and Ward, V.K. (1999). A novel capsid expression strategy for *Thosea asigna* virus (*Tetraviridae*). J. Gen. Virol. *80*, 1855–1863.

Schneemann, A., Dasgupta, R., Johnson, J.E., and Rueckert, R.R. (1993). Use of recombinant baculoviruses in synthesis of morphologically distinct viruslike particles of Flock House virus, a nodavirus. J. Virol. *67*, 2756–2763.

Schneemann, A., and Marshall, D. (1998). Specific encapsidation of nodavirus RNAs is mediated through the C-terminus of capsid precursor protein alpha. J. Virol. 72, 8738–8746.

Schneemann, A., Zhong, W., Gallagher, T.M., and Rueckert, R.R. (1992). Maturation cleavage required for infectivity of a nodavirus. J. Virol. *66*, 6728–6734.

Speir, J.A., and Johnson, J.E. (2008). Tetraviruses. Encyclopedia of Virology, Third Edition *5*, 27–37.

Speir, J.A., Taylor, D.J., Natarajan, P., Pringle, F.M., Ball, L.A., and Johnson, J.E. (2010). Evolution in action: N and C termini of subunits in related T = 4 viruses exchange roles as molecular switches. Structure *18*, 700–709.

Tang, J., Lee, K.K., Bothner, B., Baker, T.S., Yeager, M., and Johnson, J.E. (2009). Dynamics and stability in maturation of a T = 4 virus. J. Mol. Biol. *392*, 803–812.

Tang, L., Johnson, K.N., Ball, L.A., Lin, T., Yeager, M., and Johnson, J.E. (2001). The structure of Pariacoto virus reveals a dodecahedral cage of duplex RNA. Nat. Struct. Biol. *8*, 77–83.

Tang, L., Lin, C.S., Krishna, N.K., Yeager, M., Schneemann, A., and Johnson, J.E. (2002). Virus-like particles of a fish nodavirus display a capsid subunit domain organization different from that of insect nodaviruses. J. Virol. *76*, 6370–6375.

Tate, J., Liljas, L., Scotti, P., Christian, P., Lin, T., and Johnson, J.E. (1999). The crystal structure of cricket paralysis virus: the first view of a new virus family. Nat. Struct. Biol. *6*, 765–774.

Taylor, D.J., Krishna, N.K., Canady, M.A., Schneemann, A., and Johnson, J.E. (2002). Large-scale, pH-dependent, quaternary structure changes in an RNA virus capsid are reversible in the absence of subunit autoproteolysis. J. Virol. *76*, 9972–9980.

Taylor, D.J., Speir, J.A., Reddy, V., Cingolani, G., Pringle, F.M., Ball, L.A., and Johnson, J.E. (2006). Preliminary x-ray characterization of authentic providence virus and attempts to express its coat protein gene in recombinant baculovirus. Arch. Virol. *151*, 155–165.

Tihova, M., Dryden, K.A., Le, T.V., Harvey, S.C., Johnson, J.E., Yeager, M., and Schneemann, A. (2004). Nodavirus coat protein imposes dodecahedral RNA structure independent of nucleotide sequence and length. J. Virol. *78*, 2897–2905.

Tsai, B. (2007). Penetration of nonenveloped viruses into the cytoplasm. Ann. Rev. Cell Dev. Biol. *23*, 23–43.

Walukiewicz, H.E., Johnson, J.E., and Schneemann, A. (2006). Morphological changes in the T = 3 capsid of Flock House virus during cell entry. J. Virol. *80*, 615–622.

Wery, J.P., Reddy, V.S., Hosur, M.V., and Johnson, J.E. (1994). The refined three-dimensional structure of an insect virus at 2.8 Å resolution. J. Mol. Biol. *235*, 565–586.

Xie, L., Zhang, Q., Lu, X., Dai, X., Li, K., Hong, J., and Zhou, X. (2009). The three-dimensional structure of Infectious flacherie virus capsid determined by cryo-electron microscopy. Sci. China C Life Sci. *52*, 1186–1191.

Yu, X., Jin, L., and Zhou, Z.H. (2008). 3.88 Å structure of cytoplasmic polyhedrosis virus by cryo-electron microscopy. Nature *453*, 415–419.

Zhang, H., Zhang, J., Yu, X., Lu, X., Zhang, Q., Jakana, J., Chen, D.H., Zhang, X., and Zhou, Z.H. (1999). Visualization of protein–RNA interactions in cytoplasmic polyhedrosis virus. J. Virol. *73*, 1624–1629.

Zhang, J., Feng, J., Liang, Y., Chen, D., Zhou, Z.H., Zhang, Q., and Lu, X. (2001). Three-dimensional structure of the Chinese Sacbrood bee virus. Sci. China C Life Sci. *44*, 443–448.

Zhou, Z.H., Zhang, H., Jakana, J., Lu, X.Y., and Zhang, J.Q. (2003). Cytoplasmic polyhedrosis virus structure at 8 Å by electron cryomicroscopy: structural basis of capsid stability and mRNA processing regulation. Structure *11*, 651–663.

Zlotnick, A., McKinney, B.R., Munshi, S., Bibler, J., Rossmann, M.G., and Johnson, J.E. (1993). A monoclinic crystal with R32 pseudo-symmetry: a preliminary report of Nodamura virus structure determination. Acta Cryst. *49*, 580–587.

Zlotnick, A., Reddy, V.S., Dasgupta, R., Schneemann, A., Ray, W.J., Jr., Rueckert, R.R., and Johnson, J.E. (1994). Capsid assembly in a family of animal viruses primes an autoproteolytic maturation that depends on a single aspartic acid residue. J. Biol. Chem. *269*, 13680–13684.

Role of MicroRNAs as Regulators of Host–Virus Interactions

15

Sassan Asgari and Christopher S. Sullivan

Abstract

MicroRNAs (miRNAs) are small non-coding RNA molecules that play a central role in the regulation of gene expression impacting many biological processes. These include development, cancer, apoptosis, immunity, and longevity. In addition, accumulating evidence suggests that miRNAs are likely to be involved in host–virus interactions by modulating expression levels of either defence genes or virus-encoded genes. Several groups of animal viruses, as well as insect viruses, encode miRNAs that are instrumental in virus biology, including replication, pathogenesis and latency. In this chapter, we present an overview of the biogenesis of miRNAs, current approaches to the discovery of miRNAs, their mode of action and strategies for determining viral miRNA function.

Introduction

RNA interference (RNAi) is an evolutionarily conserved mechanism involved in the regulation of gene expression existing in metazoans, plants and some unicellular eukaryotes. A characteristic of RNAi is the production of small RNA molecules of about 19–25 nucleotides (nt). These are used in a variety of mechanisms in association with protein complexes that may silence endogenous genes or control the expression of exogenous molecules introduced by parasitic or pathogenic invaders, including viruses. There are two main types of small RNAs produced in RNAi pathways based on their precursors: short interfering RNAs (siRNAs) and microRNAs (miRNAs). siRNAs are often cleavage products of long exogenous double-stranded RNAs (dsRNA) generated in response to infection by RNA viruses, whereas miRNAs are processed from genome-encoded transcripts comprising stem–loop structures (Umbach and Cullen, 2009). Production of siRNAs from dsRNAs has led to the development of a powerful molecular technique, RNA silencing, to specifically silence a particular target gene by introducing artificial dsRNA homologous to the target sequence (see Chapter 16 for further details). A third class of small RNAs, Piwi RNAs (piRNAs), are expressed in animal germ cells and are associated with transcriptional gene silencing of selfish genetic elements, such as retrotransposons, preventing their spread (Nowotny and Yang, 2009). Biogenesis of these small molecules is not currently well understood.

miRNAs are small non-coding RNA molecules, ranging in size from 21–25 nt, that constitute a major class of gene-regulatory molecules. They play important roles in development, cancer, apoptosis, immunity, and longevity (Bartel, 2004; Bushati and Cohen, 2007; Lynam-Lennon *et al.*, 2009; Miska, 2005; Sonkoly and Pivarcsi, 2009; Zhao and Srivastava, 2007). In addition, miRNAs are likely to be involved in defence responses to parasites (including viruses) by modulating expression levels of either cellular defence genes or parasite genes. In 2004, the first viral encoded miRNAs were reported from the Epstein–Barr virus (EBV) belonging to the γ-herpesvirus subfamily (Pfeffer *et al.*, 2004).

Since then, members of the herpesviruses from all three subfamilies (α, β and γ) have been found to express miRNAs. For example, human viruses including herpes simplex viruses type 1 (HSV-1) and 2 (HSV-2) (α-herpesviruses) encode eight and three, respectively (Cui *et al.*, 2006), human cytomegalovirus (HCMV; a β-herpesvirus) encodes 11 (Petters and Stefanelli, 2005), EBV (a γ-herpesvirus) encodes 25 (Grundhoff *et al.*, 2006), and Kaposi's sarcoma-associated herpesvirus (KSHV; another γ-herpesvirus) encodes 12 miRNAs (Cai *et al.*, 2005). Other DNA viruses have also been shown to encode miRNAs, but a smaller number of miRNAs (commonly 1 or 2); these include human adenoviruses and polyomaviruses (Sullivan *et al.*, 2005; Xu *et al.*, 2007). Recently, a miRNA was reported from an insect ascovirus (*Heliothis virescens* ascovirus, HvAV-3e) (Hussain *et al.*, 2008) and 13 predicted precursor miRNAs (pre-miRNAs) from salivary gland hypertrophy viruses (Garcia-Maruniak *et al.*, 2009). Although many large DNA viruses have not been looked at in depth, it appears that members of the *Herpesviridae* family encode the highest number of miRNAs. It is postulated that this might be related to the biology of herpesviruses in that they have the ability to establish latent infections. Consequently, miRNAs may facilitate entry or maintenance of latent infection, or manipulation of host immune responses during this stage (Cullen, 2009). In contrast, many other types of DNA viruses have less well-understood strategies for persistent infection. Of the RNA viruses, only HIV has been suggested to encode miRNAs and these assertions remain controversial (Klase *et al.*, 2007; Lin and Cullen, 2007; Omoto and Fujii, 2005; Ouellet *et al.*, 2008).

There are few studies that have addressed the role of miRNAs in insect–virus interactions. However, because of the importance and growing impact of miRNAs to virology, the latest information on the biogenesis, mechanisms of gene regulation and resources available for miRNA studies are presented here. These may be useful for students, researchers, and academics who are interested in exploring the miRNA world in insect host–virus interactions. Inevitably, the majority of the examples are from non-insect viruses.

Biogenesis of miRNAs

The primary transcript that gives rise to a miRNA (pri-miRNA; Fig. 15.1) is generally transcribed by RNA polymerase II (pol II). miRNAs can arise from exonic or intronic regions (Bartel, 2004). Less commonly, there are examples of pri-miRNAs that are transcribed by pol III (Pfeffer *et al.*, 2005; Winter *et al.*, 2009). The pri-miRNA may comprise one (monocistronic) or multiple (polycistronic) stem-loops that are cleaved from the stem by an RNase III type endonuclease, Drosha. Drosha functions in the nucleus together with its stoichiometric cofactor, DGCR8 protein (DiGeorge syndrome critical region gene 8, also known as Pasha in *Drosophila melanogaster* and *Caenorhabditis elegans*) to generate the hairpin precursor miRNA (pre-miRNA) of about 65–70 nt (Landthaler *et al.*, 2004). The Drosha–DGCR8 complex is referred to as the microprocessor complex (Landthaler *et al.*, 2004). DGCR8/Pasha is essential for Drosha function and ensures precise cleavage at the 5′ and 3′ arms of the stem–loop structure (Han *et al.*, 2006). For efficient cleavage, some miRNAs may require additional specificity factors (reviewed in Winter *et al.*, 2009). A minority of miRNAs originating from introns do not require Drosha activity for miRNA generation as splicing releases the corresponding pre-miRNA stem–loop structures (Ruby *et al.*, 2007). This subclass of intronic miRNAs is called 'mirtrons'. So far, mirtrons have not been described as arising from viral genomes.

Once cleaved, pre-miRNAs are transported into the cytoplasm by Exportin-5 where they are further processed. The RNase III-like nuclease Dicer 1 complexed with TRBP (TAR RNA binding protein in mammals and the *loquacious* protein in *D. melanogaster*) excises the terminal loop from the pre-miRNA, yielding an imperfect, ~22 nt long RNA duplex (miRNA-miRNA*) with a 2-nt overhang at each of the 3′ ends (Fig. 15.1). Reports suggest involvement of helicases in unwinding the two strands (Meister *et al.*, 2005; Robb and Rana, 2007; Salzman *et al.*, 2007), but it does not seem to be generally essential (Gregory *et al.*, 2005; Maniataki and Mourelatos, 2005). This might be facilitated by a helicase-like domain present at the N-terminus of Dicer 1 protein (Soifer *et al.*, 2008). After separation of the two strands, one is stably incorporated into

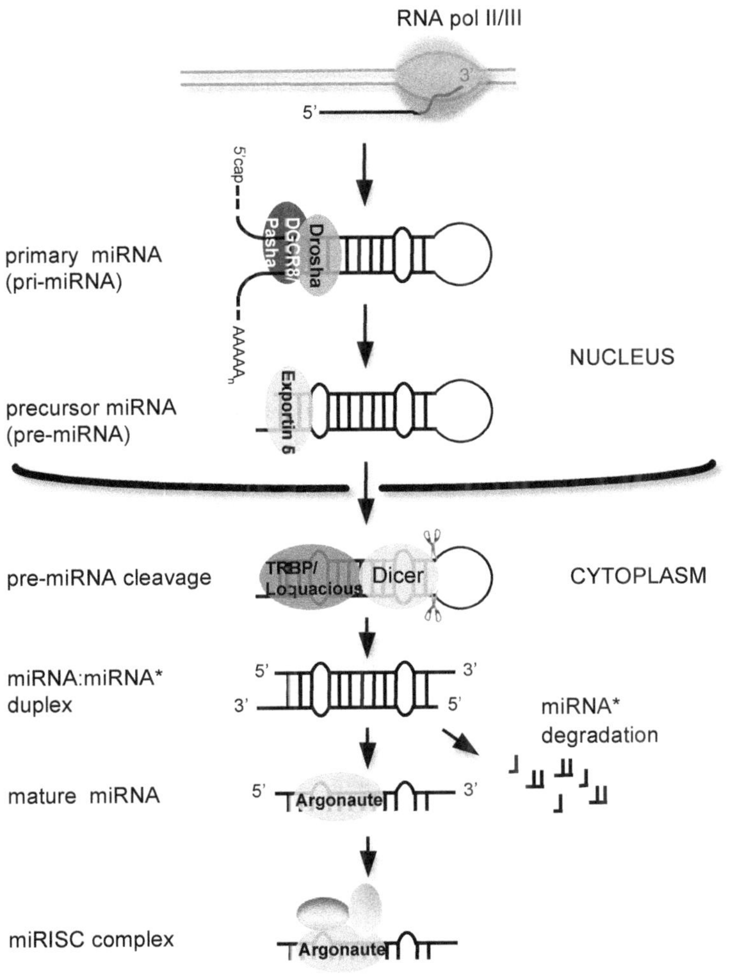

Figure 15.1 Biogenesis of miRNAs. The primary miRNA transcript (pri-miRNA), which is usually 5′-capped and 3′-polyadenylated, is produced by RNA polymerase II or III in the nucleus. pri-miRNA forms one or more hairpin loops with the stem–loop containing mismatches. The pri-miRNA is then cleaved by the RNase III enzyme Drosha and DGCR8 protein (known as Pasha in *D. melanogaster*). The excised hairpin (~65–70 nt), called precursor miRNA (pre-miRNA), is transported into the cytoplasm by Exportin 5. In the cytoplasm, pre-miRNA is further processed by another RNase II enzyme Dicer in association with the dsRNA-binding protein TRBP generating a miRNA:miRNA* duplex. The complementary strand to the mature miRNA (miRNA*) is then degraded and mature miRNA induces formation of the miRNA-RNA induced silencing complex (miRISC) in association with Argonaute protein. Subsequently, the loaded miRNA guides the miRISC to target mRNA transcript(s) (see Fig. 15.4).

RNA-induced silencing complex (RISC; guide strand), with an Argonaute (Ago) protein at its core, while the unincorporated strand (passenger strand) is degraded. However, in a subset of miRNAs both strands may become incorporated into and active within RISC. A typical viral example is a pre-miRNA encoded by the monkey polyomavirus simian virus 40 (SV40) from which two mature miRNAs are produced, one from the 5′ stem and the other from the 3′ stem of the stem–loop structure (Sullivan *et al.*, 2005). Following viral infection, cellular as well as viral miRNAs (if the virus encodes miRNA) are produced in a similar fashion, which may interact with target mRNAs transcribed from the host cell or the virus genome (Fig. 15.2).

In general, the miRNA-RISC (miRISC) complex exerts its function in down-regulating target gene expression by binding to sequences in the 3′ untranslated region (UTR) of the mRNA. The mechanism of regulation varies depending on whether the miRNA possesses complete or partial complementarity to the target sequences. In animals, the predominant mode of regulation involves translational inhibition and sometimes subsequent mRNA degradation of mRNA targets (Bushati and Cohen, 2007; Lagos-Quintana *et al.*, 2001). Ago proteins are the main effectors that reduce protein expression from the bound mRNA transcripts (Meister and Tuschl, 2004; Okamura *et al.*, 2004). Some Agos also have endonucleolytic RNase H activity. When these Agos are loaded with miRNAs possessing perfect complementarity to mRNA targets, they can direct specific cleavage of bound mRNA targets (Wang *et al.*, 2008a). While rare in animals, this mode of regulation predominates in plants and some viruses. In *Drosophila*, Ago1 has been shown to be essential for miRNA-mediated target RNA cleavage, while Ago2 functions in translational repression (Okamura *et al.*, 2004). Considering that there are no known examples

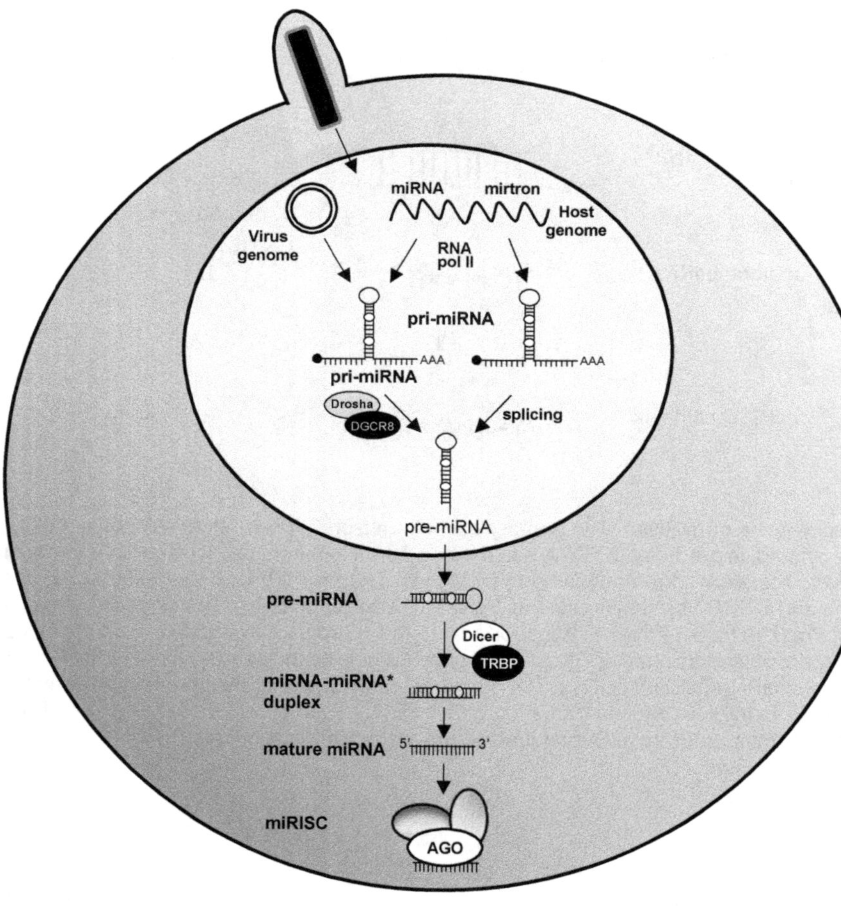

Figure 15.2 Processing of cellular and viral miRNAs. Following virus infection, cellular as well as viral miRNAs are expressed in the nucleus as primary miRNA transcripts (pri-miRNA) by the host cell RNA polymerase II. Pri-miRNA is then processed into precursor miRNA (pre-miRNA) either via cleavage by Drosha-DGCR8 or splicing. Pre-miRNA is transported into the cytoplasm where it is further processed into mature miRNA by Dicer-TRBP. Mature miRNA induces formation of miRNA–RISC complex (miRISC), which interacts with target cellular or viral mRNAs.

of viral-encoded miRNA processing or RISC components, viral miRNAs must be synthesized and processed by host cell factors similar to host miRNAs.

Regulation of miRNA expression is complex and relatively little is known about it. This regulation seems to occur at various levels, including both transcriptional and post-transcriptional levels. Temporal and tissue specific expression of both cellular and virally encoded miRNAs indicate tight regulation of miRNA synthesis. In vertebrates, abnormal expression of cellular miRNAs due to deregulation of pri-miRNA transcription is often associated with cancer, cardio-vascular diseases and developmental disorders (Chang and Mendell, 2007; Davis *et al.*, 2008). In addition, several recent studies have reported stage-specific expression of viral miRNAs (Barth *et al.*, 2008; Cai *et al.*, 2006a; Dölken *et al.*, 2007; Grey *et al.*, 2007a; Hussain *et al.*, 2008; Pfeffer *et al.*, 2004; Sullivan *et al.*, 2005). Since the majority of pri-miRNAs are transcribed by the RNA pol II, the same enzyme used for production of protein-coding messenger RNAs (mRNAs), similar transcriptional regulatory mechanisms applies to miRNA genes. Evidence of post-transcriptional regulation is accumulating and may occur at various steps of miRNA maturation. In the nucleus, this includes: miRNA-specific regulation of microprocessor complex requiring additional specificity factors; splicing of mirtrons; regulators of miRNA processing such as Lin-28, which competes with Drosha; post-transcriptional self-regulation of *Drosha/DGCR8* forming the microprocessor complex; and protection against nuclear degradation possibly by Exportin-5 (reviewed in Winter *et al.*, 2009). Examples of regulatory steps in maturation of miRNAs in the cytoplasm are: at cleavage by Dicer, its auto-regulation by an endogenous miRNA (let-7), and Lin-28 as an inhibitor of Dicer cleavage and destabilizer of pre-miRNAs. For further detailed information refer to these reviews (Ding *et al.*, 2008; Lau and MacRae, 2009; Obernosterer *et al.*, 2008; Winter *et al.*, 2009).

Discovery of viral miRNAs

Viral miRNAs have been discovered via direct cloning or with the aid of bioinformatics. For both approaches, several minimal criteria as defined by Ambros *et al.* must be met in order to ensure the small RNA of interest is a bona fide miRNA (Ambros *et al.*, 2003). First, a candidate miRNA must satisfy expression criteria, whereby it is detectable by Northern blot (or similar hybridization-based approach) or via presence in a cDNA library (ideally represented as multiple amplicons). Second, the candidate miRNA must satisfy biogenesis criteria that can include being derived from a predicted hairpin secondary structure, being evolutionarily conserved (a less common but not unprecedented characteristic of some viral miRNAs) (Cai and Cullen, 2006; Seo *et al.*, 2008), or proof of increased precursor levels in cells hypomorphic for Dicer activity. In the absence of multiple lines of evidence, claims of new viral miRNAs should be met with scepticism. For example, hairpin secondary structures are commonly predicted by RNA folding algorithms, sometimes in regions of viral genomes that do not undergo transcription. While such predictions may represent a useful starting point to identify new miRNAs, they are relatively meaningless on their own in the absence of additional supporting evidence. Similarly, viral infection can often trigger non-specific degradation of RNA, thus Northern blot analysis demonstrating the appearance of new ~22 nucleotide bands during infection, by itself, is not sufficient proof of new miRNAs. Such cases ideally require mapping the 5′ and 3′ ends of the derivative candidate miRNAs onto a predicted secondary structure hairpin precursor.

Cloning and sequencing of viral miRNAs

The first viral-encoded miRNAs were discovered by cDNA cloning – the predominant method used to identify most known host miRNAs (Pfeffer *et al.*, 2004). Total RNA, from a B cell line (B95–8) that maintains a stable latent infection with EBV, was size-fractionated on a denaturing gel, ligated to oligonucleotide linkers, reverse-transcribed, concatemerized, and sequenced by traditional Sanger sequencing methodology. These studies demonstrated that two regions of the EBV genome encoded miRNAs in a polycistronic fashion, a common feature of many host and viral miRNAs. Similar cloning strategies

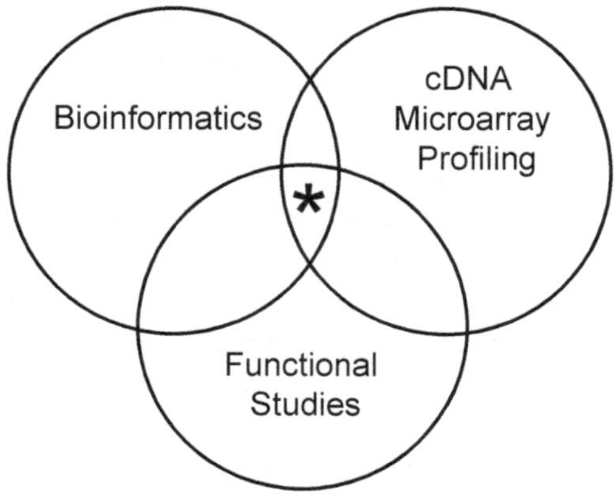

Figure 15.3 Strategies for uncovering viral miRNA function. As new viral miRNAs are discovered, uncovering their virologically relevant functions becomes the imperative challenge. Although an understanding of the function of the majority of viral miRNAs awaits future experiments, some success has already been achieved through various combinations of bioinformatics, cDNA microarray profiling and functional studies.

have contributed to the identification of miRNAs from other herpesviruses including HCMV, KSHV, and murine gammaherpesvirus-68 (MHV68) (Cai *et al.*, 2005, 2006b; Pfeffer *et al.*, 2004,2005; Samols *et al.*, 2005). While there are some disadvantages to the cloning approach (e.g. expense, biased library composition), this methodology has several advantages including the mapping of both the 5′ and 3′ ends of the miRNAs. Furthermore, in cases where the star strand derivative (miRNA*; see Fig. 15.1) of the pre-miRNA (the less abundant small RNA derived from the pre-miRNA stem arm opposite the miRNA) is also cloned, this provides very strong evidence that the cloned viral-derived small RNA is a bona fide miRNA, derived from a pre-miRNA hairpin structure.

With the advent of multiple next generation sequencing platforms (Applied_Biosystems, 2009; Margulies *et al.*, 2005), an alternative cloning strategy involves skipping the concatemerization step and directly sequencing large numbers of amplicons that each contain a single insert. This method still suffers from the same drawbacks of low-depth cDNA library sequencing – next generation platforms are expensive and have distinct library biases inherent to each methodology. However, these methods can offer several orders of magnitude more library coverage (some next generation methods offer hundreds of millions of reads compared to the thousands of reads that traditional low depth cDNA sequencing approaches top out at). Next generation sequencing methods have already contributed to the discovery of Marek's disease virus (MDV) and HSV miRNAs (Burnside *et al.*, 2006; Umbach *et al.*, 2008; Yao *et al.*, 2007), and as costs come down and the technologies become more accessible, it is likely that these methods will play a prominent future role in identifying novel viral miRNAs.

Bioinformatics approaches to identify viral miRNAs

Due to the relatively small size of viral genomes, bioinformatics approaches have contributed to the discovery of numerous viral miRNAs (Cantalupo *et al.*, 2005; Grey *et al.*, 2005; Hussain *et al.*, 2008; Pfeffer *et al.*, 2005; Seo *et al.*, 2008, 2009; Sullivan *et al.*, 2005, 2009). While such methods suffer from both false positive and false negative artefacts, they can offer the advantages of low cost and rapid discovery. Typically, these approaches first identify candidate pre-miRNA hairpin structures and then rely on further wet bench studies to confirm miRNA derivatives. In the simplest form, small regions of the viral genome can be analysed via RNA folding software and examined by eye to identify hairpin regions that resemble pre-miRNA structures. More sophisticated approaches utilize integrated secondary structure and mapping algorithms that can rely on empirical based rules or support vector machine learning approaches to determine what dictates high scoring pre-miRNA structures (Sullivan and Grundhoff, 2007).

Bioinformatics approaches have been particularly fruitful when applied to members of the polyomavirus family. These viruses have double-stranded DNA (dsDNA) genomes that are approximately 5 kilobases in size. Such small genomes allow for rapid testing of all high scoring candidate pre-miRNAs via Northern blot analysis (and subsequent cDNA cloning and PCR mapping of those candidates that score positive). So far, six different members of the polyomavirus family have been shown to encode miRNAs (Seo *et al.*, 2008, 2009; Sullivan and Grundhoff, 2007; Sullivan *et al.*, 2005, 2009). Bioinformatics has also served as the starting point for the identification of miRNAs from larger genome viruses from the ascovirus and herpesvirus families. However, for viruses with larger genomes, it may be advantageous to combine computational prediction with additional filters to help prioritize which of the potentially numerous candidate pre-miRNAs are further screened (Grundhoff *et al.*, 2006; Pfeffer *et al.*, 2005). For example, some miRNAs from the EBV and KSHV genomes were uniquely identified by combining computational prediction with custom small RNA microarray analysis (Grundhoff *et al.*, 2006). This strategy allows for greater weight being applied to candidate pre-miRNAs mapping to regions of the genome known to encode small RNA transcripts. Additional filters that can be applied include prioritizing candidates based on genomic regions of interest such as those already known to encode miRNAs (miRNAs are often polycistronic), or where antisense transcripts have already been detected (viral miRNAs are often encoded antisense to known transcripts). In summary, despite the caveat that one can never be sure that all *bona fide* miRNAs were screened for, bioinformatics (combined with additional wet bench methodology) is a useful, cost effective strategy for identifying viral miRNAs.

Mode of action of miRNAs

miRNAs regulate expression of their target genes at the post-transcriptional level by directing destabilization via cleavage and degradation or by directing translational inhibition. In the case of miRNA-mediated gene silencing, both mature miRNA and Ago proteins are important components of miRISC; the former guides the complex to target mRNA while the latter is capable of endonucleolytic cleavage activity or repression of translation. Based on the information obtained from the first few miRNAs discovered, it was generally believed that high miRNA-mRNA complementarity results in mRNA slicing and degradation while partial complementarity leads to translational repression. However, recent evidence indicates that these rules may not always apply. Several miRNAs (e.g. miR-834 and miR-H2-3p) with perfect complementary to their mRNA targets were found to have translational inhibition effect without mRNA slicing (Brodersen *et al.*, 2008; Umbach *et al.*, 2008). Instead, studies on Ago proteins have shown that different Ago proteins have differing effects on regulatory outcomes. For example, while all four mammalian somatic Ago proteins (Ago1–4) are able to repress translation (Wu *et al.*, 2008), only Ago2 is able to catalytically cleave mRNA (Meister *et al.*, 2004).

In animals, miRNAs predominantly function by targeting multiple sites in the 3′ UTR of mRNAs binding with partial sequence basepairing. For efficient repression of target messages, typically multiple miRNA binding sites are required in the 3′ UTR of the mRNA. These might be multiple copies of the same target or different ones (Brodersen and Voinnet, 2009; Doench and Sharp, 2004; van den Berg *et al.*, 2008). Recent reports suggest that some target sequences might also reside in the coding region or 5′ UTR as well (Duursma *et al.*, 2008; Hussain *et al.*, 2008; Lal *et al.*, 2008; Lytle *et al.*, 2007; Tay *et al.*, 2008).

miRNA-mRNA complementarity is widely perceived to affect the regulatory outcome (Ding *et al.*, 2008). Recruitment of miRISC may result in site specific cleavage (slicing) of mRNA (Dykxhoorn, 2007). However, even for miRNA-mediated translation repression, the seed region (2–8 nt from 5′ end) complementarity is thought to be critical for target recognition, although this has been questioned based on several miRNA–target interactions that violate the seed rule (Brodersen and Voinnet, 2009).

From these postulations, several non-mutually-exclusive models of miRNA-mediated repression mechanisms have been conceived (Fig. 15.4). For post-transcriptional repression due to mRNA degradation, mRNA bound miRISC interacts with GW182 (a 182-kDa protein associated with P/GW-bodies) and the mRNA–microribonucleoproteins–GW182 complex is delivered to processing bodies (P-bodies) where the CCR-NOT deadenylating complex begins to shorten the poly(A) tail, leading to mRNA destabilization and eventual mRNA degradation (Behm-Ansmant *et al.*, 2006). P-bodies, also known as GW or Dcp bodies, are granules in the cytoplasm of eukaryotic cells involved in mRNA catabolism (decapping, deadenylation, and mRNA degradation) and translational repression (Eulalio *et al.*, 2009). Repression of translation can occur during initiation by recruiting proteins that block eukaryotic translation initiation factor 4E (eIF4E)-m^{7}G cap interaction or repressing 60S biogenesis through eIF6 repression. A recent study also demonstrated that Ago2 is able to bind directly to m^{7}G cap (Chendrimada *et al.*, 2007; Humphreys *et al.*, 2005). Translational repression during post-initiation stages can also occur due to slowed elongation or premature termination (premature ribosome drop-off) of translation (Petersen *et al.*, 2006). In addition to sequence complementarity of miRNA-mRNA, the nucleotide sequence of the loop in the stem–loop structure may influence the processing of miRNAs. Based on experimental evidence it appears that identical or near identical mature miRNAs may have different biological roles depending on the nucleotide sequence of their pre-miRNA loop (Liu *et al.*, 2008). Finally, although the mechanisms are not fully elucidated, in some situations miRISC may actually increase protein expression (Vasudevan and Steitz, 2007; Steitz and Vasudevan, 2009).

Insect virus miRNAs

So far, HvAV-miR-1, has been the only miRNA isolated and validated from an insect virus (Hussain *et al.*, 2008). However, bioinformatics

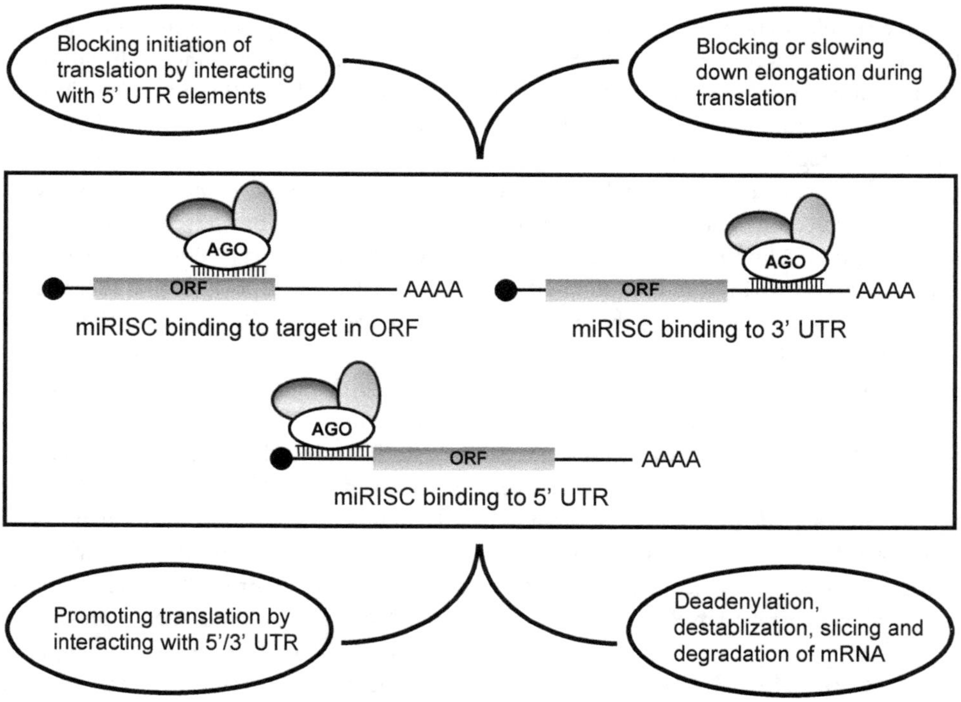

Figure 15.4 Interaction of miRNA–RISC complex with mRNA targets. The complex may interact with the 3′ UTR, 5′ UTR or ORF of target mRNAs leading to various outcomes.

analyses of several insect dsDNA viruses have predicted the presence of multiple putative miRNAs. In *Musca domestica* salivary gland hypertrophy virus (MdSGHV) and *Glossina pallidipes* salivary gland hypertrophy virus (GpSGHV), 6 and 7 pre-miRNAs were predicted, respectively (Garcia-Maruniak *et al.*, 2009). In *Autographa californica* multiple nucleopolyhedrovirus (AcMNPV) four putative miRNAs were predicated residing in the homologous regions (hr) in its genome (Kim and Burand, 2008). On the Vir-Mir Database (http://alk.ibms.sinica.edu.tw./; Li *et al.*, 2006, 2007), (a collection of predicted viral miRNA candidate hairpins), the accumulated number of predicted miRNAs for baculoviruses is 1683, ranging from 4 in *Neodiprion lecontii* NPV to 303 in *Orgyia pseudotsugata* MNPV. Apart from HvAV-miR-1, expression of these candidate insect viral miRNAs has not been experimentally confirmed, and it should be emphasized that miRNA prediction (see *Bioinformatics approaches to identify viral miRNAs*) based on sequence alone is an imperfect science and in the absence of experimental confirmation, relatively meaningless.

Insect miRNAs

In insects, miRNAs have been cloned or their expression confirmed from *D. melanogaster*, *D. pseudoobscure*, *Bombyx mori*, *Apis mellifera*, *Tribolium castaneum*, *Anopheles gambiae*, and *An. stephensi* (Behura, 2007). However, functional analyses of these miRNAs have only been carried out in *D. melanogaster*, *B. mori* and *H. zea* which established their role in cell proliferation and development (Jin *et al.*, 2004; Zhang *et al.*, 2009), neural morphogenesis (Berdnik *et al.*, 2008), hormonal signalling (Varghese and Cohen, 2007), stress resistance (Li *et al.*, 2009), fat metabolism (Xu *et al.*, 2004), cardiogenesis and muscle growth (Sokol and Ambros, 2005) and host–virus interaction (Hussain and Asgari, 2010). In *D. melanogaster* and *B. mori* a total of 110 (Lai *et al.*, 2003) and 262 (Yu *et al.*, 2008) putative miRNAs, have been identified, respectively. These numbers are, however, increasing as more miRNAs are reported. Examination of the expression profile of selected miRNAs from *D. melanogaster* and *B. mori* indicates that they are differentially expressed at various developmental stages (Aravin *et al.*, 2003; Yu *et al.*, 2008). Similarly, in the Asian malaria mosquito, *An. stephensi*, analysis of the expression profiles of 27 miRNAs confirmed distinct expression patterns of eight of those across the life cycle of the mosquito (Mead and Tu, 2008).

A substantial number of miRNAs are conserved across insects and other animals. In a rigorous bioinformatics analysis, it was found that 73 out of 139 *C. elegans* miRNAs share sequence homology with miRNAs encoded by both *Drosophila* and human genomes (Ibáñez-Ventoso *et al.*, 2008). In *An. gambiae*, 38 miRNAs were predicted to be conserved with *Drosophila* miRNAs (Lai *et al.*, 2003). In *An. stephensi*, 23 conserved and 4 new miRNAs were cloned and characterized (Mead and Tu, 2008). Most insect miRNAs appear in clusters in the genome and experimental evidence suggests that they are transcribed as polycistrons (Lai *et al.*, 2003; Mead and Tu, 2008; Winter *et al.*, 2007). However, maturation of individual miRNAs from the polycistronic pri-miRNA may be differentially regulated (Winter *et al.*, 2007).

There is very little known about the role of miRNAs in insect host–pathogen interactions and immunity. The only example in the literature, so far, examined their role in invasion of *An. gambiae* by *Plasmodium berghei*. Out of 18 miRNAs isolated from *An. gambiae*, four were affected by *Plasmodium* invasion of the midgut, three showing a decrease and one an increase in their expression levels (Winter *et al.*, 2007). Further, knock down of two key enzymes in miRNA metabolism in *An. gambiae*, Dicer1 and Ago1, led to increased sensitivity of the mosquito to *Plasmodium* infection. Although the exact function of the differentially expressed miRNAs remains to be elucidated, the results suggest that miRNAs play a role in the regulation of *Anopheles* immune system (Winter *et al.*, 2007).

Role of cellular and viral miRNAs in virus–host interactions

miRNAs produced by viruses and host cells have been shown to have significant impact on host–pathogen interactions by regulating viral and/or host cell gene expression (Cullen, 2009; Ghosh *et al.*, 2008). In addition, viruses may use cellular miRNAs for their replication, and in some cases cellular miRNAs may alter gene expression profile of the infected cell to the benefit of the virus. Despite identification of a large number of viral miRNAs, only a small number of them have been studied in detail. Among the targets of those are viral (immediate early and early proteins and DNA polymerase) as well as cellular genes (signalling molecules and pathogenicity factors) (reviewed in Cullen, 2009). Although most of scenarios and examples presented in this section are from mammalian viruses, they provide fruitful insights for those who are interested in exploring the miRNA world in insect viruses.

Virus-encoded miRNAs and their function

From evolutionary and functional points of views, miRNAs can be advantageous for viruses as regulatory elements of both host and viral genes (Cullen, 2009). Compared to protein coding genes, miRNAs are quite short, and therefore they do not occupy much space on the viral genome where space is at a premium. In addition, the small size may facilitate faster adaptation and evolution of a virus by adoption of new targets via small changes in the complementary nucleotides compared to those in a regulatory protein. One miRNA may have several targets specifically affecting several genes at the same time. Also, miRNAs are not immunogenic (Dykxhoorn, 2007); although this has not been confirmed in insects.

The first virus encoded miRNAs were reported from EBV, a human herpesvirus. One of those, miR-BART2, was shown to target the viral DNA polymerase mRNA transcript, causing its cleavage (Pfeffer *et al.*, 2004). This miRNA and some other EBV miRNAs cluster in intronic regions of the BART (BamHI-A region rightward transcript) gene (Pfeffer *et al.*, 2004). It is postulated that inhibition of viral replication by EBV may facilitate its entry into/maintenance of latent infection (Barth *et al.*, 2008). Similarly, a miRNA (HvAV-miR-1) expressed by HvAV-3e was shown to regulate the virus's own replication by targeting the DNA polymerase gene (Hussain *et al.*, 2008). Adoption of a similar mechanism to regulate virus replication by unrelated viruses suggests that virus-encoded miRNAs may function generally to regulate virus life cycles. The mature HvAV-miR-1 is produced from the coding region of the mRNA that encodes the major capsid protein (MCP). Interestingly, the mature HvAV-miR-1 is detected from 96 hours after infection whereas the primary transcript (MCP; *ORF56*) can be detected earlier in infection (from 16 hours), suggesting the presence of post-transcriptional miRNA regulatory mechanisms. Stage and tissue specific maturation of miRNAs, despite the constitutive expression of the pri-miRNA, have been reported, especially during development (Zhao and Srivastava, 2007), in cancer cells (Thomson *et al.*, 2006) and even in viral infections (Barth *et al.*, 2008; Burnside *et al.*, 2006; Cai and Cullen, 2006; Dölken *et al.*, 2007; Grey *et al.*, 2007b; Pfeffer *et al.*, 2004; Sullivan and Ganem, 2005). For most virally encoded miRNAs, the mechanism regulating the kinetics of temporal expression is not understood. When HvAV-miR-1 was ectopically expressed in insect cells followed by virus infection, virus replication was inhibited due to degradation of virally encoded DNA polymerase transcripts (*ORF1*; Hussain *et al.*, 2008).

miRNAs encoded by SV40 (SVmiRNAs) expressed late in infection were shown to target early viral mRNAs leading to their cleavage (Sullivan *et al.*, 2005). The miRNAs are produced from a single pre-miRNA stem–loop structure. Studies of a viral mutant that lacks the pre-miRNA demonstrated that the SVmiRNAs are not required for virus replication in cultured cells. Nevertheless, they may provide an advantage for the virus, as the wild type-infected cells were less sensitive to lysis by cytotoxic T cells in *in vitro* assays. This is likely due to a reduction in the accumulation of immunogenic T antigens produced by the early genes. The SVmiRNAs are conserved in two other members of polyomaviruses, the human polyomaviruses JC virus (JCV) and BK virus (BKV) (Seo *et al.*, 2008). Despite significant differences in the seed region between SVmiRNAs and those encoded by JCV and BKV, the target (the early genes) and function (the down-regulation of early gene transcript levels) of the miRNAs were found to be conserved.

Several viral-encoded miRNAs have been shown to be predominantly expressed during latency of herpesviruses. As would be expected, these are derived from transcripts transcribed during latent infection (Samols *et al.*, 2005; Umbach *et al.*, 2008). In HSV-1 and HSV-2, the miRNAs target a viral gene that plays a role in reactivation from latency and a transcription factor required for expression of most viral genes during lytic infection (Tang *et al.*, 2008; Umbach *et al.*, 2008). It would be interesting to examine insect viruses that are able to establish latent infection to find out if miRNAs may play a role in maintenance of latency. Latent/persistent virus infections have been reported from baculoviruses (Cooper *et al.*, 2003; Hughes *et al.*, 1997) and Hz-1 (Lin *et al.*, 1999), although these might be sublethal infections rather than true latent infections (Burden *et al.*, 2002).

A viral-encoded miRNA from HCMV, miR-UL112-1, targets three viral mRNAs, including an immediate early gene (known as IE1 and IE72) which is believed to be involved in transition from the latent to productive infection (Grey *et al.*, 2007a). This miRNA also inhibits viral DNA replication as premature expression of the miRNA resulted in a significant decrease in the viral DNA production.

Cellular miRNAs and their impact on viruses

Generating profiles of global expression levels of miRNAs (miRNome) may provide useful information to study their role in biological processes, including viral infection. Very few differential profiles of cellular miRNAs in response to viral infection are available. In a recent study, microarray analysis of miRNAs from a *H. zea* fat body cell line (Hz-miRs) demonstrated differential expression of cellular miRNAs in response to HvAV-3e infection (Hussain and Asgari, 2010). Several miRNAs were also down-regulated after HvAV-3e infection. In another example, changes in the global expression profile of T-lymphoma cell lines to MDV infection were determined. The results indicated general down-regulation of certain miRNAs (miR-150 and miR-223) regardless of virus aetiology but explicit down-regulation of a specific miRNA (miR-155) due to MDV infection (Godshalk *et al.*, 2008; Yao *et al.*, 2009).

The role of the miRNAs in host–virus interactions has not been determined yet. The few examples of the impact of cellular miRNAs on viruses described below are again mostly based on non-insect viruses. A cellular miRNA from *Helicoverpa zea* fat body (HzFB) cells (Hz-miR-24) is transiently down-regulated after HvAV-3e infection (24–48 hours post infection, hpi) (Hussain and Asgari, 2010). Hz-miR-24 targets were found in HvAV-3e *ORF64* and *ORF82* encoding for proteins with homology to eukaryotic RNA polymerase RPC2 and β subunits, respectively, with perfect seed region complementarity (Fig. 15.5). In the presence of synthesized Hz-miR-24 mimic or ectopic expression of the miRNA, transcript levels of *ORF64* and *ORF82* were both significantly decreased indicating a true miRNA–target interaction. Although *ORF64* and *ORF82* have not be characterized at the functional level, in general, virally encoded RNA polymerases from DNA viruses transcribe late and very late gene expression (e.g. Baculoviruses; Passarelli and Guarino, 2007), whereas immediate and early genes are transcribed by host RNA polymerase II. Considering that synthesis of Hz-miR-24, which targets the HvAV-3e RNA polymerases, is down-regulated in host cells following HvAV-3e infection, it suggests that the virus has evolved a mechanism to specifically down-regulate expression of the miRNA that can potentially affect expression of late AV genes. The mechanism by which expression of Hz-miR-24 is down-regulated still remains to be investigated.

A liver-specific miRNA (miR-122) increases the abundance of hepatitis C virus (HCV) RNA, a flavivirus. Suppression of miR-122 strongly reduces HCV RNA abundance, whereas addition of miR-122 increases RNA abundance and may also increase HCV translation in liver cell lines (Henke *et al.*, 2008). It is currently unclear if the effects are predominantly due to an increase in viral RNA abundance, translation, or both (Jopling *et al.*, 2005; Sarnow *et al.*, 2006). Interestingly, miR-122 also slightly enhances replication of West Nile virus (WNV) replicon RNA (Chang *et al.*, 2009). It was postulated that miR-122 may bind to its target in the WNV genome, similar to HCV, to regulate its replication or regulate host genes essential for replication and indirectly affects WNV replication.

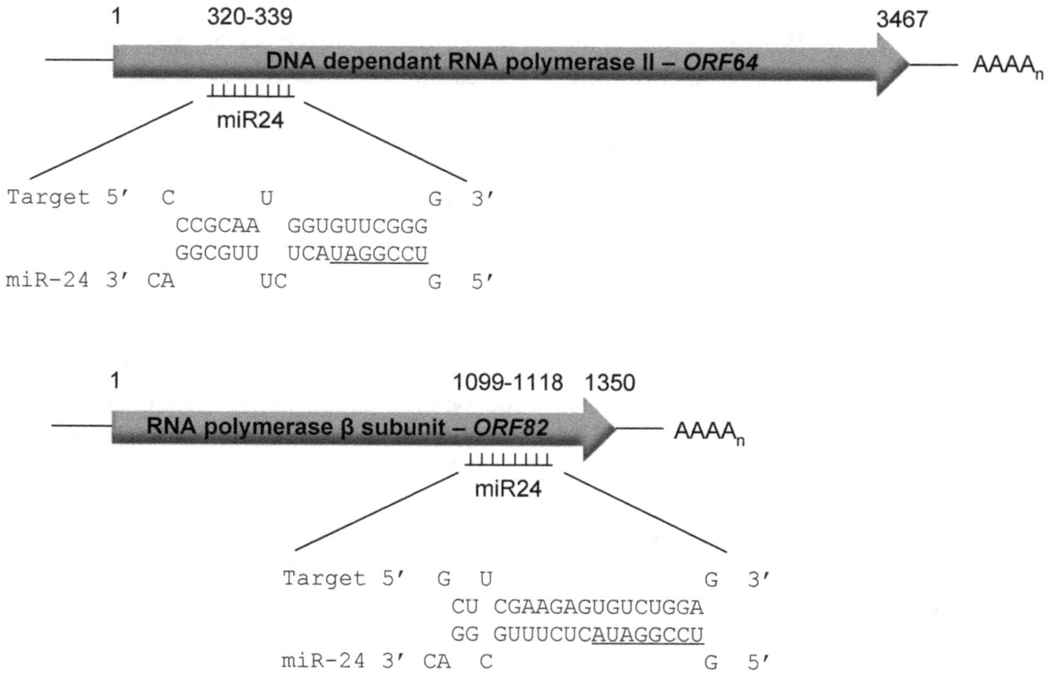

Figure 15.5 *Heliothis virescens* ascovirus (HvAV-3e) encoded DNA-dependant RNA polymerase II and RNA polymerase β subunit genes are targets of the host cell Hz-miR24 miRNA (miR24). Predicated target sequences are within the open reading frames of the genes. Seed regions (nucleotides 2–8 from the 5′ end of the miRNA) are underlined.

Conversely, expression of certain cellular miRNAs may limit virus replication (Gottwein and Cullen, 2008). In addition, some viruses alter the expression of host miRNA for their own benefit. For example, infection by EBV induces expression of a miRNA (miR-155) in primary human B lymphocytes (Lu *et al.*, 2008). The miRNA has been shown to attenuate NF-kappa B signalling and stabilizes latent virus persistence. The feasibility of harnessing endogenous miRNAs to control virus tissue tropism as a strategy for developing attenuated virus vaccines has also been shown (Barnes *et al.*, 2008). Using poliovirus, the authors demonstrated that insertion of miRNA homology sequences into a viral genome can restrict its tissue tropism, thereby preventing pathogenicity and yielding an attenuated viral strain. In addition, overexpression of miRNAs directed against viral sequences inhibits viral replication efficiently (Wang *et al.*, 2008b). Thus, the miRNA pathway can be exploited to modulate host–pathogen interactions and prevent viral replication.

Strategies for determining viral miRNA function

While there has been much success in discovering new viral miRNAs, similar to host miRNAs, an understanding of the functions of the vast majority of viral miRNAs is lacking (Fig. 15.3). So far, over 120 viral miRNAs have been described from at least 3 different virus families (Griffiths-Jones, 2006). Of these, only about 20 have defined molecular targets, and even fewer have validated functions during infection. With that said, a model is taking shape whereby viral miRNAs will regulate viral transcripts, host transcripts, or both. Most studies to date have relied on identifying the molecular targets of viral miRNAs, and deducing back from these targets the function of the miRNA. Most miRNA targets have been identified by taking advantage of the small changes in steady state levels of mRNA transcript targets that occur during ectopic expression of the miRNA (Lim *et al.*, 2005). Described below are several different types of approaches that have successfully identified miRNA targets and in some cases led to an understanding of viral miRNA function.

miRNA-mediated cleavage of antisense transcripts

The first functions identified for viral miRNAs was the autoregulation of antisense transcripts (Barth *et al.*, 2008; Pfeffer *et al.*, 2004; Sullivan *et al.*, 2005). Similar to siRNAs, miRNAs can direct cleavage of an mRNA target if they possess perfect complementarity. Although this mode of regulation is common in plants, in animals there is only a single example of a host miRNA directing cleavage of its mRNA target (Mansfield *et al.*, 2004; Yekta *et al.*, 2004). Nonetheless, multiple examples of this mode of regulation exist in animal viral miRNAs. Thus, if a viral miRNA is discovered that is encoded antisense to a known mRNA transcript, it is likely (with some exceptions, e.g. Umbach *et al.*, 2008) that it will direct cleavage and negative regulation of that transcript. For example, the EBV miR-BART2 directs cleavage of the DNA polymerase (BamHI L fragment 5, BALF5) (Barth *et al.*, 2008), and all the polyomaviral miRNAs that have been described are made late during infection and have the capacity to direct cleavage of the early transcripts.

Computational prediction

Bioinformatics approaches have been used with some success for identifying relevant miRNA targets. These approaches rely on complementarity matches between the miRNA of interest and the candidate mRNA target. Most algorithms provide extra weight to perfect complementarity between the 5′ end of the miRNA from positions 2–8 nt (the so-called seed region). Some algorithms also factor in the secondary structure of flanking regions of the predicted binding site or the predicted free energy changes that occur when the miRNA binds to the target (reviewed in Alexiou *et al.*, 2009). Other algorithms factor in evolutionary conservation of predicted miRNA binding sites. While these various programs offer some utility as a starting point for identifying miRNA function, it is worth noting that there is often little overlap between the predicted targets from each program. Thus, in the absence of any additional information, it can be challenging to determine which of the numerous predicted miRNA targets are real. This difficulty is compounded for viral miRNAs because most are not conserved with either host or other viral miRNAs, and therefore would not be predicted to have a preference for conserved binding sites. With that said, use of predictive algorithms is fast and inexpensive and therefore represents a logical starting point for understanding viral miRNA functions. Some examples of viral miRNA functions that have been uncovered via this route include the EBV miR-BART5 which targets PUMA (p53 up-regulated modulator of apoptosis which induces apoptosis) and has a role in promoting cell survival during latent infection and the ascovirus HvAV-miR-1 which targets and autoregulates the HvAV-3e DNA polymerase transcripts (Choy *et al.*, 2008; Hussain *et al.*, 2008). Once a candidate miRNA target is identified, typically, chimeric reporter constructs containing the predicted mRNA binding region are utilized to confirm direct regulation by the miRNA of interest.

Ectopic expression induces decreased steady-state levels of target transcripts

Many viral miRNA targets have been elucidated by combining ectopic expression with cDNA expression microarray profiling. miRNAs can be ectopically expressed via pol II expression vectors or via oligonucleotide mimics. The pol II vector strategy has the advantage of affordability and greater likelihood of expressing physiologically relevant levels of miRNAs. The oligonucleotide mimic strategy can be used in transient assays to drive high levels of expression. In addition, miRNA mimics are highly transfectable, even in cell types that do not efficiently take up DNA plasmids. Several targets of the KSHV miRNAs were elucidated by profiling cDNA levels from cells that ectopically express 10 of the 12 KSHV miRNAs in a stable fashion (Samols *et al.*, 2007). These targets include thrombospondin 1 (THBS-1), that normally functions in angiogenesis and cell cycle control, which may suggest that the KSHV viral miRNAs are involved in the transforming phenotype associated with this virus. Similarly, ectopic expression of a single KSHV miRNA, miR-K12–11 has been shown to decrease the steady state levels of several transcripts that may be pertinent to the pathology associated with KSHV infection (Gottwein *et al.*, 2007; McClure and Sullivan, 2008; Skalsky *et al.*, 2007). Ectopic expression of HvAV-miR-1 (a viral miRNA) and Hz-miR-24 (a cellular miRNA) also was shown to regulate transcript levels of their target genes in HvAV-3e infected cells (Hussain *et al.*, 2008; Hussain and Asgari, 2010). A variation of this approach relies on blocking the function of a viral miRNA of interest via transfection of a modified antisense oligonucleotide inhibitor. In this approach, some direct targets of the miRNA of interest will show an increase in steady-state levels. Zigelbauer and colleagues combined this strategy with other analyses to demonstrate that several KSHV-encoded miRNAs target the cellular transcript encoding BCLAF1, a protein associated with apoptosis and linked to control of lytic replication of KSHV (Ziegelbauer *et al.*, 2009). This strategy has the advantage that physiological levels of miRNAs can be examined during infection.

A caveat to using cDNA microarray analysis for identifying viral miRNA targets is that not all mRNAs targeted by miRNAs show a robust decrease in steady state levels. A recent advance used to identify targets of host miRNAs involves immunoprecipitation of the RISC-miRNA complex, followed by microarray analysis to identify associated transcripts (Chi *et al.*, 2009; Kirino and Mourelatos, 2008). This approach does not rely on decreases in steady state levels, but rather scores for positive enrichment of mRNA transcripts in the presence of ectopic expression of the miRNA of interest. While so far there have not been any reports of viral miRNA targets identified with this methodology, it seems likely that this approach will contribute much to our understanding of viral miRNA function in the near future.

'Functions first' approaches to identifying viral miRNA targets and functions

Although bioinformatics and cDNA expression methods have been useful for identifying viral miRNA targets, understanding the biologically relevant function of a viral miRNA requires further experimentation. For example, ectopic expression of a miRNA can lead to statistically significant changes in the steady state levels of more than a hundred mRNA transcripts (Giraldez *et al.*, 2006; Lim *et al.*, 2005). As many of these changes may be neutral towards viral infectivity, they are tolerated so long as the functionally relevant targets are affected. Similar problems of separating the 'wheat from the chaff' can arise from purely computational methods. Thus, lists of transcripts generated from microarray experiments and bioinformatics approaches represent a useful starting point. With luck, clear patterns of transcripts involved in a virologically relevant process will emerge from the list of targets that can inform further functional studies. This approach constitutes a 'bottom-up' approach where the repertoire of putative viral miRNA targets is used as a base to construe the major functions of the miRNA. An opposite 'top-down' approach could be taken whereby viral functions are initially screened for, which is then followed by identification of the relevant target.

Although a top-down approach has not yet been utilized in many viral miRNA studies, given the similar challenges that disparate viruses must overcome (host defences, optimization of cell cycle and metabolism for virus replication), it seems likely that 'functions-first' approaches will be a fruitful source of understanding for viral miRNAs. Recently, a functional screen was used to identify a KSHV miRNA that targets the natural killer (NK) cell ligand, MHC class I-related chain B (MICB) (Nachmani *et al.*, 2009). As part of the antiviral defence response, mammals utilize NK cells to detect cells undergoing aberrant stressors associated with viral infection. NK cells bind to ligands (such as MICB) on the surface of stressed cells to direct their destruction. Previous studies had demonstrated that an HCMV miRNA directly targets MICB for down-regulation (Stern-Ginossar *et al.*, 2007). Nachmani *et al.* (2009) utilized a flow cytometry-based assay to screen all 17 miRNA derivatives of the 12 KSHV pre-miRNAs to identify one (miR-K12-7) that acts as a regulator of MICB. This regulation occurs in spite of the fact that the HCMV miRNA and miR-K12-7 share no sequence identity. Interestingly, one of the putative miR-K12-7 binding sites possesses imperfect complementarity within the seed region. This fact, combined with the absence of a reported decrease in MICB transcript levels upon KSHV miRNA expression, may mean that a bottom-up approach would have missed this evolutionarily conserved mode of viral miRNA regulation. In the future, similar functional screens could be applied to large libraries of viral miRNAs.

Concluding remarks

Interactions between pathogens and insect hosts has been the subject of extensive investigations aimed at discovering novel and efficient approaches to reduce or disrupt pathogen transmission, improve the efficacy of microbial control agents and devise novel insect control strategies by manipulating their physiology. In particular, the immune response of insect hosts to pathogen invasion has attracted immense interest. Interestingly, phylogenomic analysis of the insect immune repertoire in various insects has revealed both conservative (mostly genes involved in recognition and signal transduction) and rapidly evolving (effector genes) features of various functional gene categories (Waterhouse *et al.*, 2007). However, the involvement of miRNA pathway in pathogen invasion and innate immune defence has not been experimentally investigated and the interaction between the host insects and pathogens, including viruses, is even less understood.

miRNAs are now recognized as a major class of gene-regulatory molecules that play important roles in various biological processes, including immunity (Miska, 2005). So far, there has been very few miRNAs reported from insect viruses, and we know little about their role in insect host–virus interactions. Given the nuclear localization of Drosha, it is unlikely that dsDNA viruses that replicate in the cytoplasm produce miRNAs. However, this does not exclude the possibility, as alternative miRNA biogenesis pathways may exist. With the increasing set of tools for detection of potential miRNAs and their targets, a variety of avenues for investigating the role of these small non-coding RNAs in host–virus interactions exist. These will further our understanding of insect virus biology as well as host antiviral immune responses.

References

Alexiou, P., Maragkakis, M., Papadopoulos, G.L., Reczko, M., and Hatzigeorgiou, A.G. (2009). Lost in translation: an assessment and perspective for computational microRNA target identification. Bioinformatics *29*, 29.

Ambros, V., Bartel, B., Bartel, D.P., Burge, C.B., Carrington, J.C., Chen, X., Dreyfuss, G., Eddy, S.R., Griffiths-Jones, S., Marshall, M., *et al.* (2003). A uniform system for microRNA annotation. RNA *9*, 277–279.

Applied_Biosystems (2009). Whole-genome discovery and profiling of small RNAs using the SOLiD™ System. BioTechniques *46*, 232–234.

Aravin, A.A., Lagos-Quintana, M., Yalcin, A., Zavolan, M., Marks, D.S., Gaasterland, T., Meyer, J., and Tuschl, T. (2003). The small RNA profile during *Drosophila melanogater* development. Insect Biochem. Mol. Biol. *37*, 3–9.

Barnes, D., Kunitomi, M., Vignuzzi, M., Saksela, K., and Andino, R. (2008). Harnessing endogenous miRNAs to control virus tissue tropism as a strategy for developing attenuated virus vaccines. Cell Host Microbe *4*, 239–248.

Bartel, D.P. (2004). MicroRNAs: Genomics, biogenesis, mechanism, and function. Cell *116*, 281–297.

Barth, S., Pfuhl, T., Mamiani, A., Ehses, C., Roemer, K., Kremmer, E., Jaker, C., Hock, J., Meister, G., and Grasser, F.A. (2008). Epstein–Barr virus-encoded microRNA miR-BART2 down-regulates the viral DNA polymerase BALF5. Nucleic Acids Res. *36*, 666–675.

Behm-Ansmant, I., Rehwinkel, J., Doerks, T., Stark, A., Bork, P., and Izaurralde, E. (2006). mRNA degradation by miRNAs and GW182 requires both CCR4:NOT deadenylase and DCP1:DCP2 decapping complexes. Genes Dev. *20*, 1885–1898.

Behura, S.K. (2007). Insect microRNAs: structure, function and evolution. Insect Biochem. Mol. Biol. *37*, 3–9.

Berdnik, D., Fan, A.P., Potter, C.J., and Luo, L. (2008). MicroRNA processing pathway regulates olfactory neuron morphogenesis. Curr. Biol. *18*, 1754–1759.

Brodersen, P., Sakvarelidze-Achard, L., Bruun-Rasmussen, M., Dunoyer, P., Yamamoto, Y.Y., Sieburth, L., and Vionnet, O. (2008). Widespread translational inhibition by plant miRNAs and siRNAs. Science *320*, 1185–1190.

Brodersen, P., and Voinnet, O. (2009). Revisiting the principles of microRNA target recognition and mode of action. Nat. Rev. Mol. Cell Biol. *10*, 141–148.

Burden, J.P., Griffiths, C.M., Cory, J.S., Smith, P., and Sait, S.M. (2002). Vertical transmission of sublethal granulovirus infection in the Indian meal moth, *Plodia interpunctella*. Mol. Ecol. *11*, 547–555.

Burnside, J., Bernberg, E., Anderson, A., Lu, C., Meyers, B.C., Green, P.J., Jain, N., Isaacs, G., and Morgan, R.W. (2006). Marek's disease virus encodes microRNAs that map to meq and the latency-associated transcript. J. Virol. *80*, 8778–8786.

Bushati, N., and Cohen, S.M. (2007). MicroRNA functions. Annu. Rev. Cell Dev. Biol. *23*, 175–205.

Cai, X., and Cullen, B.R. (2006). Transcriptional origin of Kaposi's sarcoma-associated herpesvirus microRNAs. J. Virol. *80*, 2234–2242.

Cai, X., Li, G., Laimins, L.A., and Cullen, B.R. (2006a). Human papillomavirus genotype 31 does not express detectable microRNA levels during latent or productive virus replication. J. Virol. *80*, 10890–10893.

Cai, X., Lu, S., Zhang, Z., Gonzalez, C.M., Damania, B., and Cullen, B.R. (2005). Kaposi's sarcoma-associated herpesvirus expresses an array of viral microRNAs in latently infected cells. Proc. Natl. Acad. Sci. U.S.A. *102*, 5570–5575.

Cai, X., Schäfer, A., Lu, S., Bilello, J.P., Desrosiers, R.C., Edwards, R., Raab-Traub, N., and Cullen, B.R. (2006b). Epstein–Barr virus microRNAs are evolutionarily conserved and differentially expressed. PLoS Pathog. *2*, e23.

Cantalupo, P., Doering, A., Sullivan, C.S., Pal, A., Peden, K.W., Lewis, A.M., and Pipas, J.M. (2005). Complete nucleotide sequence of polyomavirus SA12. J. Virol. *79*, 13094–13104.

Chang, J., Guo, J.T., Jiang, D., Guo, H., Taylor, J.M., and Block, T.M. (2009). Liver specific microRNA, miR-122, enhances the replication of hepatitis C virus in non-hepatic cells. J. Virol. *82*, 8215–8223.

Chang, T.C., and Mendell, J.T. (2007). microRNAs in vertebrate physiology and human disease. Annu. Rev. Genom. Hum. G. *8*, 215–239.

Chendrimada, T.P., Finn, K.J., Ji, X.J., Baillat, D., Gregory, R.I., Liebhaber, S.A., Pasquinelli, A.E., and Shiekhattar, R. (2007). MicroRNA silencing through RISC recruitment of eIF6. Nature *447*, 823–828.

Chi, S.W., Zang, J.B., Mele, A., and Darnell, R.B. (2009). Argonaute HITS-CLIP decodes microRNA–mRNA interaction maps. Nature *17*, 17.

Choy, E.Y., Siu, K.L., Kok, K.H., Lung, R.W., Tsang, C.M., To, K.F., Kwong, D.L., Tsao, S.W., and Jin, D.Y. (2008). An Epstein–Barr virus-encoded microRNA targets PUMA to promote host cell survival. J. Exp. Med. *205*, 2551–2560.

Cooper, D., Cory, J.S., Theilmann, D.A., and Myers, J.H. (2003). Nucleopolyhedroviruses of forest and western tent caterpillars: cross-infectivity and evidence for activation of latent virus in high-density field populations. Ecol. Entomol. *28*, 41–50.

Cui, C., Griffiths, A., Li, G., Silva, L.M., Kramer, M.F., Gaasterland, T., Wang, X.J., and Coen, D.M. (2006). Prediction and identification of Herpes Simplex Virus 1-encoded microRNAs. J. Virol. *80*, 5499–5508.

Cullen, B.R. (2009). Viral and cellular messenger RNA targets of viral microRNAs. Nature *457*, 421–425.

Davis, B.N., Hilyard, A.C., Lagna, G., and Hata, A. (2008). SMAD proteins control DROSHA-mediated microRNA maturation. Nature *454*, 56–61.

Ding, X.C., Weiler, J., and Großhans, H. (2008). Regulating the regulators: mechanisms controlling the maturation of microRNAs. Trend Biotechnol. *27*, 27–36.

Doench, J.G., and Sharp, P.A. (2004). Specificity of microRNA target selection in translational repression. Genes Dev. *18*, 504–511.

Dölken, L., Perot, J., Cognat, V., Alioua, A., John, M., Soutschek, J., Ruzsics, Z., Koszinowski, U., Voinnet, O., and Pfeffer, S. (2007). Mouse cytomegalovirus microRNAs dominate the cellular small RNA profile during lytic infection and show features of posttranscriptional regulation. J. Virol. *81*, 13771–13782.

Duursma, A.M., Kedde, M., Schrier, M., le Sage, C., and Agami, R. (2008). miR-148 targets human DNMT3b protein coding region. RNA *14*, 872–877.

Dykxhoorn, D.M. (2007). MicroRNAs in viral replication and pathogenesis. DNA Cell Biol. *26*, 239–249.

Eulalio, A., Huntzinger, E., Nishihara, T., Rehwinkel, J., Fauser, M., and Izaurralde, E. (2009). Deadenylation is a widespread effect of miRNA regulation. RNA *15*, 21–32.

Garcia-Maruniak, A., Abd-Alla, A.M.M., Salem, T.Z., Parker, A.G., Lietze, V.-U., van Oers, M.M., Maruniak, J.E., Kim, W., Burand, J.P., Cousserans, F., *et al.* (2009). Two viruses that cause salivary gland hypertrophy in *Glossina pallidipes* and *Musca domestica* are related and form a distinct phylogenetic clade. J. Gen. Virol. *90*, 334–346.

Ghosh, Z., Mallick, B., and Chakrabart, J. (2008). Cellular versus viral microRNAs in host–virus interaction. Nucleic Acids Res. *37*, 1035–1048.

Giraldez, A.J., Mishima, Y., Rihel, J., Grocock, R.J., Van Dongen, S., Inoue, K., Enright, A.J., and Schier, A.F. (2006). Zebrafish MiR-430 promotes deadenylation and clearance of maternal mRNAs. Science *312*, 75–79.

Godshalk, S.E., Bhaduri-McIntosh, S., and Slack, F.J. (2008). Epstein–Barr virus-mediated dysregulation of human microRNA expression. Cell Cycle *7*, 3595–3600.

Gottwein, E., and Cullen, B.R. (2008). Viral and cellular microRNAs as determinants of viral pathogenesis and immunity. Cell Host Microbe *3*, 375–387.

Gottwein, E., Mukherjee, N., Sachse, C., Frenzel, C., Majoros, W.H., Chi, J.-T.A., Braich, R., Manoharan, M., Soutschek, J., Ohler, U., *et al.* (2007). A viral microRNA functions as an orthologue of cellular miR-155. Nature *450*, 1096–1099.

Gregory, R.I., Chendrimada, T.P., Cooch, N., and Shiekhattar, R. (2005). Human RISC couples microRNA biogenesis and posttranscriptional gene silencing. Cell *123*, 631–640.

Grey, F., Antoniewicz, A., Allen, E., Saugstad, J., McShea, A., Carrington, J.C., and Nelson, J. (2005). Identification and characterization of human cytomegalovirus-encoded microRNAs. J. Virol. *79*, 12095–12099.

Grey, F., Meyers, H., White, E.A., Spector, D.H., and Nelson, J. (2007a). A human cytomegalovirus-encoded microRNA regulates expression of multiple viral genes involved in replication. PLoS Pathog. *3*, e163.

Grey, F., Meyers, H., White, E.A., Spector, D.H., and Nelson, J. (2007b). A human cytomegalovirus-encoded microRNA regulates expression of multiple viral genes involved in replication. PLoS Pathog. *3*, e163.

Griffiths-Jones, S. (2006). miRBase: the microRNA sequence database. Meth. Mol. Biol. *342*, 129–138.

Grundhoff, A., Sullivan, C.S., and Ganem, D. (2006). A combined computational and microarray-based approach identifies novel microRNAs encoded by human gamma-herpesviruses. RNA *12*, 733–750.

Han, J., Lee, Y., Yeom, K.H., Nam, J.W., Heo, I., Rhee, J.K., Sohn, S.Y., Cho, Y., Zhang, B.T., and Kim, V.N. (2006). Molecular basis for the recognition of primary microRNAs by the Drosha-DGCR8 complex. Cell *125*, 887–901.

Henke, J.I., Goergen, D., Zheng, J., Song, Y., Schüttler, C.G., Fehr, C., Jünemann, C., and Niepmann, M. (2008). MicroRNA-122 stimulates translation of hepatitis C virus RNA. EMBO J. *27*, 3300–3310

Hughes, D.S., Possee, R.D., and King, L.A. (1997). Evidence for the presence of a low-level, persistent baculovirus infection of *Mamestra brassicae* insects. J. Gen. Virol. *78*, 1801–1805.

Humphreys, D.T., Westman, B.J., Martin, D.I., and Preiss, T. (2005). MicroRNAs control translation initiation by inhibiting eukaryotic initiation factor 4E/cap and poly(A) tail function. Proc. Natl. Acad. Sci. U.S.A. *102*, 16961–16966.

Hussain, M., and Asgari, S. (2010). Functional analysis of a cellular microRNA in insect host–ascovirus interaction. J. Virol. *85*, 612–620.

Hussain, M., Taft, R.J., and Asgari, S. (2008). An insect virus-encoded microRNA regulates viral replication. J. Virol. *82*, 9164–9170.

Ibáñez-Ventoso, C., Vora, M., and Driscoll, M. (2008). Sequence relationships among *C. elegans*, *D. melanogaster* and human microRNAs highlight the extensive conservation of microRNAs in biology. PLoS ONE 3, e2818.

Jopling, C.L., Yi, M.K., Lancaster, A.M., Lemon, S.M., and Sarnow, P. (2005). Modulation of hepatitis C virus RNA abundance by a liver-specific microRNA. Science *309*, 1577–1581.

Jin, P., Zarnescu, D.C., Ceman, S., Nakamoto, M., Mowrey, J., Jongens, T.A., Nelson, D.L., Moses, K., and Warren, S.T. (2004). Biochemical and genetic interaction between the fragile X mental retardation protein and the microRNA pathway. Nat. Neurosci. 7, 113–117.

Kim, W., and Burand, J.P. (2008). Identification of putative miRNA sequences in four insect pathogenic viruses. Paper presented at: 41st Annual Meeting of the Society for Invertebrate Pathology (Warwick, UK).

Kirino, Y., and Mourelatos, Z. (2008). Site-specific cross-linking of human microRNPs to RNA targets. RNA *14*, 2254–2259.

Klase, Z., Kale, P., Winograd, R., Gupta, M.V., Heydarian, M., Berro, R., McCaffrey, T., and Kashanchi, F. (2007). HIV-1 TAR element is processed by Dicer to yield a viral micro-RNA involved in chromatin remodeling of the viral LTR. BMC Mol. Biol. *8*, 63.

Lagos-Quintana, M., Rauhut, R., Lendeckel, W., and Tuschl, T. (2001). Identification of novel genes coding for small expressed RNAs. Science *294*, 853–858.

Lai, E.C., Tomancak, P., Williams, R.W., and Rubin, G.M. (2003). Computational identification of *Drosophila* microRNA genes. Genome Biol. *4*, R42.

Lal, A., Kim, H.H., Abdelmohsen, K., Kuwano, Y., Pullmann Jr., R., Srikantan, S., Subrahmanyam, R., Martindale, J.L., Yang, X., Ahmed, F., *et al.* (2008). p16INK4a translation suppressed by miR-24. PLoS ONE 3, e1864.

Landthaler, M., Yalcin, A., and Tuschl, T. (2004). The human DiGeorge syndrome critical region gene 8 and its *D. melanogaster* homolog are required for miRNA biogenesis. Curr. Biol. *14*, 2162–2167.

Lau, P.W., and MacRae, I.J. (2009). The molecular machines that mediate microRNA maturation. J. Cell. Mol. Med. *13*, 54–60.

Li, S.C., Pan, C.Y., and Lin, W.C. (2006). Bioinformatic discovery of microRNA precursors from human ESTs and introns. BMC Genomics 7, 164.

Li, S.C., Shiau, C.K., and Lin, W.C. (2007). Vir-Mir db: prediction of viral microRNA candidate hairpins. Nucleic Acids Res. *36*, 184–189.

Li, X., Cassidy, J.J., Reinke, C.A., Fischboeck, S., and Carthew, R.W. (2009). A microRNA imparts robustness against environmental fluctuation during development. Cell *137*, 273–282.

Lim, L.P., Lau, N.C., Garrett-Engele, P., Grimson, A., Schelter, J.M., Castle, J., Bartel, D.P., Linsley, P.S., and Johnson, J.M. (2005). Microarray analysis shows that some microRNAs down-regulate large numbers of target mRNAs. Nature 433, 769–773.

Lin, C.L., Lee, J.C., Chen, S.S., Wood, H.A., Li, M.L., Li, C.F., and Chao, Y.C. (1999). Persistent Hz-1 virus infection in insect cells: Evidence for insertion of viral DNA into host chromosomes and viral infection in a latent status. J. Gen. Virol. 73, 128–139.

Lin, J., and Cullen, B.R. (2007). Analysis of the interaction of primate retroviruses with the human RNA interference machinery. J. Virol. *81*, 12218–12226.

Liu, G., Min, H., Yue, S., and Chen, C.-Z. (2008). Pre-miRNA loop nucleotides control the distinct activities of mir-181a-1 and mir-181c in early T cell development. PloS ONE 3, e3592.

Lu, F.W., A., Liu, C.G., Volinia, S., Croce, C.M., and Lieberman, P.M. (2008). Epstein–Barr virus-induced miR-155 attenuates NF-kappa B signaling and stabilizes latent virus persistence. J. Virol. *82*, 10436–10443.

Lynam-Lennon, N., Maher, S.G., and Reynolds, J.V. (2009). The roles of microRNA in cancer and apoptosis. Biol. Rev. *84*, 55–71.

Lytle, R.J., Yario, T. A., and Steitz, J. A. (2007). Target mRNAs are repressed as efficiently by microRNA-binding sites in the 5′ UTR as in the 3′ UTR. Proc. Natl. Acad. Sci. U.S.A. *104*, 9667–9672.

Maniataki, E., and Mourelatos, Z. (2005). A human, ATP-independent, RISC assembly machine fueled by pre-miRNA. Genes Dev. *19*, 2979–2990.

Mansfield, J.H., Harfe, B.D., Nissen, R., Obenauer, J., Srineel, J., Chaudhuri, A., Farzan-Kashani, R., Zuker, M., Pasquinelli, A.E., Ruvkun, G., *et al.* (2004). MicroRNA-responsive 'sensor' transgenes uncover Hox-like and other developmentally regulated patterns of vertebrate microRNA expression. Nat. Genet. *36*, 1079–1083.

Margulies, M., Egholm, M., Altman, W.E., Attiya, S., Bader, J.S., Bemben, L.A., Berka, J., Braverman, M.S., Chen, Y.J., Chen, Z., *et al.* (2005). Genome sequencing in microfabricated high-density picolitre reactors. Nature *437*, 376–380.

McClure, L.V., and Sullivan, C.S. (2008). Kaposi's sarcoma herpes virus taps into a host microRNA regulatory network. Cell Host Microbe *3*, 1–3.

Mead, E.A., and Tu, Z. (2008). Cloning, characterization, and expression of microRNAs from the Asian malaria mosquito, *Anopheles stephensi*. BMC Genomics *9*, 244.

Meister, G., Landthaler, M., Patkaniowska, A., Dorsett, Y., Teng, G., and Tuschl, T. (2004). Human Argonaute2 mediates RNA cleavage targeted by miRNAs and siRNAs. Mol. Cell *15*, 185–197.

Meister, G., Landthaler, M., Peters, L., Chen, P.Y., Urlaub, H., Lührmann, R., and Tuschl, T. (2005). Identification of novel argonaute-associated proteins. Curr. Biol. *15*, 2149–2155.

Meister, G., and Tuschl, T. (2004). Mechanisms of gene silencing by double-stranded RNA. Nature *431*, 343–349.

Miska, E.A. (2005). How microRNAs control cell division, differentiation and death. Curr. Opin. Gent. Dev. *15*, 563–568.

Nachmani, D., Stern-Ginossar, N., Sarid, R., and Mandelboim, O. (2009). Diverse herpesvirus microRNAs target the stress-induced immune ligand MICB to escape recognition by natural killer cells. Cell Host Microbe *5*, 376–385.

Nowotny, M., and Yang, W. (2009). Structural and functional modules in RNA interference. Curr. Opin. Struc. Biol. *19*, 286–293.

Obernosterer, G., Leuschner, P.J.F., Alenius, M., and Martinez, J. (2008). Post-transcriptional regulation of microRNA expression. RNA *12*, 1161–1167.

Okamura, K., Ishizuka, A., Siomi, H., and Siomi, M.C. (2004). Distinct roles for Argonaute proteins in small RNA-directed RNA cleavage pathways. Genes Dev. *18*, 1655–1666.

Omoto, S., and Fujii, Y.R. (2005). Regulation of human immunodeficiency virus 1 transcription by nef microRNA. J. Gen. Virol. *86*, 751–755.

Ouellet, D.L., Plante, I., Landry, P., Barat, C., Janelle, M.E., Flamand, L., Tremblay, M.J., and Provost, P. (2008). Identification of functional microRNAs released through asymmetrical processing of HIV-1 TAR element. Nucleic Acids Res. *36*, 2353–2365.

Passarelli, A.L., and Guarino, L.A. (2007). Baculovirus late and very late gene regulation. Curr. Drug Targets *8*, 1103–1115.

Petersen, C.P., Bordeleau, M.E., Pelletier, J., and Sharp, P.A. (2006). Short RNAs repress translation after initiation in mammalian cells. Mol. Cell *21*, 533–542.

Petters, R.M., and Stefanelli, J. (2005). Developmental arrest of endoparasitoid wasp larvae (*Nemeritis canescens* grav.) caused by an ectoparasitoid wasp (*Bracon hebetor* say). J. Exp. Zool. *225*, 459 – 465.

Pfeffer, S., Zavolan, M., Grasser, F.A., Chein, M., Russo, J.J., Ju, J., John, B., Enright, A.J., Marks, D., Sander, C., *et al.* (2004). Identification of virus-encoded microRNAs. Science *304*, 734–736.

Pfeffer, S., Zavolane, M., Grasser, F.A., Chien, M.C., Russo, J.J., Ju, J.Y., John, B., Enright, A.J., Marks, D., Sander, C., *et al.* (2005). Identification of microRNAs of the herpesvirus family. Nat. Meth. *2*, 269–276.

Robb, G.B., and Rana, T.M. (2007). RNA helicase A interacts with RISC in human cells and functions in RISC loading. Mol. Cell *26*, 523–537.

Ruby, J.G., Jan, C.H., and Bartel, D.P. (2007). Intronic microRNA precursors that bypass Drosha processing. Nature *448*, 83–86.

Salzman, D.W., Shubert-Coleman, J., and Furneaux, H. (2007). P68 RNA helicase unwinds the human let-7 microRNA precursor duplex and is required for let-7-directed silencing of gene expression. J. Biol. Chem. *282*, 32773–32779.

Samols, M.A., Hu, J., Skalsky, R.L., and Renne, R. (2005). Cloning and identification of a microRNA cluster within the latency-associated region of Kaposi's sarcoma-associated herpesvirus. J. Virol. *79*, 9301–9305.

Samols, M.A., Skalsky, R.L., Maldonado, A.M., Riva, A., Lopez, M.C., Baker, H.V., and Renne, R. (2007). Identification of cellular genes targeted by KSHV-encoded microRNAs. PLoS Pathog. *3*, e65.

Sarnow, P., Jopling, C., Norman, K., M.K., Y., and Lemon, S. (2006). Modulation of hepatitis C virus RNA by a liver-specific microRNA. FASEB J. *20*, A1336-A1336.

Seo, G.J., Chen, C.J., and Sullivan, C.S. (2009). Merkel cell polyomavirus encodes a microRNA with the ability to autoregulate viral gene expression. Virology *383*, 183–187.

Seo, G.J., Fink, L.H.L., O'Hara, B., Atwood, W.J., and Sullivan, C.S. (2008). Evolutionarily conserved function of a viral microRNA. J. Virol. *82*, 9823–9828.

Skalsky, R.L., Samols, M.A., Plaisance, K.B., Boss, I.W., Riva, A., Lopez, M.C., Baker, H.V., and Renne, R. (2007). Kaposi's sarcoma-associated herpesvirus encodes an ortholog of miR-155. J. Virol. *81*, 12836–12845.

Soifer, H.S., Sano, M., Sakurai, K., Chomchan, P., Saetrom, P., Sherman, M.A., Collingwood, M.A., Behlke, M.A., and Rossi, J.J. (2008). A role for the Dicer helicase domain in the processing of thermodynamically unstable hairpin RNAs. Nucleic Acids Res. *36*, 6511–6522.

Sokol, N.S., and Ambros, V. (2005). Mesodermally expressed *Drosophila* microRNA-1 is regulated by Twist and is required in muscles during larval growth. Genes Dev. *19*, 2343–2354.

Sonkoly, E., and Pivarcsi, A. (2009). Advances in microRNAs: implications for immunity and inflammatory diseases. J. Cell. Mol. Med. *13*, 24–38.

Steitz, J.A., and Vasudevan, S. (2009). miRNPs: versatile regulators of gene expression in vertebrate cells. Biochem. Soc. Trans. *37*, 931–935.

Stern-Ginossar, N., Elefant, N., Zimmerman, A., Wolf, D.G., Saleh, N., Biton, M., Horwitz, E., Prokocimer, Z., Prichard, M., Hahn, G., *et al.* (2007). Host immune system gene targeting by a viral miRNA. Science *317*, 376–381.

Sullivan, C.S., and Ganem, D. (2005). A virus-encoded inhibitor that blocks RNA interference in mammalian cells. J. Virol. *79*, 7371–7379.

Sullivan, C.S., and Grundhoff, A. (2007). Identification of viral microRNAs. Meth. Enzymol. *427*, 1–23.

Sullivan, C.S., Grundhoff, A.T., Tevethia, S., Pipas, J.M., and Ganem, D. (2005). SV40-encoded microRNAs regulate viral gene expression and reduce susceptibility to cytotoxic T cells. Nature *435*, 682–686.

Sullivan, C.S., Sung, C.K., Pack, C.D., Grundhoff, A., Lukacher, A.E., Benjamin, T.L., and Ganem, D. (2009). Murine Polyomavirus encodes a microRNA that cleaves early RNA transcripts but is not essential for experimental infection. Virology 387, 157–167

Tang, S., Bertke, A.S., Patel, A., Wang, K., Cohen, J.I., and Krause, P.R. (2008). An acutely and latently expressed herpes simplex virus 2 viral microRNA inhibits expression of ICP34.5, a viral neurovirulence factor. Proc. Natl. Acad. Sci. U.S.A. *105*, 10931–10936.

Tay, Y., Zhang, J., Thompson, A.M., Lim, B., and Rigoutsos, I. (2008). MicroRNAs to Nanog, Oct4 and Sox2 coding regions modulate embryonic stem cell differentiation. Nature *455*, 1124–1128.

Thomson, J.M., Newman, M., Parker, J.S., Morin-Kensicki, E.M., Wright, T., and Hammond, S.M. (2006). Extensive post-transcriptional regulation of microRNAs and its implications for cancer. Genes Dev. *20*, 2202–2207.

Umbach, J.L., Kramer, M.F., Jurak, I., Karnowski, H.W., Coen, D.M., and Cullen, B.R. (2008). MicroRNAs expressed by herpes simplex virus 1 during latent infection regulate viral mRNAs. Nature *454*, 780–783.

Umbach, J.L., and Cullen, B.R. (2009). The role of RNAi and microRNAs in animal virus replication and antiviral immunity. Genes Dev. *23*, 1151–1164.

van den Berg, A., Mols, J., and Han, J.H. (2008). RISC–target interaction: Cleavage and translational suppression. Biochem. Biophy. Acta- Gene Reg. Mech. *1779*, 668–677.

Varghese, J., and Cohen, S.M. (2007). microRNA miR-14 acts to modulate a positive autoregulatory loop controlling steroid hormone signaling in *Drosophila*. Genes Dev. *21*, 2277–2282.

Vasudevan, S., and Steitz, J.A. (2007). AU-rich-element-mediated up-regulation of translation by FXR1 and argonaute 2. Cell *128*, 1105–1118.

Wang, Y., Sheng, G., Juranek, S., Tuschl, T., and Patel, D.J. (2008a). Structure of the guide-strand-containing argonaute silencing complex. Nature *456*, 209–213.

Wang, F.-Z., Weber, F., Croce, C., Liu, C.-C., Liao, X., and Pellett, P.E. (2008b). Human cytomegalovirus infection alters the expression of cellular microRNA species that affect its replication. J. Virol. *82*, 9065–9074.

Waterhouse R.M., Kriventseva, E.V., Meister, S., Xi, Z., Alvarez, K.S., Bartholomay, L.C., Barillas-Mury C., Bian, G., Blandin, S., Christensen, B.M., *et al.* (2007). Evolutionary dynamics of immune-related genes and pathways in disease-vector mosquitoes. Science *316*, 1738–1743.

Winter, F., Edaye, S., Hüttenhofer, A., and Brunel, C. (2007). *Anopheles gambiae* miRNAs as actors of defence reaction against *Plasmodium* invasion. Nucleic Acids Res. 35, 6953–6962.

Winter, J., Jung, S., Keller, S., Gregory, R.I., and Diederichs, S. (2009). Many roads to maturity: microRNA biogenesis pathways and their regulation. Nat. Cell Biol. *11*, 228–234

Wu, L., Fan, J., and Belasco, J.G. (2008). Importance of translation and nonnucleolytic Ago proteins for on-target RNA interference. Curr. Biol. *18*, 1327–1332.

Xu, N., Segerman, B., Zhou, X., and Akusjarvi, G. (2007). Adenovirus virus-associated RNAII-derived small RNAs are efficiently incorporated into the RNA-induced silencing complex and associate with polyribosomes. J. Virol. *81*, 10540–10549.

Xu, P., Gue, M., and Hay, B.A. (2004). MicroRNAs and the regulation of cell death. Trends Genet. *20*, 617–624.

Yao, Y., Zhao, Y., Smith, L.P., Lawrie, C.H., Saunders, N.J., Watson, M., and Nair, V. (2009) Differential expression of microRNAs in Marek's disease virus-transformed T-lymphoma cell lines. J. Gen. Virol. *90*, 1551–1559.

Yao, Y., Zhao, Y., Xu, H., Smith, L.P., Lawrie, C.H., Sewer, A., Zavolan, M., and Nair, V. (2007). Marek's disease virus type 2 (MDV-2)-encoded microRNAs show no sequence conservation to those encoded by MDV-1. J. Virol. 25, 25.

Yekta, S., Shih, I.H., and Bartel, D.P. (2004). MicroRNA-directed cleavage of HOXB8 mRNA. Science *304*, 594–596.

Yu, X., Zhou, Q., Li, S.-C., Luo, Q., Cai, Y., Lin, W.-C., Chen, H., Yang, Y., Hu, S., and Yu, J. (2008). The silkworm (*Bombyx mori*) microRNAs and their expressions in multiple developmental stages. PLoS ONE 3, e2997.

Zhang, Y., Zhou, X., Ge, X., Jiang, J., Li, M., Jia, S., Yang, X., Kan, Y., Miao, X., Zhao, G., *et al.* (2009). Insect-specific microRNA involved in the development of the silkworm *Bombyx mori*. PLoS ONE 4, e4677.

Zhao, Y., and Srivastava, D. (2007). A developmental view of microRNA function. Trends Biochem. Sci. *32*, 189–197.

Ziegelbauer, J.M., Sullivan, C.S., and Ganem, D. (2009). Tandem array-based expression screens identify host mRNA targets of virus-encoded microRNAs. Nat. Genet. *41*, 130–134.

The Antiviral Role of RNA Interference 16

Michelle L. Flenniken, Mark Kunitomi, Michel Tassetto and Raul Andino

Abstract

Insects, like all living organisms, have developed defence mechanisms to resist infection. RNA interference (RNAi), a nucleic acid-based, post-transcriptional gene regulation process has recently emerged as a central pathway in antiviral defence in insects. In this chapter, we outline the role of RNAi in insect immunity and highlight research that led to its discovery as well as research aimed at understanding the mechanistic details of antiviral RNAi and the counter-measures viruses employ to modulate this immunological mechanism. As our knowledge of the pathways and mechanisms involved in insect immunity expands, so do the opportunities to employ insects as model systems to examine the general principles of RNAi and the co-evolution of hosts and their pathogens.

Introduction

All organisms have evolved sophisticated strategies to resist infectious diseases. The immune system is composed of mechanisms, which have been broadly classified as either innate (non-specific) or adaptive (pathogen-specific) immunity. Components of the innate immune system protect against a wide range of pathogens, whereas adaptive responses generate pathogen-specific immunity. The goal of the immune system is to completely eliminate or significantly reduce the number of pathogens from the host organism. Clearance of viruses is often achieved by the destruction of infected cells. However, clearance mechanisms that enable cell survival following infection have been described (Guidotti and Chisari, 2001). Studies in plants determined that one such mechanism was based on RNA-induced gene silencing, now referred to as RNA silencing or RNA interference (RNAi). This ancient pathway is conserved among species from different kingdoms (Fungi, Animalia and Plantae) and it controls gene expression at the transcriptional and post-transcriptional level (Bernstein *et al.*, 2001; Hammond *et al.*, 2001). Recent studies in insects indicate that RNAi plays a central role in antiviral immunity. In this chapter, we outline the role of RNAi in insect antiviral immunity and highlight research aimed at understanding the biological significance, biochemical mechanisms, and evolutionary consequences of this phenomenon.

The focus of this chapter is the role of RNAi in insect antiviral immunity. However, in addition to the RNAi response described herein, the insect immune response to pathogens (parasites, bacteria, fungi, viruses) includes: haemolymph coagulation, melanization, encapsulation, phagocytosis, and antimicrobial peptide production. Several signal transduction pathways, including the Toll, Imd (for immune deficiency), and JAK-STAT (for Janus kinases and signal transducers and activators of transcription) play important roles in insect immunity (see Chapter 17; Ferrandon *et al.*, 2007; Govind, 2008; Hoffmann, 2003; Lazzaro, 2008; Marmaras and Lampropoulou, 2009; Waterhouse *et al.*, 2007; Yassine and Osta, 2010). Initial characterization of the Toll pathway suggested it was primarily responsible for combating fungal and gram-positive bacterial infections whereas the Imd pathway was shown to be involved in the *Drosophila* immune response against gram-negative bacterial infections. However, further analysis suggests a larger overlap among immune responses than previously anticipated (Costa *et al.*, 2009; Dionne and Schneider, 2008; Dostert *et al.*, 2005; Kemp and Imler, 2009; Lemaitre *et al.*, 1995; Rutschmann *et al.*, 2002; Souza-Neto *et al.*, 2009; Xi *et al.*, 2008). While there has been a significant amount of research investigating the insect immune response against fungal and bacterial infections, more recent studies have revealed that these pathways also play a role in controlling virus replication. Investigations of specific virus–host interactions have demostrated a role of the Imd pathway in combating cricket paralysis virus (CrPV) infection of *Drosophila* (Costa *et al.*, 2009), and the involvement of the JAK-STAT and Toll pathways in combating Dengue virus (DENV) infection in *Aedes aegypti* mosquitoes (Souza-Neto *et al.*, 2009; Xi *et al.*, 2008). In addition, the composition of the insect microbial flora influences antiviral responses (Hedges *et al.*, 2008; Moreira *et al.*, 2009; Xi *et al.*, 2008). For example, research examining viral and bacterial (*Wolbachia*) co-infection in both *Drosophila* and *Aedes* mosquitoes indicates that insects infected with some *Wolbachia* strains are less sensitive to viral infection (Hedges *et al.*, 2008; Moreira *et al.*, 2009; Osborne *et al.*, 2009; Teixeira *et al.*, 2008). These and other studies suggest that we are just beginning to appreciate the complexity of insect immunity and antiviral defence. As our knowledge of the pathways and mechanisms involved in insect immunity expands, so do the opportunities to employ insects as model systems to examine the co-evolution of hosts and their pathogens.

Viruses and insects used to investigate the antiviral role of RNAi

As the preceding chapters discuss, insects are readily infected by a variety of viruses. In this chapter, we focus on a subset of RNA viruses that infect *Drosophila melanogaster* and select arboviruses (arthropod borne) vectored by *Aedes*, *Anopheles* and *Culex* mosquito species. These viruses were utilized in studies aimed at understanding the antiviral role of RNAi in insects. The complete list of the viruses and insects described herein can be found in Table 16.1. Understanding basic antiviral defence mechanisms will reveal the biological significance of RNAi in dictating the outcome of infections. For example, it may be that species-specific differences in RNAi antiviral defence determine whether a particular mosquito species is capable of transmitting viruses. These findings may aid in the development of strategies that decrease pathogenic arbovirus transmission to humans (Campbell *et al.*, 2008; Franz *et al.*, 2006; Sanchez-Vargas *et al.*, 2004). In addition, RNAi may prove important for protecting agriculturally important insects, such as *Apis mellifera* (Western honey bee), from viral infection (Maori *et al.*, 2009).

RNA interference (RNAi)

RNA interference (RNAi) is a post-transcriptional sequence-specific gene silencing process involved in the regulation of endogenous gene expression that was first observed in plants in 1990 (Napoli *et al.*, 1990) and in fungi where it is known as quelling (Romano and Macino, 1992). However, it was in the nematode *Caenorhabditis elegans* (Fire *et al.*, 1998), that the RNAi phenomenon was first characterized at the molecular level, work that earned its authors Craig Mello and Andrew Fire, the 2006 Nobel Prize in Physiology or Medicine. RNAi pathways were subsequently identified in insects and mammals. These discoveries have revolutionized the way we think about post-transcriptional gene regulation in eukaryotes. Questions surrounding the multiple functions and mechanism(s) of RNAi in animals continue to fuel many exciting avenues of research.

RNAi is mediated by single-stranded RNA molecules (20–30 nucleotides in length). These small RNA molecules associate with a component of the RNAi machinery, the Argonaute (Ago) protein, and target RNAs for sequence-specific silencing. In metazoans, three somewhat independent RNAi pathways have been characterized based on the origin of the small RNA guides and the proteins involved in gene silencing (Fig. 16.1) (Umbach and Cullen, 2009).

The micro-RNA (miRNA) pathway employs endogenous nuclear encoded RNAs which are processed by Drosha in the nucleus, exported into the cytoplasm by Exportin-5, and then incorporated in the cytoplasmic RNA Induced Silencing Complex (RISC) where they serve as guides to target mRNA for degradation or translational inhibition (Bartel, 2004).

Table 16.1 Insect and virus taxonomy

Insect – genus and species	Common name	Family	Order	
Aethina tumida	Small hive beetle	Nitidulidae	Coleoptera	
Harmonia axyridis	Ladybird beetle	Coccinellidae	Coleoptera	
Blatella germanica	German cockroach	Blattellidae	Dictyoptera	
Drosophila melanogaster	Vinegar, wine or fruit fly	Drosophilidae	Diptera	
Aedes (Ae.) aegyptii	Mosquito	Culicidae	Diptera	
Anopholes gambiae	Mosquito	Culicidae	Diptera	
Culex (Cx.) tritaeniorhynchus	Mosquito	Culicidae	Diptera	
Bemisia tabaci	Sweetpotato whitefly	Aleyrodidae	Hemiptera	
Rhodnius prolixus	Triatomid bug	Reduviidae	Hemiptera	
Acyrthosiphon pisum	Pea aphid	Aphididae	Hemiptera	
Oncopeltus fasciatus	Milkweed bug	Lygaeidae	Heteroptera	
Apis mellifera	Western honey bee	Apidae	Hymenoptera	
Manduca sexta	Tomato hornworm	Sphingidae	Lepidoptera	
Galleria mellonella	Wax moth	Pyralidae	Lepidoptera	
Epiphyas postvittana	Light brown apple moth	Tortricidae	Lepidoptera	
Spodoptera frugiperda	Fall armyworm	Noctuidae	Lepidoptera	
Teleogryllus oceanicus	Black field cricket	Gryllidae	Neoptera	
Virus	**Natural insect host(s)**	**Genus**	**Family**	**Genome**
Drosophila X virus (DXV)	*Drosophila melanogaster*	*Entomobirnavirus*	Birnaviridae	dsRNA
Drosophila C virus (DCV)	*Drosophila melanogaster*	*Cripavirus*	Dicistroviridae	(+) ssRNA
Cricket paralysis virus (CrPV)	*Teleogryllus oceanicus*	*Cripavirus*	Dicistroviridae	(+) ssRNA
Israeli acute paralysis virus (IAPV)	*Apis mellifera*	unclassified	Dicistroviidae	(+) ssRNA
Black queen cell virus (BQCV)	*Apis mellifera*	*Cripavirus*	Dicistroviridae	(+) ssRNA
Deformed wing virus (DWV)	*Apis mellifera*	*Iflavirus*	Iflaviridae	(+) ssRNA
Flock House virus (FHV)	*Galleria / Hordeum vulgar*	*Alphanodavirus*	Nodaviridae	(+) ssRNA
Dengue virus (DENV)	*Aedes / Culex* / human	*Flavivirus*	Flaviviridae	(+) ssRNA
West Nile virus (WNV)	*Aedes / Culex / Aves* / human	*Flavivirus*	Flaviviridae	(+) ssRNA
Yellow fever virus (YFV)	*Aedes / Culex* / human	*Flavivirus*	Flaviviridae	(+) ssRNA
Sindbis virus (SINV)	*Aedes / Culex* / human	*Alphavirus*	Togaviridae	(+) ssRNA
O'nyong-nyong virus (ONNV)	*Anopholes* / human	*Alphavirus*	Togaviridae	(+) ssRNA
Semliki Forest virus (SFV)	*Aedes* / human	*Alphavirus*	Togaviridae	(+) ssRNA
Chikungunya virus (CHIKV)	*Aedes* / human	*Alphavirus*	Togaviridae	(+) ssRNA

Another RNAi pathway, herein called the siRNA pathway, is initiated by dsRNA (virus or transposon derived), which is processed by Dicer into small interfering RNAs (siRNAs) that are in turn incorporated into RISC to guide the targeting of homologous mRNAs. Degradation of target mRNA is catalysed by Argonaute-2 (Ago-2), a RISC component shown to exhibit target mRNA slicing activity (Hutvagner and Simard, 2008; Kavi *et al.*, 2008).

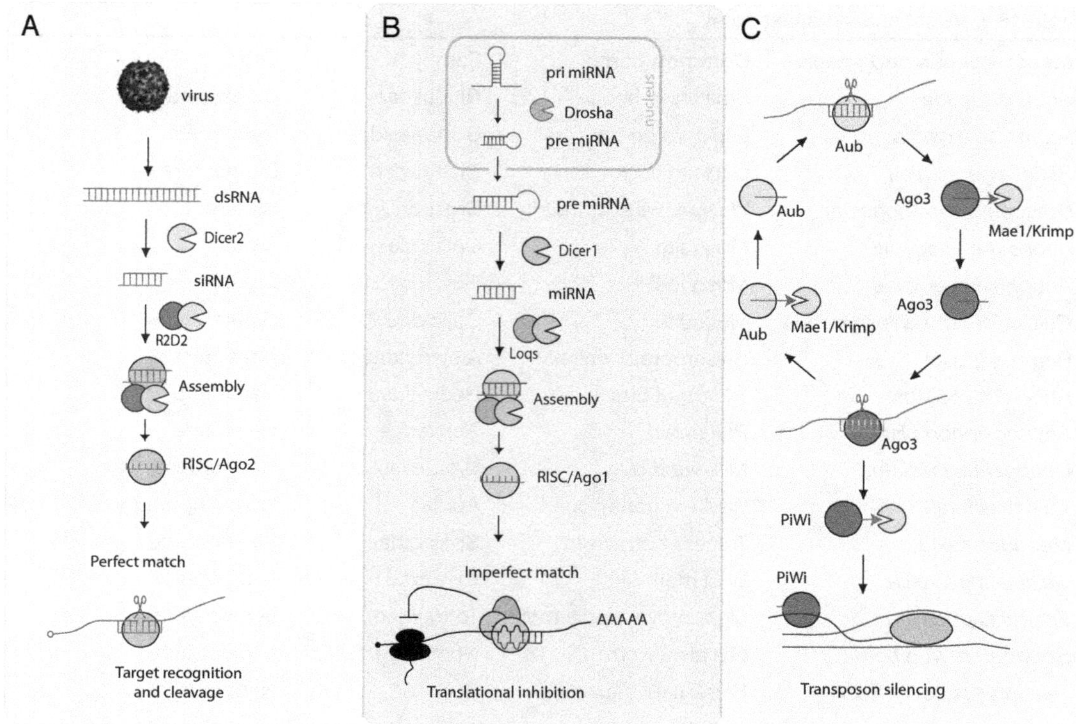

Figure 16.1 Schematic representation of three RNAi pathways found in insects. (A) The siRNAi pathway: The exogenous RNAi pathway that employs dsRNA produced by viruses (intermediates of replication or secondary structures) and processes it into small interfering RNAs (siRNAs) that direct the cleavage and degradation of viral target mRNAs. (B) The microRNA pathway: the miRNA pathway is initiated by endogenously expressed microRNA precursors (pre-miRNAs). These precursors are encoded in the genome of the cell and are expressed in the nucleus, transported to the cytoplasm and used to modulate translation and stability of target mRNAs. (C) The piRNA pathway: piRNAs are produced by annealing of sense and antisense related transposable elements. The biogenesis of piRNA is independent of Dicer, but does require the Argonaute proteins (Argonaute 3, Ago 3, Piwi and Aubergine (Aub), which is a Piwi-like protein). piRNA act to regulate chromatin structure and transcriptional silencing.

The third RNAi pathway is mediated by a group of argonaute-superfamily proteins called Piwi. Piwi-associated small RNAs (piRNAs) are slightly longer (25–30 nucleotides) than siRNAs and miRNAs (21–23 nucleotides). Piwi proteins and piRNAs are found only in the germline where they repress transposable elements (Aravin *et al.*, 2007; Brennecke *et al.*, 2008; Lin, 2007; Zhou *et al.*, 2007). In *Drosophila*, mutations in the Piwi proteins that associate with piRNA lead to sterility and loss of germline cells in both males and females.

Investigation of plant host–virus interactions demonstrated that plants employ RNAi as an antiviral strategy and in turn plant viruses encode suppressors of RNAi (VSRs) (see *Viral suppressors of RNAi* below). These discoveries led to analogous investigations in insects and the discovery of the evolutionarily conserved antiviral RNAi pathway. Three key components of the siRNA pathway, Dicer-2, R2D2, and Ago-2, have been studied for their role in insect antiviral immunity. *Drosophila* mutants lacking these proteins exhibit increased mortality in response to viral infection as compared to wild type flies (Fig. 16.2A and B). These data established that the siRNA pathway is critical to antiviral defence in *Drosophila*.

Exogenous dsRNA-derived small interfering RNAs are generated by Dicer-2

The siRNA pathway is initiated by Dicer-2 (Dcr-2), a multidomain ribonuclease of the RNase III family that processes long exogenous dsRNA molecules (such as dsRNA virus genomes or viral

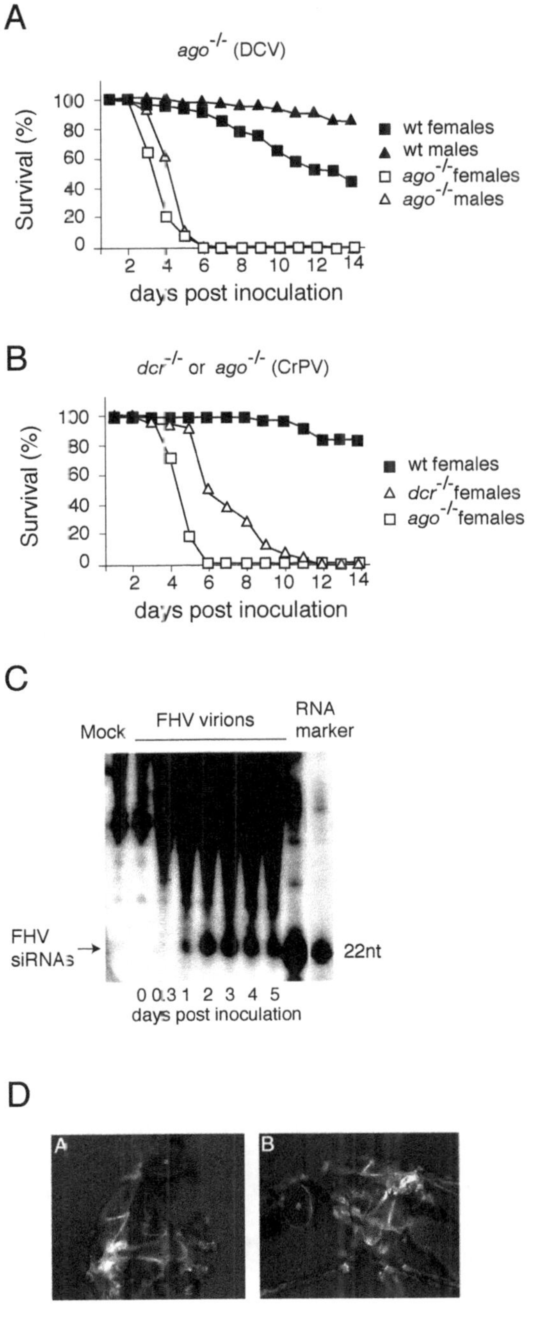

Figure 16.2 Antiviral RNAi in Insects. (A,B) *Drosophila melanogaster* RNAi mutant flies are hypersensitive to virus infection. Homozygous *ago-2*$^{-/-}$, *dcr-2*$^{-/-}$ and wild type (WT) flies were injected with (A) DCV or (B) CrPV and monitored daily for survival. Adapted from (van Rij *et al.*, 2006) with permission. (C) Induction and suppression of RNA silencing in *Drosophila* by FHV. A time course Northern blot analysis shows accumulation of FHV siRNAs. Adapted from (Li *et al.*, 2002, with permission). (D) eGFP expression at 3 days post inoculation (d.p.i.) in mosquitoes co-injected with dsRNAs and O'nyong-nyong virus-eGFP (ONNV-eGFP). (*a*) Co-injection of ONNV-eGFP and ONNV-virus specific dsRNA (dsnsP3). (*b*) Co-injection of ONNV-eGFP and non-specific dsRNA (dsβgal). Mosquitoes injected with dsnsP3 show a dramatic reduction in eGFP expression when compared with dsRNA gal-injected controls. Adapted from (Keene *et al.*, 2004) Copyright (2004, with permission) National Academy of Sciences, USA.

replication intermediates) into siRNA. Dicer-2 cleaved products are 21–23 nucleotide duplexes with an approximately 19 basepair double-stranded core flanked by a 5′-phosphate end and a two-nucleotide single-stranded overhang at their 3′ end. siRNAs are noticeably longer than most other RNase III enzyme cleavage products. siRNA size is dictated directly by Dicer structure and is an important feature of RNAi, as shorter siRNA cannot trigger silencing. Dicer's two RNase III domains act as an intermolecular dimer to form a single catalytic site where residues from each RNase III domain cleave one strand of the substrate leaving the 2-nucleotide single-stranded overhang at the 3′ hydroxyl end (Tomari and Zamore, 2005; Zhang *et al.*, 2004). Through its PAZ domain (named after the proteins Piwi, Argonaute and Zwille), Dcr-2 can then bind to the newly formed single-stranded overhang and proceeds with another round of cleavage; the distance between the PAZ domain and the catalytic site defining the specific length of siRNAs.

The RISC loading complex

Dicer-2 generated double-stranded siRNAs, which must be loaded into the RISC in order to initiate gene silencing. Alone, Dcr-2 fails to remain bound to its cleavage product and requires the association of a cofactor, R2D2, to interact with duplex siRNA. The C-terminus of R2D2 is essential for binding to Dcr-2, whereas its two dsRNA binding domains are critical for siRNA binding and RISC assembly (Liu *et al.*, 2006). Together, Dcr-2 and R2D2 facilitate loading of siRNAs into RISC. Interaction of Dcr-2 and R2D2 with RISC during loading forms the RISC loading complex (RLC). The thermodynamic stability of basepairs at the ends of the siRNA determines the orientation of the associated Dcr-2 – R2D2 complex and, ultimately, which strand is retained within RISC (Tomari *et al.*, 2004). R2D2 binds the most stable siRNA strand; conversely, Dcr-2 associates more closely with the less stable strand. Following loading into RISC, the strand associated with Dcr-2, called the 'guide' strand, is selectively retained by the Ago-containing RISC and used to target complementary RNAs for degradation. The other strand of the siRNA duplex, called the 'passenger' strand, is cleaved by Ago-2. The cleaved passenger strand disassociates from RISC in a reaction catalysed by the component 3 promoter of RISC (C3PO), another component of the RLC that is composed of two proteins: Trax and Translin (Liu *et al.*, 2009). Thus, strand selection occurs at the level of the RLC where R2D2 and Dcr-2 act as sensors for siRNA asymmetry.

The RNA-induced silencing complex (RISC)

The catalytic activity of the effector protein Ago-2 is Mg^{2+}-dependent (Schwarz *et al.*, 2004). Several biochemical studies have focused on RNA target cleavage by the programmed RISC and have unravelled sophisticated molecular mechanisms behind target recognition and cleavage site (Elbashir *et al.*, 2001; Haley and Zamore, 2004; Li *et al.*, 2007; Ma *et al.*, 2005; Ui-Tei *et al.*, 2008; Yuan *et al.*, 2005). It appears that the 5′ bases of the siRNA guide strand, also referred to as the 'seed region', play a key role in the binding affinity of RISC to RNA targets. Structurally, the seed region of the siRNA is linked to the PIWI domain of Ago-2 and exposed to the solvent in a semi-helical conformation favouring duplex formation. The conformation of the holo-RISC enables the complex to scan RNA molecules for target sites. On the other hand, the 3′ nucleotides of the guide strand bound to the PAZ domain of Ago-2 are initially less amenable to hybridization. Binding of the seed region facilitates the annealing of the guide strand 3′ end to the rest of the RNA target site. In turn, the complementarity between the 3′ nucleotides of the guide strand and the RNA target favours an A-form helix geometry at the core of the siRNA-RNA target duplex which is required for efficient catalysis by RISC (Ameres *et al.*, 2007). Cleavage occurs at the phosphodiester bond located across from nucleotides 10 and 11 of the guide strand (from the 5′ end) or between nucleotides 9 and 10 for passenger strand cleavage.

Another feature of the activated RISC is the 2′-O-methylation at the 3′ end of siRNA guide strand (Horwich *et al.*, 2007). This modification is mediated by the RNA methyltransferase DmHen1 (also termed Pimet, for its role in the methylation of Piwi RNA) using *S*-adenosyl methionine (SAM) as hydroxyl group donor. This modification has been only observed for Ago-2-loaded siRNA and occurs only after the pre-RISC has matured into the holo-RISC. In the PIWI RNAi pathway that represses transposable elements in the genome, the loss of 2′-O-methylation at the 3′ end of piRNAs (which are also loaded on Ago-2- RISC) correlates with a derepression of transposons (Forstemann *et al.*, 2007). However, the role of siRNA 2′-O-methylation in the siRNA pathway remains unknown.

RNAi as an antiviral response

RNAi interference (RNAi) plays a significant role in insect antiviral responses. Direct evidence of the antiviral role of RNAi in insects has predominantly come from investigating virus–host interactions in fruit flies and mosquitoes. The genetic malleability of *D. melanogaster* makes it an ideal model organism for investigating antiviral RNAi and determining gene-silencing mechanisms. Parallel research in non-model insects, such as mosquitoes, may lead to strategies that decrease virus transmission to human populations.

Before the discovery of the antiviral role of RNAi in insects, it was known that the insect RNAi machinery, like that of other eukaryotes, functioned to down-regulate endogenous genes in response to dsRNA (Amdam *et al.*, 2003; Araujo *et al.*, 2006; Cruz *et al.*, 2006; Ghanim *et al.*, 2007; Hughes and Kaufman, 2000; Kennerdell and Carthew, 1998; Kuwayama *et al.*, 2006; Mutti *et al.*, 2006; Paskewitz *et al.*, 2006; Turner *et al.*, 2006; Vermehren *et al.*, 2001). Subsequent experiments in cultured *Drosophila* S2 cells demonstrated that the RNAi machinery plays a role in limiting *Flock House virus* (FHV) replication (Li *et al.*, 2002). Initial evidence of an insect RNAi mediated antiviral strategy was the identification of FHV siRNAs in *Drosophila* S2 cells infected with FHV (Fig. 16.2C). Depletion of a key component of the RNAi machinery, Ago-2, from cultured *Drosophila* S2 cells prior to FHV infection resulted in the production of 2 to 3-fold more viral RNA as compared to controls (Li *et al.*, 2002). Together, these data indicate that an RNAi-mediated mechanism plays an important role in insect antiviral defence. Further examination of the host–pathogen interaction in both plants and *Drosophila* S2 cells revealed that the FHV-B2 protein is a viral encoded suppressor of RNAi (VSR) and that this protein is essential to FHV RNA accumulation in *Drosophila* cells (Li *et al.*, 2002).

Prior to the description of RNAi as an antiviral pathway in mosquitoes, promising research indicated that DENV resistance could be achieved in mosquitoes via introduction of a viral vector expressing DENV2-specific RNAs (Gaines *et al.*, 1996; Olson *et al.*, 1996). Both *Ae. albopictus* mosquito cells (C6/36) and *Ae. aegypti* mosquitoes infected with a recombinant Sindbis virus (SINV) expressing the premembrane (prM) coding region of DENV2 in either sense or antisense orientation were resistant to DENV2 challenge (Gaines *et al.*, 1996; Olson *et al.*, 1996). Additional experiments in C6/36 cells determined that an RNAi mechanism was responsible for DENV2 resistance and in addition demonstrated dsRNA induced silencing of DENV1 and Semliki Forest virus (SFV) (Adelman *et al.*, 2002; Caplen *et al.*, 2002; Sanchez-Vargas *et al.*, 2004). DENV2-specific siRNAs, a hallmark of RNAi pathway activity, were isolated from stable DENV2 prM hairpin RNA expressing C6/36 cells (FB9.1) after DENV2 challenge (Sanchez-Vargas *et al.*, 2004). Interestingly, FB9.1 cells limited both DENV2 and DENV3 replication, due to the high degree of similarity between the prM sequences of both serotypes (Sanchez-Vargas *et al.*, 2004)

Subsequent evidence that RNAi serves as an insect antiviral response came from studies of *Anopheles gambiae* mosquitoes infected with O'nyong-nyong virus (ONNV). In this work, the ONNV was engineered to express enhanced green fluorescent protein (eGFP) to facilitate *in vivo* localization and virus quantification (Keene *et al.*, 2004). The effect of the siRNA pathway on ONNV-eGFP replication and dissemination was evaluated in experiments in which the virus and either virus-specific dsRNA (dsRNA ONNV-nsP3) or non-specific dsRNA (dsRNA βgal), were co-injected in the mosquito thorax. These studies revealed that ONNV-eGFP dissemination and titre decreased in response to viral specific dsRNA, whereas non-specific dsRNA did not limit virus replication (Fig. 16.2D). Direct evidence of an RNAi mediated antiviral pathway, came from experiments in which Ago-2, the RISC component responsible for 'slicing' target RNA, was knocked-down in conjunction with ONNV-eGFP infection. These experiments resulted in increased virus spread and titre in *An. gambiae* mosquitoes with compromised RNAi pathways (Keene *et al.*, 2004). Interestingly, parallel experiments investigating the potential role of other Ago-proteins in antiviral defence indicated that *An. gambiae* Ago-3 also plays a modest role in limiting viral infection (Keene *et al.*, 2004).

Further evidence for a critical role of RNAi in insect antiviral immunity was demonstrated using mutants lacking *dcr-2*, *r2d2*, or *ago-2*, each of which demonstrated hypersensitivity to viral infection (Galiana-Arnoux *et al.*, 2006; van Rij *et al.*, 2006; Wang *et al.*, 2006; Zambon *et al.*, 2006) (Fig. 16.2A and B). Adult flies deficient in components of the RNAi machinery exhibited increased mortality when challenged with FHV, CrPV, SINV, *Drosophila C virus* (DCV), and *Drosophila X virus* (DXV) (Galiana-Arnoux *et al.*, 2006; van Rij *et al.*, 2006; Wang *et al.*, 2006; Zambon *et al.*, 2006). Northern blot analysis verified that viral-specific siRNAs were produced in response to both FHV and CrPV infection in wild type (wt) flies, but not in *dcr-2*$^{-/-}$ mutants and demonstrated increased DCV genomic RNA in *ago-2*$^{-/-}$ flies (van Rij *et al.*, 2006; Wang *et al.*, 2006). Thus, antiviral siRNAs appear to be generated by Dcr-2 and the catalytic component of RISC, Ago-2, plays an essential role in insect antiviral immunity (van Rij *et al.*, 2006). As expected from the model of the RNAi mechanism (Fig. 16.1), *r2d2*$^{-/-}$ flies were able to produce siRNAs, but since these siRNAs were not properly loaded into RISC, they could not serve as homing oligonucleotides to target viral genomes. Thus *r2d2*$^{-/-}$ mutant flies were also more susceptible to viral infection (Wang *et al.*, 2006). More recent experiments showed the ability to induce RNAi-mediated antiviral protection in *Drosophila* via virus-specific dsRNA immunization (Saleh *et al.*, 2009) (Fig. 16.3). Wild type flies infected with Sindbis-GFP virus two days after intrathoracic injection of SINV-specific dsRNA (dsSin1, dsSin2) are protected from SINV infection, whereas flies immunized with non-specific dsRNA (dsDCV) were susceptible to virus infection (Saleh *et al.*, 2009). This study also demonstrated systemic spread of RNAi-mediated antiviral protection (Saleh *et al.*, 2009) (see 'Systemic RNAi' and Fig. 16.3). Together, these studies demonstrated that RNAi acts against a broad spectrum of viruses as the insect antiviral defence system.

It is noteworthy that the antiviral role of Dcr-2 in *Drosophila* extends beyond the canonical RNAi antiviral pathway. Research examining one of the numerous *Drosophila* encoded genes with increased expression in response to DCV infection (Dostert *et al.*, 2005), *vago-2*, determined its expression was Dcr-2 dependent (Deddouche *et al.*, 2008). Although, DCV induced mortality of vago-2$^{-/-}$ flies did not exceed that of wild type flies, an increase in virus accumulation was observed in the fat body of these mutants (Deddouche *et al.*, 2008). This result resonates with the observation that Vago-2 expression is up-regulated in fat bodies of DCV infected transgenic reporter (*Vago::lacZ)* flies (Deddouche *et al.*, 2008). Double-stranded RNA is an intermediate of RNA viral replication in infected cells, but not a typical product of host gene expression. Therefore, in mammalian systems, dsRNA is recognized by pattern recognition receptors including endosomal Toll-3, and cytosolic Mda-5, LGP2 and RIG-I (Akira *et al.*, 2006). Intriguingly, phylogenetic analysis of the DExD/H-box helicase domain of Dcr-2 as compared to 389 proteins that share this domain, place it on the same phylogenetic branch as mammalian sensors of dsRNA (Deddouche *et al.*, 2008). It is fascinating that in insects Dcr-2 serves as an effector molecule within the antiviral RNAi pathway and potentially as a general sensor of viral infection. Future work in this arena may uncover addition overlap between RNAi, Toll, Imd, and JAK-STAT pathways.

The antiviral role of RNAi in the context of natural insect infections

The majority of early experiments aimed to investigate RNAi as an antiviral defence system in insects relied on intrathoracic injection of virus in order to precisely control infectious dose. While this method has been very useful, research aimed at understanding antiviral role of RNAi in the context of natural infection routes (e.g. through feeding) are critical to our understanding of the biological significance of antiviral RNAi. This work is particularly important, for understanding the role of RNAi in the context of persistent infections and arbovirus transmission.

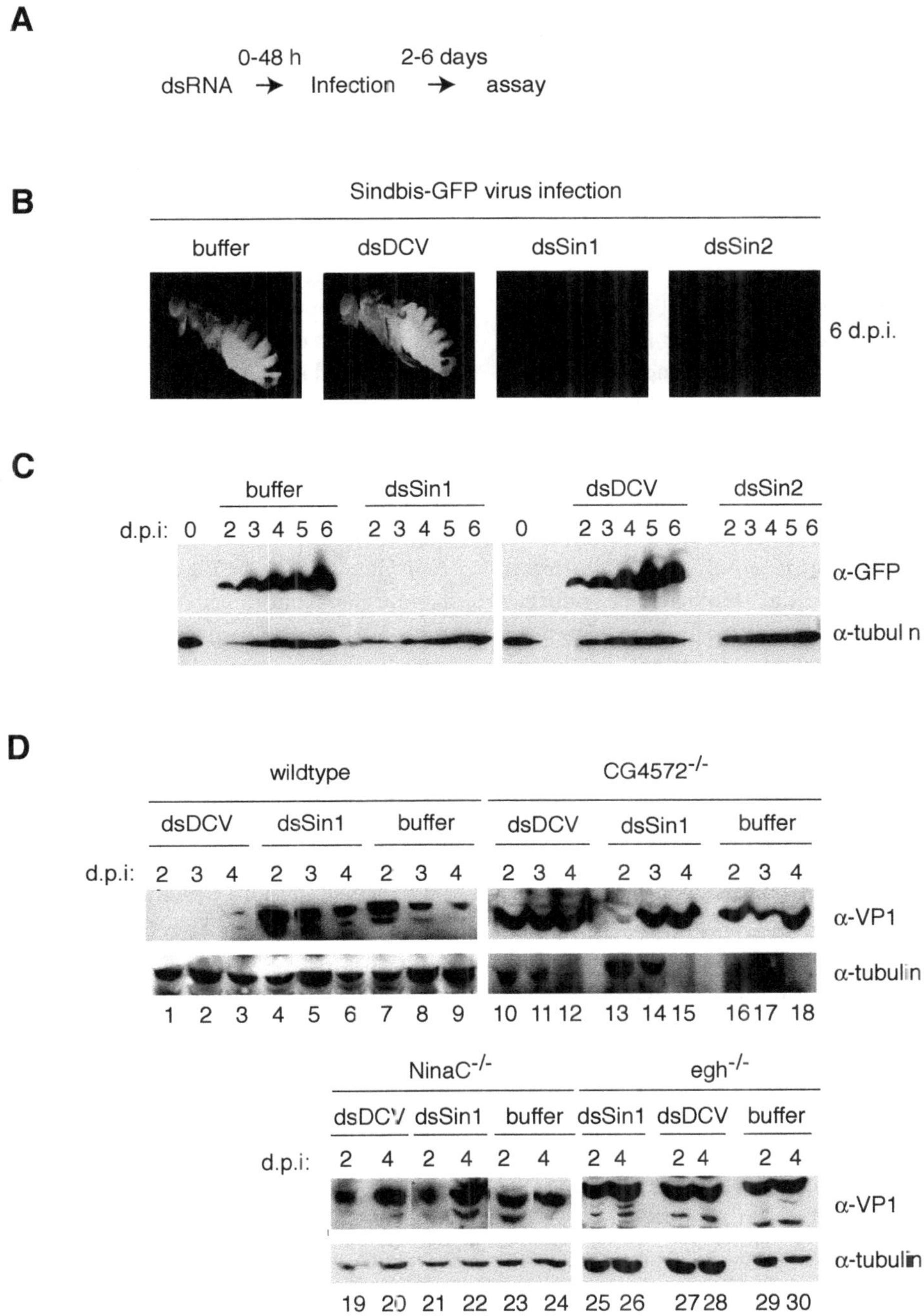

Figure 16.3 *In vivo* dsRNA immunization provides sequence-specific antiviral protection in *Drosophila*. (A) Immunization protocol. (B, C) Wild type flies infected with Sindbis-GFP viruses two days after intrathoracic injection of dsRNA against *Drosophila C virus* (dsDCV), against Sindbis virus non-structural proteins 1 and 2 (dsSin1) or Sindbis virus non-structural proteins 3 and 4 (dsSin2). Buffer: control injection. d.p.i.: days post infection. (B, C) Sindbis-GFP virus replication was monitored by GFP production. (B) Fluorescence images and (C) Western blot with an anti-GFP antibody. (D) Systemic spread of dsRNA follows virus infection and is essential for effective antiviral immunity. *Drosophila C virus* infection in wild type and homozygous mutant flies [egh^{EP804} ($egh^{-/-}$), $NinaC^{3}$ ($NinaC^{-/-}$) and $CG4572^{c05963}$ ($CG4572^{-/-}$)] inoculated with DCV specific or unrelated dsRNA. DCV replication was monitored by Western blot using an antibody directed against DCV capsid protein VP1. (Adapted from Saleh *et al*., 2009).

Arboviruses are responsible for many emerging diseases. Increased prevalence of arbovirus associated disease in naive human populations is expected as changes in global weather patterns affect the range of mosquito and bird species that harbour these pathogens (Peters, 2007).

Although mosquitoes utilize RNAi as an antiviral defence system, they still readily harbour and transmit alphaviruses (e.g. SINV, Eastern equine encephalitis virus) and flaviviruses (e.g. DENV, West Nile virus, yellow fever virus) to human populations. Interestingly, mosquito species differ in their abilities to transmit viruses; it may be that differences in the regulation and/or expression of the RNAi machinery account for vector differences (Campbell *et al.*, 2008; Cirimotich *et al.*, 2009; Sánchez-Vargas *et al.*, 2009; Sanchez-Vargas *et al.*, 2004). Increased understanding of the biological significance of antiviral RNAi may result in a significant reduction of arbovirus infections in human populations.

Two important arboviruses which have been utilized in a number of studies aimed at understanding the antiviral role of RNAi in the context of natural infections are SINV and DENV. SINV is an enveloped alphavirus with an 11.7-kb single-stranded, positive polarity RNA genome (Griffin, 2007). SINV is less pathogenic to humans (infections can result in fever, rash, and arthritis) than other alphaviruses such as Venezuelan, Western, and Eastern equine encephalitis viruses which although rare, can be deadly to humans (Griffin, 2007). SINV is transmitted by mosquitoes to human and bird hosts. DENV is a positive-sense single-stranded RNA flavivirus vectored by *Ae. aegypti* (Griffin, 2007). DENV 1–4 cause Dengue fever and Dengue haemorrhagic fever in infected individuals (Griffin, 2007). Both SINV and DENV are responsible for regional epidemic outbreaks of disease (Lindenbach *et al.*, 2007).

RNAi is important in reducing SINV titres in orally infected *Ae. aegypti* mosquitoes (Campbell *et al.*, 2008; Cirimotich *et al.*, 2009). Transient knock-down of *Ae. aegypti* RNAi proteins (Ago-2, Dcr-2, tudor staphylococcal endonuclease (TSN), a component of RISC; Caudy *et al.*, 2003) prior to SINV-eGFP infection increased viral accumulation (Campbell *et al.*, 2008). These results were consistent with earlier findings from co-injection of Ago-2 dsRNA and ONNV in *An. gambiae* (Keene *et al.*, 2004). Although, transient reduction of RNAi components in conjunction with SINV infection did not increase mosquito mortality, compromising the RNAi machinery by expressing the FHV-B2 protein from SINV constructs (SINV TE/3′ 2J/B2) did increase mosquito mortality; 83% dead by day 21 post infection compared to 11% mortality in SINV infected controls (Cirimotich *et al.*, 2009). FHV-B2 mediated suppression of RNAi in the context of SINV infection by intrathoracic injection in *Ae. aegypti* and two additional mosquito species, *Ae. albopictus* and *Culex tritaeniorhynchus*, yielded similar overall results (Cirimotich *et al.*, 2009). Differences, such as higher overall mortality of SINV TE/3′ 2J/B2 infected *Ae. aegypti* mosquitoes as compared to the other mosquito species and higher mortality of *Cx. tritaeniorhynchus* mosquitoes infected with SINV-TE/3′ 2J and SINV-TE/3′ 2J/GFP as compared with the mock infection controls, indicate that further investigation of species-specific antiviral RNAi responses is warranted (Cirimotich *et al.*, 2009).

The antiviral role of RNAi has been investigated in both wild type and transgenic strains of *Ae. aegypti* Higgs' white eye (HWE) mosquitoes infected via ingestion of a DENV containing blood meal (Franz *et al.*, 2006; Franz *et al.*, 2009; Sánchez-Vargas *et al.*, 2009). An RNAi based strategy was employed to engineer DENV2 resistant transgenic *Ae. aegypti* mosquitoes (Carb77) (Franz *et al.*, 2006). Carb77 mosquitoes express a DENV2 prM hairpin RNA under the control of a carboxypeptidase A promoter, which induces expression in midgut epithelial cells following a blood meal (Franz *et al.*, 2006). The majority of these transgenic mosquitoes limited DENV2 replication and exhibited reduced virus transmission capabilities (Franz *et al.*, 2006). Unfortunately, the DENV2 resistance phenotype of Carb77 mosquitoes diminished by the 14th generation and was gone by the 17th generation (Franz *et al.*, 2009). Investigation into this phenomenon revealed that although the transgene was not mutated, expression of the prM inverted-repeat RNA was no longer detectable (Franz *et al.*, 2009). Nonetheless, these experiments demonstrated that RNAi could efficiently prevent infection via a natural route of infection.

Detailed characterization of wild type *Ae. aegypti* mosquitoes infected with DENV2 via ingestion of an artificial blood meal revealed the prevalence of dsRNA substrates in the mosquito midgut, the generation of DENV2-specific siRNAs, and confirmed the importance of the RNAi machinery in mitigating natural DENV2 infection (Sánchez-Vargas *et al.*, 2009). Most interestingly, a series of DENV2 infection experiments comparing the outcomes of different parameters of viral infection over time from mosquitoes with reduced levels of Dcr-2, Ago-2, and R2D2, demonstrated that mosquitoes with compromised levels of the RNAi machinery exhibited decreased extrinsic incubation periods (EIP), the time from mosquito infection to transmission (Sánchez-Vargas *et al.*, 2009). The transmission competence of infected mosquitoes was assessed by analysing saliva deposited into feeding solutions for DENV2. By day 7-post ingestion (p.i.), the feeding solutions probed by mosquitoes with reduced Dcr-2 levels were 62.5% positive as compared to 11% of controls (Sánchez-Vargas *et al.*, 2009).

Together these results indicate that RNAi plays a critical role in mitigating arbovirus infection in mosquito vectors. Future research aimed at understanding the effects of tissue-specific expression of the RNAi machinery and the effects of strain dependant viral replication kinetics will provide additional insights into RNAi mediated control of insect virus infection (Campbell *et al.*, 2008; Cirimotich *et al.*, 2009; Sánchez-Vargas *et al.*, 2009).

The antiviral role of RNAi in non-model insects

Our knowledge of RNAi as an insect antiviral defence system is predominantly based on the results from studies of fruit flies and mosquitoes (order Diptera). Although, numerous reports demonstrate that a variety of non-model insects have functional RNAi pathways, the antiviral role of RNAi in these insects remains relatively underexplored. One notable exception is a study of the antiviral effects of dsRNA in *Apis mellifera* (Western honey bee) (Maori *et al.*, 2009). Many dicistroviruses and iflaviruses including Israeli acute paralysis virus (IAPV), Black queen cell virus (BQCV) and deformed wing virus (DWV), are known to infect honey bees (Chen and Siede, 2007). These honey bee viral infections can remain asymptomatic, cause paralysis, or result in death (Chen *et al.*, 2006; Chen and Siede, 2007). Recently, the US honey bee population has suffered dramatic overwinter losses (2006–2007, 32%; 2007–2008, 36%; 2008–2009, 29%) (van Engelsdorp *et al.*, 2009b). This phenomenon, known as colony collapse disorder (CCD), is defined as the death of the colony characterized by a rapid loss of adult bees relative to the number of young bees and brood, the absence of dead or dying bees in the area surrounding the hive, delayed invasion of pests (e.g. *Galleria mellonella* (wax moth) and *Aethina tumida* (small hive beetle), and delayed kleptoparasitism by neighbouring honey bees (Cox-Foster *et al.*, 2007; van Engelsdorp *et al.*, 2009a). Although no specific virus or microbe has been definitively shown to cause CCD, affected hives have an increased number of co-infections (viruses and *Nosema*, a microsporidia) compared with non-CCD affected colonies (van Engelsdorp *et al.*, 2009a). Recent results, examining the potential to utilize dsRNA as an antiviral treatment in *A. mellifera* are promising from both the applied and basic science perspectives (Maori *et al.*, 2009). Honey bees (200 bee mini-colonies) were feed-inoculated with either IAPV alone or IAPV in conjunction with either IAPV-specific dsRNA (dsRNA-IAPV) or non-specific (GFP) dsRNA (Maori *et al.*, 2009). Colony survival of IAPV inoculated mini-hives was reduced (20% at day 8 p.i.) compared to honey bee colonies that were co-fed dsRNA-IAPV and IAPV (67% at day 8 p.i.) (Maori *et al.*, 2009). Quantitative PCR analysis indicated that dsRNA-IAPV treated hives had approximately 2-logs less IAPV genomic RNA as compared to those hives fed IAPV alone (Maori *et al.*, 2009). An RNAi mediated mechanism of colony protection was indicated by the detection of IAPV specific siRNAs (21–25 bp) from honey bees treated with IAPV-dsRNA (Maori *et al.*, 2009). It will be interesting to further investigate the mechanism(s) by which exogenous dsRNA is internalized by gut epithelial cells and triggers antiviral responses. Honey bees encode a SID-1 homologue, in *C. elegans* this gene was found to participate in systemic spread of the RNAi response (see below), and endogenous gene expression has been decreased in larva by soaking them in dsRNA (Aronstein *et al.*, 2006; Kucharski *et al.*, 2008; Schluns and Crozier, 2007). Ingestion of dsRNA resulting in decreased gene expression has also been reported in other insects (e.g. *Rhodnius prolixus*, triatome bugs; *Epiphyas postvittana*, light brown apple moth) (Araujo *et al.*, 2006; Turner *et al.*, 2006). Research in *D. melanogaster* demonstrated that the endocytic pathway is critical to RNAi mediated antiviral defence (Saleh *et al.*, 2006); it will be interesting to determine the degree of conservation between dsRNA uptake mechanisms in *Drosophila* (Diptera) and *A. mellifera* (Hymenoptera).

RNA viruses, the dsRNA replication intermediate and secondary structures

The antiviral role of RNAi has been clearly established in both *Drosophila* and *Ae. aegypti*. Although the mechanistic details are not well understood. Viral-derived siRNAs (viRNAs) have been detected in several studies, but the viral substrate of Dcr-2 is not known. Genetic studies have demonstrated the requirement of Dcr-2 for antiviral RNAi and biochemical studies have shown that long dsRNA is a substrate for Dcr-2 in many species (Saleh *et al.*, 2006; Zhang *et al.*, 2002). Based on these results, two forms of viral dsRNA are logical candidate substrates for Dcr-2; viral replication intermediates and viral RNA secondary structure. Recently, high throughput sequencing technologies have been utilized to gain insight on the form of viral RNA targeted by the RNAi machinery. Several studies have analysed the production of viRNAs generated during the course of infection both in cell culture and *in vivo* using a variety of viruses of different families.

One such study determined the small RNA profile produced in *Drosophila* S2 cells infected with an FHV mutant (FHVΔB2) deficient for the RNAi suppressor protein B2 (Aliyari *et al.*, 2008). Sequence analysis of viRNAs in this study indicated a slight bias for the positive strand and a strong bias for the 5′ end of the positive strand of RNA1. The authors speculate that this profile represents dicing of the dsRNA replication intermediate of RNA1 during positive-strand synthesis. Dcr-2 targeting of this region generated equal amounts of positive and negative strand viRNAs beginning at the 5′ end of the positive strand (Aliyari *et al.*, 2008). A subsequent study analysed the small RNA profile of populations of *Drosophila* S2 cells persistently infected with FHV (Flynt *et al.*, 2009). This profile agreed with the previous study, indicating a slight bias towards the positive strand and a strong bias towards the 5′ end of the positive strand of RNA1. It was also observed that cloned viRNAs were clustered in discreet regions of the viral genome. Curiously, these viRNAs were not active against a luciferase reporter with homologous sequence engineered into the 3′ UTR (Flynt *et al.*, 2009). Both FHV studies analysed viRNA RISC association and 3′ end chemical modification. 2′O-methylation is a characteristic of siRNAs loaded into Ago-2-RISC (Horwich *et al.*, 2007). viRNAs from FHV lacking B2 were strongly associated with Ago-2 and accordingly, showed some resistance to beta-oxidation, indicating that their 3′ ends were 2′O-methylated (Aliyari *et al.*, 2008). Interestingly, the viRNAs generated during latent infection of S2 cells with wild type FHV showed a very different pattern (Flynt *et al.*, 2009). These viRNAs were very weakly associated with Ago-2 and showed no resistance to beta-oxidation. These differences are likely due to one of two major differences in these studies. First, these investigations involved different viral constructs, an FHV-B2 deficient mutant and wild type FHV, respectively. The use of an FHV-B2 deficient mutant likely explains the higher abundance of viRNAs and in turn increased RISC association. A second major difference is that one study analysed viRNAs generated during an acute infection, whereas the other analysed viRNAs generated during a latent infection.

The first high throughput sequence analysis of viRNA populations in insects came from *Ae. aegypti* mosquitoes infected with SINV via injection (Myles *et al.*, 2008). Small RNA populations were sequenced from wild type SINV or transgenic SINV expressing FHV-B2 (SINV-B2). Sequence analysis revealed reduced accumulation of viRNAs from mosquitoes infected with SINV-B2 compared to SINV

(Myles *et al.*, 2008). In this study, there was a significant bias towards viRNAs derived from the positive strand, whereas previous studies of FHV infection in *Drosophila* S2 cells indicated roughly equal ratios of positive and negative strand viRNA (Aliyari *et al.*, 2008; Flynt *et al.*, 2009). In addition, viRNAs identified in the mosquito study were enriched in equally dispersed discreet regions of the viral genome (Myles *et al.*, 2008). Likewise, sequencing results from an analogous study showed that viRNA populations produced during West Nile virus infection of *Ae. aegypti* (Brackney *et al.*, 2009) are derived from equally dispersed, discrete regions of the viral genomic strand.

Sequence analysis of viRNA populations generated during viral infection aid in understanding the mechanism of antiviral RNAi, but experimental caveats associated with this approach may affect the data. For example, the form (guide or passenger strand) of the sequenced siRNA cannot be determined by this technique. Another difficulty in determining how sequenced viRNAs relate to the original substrate is due to the unknown half-life of small RNAs. To date, viRNA sequencing methods have employed adaptor ligation coupled with polymerase chain reaction (PCR) amplification which may introduce bias.

The siRNA pathway in insects: cell and non-cell autonomous functions

Plants were the first organisms known to use RNAi as a virus resistance mechanism (Dougherty and Parks, 1995). At the turn of the new century, the RNAi machinery of insect species, such as *D. melanogaster* (Galiana-Arnoux *et al.*, 2006; Li *et al.*, 2002; van Rij *et al.*, 2006; Wang *et al.*, 2006) and the mosquito *An. gambiae* (Keene *et al.*, 2004) was also found to be required for antiviral immune response. During infection, viruses can multiply (within a few hours) and spread (within a few days) throughout the entire host organism. To efficiently control viral infection, most antiviral immune defences in multicellular organisms rely on a two step process: (i) a cell-autonomous response limiting viral lytic effects in infected cells and (ii) a non- cell-autonomous response which impedes virus spread to uninfected tissues. How are these antiviral features in multicellular organisms implemented in an RNAi-based antiviral response?

In plants, intercellular RNA trafficking is involved in development (Lucas *et al.*, 1995) and virus infection (Ding *et al.*, 2005). Plant cells are surrounded by thick cellulose cell walls preventing cell migration and limiting simple diffusion via the extracellular media. The specificity of multicellular organization in plants has facilitated the investigation of intercellular signalling and macromolecular trafficking, including siRNAs. In plants, intercellular trafficking relies heavily on intercytoplasmic channels called plasmodesmata (Kalantidis *et al.*, 2008). Although insects are devoid of such structures, insect cells are readily amenable to extracellular dsRNA exposure to trigger a sequence-specific RNAi response (Boutros *et al.*, 2004).

dsRNA uptake

The mechanism for dsRNA uptake in insect cells was investigated in *Drosophila* S2 cells by an RNAi of RNAi approach, where candidate genes are knocked down by a first exposure to an endogenous gene-specific dsRNA library and then tested for RNAi activity against an exogenous reporter gene when soaked with the corresponding dsRNA (Saleh *et al.*, 2006; Ulvila *et al.*, 2006). This approach revealed that dsRNA uptake in insect cells requires many components of the receptor mediated endocytic pathway, such as elements of clathrin coated vesicles, cytoskeleton proteins, oligomeric Golgi complex, cellular motors and vesicle mediated transport proteins (e.g. the vacuolar H^+ ATPase). Fluorochrome-labelled dsRNA molecules applied to S2 cells also confirm the transport of dsRNA to vesicles after uptake. Although these first screens failed to identify a dsRNA-specific receptor, a second RNAi experiment targeting multiple receptors, coupled with a pharmacological assay demonstrated that a combination of scavenger receptors is responsible for cellular uptake of dsRNA (Saleh *et al.*, 2006). Scavenger receptors do not have highly specific ligands but instead recognize repetitive molecular signatures and play an important role in insect immunity as they are involved in pathogen recognition and bacteria phagocytosis (Erturk-Hasdemir and Silverman, 2005).

The mechanisms of systemic RNAi in animals have been best characterized in the model organism *C. elegans* (Boisvert and Simard, 2008). In the worm, multispan transmembrane proteins thought to act as dsRNA receptors and named SID (for Systemic Interference Defective), are essential for RNAi signal spreading between cells. Interestingly, *Sid*-1 orthologues are found in some Coleoptera (e.g. *Tribolium castaneum*), Hemiptera (e.g. *Aphis gossypii*), Hymenoptera (e.g. *A. mellifera*), Lepidoptera (e.g. *Bombyx mori*) and Orthoptera (e.g. *Schistocerca americana*) but not in Diptera species (e.g. *D. melanogaster* and *An. gambiae*) (Xu and Han, 2008). Despite this difference between *D. melanogaster* and *C. elegans*, most of the endocytosis pathway components involved in dsRNA uptake in *Drosophila* are also required for RNAi spread in worms. Therefore, downstream to the dsRNA binding at the cell surface, the endocytosis pathway is evolutionarily conserved to mediate extracellular dsRNA transport to the intracellular RNAi machinery in worms and flies.

Extracellular dsRNA

The conservation across different phyla of mechanisms linking exogenous dsRNA uptake and RNAi response raises the question of the extracellular dsRNA origin. In plants and worms (Sijen *et al.*, 2001; Vaistij *et al.*, 2002), *in vivo* expression of an siRNA producing transgene that targets an endogenous gene in a subset of cells induces siRNA-specific silencing in neighbouring wild-type cells. This is not the case in the best studied insect model system, *D. melanogaster* (Roignant *et al.*, 2003). Interestingly, SID proteins are not required for systemic RNAi signal export in worm (Kaplan *et al.*, 2003) and the mechanisms by which dsRNA molecules gain access to the extracellular media have not been identified in either worms or insects.

Systemic RNAi

Several studies assessing the potential of RNAi for pest control (for review see Huvenne and Smagghe, 2009) have shown that several insect species are amenable to dsRNA feeding to induce gene-specific RNAi. This RNAi response triggered by uptake of dsRNA from the environment has been mostly observed in the digestive tract, the tissues primarily in contact with the external dsRNA molecules. This non-cell autonomous RNAi has thus been coined environmental RNAi. Moreover, *in vivo* dsRNA injections in fruit fly and mosquito can induce down-regulation of endogenous gene expression and viral replication (Dzitoyeva *et al.*, 2001; Keene *et al.*, 2004). Therefore, despite the apparent lack of intercellular spread for endogenous RNAi signals in insects, systemic RNAi can be triggered by exogenous molecules. Extracellular virus-derived RNAs (e.g. viral replication intermediates) are released from virus-infected cell lysis. Confirmation that the endocytosis pathway, involved in dsRNA uptake, plays a role in antiviral systemic RNAi was demonstrated by investigating the outcome of viral infection in *Drosophila* mutants deficient in endocytic components (Saleh *et al.*, 2009). Viruses accumulated at an increased rate in *Drosophila* deficient for genes involved in dsRNA uptake and induced host death at normally sublethal dose (Saleh *et al.*, 2009). Wild-type flies accumulated less virus than endocytosis pathway mutants (Saleh *et al.*, 2009) (Fig. 16.3). In accordance with their hypersensitivity to virus infection, dsRNA uptake mutant flies are unable to mount an RNAi-based antiviral immune response when injected with virus-specific dsRNA prior to or in parallel with virus infection (Fig. 16.3D). Notably, dsRNA uptake mutant flies maintain cell-autonomous RNAi and mount proper immune responses against bacterial pathogens. Lastly, during infection with a reporter gene-expressing virus, a virus-derived RNAi signal can spread to uninfected cells and specifically silence the same host-encoded reporter gene (Saleh *et al.*, 2009).

These data have led to a model for systemic RNAi in *Drosophila* immunity where upon viral infection, virus-derived dsRNAs are first released from dead cells due to the virus lytic effects or through a yet unknown host secretion process, spread to neighbouring cells and possibly more distal tissues where they are internalized via endocytosis and subsequently used to prime the RNAi machinery of the receiving cell against the virus (Fig. 16.4) (Saleh *et al.*, 2009).

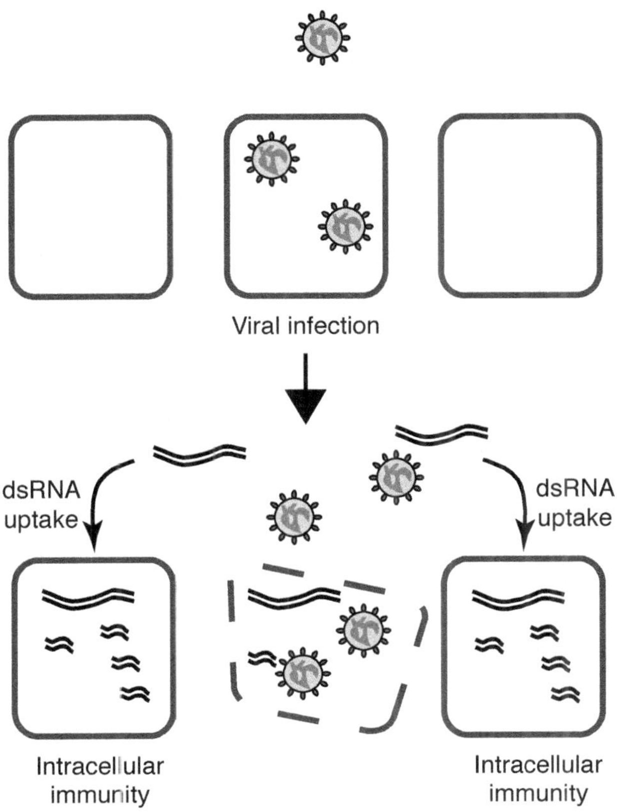

Figure 16.4 A model for systemic RNAi viral immunity in *Drosophila melanogaster*. Upon viral infection, virus-specific dsRNAs (e.g. replication intermediates or RNA structures require for replication) accumulate during the initial rounds of virus replication. Following cell death or lysis, dsRNAs are released and spread locally and/or systemically. dsRNA is then taken up and processed by uninfected cells to protect them from subsequent infection, thereby preventing virus propagation. Adapted from Saleh *et al.* (2009).

Viral suppressors of RNAi

In insects, antiviral RNAi provides robust protection against a wide variety of viruses; however, many viruses encode suppressors of RNAi to modulate this antiviral response. Viral suppressors of RNAi (VSR) have independently evolved in many different virus families, indicating the evolutionary pressure of the RNAi system. Co-evolution of insect antiviral immunity and viral suppression/evasion has resulted in rapid evolution of the *Drosophila* RNAi machinery (Obbard *et al.*, 2006). Here we review three well-characterized suppressors of insect antiviral RNAi: the B2 protein of FHV (*Nodaviridae, Alphanodavirus*), DCV-1A of DCV (*Dicistroviridae, Cripavirus*), and CrPV-1A of CrPV (*Dicistroviridae, Cripavirus*).

FHV-B2

The FHV B2 protein is a 106 amino acid dsRNA binding protein encoded by RNA1 of the bipartite positive-stranded RNA viral genome. Discovery of FHV-B2-mediated RNAi suppression in *Drosophila* was simultaneous with the discovery of insect antiviral RNAi (Li *et al.*, 2002). Studies in FHV infected *Drosophila* S2 cells determined that FHV-B2 was required for viral accumulation. However, in *Drosophila* S2 cells with transient knockdown of core RNAi components (e.g. Ago-2 and Dcr-2), FHV-B2 was not required for viral accumulation. These results in combination with experiments which demonstrated the ability of FHV-B2 to suppress RNA silencing in plants, classified this protein as a bonafide suppressor of RNAi.

Subsequent biochemical and structural studies revealed that FHV-B2 encodes a non-canonical dsRNA binding domain that binds to both long dsRNA and siRNAs with high affinity (Chao *et al.*, 2005). Expression of FHV-B2 *in vitro* inhibits Dcr–2 cleavage of long dsRNA into siRNAs (Chao *et al.*, 2005; Lingel and Sattler, 2005). The dsRNA binding domain of FHV-B2 is located in the N-terminal region (Chao *et al.*, 2005). Structural studies (X-ray crystallography in combination with NMR) indicate that B2 acts as a homodimer in solution (Chao *et al.*, 2005; Lingel and Sattler, 2005). Each monomer consists of three α-helices. The first two helices (α1 and α2) fold into a helix-turn-helix hairpin structure, whereas the third, shorter α helix (α3) packs perpendicular to α1 and α2. The homodimer is formed by a head–to-tail interaction of two of the monomers, involving interactions spanning all three helices. In the protein–RNA complex, the antiparallel α2α2′ helices of the homodimer form an extended RNA binding surface parallel to A-form dsRNA that covers two minor grooves and the intervening major groove (Chao *et al.*, 2005). It is widely assumed that RNAi suppressors which bind dsRNA protect the viral substrate from Dcr-2, although it is not clear what form of viral RNA Dcr-2 recognizes *in vivo*. Recent work analysing the mechanism of FHV-B2 has shown that in the absence of FHV-B2 small RNAs are generated from the 5′ end of RNA1 (Aliyari *et al.*, 2008). These virally derived siRNAs (viRNAs) are equally distributed from the positive and negative strands indicating that the viral replicative double-stranded intermediate is targeted by Dcr-2. In addition, FHV-B2 co-immunoprecipitates with the RdRp of the virus (Aliyari *et al.*, 2008). Therefore, the authors hypothesized that FHV-B2 localizes to the 5′ end of nascent positive-strand RNA during viral replication, preventing Dcr-2 from targeting this potential dsRNA substrate (Aliyari *et al.*, 2008). Although more studies are needed, this work presented some very enticing evidence for a mechanism of RNAi suppression by a dsRNA binding protein and has provided some evidence for replicative dsRNA intermediates as a viral substrate for Dicer.

Binding of dsRNA may not be the entire story for FHV-B2. In armyworm (*Spodoptera frugiperda*) cell culture (Sf21 cells), FHV-B2 was shown to interact with the PAZ domain of a Dicer-like protein (Singh *et al.*, 2009). In this system FHV-B2 co-fractionated with this Dicer-like protein and immunoprecipitated it from cell extract. In addition, the FHV-B2 interacted with recombinant PAZ domain from both Dcr-1 and Dcr-2 of *D. melanogaster*. The interaction of FHV-B2 and Dicer is intriguing; it may be that FHV-B2 is a dual functional suppressor of RNAi. However, the Dicer-like protein in Sf21 cells requires further characterization and the interaction of FHV-B2 and full-length *Drosophila* Dicer has not been demonstrated. FHV-B2 is a tremendous tool for investigating both antiviral RNAi and viral suppression of RNAi.

DCV-1A

DCV was originally isolated from wild populations of *Drosophila* and is one of the most studied natural pathogens of *Drosophila* (see Chapter 9). DCV is a positive-stranded RNA virus in the *Dicistroviridae* family, which is characterized by two internal ribosomal entry sites (IRES) regulating the translation of two open reading frames. DCV-1A is a suppressor of RNAi encoded in 5′ end of the first ORF (van Rij *et al.*, 2006). DCV-1A was discovered utilizing a *Drosophila* S2 cell RNAi reporter assay (van Rij *et al.*, 2006). DCV infected S2 cells failed to silence a luciferase reporter when treated with luciferase-specific dsRNA. This suppression of RNAi was mapped to the first 99 amino acids of the N-terminal region of the 5′ ORF and named DCV-1A (van Rij *et al.*, 2006). The C-terminal region of DCV-1A contains homology to the canonical dsRNA binding domain (dsRBD) found in many cellular proteins including Dcr-2 and R2D2. Crystal structures and NMR data from several dsRBD family members display a α1β1β2β3α2 fold, in which the two α helices lie on either side and pack against a three-stranded antiparallel β-sheet (Ryter and Schultz, 1998). dsRBDs are predicted to contact A-form dsRNA at three key regions. The α2 helix is thought to interact with the dsRNA major groove whereas the α1 helix and the loop in between the β1 and β2 sheets are believed to contact the flanking two minor grooves. Mutational analysis determined that residues structurally predicted to contact dsRNA are critical to dsRNA binding and for the ability of DCV-1A to suppress RNAi (van Rij *et al.*, 2006). In addition, it was established that DCV-1A displays a strong bias for binding long dsRNA and has only a weak affinity for 21 bp siRNAs *in vitro* (van Rij *et al.*, 2006), in contrast to FHV-B2 which binds strongly to both long dsRNA and siRNAs. Addition of DCV-1A to S2 cell extracts also inhibits Dcr-2 cleavage of long dsRNA in a similar fashion to FHV-B2. Based on these *in vitro* experiments, it is believed that DCV-1A acts to suppress RNAi by binding to and protecting a virally formed substrate of Dicer.

The identification of two independently evolved dsRNA binding proteins (FHV-B2 and DCV-1A) that act as suppressors of RNAi suggests that this may be a general strategy among viruses to suppress RNAi. Indeed, in plants a plethora of dsRNA binding proteins have been shown to act as suppressors of RNAi (Merai *et al.*, 2006). However, the mechanisms behind this type of suppression remain largely unexplored. Although many speculate that these dsRNA binding proteins act to protect viral dsRNA from being targeted by Dicer, there is little *in vivo* evidence.

CrPV-1A

CrPV is the most closely related member of DCV in the *Dicistroviridae* family. CrPV infected flies deficient in components of the RNAi machinery also exhibit increased viral RNA loads and an increased mortality rate. However, CrPV displays a much higher rate of mortality than DCV in orally inoculated flies and in cell culture (Manousis and Moore, 1987). CrPV also encodes a suppressor of RNAi at the N-terminus of its 5′ ORF (CrPV-1A) (Nayak *et al.*, 2010; Wang *et al.*, 2006). However, alignment of DCV and CrPV in this suppressor-encoding region revealed no significant homology as compared to downstream amino acid sequences. CrPV-1A suppresses RNAi initiated by both long dsRNA and siRNAs; but, unlike FHV-B2 or DCV-1A, CrPV-1A is unable to bind dsRNA or suppress Dcr-2 activity *in vitro* (Nayak *et al.*, 2010). These data are supported by the detection of CrPV specific viRNA accumulation by Northern Blot (Wang *et al.*, 2006), indicating that CrPV does not suppress RNAi at the Dicer processing stage. It was demonstrated that CrPV-1A specifically interacts with Ago-2 within assembled holo-RISC, without modifying RISC composition, and that this interaction prevents RISC cleavage of target mRNAs (Nayak *et al.*, 2010).

This novel mechanism of RNAi suppression was shown to affect both strength of RNAi suppression and outcome of infection (Nayak *et al.*, 2010). CrPV-1A is a more potent suppressor of RNAi than DCV-1A in tests using a luciferase reporter targeted by long dsRNA in cell culture and *in vivo* using transgenic flies expressing a hairpin targeting the *white* gene and measuring the changes in pigmentation of the eye (Nayak *et al.*, 2010). It was hypothesized that the efficient suppression of RNAi by CrPV-1A stems from a direct and specific inhibition of the catalytic component of RISC, Ago-2 (Nayak *et al.*, 2010). This strategy of RNAi suppression is significantly more potent than DCV-1A suppression of RNAi through binding of long dsRNA. These distinct strategies employed by viruses to repress the RNAi machinery are correlated with the differences in viral pathogenesis between DCV and CrPV (Nayak *et al.*, 2010). To test the effect of suppressor potency on the outcome of infection, recombinant Sindbis virus was engineered to express either DCV-1A, CrPV-1A, or a non-functional truncated CrPV-1A (CrPV108) as a control (SINV-DCV, SINV-CrPV, or SINV-CrPV108, respectively). SINV-CrPV was shown to cause a dramatic increase in mortality of *Drosophila* following intrathoracic injection when compared to SINV-DCV or SINV-CrPV108 (Nayak *et al.*, 2010).

Differences in viral RNAi suppressor potency and mechanism affect the outcome of infection. The differences in the co-evolution of specific virus–host pairs present a very interesting area of research. To date, there are no characterized arbovirus suppressors of RNAi. The recent generation of a transgenic mosquito sensor for RNAi functionality may help reveal VSRs, as well as other RNAi modulating proteins (Adelman *et al.*, 2008).

Virus–host interactions; RNAi and persistent infections

Viruses must infect cells and commandeer the host cellular machinery in order to replicate and package their genomes into new viral capsids. On the other hand, the host organism employs several immunologic defence mechanisms in order to avoid and combat viral infection. These goals appear mutually exclusive; however, many experiments in *Drosophila* have demonstrated that host and virus have evolved to strike a balance between viral replication and host defence.

DCV was originally isolated from several stocks of *D. melanogaster*. Although DCV causes a high degree of mortality when injected into or fed to naive flies (Galiana-Arnoux *et al.*, 2006; van Rij *et al.*, 2006; Wang *et al.*, 2006), the progeny of *Drosophila* that survive this acute infection show little deleterious effects from harbouring the virus (Thomas-Orillard, 1984). Populations of *D. melanogaster* that are persistently infected show increased fecundity and decreased time of development; however, mortality of larva is slightly higher in DCV-infected populations when compared to uninfected populations (Thomas-Orillard, 1984).

Persistent infection of *Drosophila* has also been studied in cell culture revealing some of the key molecular components involved. Although acute FHV infection of *Drosophila* S2 cells causes widespread apoptosis, in many laboratories S2 cell lines are persistently infected with FHV and exhibit only mild cytopathic effect (Settles and Friesen, 2008). In order to explore this phenomenon, the FHV genome was sequenced to determine if deleterious mutations in the viral genome were responsible for persistent infection. One hypothesis was that mutations within the FHV-B2 suppressor-encoding region, which is required for rapid accumulation of viral RNA, might account for reduced FHV pathogenicity. However, no obvious mutations were found. In addition, FHV isolated from persistently infected S2 cells lead to acute infections in uninfected S2 cells, indicating the virus was completely viable (Flynt *et al.*, 2009). RNAi appears to play a role in maintenance of latent infection, as knockdown of host RNAi machinery results in an increase in viral RNA accumulation; Dcr-2 knockdown had the greatest effect on viral replication. Deep sequencing of small RNA populations from persistently infected S2 cells revealed viral specific siRNAs were generated from equally dispersed discreet regions of both the positive and negative virus strands (Flynt *et al.*, 2009). However, these siRNAs failed to silence a luciferase reporter with complimentary sequence to the most abundant siRNA populations (Flynt *et al.*, 2009). The generation of viRNAs that cannot target complimentary RNAs, combined with the strong effect specifically of Dcr-2 knockdown in persistently infected cells, suggests the balance of persistent infection is maintained at the production and loading of virally derived siRNAs into RISC. These cases of persistent infection illustrate an exquisite equilibrium between the host RNAi machinery and the viral RNAi suppressor, a balance that determines the pathogenic outcome of infection, virus survival, and virus spread. The virus and host mechanisms responsible for maintaining the delicate balance achieved in persistent infections in which both parties thrive is a fascinating avenue of research.

Concluding remarks

Co-evolution of virus and host is one of the most interesting aspects of biology. In this chapter, we have focused on the recently discovered role of RNAi in insect antiviral immunity. Although RNAi is an ancient pathway, conserved among species from different kingdoms (Fungi, Animalia, Plantae), we are just beginning to understand the biological implications and mechanistic details of this pathway in the context of viral infection. As our understanding of the pathways and mechanisms involved in insect immunity expands, so do the opportunities to employ insects as model systems to examine the general principles of RNAi and the co-evolution of hosts and their pathogens. In addition, understanding fundamental mechanisms of insect immune systems may have broad societal impact (e.g. decreasing the global incidence of arbovirus-associated diseases and/or the development of treatments for agriculturally important insects).

References

Adelman, Z.N., Anderson, M.A.E., Morazzani, E.M., and Myles, K.M. (2008). A transgenic sensor strain for monitoring the RNAi pathway in the yellow fever mosquito, *Aedes aegypti*. Insect Biochem. Mol. Biol. *38*, 705–713.

Adelman, Z.N., Sanchez-Vargas, I., Travanty, E.A., Carlson, J.O., Beaty, B.J., Blair, C.D., and Olson, K.E. (2002). RNA silencing of dengue virus type

2 replication in transformed C6/36 mosquito cells transcribing an inverted-repeat RNA derived from the virus genome. J. Virol. 76, 12925–12933.

Akira, S., Uematsu, S., and Takeuchi, O. (2006). Pathogen recognition and innate immunity. Cell *124*, 783–801.

Aliyari, R., Wu, Q.F., Li, H.W., Wang, X.H., Li, F., Green, L.D., Han, C.S., Li, W.X., and Ding, S.W. (2008). Mechanism of induction and suppression of antiviral immunity directed by virus-derived small RNAs in *Drosophila*. Cell Host Microbe 4, 387–397.

Amdam, G.V., Simoes, Z.L., Guidugli, K.R., Norberg, K., and Omholt, S.W. (2003). Disruption of vitellogenin gene function in adult honeybees by intra-abdominal injection of double-stranded RNA. BMC Biotechnol. 3, 1.

Ameres, S.L., Martinez, J., and Schroeder, R. (2007). Molecular basis for target RNA recognition and cleavage by human RISC. Cell *130*, 101–112.

Araujo, R.N., Santos, A., Pinto, F.S., Gontijo, N.F., Lehane, M.J., and Pereira, M.H. (2006). RNA interference of the salivary gland nitrophorin 2 in the triatomine bug *Rhodnius prolixus* (Hemiptera: Reduviidae) by dsRNA ingestion or injection. Insect Biochem. Mol. Biol. *36*, 683–693.

Aravin, A.A., Hannon, G.J., and Brennecke, J. (2007). The Piwi-piRNA pathway provides an adaptive defense in the transposon arms race. Science *318*, 761–764.

Aronstein, K., Pankiw, T., and Saldivar, E. (2006). SID-I is implicated in systemic gene silencing in the honey bee. J. Api. Res. *45*, 20–24.

Bartel, D.P. (2004). MicroRNAs: genomics, biogenesis, mechanism, and function. Cell *116*, 281–297.

Bernstein, E., Caudy, A.A., Hammond, S.M., and Hannon, G.J. (2001). Role for a bidentate ribonuclease in the initiation step of RNA interference. Nature *409*, 363–366.

Boisvert, M.E., and Simard, M.J. (2008). RNAi pathway in *C. elegans*: the argonautes and collaborators. Curr. Top. Microbiol. Immunol. *320*, 21–36.

Boutros, M., Kiger, A.A., Armknecht, S., Kerr, K., Hild, M., Koch, B., Haas, S.A., Paro, R., and Perrimon, N. (2004). Genome-wide RNAi analysis of growth and viability in *Drosophila* cells. Science *303*, 832–835.

Brackney, D.E., Beane, J.E., and Ebel, G.D. (2009). RNAi targeting of West Nile Virus in mosquito midguts promotes virus diversification. PLoS Pathog. *5*, e1000502.

Brennecke, J., Malone, C.D., Aravin, A.A., Sachidanandam, R., Stark, A., and Hannon, G.J. (2008). An epigenetic role for maternally inherited piRNAs in transposon silencing. Science *322*, 1387–1392.

Campbell, C.L., Keene, K.M., Brackney, D.E., Olson, K.E., Blair, C.D., Wilusz, J., and Foy, B.D. (2008). *Aedes aegypti* uses RNA interference in defense against Sindbis virus infection. BMC Microbiol. *8*, 47.

Caplen, N.J., Zheng, Z.L., Falgout, B., and Morgan, R.A. (2002). Inhibition of viral gene expression and replication in mosquito cells by dsRNA-triggered RNA interference. Mol. Ther. *6*, 243–251.

Caudy, A.A., Ketting, R.F., Hammond, S.M., Denli, A.M., Bathoorn, A.M., Tops, B.B., Silva, J.M., Myers, M.M., Hannon, G.J., and Plasterk, R.H. (2003). A micrococcal nuclease homologue in RNAi effector complexes. Nature *425*, 411–414.

Chao, J.A., Lee, J.H., Chapados, B.R., Debler, E.W., Schneemann, A., and Williamson, J.R. (2005). Dual modes of RNA-silencing suppression by Flock House virus protein B2. Nat. Struct. Mol. Biol. *12*, 952–957.

Chen, Y.P., Pettis, J.S., Collins, A., and Feldlaufer, M.F. (2006). Prevalence and transmission of honeybee viruses. Appl. Environ. Microbiol. *72*, 606–611.

Chen, Y.P., and Siede, R. (2007). Honey bee viruses. Adv. Virus Res. *70*, 33–80.

Cirimotich, C.M., Scott, J.C., Phillips, A.T., Geiss, B.J., and Olson, K.E. (2009). Suppression of RNA interference increases alphavirus replication and virus-associated mortality in *Aedes aegypti* mosquitoes. BMC Microbiol. *9*, 49.

Costa, A., Jan, E., Sarnow, P., and Schneider, D. (2009). The Imd pathway is involved in antiviral immune responses in *Drosophila*. PLoS ONE *4*, e7436.

Cox-Foster, D.L., Conlan, S., Holmes, E.C., Palacios, G., Evans, J.D., Moran, N.A., Quan, P.L., Briese, T., Hornig, M., Geiser, D.M., *et al.* (2007). A metagenomic survey of microbes in honey bee colony collapse disorder. Science *318*, 283–287.

Cruz, J., Mane-Padros, D., Belles, X., and Martin, D. (2006). Functions of the ecdysone receptor isoform-A in the hemimetabolous insect *Blattella germanica* revealed by systemic RNAi *in vivo*. Dev Biol. *297*, 158–171.

Deddouche, S., Matt, N., Budd, A., Mueller, S., Kemp, C., Galiana-Arnoux, D., Dostert, C., Antoniewski, C., Hoffmann, J.A., and Imler, J.L. (2008). The DExD/H-box helicase Dicer-2 mediates the induction of antiviral activity in *Drosophila*. Nat. Immunol. *9*, 1425–1432.

Ding, B., Itaya, A., and Zhong, X. (2005). Viroid trafficking: a small RNA makes a big move. Curr. Opin. Plant Biol. *8*, 606–612.

Dionne, M.S., and Schneider, D.S. (2008). Models of infectious diseases in the fruit fly *Drosophila melanogaster*. Dis. Model Mech. *1*, 43–49.

Dostert, C., Jouanguy, E., Irving, P., Troxler, L., Galiana-Arnoux, D., Hetru, C., Hoffmann, J.A., and Imler, J.L. (2005). The Jak-STAT signaling pathway is required but not sufficient for the antiviral response of *Drosophila*. Nat. Immunol. *6*, 946–953.

Dougherty, W.G., and Parks, T.D. (1995). Transgenes and gene suppression: telling us something new? Curr. Opin. Cell Biol. *7*, 399–405.

Dzitoyeva, S., Dimitrijevic, N., and Manev, H. (2001). Intra-abdominal injection of double-stranded RNA into anesthetized adult *Drosophila* triggers RNA interference in the central nervous system. Mol. Psychiatr. *6*, 665–670.

Elbashir, S.M., Martinez, J., Patkaniowska, A., Lendeckel, W., and Tuschl, T. (2001). Functional anatomy of siRNAs for mediating efficient RNAi in *Drosophila melanogaster* embryo lysate. EMBO J. *20*, 6877–6888.

Erturk-Hasdemir, D., and Silverman, N. (2005). Eater: a big bite into phagocytosis. Cell *123*, 190–192.

Ferrandon, D., Imler, J.L., Hetru, C., and Hoffmann, J.A. (2007). The *Drosophila* systemic immune response:

sensing and signalling during bacterial and fungal infections. Nat. Rev. Immunol. 7, 862–874.

Fire, A., Xu, S., Montgomery, M.K., Kostas, S.A., Driver, S.E., and Mello, C.C. (1998). Potent and specific genetic interference by double-stranded RNA in *Caenorhabditis elegans*. Nature *391*, 806–811.

Flynt, A., Liu, N., Martin, R., and Lai, E.C. (2009). Dicing of viral replication intermediates during silencing of latent *Drosophila* viruses. Proc. Natl. Acad. Sci. U.S.A. *106*, 5270–5275.

Forstemann, K., Horwich, M.D., Wee, L., Tomari, Y., and Zamore, P.D. (2007). *Drosophila* microRNAs are sorted into functionally distinct argonaute complexes after production by Dicer-1. Cell *130*, 287–297.

Franz, A.W., Sanchez-Vargas, I., Adelman, Z.N., Blair, C.D., Beaty, B.J., James, A.A., and Olson, K.E. (2006). Engineering RNA interference-based resistance to dengue virus type 2 in genetically modified *Aedes aegypti*. Proc. Natl. Acad. Sci. U.S.A. *103*, 4198–4203.

Franz, A.W., Sanchez-Vargas, I., Piper, J., Smith, M.R., Khoo, C.C., James, A.A., and Olson, K.E. (2009). Stability and loss of a virus resistance phenotype over time in transgenic mosquitoes harbouring an antiviral effector gene. Insect Mol. Biol. *18*, 661–672.

Gaines, P.J., Olson, K.E., Higgs, S., Powers, A.M., Beaty, B.J., and Blair, C.D. (1996). Pathogen-derived resistance to dengue type 2 virus in mosquito cells by expression of the premembrane coding region of the viral genome. J. Virol. *70*, 2132–2137.

Galiana-Arnoux, D., Dostert, C., Schneemann, A., Hoffmann, J.A., and Imler, J.-L. (2006). Essential function *in vivo* for Dicer-2 in host defense against RNA viruses in *Drosophila*. Nat. Immunol. 7, 590–597.

Ghanim, M., Kontsedalov, S., and Czosnek, H. (2007). Tissue-specific gene silencing by RNA interference in the whitefly *Bemisia tabaci* (Gennadius). Insect Biochem. Mol. Biol. *37*, 732–738.

Govind, S. (2008). Innate immunity in *Drosophila*: Pathogens and pathways. Insect Science *15*, 29–43.

Griffin, D.E. (2007). Alphaviruses. In Field's Virology, D.M. Knipe, and P.M. Howley, eds. (Philadelphia, PA USA, Lippincott, Williams & Wilkins).

Guidotti, L.G., and Chisari, F.V. (2001). Noncytolytic control of viral infections by the innate and adaptive immune response. Annu. Rev. Immunol. *19*, 65–91.

Haley, B., and Zamore, P.D. (2004). Kinetic analysis of the RNAi enzyme complex. Nat. Struct. Mol. Biol. *11*, 599–606.

Hammond, S.M., Boettcher, S., Caudy, A.A., Kobayashi, R., and Hannon, G.J. (2001). Argonaute2, a link between genetic and biochemical analyses of RNAi. Science *293*, 1146–1150.

Hedges, L.M., Brownlie, J.C., O'Neill, S.L., and Johnson, K.N. (2008). *Wolbachia* and virus protection in insects. Science *322*, 702.

Hoffmann, J.A. (2003). The immune response of *Drosophila*. Nature *426*, 33–38.

Horwich, M.D., Li, C., Matranga, C., Vagin, V., Farley, G., Wang, P., and Zamore, P.D. (2007). The *Drosophila* RNA methyltransferase, DmHen1, modifies germline piRNAs and single-stranded siRNAs in RISC. Curr. Biol. *17*, 1265–1272.

Hughes, C.L., and Kaufman, T.C. (2000). RNAi analysis of deformed, proboscipedia and sex combs reduced in the milkweed bug *Oncopeltus fasciatus*: novel roles for Hox genes in the hemipteran head. Development *127*, 3683–3694.

Hutvagner, G., and Simard, M.J. (2008). Argonaute proteins: key players in RNA silencing. Nat. Rev. Mol. Cell Biol. *9*, 22–32.

Huvenne, H., and Smagghe, G. (2009). Mechanisms of dsRNA uptake in insects and potential of RNAi for pest control: A review. J. Insect Physiol. *56*, 227–235.

Kalantidis, K., Schumacher, H.T., Alexiadis, T., and Helm, J.M. (2008). RNA silencing movement in plants. Biol. Cell. *100*, 13–26.

Kaplan, C.D., Laprade, L., and Winston, F. (2003). Transcription elongation factors repress transcription initiation from cryptic sites. Science *301*, 1096–1099.

Kavi, H.H., Fernandez, H., Xie, W.W., and Birchler, J.A. (2008). Genetics and biochemistry of RNAi in *Drosophila*. RNA Interference *320*, 37–75.

Keene, K.M., Foy, B.D., Sanchez-Vargas, I., Beaty, B.J., Blair, C.D., and Olson, K.E. (2004). RNA interference acts as a natural antiviral response to O'nyong-nyong virus (Alphavirus; Togaviridae) infection of *Anopheles gambiae*. Proc. Natl. Acad. Sci. U.S.A. *101*, 17240–17245.

Kemp, C., and Imler, J.L. (2009). Antiviral immunity in *Drosophila*. Curr. Opin. Immunol. *21*, 3–9.

Kennerdell, J.R., and Carthew, R.W. (1998). Use of dsRNA-mediated genetic interference to demonstrate that frizzled and frizzled 2 act in the wingless pathway. Cell *95*, 1017–1026.

Kucharski, R., Maleszka, J., Foret, S., and Maleszka, R. (2008). Nutritional control of reproductive status in honeybees via DNA methylation. Science *319*, 1827–1830.

Kuwayama, H., Yaginuma, T., Yamashita, O., and Niimi, T. (2006). Germ-line transformation and RNAi of the ladybird beetle, *Harmonia axyridis*. Insect Mol. Biol. *15*, 507–512.

Lazzaro, B.P. (2008). Natural selection on the *Drosophila* antimicrobial immune system. Curr. Opin. Microbiol. *11*, 284–289.

Lemaitre, B., Kromer-Metzger, E., Michaut, L., Nicolas, E., Meister, M., Georgel, P., Reichhart, J.M., and Hoffmann, J.A. (1995). A recessive mutation, immune deficiency (imd), defines two distinct control pathways in the *Drosophila* host defense. Proc. Natl. Acad. Sci. U.S.A. *92*, 9465–9469.

Li, F., Pallan, P.S., Maier, M.A., Rajeev, K.G., Mathieu, S.L., Kreutz, C., Fan, Y., Sanghvi, J., Micura, R., Rozners, E., *et al.* (2007). Crystal structure, stability and *in vitro* RNAi activity of oligoribonucleotides containing the ribo-difluorotoluyl nucleotide: insights into substrate requirements by the human RISC Ago2 enzyme. Nucleic Acids Res. *35*, 6424–6438.

Li, H., Li, W.X., and Ding, S.W. (2002). Induction and suppression of RNA silencing by an animal virus. Science *296*, 1319–1321.

Lin, H. (2007). piRNAs in the germ line. Science *316*, 397.

Lindenbach, B.D., Thiel, H., and Rice, C.M. (2007). Flaviviridae: The viruses and their replication. In Field's Virology, D.M. Knipe, and P.M. Howley, eds. (Philadelphia, PA USA, Lippincott, Wiilliams & Wilkins).

Lingel, A., and Sattler, M. (2005). Novel modes of protein-RNA recognition in the RNAi pathway. Curr. Opin. Struct. Biol. *15*, 107–115.

Liu, X., Jiang, F., Kalidas, S., Smith, D., and Liu, Q. (2006). Dicer-2 and R2D2 coordinately bind siRNA to promote assembly of the siRISC complexes. RNA *12*, 1514–1520.

Liu, Y., Ye, X., Jiang, F., Liang, C., Chen, D., Peng, J., Kinch, L.N., Grishin, N.V., and Liu, Q. (2009). C3PO, an endoribonuclease that promotes RNAi by facilitating RISC activation. Science *325*, 750–753.

Lucas, W.J., Bouchepillon, S., Jackson, D.P., Nguyen, L., Baker, L., Ding, B., and Hake, S. (1995). Selective trafficking of knotted1 homeodomain protein and its messenger RNA through plasmodesmata. Science *270*, 1980–1983.

Ma, J.B., Yuan, Y.R., Meister, G., Pei, Y., Tuschl, T., and Patel, D.J. (2005). Structural basis for 5′-end-specific recognition of guide RNA by the A. fulgidus Piwi protein. Nature *434*, 666–670.

Manousis, T., and Moore, N.F. (1987). Cricket paralysis virus, a potential control agent for the olive fruit fly, *Dacus oleae Gmel*. Appl. Environ. Microbiol. *53*, 142–148.

Maori, E., Paldi, N., Shafir, S., Kalev, H., Tsur, E., Glick, E., and Sela, I. (2009). IAPV, a bee-affecting virus associated with Colony Collapse Disorder can be silenced by dsRNA ingestion. Insect Mol. Biol. *18*, 55–60.

Marmaras, V.J., and Lampropoulou, M. (2009). Regulators and signalling in insect haemocyte immunity. Cell. Signal. *21*, 186–195.

Merai, Z., Kerenyi, Z., Kertesz, S., Magna, M., Lakatos, L., and Silhavy, D. (2006). Double-stranded RNA binding may be a general plant RNA viral strategy to suppress RNA silencing. J. Virol. *80*, 5747–5756.

Moreira, L.A., Iturbe-Ormaetxe, I., Jeffery, J.A., Lu, G., Pyke, A.T., Hedges, L.M., Rocha, B.C., Hall-Mendelin, S., Day, A., Riegler, M., *et al.* (2009). A *Wolbachia* symbiont in *Aedes aegypti* limits infection with dengue, Chikungunya, and *Plasmodium*. Cell *139*, 1268–1278.

Mutti, N.S., Park, Y., Reese, J.C., and Reeck, G.R. (2006). RNAi knockout of a salivary transcript leading to lethality in the pea aphid, *Acyrthosiphon pisum*. J. Insect Sci. *6*, 1–7.

Myles, K.M., Wiley, M.R., Morazzani, E.M., and Adelman, Z.N. (2008). Alphavirus-derived small RNAs modulate pathogenesis in disease vector mosquitoes. Proc. Natl. Acad. Sci. U.S.A. *105*, 19938–19943

Napoli, C., Lemieux, C., and Jorgensen, R. (1990). Introduction of a chimeric chalcone synthase gene into petunia results in reversible co-suppression of homologous genes in trans. Plant Cell *2*, 279–289.

Nayak, A., Berry, B., Acevedo, A., Deng, C., Kruchinsky, A., Gross, J., Antoniewski, C., and Andino, R. (2010). An animal virus RNAi suppressor that directly blocks catalysis by Argonaute. Nat. Struct. Mol. Biol. *17*, 547–554

Obbard, D.J., Jiggins, F.M., Halligan, D.L. and Little, T.J. (2006). Natural selection drives extremely rapid evolution in antiviral RNAi genes. Curr. Biol. *16*, 580–585.

Olson, K.E., Higgs, S., Gaines, P.J., Powers, A.M., Davis, B.S., Kamrud, K.I., Carlson, J.O., Blair, C.D., and Beaty, B.J. (1996). Genetically engineered resistance to dengue-2 virus transmission in mosquitoes. Science *272*, 884–886.

Osborne, S.E., Leong, Y.S., O'Neill, S.L., and Johnson, K.N. (2009). Variation in antiviral protection mediated by different *Wolbachia* strains in *Drosophila simulans*. PLoS Pathog. *5*, e1000656.

Paskewitz, S.M., Andreev, O., and Shi, L. (2006). Gene silencing of serine proteases affects melanization of Sephadex beads in *Anopheles gambiae*. Insect Biochem. Mol. Biol. *36*, 701–711.

Peters, C.J. (2007). Emerging viral diseases In Field's Virology, D.M. Knipe, and P.M. Howley, eds. (Philadelphia, PA USA, Lippincott, Williams & Wilkins).

Roignant, J.Y., Carre, C., Mugat, B., Szymczak, D., Lepesant, J.A., and Antoniewski, C. (2003). Absence of transitive and systemic pathways allows cell-specific and isoform-specific RNAi in *Drosophila*. RNA *9*, 299–308.

Romano, N., and Macino, G. (1992). Quelling – Transient inactivation of gene-expression in *Neurospora crassa* by transformation with homologous sequences. Mol. Microbiol. *6*, 3343–3353.

Rutschmann, S., Kilinc, A., and Ferrandon, D. (2002). Cutting edge: the toll pathway is required for resistance to gram-positive bacterial infections in *Drosophila*. J. Immunol. *168*, 1542–1546.

Ryter, J.M., and Schultz, S.C. (1998). Molecular basis of double-stranded RNA–protein interactions: structure of a dsRNA-binding domain complexed with dsRNA. EMBO J. *17*, 7505–7513.

Saleh, M.-C., Tassetto, M., van Rij, R.P., Goic, B., Gausson, V., Berry, B., Jacquier, C., Antoniewski, C., and Andino, R. (2009). Antiviral immunity in *Drosophila* requires systemic RNA interference spread. Nature *458*, 346–350.

Saleh, M.C., van Rij, R.P., Hekele, A., Gillis, A., Foley, E., O'Farrell, P.H., and Andino, R. (2006). The endocytic pathway mediates cell entry of dsRNA to induce RNAi silencing. Nat. Cell Biol. *8*, 793–802.

Sánchez-Vargas, I., Scott, J.C., Poole-Smith, B.K., Franz, A.W.E., Barbosa-Solomieu, V., Wilusz, J., Olson, K.E., and Blair, C.D. (2009). Dengue virus type 2 infections of *Aedes aegypti* are modulated by the mosquito's RNA interference pathway. PLoS Pathog. *5*, e1000299.

Sanchez-Vargas, I., Travanty, E., Keene, K., Franz, A., Beaty, B., Blair, C., and Olson, K. (2004). RNA interference, arthropod-borne viruses, and mosquitoes. Virus Res. *102*, 65–74.

Schluns, H., and Crozier, R.H. (2007). Relish regulates expression of antimicrobial peptide genes in the honeybee, *Apis mellifera*, shown by RNA interference. Insect Mol. Biol. *16*, 753–759.

Schwarz, D.S., Tomari, Y., and Zamore, P.D. (2004). The RNA-induced silencing complex is a Mg2+-dependent endonuclease. Curr. Biol. *14*, 787–791.

Settles, E.W., and Friesen, P.D. (2008). Flock house virus induces apoptosis by depletion of *Drosophila* inhibitor-of-apoptosis protein DIAP1. J. Virol. *82*, 1378–1388.

Sijen, T., Fleenor, J., Simmer, F., Thijssen, K.L., Parrish, S., Timmons, L., Plasterk, R.H., and Fire, A. (2001). On the role of RNA amplification in dsRNA-triggered gene silencing. Cell *107*, 465–476.

Singh, G., Popli, S., Hari, Y., Malhotra, P., Mukherjee, S., and Bhatnagar, R.K. (2009). Suppression of RNA silencing by Flock house virus B2 protein is mediated through its interaction with the PAZ domain of Dicer. FASEB J. *23*, 1845–1857.

Souza-Neto, J.A., Sim, S., and Dimopoulos, G. (2009). An evolutionary conserved function of the JAK-STAT pathway in anti-dengue defense. Proc. Natl. Acad. Sci. U.S.A. *106*, 17841–17846.

Teixeira, L., Ferreira, A., and Ashburner, M. (2008). The bacterial symbiont *Wolbachia* induces resistance to RNA viral infections in *Drosophila melanogaster*. PLoS Biol. *6*, e2.

Thomas-Orillard, M. (1984). Modifications of mean ovariole number, fresh weight of adult females and developmental time in *Drosophila melanogaster* induced by *Drosophila* C virus. Genetics *107*, 635–644.

Tomari, Y., Matranga, C., Haley, B., Martinez, N., and Zamore, P.D. (2004). A protein sensor for siRNA asymmetry. Science *306*, 1377–1380.

Tomari, Y., and Zamore, P.D. (2005). Perspective: machines for RNAi. Genes Dev. *19*, 517–529.

Turner, C.T., Davy, M.W., MacDiarmid, R.M., Plummer, K.M., Birch, N.P., and Newcomb, R.D. (2006). RNA interference in the light brown apple moth, *Epiphyas postvittana* (Walker) induced by double-stranded RNA feeding. Insect Mol. Biol. *15*, 383–391.

Ui-Tei, K., Naito, Y., Nishi, K., Juni, A., and Saigo, K. (2008). Thermodynamic stability and Watson-Crick basepairing in the seed duplex are major determinants of the efficiency of the siRNA-based off-target effect. Nucleic Acids Res. *36*, 7100–7109.

Ulvila, J., Parikka, M., Kleino, A., Sormunen, R., Ezekowitz, R.A., Kocks, C., and Ramet, M. (2006). Double-stranded RNA is internalized by scavenger receptor-mediated endocytosis in *Drosophila* S2 cells. J. Biol. Chem. *281*, 14370–14375.

Umbach, J.L., and Cullen, B.R. (2009). The role of RNAi and microRNAs in animal virus replication and antiviral immunity. Genes Dev. *23*, 1151–1164.

Vaistij, F.E., Jones, L., and Baulcombe, D.C. (2002). Spreading of RNA targeting and DNA methylation in RNA silencing requires transcription of the target gene and a putative RNA-dependent RNA polymerase. Plant Cell *14*, 857–867.

van Rij, R.P., Saleh, M.-C., Berry, B., Foo, C., Houk, A., Antoniewski, C., and Andino, R. (2006). The RNA silencing endonuclease Argonaute 2 mediates specific antiviral immunity in *Drosophila melanogaster*. Genes Dev. *20*, 2985–2995.

van Engelsdorp, D., Evans, J.D., Saegerman, C., Mullin, C., Haubruge, E., Nguyen, B.K., Frazier, M., Frazier, J., Cox-Foster, D., Chen, Y., *et al.* (2009a). Colony collapse disorder: a descriptive study. PLoS ONE *4*, e6481.

van Engelsdorp, D., Hayes, J., and Pettis, J. (2009b). Preliminary Results: A Survey of Honey Bee Colonies Losses in the U.S. Between September 2008 and April 2009. (Apiary Inspectors of America (AIA) and USDA-ARS Beltsville Honey Bee Lab).

Vermehren, A., Qazi, S., and Trimmer, B.A. (2001). The nicotinic alpha subunit MARA1 is necessary for cholinergic evoked calcium transients in *Manduca neurons*. Neurosci. Lett. *313*, 113–116.

Wang, X.-H., Aliyari, R., Li, W.-X., Li, H.-W., Kim, K., Carthew, R., Atkinson, P., and Ding, S.-W. (2006). RNA interference directs innate immunity against viruses in adult *Drosophila*. Science *312*, 452–454.

Waterhouse, R.M., Kriventseva, E.V., Meister, S., Xi, Z., Alvarez, K.S., Bartholomay, L.C., Barillas-Mury, C., Bian, G., Blandin, S., Christensen, B.M., *et al.* (2007). Evolutionary dynamics of immune-related genes and pathways in disease-vector mosquitoes. Science *316*, 1738–1743.

Xi, Z., Ramirez, J.L., and Dimopoulos, G. (2008). The *Aedes aegypti* toll pathway controls dengue virus infection. PLoS Pathog. *4*, e1000098.

Xu, W.N., and Han, Z.J. (2008). Cloning and phylogenetic analysis of sid-1 like genes from aphids. J. Insect Sci. *8*, 6.

Yassine, H., and Osta, M.A. (2010). *Anopheles gambiae* innate immunity. Cell. Microbiol. *12*, 1–9.

Yuan, Y.R., Pei, Y., Ma, J.B., Kuryavyi, V., Zhadina, M., Meister, G., Chen, H.Y., Dauter, Z., Tuschl, T., and Patel, D.J. (2005). Crystal structure of *A. aeolicus* argonaute, a site-specific DNA-guided endoribonuclease, provides insights into RISC-mediated mRNA cleavage. Mol. Cell *19*, 405–419.

Zambon, R.A., Vakharia, V.N., and Wu, L.P. (2006). RNAi is an antiviral immune response against a dsRNA virus in *Drosophila melanogaster*. Cell. Microbiol. *8*, 880–889.

Zhang, H., Kolb, F.A., Brondani, V., Billy, E., and Filipowicz, W. (2002). Human Dicer preferentially cleaves dsRNAs at their termini without a requirement for ATP. EMBO J. *21*, 5875–5885.

Zhang, H., Kolb, F.A., Jaskiewicz, L., Westhof, E., and Filipowicz, W. (2004). Single processing center models for human Dicer and bacterial RNase III. Cell *118*, 57–68.

Zhou, X., Liao, Z., Jia, Q., Cheng, L., and Li, F. (2007). Identification and characterization of Piwi subfamily in insects. Biochem. Biophys. Res. Commun. *362*,126–131.

Antiviral Responses in Insects: Apoptosis and Humoral Responses

17

Rollie J. Clem, Holly J.R. Popham and Kent S. Shelby

Abstract

Insects are subject to infection by many different kinds of DNA and RNA viruses. These include viruses that are pathogenic to insects, as well as vertebrate pathogens that are vectored by insects. Although the study of antiviral responses in insects has lagged behind studies of responses to other types of pathogens, progress in this area has begun to rapidly accelerate over the past several years. Within the field of insect pathology, antiviral responses in insects have traditionally been categorized as falling into one of three types: physical barriers, cellular immunity, and humoral immunity. However, it has become clear in recent years that a fourth type of response exists, called intracellular immunity. Intracellular immunity is particularly relevant to virus infection because it operates within infected cells, and includes responses such as RNA interference and, as we propose here, apoptosis. In this chapter we discuss the current understanding of antiviral responses in insects, focusing mainly on two of the best understood types of responses, apoptosis and humoral immunity.

Introduction

Insect innate immunity has traditionally been described as consisting of three broad immune functions: physical barriers to infection (such as cuticle), cellular immunity, and humoral immunity. Cellular immunity is characterized by direct interactions between invaders and circulating haemocytes (Lavine and Strand, 2002). Circulating haemocytes are recruited to infections to phagocytose small microbes, form nodules of large numbers of microbes, and encapsulate larger invaders. Humoral immunity is characterized by the prophenoloxidase system and the induced biosynthesis of antibacterial proteins (Hultmark, 2003). Acting in concert, humoral and cellular responses halt microbial infection and clear, or wall off, the resulting debris from the haemocoel by enveloping them in melanin and melanized haemocytes (Trudeau *et al.*, 2001).

In addition to these three classical types of innate immunity, a fourth type of innate immunity has been described, called intracellular immunity (Fig. 17.1). Intracellular immunity includes responses which are generated by the infected cell and affect intracellular parasites within that cell. This includes cellular responses to virus infection such as modifications of viral nucleic acids (Zheng and Peterlin, 2005) and responses involving small RNAs, including microRNAs (miRNAs), small interfering RNAs (siRNAs), and piwi-interacting RNAs (piRNAs) (Fire, 2005). We propose that apoptosis should also be included in the category of intracellular immunity. Because viruses are obligate intracellular parasites, intracellular immune functions are of particular importance in combating viral infections. In this chapter, we will focus mainly on two of the best understood types of antiviral responses in insects, apoptosis and humoral immunity. We will also briefly discuss the evidence suggesting that cellular immune functions, such as phagocytosis and encapsulation of infected cells, may play important roles in antiviral immunity in insects. Meanwhile, the important roles of small RNAs in

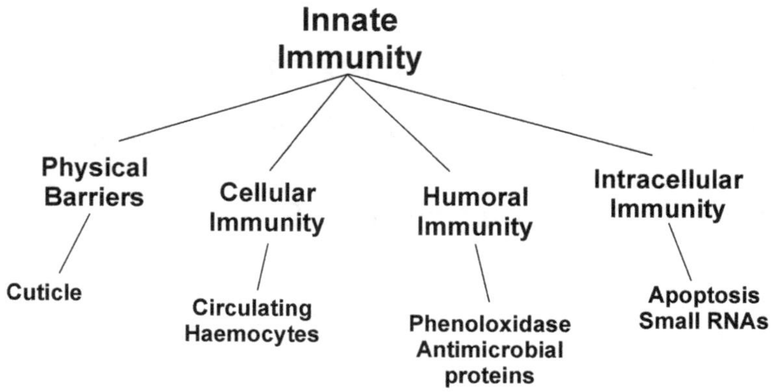

Figure 17.1 The four types of innate immunity in insects. We propose that apoptosis should be considered as a type of intracellular immunity, since viral induction of apoptosis directly affects the virus replicating within the cell where apoptosis is induced.

antiviral responses will be covered in more depth in separate chapters in this volume (see Chapters 15 and 16).

Apoptosis

Apoptosis is a type of programmed cell death that is characterized by morphological characteristics including plasma membrane blebbing and nuclear condensation and fragmentation. During apoptosis, chromatin also becomes condensed, and the DNA is digested by endonucleases into oligonucleosomal length fragments. As the cells fragment, their contents remain in membrane-bound fragments called apoptotic bodies, and *in vivo*, apoptotic bodies and cells are rapidly phagocytosed by other cells. A core apoptosis pathway exists in metazoans which involves the activation of caspases, a family of cysteine proteases. Upon receiving an apoptotic signal, initiator caspases undergo autoactivation through dimerization, and then cleave and activate effector caspases, which dismantle the cell (Fig. 17.2).

Apoptosis was first suggested to be able to function as an antiviral response in 1985 (Clouston and Kerr, 1985). It was shortly after this that evidence in support of this hypothesis began to emerge from both mammalian and insect systems. Apoptosis has been recognized as a part of antiviral immunity in insects since baculoviruses were first shown to inhibit apoptosis in the early 1990s, and subsequent experiments have demonstrated that apoptosis can be an effective defence against baculovirus infection. To date, the baculovirus–lepidopteran system remains by far the best-characterized system for understanding the role of apoptosis in antiviral immunity in insects. Several other families of DNA viruses that infect insects have also been shown to encode anti-apoptotic proteins. Because of this, apoptosis is presumed to be an important defence against these other DNA viruses, but the role of apoptosis in defence against other DNA viruses in insects has not yet been examined directly. On the other hand, whether or not apoptosis can be an effective antiviral defence against RNA viruses in insects is still largely an open question.

How do baculoviruses induce apoptosis?

The *Baculoviridae* is a large and diverse family of DNA viruses, with over 50 officially recognized species in two genera, and several hundred reported isolates (see Chapter 2 in this volume). Baculoviruses tend to be highly host-specific. The molecular basis for this exquisite host range specificity is poorly understood, but it appears to involve multiple mechanisms, including apoptosis, that can each contribute towards determining compatibility, depending on the virus–host combination.

A number of baculoviruses have been shown to induce apoptosis in insect cells (reviewed in Clem, 2007). In some cases, apoptosis is only evident when a mutation has occurred that inactivates an anti-apoptotic gene encoded by

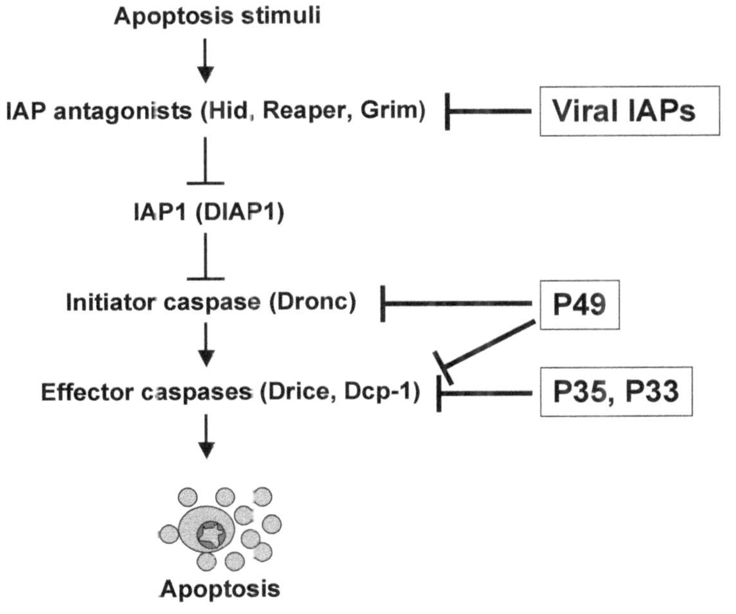

Figure 17.2 The core apoptosis pathway in insects, based on knowledge gained from studying *D. melanogaster*. The names of *D. melanogaster* proteins involved at each step in the pathway are shown in parentheses. Insect virus proteins that inhibit apoptosis are boxed, and the steps where they act are indicated.

the virus (or after silencing of the anti-apoptotic gene by RNAi). This is the case for *Autographa californica* multiple nucleopolyhedrovirus (AcMNPV) infection in cells from either permissive (Clem *et al.*, 1991; da Silveira *et al.*, 2005) or non-permissive hosts (Clem *et al.*, 1991; Du and Thiem, 1997; Wright *et al.*, 2005; Zoog *et al.*, 2002); for *Orgyia pseudotsugata* MNPV infection of *Lymantria dispar* cells (Means *et al.*, 2003), for *Bombyx mori* NPV infection of *B. mori* cells (Kamita *et al.*, 1993), and for *Anticarsia gemmatalis* MNPV infection of *A. gemmatalis* cells (Carpes *et al.*, 2009; Silveira *et al.*, 1999). These mutant viruses do not always induce apoptosis in all cell lines or hosts, however (Clem *et al.*, 1991; de Castro and Ribeiro, 2001; Griffiths *et al.*, 1999). In other examples, apoptosis is observed during a non-permissive infection involving the wild type virus. For example, wild type AcMNPV induces apoptosis in cells from *Spodoptera littoralis*, *Spodoptera litura*, and *Choristoneura fumiferana* (Chejanovsky and Gershburg, 1995; Palli *et al.*, 1996; Zhang *et al.*, 2002). Several other baculoviruses have also been shown to induce apoptosis during non-permissive infections (Clem, 2007). The extreme variability of the response to infection is illustrated by the fact that, in some cases, different responses have even been reported in different cell lines derived from the same species (Chejanovsky and Gershburg, 1995; Griffiths *et al.*, 1999; Palli *et al.*, 1996). Thus, whether or not apoptosis occurs is highly dependent on both the virus and the host species and/or cell line being infected.

The mechanisms involved in apoptosis induction have so far only been studied for AcMNPV infection in *Spodoptera frugiperda* cells. Early studies established three potential apoptotic signals that all occur during the same phase of infection, in an interdependent fashion: (1) inhibition of host RNA and protein synthesis, (2) induction of viral DNA synthesis, and (3) initiation of late viral gene expression (Clem, 2001). All three of these events occur during the transition from early to late viral gene expression, which in the case of AcMNPV infection in *S. frugiperda* cells is at around 6 hours post infection. Inhibition of viral DNA synthesis by drugs or mutations in key viral genes prevents initiation of apoptosis, suggesting that this may be the trigger (Clem and Miller, 1994; LaCount and Friesen, 1997). However, viral DNA synthesis is required for late viral gene expression, and also for shut-off of host RNA and protein synthesis. Since treatment with drugs that inhibit RNA or protein synthesis also induces apoptosis (Clem and Miller, 1994), presumably because of the short half-life of the apoptosis inhibitor protein IAP1 (Muro *et al.*, 2002), it has been difficult to disentangle the roles of these processes in triggering apoptosis. Recently, it was shown that silencing AcMNPV genes involved in DNA replication, but not those required specifically for late gene expression (non-replicative expression factors), inhibited apoptosis during infection, and also prevented host shut-off (Schultz and Friesen, 2009). Thus, it appears that apoptosis induction by AcMNPV in *S. frugiperda* may be due to a cellular DNA damage-type response, which is triggered by viral DNA replication and arrests cellular RNA and protein synthesis. Whether this is also the case in other baculovirus–host combinations remains to be determined. It has also been suggested that abnormally increased or extended production of budded virus can trigger apoptosis, since an FP (few polyhedra) mutant of AcMNPV that also lacks a functional apoptosis inhibitor causes late apoptotic-like death in *Trichoplusia ni* cells that are normally resistant to apoptosis (Kelly *et al.*, 2008).

How do baculoviruses inhibit apoptosis?

Baculoviruses express proteins that are able to inhibit apoptosis of their host cells. Extensive research has been carried out on two families of baculovirus anti-apoptotic proteins, the P35 family, and the IAP (Inhibitor of Apoptosis) family. Nearly all of the baculovirus genomes that have been sequenced to date (more than 40) contain homologues of P35, IAP, or both, indicating the importance of these genes in baculovirus evolution.

The first P35 protein discovered, AcMNPV P35, is the one that has been most thoroughly characterized (reviewed in Means and Clem, 2008). P35-related proteins have also been reported in a relatively small number of other baculoviruses, and in an entomopoxvirus. Given their limited distribution in a small number of fairly divergent viruses, it is likely that viruses acquired ancestral *p35* genes from their hosts more than once (Means and Clem, 2008). However, genes with demonstrable homology to *p35* have not been identified in any cellular genomes to date. The AcMNPV *p35* gene is transcribed both early and late during infection (Friesen and Miller, 1987). The P35 protein is a suicide inhibitor of caspases, the proteases that play a central role in apoptosis. Following an apoptotic signal, initiator caspases are activated first, which then activate effector caspases, leading to death of the cell (Fig. 17.2). P35 is a more effective inhibitor of effector caspases than initiator caspases. The mechanism of caspase inhibition by P35 is well understood, and involves cleavage of P35 by the caspase, followed by formation of a thioester bond between P35 and the caspase (Eddins *et al.*, 2002; Xu *et al.*, 2001). This results in essentially permanent inactivation of the caspase. Relatives of P35, called P49, have been found in four baculoviruses to date, and P49 is also cleaved by caspases and probably functions through a similar mechanism (Pei *et al.*, 2002; Zoog *et al.*, 2002). However, P49 is able to inhibit both initiator and effector caspases (Jabbour *et al.*, 2002; Pei *et al.*, 2002; Zoog *et al.*, 2002). The most distantly related gene in this family, P33 from *Amsacta moorei* entomopoxvirus (AmEPV), is also a functional apoptosis inhibitor and appears to be similar to P35 in its range of caspase targets (Means *et al.*, 2007).

Unlike *p35* genes, which are found in a small subset of baculoviruses, the *iap* genes are widespread in baculoviruses, and in addition have also been found in other insect DNA viruses, including entomopoxviruses, iridoviruses, and ascoviruses, among others. Interestingly, *iap* genes have not been found in viruses that exclusively infect mammals, but they are highly conserved in the cellular genomes of both insects and higher animals, where they also are involved in regulating apoptosis, as well as other cellular processes (Rumble and Duckett, 2008). All IAP proteins contain 1–3 copies of a sequence called a BIR (baculovirus IAP repeat) domain, a type of zinc-binding domain approximately 70 amino acids in length. BIR domains contain a number of highly conserved amino acids, including three cysteines and one histidine which mediate contacts with a single zinc atom. BIR domains function by mediating interactions with other proteins. In addition, many IAP proteins contain a C-terminal RING domain, which confers E3 ubiquitin ligase activity. The first IAP proteins were identified in the baculoviruses *Cydia pomonella* granulovirus and OpMNPV (Birnbaum *et al.*, 1994; Crook *et al.*, 1993). Among the numerous baculovirus IAP proteins, Op-IAP3 has been most extensively studied. While some cellular IAP proteins have been shown to be able to bind and directly inhibit caspases, as well as induce the degradation of caspases through the ubiquitin-proteasome pathway, baculovirus IAP proteins do not appear to work in this manner. Instead, Op-IAP3 appears to function by inactivating IAP antagonists, cellular proapoptotic proteins that inhibit cellular IAP proteins (Fig. 17.2) (Wilkinson *et al.*, 2004; Wright *et al.*, 2005).

Many baculoviruses contain more than one *iap* gene, and viruses that contain *p35* homologues also contain *iap* genes. This raises the question of why these viruses have retained these multiple genes. In the one case where it has been examined, only one of the four *iap* genes from *Epiphyas postvittana* NPV (EppoNPV) was able to inhibit apoptosis in the context tested, while a second gene delayed apoptosis (Maguire *et al.*, 2000). Similarly, the *iap-1* gene from AcMNPV was found not to be a functional apoptosis inhibitor. One possibility is that some of the IAP proteins have alternative functions, which would not be surprising since there is evidence that cellular IAPs also function in other pathways besides apoptosis (Rumble and Duckett, 2008). In addition, there are also many examples of mammalian viruses that contain multiple apoptosis inhibitors. Thus it may be advantageous for viruses to retain multiple genes that inhibit apoptosis stimulated by alternative pathways, or that inhibit apoptosis in alternative tissue types or hosts.

Potential effects of apoptosis on viruses

In what tissues would apoptosis be expected to play an important role in antiviral immunity in insects, and exactly how would apoptosis limit virus replication and spread? Most insects become infected by viruses through feeding on virus-contaminated material, and so the midgut epithelium is usually the first tissue that becomes infected. Baculoviruses typically need to penetrate the midgut in order to establish a fatal infection, although there are some examples where a successful infection is midgut-limited. In some cases, midgut penetration is sufficient to establish a fatal infection; that is, once the baculovirus penetrates the midgut, the insect has no effective defence against infection (for example, AcMNPV infection in *Trichoplusia ni*; Washburn *et al.*, 1998), but in other cases, additional immune mechanisms can operate (Clarke and Clem, 2002; Haas-Stapleton *et al.*, 2003). In the case of arboviruses, dissemination of the infection from the midgut to the salivary glands is often required for the arbovirus to be transmitted to a new host. Given the importance of the midgut in the infection process, it would seem to be a good place for the host to make a defensive stand, in order to prevent virus penetration into the haemocoel. Indeed, there are a small number of published reports of apoptosis occurring in response to virus infection in midgut or midgut cell cultures (Brooks *et al.*, 2002; Chikhalya *et al.*, 2009; Dougherty *et al.*, 2006; Vaidyanathan and Scott, 2006; Weaver *et al.*, 1988, 1992). In some cases, apoptosis has been correlated with resistance to infection (Chikhalya *et al.*, 2009; Vaidyanathan and Scott, 2006). Sloughing of baculovirus-infected midgut cells into the gut lumen has long been associated with resistance to infection, and a recent report indicates that this is an apoptotic mechanism (Chikhalya *et al.*, 2009). Furthermore, inhibiting apoptosis in a primary midgut cell culture allowed efficient virus replication *in vitro* (Dougherty *et al.*, 2006). Recently obtained results also indicate that apoptosis occurs in *Spodoptera exigua* midgut following oral infection with *p35* mutant AcMNPV (J.C. Means, T.E. Clarke, L. Heaton, A.L. Passarelli, and R.J. Clem, unpublished).

Another tissue where apoptosis could be important in limiting virus infection is tracheal epithelium, since it has been reported that baculoviruses utilize this tissue in order to penetrate the midgut basal lamina (Engelhard *et al.*, 1994). Apoptosis is evident in this tissue following infection of *S. frugiperda* by *p35* mutant AcMNPV by budded virus injection into the haemocoel, and in *S. exigua* following either intrahaemocoelic or oral infection (J.C. Means, T.E. Clarke, L. Heaton, A.L. Passarelli, and R.J. Clem, unpublished). Haemocytes could also be an important site of defence, given their reported role in spreading infection. Haemocyte apoptosis has been reported following infection of *S. litura* larvae with wild type AcMNPV (Zhang *et al.*, 2002), or infection of *A. gemmatalis* larvae with *p35* mutant AcMNPV (da Silveira *et al.*, 2005). Finally, arboviruses that are transmitted through the saliva of biting insects must replicate in salivary glands, and a few cases of apoptosis have been reported in this tissue following arbovirus infection (Bowers *et al.*, 2003; Girard *et al.*, 2005, 2007; Mims *et al.*, 1966).

Exactly how would apoptosis be expected to limit virus infection? One can envision several possible mechanisms (Fig. 17.3). In addition to sloughing of infected cells into the lumen of midgut or salivary gland, another way would be through reducing the amount of virus replication, as is seen when *p35* mutant AcMNPV replicates in *S. frugiperda* cells. Obviously, decreased production of budded virus would be expected to reduce or at least slow virus spread in the host. Another mechanism could involve phagocytosis or encapsulation of infected apoptotic cells by haemocytes. Apoptotic cells are known to produce 'find me' and 'eat me' signals in the early stages of death, which instruct phagocytic cells to migrate towards, engulf and digest the dying cells (Elliott *et al.*, 2009; Nakanishi *et al.*, 2009). Engulfment of the infected cell would be expected to not only further limit virus replication, but also to destroy (through lysosomal digestion) any intracellular virus remaining in the cell. Encapsulation and melanization of apoptotic tissue by haemocytes would also be expected to inactivate and limit spread of virus. Finally, viral proteins or DNA may be damaged by caspases or caspase-activated nucleases during apoptosis.

The effects of apoptosis on baculovirus replication and infectivity

The best-studied system for understanding the effects of apoptosis on virus infection in insects is that of baculovirus infection of lepidopteran hosts. Triggering of apoptosis during baculovirus infection can have severe negative consequences for the virus, although the severity of the consequences varies depending on the virus/host combination. The most extensively studied example is *p35* mutant AcMNPV infection of *S. frugiperda* cells and larvae. Mutant AcMNPV lacking *p35* replicates poorly in *S. frugiperda* cells. Yields of budded virus are reduced by 100-fold or more, while occluded virus is not produced at all (Clem and Miller, 1993; Hershberger *et al.*, 1992). When *p35* mutant AcMNPV is used to infect *S. frugiperda* larvae, the effects are also dramatic. Compared to wild type AcMNPV, *p35* mutant virus is more than 1000-fold less infectious when infection is by injection of budded virus, and 25-fold less infectious by feeding occluded virus (Clem and Miller, 1993). Three key pieces of evidence indicate that these defects are due to apoptosis. First, extensive apoptosis occurs in tracheal epithelium, fat body, and muscle of *p35* mutant-infected, but not wild type-infected, *S. frugiperda* larvae following injection of budded virus (Clarke and Clem,

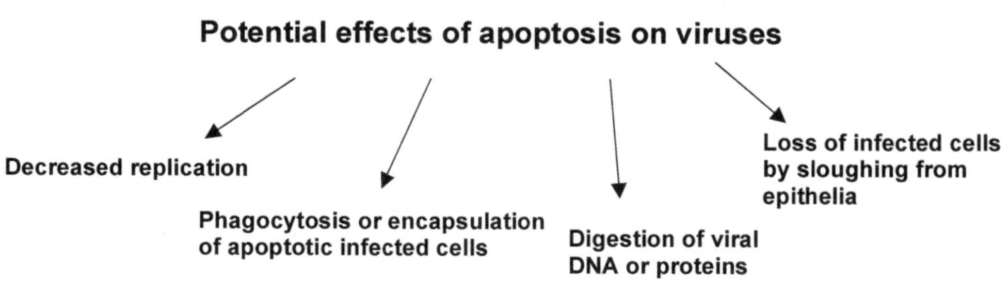

Figure 17.3 Some potential effects that apoptosis can have on the ability of viruses to cause infections *in vivo*. Each of these is supported by evidence, as discussed in the text.

2003). Second, the replication and infectivity of *p35* mutant AcMNPV is normal in *T. ni* cells and larvae, where apoptosis is not observed (Clem and Miller, 1993). Third, the defects associated with *p35* mutant infection in *S. frugiperda* cells and larvae are rescued by expression of a different anti-apoptotic gene, indicating that they are due to apoptosis and not some other function of *p35* in replication (Clem *et al.*, 1994). Since *S. frugiperda* is highly resistant to oral infection by AcMNPV at later instars, histological examination of *p35* mutant-infected midgut cells has not been reported in this species. However, apoptosis is observed in *S. exigua* midgut epithelial cells following oral infection with *p35* mutant virus (J.C. Means, T.E. Clarke, L. Heaton, A.L. Passarelli, and R.J. Clem, unpublished).

The effectiveness of apoptosis as a defence against baculovirus infection appears to vary depending on the virus/host combination. While virus replication is drastically reduced when *p35* mutant AcMNPV is used to infect *S. frugiperda* cells, replication of *p35* mutant BmNPV is not affected in *B. mori* cells or larvae compared to wild type BmNPV, despite the fact that a portion of the infected cells undergo apoptosis *in vitro* (Kamita *et al.*, 1993). Whether this is due to the presence of another anti-apoptotic gene in the BmNPV genome has not been explored, but the BmNPV P35 protein has been shown to be a less effective caspase inhibitor than AcMNPV P35 (Morishima *et al.*, 1998). Likewise, little effect on virus production was reported with an *iap-3* mutant of AgMNPV in *A. gemmatalis* cells (de Castro and Ribeiro, 2001). *In vivo*, this AgMNPV mutant was found to be only 5-fold less infectious than wild type AgMNPV in *A. gemmatalis* larvae (da Silveira *et al.*, 2007). In one case where wild type AcMNPV induces apoptosis in a non-permissive species (*S. littoralis*), it has been shown that inhibiting apoptosis alone does not allow permissive infection (Gershburg *et al.*, 1997). Thus, apoptosis may sometimes be just one of the reasons why a particular virus is not able to infect a particular host.

While the infectivity of *p35* mutant AcMNPV by intrahaemocoelic injection of budded virus is reduced more than 1,000-fold compared to wild type in *S. frugiperda* larvae, a less severe but still significant 60-fold reduction is seen in *S. exigua* larvae (J.C. Means, T. Clarke, L. Heaton, and R.J. Clem, unpublished results). Interestingly, the infectivity of wild type AcMNPV is also reduced in *S. frugiperda* larvae compared to *S. exigua* larvae under the same conditions, suggesting that there are other immune mechanisms that limit AcMNPV infection in *S. frugiperda*. Thus, apoptosis may be most effective when it acts synergistically with other immune mechanisms.

A novel effect of apoptosis on baculovirus replication was recently revealed by studying the replication of *p35* mutant AcMNPV in TN-368 cells from *T. ni* (Bryant and Clem, 2009), and is summarized in Fig. 17.4. While TN-368 cells are highly resistant to apoptosis, and do not undergo apoptosis when infected with *p35* mutant AcMNPV, a low amount of caspase activity was observed in cells infected with *p35* mutant AcMNPV compared to none in cells infected with wild type AcMNPV. Interestingly, the *p35* mutant virus produced in TN-368 cells was less stable and less infectious than wild-type virus, suggesting that it was somehow defective. This infectivity defect was shown to involve either viral attachment or entry, and could be rescued by treating the cells that were used to prepare the stocks of *p35* mutant virus with a chemical caspase inhibitor. Thus, it appears that caspase activation can have a detrimental effect on baculovirus infectivity, even if the amount of caspase activity is relatively low and apoptosis does not occur. Whether this negative effect is directly due to digestion of baculovirus proteins by caspases, or to indirect effects such as caspase cleavage of host proteins required for efficient virus assembly, is not known.

The effects of apoptosis on other insect viruses

While a great deal of information has been accumulated concerning the effectiveness of apoptosis against baculovirus infection in insects, much less is known about the role of apoptosis in immunity against other families of insect viruses.

As mentioned above, IAP homologues have been identified in several other DNA virus families that infect insects or other arthropods. Examples include an entomopoxvirus (Li *et al.*, 2005), an ascovirus (Hussain and Asgari, 2008), an iridovirus (Ince *et al.*, 2008), African

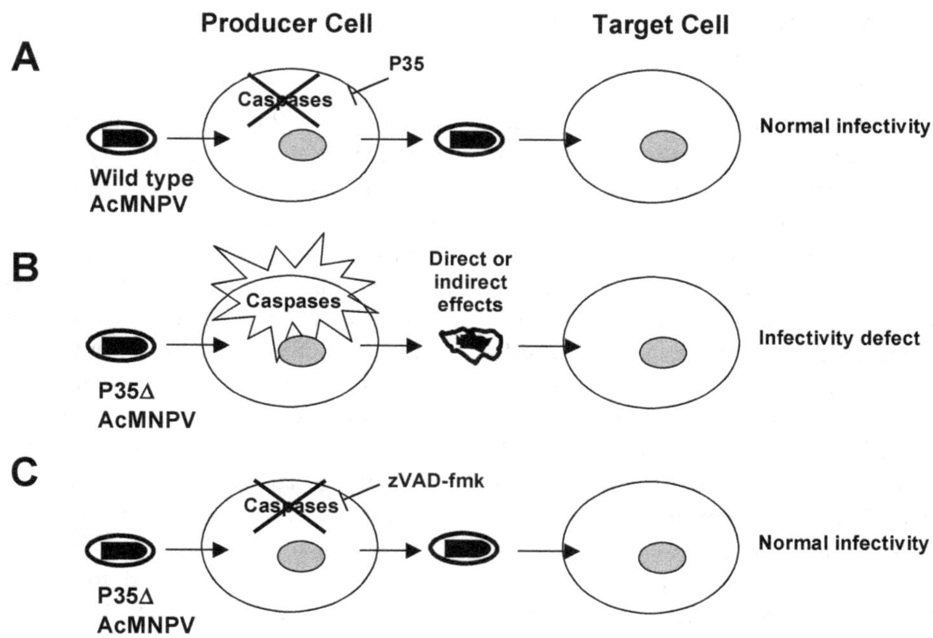

Figure 17.4 Unrestricted caspase activity in an infected cell can negatively affect the infectivity of baculovirus virions that bud from that cell (Bryant and Clem, 2009). (A) There is very little caspase activity in TN-368 cells from *T. ni* infected with wild type AcMNPV (producer cells) because of the expression of the caspase inhibitor P35 by the virus. The budded virions produced by the producer cells have normal infectivity when used to infect other cells (target cells). (B) Infection of TN-368 cells with a *p35* mutant of AcMNPV results in a low but appreciable level of caspase activity, and the budded virions that are produced have defects in attachment or entry when used to infect target cells. As discussed in the text, it is not known whether these effects are direct or indirect. (C) The infectivity defect seen with the *p35* mutant can be rescued by treatment of the producer cells with the chemical caspase inhibitor zVAD-fmk.

swine fever virus (which in addition to infecting swine also has an obligate replication stage in ticks) (Chacon *et al.*, 1995), nudiviruses (Wang and Jehle, 2009), and the currently unclassified salivary gland hypertrophy virus (Garcia-Maruniak *et al.*, 2008). Some of these genes have been shown to be functional apoptosis inhibitors. In only one case has the effect of deleting these genes been examined on virus replication; an iap deletion mutant of AmEPV exhibited 10-fold reduction of virus progeny compared to wild type (Li *et al.*, 2005). AmEPV also expresses the active P35 homologue P33 (Means *et al.*, 2007), which could influence these results. However, anti-apoptosis genes are prevalent in these other families of insect DNA viruses, and the viruses exhibit many overall similarities with baculoviruses in terms of replication strategies and pathogenesis, so it is not unreasonable to expect that apoptosis also probably plays a role in defence against them.

An interesting case is the ascoviruses, which actually seem to use an abbreviated form of apoptosis as part of their replication strategy. These viruses cause cleavage of the infected cells into vesicles that contain progeny virus particles. This cellular fragmentation is morphologically similar to apoptosis, and it has been shown that an ascovirus encodes a caspase that assists in this process (Bideshi *et al.*, 2005). It has been suggested that ascoviruses induce the early stages of apoptosis to aid in their dispersal, but then inhibit the process at a later step to allow complete virus replication (see Chapter 1 for further details).

Whether apoptosis is a defence against RNA viruses is another matter. RNA viruses do not typically encode proteins that inhibit apoptosis, for the simple reason that they do not need to. In mammals, RNA viruses often normally cause apoptosis, and apoptosis seems to have little consequence on virus replication. For example, arboviruses such as dengue and West Nile cause lytic infections in mammalian cells, and yet still replicate to high titres. This has been attributed to the rapid replication of RNA viruses, and the lack of a requirement for inducing S-phase in the host cell. However, in mosquitoes, arboviruses typically cause persistent infections which lack cytopathology. Thus, there has been a long-standing question of whether arbovirus infection has negative consequences for mosquito vectors. However, there have been scattered reports of cytopathology similar to apoptosis in mosquitoes infected with various arboviruses (Bowers *et al.*, 2003; Girard *et al.*, 2005; Mims *et al.*, 1966; Vaidyanathan and Scott, 2006; Weaver *et al.*, 1988). Similar to the situation with baculoviruses, it seems that whether apoptosis occurs likely depends on both the virus and host. In most cases, investigators have looked for cytopathology in situations where the mosquito is a competent vector for the virus in question. One would actually not expect to see incompatibility in this type of situation. It is also possible that some arboviruses have evolved as yet undiscovered mechanisms to inhibit apoptosis in their insect vectors. Recent evidence indicates that inhibiting the RNAi response in mosquitoes, which allows higher levels of arbovirus replication, also causes cytolytic infection in cultured mosquito cells and high mortality in infected mosquitoes (Cirimotich *et al.*, 2009). Thus, the role of apoptosis in defence against arboviruses may be underappreciated, but more work is needed to determine whether this is the case.

On the other hand, the insect-pathogenic RNA viruses might be expected to have more severe cytopathic effects in their hosts. In some cases, RNA viruses that are midgut-restricted cause extensive cell sloughing in the midgut, which may be part of their strategy for dispersal to new hosts, as it leads to excretion of the progeny virus (Bong and Sikorowski, 1991; Choi *et al.*, 1989; Inoue and Miyagawa, 1978; Yamaguchi, 1979). In one case, the cell sloughing caused by *Helicoverpa armigera* stunt virus (*Tetraviridae*) has been shown to be due to apoptosis (Brooks *et al.*, 2002). Flock house virus (*Nodaviridae*), which infects insects, fish, and mammals, has been shown to induce apoptosis in *Drosophila* cells, and inhibition of apoptosis by chemical means had no effect on virus replication, suggesting that apoptosis may not be an effective defence in this case (Settles and Friesen, 2008). Much more work is needed in this area, but it appears that during the evolutionary arms race between viruses and insects, some viruses, including ascoviruses and some RNA viruses, instead of inhibiting apoptosis, may have actually incorporated apoptosis into their replication strategies.

Humoral immunity

In addition to the apoptotic antiviral defence responses discussed above, insects possess humoral and cell-mediated defences such as melanization, encapsulation and phagocytosis [first observed by Elie Metchnikoff (1845–1916)]. Within the haemocoel, constitutively expressed and inducible non-cellular defences, such as antimicrobial peptides, lysozymes, proteases and phenoloxidase attack and inhibit growth of entomopathogens (Broderick *et al.*, 2009). In practice, however, the cellular and non-cellular defences are closely coordinated so as to focus the attack on pathogens, and to limit tissue damage (Rolff and Reynolds, 2009). In both vertebrates and invertebrates, pathogens are first recognized via highly conserved regulatory pathways and effectors as 'non-self' by their distinctive biopolymers such as cell wall lipopolysaccharides, peptidoglycans, or β-1,3-glucans, known as pathogen-associated molecular polymers (PAMPs; Matzinger, 2009). In insects, receptors of these non-self polymers, PAMP receptors (PRRs) act as sentinels of infection and are present on the surfaces of haemocytes, various internal organs such as the fat body, and throughout the haemolymph (Schmidt *et al.*, 2008). Detection of the polymers initiates a cascade of intricate interactions between humoral proteins, haemocytes and fat body cells coordinating an attack on the invader, subsequent removal or entombment of debris, and repair of damaged tissues.

Many antimicrobial peptides have been identified in response to bacterial infection (Broderick *et al.*, 2009). The initial humoral response to microbial incursion includes *toll* and *imd* signalling pathway up-regulation and production of antimicrobial peptides, including lysozyme. Many studies in *Drosophila melanogaster* suggest that *toll* and *imd* signalling can be activated by bacteria (De Gregorio *et al.*, 2001; Meng *et al.*, 1999), fungal (Tryselius *et al.*, 1992), or viral infection (Zambon *et al.*, 2005). Both Toll and imd pathways involve the Rel family of transcription factors that are similar to human NF-κB transcription factors. More than 30 *D. melanogaster* antimicrobial peptide genes have been identified and classified into seven families: antifungal antimicrobial peptides (drosomycin and metchnikowin), anti-Gram-negative antimicrobial peptides (attacins, diptericins, drosocins), and anti-Gram-positive antimicrobial peptides (defensins) (Khush and Lemaitre, 2000). In contrast, however, the genes responsible for humoral antiviral responses of insects appear to be unique, and involve opsonization and melanization reactions. Some of the same genes may be involved in the antiviral response but it is as yet unclear to what degree they are involved. Below we outline some of the proteins believed to have a role in antiviral response.

Phenoloxidase

A major component of the insect innate immune response is the zymogen prophenoloxidase (PPO) which is present at high concentration in insect plasma and is vital for the immune response against microbes, filariae, and parasitoids (Kanost and Gorman, 2008; Nappi and Christensen, 2005). Inhibition of PPO activation or other enzymes involved in melanization increased the susceptibility to polydnavirus infection or to parasitization in *Heliothis virescens* (Shelby *et al.*, 2000) but not in *C. fumiferana* (Doucet *et al.*, 2008). Injection of *S. frugiperda* with *Hyposoter didymator* ichnovirus increased transcription of PPO subunits, though injection of *Junonia coenia* densonucleosis virus did not (Barat-Houari *et al.*, 2006). *Micropolitis demolitor* bracovirus produced an epidermal growth factor-like motif protein that inhibits the phenoloxidase (PO) cascade likely by inhibiting the PPO-activating proteinase (Beck and Strand, 2007). *Helicoverpa zea* haemocytes in conjunction with the melanization response formed the lepidopteran's resistance to fatal infection by AcMNPV (Trudeau *et al.*, 2001). PO is believed to have a minor antiviral role in the shrimp, *Marsupenaeus japonicus*, compared to the role of apoptosis and phagocytosis (Wang and Zang, 2008).

A direct role for PO in antiviral immune responses was suggested by the observations that *H. virescens* plasma PO exhibits *in vitro* antiviral activity against several vertebrate viruses (including herpes simplex virus types 1 and 2, vesicular stomatitis virus, parainfluenza-3, coxsackie B3, Sindbis virus, and HIV-1) (Ourth, 2004). Of these, vesicular stomatitis virus and Sindbis virus also infect insects or insect cells.

Activated plasma PO from *H. virescens* directly inhibits baculoviral infections in lepidopteran cells *in vitro* (Popham *et al.*, 2004). Addition of high levels of ascorbic acid to *H. virescens* larval diet abolished this antiviral activity, indicating that dietary ascorbic acid levels may directly inhibit PO-mediated melanization of baculoviruses *in vivo* (Shelby and Popham, 2006). Conversely, the antiviral activity of plasma from *H. virescens* larvae increased when larvae were reared on Se-supplemented diet (Shelby and Popham, 2007). *Pieris brassicae* larvae fed artificial diet without added ascorbic acid beginning at the second instar exhibited no increased mortality to subsequent granulovirus infection (David *et al.*, 1972). Topical treatment with ascorbic acid completely inhibited tomato and cotton foliar peroxidase activity, which significantly enhanced baculoviral mortality of *H. virescens* larvae feeding on the treated tomato plants, relative to untreated foliage (Hoover *et al.*, 1998). Other plasma antiviral effectors beyond PO may also be present as the heat-inactivated haemolymph (i.e. without PO activity) collected from the saturnid moth *Lonomia obliqua* was found to inactivate measles, influenza and polio viruses (Greco *et al.*, 2009). Thus, the presence of soluble factors with antiviral activity in the plasma of insects has been demonstrated, however the contribution of these to susceptibility *in vivo* has yet to be shown.

Humoral responses to DNA viruses

Antiviral responses to baculovirus and densovirus are the subject of much current investigation. A great deal of information is available on how baculovirus genes affect host function (Thiem, 2009), but conversely far less is known about the host response against infection. For baculovirus infections, epithelial barriers, such as the peritrophic matrix and the midgut cells lining the lumen of the digestive tract, are the primary determinants of resistance to baculovirus infections (Clem, 2005). Following per os inoculation, one or more midgut cells become infected and the infection appears to spread from these initial foci to tracheoblasts. The virus obtains access to the haemocoel resulting in a fatal infection (Kirkpatrick *et al.*, 1998). More recent research is now revealing immune reactions to viruses at the level of midgut actions and actions within the haemocoel.

The ability to clear viral infection from haemolymph has been recorded in some virus/host combinations (Fuxa, 1993). Similarly, lepidopteran larval haemocytes recognize and encapsulate AcMNPV-infected tracheoblasts (Washburn *et al.*, 1996; Rivkin *et al.*, 2006). Factors injected into the host by the female parasitic wasp during oviposition (venoms, and symbiotic polydnaviruses) inhibit humoral and cell-mediated immune responses against the wasp eggs and larvae, including melanization catalysed by PO (Escribano *et al.*, 2000; Shelby *et al.*, 2000; Ibrahim and Kim, 2008; Bae and Kim, 2009). Parasitized larvae are subsequently more susceptible to opportunistic microbial and baculoviral infections (Rivkin *et al.*, 2006), which supports the hypothesis that PO plays an important role in resistance to baculoviral infection. A generalized suppression of immune reactions caused by parasitoid wasp oviposition (and their symbiotic polydnaviruses) prevents the clearance of an otherwise non-fatal baculovirus infection.

More recently, several host proteins have been identified which appear to possess antiviral activity (Table 17.1). Using suppression subtractive hybridization of midgut gene expression, gene fragments encoding a protein named REPAT were identified in *S. exigua* larvae to increase in expression in response to baculovirus infection (Herrero *et al.*, 2007). Several proteins with activity against BmNPV are secreted into the digestive juice of *B. mori* larvae, including a serine protease (BmSP-2), a lipase (BmLipase-1), and finally the red fluorescent proteins of mulberry fed larvae via a light-dependent mechanism (Yao *et al.*, 2006).

Baculovirus infection up-regulated the expression of haemolin, a lepidopteran-specific immunoglobulin protein, in the Chinese oak silkmoth *Antheraea pernyi* and the role of haemolin in immunity was confirmed by RNAi of haemolin, which accelerated death of *Antheraea pernyi* NPV-infected pupae (Hirai *et al.*, 2004; Table 17.1). A bracoviral protein from the wasp *Cotesia congregata* binds to *Manduca sexta* haemolin and inhibits the normal Haemolin functions of binding LPS and agglutination of bacteria (Lambropoulou *et al.*, 2008), though polydnavirus infection did not up-regulate a haemolin transcript in a *S. frugiperda* EST library (Barat-Houari *et al.*, 2006). Although haemolin transcription is strongly induced by bacterial infection in *H. virescens,* its expression is not affected by per os baculovirus infection (Terenius *et al.*, 2009).

Haemocytes are key sentinels and actors in the insect immune response against viral pathogens such as baculoviruses (Sparks *et al.*, 2008). Noctuid haemocytes may recognize and encapsulate baculovirus-infected cells (Washburn *et al.*, 1996; Washburn *et al.*, 2000). Using an EST library generated from *H virescens* haemocytes, several transcripts were identified that are believed to be involved in antiviral response (Shelby and Popham, 2009) including an orthologue to the BmNPV-induced *B. mori* Sop2 protein (suppressor of profilin 2, Xu *et al.*, 2005), a *B. mori* secreted protein that is acidic and rich in cysteine, and is induced by BmNPV or densonucleosis infection of *B. mori* midgut (Bao *et al.*, 2008). Other identified transcripts include orthologues of BmNPV-inducible paralytic peptide binding protein of *B. mori* (Hu *et al.*, 2006), *B. mori* humoral lectin haemocytin (Kotani *et al.*, 1995), a scolexin-B transcript, and the orthologue of the *S. frugiperda* scavenger receptor (Table 17.1).

B. mori larval susceptibility to different densovirus strains has recently been reviewed by

Table 17.1 Selected antiviral or antiviral response host proteins

Name	Host Insect	Virus	Reference
Abc transporter	*Bombyx mori*	BmNPV	Sagisaka *et al.* (2009)
BmLipase-1	*Bombyx mori*	BmNPV	Yao *et al.* (2006)
BmSP-2	*Bombyx mori*	BmNPV	Yao *et al.* (2006)
BmToll3	*Bombyx mori*	BmNPV	Sagisaka *et al.* (2009)
Carboxyesterase	*Epiphyas postvittana*	EppoNPV	Gatehouse *et al.* (2009)
CG12780	*D. melanogaster*	DCV[1]	Dostert *et al.* (2005)
CG9080	*D. melanogaster*	DCV	Dostert *et al.* (2005)
Gelsolin	*Epiphyas postvittana*	EppoNPV	Gatehouse *et al.* (2009)
Hdd1	*Heliothis virescens*	CsIV[2]	Nusawardani *et al.* (unpublished)
Haemocytin	*Bombyx mori*	BmNPV	Kotani *et al.* (1995)
Immunolectin a	*Heliothis virescens*	CsIV	Nusawardani *et al.* (unpublished)
Integrin	*Bombyx mori*	BmNPV	Sagisaka *et al.* (2009)
JH epoxide hydrolase	*Epiphyas postvittana*	EppoNPV	Gatehouse *et al.* (2009)
Lysozyme	*Heliothis virescens*	CsIV	Nusawardani *et al.* (unpublished)
Mucin like	*Heliothis virescens*	CsIV	Nusawardani *et al.* (unpublished)
nid-1	*Bombyx mori*	BmDNV-1	Kadono-Okuda (2009)
nsd-1	*Bombyx mori*	BmDNV-1	Kadono-Okuda (2009)
nsd-2	*Bombyx mori*	BmDNV-2	Ito *et al.* (2008)
nsd-Z	*Bombyx mori*	BmDNV-Z	Ito *et al.* (2008)
Paralytic peptide binding protein	*Bombyx mori*	BmNPV	Hu *et al.* (2006)
Pherokines	*D. melanogaster*	DCV	Sabatier *et al.* (2003)
Red fluorescent protein(s)	*Bombyx mori*	BmNPV	Yao *et al.* (2006)
REPAT	*Spodoptera exigua*	AcMNPV	Herrero *et al.* (2007)
Sop2	*Bombyx mori*	BmNPV	Xu *et al.* (2005)
Tetraspannin	*Bombyx mori*	BmNPV	Sagisaka *et al.* (2009)
toll	*Heliothis virescens*	CsIV	Nusawardani *et al.* (unpublished)
Vir-1	*D. melanogaster*	DCV	Dostert *et al.* (2005)

[1]*Drosophila* C virus.

[2]*Campoletis sonorensis* ichnovirus.

Kadono-Okuda (2009). Larval *B. mori* resistance to the type 1 densovirus BmDNV-1 is conferred by two genes, the dominant non-susceptibility to BmDNV-1 gene (*nid-1*) and the recessive non-susceptibility to BmDNV-1 gene (*nsd-1*), both of unknown function (Kadono-Okuda, 2009; Table 17.1). Resistance to the type 2 and Z densoviruses is conferred by the recessive *nsd*-2 and by *nsd*-Z genes, respectively (Ito *et al.*, 2008). Detailed mapping and analysis of the *nsd*-2 gene product indicated that a C-terminal deletion in a putative silkworm midgut amino acid transporter confers complete resistance against infection by BmDNV-2 (Ito *et al.*, 2008).

Identification of host genes responding to viral challenge has recently been explored in several microarray studies. *Drosophila* C virus challenge induced 140 genes in *D. melanogaster*, including vir-1, CG12780 and CG9080 (Dostert *et al.*, 2005; Roxtrom-Lindquist *et al.*, 2004; Table 17.1). Following per os infection of the light-brown apple moth, *E. postvittana*, with EppoNPV, microarray and further qPCR identified 21 insect genes that were strongly up-regulated and 31 that were strongly down-regulated. Many of these genes are involved in apoptosis, juvenility, actin rearrangement, transcription and translation (Gatehouse *et al.*, 2009). A microarray of gene expression in haemocytes and fat body following injection of *S. frugiperda* larvae with *Hyposoter didymator* ichnovirus found 4% of the 1750 studied genes to be altered in expression (Barat-Houari *et al.*, 2006). These included eight immune-related genes, namely both PPO subunits, prophenoloxidase activating enzyme, galectin, a novel scavenger receptor, immulectin-2, lysozyme, and calreticulin (Barat-Houari *et al.*, 2006). Three gene products, a fatty acid binding protein, actin A3, and heat shock protein 60, were also identified as being differentially expressed in *B. mori* strains that are either highly resistant or susceptible to *Bm*NPV infection (Bao *et al.*, 2009). Further quantitative polymerase chain reaction (qPCR) quantitation of actin A3 highlighted much lower expression of the gene in susceptible larvae than in resistant larvae (Bao *et al.*, 2009). The transcriptional profile of host genes in the silkworm cell line during the early phase of BmNPV infection found 35 genes significantly up-regulated and 17 genes significantly down-regulated. Further quantification of the levels of mRNA expression by quantitative reverse transcriptase PCR confirmed that the expression of 13 (such as BmEts and BmToll10-3) genes significantly increased and seven genes (such as Hsp20-1) significantly decreased after BmNPV infection (Sagisaka *et al.*, 2009).

Beyond antiviral responses in insects, insect viruses are being used to study the antiviral response in mammals (Hu, 2006). In the transduction of baculoviruses for gene therapy in humans, the mammalian antiviral response has been studied extensively with an eye to the safety of use of an insect virus in humans (Boulaire *et al.*, 2009; Chen *et al.*, 2009). Additionally, baculoviruses have recently been used to induce the innate immune response of mice and chickens to protect the animals from infection by other viruses (Abe *et al.*, 2005; Niu *et al.*, 2008).

Humoral responses to RNA viruses

Owing to the plethora of positive-, negative-, and double-stranded RNA (dsRNA) viruses infecting *Drosophila* and mosquitoes, much of the recent progress in antiviral innate immunity of insects has accumulated with these pathosystems, and has been thoroughly reviewed (see Huszar and Imler, 2008; Cherry and Silverman, 2006; Sparks *et al.*, 2008; Fragkoudis *et al.*, 2009; Chen and Siede, 2007). Therefore, in this chapter we will summarize specific activation of the response and host-specific proteins up-regulated by infection. The highly conserved RNAi multi-protein complex is activated by the detection of dsRNA viral genomes or replication intermediates in insect cells (Kemp and Imler, 2009). In brief, the protein Dicer-2 binds to and cleaves the viral dsRNA into small interfering RNAs (siRNA) which enter into the RNA-induced silencing complex (RISC). The siRNA-primed RISC seeks out and binds to complementary RNA sequences which are then cleaved into inactive fragments by argonaute, thus destroying viral RNA within the cell and preventing translation, replication and infection (Imler and Eleftherianos, 2009; also see Chapter 16 for more details). Viral dsRNA taken up by *Drosophila* cells activates antiviral immunity (Saleh *et al.*, 2009). Proteins in the RNAi pathway suppress infection by the RNA viruses dengue virus, O'nyong-nyong virus, West Nile virus, vesicular stomatitis virus, *Drosophila* C virus, Flock House virus, and Sindbis virus, but not by the double-stranded DNA vaccinia virus (Sabin *et al.*, 2009).

Activation of the RNAi defence may be mediated via the *toll* and *imd* pathways. In *D. melanogaster*, the *toll* signalling pathway is activated by *Drosophila* X virus (Zambon *et al.*, 2005). *D. melanogaster* resistance to cricket paralysis virus (Costa *et al.*, 2009) and Sindbis virus (Avadhanula *et al.*, 2009) involves the *imd* pathway, while defence against *Drosophila* C virus requires the JAK-STAT signalling pathway (Dostert *et al.*, 2005). *Drosophila* sigma virus seemed not to activate any of the three signalling pathways (Carpenter *et al.*, 2009); this is despite the finding of a *toll* pathway intermediate resistance locus called *ref*(2)P (Carre-Mlouka *et al.*, 2007). Clearance of a Nora virus infection was not dependent upon any RNAi pathway, or activation of signalling pathways (Habayeb *et al.*, 2009). Transcript profiling of RNA virus-infected insects has revealed a limited subset of differentially regulated proteins. JAK-STAT, *toll*, and *imd* signalling pathway components are generally up-regulated following infection, as are antimicrobial peptides and proteins in some, but not all virus–host systems (Xi *et al.*, 2008; Souza-Neto *et al.*, 2009; Carpenter *et al.*, 2009; Barat-Houari *et al.*, 2006).

New methods

Susceptibility of insects to virus infection has long fascinated researchers, attracting much effort using a variety of techniques (Sparks *et al.*, 2008; Imler and Eleftherianos, 2009). The field is paradoxically limited however by the extremely wide array of host insect species and viruses under investigation. Until quite recently scant proteomic or genomic resources were available for application to the problem and thus progress has lagged behind that achieved with mammalian/viral systems. Application of techniques such as fluorescent differential display enabled identification of induced *Bm*Sop2 (Xu *et al.*, 2005) and *Bm*PP (Hu *et al.*, 2006) proteins in infected *B. mori* larvae, while suppression/subtraction hybridization gene expression profiling was employed by Bao *et al.* (2009) to elucidate differences between BmNPV-susceptible and -non-susceptible strains of *B. mori*. Selection and genetic mapping have identified host susceptibility loci for *Drosophila* sigma virus (Carre-Mlouka *et al.*, 2007), densovirus (Kadono-Okuda, 2009), granulovirus (Asser-Kaiser *et al.*, 2007), and baculovirus (Yao *et al.*, 2003). Completion of the *B. mori* genome (International Silkworm Genome Consortium, 2008) and development of rapid SNP genotyping methods will speed positional cloning of susceptibility loci to viruses infecting this insect (Kadono-Okuda, 2009; Goldsmith, 2009). As noted in a previous section, numerous labs have successfully adapted the oligonucleotide microarray approach to identify candidate insect antiviral response factors and other innate immunity proteins (Tsai *et al.*, 2008; Johnson *et al.*, 2009; Gatehouse *et al.*, 2009; Barat-Houari *et al.*, 2006; T. Nusawardani, K.S. Shelby, H.J.R. Popham, M. Suzuki, and B.A. Webb, unpublished). In this section, we wish to highlight promising newer approaches to gene candidate analyses such as gene silencing and next-generation sequencing.

Sequential silencing of selected individual baculovirus genes by RNAi *in vitro* and *in vivo* can result in host survival (Kanginakudru *et al.*, 2007; Isobe *et al.*, 2004). However, for host species with significant transcriptome or genome sequence available, an ingenious inversion of the microarray concept allowing direct phenotype determination on virus infected cells, *in vitro* genome-wide RNAi, has been introduced (Cherry, 2009a). In this approach, *D. melanogaster* cells pre-treated with dsRNA designed to knock down a single *D. melanogaster* transcript per well were infected with viruses expressing fluorescent reporter genes. Quantitation of fluorescence thus enabled a high throughput knock down screen covering the entire *D. melanogaster* transcriptome. Using this approach, genes were identified that act in various cell processes such as endocytosis and autophagy that are involved in host resistance to RNA viruses (Cherry, 2009b; Cherry *et al.*, 2005; Cherry and Perrimon, 2004; Sabin *et al.*, 2009; Shelly *et al.*, 2009). The approach has since been adopted to study other cell–virus associations such as influenza (Hao *et al.*, 2008), dengue virus (Sessions *et al.*, 2009), and West Nile virus (Krishnan *et al.*, 2008). The genome-wide RNAi technique was even successfully extended into transgenic *Drosophila* larvae to identify entire suites of genes involved in oral *Serracia marcescens* infection (Cronin *et al.*, 2009). A similar approach to screening mosquito or lepidopteran cells would undoubtedly identify the majority of host responses to a number of important viruses.

Next-generation sequencing platforms will allow a faster and wider window on virus–host interactions (Morozova *et al.*, 2009). These extremely high-throughput sequencing techniques, unlike micorarrays, do not rely on oligomers predesigned from known transcripts; and in addition to directly quantifying changes in expression levels, have the further advantage of deeper sampling of novel expressed sequences

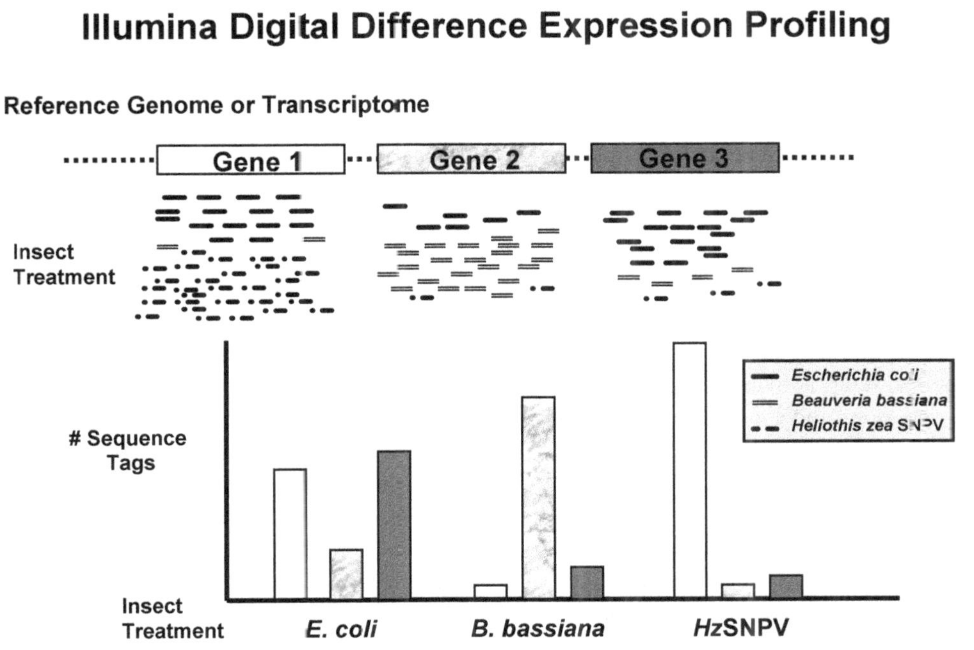

Figure 17.5 Identification of susceptibility or host response factors through digital expression profiling using short tag next-generation high throughput sequencing. Transcripts differentially expressed under different treatment regimes, e.g. baculoviral (HzSNPV; broken black line), bacterial (*Escherichia coli*; black line) or fungal (*Beauveria bassiana*; grey line) are extracted from tissues and used to generate short sequence tags with high throughput next generation methods (e.g. Illumina or SOLiD). Tags are aligned with a reference genome or transcriptome sequence. In the example shown, transcripts of Gene 1 are induced by HzSNPV infection and result in production of a greater number of sequence tags (broken black line) while transcripts of Genes 2 and 3 are less affected. The count of tags aligned with the known sequence of the gene is a quantitative measure of expression.

lacking known homologues, allelic variation, alternative splicing variants, and unannotated regions. This is a decisive advantage in insect virology where only a few well-annotated host reference genomes among the wide range of target species are currently available for transcript profiling, e.g. *Drosophila* spp., silkworm, mosquitoes, honey bee, red flour beetle, pea aphid and others for which only ESTs have been deposited in public databases. Next-generation sequencing techniques enable cost-effective genome-wide digital expression profiling approaches that avoid the biases introduced by library construction and *in vivo* cloning. The 454 (Roche) high throughput sequencing platform currently provides more easily assembled, longer reads, while in contrast assembly of the far higher throughput, short read sequencing platforms Illumina (http://www.illumina.com/), SoLiD (http://www3.appliedbiosystems.com) and Helicos (Helicos BioSciences; http://www.helicosbio.com/) are currently limited by computational methods, although these methods are rapidly improving. Digital difference profiling can be used to capture the entire transcriptional host response to an entomopathogen (Fig. 17.5). Short tags generated by the above technologies could allow the quantitative, simultaneous measurement of the entire interleaving transcriptomic responses of both the host and the entomopathogen during the course of an infection (e.g. Yang *et al.*, 2010; Legendre *et al.*, 2010). Elimination of PCR amplification bias and direct sequencing of RNA promises to increase throughput and specificity even further (Mamanova *et al.*, 2010).

Concluding remarks

Prior to a few years ago, relatively little was known about antiviral responses in insects. Over the past few years, however, progress in this area has dramatically accelerated, with the advent of genetic studies of antiviral responses using *Drosophila*, and studies of the role of RNAi and other pathways in disease vectors. Functional genomics tools such as genetic manipulation, RNAi, microarrays and deep sequencing strategies are allowing rapid progress to be made in understanding antiviral immunity in both model and non-model organisms. A singularly promising application of next-generation methods is metagenomics, i.e. simultaneous sampling and identification of all associated commensal, symbiotic and pathogenic organisms within a host or ecosystem sample. This approach was recently used to confirm the association of Israeli acute paralysis virus with colony collapse disorder of *Apis mellifera* (Cox-Foster *et al.*, 2007). With application of this powerful arsenal of newer methods to entomopathogens, their hosts and associated microbiota, we expect a bewildering and exciting expansion of our knowledge in the very near future.

Apoptosis has been known to be an antiviral response in insects for some time, but many questions remain. So far, the ability of apoptosis to limit virus infection in insects has exclusively been studied using baculoviruses in lepidopteran larvae. Whether apoptosis can be an effective defence against other families of DNA viruses, and in particular, against RNA viruses, in lepidoptera and other insects remains to be determined. Even within the baculovirus-Lepidoptera system, the effectiveness of apoptosis appears to vary widely depending on the particular virus and host in question. Further investigations into the mechanisms of apoptosis induction by baculoviruses may aid in determining why apoptosis results in some virus–host combinations, but not in others. The exact mechanisms involved in the protective effects of apoptosis against virus infection *in vivo* are still vague, and more study is needed in this area. Synergistic interactions between apoptosis and other types of antiviral responses are likely, and together these types of interactions undoubtedly govern the outcome of infection.

While apoptosis is likely an ancient type of intracellular immunity against viruses, some viruses appear to have co-opted apoptosis as part of their replication strategies. These include the DNA-containing ascoviruses, which utilize the membrane blebbing mechanisms of apoptosis to aid in their replication and dispersal, and insect pathogenic RNA viruses, which may utilize apoptosis in the gut to facilitate horizontal infection. Studies of these types of interactions will broaden our perspective of how apoptosis influences virus infection in insects.

We do not yet know whether apoptosis affects arboviruses, which do not appear to normally cause cytopathic effects in their insect vectors, despite the damage they do in vertebrate hosts. Whether arboviruses actively inhibit apoptosis in their insect vectors, or simply avoid stimulating apoptosis in the first place, is not yet known. However, interesting recent results indicate that inhibition of RNAi in infected mosquitoes, which allows higher levels of replication, causes cell death and increased mortality in the infected mosquitoes (Cirimotich *et al.*, 2009). Thus, a fine balance has apparently evolved between arboviruses and their mosquito vectors, in which virus replication is inhibited by RNAi just enough so that it is not cytolytic, but not so much that virus dissemination to new hosts is prevented. This leads to the obvious question of whether ectopically triggered apoptosis could provide protection against arbovirus infection in insect vectors. It is possible that strategies involving manipulation of apoptosis could be incorporated into future control strategies for arboviral diseases.

Acknowledgements

Mention of trade names or commercial products in this article is solely for the purpose of providing specific information and does not imply recommendation or endorsement by the US Department of Agriculture.

References

Abe, T., Hemmi, H., Miyamoto, H., Moriishi, K., Tamura, S., Takaku, H., Akira, S., and Matsuura,Y. (2005). Involvement of the toll-like receptor 9 signaling pathway in the induction of innate immunity by baculovirus. J. Virol. 79, 2847–2858.

Asser-Kaiser, S., Fritsch, E., Undorf-Spahn, K., Kienzle, J., Eberle, K.E., Gund, N.A., Reineke, A., Zebitz, C.P.W., Heckel, D.G., and Jehle, J.A. (2007). Rapid emergence of baculovirus resistance in Codling moth due to dominant, sex-linked inheritance. Science *317*, 1916–1918.

Avadhanula, V., Weasner, B.P., Hardy, G.G., Kumar, J.P., and Hardy, R.W. (2009). A novel system for the launch of alphavirus RNA synthesis reveals a role for the IMD pathway in arthropod antiviral response. PLoS Pathog. *5*, e1000582.

Bae, S., and Kim, Y. (2009). IkB genes encoded in *Cotesia plutellae* bracovirus suppress an antiviral response and enhance baculovirus pathogenicity against the diamondback moth, *Plutella xylostella*. J. Invertebr. Pathol. *102*, 79–87.

Bao, Y.Y., Li, M.-W., Zhao, Y.-P., Ge, J.-Q., Wang, C.-S., Huang, Y.-P., and Zhang, C.-X. (2008). Differentially expressed genes in resistant and susceptible *Bombyx mori* strains infected with a densonucleosis virus. Insect Biochem. Mol. Biol. *38*, 853–861.

Bao, Y.Y., Tang, X.-D., Lv, Z.-Y., Wang, X.-Y., Tian, C.-H., Xu, Y. P., and Zhang, C.-X. (2009). Gene expression profiling of resistant and susceptible *Bombyx mori* strains reveals nucleopolyhedrovirus-associated variations in host gene transcript levels. Genomics *94*, 138–145.

Barat-Houari, M., Hilliou, F., Jousset, F.-X., Sofer, L., Deleury, E., Rocher, J., Ravallec, M., Galibert, L., Delobel, P., Feyereisen, R., Fournier, P., and Volkoff, N. (2006). Gene expression profiling of *Spodoptera frugiperda* haemocytes and fat body using cDNA microarray reveals polydnavirus-associated variations in lepidopteran host genes transcript levels. BMC Genomics *7*, 160.

Beck, M.H., and Strand, M.R. (2007). A novel polydnavirus protein inhibits the insect prophenoloxidase activation pathway. Proc. Natl. Acad. Sci. U.S.A. *104*, 19267–19272.

Bideshi, D.K., Tan, Y., Bigot, Y., and Federici, B.A. (2005). A viral caspase contributes to modified apoptosis for virus transmission. Genes Dev. *19*, 1416–1421.

Birnbaum, M.J., Clem, R.J., and Miller, L.K. (1994). An apoptosis-inhibiting gene from a nuclear polyhedrosis virus encoding a peptide with cys/his sequence motifs. J. Virol. *68*, 2521–2528.

Bong, C.F.J., and Sikorowski, P.P. (1991). Histopathology of cytoplasmic polyhedrosis virus (Reoviridae) infection in corn earworm, *Helicoverpa zea* (Boddie), larvae (*Insecta: Lepidoptera: Noctuidae*). Can. J. Zool. *69*, 2127–2130.

Boulaire, J., Zhao, Y., and Wang, S. (2009). Gene expression profiling to define host response to baculoviral transduction in the brain. J. Neurochem. *109*, 1203–1214.

Bowers, D.F., Coleman, C.G., and Brown, D.T. (2003). Sindbis virus-associated pathology in Aedes albopictus (Diptera: Culicidae). J. Med. Entomol. *40*, 698–705.

Broderick, N.A., Welchman, D.P., and Lemaitre, B. (2009). Recognition and response to microbial infection in *Drosophila*. In Insect Infection and Immunity: Evolution, Ecology, and Mechanisms, J. Rolff and S.E. Reynolds, eds (New York, Oxford University Press), pp. 13–33.

Brooks, E.M., Gordon, K.H.J., Dorrian, S.J., Hines, E.R., and Hanzlik, T.N. (2002). Infection of its lepidopteran host by the *Helicoverpa armigera stunt virus* (*Tetraviridae*). J. Invertebr. Pathol. *80*, 97–111.

Bryant, B., and Clem, R.J. (2009). Caspase inhibitor P35 is required for the production of robust baculovirus virions in *Trichoplusia ni* TN-368 cells. J. Gen. Virol. *90*, 654–661.

Carpenter, J., Hutter, S., Baines, J.F., Roller, J., Saminadin-Peter, S.S., Parsch, J., and Jiggins, F.M. (2009). The transcriptional response of *Drosophila melanogaster* to infection with the sigma virus (Rhabdoviridae). PLoS One *4*, e6838.

Carpes, M.P., Nunes, J.F., Sampaio, T.L., Castro, M.E., Zanotto, P.M., and Ribeiro, B.M. (2009). Molecular analysis of a mutant *Anticarsia gemmatalis* multiple nucleopolyhedrovirus (AgMNPV) shows an interruption of an inhibitor of apoptosis gene (*iap-3*) by a new class-II piggyBac-related insect transposon. Insect Mol. Biol. *18*, 747–757.

Carre-Mlouka, A., Gaumer, S., Gay, P., Petitjean, A.M., Coulondre, C., Dru, P., Bras, F., Dezelee, S., and Contamine, D. (2007). Control of sigma virus multiplication by the ref(2)P gene of *Drosophila melanogaster*: An *in vivo* study of the PB1 domain of Ref(2)P. Genetics *176*, 409–419.

Chacon, M.R., Almazan, F., Nogal, M.L., Vinuela, E., and Rodriguez, J.F. (1995). The African swine fever virus IAP homolog is a late structural polypeptide. Virology *214*, 670–674.

Chejanovsky, N., and Gershburg, E. (1995). The wild-type *Autographa californica* nuclear polyhedrosis virus induces apoptosis of *Spodoptera littoralis* cells. Virology *209*, 519–525.

Chen, G.-Y., Shiah, H.-C., Su, H.-J., Chen, C.-Y., Chuang, Y.-J., Lo, W.-H., Huang, J.-L., Chuang, C.-K., Hwang, S.-M., and Hu, Y.-C. (2009). Baculovirus transduction of mesenchymal stem cells triggers the toll-like receptor 3 pathway. J. Virol. *83*, 10548–10556.

Chen, Y.P., and Siede, R. (2007). Honey bee viruses. In Advances in Virus Research, K.Maramorosch, A.J.Shatkin, and F.A.Murphy, eds. (New York, NY: Elsevier), pp. 33–80.

Chen, Y.P., and Siede, R. (2007). Honey bee viruses. Adv. Virus Res. *70*, 33–80.

Cherry, S. (2009a). What have RNAi screens taught us about viral–host interactions? Curr. Op. Microbiol. *12*, 446–452.

Cherry, S. (2009b). VSV infection is sensed by *Drosophila*, attenuates nutrient signaling, and thereby activates antiviral autophagy. Autophagy *5*, 1062–1063.

Cherry, S., Doukas, T., Armknecht, S., Whelan, S., Wang, H., Sarnow, P., and Perrimon, N. (2005). Genome-wide RNAi screen reveals a specific sensitivity of IRES-containing RNA viruses to host translation inhibition. Genes Dev. *19*, 445–452.

Cherry, S., and Perrimon, N. (2004). Entry is a rate-limiting step for viral infection in a *Drosophila melanogaster* model of pathogenesis. Nat. Immunol. *5*, 81–87.

Cherry, S., and Silverman, N. (2006). Host–pathogen interactions in *Drosophila*: new tricks from an old friend. Nat. Immunol. *7*, 911–917.

Chikhalya, A., Luu, D.D., Carrera, M., De La Cruz, A., Torres, M., Martinez, E.N., Chen, T., Stephens, K.D., and Haas-Stapleton, E.J. (2009). Pathogenesis of *Autographa californica* multiple nucleopolyhedrovirus in fifth-instar *Anticarsia gemmatalis* larvae. J. Gen. Virol. *90*, 2023–2032.

Choi, H.K., Kobayashi, M., and Kawase, S. (1989). Changes in infectious flacherie virus specific polypeptides and translatable m-RNA in the midgut of the silkworm, *Bombyx mori*, during larval molt. J. Invertebr. Pathol. *53*, 128–131.

Cirimotich, C.M., Scott, J.C., Phillips, A.T., Geiss, B.J., and Olson, K.E. (2009). Suppression of RNA interference increases alphavirus replication and virus-associated mortality in *Aedes aegypti* mosquitoes. BMC Microbiol. *9*, 49. doi:10.1186/1471-2180-9-49.

Clarke, T.E., and Clem, R.J. (2002). Lack of involvement of haemocytes in the establishment and spread of infection in *Spodoptera frugiperda* larvae infected with the baculovirus *Autographa californica* M nucleopolyhedrovirus by intrahemocoelic injection. J. Gen. Virol. *83*, 1565–1572.

Clarke, T.E., and Clem, R.J. (2003). *In vivo* induction of apoptosis correlating with reduced infectivity during baculovirus infection. J. Virol. *77*, 2227–2232.

Clem, R.J. (2001). Baculoviruses and apoptosis: the good, the bad, and the ugly. Cell Death Differ. *8*, 137–143.

Clem, R.J. (2005). The role of apoptosis in defense against baculovirus infection in insects. Curr. Top. Microbiol. Immunol. *289*, 113–129.

Clem, R.J. (2007). Baculoviruses and apoptosis: A diversity of genes and responses. Curr. Drug Targets *8*, 1069–1074.

Clem, R.J., Fechheimer, M., and Miller, L.K. (1991). Prevention of apoptosis by a baculovirus gene during infection of insect cells. Science *254*, 1388–1390.

Clem, R.J., and Miller, L.K. (1993). Apoptosis reduces both the *in vitro* replication and the *in vivo* infectivity of a baculovirus. J. Virol. *67*, 3730–3738.

Clem, R.J., and Miller, L.K. (1994). Control of programmed cell death by the baculovirus genes *p35* and *iap*. Mol. Cell. Biol. *14*, 5212–5222.

Clem, R.J., Robson, M., and Miller, L.K. (1994). Influence of infection route on the infectivity of baculovirus mutants lacking the apoptosis-inhibiting gene *p35* and the adjacent gene *p94*. J. Virol. *68*, 6759–6762.

Clouston, W.M., and Kerr, J.F.R. (1985). Apoptosis, lymphocytotoxicity and the containment of viral infections. Med. Hypoth. *18*, 399–404.

Costa, A., Jan, E., Sarnow, P., and Schneider, D. (2009). The Imd pathway is involved in antiviral immune responses in *Drosophila*. PLoS One *4*, e7436.

Cox-Foster, D.L., Conlan, S., Holmes, E.C., Palacios, G., Evans, J.D., Moran, N.A., Quan, P.L., Briese, T., Hornig, M., Geiser, D.M., Martinson, V., van Engelsdorp, D., Kalkstein, A.L., Drysdale, A., Hui, J., Zhai, J., Cui, L., Hutchison, S.K., Simons, J.F., Egholm, M., Petris, J.S., and Lipkin, W.I. (2007). A metagenomic survey of microbes in honey bee colony collapse disorder. Science *318*, 283–287.

Cronin, S.J.F., Nehme, N.T., Limmer, S., Liegeois, S., Pospisilik, J.A., Schramek, D., Leibbrandt, A., de Matos Simoes, R., Gruber, S., Puc, U., Ebersberger, I., Zoranovic, T., Neely, G.G., von Haeseler, A., Ferandon, D., and Penninger, J.M. (2009). Genome-wide RNAi screen identifies genes involved in intestinal pathogenic bacterial infection. Science *325*, 340–343.

Crook, N.E., Clem, R.J., and Miller, L.K. (1993). An apoptosis-inhibiting baculovirus gene with a zinc finger-like motif. J. Virol. *67*, 2168–2174.

da Silveira, E.B., Cordeiro, B.A., Ribeiro, B.M., and Bao, S.N. (2005). *In vivo* apoptosis induction and reduction of infectivity by an *Autographa californica* multiple nucleopolyhedrovirus *p35*⁻ recombinant in haemocytes from the velvet bean caterpillar *Anticarsia gemmatalis* (Hubner) (*Lepidoptera: Noctuidae*). Res. Microbiol. *156*, 1014–1025.

da Silveira, E.B., Cordeiro, B.A., Ribeiro, B.M., de Castro, M.E., Soares, E.F., and Bao, S.N. (2007). An *Anticarsia gemmatalis* multiple nucleopolyhedrovirus mutant, vApAg, induces haemocytes apoptosis *in vivo* and displays reduced infectivity in larvae of *Anticarsia gemmatalis* (Hubner) (*Lepidoptera: Noctuidae*). Virus Res. *130*, 182–192.

David, W.A.L., Ellaby, S., and Taylor, G. (1972). The effect of reducing the content of certain ingredients in a semisynthetic diet on the incidence of granulosis virus disease in *Pieris brassicae*. J. Invertebr. Pathol. *20*, 332–340.

de Castro, M.E.B., and Ribeiro, B.M. (2001). Production of viral progeny in insect cells undergoing apoptosis induced by a mutant *Anticarsia gemmatalis* nucleopolyhedrovirus. Microbiol. Res. *156*, 369–376.

De Gregorio, E., Spellman, P.T., Rubin, G.M., and Lemaitre, B. (2001). Genome-wide analysis of the Drosophila immune response by using oligonucleotide microarrays. Proc. Natl. Acad. Sci. U.S.A. *98*, 12590–5.

Dostert, C., Jouanguy, E., Irving, P., Troxler, L., Galiana-Arnoux, D., Hetru, C., Hoffmann, J.A., and Imler, J.L. (2005). The Jak-STAT signaling pathway is required but not sufficient for the antiviral response of *Drosophila*. Nat. Immunol. *6*, 946–953.

Doucet, D., Beliveau, C., Dowling, A., Simard, J., Feng, Q., Krell, P.J., and Cusson, M. (2008). Prophenoloxidases 1 and 2 from the Spruce budworm, *Choristoneura fumiferana*: molecular cloning and assessment of transcriptional regulation by a polydnavirus. Arch. Insect Biochem. Physiol. *67*, 188–201.

Dougherty, E.M., Narang, N., Loeb, M.J., Lynn, D.E., and Shapiro, M. (2006). Fluorescent brightener inhibits apoptosis in baculovirus-infected gypsy moth larval midgut cells *in vitro*. Biocontrol Sci. Technol. *16*, 157–168.

Du, X., and Thiem, S.M. (1997). Responses of insect cells to baculovirus infection: protein synthesis shutdown and apoptosis. J. Virol. *71*, 7866–7872.

Eddins, M.J., Lemongello, D., Friesen, P.D., and Fisher, A.J. (2002). Crystallization and low-resolution structure of an effector–caspase/P35 complex: similarities and differences to an initiator–caspase/P35 complex. Acta Cryst. *D58*, 299–302.

Elliott, M.R., Chekeni, F.B., Trampont, P.C., Lazarowski, E.R., Kadl, A., Walk, S.F., Park, D., Woodson, R.I., Ostankovich, M., Sharma, P., Lysiak, J.J., Harden, T.K., Leitinger, N., and Ravichandran, K.S. (2009). Nucleotides released by apoptotic cells act as a find-me signal to promote phagocytic clearance. Nature *461*, 282–286.

Engelhard, E.K., Kam-Morgan, L.N.W., Washburn, J.O., and Volkman, L.E. (1994). The insect tracheal system: A conduit for the systemic spread of *Autographa californica* M nuclear polyhedrosis virus. Proc. Natl. Acad. Sci. U.S.A. *91*, 3224–3227.

Escribano, A., Williams, T., Goulson, D., Cave, R.D., Chapman, J.W., and Caballero, P. (2000). Effect of parasitism on a nucleopolyhedrovirus amplified in *Spodoptera frugiperda* larvae parasitized by *Campoletis sonorensis*. Entomol. Exp. Appl. *97*, 257–264.

Fire, A. (2005). Nucleic acid structure and intracellular immunity: some recent ideas from the world of RNAi. Q. Rev. Biophys. *38*, 303–300

Fragkoudis, R., Attarzadeh-Yazdi, G., Nash, A.A., Fazakerley, J.K., and Kohl, A. (2009). Advances in dissecting mosquito innate immune responses to arbovirus infection. J. Gen. Virol. *90*, 2061–2072.

Friesen, P.D., and Miller, L.K. (1987). Divergent transcription of early 35 and 94-kilodalton protein genes encoded by the *Hind*-III K genome fragment of the baculovirus *Autographa californica* nuclear polyhedrosis virus. J. Virol. *61*, 2264–2272.

Fuxa, J.R. (1993). Insect resistance to viruses. In *Parasites and Pathogens of Insects, Vol. 2, Pathogens*, N.E.Beckage, S.N.Thompson, and B.A.Federici, eds. (New York: Academic Press), pp. 197–209.

Garcia-Maruniak, A., Maruniak, J.E., Farmerie, W., and Boucias, D.G. (2008). Sequence analysis of a non-classified, non-occluded DNA virus that causes salivary gland hypertrophy of *Musca domestica*, MdSGHV. Virology 377, 184–196.

Gatehouse, H.S., Poulton, J., Markwick, N.P., Gatehouse, L.N., Ward, V.K., Young, V.L., Luo, Z., Schaffer, R., and Christeller, J.T. (2009). Changes in gene expression in the permissive larval host lightbrown apple moth (*Epiphyas postvittana,* Tortricidae) in response to EppoNPV (Baculoviridae) infection. Insect Mol. Biol. *18*, 635–648.

Gershburg, E., Rivkin, H., and Chejanovsky, N. (1997). Expression of the *Autographa californica* nuclear polyhedrosis virus apoptotic suppressor gene *p35* in nonpermissive *Spodoptera littoralis* cells. J. Virol. *71*, 7593–7599.

Girard, Y.A., Popov, V., Wen, J., Han, V., and Higgs, S. (2005). Ultrastructural study of West Nile virus pathogenesis in *Culex pipiens quinquefasciatus* (Diptera: Culicidae). J. Med. Entomol. 42, 429–444.

Girard, Y.A., Schneider, B.S., McGee, C.E., Wen, J., Han, V.C., Popov, V., Mason, P.W., and Higgs, S. (2007). Salivary gland morphology and virus transmission during long-term cytopathologic West Nile virus infection in *Culex* mosquitoes. Am. J. Trop. Med. Hyg. *76*, 118–128.

Goldsmith, M.R. (2009). Recent progress in silkworm genetics and genomics. In Molecular Biology and Genetics of the Lepidoptera, M.R.Goldsmith and F.Marec, eds. (Boca Raton, FL: CRC Press), pp. 25–48.

Greco, K.N., Mendonca, R.M.Z., Moraes, R.H.P., Mancini, D.A.P., and Mendonca, R.Z. (2009). Antiviral activity of the haemolymph of *Lonomia obliqua* (Lepidoptera: Saturniidae). Antiviral Res. *84*, 84–90.

Griffiths, C.M., Barnett, A.L., Ayres, M.D., Windass, J., King, L.A., and Possee, R.D. (1999). *In vitro* host range of *Autographa californica* nucleopolyhedrovirus recombinants lacking functional *p35, iap1* or *iap2*. J. Gen. Virol. *80*, 1055–1066.

Haas-Stapleton, E., Washburn, J.O., and Volkman, L.E. (2003). Pathogenesis of *Autographa californica* M nucleopolyhedrovirus in fifth instar *Spodoptera frugiperda*. J. Gen. Virol. *84*, 2033–2040.

Habayeb, M.S., Ekstrom, J.-O., and Hultmark, D. (2009). Nora virus persistent infections are not affected by the RNAi machinery. PLoS One *4*, e5731.

Hao, L., Sakurai, A., Watanabe, T., Sorensen, E., Nidom, C.A., Newton, M.A., Ahlquist, P., and Kawaoka, Y. (2008). *Drosophila* RNAi screen identifies host genes important for influenza virus replication. Nature *454*, 890–893.

Herrero, S., Ansems, M., Van Oers, M.M., Vlak, J.M., Bakker, P.L., and de Maagd, R.A. (2007). REPAT, a new family of proteins induced by bacterial toxins and baculovirus infection in *Spodoptera exigua*. Insect Biochem. Mol. Biol. *37*, 1109–1118.

Hershberger, P.A., Dickson, J.A., and Friesen, P.D. (1992). Site-specific mutagenesis of the 35-kilodalton protein gene encoded by *Autographa californica* nuclear polyhedrosis virus: cell line-specific effects on virus replication. J. Virol. *66*, 5525–5533.

Hirai, M., Terenius, O., and Faye, I. (2004). Baculovirus and dsRNA induce Hemolin, but not antibacterial activity, in *Antheraea pernyi*. Insect Mol. Biol. *13*, 399–405.

Hoover, K., Kishida, K.T., Digiorgio, L.A., Workman. J., Alaniz, S.A., Hammock, B.D., and Duffey, S.S. (1998). Inhibition of baculoviral disease by plant mediated peroxidase activity and free radical generation. J. Chem. Ecol. *24*, 1949–2001.

Hu, Y.C. (2006). Baculovirus vectors for gene therapy. Adv. Virus Res. *68*, 287–320.

Hu, Z.G., Chen, K.P., Yao, Q., Gao, G.T., Xu, J.P., and Chen, H.Q. (2006). Cloning and Characterization of *Bombyx mori PP-BP*, a Gene Induced by Viral Infection. Acta Genetica Sinica *3*, 975–983.

Hultmark, D. (2003). *Drosophila* immunity: paths and patterns. Curr. Opin. Immunol. *15*, 12–19.

Hussain, M., and Asgari, S. (2008). Inhibition of apoptosis by *Heliothis virescens* ascovirus (HvAV-3e): characterization of orf28 with structural similarity to inhibitor of apoptosis proteins. Apoptosis *13*, 1417–1426.

Huszar, T., and Imler, J.-L. (2008). *Drosophila* viruses and the study of antiviral host-defense. In Advances in Virus Research, Elsevier), pp. 227–265.

Ibrahim, A.M., and Kim, Y. (2008). Transient expression of protein tyrosine phosphatases encoded in *Cotesia plutellae* bracovirus inhibits insect cellular immune responses. Naturwissenschaften *95*, 25–32.

Imler, J.-L., and Eleftherianos, I. (2009). *Drosophila* as a model for studying antiviral defences. In Insect Infection and Immunity: Evolution, Ecology, and Mechanisms, J. Rolff and S.E. Reynolds, eds. (Oxford University Press), pp. 49–68.

Ince, I.A., Westenberg, M., Vlak, J.M., Demirbag, Z., Nalcacioglu, R., and van Oers, M.M. (2008). Open reading frame 193R of Chilo iridescent virus encodes a functional inhibitor of apoptosis (IAP). Virology *376*, 124–131.

Inoue, H., and Miyagawa, M. (1978). Regeneration of midgut epithelial cells in the silkworm, *Bombyx mori*, infected with viruses. J. Invertebr. Pathol. *32*, 373–380.

International Silkworm Genome Consortium (2008). The genome of a lepidopteran model insect, the silkworm, *Bombyx mori*. Insect Biochem. Mol. Biol. *38*, 1036–1045.

Isobe, R., Kojima, K., Matsuyama, T., Quan, G.-X., Kandra, T., Tamura, T., Sahara, K., Asano, S.-I., and Bando, H. (2004). Use of RNAi technology to confer enhanced resistance to BmNPV on transgenic silkworms. Arch. Virol. *149*, 1931–1940.

Ito, K., Kidokoro, K., Sezutsu, H., Nohata, J., Yamamoto, K., Kobayashi, I., Uchino, K., Kalyebi, A., Eguchi, R., Hara, W., Tamura, T., Katsuma, S., Shimada, T., Mita, K., and Kadono-Okuda, K. (2008). Deletion of a gene encoding an amino acid transporter in the midgut membrane causes resistance to a *Bombyx* parvo-like virus. Proc. Natl. Acad. Sci. U.S.A. *105*, 7523–7527.

Jabbour, A.M., Ekert, P.G., Coulson, E.J., Knight, M.J., Ashley, D.M., and Hawkins, C.J. (2002). The p35 relative, p49, inhibits mammalian and *Drosophila* caspases including DRONC and protects against apoptosis. Cell Death Differ. *9*, 1311–1320.

Johnson, R.M., Evans, J.D., Robinson, G.E., and Berenbaum, M.R. (2009). Changes in transcript abundance relating to colony collapse disorder in honey bees (*Apis mellifera*). Proc. Natl. Acad. Sci. U.S.A. *106*, 14790–14795.

Kadono-Okuda, K. (2009). Densovirus resistance in *Bombyx mori*. In Molecular Biology and Genetics of the Lepidoptera, M.R.Goldsmith and F.Marec, eds. (Boca Raton, FL: CRC Press), pp. 337–348.

Kamita, S.G., Majima, K., and Maeda, S. (1993). Identification and characterization of the p35 gene of *Bombyx mori* nuclear polyhedrosis virus that prevents virus-induced apoptosis. J. Virol. *67*, 455–463.

Kanginakudru, S., Royer, C., Edupalli, S.V., Jalabert, A., Mauchamp, B., Chandrashekaraiah, Prasad, S.V., Chavancy, G., Couble, P., and Nagaraju, J. (2007). Targeting ie-1 gene by RNAi induces baculoviral resistance in lepidopteran cell lines and in transgenic silkworms. Insect Mol. Biol. *16*, 635–644.

Kanost, M.R., and Gorman, M.J. (2008). Phenoloxidases in insect immunity. In Insect Immunology, N.Beckage, ed. (New York: Academic Press), pp. 69–96.

Kelly, B.J., Chapple, S.D.J., Allen, C., Pritchard, C., King, L.A., and Possee, R.D. (2008). Extended budded virus formation and induction of apoptosis by an AcMNPV *FP-25/p35* double mutant in *Trichoplusia ni* cells. Virus Res. *133*, 157–166.

Kemp, C., and Imler, J.-L. (2009). Antiviral immunity in *Drosophila*. Curr. Op. Microbiol. *21*, 3–9.

Khush, R.S., and Lemaitre, B.A. (2000). Genes that fight infection: what the Drosophila genome says about animal immunity. Trends Gen. *16*, 442–449.

Kirkpatrick, B.A., Washburn, J.O., and Volkman, L.E. (1998). AcMNPV pathogenesis and developmental resistance in fifth instar *Heliothis virescens*. J. Invertebr. Pathol. *72*, 63–72.

Kotani, E., Yamakawa, M., Iwamoto, S., Tashiro, M., Mori, H., Sumida, M., Matsubara, F., Taniai, K., Kadono-Okuda, K., and Kato, Y. (1995). Cloning and expression of the gene of hemocytin, an insect humoral lectin which is homologous with the mammalian von Willebrand factor. Biochim. Biophys. Acta *1260*, 245–258.

Krishnan, M.N., Ng, A., Sukumaran, B., Gilfoy, F.D., Uchil, P.D., Sultana, H., Brass, A.L., Adametz, R., Tsui, M., Qian, F., Montgomery, R.R., Lev, S., Mason, P.W., Koski, R.A., Elledge, S.J., Xavier, R.J., Agaisse, H., and Fikrig, E. (2008). RNA interference screen for human genes associated with West Nile virus infection. Nature *455*, 242–247.

LaCount, D.J., and Friesen, P.D. (1997). Role of early and late replication events in induction of apoptosis by baculoviruses. J. Virol. *71*, 1530–1537.

Lambropoulou, V., Douris, V., Stefanou, D., Magrioti, C., Swevers, L., and Iatrou, K. (2008). Endoparasitoid wasp bracovirus-mediated inhibition of hemolin function and lepidopteran host immunosuppression. Cell Microbiol. *10*, 2118–2128.

Lavine, M.D., and Strand, M.R. (2002). Insect haemocytes and their role in immunity. Insect Biochem. Mol. Biol. *32*, 1295–1309.

Legendre, M. Audic, S. Poirot, O., Hingamp, P. Seltzer, V., Byrne, D. Lartigue, A. Lescot, M. Bernadec, A., Povlain, J. *et al.* (2010). mRNA deep sequencing reveals 75 new genes and a complex transcriptional landscape in Mimivirus. Genome Res. *20*, 664–674.

Li, Q., Liston, P., and Moyer, R.W. (2005). Functional analysis of the inhibitor of apoptosis (*iap*) gene carried by the entomopoxvirus of *Amsacta moorei*. J. Virol. *79*, 2335–2345.

Maguire, T., Harrison, P., Hyink, O., Kalmakoff, J., and Ward, V.K. (2000). The inhibitors of apoptosis of *Epiphyas postvittana* nucleopolyhedrovirus. J. Gen. Virol. *81*, 2803–2811.

Mamanova, L. Andrews, R.M., James, K.D. Sheridan, E.M., Ellis, P.D., Langford, C.F., Ost, T.W.B., Collins, J.E., and Turner, D.J. (2010). FRT-seq: amplification-free, strand-specific transcriptome sequencing. Nature Methods *7*, 130–134.

Matzinger, P. (2009). Friendly and dangerous signals: is the tissue in control? Nature Immunol. *8*, 11–13.

Means, J.C., and Clem, R.J. (2008). Evolution and function of the p35 family of apoptosis inhibitors. Future Virol. *3*, 383–391.

Means, J.C., Muro, I., and Clem, R.J. (2003). Silencing of the baculovirus Op-*iap3* gene by RNA interference reveals that it is required for prevention of apoptosis during Orygia pseudotsugata *M* nucleopolyhedrovirus infection of Ld652Y cells. J. Virol. *77*, 4481–4488.

Means, J.C., Penabaz, T., and Clem, R.J. (2007). Identification and functional characterization of *AMVp33*, a novel homolog of the baculovirus caspase inhibitor *p35* found in *Amsacta moorei* entomopoxvirus. Virology *358*, 436–447.

Meng, X., Khanuja, B.S., and Ip, Y.T. (1999). Toll receptor-mediated Drosophila immune response requires Dif, an NF-kappaB factor. Genes Dev. *13*, 792–7.

Mims, C.A., Day, M.F., and Marshall, I.D. (1966). Cytopathic effect of Semliki Forest virus in the mosquito *Aedes aegypti*. Am. J. Trop. Med. Hyg. *15*, 775–784.

Morishima, N., Okano, K., Shibata, T., and Maeda, S. (1998). Homologous p35 proteins of baculoviruses show distinctive anti-apoptotic activities which correlate with the apoptosis-inducing activity of each virus. FEBS Lett. *427*, 144–148.

Morozova, O., Hirst, M., and Marra, M.A. (2009). Applications of new sequencing technologies for transcriptome analysis. Annu. Rev. Genomics Hum. Genet. *10*, 135–151.

Muro, I., Hay, B.A., and Clem, R.J. (2002). The Drosophila DIAP1 protein is required to prevent accumulation of a continuously generated, processed form of the apical caspase DRONC. J. Biol. Chem. *277*, 49644–49650.

Nakanishi, Y., Henson, P.M., and Shiratsuchi, A. (2009). Pattern recognition in phagocytic clearance of altered self. Adv. Exp. Med. Biol. *653*, 129–138.

Nappi, A.J., and Christensen, B.M. (2005). Melanogenesis and associated cytotoxic reactions: Applications to insect innate immunity. Insect Biochem. Molec. Biol. *35*, 443–449.

Niu, M., Han, Y., and Li, W. (2008). Baculovirus up-regulates antiviral systems and induces protection against infectious bronchitis virus challenge in neonatal chicken. Int. Immunopharm. *8*, 1609–1615.

Ourth, D.D. (2004). Antiviral activity against human immunodeficiency virus-1 *in vitro* by myristoylated-peptide from *Heliothis virescens*. Biochem. Biophys. Res. Commun. *320*, 190–196.

Palli, S.R., Caputo, G.F., Sohi, S.S., Brownwright, A.J., Ladd, T.R., Cook, B.J., Primavera, M., Arif, B.M., and Retnakaran, A. (1996). *Cf*MNPV blocks *Ac*MNPV-induced apoptosis in a continuous midgut cell line. Virology *222*, 201–213.

Pei, Z., Reske, G., Huang, Q., Hammock, B.D., Qi, Y., and Chejanovsky, N. (2002). Characterization of the apoptosis suppressor protein P49 from the *Spodoptera littoralis* nucleopolyhedrovirus. J. Biol. Chem. *277*, 48677–48684.

Popham, H.J.R., Shelby, K.S., Brandt, S.L., and Coudron, T.A. (2004). Potent virucidal activity against HzSNPV in larval *Heliothis virescens* plasma. J. Gen. Virol. *85*, 2225–2261.

Rivkin, H., Kroemer, J.A., Bronshtein, A., Belausov, E., Webb, B.A., and Chejanovsky, N. (2006). Response of immunocompetent and immunosuppressed *Spodoptera littoralis* larvae to baculovirus infection. J. Gen. Virol. *87*, 2217–2225.

Rolff, J., and Reynolds, S.E. (2009). Introducing insect infection and immunity. In Insect Infection and Immunity: Evolution, Ecology, and Mechanisms, J.Rolff and S.E.Reynolds eds (Oxford University Press), pp. 1–12.

Roxtrom-Lindquist, K., Terenius, O., and Faye, I. (2004). Parasite-specific immune response in adult *Drosophila melanogaster*: a genomic study. EMBO Rep. *5*, 207–212.

Rumble, J.M., and Duckett, C.S. (2008). Diverse functions within the IAP family. J. Cell Sci. *121*, 3505–3507.

Sabatier, L., Jouanguy, E., Dostert, C., Zachary, D. Dimascq, J.-L., Bulet, P., and Imler, J.L. (2003). Pherokine-2 and -3: Two *Drosophila* molecules related to pheromone/odor-binding proteins induced by viral and bacterial infections. Eur. J. Biochem. *270*, 3398–3407.

Sabin, L.R., Zhou, R., Gruber, J.J., Lukinova, N., Bambina, S., Berman, A., Lau, C.-K., Thompson, C.B., and Cherry, S. (2009). Ars2 regulates both miRNA- and siRNA-dependent silencing and suppresses RNA virus infection in *Drosophila*. Cell *138*, 340–351.

Sagisaka, A., Fujita, K., Nakamura, Y., Ishibashi, J., Noda, H., Imanishi, S., Mita, K., Yamakawa, M., and Tanaka, H. (2009). Genome-wide analysis of host gene expression in the silkworm cells infected with *Bombyx mori* nucleopolyhedrovirus. Virus Res. *147*, 166–175.

Saleh, M.-C., Tassetto, M., van Rij, R.P., Goic, B., Gausson, V., Berry, B., Jacquier, C., Antoniewski, C., and Andino, R. (2009). Antiviral immunity in *Drosophila* requires systemic RNA interference spread. Nature *458*, 346–351.

Schmidt, O., Theopold, U., and Beckage, N.E. (2008). Insect and vertebrate immunity: key similarities versus differences. In Insect Immunology, N. Beckage, ed. (New York: Academic Press), pp. 1–23.

Schultz, K.L.W., and Friesen, P.D. (2009). Baculovirus DNA replication-specific expression factors trigger apoptosis and shutoff of host protein synthesis during infection. J. Virol. *83*, 11123–11132.

Sessions, O.M., Barrows, N.J., Souza-Neto, J.A., Robinson, T.J., Hershey, C.L., Rodgers, M.A., Ramirez, J.L., Dimoupoulos, G., Yang, P.L., Pearson, J.L., and Garcia-Blanco, M.A. (2009). Discovery of insect and human dengue virus host factors. Nature *458*, 1047–1050.

Settles, E.W., and Friesen, P.D. (2008). Flock house virus induces apoptosis by depletion of *Drosophila* Inhibitor-of-apoptosis protein DIAP1. J. Virol. *82*, 1378–1388.

Shelby, K.S., Adeyeye, O.A., Okot-Kotber, B.M., and Webb, B.A. (2000). Parasitism-linked block of host plasma melanization. J. Invertebr. Pathol. *75*, 218–225.

Shelby, K.S., and Popham, H.J.R. (2006). Plasma phenoloxidase of larval *Heliothis virescens* is virucidal. J. Insect Sci. *6*, 13.

Shelby, K.S., and Popham, H.J.R. (2007). Dietary selenium supplementation elevates HzSNPV resistance of *Heliothis virescens* larvae. J. Invertebr. Pathol. *95*, 77–83.

Shelby, K.S., and Popham, H.J.R. (2009). Analysis of ESTs generated from immune-stimulated haemocytes of larval *Heliothis virescens*. J. Invertebr. Pathol. *101*, 86–95.

Shelly, S., Lukinova, N., Bambina, S., Berman, A., and Cherry, S. (2009). Autophagy is an essential component of Drosophila immunity against vesicular stomatitis virus. Immunity *30*, 588–598.

Silveira, E.B., Ribeiro, B.M., and Bao, S.N. (1999). Morphological studies of apoptosis in insect cells infected with vApAg, an *Anticarsia gemmatalis* nucleopolyhedrovirus mutant. J. Submicrosc. Cytol. Pathol. *31*, 543–554.

Souza-Neto, J.A., Sim, S., and Dimoupoulos, G. (2009). An evolutionary conserved function of the JAK-STAT pathway in anti-dengue defense. Proc. Natl. Acad. Sci. U.S.A. *106*,17841–17846.

Sparks,W.O., Bartholomay,L.C., and Bonning, B.C. (2008). Insect Immunity to Viruses. In Insect Immunology, N.E.Beckage, ed. (New York: Elsevier), pp. 209–242.

Terenius, O., Popham, H.J.R., and Shelby, K.S. (2009). Bacterial, but not baculoviral infections stimulate Hemolin expression in noctuid moths. Dev. Comp. Immunol. *33*, 1176–1185.

Thiem, S. (2009). Baculovirus genes affecting host function. *In Vitro* Cell. Dev. Biol. Anim. *45*, 111–126.

Trudeau, D., Washburn, J.O., and Volkman, L.E. (2001). Central role of haemocytes in *Autographa californica* M nucleopolyhedrovirus pathogenesis in *Heliothis virescens* and *Helicoverpa zea*. J. Virol. *75*, 996–1003.

Tsai, C.W., McGraw, E.A., Ammar, E.-D., Dietzgen, R.G., and Hogenhout, S.A. (2008). *Drosophila melanogaster* mounts a unique immune response to the rhabdovirus *Sigma* virus. App. Environ. Microbiol. *74*, 3251–3256.

Tryselius, Y., Samakovlis, C., Kimbrell, D. A., and Hultmark, D. (1992). CecC, a cecropin gene expressed during metamorphosis in Drosophila pupae. Eur. J. Biochem. *204*, 395–399.

Vaidyanathan, R., and Scott, T.W. (2006). Apoptosis in mosquito midgut epithelia associated with West Nile virus infection. Apoptosis *11*, 1643–1651.

Wang, Y., and Jehle, J.A. (2009). Nudiviruses and other large, double-stranded circular DNA viruses of invertebrates: new insights on an old topic. J. Invertebr. Pathol. *101*, 187–193.

Wang, W., and Zhang, X. (2008). Comparison of antiviral efficiency of immune responses in shrimp. Fish Shellfish Immunol. *25*, 522–527.

Washburn, J.O., Haas-Stapleton, E.J., Tan, F.F., Beckage, N.E., and Volkman, L.E. (2000). Co-infection of *Manduca sexta* larvae with polydnavirus from *Cotesia congregata* increases susceptibility to fatal infection by *Autographa californica* M Nucleopolyhedrovirus. J. Insect. Physiol. *46*, 179–190.

Washburn, J.O., Kirkpatrick, B.A., Haas-Stapleton, E., and Volkman, L.E. (1998). Evidence that the stilbene-derived optical brightener M2R enhances *Autographa californica* M nucleopolyhedrovirus infection of *Trichoplusia ni* and *Heliothis virescens* by preventing sloughing of infected midgut epithelial cells. Biol. Control *11*, 58–69.

Washburn, J.O., Kirkpatrick, B.A., and Volkman, L.E. (1996). Insect protection against viruses. Nature *383*, 767.

Weaver, S.C., Lorenz, L.H., and Scott, T.W. (1992). Pathologic changes in the midgut of *Culex tarsalis* following infection with Western equine encephalomyelitis virus. Am. J. Trop. Med. Hyg. *47*, 691–701.

Weaver, S.C., Scott, T.W., Lorenz, L.H., Lerdthusnee, K., and Romoser, W.S. (1988). Togavirus-associated pathologic changes in the midgut of a natural mosquito vector. J. Virol. *62*, 2083–2090.

Wilkinson, J.C., Wilkinson, A.S., Scott, F.L., Cosmos, R.A., Salvesen, G.S., and Duckett, C.S. (2004). Neutralization of Smac/Diablo by inhibitors of apoptosis (IAPs): A caspase-independent mechanism for apoptotic inhibition. J. Biol. Chem. *279*, 51082–51090.

Wright, C.W., Means, J.C., Penabaz, T., and Clem, R.J. (2005). The baculovirus anti-apoptotic protein Op-IAP does not inhibit *Drosophila* caspases or apoptosis in *Drosophila* S2 cells and instead sensitizes S2 cells to virus-induced apoptosis. Virology *335*, 61–71.

Xi, Z., Ramirez, J.L., and Dimopoulos, G. (2008). The *Aedes aegypti* toll pathway controls dengue virus infection. PLoS Pathogens *4*, e1000098.

Xu, G., Cirilli, M., Huang, Y., Rich, R.L., Myszka, D.G., and Wu, H. (2001). Covalent inhibition revealed by the crystal structure of the caspase–8/p35 complex. Nature *410*, 494–497.

Xu, J.P., Chen, K.P., Yao, Q., Liu, M.H., Gao, G.T., and Zhao,Y. (2005). Identification and characterization of an NPV infection-related gene BMsop2 in *Bombyx mori* L. J. Appl. Entomol. *129*, 425–431.

Yamaguchi, K. (1979). Natural recovery of the fall web worm, *Hyphantria cunea*, to infection by a cytoplasmic-polyhedrosis virus of the silkworm *Bombyx mori*. J. Invertebr. Pathol. *33*, 126–128.

Yang, Z., Bruno, D.P., Martens, C.A., Porcella, S.F., and Moss, B. (2010). Simultaneous high-resolution analysis of vaccinia virus and host cell transcriptomes by deep RNA sequencing. Proc. Natl. Acad. Sci. U.S.A. *107*, 11513–11518.

Yao, Q., Li, M.W., Wang, Y., Wang, W.B., Lu, J., Dong, Y., and Chen, K.P. (2003). Screening of molecular markers for NPV resistance in *Bombyx mori* L. (Lep., Bombycidae). J. Appl. Entomol. *127*, 134–136.

Yao, H.-P., Wu, X.-F., and Gokulamma, K. (2006). Antiviral activity in the mulberry silkworm, *Bombyx mori* L. J. Zhejiang Univ. Science A *7(Suppl. II)*, 350–356.

Zambon, R.A., Nandakumar, M., Vakharia, V.N., and Wu, L.P. (2005). The Toll pathway is important for an antiviral response in *Drosophila*. Proc. Natl. Acad. Sci. U.S.A. *102*, 7257–7262.

Zhang, P., Yang, K., Dai, X., Pang, Y., and Su, D. (2002). Infection of wild-type *Autographa californica* multicapsid nucleopolyhedrovirus induces *in vivo* apoptosis of *Spodoptera litura* larvae. J. Gen. Virol. *83*, 3003–3011.

Zheng, Y.H., and Peterlin, B.M. (2005). Intracellular immunity to HIV-1: newly defined retroviral battles inside infected cells. Retrovirology *2*, 25.

Zoog, S.J., Schiller, J.J., Wetter, J.A., Chejanovsky, N., and Friesen, P.D. (2002). Baculovirus apoptotic suppressor P49 is a substrate inhibitor of initiator caspases resistant to P35 *in vitro*. EMBO J. *21*, 5130–5140.

The Ecology of Baculoviruses

18

Jenny S. Cory

Abstract

Ecological studies involving insect viruses have centred on baculoviruses, partly because they are associated with population declines of some insect species, and also because they are highly pathogenic to insects, making them ideal candidates for pest control. Recent research has focussed on four main areas: (i) the influence of host condition on resistance to viral infection, (ii) the role and maintenance of baculovirus diversity, (iii) the prevalence of covert infections and (iv) the elucidation of patterns of host resistance in field populations. Tritrophic interactions, either via direct effects of plant secondary chemicals or through nutritionally mediated changes in host immunity, can have a significant impact on baculovirus efficacy. Variation within baculovirus populations appears to be ubiquitous, and mixed genotype infections apparently act to generate higher levels of pathogenicity. Covert infections are increasingly being shown to be common in field populations of Lepidoptera but their importance in generating overt baculovirus infections is still unclear. Field studies on forest insects indicate that host resistance varies with fluctuating host density and condition. Synthesis of the impacts of host condition on susceptibility, the role of genetic variability in infection, and of the relationship between overt and covert infection, will promote understanding of the ecological interactions between baculoviruses and natural host populations.

Introduction

The ecological literature is rich with examples of studies on interactions between insect populations and their parasitoids and predators, but information on their relationship with pathogens lags well behind in mainstream ecology, even though insects are attacked by a wide range of pathogens. This omission in part pertains to the fact that early interest in insect pathogens was stimulated by their potential role as pest control agents. The difficulty in finding and identifying pathogens was also a major factor, which is why many of the early studies concentrated on diseases with obvious symptoms, such as baculoviruses. Even now, ecological studies of insect viruses are primarily concerned with host survival or host population dynamics, while ignoring the impact on the pathogen population.

Insects are attacked by a wide range of viruses; however, in this chapter I focus on the insect baculoviruses, the group for which most information is known and recent progress has been made in unravelling their ecology. There are several earlier reviews on baculovirus ecology (Cory *et al.*, 1997; Cory and Myers, 2003; Fuxa, 2004); these reviews provide greater background to baculovirus biology (see also Chapter 2) and the topics discussed here. Recently, molecular tools have allowed a shift of attention from host to pathogen populations. Although it has been known for many years that viruses can be very genetically diverse, the role of this diversity in the host–pathogen relationship has largely been ignored (Cory and Myers, 2003). Similarly, assumptions concerning mechanisms for virus persistence can now be scrutinized using a selection of molecular tools to test theories of virus maintenance and persistence in the field. The current review primarily considers recent advances in knowledge.

Individual level effects – factors that alter host condition and virus fitness

As a result of the interest in insect viruses as potential microbial control agents, the greatest attention has centred on viruses that kill their host relatively rapidly. Other viruses that produce chronic infections exist, but they have received very little attention, in part due to the difficulty of isolating and identifying them. In this virus-centred chapter, I focus on those factors that influence virus fitness at the individual level, i.e. what modulates the likelihood and impact of virus infection in an individual insect. Simple host-pathogen models indicate parameters that are likely to influence virus fitness (Cory *et al.*, 1997). The capacity to infect hosts is obviously crucial, but also important is the number of progeny produced by the virus and their transmission potential. For baculoviruses, horizontal transmission to a new susceptible host usually occurs when the infected host dies and releases occlusion bodies (unless cannibalism occurs prior to this). Thus virus-induced mortality is the end point of infection and the starting point for transmission. If maximizing virus yield increases the likelihood of transmission, then it is an advantage for the baculovirus to keep the host alive and growing as long as possible as a negative correlation between the speed of kill and the quantity of virus progeny produced occurs. A key question is therefore what determines or modulates virus virulence (degree of harm to the host) and the rate of virus replication (speed of kill)?

One of the emerging areas in host-parasite ecology is the role of host condition or vigour in determining susceptibility. In natural populations, hosts are likely to be genetically and phenotypically variable. Given that many bioassays of insect viruses are carried out in inbred insect cultures that are fed *ad libitum*, the results generated do not necessarily accurately represent the more complex field situation. We might predict that increasing the genetic diversity of the host species will decrease the efficacy of a microbial control agent (or at least its variance), whereas if host condition is not optimal in the field, the impact of the pathogen may be increased. Host variation in susceptibility in field populations has only been investigated in a few instances, although significant differences have been found and could be important for host-pathogen dynamics (see Cory *et al.*, 1997, for a review and below; Dwyer *et al.*, 2000). Recent advances in plant–mediated interactions and the impact of host condition on baculoviruses are discussed below.

The role of insect nutrition

How does variation in food quality influence interactions between viruses and their insect hosts? It is often assumed that insects eating poor quality food will be more vulnerable to pathogens. This is assumed for several reasons; first, poor-quality diet is correlated with a reduction in immune function; second, slower insect growth may increase exposure to pathogens and allow more time for infection to proceed and; third, if the insects compensate for poor quality of food by eating more, the likelihood of ingesting virus occlusion bodies (OBs) will increase. As baculoviruses apparently only infect and multiply in larval tissue, larval food, in terms of both nutritional quality and the presence or concentration of secondary chemicals, will influence the infection process. The nutritional ecology of insect-pathogen infection has not received much attention; most of the focus of plant-mediated effects has been directed at the role of secondary chemicals when ingested with the viral OBs. However, several frameworks for the study of nutritional ecology of insects exist (Raubenheimer *et al.*, 2009) which could be readily incorporated into studies of insect viruses.

For viruses such as baculoviruses that must be ingested to initiate infection, the initial barriers to infection occur within the midgut. The first physical barrier the viruses will encounter is the peritrophic matrix. Interestingly, recent research by Plymale *et al.* (2008) has shown that plant-mediated alteration to the peritrophic matrix influences insect resistance to baculoviruses. *Heliothis virescens* larvae that were fed on a variety of host plants had thicker peritrophic matrices than insects fed on diet. In addition, in a comparison of larvae challenged on cotton versus diet, cotton-fed insects had fewer foci of infection and lower resulting mortality and this was related to peritrophic matrix width. This study opens up a whole new area of possibilities for investigation of plant-mediated effects and pathogen resistance in insects.

The nutritional content of an insect's diet is likely to affect its disease resistance through influences on its immune function. Several studies have looked at the influence of starvation and clearly show that food deprivation reduces immune responsiveness (Siva-Jothy and Thompson, 2002; Yang *et al.*, 2007) and that immune activation has a cost in the absence of compensatory feeding (Moret and Schmidt-Hempel, 2000). The effect of more subtle changes in diet on immune function is less clear, and currently a rapidly expanding literature on ecological immunology investigates variation in immunity in the context of costs and trade-offs (Sadd and Schmid-Hempel, 2009).

Two aspects of immune responsiveness are, innate immunity, which is essentially the standing level of immunity maintained by the insect, and the induced immunity that occurs after pathogen challenge. Mounting an immune response should be energetically costly and thus differences in reserves available to mount or maintain this response are going to be directly related to diet. It is not clear which antiviral responses (see Chapters 16 and 17) are important, however. This is a rapidly moving field in terms of molecular immunology in relation to model organisms such as *Drosophila* spp. and beyond the realms of this review. The relevant immunological targets for ecological studies, however, are not clear. Evolutionary ecologists studying the costs of disease resistance and other trade-offs within the immune system, tend to utilize a suite of tests to estimate immunocompetence, including phenoloxidase (PO) assays, anti-bacterial assays, haemocyte counts and encapsulation assays, but the relevance of each of these for measuring the resistance to the pathogen of choice is less clear. Whatever the mechanism, the ultimate test is whether variation in insect nutrition alters the outcome of virus challenge. An interesting study of *Spodoptera littoralis* larvae showed that when they were given diets that varied in their protein and digestible carbohydrate content *after* challenge with *S. littoralis* nucleopolyhedrovirus (SlNPV), the insects that ate more protein had a better survival rate (Lee *et al.*, 2006) Moreover, the diet choices altered various immune measures: encapsulation and lysozyme-like antimicrobial activity were strongly reduced on the lower-protein diet, PO levels also declined, but total haemocyte numbers remained the same. One of the most intriguing results from this study is what happened when the larvae were given a choice of a diet optimal for growth or a high protein diet. Insects that survived baculovirus challenge ate relatively more protein than either the unchallenged controls or the insects that died of infection (Lee *et al.*, 2006). These results indicate that demand for protein increases after infection, resulting in a shift in the normal feeding regime to compensate. It also suggests that it might be possible for insects to self-medicate to increase their likelihood of survival. While such results are interesting, further studies on the effect of diet on disease are needed, in addition to investigation into causal relationships linking changes in immunity and pathogen resistance. Future research should also consider the effect of stress factors, such as food depletion, on the expression of diseases that under 'normal' laboratory conditions might be considered as lacking in virulence and impact. Expression of disease is likely to be context dependent.

Plant secondary chemicals

While controlled studies help to separate out the effects of nutrients from other plant components, the effects of plant secondary chemicals also need to be considered, particularly as related to whole plants and the field situation. Considerable work has looked at the direct effects of secondary chemicals on baculoviruses (Duffey *et al.*, 1995; Cory and Hoover, 2006). Mechanistically, numerous plant secondary chemicals, including phenolics, tannins, peroxidases and alkaloids, have been shown to have a direct impact on the virus infection process via the formation of OB-phenolic aggregates in the midgut, damage to the occlusion-derived virus or virus receptors, oxidative stress and alteration to the peritrophic matrix (reviewed in Cory and Hoover, 2006). Plant-mediated effects on virus-induced mortality have received the most attention and tend to show that allelochemicals inhibit virus infection, particularly when ingested at the same time as the virus (Cory and Hoover, 2006). This effect is particularly strong in plants, such as cotton, that have multiple inhibitory mechanisms, including alkaline leaf exudates as well as secondary chemicals (Young *et al.*, 1977; Hoover *et al.*, 2000). Where it has been studied, other aspects of the insect–virus relationship, primarily speed of kill and virus yield, are also affected by host plant (Ali *et al.*, 2002; Raymond *et al.*, 2002; Raymond and Hails, 2007). In addition, host plants can influence the virus–insect relationship in other ways, such as by plant architecture altering virus persistence via shading effects (Raymond *et al.*, 2005) or possibly by the plant surface structure (e.g. presence or absence of trichomes) altering insect mobility and thus virus encounter rate.

Many of the studies focus on constitutive chemicals, that is, those that are always present in the plants; however, it is well established that plants respond to herbivore attack by induction of increased levels of secondary chemicals. For chewing insects, this is primarily mediated by the jasmonate-dependent wound response pathway (Thaler *et al.*, 2002a; Smith *et al.*, 2009). Few studies have looked specifically at induction of allelochemicals, and where it has been investigated, the outcome is not clear-cut. This is perhaps not surprising given that induced responses, either rapid or delayed, have been recorded as having positive, negative or neutral effects on insect herbivores (Nykänen and Koricheva, 2004). Results can also be complicated by differing methodologies, a problem in comparing many tritrophic entomopathogen studies. For example, two studies investigated the effect of the chemical induction by prior larval feeding on NPV infection of *Helicoverpa zea* on tomato; however, the outcome was not the same, with levels of infection being reduced by chemical induction in one study (Hoover *et al.*, 1998) and increased in the other (Ali *et al.*, 1998). In the first experiment, the insects were exposed to the chemicals only when they were infected, whereas in the second experiment they were reared on the plants continuously and so differences in growth, as well as other possible effects on insect condition, would also be involved.

One of the most important questions regarding constitutive and induced food plant effects is what is the balance between influences on the herbivore and effects on its natural enemy? Could high levels of plant secondary chemicals represent a refuge from pathogen attack for the herbivore? This question has received most attention in the gypsy moth, *Lymantria dispar*–NPV system. Gypsy moth can feed on a wide range of deciduous tree species some of which have been shown to respond to larval feeding by inducing secondary chemicals, particularly tannins, chemicals that reduce gypsy moth growth and fecundity (Rossiter *et al.*, 1988). NPV-induced mortality in gypsy moth varies with plant species and it has also been shown to relate to levels of hydrolysable tannins (Keating *et al.*, 1988). Bioassay of field-collected foliage showed a correlation between increasing defoliation levels and reduced NPV-induced mortality. This suggested that the induction of secondary chemicals provides a partial refuge from disease for the herbivore (Hunter and Schultz, 1993). However, later field experiments found no differences in virus transmission later in the season, in part because measurable levels of induction could not be found (D'Amico *et al.*, 1998; Dwyer *et al.*, 2005).

Plants also respond to phytopathogen attack by increasing levels of certain chemicals, a process known as systemic acquired resistance (SAR). In fact, plants can be protected again pathogens by the application of an elicitor, usually a functional analogue of salicylic acid, that essentially inoculates against infection (Smith *et al.*, 2009). There is also evidence, however, for antagonistic effects between the jasmonate pathway and SAR (Thaler *et al.*, 2002b), making it difficult to predict the outcome when the plant is attacked by a diverse array of natural enemies. Plymale *et al.* (2007) investigated whether using an elicitor of SAR on cotton was compatible with herbivore control using a baculovirus. SAR application significantly increased levels of a variety of chemicals including, condensed tannin and foliar peroxidase; however, it did not alter the pupal weight of *Heliothis virescens* nor the levels of mortality after challenge with *Autographa californica* multiple nucleopolyhedrovirus (AcMNPV).

What is less clear is whether these chemicals cross the midgut and continue to affect virus infection directly, or cause indirect effects via alteration of host physiology and the cells' permissiveness to infection. This response will, in part, depend on whether an insect excretes, detoxifies or sequesters the chemicals. A recent study with the nymphalid butterfly *Junonia coenia*, which sequesters iridoid glycosides, has shown that larvae fed on diets with high levels of glycosides have reduced immune function, as indicated by reduced melanization of latex beads (Smilanich *et al.*, 2009). The authors suggested that this might indicate a conflict for the insect between defence against predators and protection against parasites. Can these controlled dietary studies on changes in insect immunity be linked to actual foliar quality? Several studies have investigated how host plants affect insect immunity; Klemola *et al.* (2007a) compared the effect of 'high-' and 'low'-quality birch, *Betula pubescens*, foliage on a number of immune parameters of the autumnal moth, *Epirrita autumnata*. Foliage quality was based on the relative growth rate of the larvae on the trees in the previous year. They found that larvae reared on low quality food produced pupae that had a greater encapsulation response to implanted nylon filament than pupae of larvae reared on high-quality food. Levels of haemolymph phenoloxidase did not vary, however. Ojala *et al.* (2005) compared various measures of immune function of the moth *Parasemia plantaginis* reared on different plant species, and on mixtures of plants. They found that the capacity to encapsulate an imbedded foreign object varied with diet, but was not correlated with fitness parameters such as pupal mass, growth rate and survival, as might have been expected.

In looking at how disease resistance relates to food quality, McVean *et al.* (2002) found that *Plodia interpunctella* larvae reared on low-quality diet were more resistant to challenge with granulovirus. Shikano *et al.* (2010) took this type of study a stage further and compared both immune parameters (haemolymph protein, phenoloxidase and haemocyte counts) and susceptibility to an NPV of *Trichoplusia ni* reared on a 'good' host plant, broccoli, and a 'bad' food plant, cucumber. Plant quality was defined by superior growth rate and survival. Insects were reared on the plants up until the point of infection, but the virus dose was given on artificial diet to rule out the direct effects of host plant chemistry at the point of infection in the midgut. Virus mortality was significantly higher on the lower quality food plant. Insects reared on broccoli had a higher protein concentration in the haemolymph, indicating better condition and they also had a greater number of haemocytes. Insects reared on cucumber had higher levels of phenoloxidase measured as both absolute and per unit weight of protein (Shikano *et al.*, 2010). These results clearly show that host plant quality can influence the immune response of an insect and that diet prior to virus challenge affects the susceptibility to viral infection. Thus, while trophic cascades can influence insect pathogens, identifying the precise mechanisms involved requires further research.

Population-level effects

Population-level effects usually refer to the influence of the virus on the host population. As obligate parasites, the dynamics of the virus are necessarily linked to that of their host; however, virus populations also have a structure that can vary, and recent research has started to investigate this at both the within and between host levels.

Within-host dynamics and virus population structure

Virus variation

Each host represents a discrete patch of resource for the virus, and depending on infection biology, the virus population in a host individual has the opportunity for numerous cycles of replication. Many studies have compared baculovirus isolates from different geographical regions, usually by using restriction fragment length polymorphism (RFLP) analysis of the DNA, and have found that genetic variation is common. When these isolates are probed further, such as by *in vitro* or *in vivo* cloning, they are often shown to be mixtures of genetically distinct variants or genotypes with a variety of deletions and other differences (Cory *et al.*, 1997; Muñoz *et al.*, 1998; Simón *et al.*, 2004). This high level of variation could be present because infection initially results from a mixture of genetically distinct genotypes or it could be generated though recombination, point mutations and DNA insertions and deletions during the infection process (Cory and Myers, 2003). Different virus genotypes must co-infect the same cell for recombination to occur. In NPVs, genetically distinct variants could be co-occluded in the same OB or virion, which would increase the likelihood of recombination and make retention and generation of variation likely (Clavijo *et al.*, 2010). In contrast, granuloviruses (GVs) only occlude a single virion, and while molecular variation has been found in GVs, it appears to be at a lower level than NPVs (e.g. Eberle *et al.*, 2009). Studies using mixtures of occlusion negative (i.e. minus the polyhedrin gene and thus unable to make OBs) and occlusion positive viruses strongly suggest that the two NPV variants can be co-occluded in the same OB (Bull *et al.*, 2001). Clavijo *et al.* (2010) go one step further and analyse the number of genotypes present in single virion plaques, and show that mixtures are common. Furthermore, a combination of passaging studies followed by population genetic modelling indicated that approximately four virus genomes infect individual cells at high doses (Bull *et al.*, 2001). This supports the interpretation that co-infection likely takes place. The likelihood of mixed infections appears to be condition dependent, however, with later instars and less permissive host species being more likely to succumb to multiple genotype infections (Zwart *et al.*, 2009a).

More relevant to the ecology and evolution of baculoviruses is whether these molecular variants actually change the biological characteristics of the genotypes. Demonstrating differences between individual variants will depend on what parameter is studied; a negative result does not necessarily mean that there are no phenotypic differences. Individual variants obtained by *in vivo* techniques have been shown to differ in terms of pathogenicity, speed of kill and productivity (Hodgson *et al.*, 2001; Cory *et al.*, 2005). These are all factors that are likely to influence host–virus dynamics; however, the role of baculovirus variation and why it is maintained has only recently been addressed.

Maintenance of variation and mixed infections

Why is there so much variation in insect baculoviruses? As baculoviruses reproduce asexually, it might be predicted that a single genotype with superior fitness would come to dominate in the competition to exploit an individual host. However, this does not appear to be the case. Theoretically, genetic variation could be maintained in numerous ways e.g. through trade-offs, differential selection or facilitation among genotypes (Hodgson *et al.*, 2003). If traits are correlated such that different phenotypes have equal fitness, they can coexist. For example, for baculoviruses speed of kill and the production of OBs are correlated (fast kill-low yield versus slow kill-high yield). This trade-off is commonly seen in bioassay studies involving a single virus, but few studies have compared different natural genotypes. Those that have done so have not found significant correlations. This could in part be due to the small number of genotypes compared (Hodgson *et al.*, 2001).

The ecology of the host can also provide the opportunity for differential selection of virus genotypes under different conditions, promoting the persistence of numerous genotypes. One area where this possibility has been investigated is in tritrophic interactions. Many lepidopteran larvae feed on more than one species of host plant and if specific virus genotypes perform differently on these host plants, differential selection can result. An example is seen with the pine beauty moth, *Panolis flammea*, and its NPV; in the UK *P. flammea* feeds on the native Scots pine, *Pinus sylvestris*, but also on the introduced lodgepole pine, *P. contorta*. When *P. flammea* larvae were reared on the two pine species and then infected with two *P. flammea* NPV variants, pathogenicity varied differentially depending on the specific pairing, with one variant producing higher levels of mortality on one pine species and the other variant producing higher mortality on the other (Hodgson *et al.*, 2002). Similar differences were also seen for speed of kill. This result implies that the viruses adapt to different host plants to increase their fitness. As discussed above, baculoviruses have an intimate relationship with the plant species that their host feeds on and plant chemistry can have an impact (often negative) on the outcome of baculovirus infection. This negative impact could select for particular virus genotypes that are better able to cope with particular plant chemistries. Some evidence for adaptation to host plant has been found in baculoviruses. The western tent caterpillar *Malacosoma californicum pluviale* feeds on numerous deciduous plants and is often found in discrete populations dominated by one plant species. Cory and Myers (2004) collected *M.c. pluviale* NPV from two sites; one where alder was the main host food plant and another where the dominant plant was wild rose, and then bioassayed the two viral isolates on the same two host plant species using a common stock of insects reared initially on alder. They found that each isolate performed best in terms of speed of kill on the host plant that it had been originally isolated from. This implies that viral types had adapted to that plant.

Other ecological or environmental factors could also result in differential performance and selection, including sensitivity to ultraviolet irradiation (UV), temperature, host life span and alternative hosts. For baculoviruses that are capable of infecting more than one host species, alternative hosts present a very plausible option for the generation of diversity. Numerous laboratory studies have shown that passaging a wild type baculovirus isolate composed of a mixture of variants in an alternative host, can alter both its DNA profile and its pathogenicity (ability to kill) for a particular host species (e.g. Tompkins *et al.*, 1988; Kolodny-Hirsch and van Beek, 1997). Studies using wild type (mixed) *P. flammea* NPV demonstrated that the proportions of various polymorphic loci in the genome changed after passage through different susceptible species of Lepidoptera, indicating selection of different genotypes in the different hosts (Hitchman *et al.*, 2007). This certainly indicates a possible mechanism for maintaining diverse genotypes; however, in all these cases the various genotypes also need to disperse between sites of selection to be maintained within the population as whole.

One of the most interesting mechanisms for maintaining diversity is the possibility that mixed pathogen infections are advantageous in some way. Several recent studies have shown that pairs of NPV genotypes are more pathogenic than either of the individual genotypes alone, although the outcome depends on the combination of genotypes tested (Hodgson *et al.*, 2004; Simón *et al.*, 2005; Redman and Cory, unpublished data). In all of these examples, the mixed wild-type virus was always more pathogenic than any of the individual variants, suggesting a positive interaction between them. Other parameters such as speed of kill and virus yield are also influenced by the genetic composition of the virus population (Simón *et al.*, 2008). High levels of mixed infections can also promote the evolution of cheating genotypes (Hodgson *et al.*, 2003). Viruses with smaller genomes are likely to have a replication advantage over complete genomes and can utilize the gene products from co-infecting strains.

Detailed analyses of the population genetic structure of the NPV of the fall armyworm *S. frugiperda* have clearly shown that the natural NPV populations include deletion mutants that are not capable of infection on their own and are therefore dependent on complete genotypes (Simón *et al.*, 2004). Presence of these deletion genotypes has different effects; in *S. exigua* MNPV isolate, they reduced overall pathogenicity (Muñoz and Caballero, 2000), whereas in the *S. frugiperda* MNPV study certain combinations were found to have positive effects (Lopez-Ferber *et al.*, 2003). A mixture of a complete SfMNPV genotype and a deletion variant that is unable to be transmitted alone, increased pathogenicity, but only when the two viruses were co-occluded and the mutant occurred at intermediate levels (López-Ferber *et al.*, 2003). The main factor involved in the lack of infectivity of the SfMNPV mutants was the deletion of one or more *pif* genes that are responsible for oral infectivity (Kikhno *et al.*, 2002; Simón *et al.*, 2004). The presence of non-infective mutants is likely to be frequency dependent, as their presence at either high or low levels will reduce the fitness of the population. In the SfMNPV study, the largest increase in pathogenicity was found at a ratio of 3 complete to 1 defective genotype (López-Ferber *et al.*, 2003). In a follow-up study, in which various ratios of the same two genotypes were passaged though larvae for several generations, it was found that the populations converged on a ratio of approximately 80% complete to 20% defective virus, very similar to the 78% to 22% ratio seen in the wild-type population (Simón *et al.*, 2006).

The mechanisms behind the apparent synergism between genotypes is still unclear; in vertebrates the enhanced virulence seen with multiple strains of malaria parasites, for example, has been explained by effects on the immune system, and the different response elicited by individual strains (Read and Taylor, 2001). This type of response would require strain specific immunity in invertebrate systems. This is not considered to be likely, although limited evidence in its support is found from other invertebrates and parasites (Carius *et al.*, 2001). Some type of differential tissue tropism among the different baculovirus strains may also be possible, although given that most tissues are infected in a normal baculovirus infection in lepidopterans, this seems unlikely. Multiple genotypes may also cooperate in some way through sharing of gene products; it is relatively straightforward to see how a deletion mutant might benefit in this way, but it is as yet unclear what the more complete genotypes lack that the deletion mutant can contribute. Presence or absence of genes is not difficult to ascertain for baculoviruses and differences are more likely to be more subtle changes in gene expression. However, in terms of biological control using baculoviruses, the message is clear, single virus clones are less likely to be effective than wild type mixtures. In addition, local isolates may possess variants adapted to deal with local conditions and diverse virus populations will have the capacity to adapt and perform under different conditions.

What is not clear from current studies is the importance of virus variation at different scales. Studies that have compared virus variation in individual larva, without further amplification, have showed considerable variation (Laitinen *et al.*, 1996; Cooper *et al.*, 2003). RFLP profiles of baculovirus DNA extracted from individual larvae frequently show submolar bands indicative of a multiple variant infection, and cloning studies have shown that dozens of genetic variants can be isolated from a single host (Cory *et al.*, 2005), but what scale is the most relevant for virus transmission and maintaining genetic variation in the virus population? Multigenerational laboratory experiments indicate that the outcome from simply passaging two variants over several insect generations can result in different end points, depending on whether the inoculum was derived from a single or multiple larvae (Zwart *et al.*, 2009b, 2010). Variants are more likely to be retained in the population if the inoculum originates from multiple larvae, although this depends on the relative fitness of the isolates. It is difficult to predict which situation is the more realistic in the field. Transmission of virus in the field is poorly understood but can be horizontal between individuals or vertical, from parents to offspring (Table 18.1). For horizontal transmission, the viral dose ingested could involve OBs from one or several dead hosts and thus vary in the number of variants present. In addition, within-host infection processes via the production and spread of budded virus and followed by OB production are likely to expose the variants to very different selection pressures compared to between-host horizontal transmission that depends on the infectivity of the OB. Between-host transmission will also be influenced by other factors, such as the ability of the OB to persist outside the host and insect behaviour. Understanding the source of infective material and how this affects virus diversity is clearly important if we want to understand the transmission process, as well as virus persistence outside the host.

Host population dynamics and virus persistence

How do insect viruses affect the dynamics of their hosts?

Host population studies have mostly centred on the baculoviruses of Lepidoptera, particularly cyclic, temperate forest species, although there are some data on sawflies in forests (reviewed in Cory and Myers, 2003). However, the impact of viral infection on host populations has not been extensively studied. Viral epizootics have primarily been observed in two families of forest Lepidoptera, the Lymantriidae and Lasiocampidae, whereas baculovirus epizootics also appear to be common in some species of migratory, agricultural pests such as armyworms (e.g. *Spodoptera* spp.), found in warmer climates

Table 18.1 Types of transmission that potentially influence the ecology of baculoviruses in the field

Horizontal	Within generation	Through contamination of foliage by cadavers following the death of larvae
	Between generations	Through environmental contamination and spread of OBs over time to foliage, bark and soil
Vertical	Between parents and offspring – overt	Passage of an active virus infection (or viable virus DNA) carried on or within eggs
	Between parents and offspring – covert	Passage of a persistent infection within the eggs

(e.g. Browne and Swaine, 1965; Rose *et al.*, 2000). Although baculoviruses have been isolated from a wide range of host species, not all species of Lepidoptera experience baculovirus epizootics and this feature appears to be system dependent (Fuxa, 2004). The ecology of baculoviruses in armyworm populations is particularly intriguing as these species often live in habitats with high UV levels that would appear to be hostile to baculovirus persistence. In addition, the host is often highly migratory, such that it is continually moving away from sources of inocula and these species also frequently exhibit large fluctuations in population density.

The best-studied armyworm system is the African armyworm, *Spodoptera exempta*; NPV is the major mortality factor in this species. NPV levels are often low early in the season, but build up from one generation to the next and can reach infection levels approaching 100% late in the season (Rose *et al.*, 2000). The African armyworm has two phases, gregarious and solitary and these respond differently to the virus. The solitary phase occurs at low density, is hard to find and is more susceptible to the NPV (Reeson *et al.*, 1998); however, in reality these insects are less likely to come into contact with NPV OBs than the gregarious form. Towards the end of the year and coinciding with the early rains, insects start to converge as a result of wind patterns or topography and primary outbreaks of moths start to form (Rose *et al.*, 2000). These are the darker, gregarious insects that initiate outbreaks. Recent work has shown that *S. exempta* populations in Tanzania carry high levels of persistent infection (Vilaplana *et al.*, 2010), which is one means whereby the virus could persist during the solitary, low density phase. However, the mechanism by which overt viral infection is introduced back into the gregarious populations is unknown, as is the population dynamics of the virus over multiple generations.

Longer term population dynamics and virus infection have only been studied in forest insects. The dynamics of some species of forest caterpillars are unstable but tend to show regular population cycles, such as those observed for gypsy moth, *L. dispar*, and tent caterpillars, e.g. *M. c. pluviale* (Liebhold *et al.*, 2000; Myers, 2000), and the population decline of these species is often associated with high levels of NPV infection. Research on these systems has not only considered disease prevalence and change with host population density, but also the patterns that underlie the host dynamics and the mechanisms that might drive the cycles. Virus transmission tends to be non-linear with host density and is more efficient at low host density than at high (Beisner and Myers, 1999; Fenton *et al.*, 2002). Levels of infection can increase very rapidly between host generations and then diminish quickly as host populations decline (Cory and Myers, 2009). Recent studies have shown a high level of variation in susceptibility to NPV infection among families in western tent caterpillar populations (Cory and Myers, 2009), although the factors that maintain this variation are not clear.

One obvious underlying pattern in western tent caterpillar population dynamics is a decrease in fecundity after the population peaks. If reduced fecundity is related to virus infection, it could be caused by two mechanisms; a cost resulting from selection for virus resistance or an effect of sublethal infection in surviving, virus-exposed insects. Laboratory studies have shown that insects that survive virus challenge have reduced fecundity (Rothman and Myers, 1994); however, the mechanism behind this finding is not clear. Comparison of the disease resistance of different families of *M.c. pluviale* collected from field populations with known histories showed no clear relationship with fecundity (Cory and Myers, 2009). This indicates that reduced fecundity is not likely to be a result of a cost of NPV resistance in tent caterpillars but is more likely to be associated with sublethal infections or food limitation. In addition, although there was a small increase in virus resistance after epizootics in field populations, they also became slightly more resistant to viral infection at peak densities, even before any potential selection from the viral epizootic could occur (Cory and Myers, 2009). These results indicate that other mechanisms appear to be acting to reduce insect fecundity and alter the susceptibility of western tent caterpillars to disease.

Other factors could be related to insect population eruptions and viral epizootics. As discussed above, food quality or availability can alter the response of an insect to pathogen challenge. Peak populations of forest caterpillars are likely to experience some food limitation, and possibly an increase in secondary chemicals of their food plants via induced effects. Thus, it is relevant to know whether this might influence their population dynamics. Recent interest has focussed on how immunity, in combination with foliage quality, might change with densities of populations of the autumnal moth, *Epirrita autumnata* (Haukioja, 2005). Kapari *et al.* (2006) showed that induced plant defence was associated with an increase in immunocompetence in increasing populations of *E. autumnata*; however, other studies did not find any relationship between immune function and either the level of parasitism or population fluctuations (Klemola *et al.*, 2007b). While viral infection is known to occur in this species, diseases are apparently not a major mortality factor.

Although many mathematical models have been developed to look at interactions between viral disease and population dynamics of forest Lepidoptera, most of which show some cyclic dynamics (e.g. Liu *et al.*, 2007; Dwyer *et al.*, 2004), the actual interactions between host density and viral infection in natural populations are still poorly understood. It is interesting that for forest Lepidoptera the factors that are likely to change at high host density, such as food availability, food quality (including secondary chemicals) and crowding, all apparently increase disease resistance (at least in the laboratory) rather than reduce it, but even so, virus epizootics occur.

How do viruses persist from one host generation to the next?

In multivoltine species, where several susceptible larval generations may overlap, it is likely that baculoviruses can persist via horizontal transmission from infected cadavers to healthy larvae, or via environmental contamination following the release of OBs that persist on the foliage or the soil (Table 18.1). Environmental contamination can also initiate infection in the next generation of hosts, for example, by contamination of egg batches laid on bark. Indeed, horizontal transmission has always been assumed to be the most common route of baculovirus transmission, as well as the means by which baculovirus infection is acquired following the application of baculovirus-based insecticides for pest control. Baculovirus OBs have been shown to persist for considerable periods of time outside their hosts, years in some cases, if they are protected from UV such as by being in the soil (Fuxa, 2004). Several studies have shown that viable virus can be reintroduced into populations from virus reservoirs (Fuxa, 2004), and highlight, in particular, the importance of insect behaviour in facilitating this process (Richards *et al.*, 1999).

Explaining viral transmission and infection is particularly challenging in situations where host populations undergo large fluctuations in density over time, or the host is migratory or highly mobile and does not revisit sites where epizootics have previously taken place. Environmental conditions such as high UV levels that reduce environmental persistence will also have a strong influence on transmission. In these situations it is thought likely that the virus persists via vertical transmission, that is, transmission from parent to young (Table 18.1). Vertical transmission has long been discussed in the baculovirus literature (Cory *et al.*, 1997) but it is only recently that we have had the tools to start investigating this phenomenon in more detail. Vertically transmitted virus can cause lethal infection in the next generation of larvae. This is likely to result from external contamination of the eggs by the female, an effect that can be difficult to separate from environmental contamination of the eggs from OBs on bark or foliage. Levels of vertical transmission that cause lethal disease can be high (>50%) and appear to vary between insect–virus combinations. Some evidence suggests that migratory species, such as those in the genus *Spodoptera*, have a particular facility for vertical transmission (Kukan, 1999). External contamination of the eggs (or transovum transmission) can usually be demonstrated by surface-sterilizing a portion of the egg batch and comparing the resulting level of virus mortality of larvae hatching from the unsterilized portion. However, surface-sterilization does not always totally eliminate infection, implying that overt infection can also be carried inside the egg (transovarially) at low levels (<5%) (Kukan, 1999).

Insects that show density-dependent phase change can respond differently to baculovirus challenge. It has already been mentioned that the darker, gregarious form of the African armyworm, *S. exempta*, is more resistant to NPV disease, than the lighter, solitary phase (Reeson *et al.*, 1998). This also results in reduced virus transmission in gregarious insects in the field (Reeson *et al.*, 2000). If this comparison is taken further and the survivors of virus challenge are mated and the effects on their progeny observed, it is found that overt NPV infection in the next generation is increased in virus-challenged insects and this increase is 2–3 times greater in insects that were reared solitarily (Vilaplana *et al.*, 2008). This result was unrelated to the sex of the parent or whether the eggs were surface-sterilized, indicating that transmission was transovarial. The mechanism behind this difference is unknown but might be related to the greater resistance of the gregarious insects as larvae.

The type of vertical transmission that has attracted most attention is that involving a latent form. There are numerous reports of apparently healthy larvae suddenly succumbing to baculovirus infection without obvious exposure to virus. However, it is only with the more recent use of molecular techniques that make it possible to determine if insects are carrying covert virus. Earlier studies on the cabbage moth, *Mamestra brassicae,* found sublethal levels of *M. brassicae* NPV in all life stages (Hughes *et al.*, 1993). Further studies showed that RNA transcripts were being produced, implying that the virus was not truly latent, but was better described as being persistent with low levels of replication or covert (Table 18.1) (Hughes *et al.*, 1997). Initial studies were carried out on a laboratory population and thus gave no indication of the prevalence of persistent infections in field populations. Burden *et al.* (2003) addressed this issue and found high levels of persistent infection in field populations of *M. brassicae* in the UK as well.

Recent work on *S. exempta* populations in Tanzania have shown very high levels of covert NPV infection in field populations; 97% of insects collected directly from the field were positive for *S. exempta* NPV DNA as adults and 60% of these were positive for RNA, indicating some level of transcription (Vilaplana *et al.*, 2010). These high levels of viral DNA in the adults were maintained when the insects were reared in the laboratory. *S. exempta* is a highly migratory species, often travelling large distances between generations. NPV epizootics frequently occur at the end of the larval season and it is not clear where the virus that initiates these epizootics originates. As mentioned above, *S. exempta* goes through both solitary and gregarious phases; the NPV epizootics occur in the damaging gregarious phase but during the solitary phase population densities are too low to maintain the virus by horizontal transmission. Vertical transmission therefore seems a likely route for maintaining NPV in this species. It may well be that further studies will show that persistent infection is very common in lepidopteran populations, and it will be interesting to see if persistent infections show the variation so prevalent in horizontally transmitted disease.

The big questions with covert infections are whether they actually play a role in the development of baculovirus epizootics, or at least, the reintroduction of active virus into the insect population, and if so, how they are triggered into an active form. Numerous reports, often anecdotal, suggest that factors such as crowding, diet, temperature, relative humidity, and the introduction of a second pathogen trigger covert infections (Cory *et al.*, 1997). However, finding a consistent trigger continues to prove elusive. Several theoretical studies have investigated the conditions under which covert infection might have evolved. The simplest approach is that reactivation of persistent virus is a stochastic process that is essentially a bet-hedging strategy. Models of this type have been developed for herpes virus that has a latent strategy (Stumpf *et al.*, 2002). Thus, the virus reactivates at intervals without a specific trigger or host density, and under some conditions this strategy will outcompete non-persistent forms of the virus over time (see Vilaplana *et al.* (2010) for further discussion of nature and evolution of covert infections). Another recent theoretical study examined the evolution of covert infection strategies, and demonstrated that covert infection could possibly evolve when host population density and thus transmission rates, are highly variable (Sorrell *et al.*, 2009). However, this model predicts much lower levels of covert infection than the very high frequency recorded in insect populations, such as the African armyworm, indicating that additional factors that were not taken into consideration are likely to be important.

Community-level effects

Insect viruses (and other entomopathogens) tend to be studied in isolation under optimal conditions; however, outside the laboratory they may well have to share their host with other natural enemies. In a pest control context, they could be applied as part of an integrated pest management strategy that includes parasitoids and other pathogens. In natural populations they are likely to interact with a whole range of other micro- and macro-organisms, including sublethal infections of other microorganisms. From an applied perspective, we want to know whether insect viruses compete directly with other natural enemies and whether this has detrimental effects on pest suppression. In natural populations, we want to understand how natural enemies interact and how this influences the population dynamics of their hosts and the functioning of the community that they belong to.

Within-host interactions

An individual host is essentially a discrete nutrient source for a pathogen to infect and colonize to maximize their fecundity. How do insect viruses respond when they have to compete for this resource? There have been numerous laboratory studies which have investigated the interaction between viruses and parasitoids, and some which have looked at multipathogen infections. The competition between parasitoids and pathogens is not symmetrical: when a parasitoid competes with a pathogen it is more likely to be a life or death struggle for the parasitoid because if it does not complete its development it will not survive. Pathogens, however, are more likely to suffer reduced yield if in competition with parasitoids. One of the most complete reviews of host–parasitoid-pathogen interaction was carried out by Brooks (1993), who took a mechanistic approach, detailing both the potential negative and positive effects of co-infection. For the parasitoid, detrimental effects included, premature death of the host, toxic substances, reduced ovipositional attractiveness, pathogen replication and reduction in host numbers. Positive effects primarily benefited the pathogen and included parasitoid-mediated transmission of infection. Overall, the outcome tended to be predictable in terms of the parasitoid, such that the parasitoid rarely survived if the pathogen had infected first because it could not complete its development (Brooks, 1993). When the parasitoid preceded the pathogen the outcome tended to depend on the period between parasitism and infection and pathogen dose, although parasitism has been shown to decrease susceptibility to disease (Cory *et al.*, 1997). There are few data on how competition affects the pathogen; however, Escribano *et al.* (2000) showed that mean viral yield was reduced in *S. frugiperda* parasitized by an endoparasitoid, *Chelonus sonorensis*, at larger instars, but not earlier ones. More intriguingly, there was also some indication that parasitism also affected the genetic composition of the NPV (Escribano *et al.*, 2000).

A number of studies have investigated mixed infections with baculoviruses, usually a combination of NPVs and GVs; however, competition with other pathogen groups has received less attention. As with the parasitoid interactions, most focus has been on the effect on mortality, in terms of whether the interaction was deemed to be synergistic, antagonistic or neutral, but with little information on the effect on the individual pathogens. An area that has received attention is the observed ability of GVs, via enhancin, to increase the potency of NPV infections, following on from the work of Tanada in 1959 (Lepore *et al.*, 1996). These studies have primarily concerned the development of more effective pest control strategies (e.g. Guo *et al.*, 2002) and identifying the mechanism of action of the enhancin protein (Wang and Granados, 1997) rather than elucidating ecological interactions. However, despite the positive effect of enhancins, NPV and GV infections can also interfere with each other (Hackett *et al.*, 2000) and the outcome is likely to be related to the relative infectivity of each virus to the host, in addition to timing and dose. A more detailed investigation of NPV interaction with a non-baculovirus is the study of Ishii *et al* (2002), in which competition between an NPV and an entomopoxvirus (EPV) was followed in the smaller tea tortrix, *Adoxophyes honmai*. In simultaneous mixed infections, the more rapid acting NPV out-competed the EPV, with no loss in yield. When the viruses were introduced sequentially, the EPV could out-compete the NPV if it was introduced sufficiently early with no loss in yield, whereas the production of NPV OBs was significantly reduced. These results make it clear that both the sequence and timing of mixed infection alter the outcome, but extrapolating this to the patterns observed in the field is more difficult.

Most studies on multispecies interactions have limitations as they rarely investigate the costs of competition for both players. They often record overall mortality and parasitoid survival in single versus mixed infections, but they usually do not estimate costs to the pathogen, such as changes in pathogen yield and production per infecting pathogen. Additional costs could include the subsequent oviposition success and fecundity in the parasitoid. The lack of data on costs makes it hard to conclude anything about the longer-term outcome of competition between the species. One approach is that adopted by Hodgson *et al.* (2004) who investigated the competition between two NPV genotypes by not only looking at overall OB yield, but also by comparing yield per infecting OB for each genotype in single and mixed infections. In this way, they demonstrated that there was no cost to either genotype to sharing a host; total yield per genotype was reduced in mixed infections, as might be predicted, but yield per infecting OB was unaffected and in one case increased in the presence of a second genotype. Detailed data on the outcome of mixed infections needs to be generated and analysed within an ecological framework if we are to understand the impact of mixed infections at both the individual and the population level.

Population level interactions

A small number of studies have investigated the population level effects of multispecies interactions involving entomopathogens, although there may well be data on short-term interactions within the biological control literature. A halfway house for studying mixed species interactions is to use microcosms in which contained and environmentally controlled host populations can be monitored over set time periods. Most microcosm studies have involved stored product pests such as the Indian meal moth, *Plodia interpunctella*, because they are easy to rear in small containers. In isolation, *P. interpunctella* exhibits population cycles of approximately one generation in length that are thought to be driven primarily by asymmetrical larval competition. These microcosm studies have shown, (i) that the population cycles increase slightly and host population density decreases when *P. interpunctella* GV (PiGV) is added (Sait *et al.*, 1994), (2) population cycles become multigenerational but eventually go extinct when a parasitoid is introduced to the system (Begon *et al.*, 1996) and (3) that the outcome of multispecies interactions depends on what order the participants are introduced into the microcosm (Sait *et al.*, 2000). A more recent experiment has investigated the interaction between PiGV and PiGV plus *Ephestia cautella* NPV in exactly the same set up. The study showed that the introduction of the second pathogen only resulted in very subtle changes in the population dynamics of the host, which had shorter cycles and greater abundance compared to the PiGV alone populations, so in essence, addition of a second pathogen reversed the effects of adding PiGV alone (Griffiths *et al.*, 2003). These studies indicate that multispecies dynamics are difficult to predict from single species interactions, and the relevance of microcosm studies to the more complex field situation remains unknown.

Concluding remarks

Molecular techniques have allowed much progress to be made in describing high levels of variation in baculoviruses and discovering the widespread occurrence of covert infection in some species of Lepidoptera. Detailed studies have demonstrated complex interactions between viruses and both food plants of their hosts and the role of nutrients as an influence on host susceptibility to virus infection. The role of virus infection in field populations of lepidopteran hosts however, still remains a 'black box'. Epizootics are only reported to occur regularly in a modest number of forest caterpillars and agricultural pests. How virus is maintained in populations of these hosts that either fluctuate in density or migrate long distances, and thus are often rare, remains a puzzle. The key to this puzzle will be found through determining (1) the prevalence of covert infection in host populations for which population dynamics are well studied, (2) what factors facilitate the transition from covert to overt infection, and (3) the factors that impact host condition and disease expression in the field. Although virulence of baculoviruses varies with the levels of molecular variation and the food plants of the hosts, whether this influences the dynamics of host–virus interactions in field populations remains to be determined. Because the ecology of viruses is so intimately related to the ecology of their hosts, research in this area must involve an understanding of the ecological arena in which both reside.

Acknowledgements

I am very grateful to Judy Myers for her comments on the manuscript and help with editing, and NSERC for funding.

References

Ali, M.I., Felton, G.W., Meade, T., and Young, S.Y. (1998). Influence of interspecific and intraspecific host plant variation on the susceptibility of heliothines to a baculovirus. Biol. Control *12*, 42–49.

Ali, M.I., Young, S.Y., Felton, G.W., and McNew, R.W. (2002). Influence of host-plant on occluded virus productivity and lethal infectivity of a baculovirus. J. Invertebr. Pathol. *81*, 158–165.

Begon, M., Sait, S.M., and Thompson, D.J. (1996). Predator-prey cycles with period shifts between two- and three-species systems. Nature *381*, 311–315.

Beisner, B.E., and Myers, J.H. (1999). Population density and transmission of virus in experimental populations of western tent caterpillar (Lepidoptera: Lasiocampidae). Environ. Entomol. *28*, 1107–1113.

Brooks, W. (1993). Host–parasitoid-pathogen interactions. In Parasites and Pathogens of Insects, volume 2, N.E. Beckage, S.N. Thompson and B.A. Federici, eds (San Diego, USA: Academic Press) pp. 231–272.

Browne, E.S, and Swaine, G. (1965). Virus disease of African armyworm *Spodoptera exempta* (Wlk). B. Entomol. Res. *56*, 95–116.

Bull, J.C., Godfray, H.C.J., and O'Reilly, D.R. (2001). Persistence of an occlusion-negative recombinant nucleopolyhedrovirus in *Trichoplusia ni* indicates high multiplicity of cellular infection. Appl. Environ. Microb. *67*, 5204–5209.

Burden, J.P., Nixon, C.P., Hodgkinson, A.E., Possee, R.D., Sait, S.M., King, L.A., and Hails, R.S. (2003). Covert infection as a mechanism for long term persistence of baculoviruses. Ecol. Lett. *6*, 524–531.

Carius, H.J., Little, T.J., and Ebert, D. (2001). Genetic variation in a host–parasite association: potential for coevolution and frequency dependent selection. Evolution *55*, 1136–1145.

Clavijo, G., Williams, T., Muñoz, D., Caballero, P., and Lopez-Ferber, M. (2010). Mixed genotype transmission bodies and virions contribute to the maintenance of diversity in an insect virus. P. Roy. Soc. B Biol. Sci. *277*, 943–951.

Cooper, D., Cory, J.S., and Myers, J.H. (2003). Hierarchical spatial structure of genetically variable nucleopolyhedroviruses infecting cyclic populations of western tent caterpillars. Mol. Ecol. *12*, 881–890.

Cory, J.S., Green, B.M., Paul, R.K., and Hunter-Fujita, F.R. (2005). Genotypic and phenotypic diversity of a baculovirus population within an individual insect host. J. Invertebr. Pathol. *89*, 101–111.

Cory, J.S., Hails, R.S., and Sait, S.M. (1997). Baculovirus ecology. In The Baculoviruses, L.K. Miller, ed. (New York: Plenum Press), pp.301–339.

Cory, J.S., and Hoover, K. (2006). Plant-mediated effects in insect–pathogen interactions. Trends Ecol. Evol. *21*, 278–286.

Cory, J.S., and Myers, J.H. (2003). The ecology and evolution of insect baculoviruses. Annu. Rev. Ecol. Evol. S. *34*, 239–272.

Cory, J.S., and Myers, J.H. (2004). Adaptation in an insect-host plant pathogen interaction. Ecol. Lett. *7*, 632–639.

Cory, J.S., and Myers, J.H. (2009). Within and between population variation in disease resistance in cyclic populations of western tent caterpillars: a test of the disease defence hypothesis. J. Anim. Ecol. *78*, 646–655.

D'Amico, V., Elkinton, J.S., Dwyer, G., Willis, R.B., and Montgomery, M.E. (1998). Foliage damage does not affect within-season transmission of an insect virus. Ecology *79*, 1104–1110.

Duffey, S.S., Hoover, K., Bonning, B.C., and Hammock, B.D. (1995). The impact of host-plant on the efficacy of baculoviruses. In Reviews in Pesticide Toxicology, M Roe and R Kuhr, eds), (Raleigh, North Carolina: CTI Toxicology Communications), pp. 137–275.

Dwyer, G., Dushoff, J., Elkinton, J.S., and Levin, S.A. (2000). Pathogen-driven outbreaks in forest defoliators revisited: building models from experimental data. Am. Nat. *156*, 105–120.

Dwyer, G., Dushoff, J., and Yee, S.H. (2004). The combined effects of pathogens and predators on insect outbreaks. Nature *430*, 341–345.

Dwyer, G., Firestone, J., and Stevens, T.E. (2005) Should models of disease dynamics in herbivorous insects include the effects of variability of host-plant foliage quality? Am. Nat. *165*, 16–31.

Eberle, K.E., Sayed, S., Rezapanah, M., Shojai-Estabragh, S., and Jehle, J.A. (2009). Diversity and evolution of the *Cydia pomonella* granulovirus. J. Gen. Virol. *90*, 662–671.

Escribano, A., Williams, T. Goulson, D., Cave, R.D. Chapman, J.W., and Caballero, P. (2000). Effect of parasitism on a nucleopolyhedrovirus amplified in *Spodoptera frugiperda* larvae parasitized by *Chelonus sonorensis*. Ent. Expl. Appl. *97*, 257–264.

Fenton, A., Fairbairn, J.P., Norman, R., and Hudson, P.J. (2002). Parasite transmission: reconciling theory and reality. J. Anim. Ecol. *71*, 893–905.

Fuxa, J.R. (2004). Ecology of insect nucleopolyhedroviruses. Agr. Ecosyst. Environ. *103*, 27–43.

Griffiths, C.M., McVean, R., Benmayor, R., Sait, S.M., Thompson, D.J., Possee, R.D., Cory, J.S., and Begon, M. (2003). Host dynamics and pathogen variation in insect–baculovirus interactions. In Genes in the Environment. R.S. Hails, J. Beringer, H.C.J. Godfray, eds, (Oxford, UK: Blackwell Science Ltd), pp. 281–305.

Guo, H.F., Fang, J.C., Wang, J.P., Zhong, W.F., and Liu, B.S. (2002). Interaction of *Xestia c-nigrum* granulovirus with peritrophic matrix and *Spodoptera litura* nucleopolyhedrovirus and *Spodoptera litura*. J. Econ. Entomol. *100*, 20–25.

Hackett, K.J., Boore, A., Deming, C., Buckley, E., Camp, M., and Shapiro, M. (2000). *Helicoverpa armigera* granulovirus interference with progression of *H. zea* nucleopolyhedrovirus disease in *H. zea* larvae. J. Invertebr. Pathol. *75*, 99–106.

Haukioja, E. (2005). Plant defences and population fluctuations of forest defoliators: mechanism-based scenarios. Ann. Zool. Fennici *42*, 313–325.

Hitchman, R.B., Hodgson, D.J., King, L.A., Hails, R.S., Cory, J.S., and Possee, R.D. (2007). Host mediated selection of pathogen genotypes as a mechanism for the maintenance of baculovirus diversity in the field. J. Invertebr. Pathol. *94*, 153–162.

Hodgson, D.J., Hitchman, R.B., Vanbergen, A.J., Hails, R.S., Hartley, S.E., Possee, R.D., Watt, A D., and Cory, J.S. (2003). The existence and persistence of genotypic variation in nucleopolyhedrovirus populations. In Genes in the Environment, R.S. Hails, J. Beringer, H.C.J. Godfray, eds, (Oxford, UK: Blackwell Science Ltd), pp.258–280.

Hodgson, D.J., Hitchman, R.B., Vanbergen, A.J., Hails, R.S., Possee, R.D., and Cory, J.S. (2004). Host ecology determines the relative fitness of virus genotypes in mixed-genotype nucleopolyhedrovirus infections. J. Evolution. Biol. *17*, 108–1025.

Hodgson, D.J., Vanbergen, A.J., Watt, A.D., Hails, R.S., and Cory, J.S. (2001). Phenotypic variation between naturally co-existing genotypes of a lepidopteran baculovirus. Evol. Ecol. Res. *3*, 687–701.

Hodgson, D.J., Vanbergen, A.J., Watt, A.D., Hartley, S.E., Hails, R.S., and Cory, J.S. (2002). Differential selection of baculovirus genotypes mediated by different species of host food plant. Ecol. Lett. *5*, 512–518.

Hoover, K., Stout, M.J., Alaniz, S.A., Hammock, B.D., and Duffey, S.S. (1998). Influence of induced plant defences in cotton and tomato on the efficacy of baculoviruses on noctuid larvae. J. Chem. Ecol. *24*, 253–271.

Hoover, K., Washburn, J.O., and Volkman, L.E. (2000). Midgut-based resistance of *Heliothis virescens* to baculovirus infection mediated by phytochemicals in cotton. J. Insect Physiol. *46*, 999–1007.

Hughes, D.S. Possee, R.D., and King, L.A. (1993). Activation and detection of latent baculovirus resembling *Mamestra brassicae* nuclear polyhedrosis virus in *M. brassicae* insects. Virology *194*, 604–615.

Hughes, D.S., Possee, R.D., and King, L.A. (1997). Evidence for the presence of a low-level, persistent baculovirus infection of *Mamestra brassicae* insects. J. Gen. Virol. *78*, 1801–1805.

Hunter, M.D., and Schultz, J.C. (1993). Induced plant defenses breached? Phytochemical induction protects an herbivore from disease. Oecologia *94*, 195–203.

Ishii, T., Takatsuka, J., Nakai, M., and Kunimi, Y. (2002). Growth characteristics and competitive ability of a nucleopolyhedrovirus and an entomopoxvirus in larvae of the smaller tea tortrix, *Adoxophyes honmai* (Lepidoptera: Tortricidae). Biol. Control *23*, 96–105.

Kapari, L., Haukioja, E., Rantala, M.J., and Ruuhola, T. (2006). Defoliating insect immune defense interacts with induced plant defence during a population outbreak. Ecology *87*, 291–296.

Keating, S.T., Yendol, W.G., and Schultz, J.C. (1988). Relationship between susceptibility of gypsy moth larvae (Lepidoptera: Lymantriidae) to a baculovirus and host plant constituents. Environ. Entomol. *17*, 942–958.

Kikhno, I., Gutiérrez, S., Crozier, L., Crozier, G., and López-Ferber, M. (2002). Characterization of *pif*, a gene required for the per os infectivity of *Spodoptera littoralis* nucleopolyhedrovirus. J. Gen. Virol. *83*, 3013–3022.

Klemola, N., Klemola, T., Rantala, M.J., and Ruuhola, T. (2007a). Natural host plant quality affects immune defence of an insect herbivore. Entomol. Exp. Appl. *123*, 167–176.

Klemola, T., Klemola, N., Andersson, T., and Ruohomäki, K. (2007b). Does immune function influence population fluctuations and level of parasitism in the cyclic geometrid moth? Popul. Ecol. *49*, 165–178.

Kolodny-Hirsch, D.M., and van Beek, N.A.M. (1997). Selection of a morphological variant of *Autographa californica* nuclear polyhedrosis virus with increased virulence following serial passage in *Plutella xylostella*. J. Invertebr. Pathol. *69*, 205–211.

Kukan, B. (1999). Vertical transmission of nucleopolyhedroviruses in insects. J. Invertebr. Pathol. *74*, 103–111.

Laitinen, A.M., Otvos, I.S., and Levin, D.B. (1996). Genotypic variation among wild isolates of Douglas-fir tussock moth (Lepidoptera: Lymantriidae) nuclear polyhedrosis virus. J. Econ. Entomol. *89*, 640–647.

Lee, K.P., Cory, J.S., Wilson, K., Raubenheimer, D., and Simpson, S.J. (2006). Flexible diet choice offsets protein costs of pathogen resistance in a caterpillar. P. Roy. Soc. B Biol. Sci. *273*, 823–829.

Lepore, L.S., Roelvink, P.R., and Granados, R. (1996). Enhancin, the granulosis virus protein that facilitates nucleopolyhedrovirus (NPV) infections, is a metalloprotease. J. Invertebr. Pathol. *68*, 131–140.

Liebhold, A., Elkinton, J., Williams, D., and Muzika, R.M. (2000). What causes outbreaks of gypsy moth in North America? Popul. Ecol. *42*, 257–266.

Liu, W.-C., Bonsall, M.B., and Godfray, H.C.J. (2007). The form of host density-dependence and the likelihood of host-pathogen cycles in forest-insect systems. Theor. Popul. Biol. *72*, 86–95.

López-Ferber, M., Simón, O., Williams, T., and Caballero, P. (2003). Defective or effective? Mutualistic interactions between virus genotypes. P. Roy. Soc. B Biol. Sci. *270*, 2249–2255.

McVean, R.I.K., Sait, S.M., Thompson, D.J., and Begon, M. (2002). Dietary stress reduces susceptibility of *Plodia interpunctella* to infection by a granulovirus. Biol. Control *25*, 81–84.

Moret, Y., and Schmidt-Hempel, P. (2000). Survival for immunity: the price of immune system activation for bumblebee workers. Science *290*, 1166–1168.

Muñoz, D., and Caballero, P. (2000). Persistence and effects of parasitic genotypes in a mixed population of *Spodoptera exigua* nucleopolyhedrovirus. Biol. Control *19*, 259–264.

Muñoz, D., Castillejo, J.I., and Caballero, P. (1998). Naturally occurring deletion mutants are parasitic genotypes in a wild-type nucleopolyhedrovirus population of *Spodoptera exigua*. Appl. Environ. Microb. *64*, 4371–4377.

Myers, J.H. (2000). Population fluctuations of the western tent caterpillar in southwestern British Columbia. Popul. Ecol. *42*, 231–241.

Nykänen, H., and Koricheva, J. (2004). Damage-induced changes in woody plants and their effect on insect herbivore performance: a meta-analysis. Oikos *104*, 247–268.

Ojala, K., Julkunen-Tiitto, R., Lindström, L., and Mappes, J. (2005). Diet affects the immune defence and life-history traits of an arctiid moth *Parasemia plantaginis*. Evol. Ecol. Res. *7*, 1153–1170.

Plymale, R.C., Felton, G.W., and Hoover, K. (2007). Induction of systemic acquired resistance in cotton foliage does not adversely affect the performance of an entomopathogen. J. Chem. Ecol. *33*, 1570–1581.

Plymale, R.C., Grove, M.J., Cox-Foster, D., Ostiguy, N., and Hoover, K. (2008). Plant-mediated alteration of the peritrophic matrix and baculovirus infection in lepidopteran larvae. J. Insect Physiol. *54*, 737–749.

Raubenheimer, D., Simpson, S.J., and Mayntz, D. (2009). Nutrition, ecology and nutritional ecology: toward and integrated framework. Funct. Ecol. *23*, 4–16.

Raymond, B., and Hails, R.S. (2007). Variation in plant resource quality and the transmission and fitness of the winter moth, *Operophtera brumata* nucleopolyhedrovirus. Biol. Control *41*, 237–245.

Raymond, B., Hartley, S.E., Cory, J.S., and Hails, R.S. (2005). The role of food plant and pathogen-induced behaviour in the persistence of a nucleopolyhedrovirus. J. Invertebr. Pathol. *88*, 49–57.

Raymond, B., Vanbergen, A., Pearce, I., Hartley, S.E., Cory, J.S., and Hails, R.S. (2002). Host plant species can influence the fitness of herbivore pathogens: the winter moth and its nucleopolyhedrovirus. Oecologia *131*, 533–541.

Read, A.F., and Taylor, L.H. (2001). The ecology of genetically diverse infections. Science *292*, 1099–1102.

Reeson, A.F., Wilson, K., Cory, J.S., Hankard, P., Weeks, J.M., Goulson, D., and Hails, R.S. (2000). Effects of phenotypic plasticity on pathogen transmission in the field in a Lepidoptera-NPV system. Oecologia *124*, 373–380.

Reeson, A.F., Wilson, K., Gunn, A., Hails, R.S., and Goulson, D. (1998). Baculovirus resistance in the noctuid *Spodoptera exempta* is phenotypically plastic and responds to population density. P. Roy. Soc. B Biol. Sci. *265*, 1787–1791.

Richards, A., Cory, J., Speight, M., and Williams, T. (1999). Foraging in a pathogen reservoir can lead to local host population extinction: a case study of Lepidoptera–virus interaction. Oecologia *118*, 29–38.

Rose, D.J.W., Dewhurst, C.F., and Page, W.W. (2000). The African armyworm handbook. 2nd edition. (Greenwich, UK: Natural Resources Institute).

Rossiter, M.C., Schultz, J.C., and Baldwin, I.T. (1988). Relationships among defoliation, red oak phenolics, and gypsy moth growth and reproduction. Ecology *69*, 267–277.

Rothman, L.D., and Myers, J.H. (1994). Nuclear polyhedrosis virus treatment effect on reproductive potential of western tent caterpillar (Lepidoptera: Lasiocampidae). Environ. Entomol. *23*, 864–869.

Sadd, B.M., and Schmid-Hempel, P. (2009). Principles of ecological immunity. Evol. Appl. *2*, 113–121.

Sait, S.M., Begon, M., and Thompson, D.J. (1994). Long-term population dynamics of the Indian meal moth *Plodia interpunctella* and its granulosis virus. J. Anim. Ecol. *63*, 861–870.

Sait, S.M., Liu, W.-C., Thompson, D.J., Godfray, H.C.J., and Begon, M. (2000). Invasion sequence affects predator-prey dynamics in a multi–species interaction. Nature *405*, 448–450.

Shikano, I., Ericsson, J.D., Cory, J.S., and Myers, J.H. (2010). Indirect plant-mediated effects on insect immunity and disease resistance in a tritrophic system. Basic Appl. Ecol. *11*, 15–22.

Simón, O., William, T., Caballero, P., and López-Ferber, M. (2006). Dynamics of deletion genotypes in an experimental virus population. P. Roy. Soc. B Biol. Sci. *273*, 783–790.

Simón, O., Williams, T., López-Ferber, M., and Caballero, P. (2004). Genetic structure of a *Spodoptera frugiperda* nucleopolyhedrovirus population: high prevalence of deletion genotypes. Appl. Environ. Microb. *70*, 5579–5588.

Simón, O., Williams, T., López-Ferber,M., and Caballero, P. (2005). Functional importance of deletion mutant genotypes in an insect nucleopolyhedrovirus population. Appl. Environ. Microb. *71*, 4254–4262.

Simón, O., Williams, T., López-Ferber, M., Taulemesse, J.-M., and Caballero, P. (2008). Population genetic structure determines speed of kill and occlusion body production in *Spodoptera frugiperda* multiple nucleopolyhedrovirus. Biol. Control *44*, 321–330.

Siva-Jothy, M.T., and Thompson, J.J.W. (2002). Short-term nutrient deprivation affects immune function. Physiol. Entomol. *27*, 206–212.

Smilanich, A.M., Dyer, L.A., Chambers, J.Q., and Bowers, M.D. (2009). Immunological cost of chemical defence and the evolution of herbivore diet breadth. Ecol. Lett. *12*, 612–621.

Smith, J.L., de Moraes, C.M., and Mescher, M.C. (2009). Jasmonate- and salicylate-mediated plant defense responses to insect herbivores, pathogens and parasitic plants. Pest Manag. Sci. *65*, 497–503.

Sorrell, I., White, A., Pedersen, A.B., Hails, R.S., and Boots, M. (2009). The evolution of covert, silent infection as a parasite strategy. P. Roy. Soc. B Biol. Sci. *276*, 2217–2226.

Stumpf, M.P.H., Laidlaw, Z., and Jansen, V.A.A. (2002). Herpes viruses hedge their bets. Proc. Natl. Acad. Sci. U.S.A. *99*, 15234–15237.

Thaler, J.S., Farad, M.A., Paré, P.W., and Dicke, M. (2002a). Jasmonate-deficient plants have reduced direct and indirect defences against herbivores. Ecol. Lett. *5*, 764–774.

Thaler, J.S., Fidantsef, A.L., and Bostock, R.M. (2002b). Antagonism between jasmonate- and salicylate-mediated induced plant resistance: effects of concentration and timing of elicitation on defense-related proteins, herbivore and pathogen performance in tomato. J. Chem. Ecol. *28*, 1131–1159.

Tompkins, G.J., Dougherty, E.M., Adams, J.R., and Diggs, D. (1988). Changes in the virulence of nuclear polyhedrosis viruses when propagated in alternate noctuid (Lepidoptera: Noctuidae) cell lines and hosts. J. Econ. Entomol. *81*, 1027–1032.

Vilaplana, L., Redman, E.M., Wilson, K., and Cory, J.S. (2008). Density-related variation in vertical transmission of a virus in the African armyworm. Oecologia *155*, 237–246.

Vilaplana, L., Wilson, K., Redman, E.M., and Cory, J.S. (2010). Pathogen persistence in migratory insects: high levels of vertically transmitted virus infection in field populations of the African armyworm. Evol. Ecol. *24*, 147–160.

Wang, P., and Granados, R.R. (1997). An intestinal mucin is the target substrate for a baculovirus enhancin. Proc. Natl. Acad. Sci. U.S.A. *94*, 6977–6982.

Yang, S., Ruuhola, T., and Rantala, M.J. (2007). Impact of starvation on immune defense and other life-history traits of an outbreaking geometrid, *Epirrita autumnata*; a possible causal trigger for the crash phase of population cycle. Ann. Zool. Fennici *44*, 89–96.

Young, S.Y., Yearian, W.C., and Kim, K.S. (1977). Effect of dew from cotton and soybean foliage on activity of *Heliothis* nuclear polyhedrosis virus. J. Invertebr. Pathol. *29*, 105–111.

Zwart, M.P., Hemerik, L., Cory, J.S., de Visser, J.A.G.M., Bianchi, F.J.J.A., van Oers, M., Vlak, J.M., Hoekstra, R., and van der Werf, W. (2009a). An experimental test of the independent action hypothesis in virus-insect pathosystems. P. Roy. Soc. B Biol. Sci. *276*, 233–2242.

Zwart, M.P., van der Werf, W., Georgievska, L., van Oers, M., Vlak, J.M., and Cory, J.S. (2010). Mixed genotype infections of *Trichoplusia ni* larvae with *Autographa californica* multicapsid nucleopolyhedrovirus: speed of action and persistence of a recombinant in serial passage. Biol. Control *52*, 77–83.

Zwart, M.P., van der Werf, W., van Oers, M., Hemerik, L., van Lent, J.M.V., de Visser J.A.G.M., Vlak, J.M., and Cory, J.S. (2009b). Mixed infections and the competitive fitness of faster-acting genetically modified viruses. Evol. Appl. *2*, 209–221.

Index

R

S

T

U

V

W

X

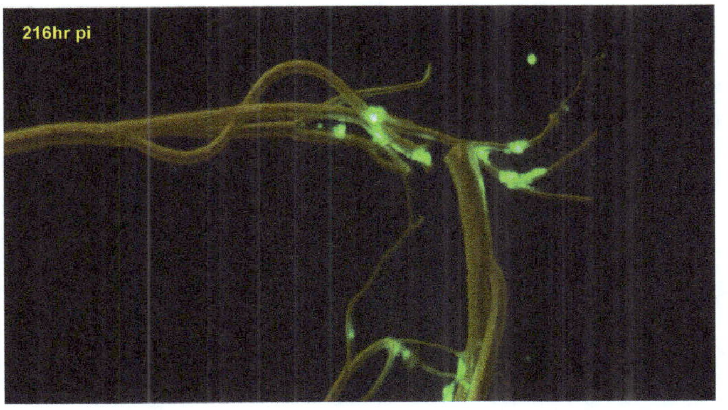

Plate 4.1 Infected haemocytes in the proximity of uninfected trachea. Gypsy moth larvae were inoculated with infected *Lymantria dispar* (Ld) cells and at various times thereafter, tissues were dissected out and examined for GFP expression from the spheroidin promoter, which indicated infection. p.i., post infection.

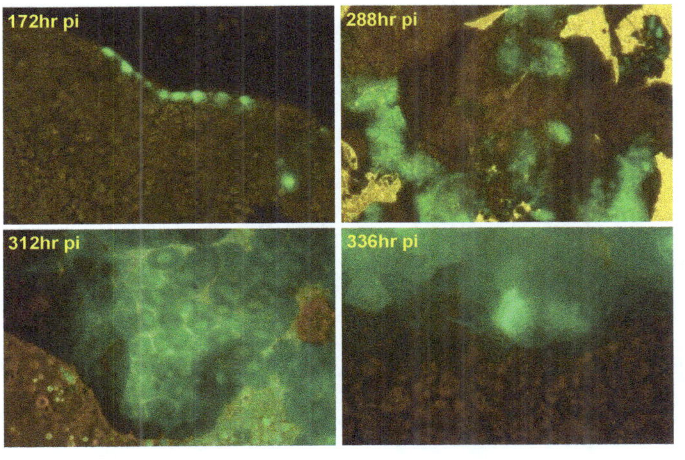

Plate 4.2 Infection of fat body tissue at various times after inoculation. Same procedure as in Plate 4.1. p.i., post infection.

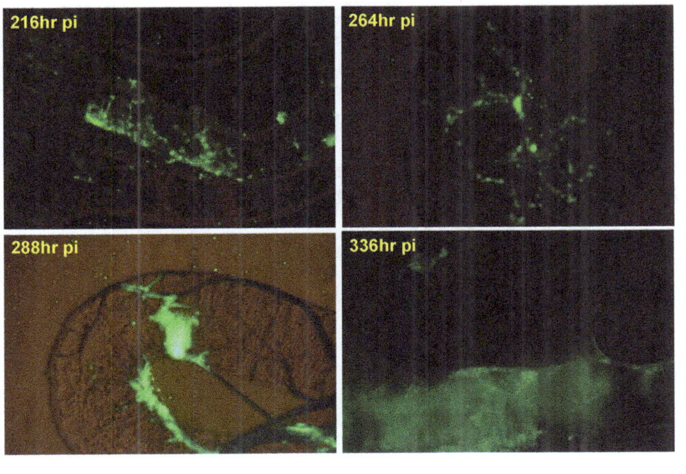

Plate 4.3 Infection in the silk gland. Same procedure as in Plate 4.1. pi, post infection.

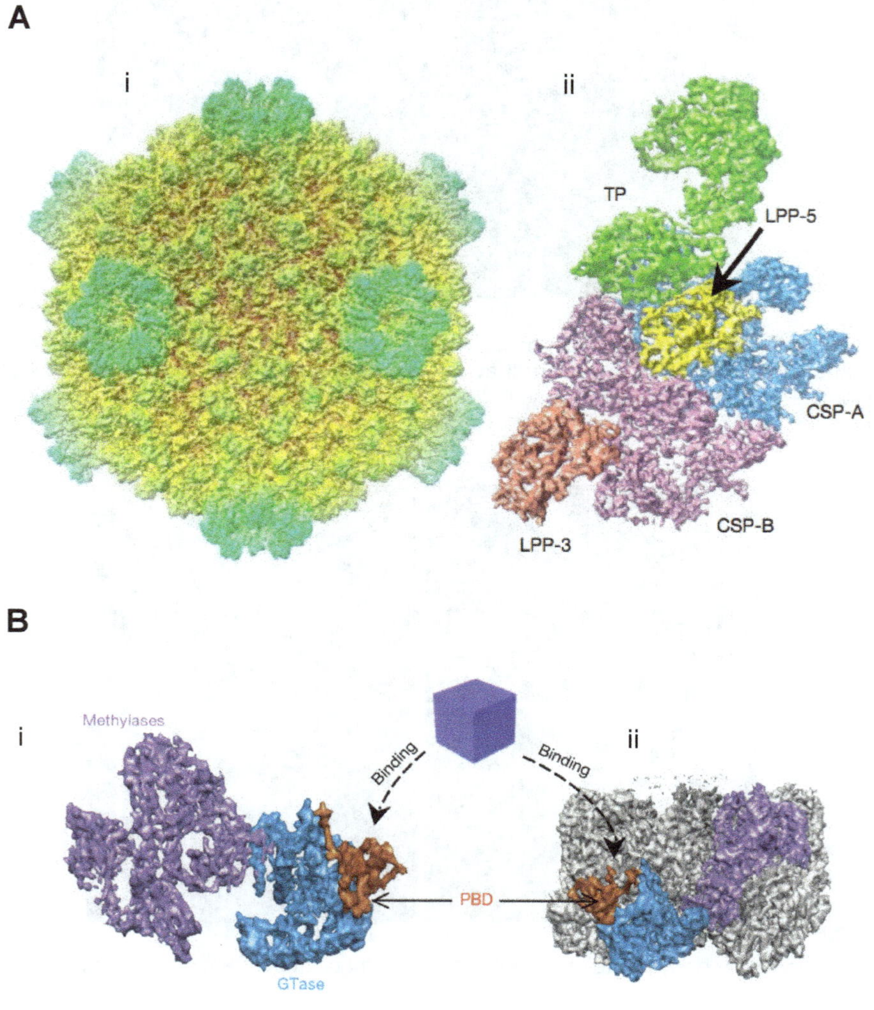

Plate 13.2 BmCPV 3.88 Å resolution electron density map derived from 12,814 cryo-electron microscopy single particle images using icosahedral averaging. (A) (i) The whole particle viewed down a two-fold symmetry axis. Eight of the twelve turret projects are visible in this orientation. The colours correspond to radial distance. (ii) The icosahedral asymmetric unit, which is repeated 60 times to generate the whole particle. The asymmetric unit contains one TP molecule, two CSP molecules (CSP-A and CSP-B) and 2 LPP molecules (LPP-3 and LPP-5). (B) Turret protein (TP, VP3) electron density. (i) A single TP protein, showing the GTase and methylase domains. (ii) Turret (like the eight turrets visible in A) containing five TP proteins. The location of the N-terminal 79 residue polyhedrin binding domain (PBD) is indicated.

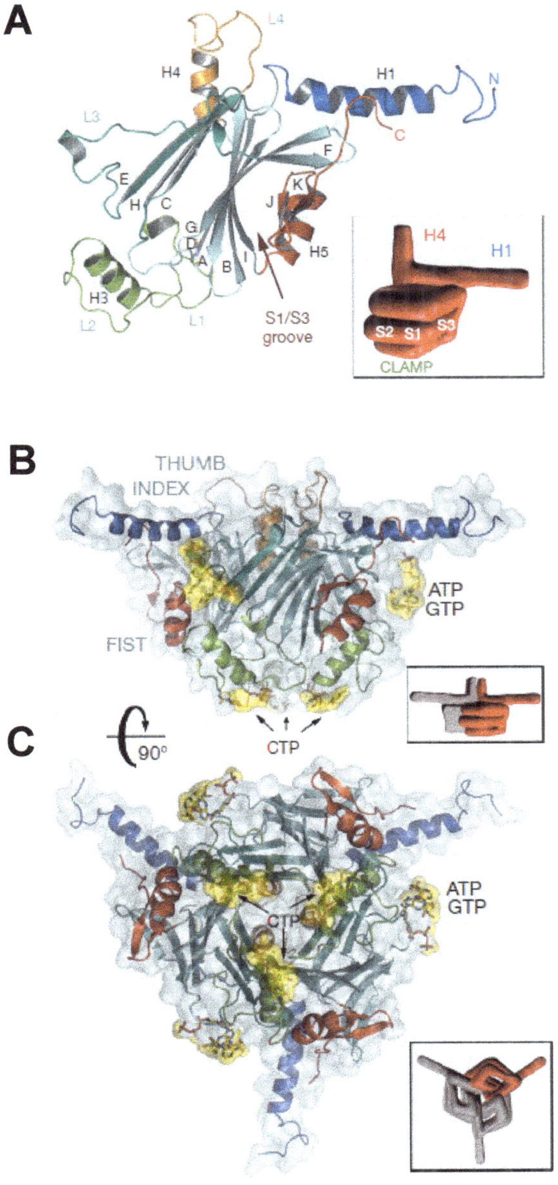

Plate 13.3 Structure of BmCPV polyhedrin. (A) Polyhedrin has the shape of a left hand (inset). The finger (H1, blue), thumb (H4, orange), fist (S1–S2, cyan and S3, red) and clamp (H3, green) all contribute to the organization of polyhedra. (B, C) The polyhedrin trimer in two orthogonal views with bound nucleotides shown in a yellow surface. The insets represent trimer models with one of the subunits highlighted in red.

A

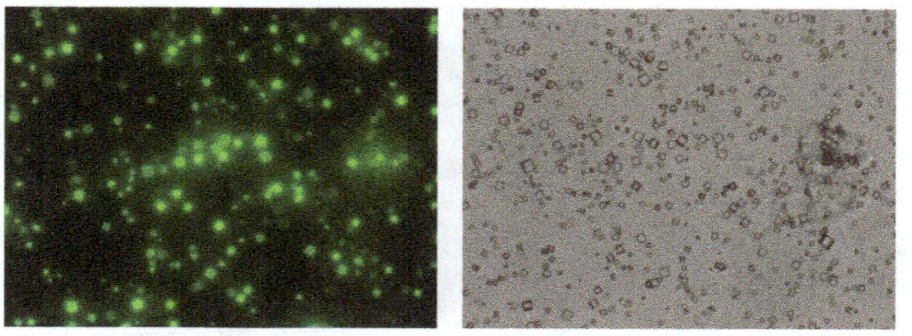

B

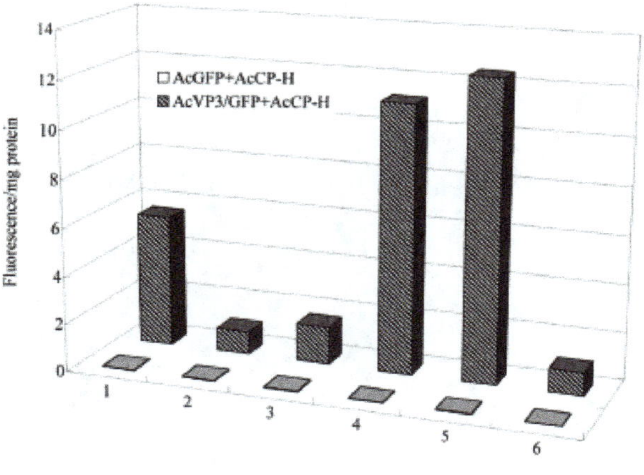

Plate 13.4 Immobilization of EGFP into the BmCPV polyhedra. (A) The green fluorescence from the polyhedra by double infection with AcVP3/GFP and AcCP-H. The green fluorescence under the irradiation of UV (left) and micrograph of BmCPV polyhedra (right). (B) Measurement of green fluorescence of the VP3-GFP chimera in BmCPV polyhedra. Polyhedra obtained by double infection with AcGFP and AcCP-H or with AcVP3/GFP and AcCP H were resuspended in distilled water, acetate buffer (pH 4.0), and carbonate buffer (pH 11.0). After resuspension in acetate buffer, the pH was adjusted to 6.0, 10.0, and 12.5 by the addition of 5 N NaOH. Fluorescence intensity was represented as fluorescence emission at 510 nm per milligram of protein which was illuminated with 475 nm light. Column; 1, distilled water; column 2, pH 4; column 3, pH 6; column 4, pH 10; column 5, pH 11; column 6, pH 12.5.

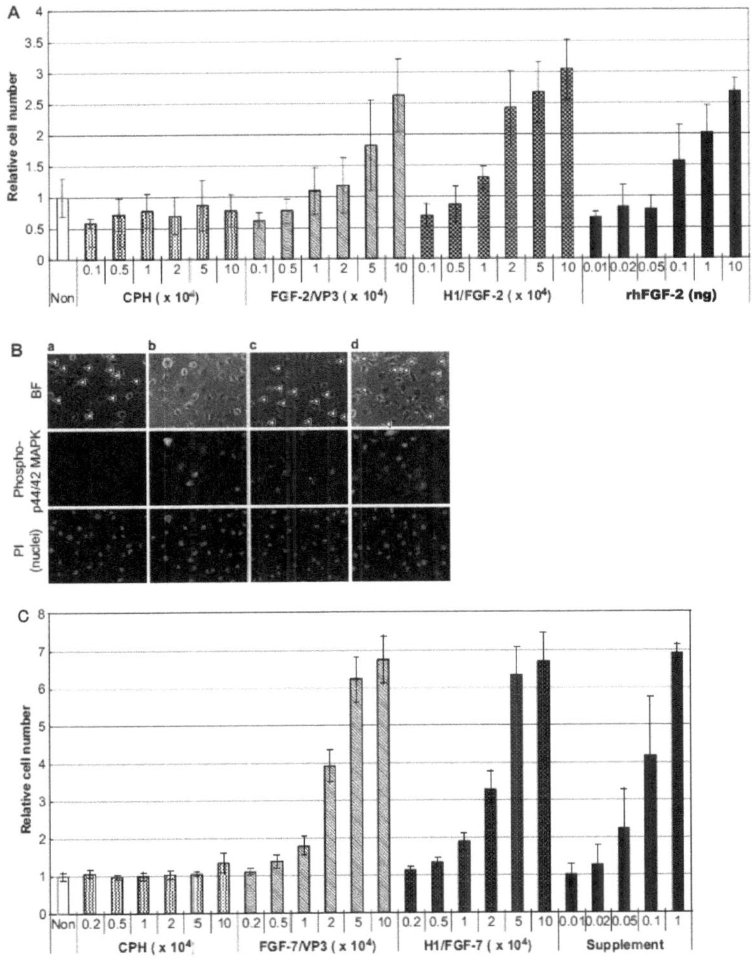

Plate 13.5 Effects of FGF-2 polyhedra on the proliferation of NIH/3T3 cells in culture. (A) Proliferation of NIH/3T3 cells by FGF-2 polyhedra. NIH3T3 cells were seeded onto a 96-well cell culture plate at a density of 4,000 cells per well and cultured overnight in DMEM medium containing 10% FCS followed by 6 hours' culture in α-MEM medium without FCS to starve the cells. Cells were stimulated with CP-H polyhedra and FGF-2 polyhedra (FGF-2/VP3 and H1/FGF-2) or recombinant human FGF-2 (rhFGF-2). Number of polyhedra or the dose of rhFGF-2 is indicated on the horizontal axis. After 5 days, the growth rate of respective cell group was determined by CCK assay using a microplate reader. Cell proliferation is reported relative to the negative control arbitrarily set to 1. The mean value of triplicate samples and its standard deviation (SD) are shown as a bar and an error bar, respectively. (B) Immunocytochemistry of phosphorylated MAPK in NIH3T3 cells. The serum-starved NIH3T3 cells on 96 well culture plate were stimulated with CP-H polyhedra (a), 5 ng of rhFGF-2 (b), FGF-2/VP3 polyhedra (c) or H1/FGF-2 polyhedra (d) (each 2.5 × 104 polyhedra) for 15 min. The cells were then fixed and nuclei were stained with propidium iodide (PI). Furthermore, the intracellular phosphorylated MAP kinase was detected by immunocytochemistry using a primary anti-phospho p44/42 MAP kinase antibody and a secondary antibody conjugated with Alexa 488 fluorescent dye. The phosphorylated cells were detected under a fluorescent microscope. Each image indicates phosphorylated p44/42 MAP kinase (green) and nuclei (red) and, in BF, polyhedra are identified with white triangles. (C) Culture of keratinocytes with FGF-7 polyhedra. NHEK at passage 3 with DK-SFM containing the growth supplement were seeded at a density of 900 cells/well and cultured for 5 days at 37°C under 5% CO2 in air. NHEK cells were grown with or without a commercial growth supplement, and with CP-H polyhedra or FGF-7 polyhedra (FGF-7/VP3 and H1/FGF-7) at concentrations of 0.2 to 10 x 104 polyhedra per well. The rate of cell proliferation was detected by CCK assay for 4 hours at 37°C under 5% CO_2 in air and measure of absorbance at 450 nm. Cell proliferation is reported relative to the negative control arbitrarily set to 1. The mean value of triplicate samples and its standard deviation (SD) are shown as a bar and an error bar, respectively.

www.ingramcontent.com/pod-product-compliance
Lightning Source LLC
LaVergne TN
LVHW062214200926
848484LV00006B/1389